FLUIDIZATION V

Proceedings of the
Fifth Engineering Foundation Conference
on Fluidization

May 18-23, 1986
Elsinore, Denmark

Knud Østergaard and Ansgar Sørensen, Editors
The Technical University of Denmark
Lyngby, Denmark

PUBLISHED BY

Engineering Foundation
New York, New York

Harold A. Comerer, *Director*

Library of Congress Catalog Card Number: 86-80478
ISBN 0-8169-0356-5

FIFTH ENGINEERING FOUNDATION CONFERENCE ON FLUIDIZATION

Executive Committee

K. Østergaard (Chairman), A. Sørensen (Vice-Chairman),
M.A. Bergougnou, J.F. Davidson, J.R. Grace, D.L. Keairns, D. Kunii,
L.S. Leung, H. Littman, J.M. Matsen, H.A. Comerer

Co-Sponsors

The Danish Society of Chemical Engineers
The European Federation of Chemical Engineering (Event No. 335)

Financial Sponsors

The Engineering Foundation and the following Danish authorities, corporations, and institutions:

Aalborg Værft A/S, Laurits Andersens Fond,
Burmeister & Wain Energi A/S, A/S Cheminova,
Danish Ministry of Education, Danish Ministry of Energy,
Danish Engineering Research Council, Dansk Esso Fond,
De Danske Sukkerfabrikker A/S, Elkraft, Elsam, Faxe Kalk,
Grindsted Products A/S, Otto Mønsteds Fond, Novo Industri A/S,
F.L. Smidth & Co. A/S' Gavefond, Superfos a/s,
Thomas B. Thriges Fond, Haldor Topsøe A/S, Tuborgfondet.

CONTENTS

IV. Particle Motion; Mixing; Circulation

V. Particle Motion; Segregation

VI. Spouted Beds; Risers; Standpipes

VII. Turbulent Fluidization; Entrainment; Fast Fluidization

VIII. Three-Phase Fluidization

IX. Heat and Mass Transfer

X. Fluid Bed Reactors

XI. Fluid Bed Gasifiers

XII. Fluid Bed Combustion

XIII. Fluid Bed Applications in the Process Industries

PREFACE

This volume contains the papers to be presented in formal sessions at the Fifth Engineering Foundation Conference on Fluidization in Elsinore, Denmark, on May 18–23, 1986.

Fluidization is an important industrial technology for the handling of solids in the form of powders and particles. It is, furthermore, a complex process of intrinsic academic interest. It has, for these reasons, been the subject of much academic and industrial research and development, and many conferences have been held in the past to present and discuss advances in this field.

Outstanding in the history of fluidization research are six previous international conferences: 1967 in Eindhoven (Holland), 1973 in Toulouse (France), 1975 in Asilomar (California), 1978 in Cambridge (England), 1980 in Henniker (New Hampshire), and 1983 in Kashikojima (Japan). The last four were organized and sponsored by the Engineering Foundation, New York.

The International Conference in Elsinore will thus be the fifth in the series of Engineering Foundation Conferences on fluidization. Its purpose will be to discuss the latest advances and applications in fluidization. A particular objective will be to bring together industrial and academic workers in order to promote the application of fundamental work to industrial problems and to stimulate fundamental work in fields of practical importance. This purpose is well-reflected in the 84 papers contained in this volume.

During preparations for the conference we have received valuable assistance from members of the Executive Committee, and in particular from Harold A. Comerer, Director of the Engineering Foundation. This is acknowledged with gratitude.

We are also grateful to the many referees who agreed to evaluate papers submitted for the Conference. Their assistance in the selection and editing of papers has been invaluable and has very markedly contributed to the high quality of the papers accepted for presentation. The referees are listed on pages 679–683 of this volume.

Our special thanks are due to Mrs. H. Wolff who most efficiently carried out all the secretarial work connected with the conference and who thus, for example, took care of the quite voluminous correspondence with authors and referees.

Finally, we wish to acknowledge the generous financial support received from the Engineering Foundation and from the many Danish sources listed on page iii. This support has decisively influenced the quality of the Conference.

Knud Østergaard   Ansgar Sørensen
Lyngby, Denmark
February 1986

HYDRODYNAMICS OF FLUIDIZATION OF SINGLE AND BINARY SIZE PARTICLES: SUPERCOMPUTER MODELING

by

Dimitri Gidaspow,
M. Syamlal and Y. Seo
Department of Chemical Engineering
IIT, Chicago, IL 60616

ABSTRACT

Computed formation, growth and bursting of a bubble in a two dimensional fluidized bed with a jet compared well with observations. Supercomputer generated movies of mixing of two sizes of particles also agreed with experiments qualitatively.

SCOPE

Fluidized beds are ideal for gasification and combustion of coal due to high rates of heat and mass transfer and solids mobility. However, one of the largest concerns when using fluidized beds to commercialize many chemical processes is scale-up. We believe this is due to the absence of an experimentally verified hydrodynamic theory that can describe the complicated transient gas and solid motion in a fluid bed. During the past few years several organizations began to develop hydrodynamic computer models that promise to be predictive in many respects. [1,2]

Our previous computer model calculated the pressure, the void fraction and the velocities of a single size solid and of the gas. Computed time averaged porosity distributions in two dimensional beds with a jet agreed with our measurements without the use of any fitted parameters. [3] Photographically determined bubble sizes compared well with the predicted sizes, [4]. Calculated gas velocity distributions also agreed with the experimental values measured at Westinghouse in a semi-circular bed with a jet. [5] Our model also predicted the high values of the heat transfer coefficients in a fluidized bed without the use of any turbulence in the model. [6] It was applied to predict gas compositions in the U-GAS process reactor [7].

CONCLUSIONS AND SIGNIFICANCE

The IIT hydrodynamic model was generalized to compute velocity fields for a mixture of particles of various sizes and densities. Mixing and segregation can be computed. The ultimate utility of such a model is for the design of fluidized bed reactors, dryers, etc. from first principles: from conservation of mass, species, momentum and energy for each phase with a minimum of empirical input consisting of drag relations, stresses, effective conductivities and reaction rates.

MATHEMATICAL MODEL

The differential equations are written in a form suggested by Soo [8]. They are a generalization of the single particle size model of Gidaspow and Ettehadieh [3]. The equations solved in the computer code developed by Syamlal [9] are the momentum equations for size k

$$\frac{\partial}{\partial t}(\rho_k \varepsilon_k \vec{v}_k) + \nabla \cdot (\varepsilon_k \rho_k \vec{v}_k \vec{v}_k) =$$

$$- \varepsilon_k \nabla P + \rho_k \varepsilon_k g - G_k \nabla \varepsilon_k + \sum_{l=1}^{N} K_{kl}(\vec{v}_l - \vec{v}_k). \qquad (1)$$

supplemented by the continuity equations,

$$\frac{\partial}{\partial t}(\rho_k \varepsilon_k) + \nabla \cdot (\varepsilon_k \rho_k \vec{v}_k) = 0 \qquad (2)$$

In Equations (1) and (2) ε_k is the volume fraction of phase k, ρ_k is the density, v_k the velocity, P the pressure, G_k is an elastic modulus for phase k arising from a force that resists compression and $K_{k\ell}$ is a friction coefficient for size k between ℓ "phases" in the sense used by S. L. Soo [8]. Figure 1 summarizes the constitutive relations used.

SINGLE PARTICLE SIZE FLUIDIZATION

The NSF sponsored Minnesota Cray-I was used to compute fluidization of 800 μm glass beads using a fine computational mesh corresponding to our experiment described in our Chem. Eng. Communications paper [4]. Figure 2 shows the computed formation, growth and bursting of the first bubble formed in a two-dimensional fluidized bed initially at minimum fluidization of 0.536 m/sec. At zero time a jet of a constant velocity of 5.77 m/sec was turned on. The jet width was 1.27 cm. The bed width was 40 cm. The initial height at minimum fluidization was 28 cm. The particle diameter was 800 μm with a density of 2.42 g/cm^3. In the computation we assumed symmetry and used a radial grid size of 0.635 cm and an axial grid size of 1.08 cm with a time step of 1. x 10^{-4} sec.

An experiment duplicating this computer run was also conducted. The observed bubble shape closely resembles the computed shapes shown in Figure 1. However, the experimental bubble size was about 30% smaller, probably due to the effect of the walls for this large bubble. For example, at 0.32 sec. the experimental bubble was 14.0 x 14.6 cm vs. 15.2 x 20.0 cm shown in Figure 2.

Figure 3 shows the computed gas flux into the bubble at its point of detachment. This figure was constructed based on the computed void fractions and velocities around the bubble periphery of a void fraction of roughly 0.8. Since the bubble acts as a short-circuit for gas flow, the flow into the bubble is much larger than that supplied by the jet alone. The jet flow is 577 cm/sec x 1.27 cm = 732 cm^2/sec compared to an inflow of 2426 versus an outflow of 1649. The computed "leakage" is clearly more complex than that used in the literature by P.N. Row and others [10]. For the computed bubble size of 98 cm^3 and the frequency of 1/0.4 sec. the conventional "leakage" is 98 cm^3/732 cm^3 x 0.4 = 0.33, a value that compares well to others, when one makes the same assumptions.

Supercomputer movies of the gas and solid velocity vectors give other details of the flow patterns. Syamlal [9] displays some of these. Bubble bursts can be compared to experimental results of Peters, et al. [11].

MIXING OF TWO PARTICLE SIZES

The computer code developed by Syamlal [9] for multiparticle size hydrodynamics was used to simulate mixing of two sizes of particles in our fluidized bed with a jet. As an aid in understanding color computer movies of volume fractions of the two sizes of the particles and vector plots of the velocities of the solids and the gas were made. In the experiment and in the simulation the large bottom layer particles were blue. Initially there was a 14.6 cm layer of 820 μm blue Ballotini at the bottom of the bed with a layer of 241 μm white glass beads on top. Gas was flowing at a velocity of 12.2 cm/sec. At time zero a central jet was turned on at a velocity of 5 m/sec and the grid flow was increased to 50.6 cm/sec. The simulation took 2 hours on Cray-I and corresponds to about 0.5 sec of real mixing time.

Figure 4 shows the density plots of the large (on the left) and the small particles (on the right) at 0.08 and at 0.4 seconds. At both times we see a bubble in the large particle layer but not in the small one. Instead we see a large bed expansion of the fine particle layer. Its height is close to that observed in the experiment. At 0.4 sec we see that the fines mixed with the larger particles. A more complete mixing takes longer. Further details such as motion in the wake seen in the computer movies, are given by Syamlal [9].

ACKNOWLEDGEMENT

This study was supported by the National Science Foundation under Grant No. CPE 8209290.

NOTATION

C_{Dk} Drag coefficient of phase k
d_k Characteristic diameter of particulate phase k (cm)
g Acceleration due to gravity (cm/sec^2)
G_k Solid stress modulus (gm/sec^2)
K_{kl} Generalized drag coefficient between phase k and phase l (gm/cm^3s)
N Total number of phases or particle sizes
P Gas Pressure, Pa

Re_k Reynolds number for particulate phase k
t Time(s)
u_k Velocity of phase k in r direction (cm/s)
v_k Velocity of phase k in z direction (cm/s)
$\vec{v}_k$ Velocity vector (cm/s)

Greek Symbols

ε_k Volume fraction of phase k
Φ_k Volume fraction of solids phase k at maximum packing
ϕ_k Characteristic shape factor of solids phase k
μ_1 Gas viscosity (gm/cms)
ρ_k Density of phase k (gm/cm^3)

LITERATURE CITED

1. Pritchett, et al., AIChE Symp. No. 176, 74, 134-148. (1978).
2. Klein, H.H. and Scharff, M.F., ASME Paper 82-FE-6 (1982).
3. Gidaspow, D. and B. Ettehadieh, I&EC Fund. 22, 193-201 (1983).
4. Gidaspow, D., Y.C. Seo, B. Ettehadieh, Chem. Eng. Comm. 22, 253-273 (1983).
5. Ettehadieh, B., D. Gidaspow and R.W. Lyczkowski, AIChE Journal 30, 529-36 (1984).
6. Syamlal, M. and D. Gidaspow, AIChE Journal 31, 127-135 (1985).
7. Gidaspow, D., B. Ettehadieh and J. Bouillard, AIChE Symp. No. 241, 80, 57-64 (1985).
8. Soo, S.L., "Fluid Dynamics of Multiphase Systems" Blaisdell, 1967.
9. Syamlal, M., "Multiphase Hydrodynamics of Gas-Solid Flow" Ph.D. Thesis, Illinois Institute of Technology, Chicago, IL (Aug. 1985).
10. Yates, J.G., P.N. Rowe and D.J. Cheesman, AIChE Journal, 30, 890-894 (1984).
11. Peters, M.H., L.-S. Fan and T.L. Sweeney, Chem. Eng. Sci. 38, 481-485 (1983).

Gas–Particle Friction

$$K_{1k} = K_{k1} = \frac{150(1-\epsilon_1)\epsilon_k\mu_1}{\epsilon_1(d_k\phi_k)^2} + \frac{1.75\rho_1|\mathbf{v}_1-\mathbf{v}_k|\epsilon_k}{(d_k\phi_k)} \qquad 0.2 \le \epsilon_1 < 0.8$$

$$= \frac{3}{4}C_{Dk}\frac{\epsilon_1|\mathbf{v}_1-\mathbf{v}_k|\rho_1\epsilon_k}{(d_k\phi_k)}f(\epsilon_1) \qquad 0.8 \le \epsilon_1 \le 1.0$$

where,

$$f(\epsilon_1) = \epsilon_1^{-2.65},$$

$$C_{Dk} = \frac{24}{Re_k}\left(1+0.15Re_k^{0.687}\right) \qquad Re_k < 1000$$

$$= 0.44, \qquad Re_k \ge 1000$$

and

$$Re_k = \frac{d_k|\mathbf{v}_1-\mathbf{v}_k|\rho_1\epsilon_1}{\mu_1}$$

Solids Stress

$$G_k = 0 \qquad \text{for } k=1$$

$$= g_1\epsilon_k\, exp\left[g_2\left(\epsilon^* - \epsilon_1\right)\right] \qquad \text{for } k = 2,\ldots,N$$

Particle-Particle Friction

$$\underset{k,l\neq 1}{K_{kl}} = F(1+e)\frac{\rho_k\rho_l\epsilon_k\epsilon_l(d_k+d_l)^2}{\rho_k d_k^3 + \rho_l d_l^3}|\mathbf{v}_k - \mathbf{v}_l|$$

where

$$F = \frac{3\epsilon_{kl}^{\frac{1}{3}} + (\epsilon_k+\epsilon_l)^{\frac{1}{3}}}{4\left(\epsilon_{kl}^{\frac{1}{3}} - (\epsilon_k+\epsilon_l)^{\frac{1}{3}}\right)}$$

$$\epsilon_{kl} = [(\Phi_k - \Phi_l) + (1-a)(1-\Phi_k)\Phi_l][\Phi_k + (1-\Phi_k)\Phi_l]\frac{X_k}{\Phi_k} + \Phi_l$$

$$\text{for} X_k \le \frac{\Phi_k}{\Phi_k + (1-\Phi_k)\Phi_l}$$

$$= (1-a)[\Phi_k + (1-\Phi_k)\Phi_l](1-X_k) + \Phi_k$$

$$\text{for} X_k \ge \frac{\Phi_k}{\Phi_k + (1-\Phi_k)\Phi_l}$$

$$a = \sqrt{\frac{d_l}{d_k}} \qquad \text{when} d_k \ge d_l$$

$$X_k = \frac{\epsilon_k}{\epsilon_k + \epsilon_l}$$

Φ_k, Φ_l – volume fractions at maximum packing

Figure 1. Constitutive Relations.

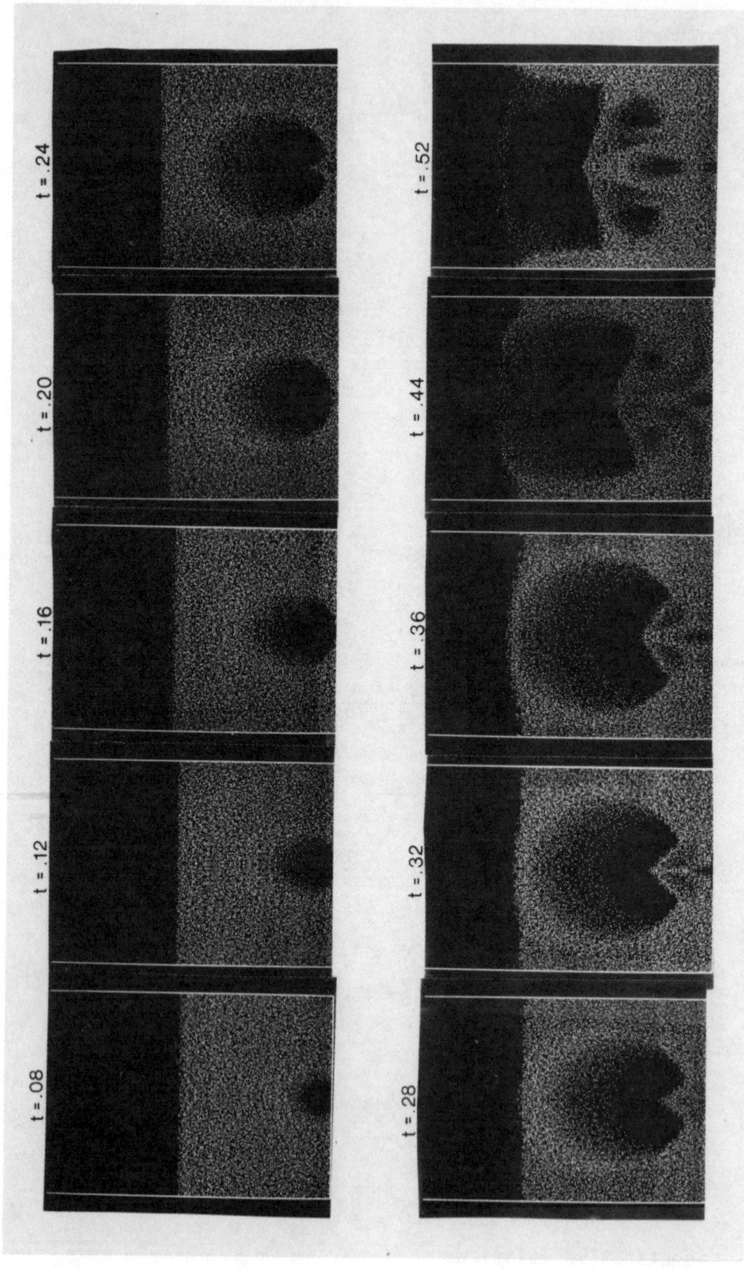

Figure 2. Computed Formation, Growth and Bursting of First Bubble in a 2-Dimensional Fluidized Bed with a Jet.

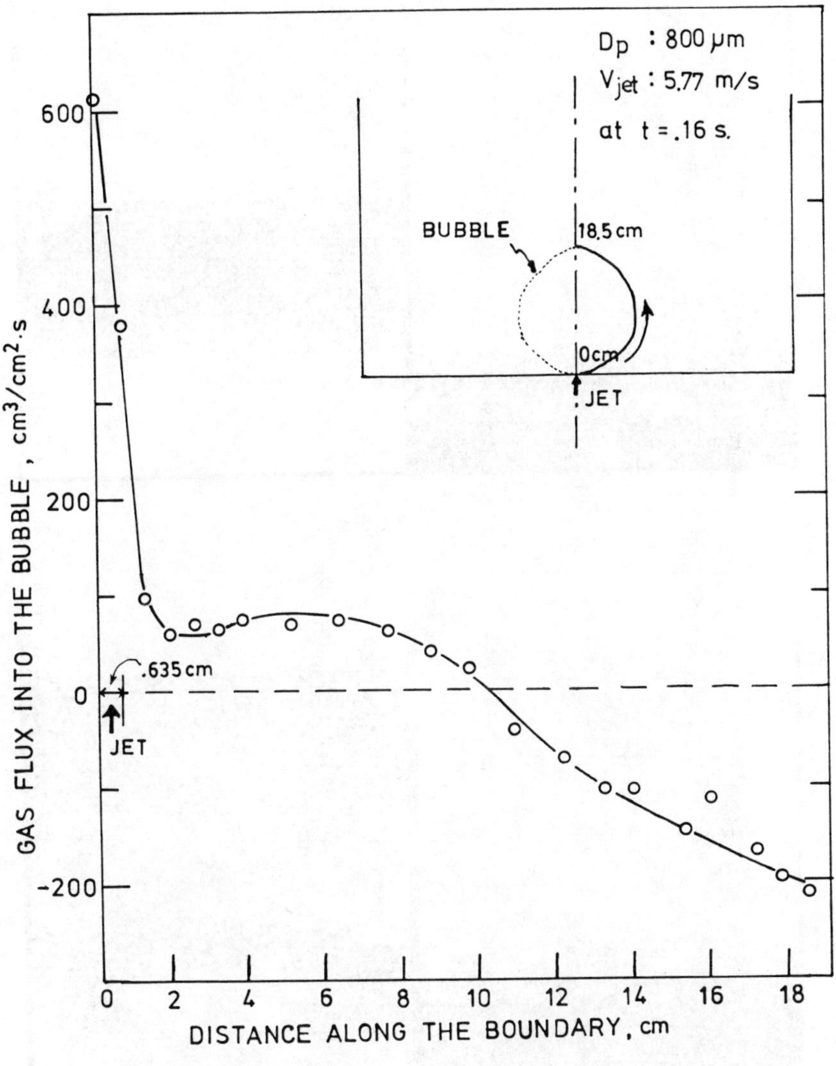

Figure 3. Computed Gas Flux into a Bubble at its Point of Detachment Corresponding to Conditions of Figure 2.

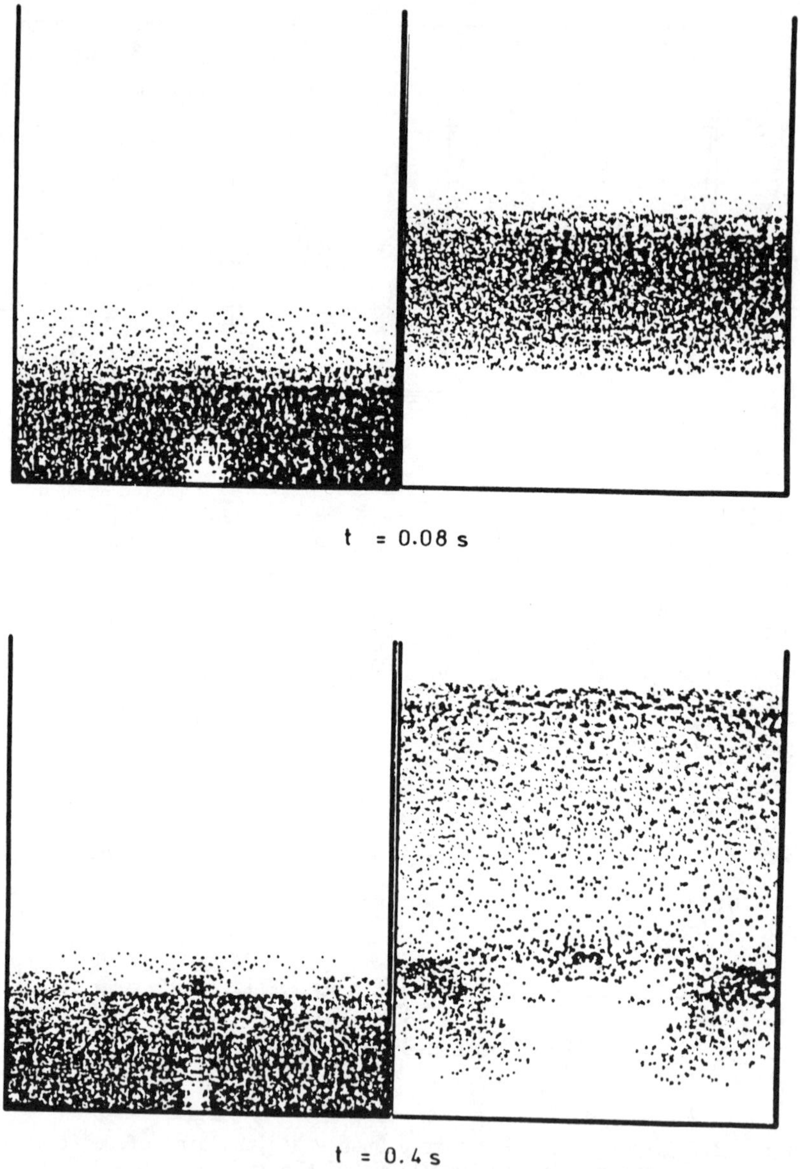

Figure 4. Mixing of Two Segregated Layers of Ballotini (ρ_p=2.94 g/cm^3 d_p=820 μm) and glass beads (ρ_p=2.42, d_p=241 μm) with a central jet at a velocity of 5 m/sec and grid flow stepped up from 12.2 to 50.6 cm/sec.

THE APPLICATION OF A PARTICULATE MODEL TO FLUIDISATION DYNAMICS

P.U. Foscolo°, L.G. Gibilaro+, S.P. Waldram+, R. Di Felice+

° Dipartimento di Chimica, Ingegneria Chimica e Materiali, University of L'Aquila, 67100 L'Aquila, Italy

+ Department of Chemical and Biochemical Engineering, University College London, London WC1E 7JE, UK

ABSTRACT

A one dimensional hydrodynamic model of a fluidised bed is derived from particle force balance considerations. It yields both the general two phase stability criterion of Wallis in explicit form, and numerical solutions relating to the growth and decay of voidage perturbations that are in accord with reported experimental observations.

SCOPE

In previous papers we have developed a general expression for the hydraulic interactions between particles and fluids (1). This applies to fixed and fluidised beds of particles for all voidages and flow regimes and has enabled friction factor relations to be specified in closed form. It has also led to the development of a unified model for particulate expansion of fluidised beds and flow in fixed porous media (2), and has been the key factor in defining a specific form of the predictive criterion for transition between particulate and aggregate fluidisation (3). This criterion was due to Wallis (4), and proposes that the stability of two phase flow is governed entirely by the relative magnitude of the velocity of propagation of a concentration disturbance within the bed u_ε (the continuity wave) and the elastic (or dynamic) longitudinal wave velocity u_e.

In this paper we employ one dimensional continuity and momentum equations for a fluidised bed using the specific form of drag force on a single particle proposed in (3). Previous, and more complex, models have used drag force expressions appropriate only to the creeping flow regime thereby limiting applications to conditions under which fluidised beds seldom operate (5,6): this work has in general concentrated on the mathematical and computing aspects rather than the formulation of system equations and it is therefore not altogether surprising that known patterns and trends in bed behaviour have not been faithfully reproduced (7,8).

The activity in this field of research is well reviewed by Rietema (9).

In our approach the array of particles is viewed as an elastic structure which will respond to changes in fluid velocity within it; the fluid is the medium through which forces are transmitted from a particle to its adjacent neighbours. Continuity and momentum equations are therefore formulated for the particles alone in terms of the physical variables of the system, the void fraction and the particle slip velocity. We consider an idealised bed in the sense that wall effects are neglected, the fluidising medium is Newtonian and incompressible and the particles are monosized spheres. Inertial effects are ignored, an assumption which has been shown to be reasonable for a specific fluidised bed (10). Our one dimensional model assumes that representative fluid and particle velocities may be expressed as the functions of bed height and time alone.

By imposing a small disturbance of known form and seeing whether this grows or decays with time, a general stability criterion is developed. This is identical to the condition proposed by Wallis and referred to above. If the imposed disturbance grows this leads to the development of a shock front that represents a transition from particulate to aggregate fluidisation. The variation with time of voidage within the bed following imposed disturbances has been studied by solving the model equations of continuity and momentum by the method of characteristics (12).

CONCLUSIONS AND SIGNIFICANCE

The proposed model of a fluidised bed is comparatively simple and yet faithfully reproduces aspects of fluidised bed behaviour which more complex models are unable to do. It predicts that bubbles will form in a bed whenever the velocity at which voidage disturbances propagate is greater than the elastic wave velocity for the bed itself. This stability criterion has previously been tested (3) and is in excellent agreement with reported experiments. Preliminary results indicate that after the onset of bubbling dense phase voidages do remain at essentially the value corresponding to the minimum fluidisation velocity thus giving qualitative support to the two phase theory of fluidisation (13).

MODEL DEVELOPMENT

The drag force exerted on a particle by a fluid is function of the 'slip velocity', that is the relative velocity between the fluid and particle. Following previous work, (3), we have

$$F_d = F_{dt} \left(\frac{u_o - u_p}{u_t}\right)^{4.8/n} \varepsilon^{-3.8} \tag{1}$$

If the particles are in a state of dynamic equilibrium then from (2)

$$F_d = F_{dt}\,\varepsilon \qquad \left(\frac{u_o - u_p}{u_t}\right) = \varepsilon^n \tag{2}$$

Equation (2) is a general voidage-velocity relationship which for the case of stationary particles reduces to the well known Richardson - Zaki equation (11). By formulating the equations of motion for an element of bed height Δz, see Figure 1, we obtain:

a) The Continuity Equation

$$\frac{\partial}{\partial z}\{(1-\varepsilon)\,u_p\} = \frac{\partial \varepsilon}{\partial t} \tag{3}$$

b) The Momentum Balance Equation

$$(1-\varepsilon)\,\rho_p u_p \frac{\partial u_p}{\partial z} + (1-\varepsilon)\,\rho_p \frac{\partial u_p}{\partial t} = \Sigma f_i \tag{4}$$

The right hand side of Equation (4) will account for all forces acting on the particles, that is particle weight, $-(1-\varepsilon)\rho_p g$, flow resistance, $6(1-\varepsilon)F_d/\pi d^3$, pressure gradient force due to the bed density, $(1-\varepsilon)\bar{\rho}g$, and the force due to the net stress gradient resulting from the elastic property of the bed structure

$$-\frac{\partial P'}{\partial z} = -\frac{\partial P'}{\partial \rho_b}\frac{\partial \rho_b}{\partial z} = u_e^2\,\rho_p\frac{\partial \varepsilon}{\partial z} \tag{5}$$

Incorporating all these into Equation (4), we have

$$\rho_p\left(u_p\frac{\partial u_p}{\partial z} + \frac{\partial u_p}{\partial t}\right) = \frac{u_e^2\,\rho_p}{(1-\varepsilon)}\frac{\partial \varepsilon}{\partial z} + \frac{6\,F_d}{\pi\,d^3} - (\rho_p - \bar{\rho})\,g \tag{6}$$

Equations (3) and (6) thus form a pair of non-linear first order partial differential equations in u_p and ε. For some purposes it is convenient to express these in dimensionless form by introducing new variables $\hat{z}$, $\hat{t}$, $\hat{u}_p$ and $\hat{u}_e$ as defined in the notation.

CORRESPONDENCE WITH THE STABILITY CRITERION DUE TO WALLIS

Consider a uniform fluidised bed with voidage ε_o and superficial fluid velocity u_o and imagine that the time averaged position of particles is stationary (u_p=0). Let a small voidage perturbation $\varepsilon = \varepsilon_o + \varepsilon'$ be imposed on the bed: substituting this expression into (6) and taking the derivative with respect to z results in terms involving $\partial u_p/\partial z$: these may be eliminated using Equation (3). After linearization and further manipulation we follow the approach of Wallis (4) and obtain

$$\frac{\partial^2 \varepsilon'}{\partial t^2} - u_e^2\frac{\partial^2 \varepsilon'}{\partial z^2} + B\left(\frac{\partial \varepsilon'}{\partial t} + u_\varepsilon\frac{\partial \varepsilon'}{\partial z}\right) = 0 \tag{7}$$

where

$$B = \frac{-6}{\pi d^3}\left(\frac{1}{\rho_p}\frac{\partial F}{\partial u_p}\right) \qquad u_\varepsilon = -(1-\varepsilon)\frac{\partial u_p}{\partial \varepsilon}\bigg|_{F=0} \tag{8}$$

In Equation (7) B is a positive quantity which represents "damping". If the imposed voidage disturbance is assumed to have the form

$$\varepsilon' = e^{\{at + i\omega(t - z/v)\}} \tag{9}$$

we have a wave with velocity v and frequency $\omega/2\pi$. This will grow or decay depending on the sign of a. By definition a and ω are both real.

On differentiating Equation (9), substituting into (7) and separating real and imaginary parts we obtain

$$a = \frac{B}{2}\left(\frac{u_\varepsilon}{v} - 1\right) \qquad \omega^2 = \frac{B^2}{4}\left(\frac{u_\varepsilon^2 - v^2}{v^2 - u_e^2}\right) \tag{10}$$

For $a<0$ a decaying voidage disturbance results and as $\omega^2>0$ then $u_\varepsilon \leqq u_e$. This is identical to Wallis' stability criterion (4). In (3) explicit expressions for u_ε and u_e are derived and this criterion is tested against a variety of experimental data for the onset of bubbling in fluidised beds and found to be in excellent agreement with reported observations.

SOLUTION OF THE SYSTEM EQUATIONS

The dimensionless forms of Equations (3) and (6) constitute a pair of hyperbolic partial differential equations which may be solved by the method of characteristics (12). For unstable beds solutions will only be possible up to the development of a shock front which represents the formation of a bubble. Consider particles of density 1500 kg/m^3 fluidised by air at atmospheric conditions and let a voidage disturbance of $\alpha \cdot \sin 2\pi\hat{z}/100$ be superimposed on a uniformly expanded bed of voidage ε_o. It is also assumed that at time t=0 Equation (2) holds at any bed height.

Case 1. Particles of diameter 610 μm, α=0.01

a) ε_o=0.75. Figure 2 shows the first family of characteristic curves for this situation: note that the first and the last characteristics are of the same form showing the periodicity of the imposed disturbance. A shock front is clearly developing as illustrated by the crossing of some characteristics among each other. The dashed lines give the locus of points where u_p and ε are successively calculated. Figure 3 shows a picture of voidage variation with distance at specific times for an unbounded bed: to the left the initial imposed voidage disturbance and to the right the form to which it has grown after approximately 0.13 sec, corresponding to $\hat{t}$=280. In the computer program a lower limit for ε has been introduced: the minimum fluidisation voidage chosen arbitrarily as 0.4.

b) ε_o=0.45. The uniform expansion is clearly unstable also in this case, as shown on Figure 4, although the shock front has not yet been reached at t=0.13 sec.

In both situations a) and b) some portions of the bed are at the minimum voidage, whilst other regions are at voidages as high as 0.85 and 0.60 respectively. These numerical results agree qualitatively with the two phase theory of fluidisation which states that fluid flow in excess of that required for minimum fluidisation will pass through the bed as bubbles (13).

Case 2. Particles of 61 μm, α=0.02

a) ε_o=0.45. By way of contrast consider a class A powder on the Geldart classification (14), which displays significant particulate expansion prior to the onset of aggregate behaviour: this transition would be anticipated at a voidage of 0.53 so that with ε_o=0.45 a stable bed should result. Figures 5 and 6 illustrate exactly this. In Figure 5 the characteristics are essentially straight and parallel and correspond to the bed voidages shown in Figure 6. In this case the presence of the distributor plate at $\hat{z}$=0 is taken into account by considering the boundary condition $u_p(\hat{z}=0)=0$ at any time (the first layer of particles is always attached at the distributor). It is clear that the imposed perturbation has been displaced upwards leaving behind an uniformly expanded bed at ε_o=0.45.

b) ε_o=0.75. In this case the voidage is in excess of that corresponding to the onset of aggregate fluidisation and instability is anticipated. Figure 7 confirms this and shows, as in case 1.a), the crossing of two characteristics of the same family. Figure 8 shows how the initial disturbance will have developed in 0.1 sec in an unbounded bed. It is noticeable that the densest part of the bed reaches the minimum fluidisation voidage, i.e. a lower value than the minimum bubbling voidage: this agrees with the experimental observation of a dense phase contraction as the fluidisation velocity is increased above the value corresponding to the onset of bubbles (15).

In principle a wide range of systems can be simulated using the techniques we have reported and a voidage disturbance with any shape can be imposed on the bed. Results of further simulations will be presented.

ACKNOWLEDGMENTS

This work was supported by the Consiglio Nazionale delle Ricerche (Piano Finalizzato Energetica 2) and also by the Science and Engineering Research Council (Specially Promoted Programme in Particulate Technology).

NOTATION

a defined in Eqn (9)
B defined in Eqn (8)
d particle diameter
f_i defined in Eqn (4)
F $F_d - \pi d^3 \varepsilon(\rho_p - \bar{\rho})g/6$
F_d drag force on a single particle

F_{dt} drag force at terminal velocity
g acceleration due to gravity
n Richardson-Zaki parameter
P' net stress
t time
$\hat{t}$ $t\,u_o/\,d$
u_ε continuity wave velocity
u_e elastic wave velocity
$\hat{u}_e$ $u_e\,/\,u_o$
u_o superficial velocity of fluid approaching the fluidised bed
u_p particle velocity
$\hat{u}_p$ $u_p\,/\,u_o$
u_t terminal settling velocity
v voidage wave velocity
z bed height
$\hat{z}$ $z\,/\,d$
α amplitude of voidage disturbance
ε voidage
ε' imposed voidage disturbance
ε_o mean value of imposed voidage disturbance
$\bar{\rho}$ suspension density
ρ_b particle bed density $\{(1-\varepsilon)\rho_p\}$
ρ_p particle density
ω voidage wave frequency

LITERATURE CITED

1. Gibilaro,L.G.,Di Felice,R.,Waldram,S.P.,Foscolo,P.U.,Chem.Engng Sci. In press

2. Foscolo,P.U.,Gibilaro,L.G.,Waldram,S.P.,Chem.Engng Sci.,38,1251(1983)

3. Foscolo,P.U.,Gibilaro,L.G.,Chem.Engng Sci.,39,1667(1984)

4. Wallis,G.B.,One-dimensional two-phase flow,McGraw Hill,New York(1969)

5. Fanucci,J.B.,Ness,N.,Yen,R.H.,J.Fluid Mech.,94,353(1979)

6. Needham,D.J.,Merken,J.H.,J.Fluid Mech.,131,427(1983)

7. Needham,D.J.,Merken,J.H.,J.of Engng Math.,18,119(1984)

8. Ettehadieh,E.,Gidaspow,D.,Lyczkowski,R.W.,AIChE J.,30,529(1984)

9. Rietema,K.,Chem.Engng Sci.,37,1125(1982)

10. Gibilaro,L.G.,Waldram,S.P.,Foscolo,P.U.,Chem.Engng Sci.,39,607(1984)

11. Richardson,J.F.,Zaki,W.N.,Trans.Instn Chem.Engrs,32,35(1954)

12. Aris,R.,Amundson,N.R.,Mathematical Methods in Chemical Engineering, Vol.2,Prentice-Hall,New Jersey(1973)

13. Toomy,R.D.,Johnstone,H.F.,Chem.Engng Prog.,48,220(1952)

14. Geldart,D.,Powder Technol.,7,285(1973)

15. Barreto,J.F.,Yates,J.G.,Rowe,P.N.,Chem.Engng Sci.,38,345(1983)

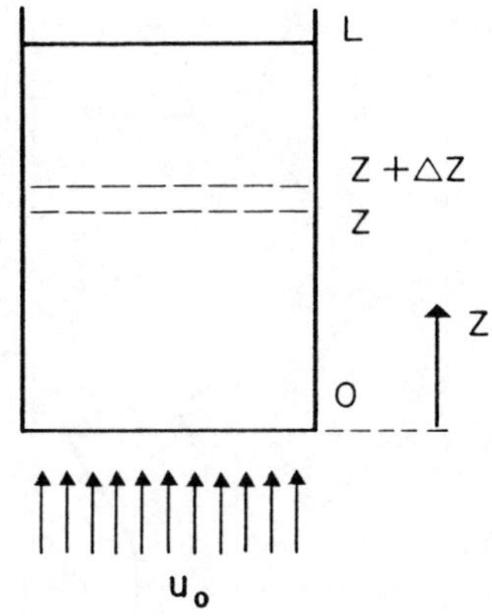

Figure 1. The fluidised bed

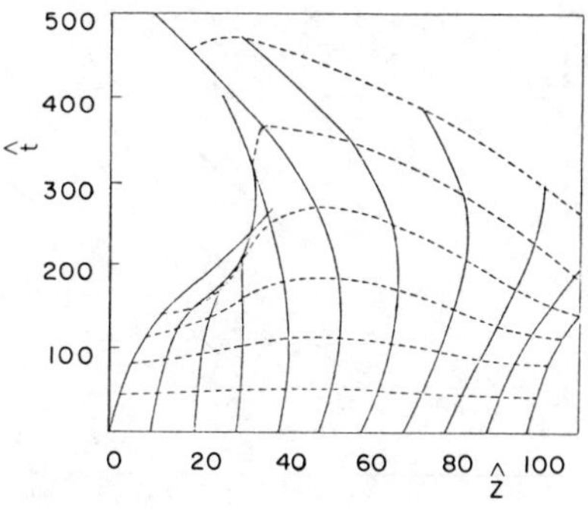

Figure 2. Case 1.a):610 µm particles, ε_o=0.75 - Characteristic curves of the first family

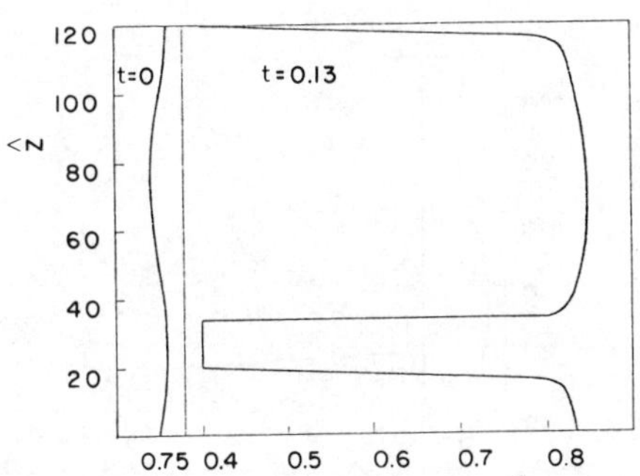

Figure 3. Case 1.a):610 µm particles, ε_o=0.75 - Initial imposed voidage disturbance and the form to which it has grown after 0.13 sec

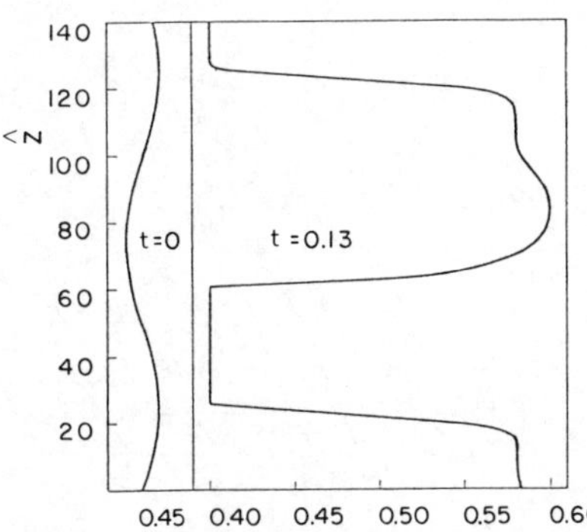

Figure 4. Case 1.b):610 µm particles, ε_o=0.45 - Initial imposed voidage disturbance and the form to which it has grown after 0.13 sec

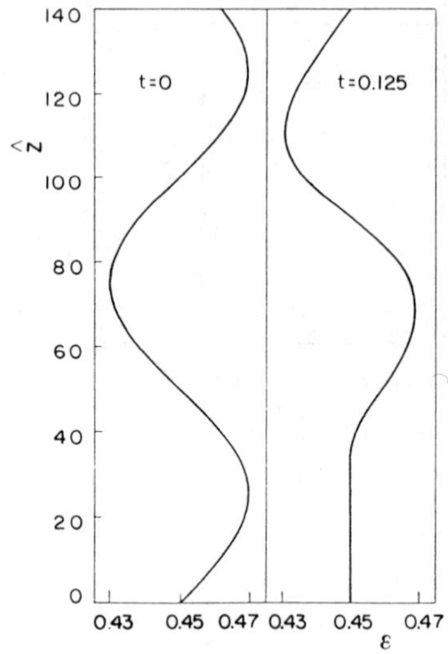

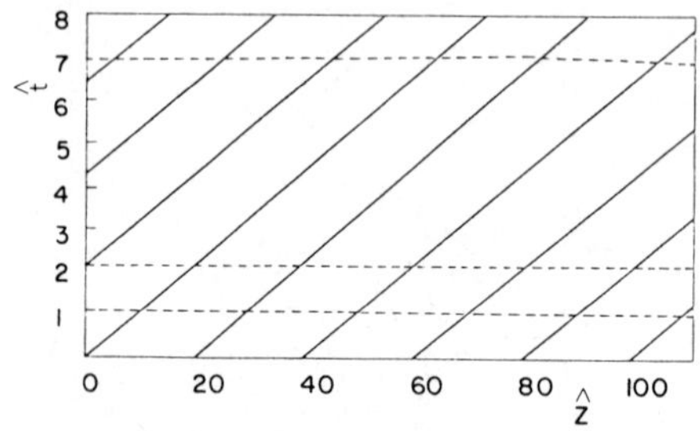

Figure 5. Case 2.a):61 μm particles, ε_o=0.45 - Characteristic curves of the first family

Figure 6. Case 2.a):61 μm particles ε_o=0.45 - Initial voidage disturbance and the form to which it has developed after 0.125 sec

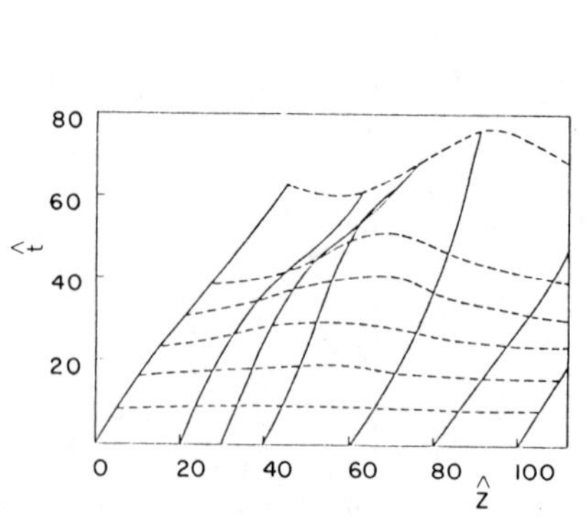

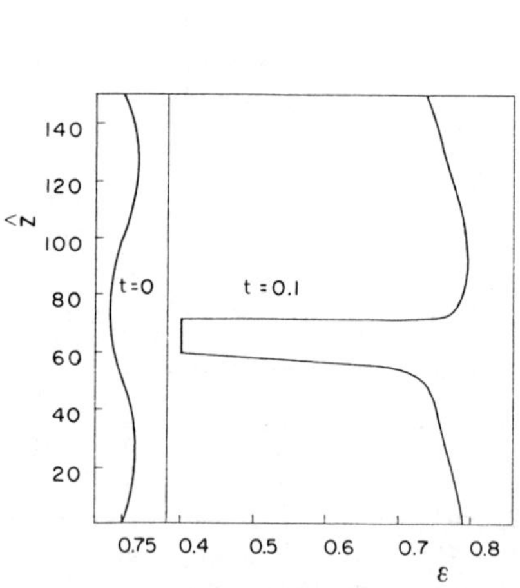

Figure 7. Case 2.b):61 μm particles, ε_o=0.75 - Characteristic curves of the first family

Figure 8. Case 2.b):61 μm particles, ε_o=0.75 - Initial imposed voidage disturbance and the form to which it has grown after 0.1 sec

STATISTICAL MECHANICS OF FLUIDIZED PARTICLES

Michael H. Peters
Department of Chemical Engineering
Rensselaer Polytechnic Institute
Troy, N.Y. 12180

ABSTRACT

In the present work, a statistical mechanical derivation of the particle phase momentum conservation equation for fluidized particles is given. The diffusion flux of particle momentum is shown to be composed of "kinetic" and "configurational" quantities which depend upon the singlet and pair distribution functions for the particles. Peculiarities of fluidized particles relevant to the evaluation of such distribution functions are also discussed.

SCOPE

Previous treatments of the transport phenomena of fluid-particle systems in which the particle phase is highly concentrated (say, a volume fraction greater than 0.1) have been based upon well-known volume-averaging techniques. This technique has many shortcomings, the most important of which is the inherent unresolved nature of the property diffusion flux vectors. Such shortcomings can be overcome by statistical mechanical treatments of the fluid-particle suspension.

CONCLUSIONS AND SIGNIFICANCE

The "closure" hypothesis employed in the particle phase momentum conservation equation for fluidized particles based upon the division of the momentum diffusion flux vector into kinetic and configurational terms are shown to be, in part, justified by statistical mechanical formulations. Expressions for the diffusion flux vectors rely upon knowledge of the singlet and pair distribution functions for the fluid molecules and particles, which are, in principle, obtainable from solutions to the appropriate contracted forms of the Liouville equation. Uncommon and complicating features of fluidized particles include non-Maxwellian behavior in particle velocity distributions, and intermediate particle Reynolds number flows.

INTRODUCTION

The popular volume-averaging techniques used to derive the two-phase transport equations for fluidized particles (see, e.g., Anderson and Jackson (1)) suffer from the drawback that the diffusion flux vectors of a property ψ_i of the fluid or particle phase remain unresolved. Proper closure of the transport equations demands that the diffusion flux of the property ψ_i (mass, momentum, or energy flux in relative coordinates) be expressed in terms of the relevant macroscopic dependent variables, such as mass density, velocity, and temperature. Even in the case of pure fluids, mathematical elaboration of the diffusion flux vectors based upon purely deterministic reasoning is formidable, and relys heavily upon reasonably established experimental procedures. In dispersed phase systems, purely deterministic reasoning could easily lead to gross inadequacies in predicting diffusion flux relationships as evident, for example, by the complicated rheological behavior of colloidal systems (see, e.g., the review by Hirtzel and Rajagopalan (2)). Furthermore, experimental methods in dispersed phase systems are hindered by the unknown intrusive nature of traditional apparatus (see, e.g., the review by Schügerl (3)). Although, recently studied nonintrusive techniques provide an attractive means of determining transport properties, provided that the relevant functional relationships for the diffusion flux vectors are properly written (see, e.g., Homsy, et al. (4)).

An alternative theoretical approach to the transport phenomena of fluidized particles, with demonstrated success in not only pure fluids, but polymeric and colloidal solutions as well, is based upon statistical mechanical formulations. Briefly, in statistical mechanical theories one seeks to determine the probability of finding one or more particles and fluid molecules with a given set of coordinates and momenta (called "phase space"). There are, of course, many advantages to such an approach. For example, the ready determination of the transport equations from the phase-space conservation equation of Liouville, based upon a system ensemble. This negates the use of ill-defined volume-averaging techniques, and can accomodate "rarefied" particle dynamics, analogous to rarefied gas dynamics, in which continuum hypotheses breakdown.

The potential ability of statistical mechanical theories applied to fluidized particles is evident from diffusion flux vector "modelling" based upon dense gas kinetic theory formalism (see, e.g., Lun et al. (5); Peters and Nadkarni (6)). For example, the total particle phase pressure tensor (i.e., the diffusion flux of particle momentum), appearing in the particle phase momentum balance equation, may be written as the sum of two components (also see Rietema (7)): (i) a particle configurationally dependent quantity, due to the force exerted by the surrounding fluid phase, and represented by the fluid pressure tensor, and (ii) a kinetic component, dependent upon the local particle velocity distribution function, in particular, the variance in that distribution. This latter quantity is directly related to the mean peculiar kinetic energy of the particles, and has been termed the "granular temperature", albeit it is not a thermometric quantity. In the simplest approach the kinetic contribution may be "modelled" by the assumption of hard collisions between particles moving in a vacuum; thus, introducing a particle phase "energy" equation which describes the macroscopic variations in the granular temperature (Lun et al. (5)). Although the above formalism is not rigorous, calculations of the granular temperature variations associated with the steady motion of a single bubble (based upon singular perturbation solutions to the energy equation) yield theoretical particle phase viscosity coefficients in remarkably close agreement with experimental measurements (in the range 2 to 10 poise) (Peters and Nadkarni (6)).

Although the above mentioned studies indicate some potential in dense gas kinetic theory formalism for certain types of problems, a rigorous analysis of the problem, including the conditions under which simpler approaches are applicable, is contained within the framework of statistical mechanics. As an example of statistical mechanical formulations of particle phase transport, in the next section we consider the results of such an approach for the particle phase momentum conservation equation, including the expressions for the pressure tensor briefly discussed above.

STATISTICAL MECHANICAL THEORY OF PARTICLE PHASE MOMENTUM TRANSPORT

Following Curtiss et al. (8) the fluid and particle phase transport equations can be rigorously derived from the Liouville equation, written for any dynamical variable B as

$$\frac{\partial}{\partial t} \langle B \rangle = \langle LB \rangle \tag{1}$$

where **L** is the Liouville operator, and the angular brackets denote an ensemble average taken over N particles and M fluid molecules. The momentum equation for the particles can be obtained by taking as the dynamic variable the total particle momentum in the system, i.e.,

$$B = \sum_{\beta=1}^{N} \underset{\sim}{p}_c^{\beta} \, \delta(\underset{\sim}{r}_c^{\beta} - \underset{\sim}{r}) \tag{2}$$

where $\underset{\sim}{p}_c^{\beta}$ is the momenta of particle β with position coordinate $\underset{\sim}{r}_c^{\beta}$, and δ is the "delta" function. The quantity B, therefore, represents the total particle momentum at position $\underset{\sim}{r}$ for one system of the ensemble. Substitution of Eq. (2) into Eq. (1) gives, after some manipulation, the momentum conservation equation for the particles as

$$\frac{\partial}{\partial t}(\rho_p \underset{\sim}{v}_p) + \frac{\partial}{\partial \underset{\sim}{r}} \cdot (\rho_p \underset{\sim}{v}_p \underset{\sim}{v}_p) = - \frac{\partial}{\partial \underset{\sim}{r}} \cdot \underset{\approx}{\Pi}_k^p + \underset{\sim}{S}_{pp} + \underset{\sim}{S}_{ps} \tag{3}$$

where,

$$\underset{\approx}{\Pi}_k^p = m_p \int \underset{\sim}{V}_p \underset{\sim}{V}_p f_p (\underset{\sim}{r},\underset{\sim}{p},t)d\underset{\sim}{p} \tag{4}$$

$$\underset{\sim}{S}_{pp} = \int \underset{\sim}{F}_{pp} \psi_{pp} (\underset{\sim}{r},\underset{\sim}{r}'_b,t)d\underset{\sim}{r}'_b \tag{5}$$

$$\underset{\sim}{S}_{ps} = \int \underset{\sim}{F}_{ps} \psi_{ps} (\underset{\sim}{r},\underset{\sim}{r}_a,t)d\underset{\sim}{r}_a \tag{6}$$

In the above equations $\underset{\sim}{V}_p$ is the peculiar velocity of a particle (i.e., the velocity relative to the mass average velocity), f_p is the singlet particle distribution function, ψ_{pp} is the pair distribution function for the particles in configuration space, ψ_{ps} is the pair distribution function for one fluid molecule and one particle in configuration space, $\underset{\sim}{F}_{pp}$ is the net force exerted on a particle at position $\underset{\sim}{r}$ due to a particle at position $\underset{\sim}{r}'_b$, and F_{ps} is the net force exerted on a particle at position $\underset{\sim}{r}$ due to a fluid molecule at position $\underset{\sim}{r}_a$ (for further details see Curtiss et al. (8)). The previously proposed division of the total particle phase pressure tensor into a kinetic contribution (as in Eq. (4) above) and a configurational contribution (as in (Eq. (5) above) appears reasonable (for example, compare Eqs. (3)-(6) above to Eq. (33) of Rietema and Van den Akker (9)).

Ultimately, a complete elaboration of the expressions for $\underset{\approx}{\Pi}_k^p$, $\underset{\sim}{S}_{pp}$, and $\underset{\sim}{S}_{ps}$ requires specification of both the singlet and pair distribution functions appearing in Eqs. (4)-(6), which are obtainable from solutions to the appropriate contracted forms of the Liouville equation, Eq. (1), along with so-called closure "ansatz" conditions. For example, in the case of the "hard" sphere approximation employed in kinetic theory models mentioned earlier, the singlet distribution function, f_p, is given by the Chapman-Enskog expansion which ultimately yields the familiar Newtonian form for the pressure tensor $\underset{\approx}{\Pi}_k^p$ in Eq. (4). It is also noteworthy that this non-Maxwellian behavior of the velocity-space distribution function for the particles is uncommon to many studies on solutions to contracted Liouville equations, such as those which arise in polymeric and colloidal systems. Theoretical treatments of the fluid molecule-particle interaction force $\underset{\sim}{S}_{ps}$, based upon fluid molecule and particle distribution functions, have been, so far at least, developed only for sufficiently dilute particle systems in which hydrodynamic interactions are negligible. Another complicated and characteristic feature, particularly of gas fluidized particles, is intermediate particle Reynolds number behavior, for which theoretical hydrodynamic treatments have not been developed; although, it is most likely that such "intractable" conditions are responsible for the evidently successful results of dense gas kinetic theory formalism. In any event, the above-mentioned peculiarities for fluidized particles pose some rather interesting theoretical challenges for statistical mechanical descriptions.

LITERATURE CITED

1. Anderson, T.B. and Jackson, R., Ind. Eng. Chem. Fundls., 4, 527 (1967).

2. Hirtzel, C.S. and Rajagopalan, R., Colloidal Phenomena. Advanced Topics, Noyes Publications, New Jersey (1985).

3. Schügerl, K., in Fluidization (J. Davidson and D. Harrison, eds.), Academic Press (1971).

4. Homsy, G.M., El-Kaissy, M.M., and Didwania, A., Int. J. Multiphase Flow, 6, 305 (1980).

5. Lun, C.K.K., Savage, S.B., Jeffrey, D.J., and Chepurniy, N., J. Fluid Mech., 140, 223 (1984).

6. Peters, M.H. and Nadkarni, A.R., "Asymptotic Solutions to the Particle Phase Energy Balance Equation for Single Bubble Motion in a Fluidized Bed", paper presented at the 16th Annual Fine Particle Society Meeting, Miami, Fla. (1985).

7. Rietema, K., Chem. Eng. Sci., 37, 1125 (1982).

8. Curtiss, C.F., Bird, R.B., and Hassager, O., Adv. Chemical Phys., 35, 31 (1976).

9. Rietema, K., and Van den Akker, H.E.A., Int. J. Multiphase Flow, 9, 21 (1983).

A NUMERICAL STUDY OF THE EFFECT OF OPERATING CONDITIONS ON THE MINIMUM FLUIDIZATION VELOCITY

Atul Padhye
West Virginia University,
Morgantown, West Virginia 26506

Thomas J. O'Brien
U.S. Department of Energy
Morgantown, West Virginia 26505

ABSTRACT: Mass and momentum conservation equations for gas-solid flows are solved numerically to determine the effect of the operating conditions on the minimum fluidization velocity of gas fluidized beds of Geldart Group B and Group D particles. Effect of bed operating temperature and pressure are studied. The numerical predictions are compared with the engineering correlations and experimental observations.

SCOPE: Coarse particle fluidized beds, operating at high pressure and temperature, are being considered for coal gasification. The minimum fluidization velocity U_{mf} is one of the basic parameters of any fluidized bed. Engineering correlations are used for estimating effect of operating conditions on U_{mf}. Unfortunately, most of these correlations were based on the experimental data for fine particle fluidized beds relevant to catalytic cracking. The object of this study was to use numerical techniques for predicting these effects, and to validate the drag function to be used in numerical simulation of more detailed properties of fluidized beds for which reliable engineering correlations are not available.

CONCLUSIONS AND SIGNIFICANCE: Numerically, it is shown that, for fluidized beds of Geldart Group B particles, the minimum fluidization velocity decreases with increasing operating pressure and temperature. On the other hand, the minimum fluidization velocity increases with increasing pressure for the beds of Geldart Group D particles. The numerical calculations, unlike traditional engineering correlations, do not need a priori knowledge of minimum fluidization void fraction to predict the minimum fluidization velocity, and hence, are more versatile in their predictive capability. Also, since the basic theory behind the numerical procedure is more general, it is not restricted to any specific particle size category. Further, these calculations demonstrate the reliability of the drag function formulation, which is critical in simulating more detailed properties of fluidized beds, for which reliable engineering correlations are not available.

INTRODUCTION

A theoretical formulation was developed by Anderson and Jackson (1) for the dynamics of fluidized-beds, based on a continuum description of the gas and the solids phases. Both of these phases, solid and gas, are represented as interpenetrating Eulerian fields coupled by a "drag" coefficient which allows for momentum exchange between the two phases.

A set of coupled partial differential equations for the motion of these phases was derived, based on the conservation of mass and momentum. These equations have been the basis for the development of two numerical codes for the simulation of fluidized-bed dynamics, CHEMFLUB [Blake (2)], the code used in the work discussed in this paper and a code developed at the Illinois Institute of Technology [Gidaspow (9)] as a variation of the K-FIX code [Rivard (12)]. In the development of the CHEMFLUB code, several approximations were made, in addition to those of the continuum theory. A complete discussion of the actual set of equations encoded in CHEMFLUB can be found in Blake, et al. (2). Using this basic code, with some modifications (particularly to the numerical representation of the drag coefficient) this computational study explores the effects of operating conditions on the calculated minimum fluidization velocity U_{mf} in order to demonstrate the validity of this code in predicting this most fundamental fluidization property. The effects of temperature, pressure, and particle size were examined, and the computed predictions compared with experimental data and engineering correlations. The close relationship between these correlations and the formulation of the drag coefficient was explored in order to verify this constitutive function. Because of the complexity of these calculations, they would certainly not be used merely to determine U_{mf}. However, this study demonstrates the reliability of the drag coefficient, which is essential in predicting details of all other aspects of fluidization, for which there are few correlations.

MATHEMATICAL FORMULATION

Although the gas-solid continuum theory provides a very general set of equations representing the fundamental conservation laws, the developers of the CHEMFLUB code assumed that the gas phase inertia is negligible compared with that of the solid phase, which resulted in an algebraic representation of the gas phase momentum equation,

$$\bar{\nabla}P = \frac{B(\theta)}{1 - \theta}(\bar{U}_s - \bar{U}_g). \tag{1}$$

This implies that the gas field adjusts instantaneously to the solids flow field and the pressure variations. This assumption greatly simplifies the numerical solution of the remaining coupled partial differential equations. Within the limits of these assumptions, this equation is further affected by the accuracy of the representation of $B(\theta)$.

A revised formulation of this coefficient based on the experimental results of Richardson (11) and the analysis of Blake (2) was developed [Padhye (10)], which brought the calculated values of $\bar{U}_{mf}$ into agreement with experiment. Richardson measured the drag force on a bed of particles and its variation with voidage, deducing a relationship between the friction factor $\psi = R/(\rho_g|\bar{U}_g|^2)$ and a modified particle Reynolds number. Richardson's experiments showed that the friction factor was a function only of the Reynolds number. This relationship was fitted [Padhye (10)] to the form

$$\psi = C\ Re^{(m-1)} \tag{2}$$

Following the approach used by Blake, et al. (2), for relating the pressure gradient to the experimentally determined friction factor and the particle Reynolds number, yields the form used for the drag coefficient.

$$B(\theta) = \frac{\theta^2}{(1 - \theta)} \mu_g S^2 \ (Re\psi) \tag{3}$$

The pressure gradient Equation (1) can then be written as

$$\nabla p = \frac{\theta^{1+m}}{C(1 - \theta)^{1+m}} S^{1+m} \mu_g^m \rho_g^{1-m} |\bar{U}_s - \bar{U}_g|^{2-m} \tag{4}$$

This equation shows that the gas pressure drop and, hence, U_{mf} depends on the fluidizing gas density and viscosity. This can be shown directly by recognizing that for $U_g = U_{mf}$, $U_s = 0$. The resulting equation can be solved for U_{mf} as

$$U_{mf} = \left[\frac{C(1 - \theta)^{1+m}}{\theta^m S^{1+m}}\right]^{\frac{1}{2-m}} \left(\frac{\rho_s - \rho_g}{\mu_g^m \rho_g^{1-m}}\right)^{\frac{1}{2-m}} \tag{5}$$

Since the value of the exponent m is always positive and less than unity [Padhye (10)], it can be seen that an increase in either the gas density or the gas viscosity will decrease U_{mf}. It is known that the gas viscosity is virtually independent of the gas pressure, so that, the effect of pressure on U_{mf} is only through the gas density. Hence, an increase in pressure leads to decrease in the minimum fluidization velocity. On the other hand, the effect of temperature on the minimum fluidization velocity is twofold. First, the density of the fluidizing gas is inversely proportional to its absolute temperature. Hence, increasing gas temperature will tend to increase the minimum fluidization velocity through its density effect. Second, the gas viscosity increases with an increase in temperature, and hence, increasing the gas temperature will tend to decrease the minimum fluidization velocity through its viscosity effect. The actual effect of operating temperature on the minimum fluidization velocity is a net result of these two opposing trends.

NUMERICAL SIMULATION

The CHEMFLUB code, with the modified drag coefficient [Padhye (10)], was used to simulate a circular bed 14 cm in diameter and 14 cm in depth. The exit pressure at the top of the bed was held at desired constant value. A uniform gas mass flow rate at the bottom of the bed was varied to achieve different superficial gas velocities. The temperature was held constant at assumed bed operating value. Since effect of operating pressure and temperature on the minimum fluidization velocity is through the gas density and gas viscosity, it was important to use appropriate values of the air density and air viscosity. The gas density is calculated using ideal gas equation of state. The gas viscosity is calculated using Sutherland's viscosity law [White (16)].

In the first simulation the bed was made of spherical ballotini particles, 460 microns in diameter and 2.48 g/cm^3 in density, Group B particles [Geldart (7)]. Rowe and Partridge (13), in their experiments with similar particles, had observed a bed voidage of 0.42 at minimum fluidization. Hence, the bed solidity was taken to be initially 0.58. In the second set of simulations, spherical sand particles 2,320 microns in diameter, and 2.61 g/cm^3 in density were used as bed material, Geldart Group D particles. Botterill, et al. (3), in their experiments with similar particles, had observed a bed voidage of 0.44 at minimum fluidization. Hence, in the simulation, initial solidity was taken to be 0.56. Numerical details on the simulation can be found in Padhye (10).

Experimentally, U_{mf} is determined by measuring the variation of the total gas pressure drop with superficial gas velocity. The pressure drop increases with increasing superficial gas velocity until it reaches its maximum value, the buoyant weight of the bed. After this, the pressure drop remains constant and independent of the gas velocity. In general, the transition from increasing pressure drop to constant pressure drop is not abrupt, and the minimum fluidization velocity is determined as the velocity at a point of intersection of tangents to the sloping and flat parts of the curve. Numerically, the minimum fluidization velocity was determined in a similar fashion. Since the computer code simulates the transient behavior of the bed, the calculated pressure drop is a function of time, a steady state, for superficial gas velocities below U_{mf}, or a quasi-periodic state, for superficial gas velocities above the U_{mf}. A time averaged pressure drop was calculated from the steady/periodic portion of the pressure versus time curve. Once the time averaged gas pressure distribution was calculated, the pressure at the top and bottom of the bed was averaged over the cross sectional area; the pressure drop being the difference between these two values.

Figure 1 is a plot of U_{mf} as a function of pressure at an operating temperature of 293 K. The bed material is 460 μ ballotini. It can be seen that at a pressure of 1 atmosphere U_{mf} was approximately 22.5 cm/sec; Rowe and Partridge (13) had experimentally found 18.5 cm/sec. This discrepancy can be explained by the fact that in the numerical simulation the bed consisted of a single particle size, whereas, in the experiments, the bed material consisted of blend of particle sizes with an average particle size of 460 microns. Some of the particles were not even of Geldart B type, and could explain the lower value. As can be observed from this figure, U_{mf} decreases with increasing pressure. At a pressure of 15 atmospheres, the U_{mf} is 11 cm/sec, approximately 50 percent lower than at 1 atmosphere. At pressures higher than 15 atmospheres, U_{mf} was virtually independent of the operating pressure, as was confirmed by less rigorous calculations.

Figure 2 shows U_{mf} as a function of temperature. The bed pressure in this simulation was held constant at 1 atmosphere. It can be seen that U_{mf} decreases from the value of 22.5 cm/sec at 293 K to a value of 10 cm/sec at 1,500 K, a reduction of about 51 percent. At temperatures higher than 1,500 K, U_{mf} was found to be virtually independent of the temperature.

Figure 3 is a plot of U_{mf} as a function of temperature for a bed of 2,320 micron sand particles. The U_{mf} increases with the temperature up to about 1,200K. Any further increase in the temperature leads to a decrease in U_{mf}. Experimentally, the same trend was observed by Botterill, et al. (3). The experimentally determined U_{mf} values are also plotted in Figure 7.

A typical fluidized-bed coal gasifier operates at about 1,200 K. The pressures in these gasifiers range from 1 to 30 atmospheres. Figure 4 shows the variation of U_{mf} over this pressure range, at 1,200 K, for Geldart Group B particles. It can be seen that U_{mf} decreases with an increase in pressure, but the reduction is only of the order of 25 percent, instead of 50 percent as predicted at 293 K.

Several engineering correlations are available for predicting U_{mf}. All of these equations are derived by equating the buoyant weight of the bed to the experimentally determined correlation for the observed pressure gradient. One such commonly used engineering correlation, given by Wen and Yu (15) and derived from the Ergun (6) correlation for the pressure gradient, can be written as

$$Re_{mf} = (33.7^2 + 0.0408\ Ga)^{1/2} - 33.7 \tag{6}$$

Where Re_{mf} is the particle Reynolds number defined as

$$Re_{mf} = \rho_g |\bar{U}_g|_{mf} d_p / \mu_g \tag{7}$$

and Ga is the Galileo number given by

$$Ga = \rho_g(\rho_s - \rho_g) g d_p^3 / \mu_g^2 \tag{8}$$

Figure 1 shows the variation of U_{mf} with operating pressure as predicted by this correlation. As can be observed from this figure, the U_{mf} at 1 atmosphere is about 17 cm/sec, and that at 15 atmospheres is about 10 cm/sec, a 41 percent decrease. The engineering correlation under predicts the minimum fluidization velocity when compared to the numerical simulation, but predicts the same trend for the effect of operating pressure on the minimum fluidization velocity.

Experimentally the same trend was observed by Sobreiro and Monteiro (14) and Chitester (4). In their experiments, Sobreiro and Monteiro measured U_{mf} for a bed of ballotini particles, using nitrogen as the fluidizing gas. Three different particle sizes (125, 177, and 250 microns) were tested. The operating pressure was varied from 1 atmosphere to approximately 35 atmospheres. It was observed that the change in operating pressure had negligible effect on U_{mf} for the bed of small particles, but the effect was pronounced for the bed of large particles. For 250 micron particles U_{mf} decreased from 5.18 cm/sec at 1 atmosphere to 4.62 cm/sec at 15 atmospheres, a decrease of about 11 percent. On the other hand, there was virtually no change in U_{mf}, over the entire range of operating pressures, for the bed of 125 micron particles. Also, it should be noted that Wen and Yu correlation underpredicted the minimum fluidization velocity for all of these experiments.

CONCLUDING REMARKS

1. The numerical method can be used for predicting U_{mf}, and the predictions compare well with experiments and other correlations.
2. For the air fluidized beds of Geldart Group B particles, the U_{mf} decreases with increasing pressure. For the particles considered in this study, the decrease in the minimum fluidization velocity was of the order of 50 percent over the operating pressure range of 1 to 15 atmospheres. This prediction compares well with the experimental observations.
3. For Geldart Group B particles, the U_{mf} decreases with increasing bed temperature. This decrease was of the order of 50 percent over the temperature range of 293 K to 1,500 K. This trend also compares well with the experimental observations.
4. For Geldart Group D particles, the minimum fluidization velocity first increased and then decreased with increasing temperature. However, this variation was not large and amounted to only a few percent of the actual velocity.
5. At high temperature, for Geldart Group B particles, U_{mf} decreased with pressure. The decrease amounted to about 25 percent over the range of 1 to 30 atmospheres.

NOMENCLATURE

B(θ)	Drag coefficient, Equation (1)	S	Specific surface of a particle
C	Constant in Equation (3)	$\bar{U}$	Velocity
d	Diameter	$\bar{\nabla}$	Gradient vector
Ga	Galileo number	μ	Viscosity
g	Acceleration due to gravity	Ψ	Friction factor
m	Exponent in Equation (3)	ρ	Density
P	Gas pressure	θ	Solidity
R	Drag force per unit surface	ϕ	Sphericity
Re	Reynolds number		

SUBSCRIPTS

g	Gas property
mf	At minimum fluidization velocity
p	Particle property
s	Solid property

LITERATURE CITED

1. Anderson, T. B., and R. Jackson (1967). "Fluid Mechanics Description of Fluidized Beds," Ind. Eng. Chem. Fund., 6, pp. 527.

2. Blake, T. R., S. K. Garg, H. B. Levine, and J. W. Pritchett (1981). "Computer Modeling of Coal Gasification Reactors," Department of Energy Report No. FE-1770-75 (Volume 2), Final Program Report, June 1975 - November 1980.

3. Botterill, J. S. M., U. Teoman, and K. R. Yuregir (1982). "The Effect of Operating Temperature on the Velocity of Minimum Fluidization, Bed Voidage, and General Behavior," Powder Technology, 31, pp. 101-110.

4. Chitester, D. (1982). "PETC Advanced Research and Technology Development, Surface Gasification, High-Pressure Fluidization Cold Model Research," Second Annual Advanced Gasification Projects Contractor's Meeting Proceedings, Ed., M. R. Ghate, U.S. DOE/METC Report DOE/METC/82-51, pp. 109-111.

5. DelleDonne, M. J., P. Nicoletti, and T. J. O'Brien (1984). "Correlation Function Analysis of a Fluidized Bed with a Weak Central Jet." In Gas-Solid Flows: Proceedings of the 1984 Energy Sources Technology Conference, ed. by J. T. Jurewicz, FED-Vol. 10, pp. 97-106. New York: American Society of Mechanical Engineers.

6. Ergun, S. (1952). "Fluid Flow Through Packed Columns," Chemical Engineering Progress, 48 (2), pp. 89-94.

7 Geldart, D. (1973). "Types of Gas Fluidization," Powder Technology, 7, pp. 285-292.

8. Geldart, D., and A. R. Abrahamsen (1977). "Homogeneous Fluidization of Fine Powders Using Various Gases and Pressures," Powder Technology, 19, pp. 133-136.

9. Gidaspow, D., and B. Ettehadieh (1983). "Fluidization in Two-Dimensional Beds with a Jet. 2, Hydrodynamic Modeling," I&EC Fundamentals, Vol. 22, No. 2, pp. 193-201.

10. Padhye, A., T. J. O'Brien (1985). "Numerical Prediction of Effect of Operating Pressure on Minimum Fluidization Velocity in Gas Fluidized Beds," AIChE Summary National Meeting, Seattle, Paper No. 44c.

11. Richardson, J. F. (1971). "Incipient Fluidization and Particulate Systems," in Fluidization, Ed., Davidson and Harrison, Academic Press, London.

12. Rivard, W. C., and M. D. Torrey (1977). "K-FIX: A Computer Program for Transient Two-Dimensional, Two-Fluid Flow," Los Alamos Scientific Laboratory Report LA-NUREG-6623, Los Alamos, New Mexico.

13. Rowe, P. N., and B. A. Partridge (1965). "An X-ray Study of Bubbles in Fluidized Beds," Trans. Instn. Chem. Engrs., 43, pp. T157-T175.

14. Sobreiro, L. E. L., and J. L. F. Monteiro (1982). "The Effect of Pressure on Fluidized-Bed Behavior," Powder Technology, 33, pp. 95-100.

15. Wen, C. Y., and Y. H. Yu (1966). "Chemical Engineering Progress Symposium Series," No. 62, 62, pp. 100.

16. White, F. M. (1974). "Viscous Fluid Flow," Chapter 1, McGraw Hill, New York.

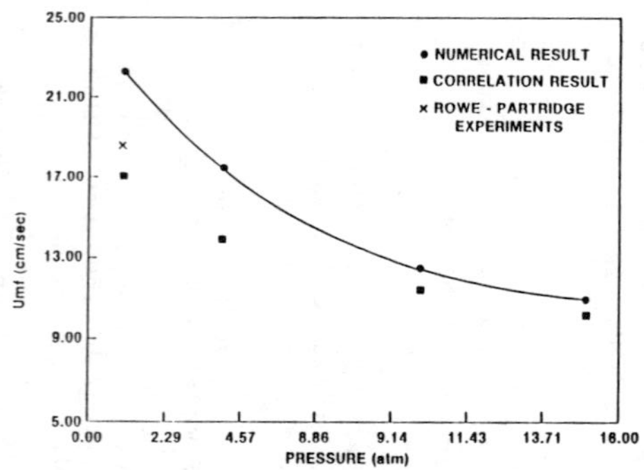

Figure 1. Calculated Minimum Fluidization Velocity as a Function Operating Pressure (Bed Material: 460 Micron Ballotini)

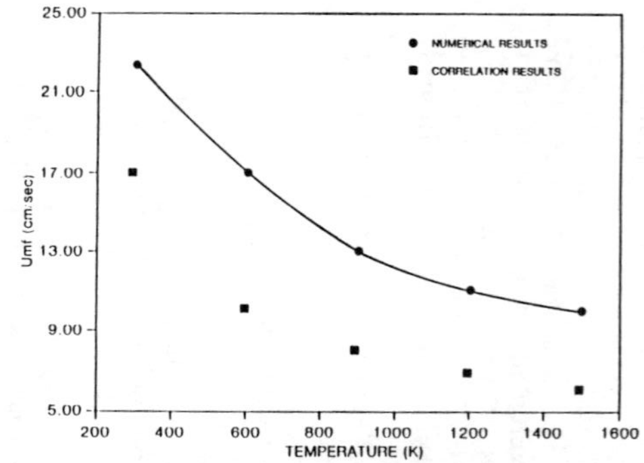

Figure 2. Calculated Minimum Fluidization Velocity as a Function of Operating Temperature (Bed Material: 460 Micron Ballotini, Operating Pressure: 1 atmosphere)

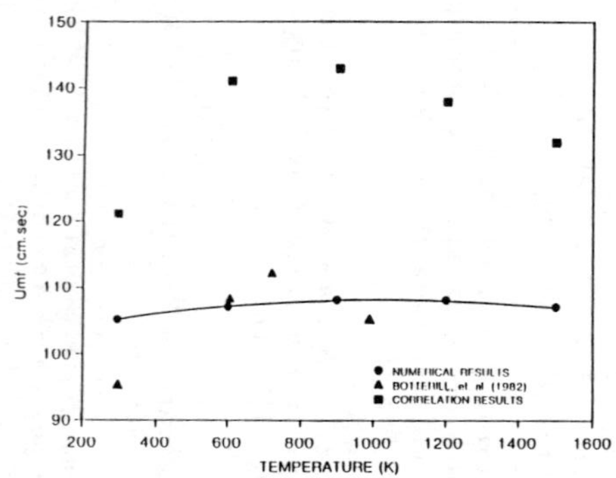

Figure 3. Calculated Minimum Fluidization Velocity as a Function of Operating Temperature (Bed Material: 2320 Micron Sand, Operating Pressure: 1 atmosphere)

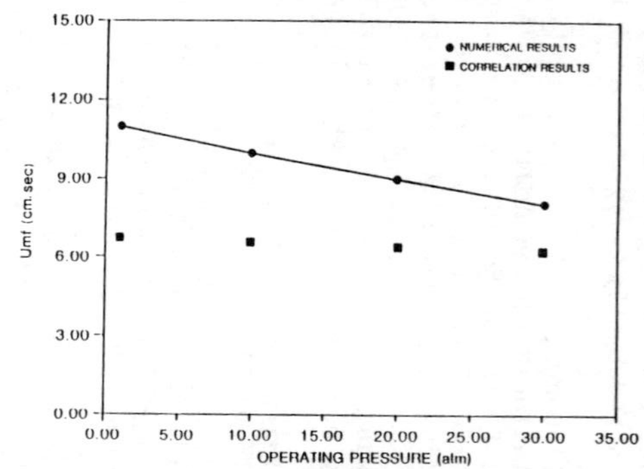

Figure 4. Calculated Minimum Fluidization Velocity as a Function of Operating Pressure at a Typical Gasifier Operating Temperature: 1200K)

TEST OF THE SCALING RELATIONSHIPS FOR FLUID-BED DYNAMICS

R. A. Newby and D. L. Keairns
Westinghouse Electric R&D Center
Pittsburgh, PA 15235

ABSTRACT

Experiments have been conducted to test the validity of scaling rules for fluidized beds. The results show correspondence between the bubble dynamics of two geometrically similar beds with respect to spatial bubble patterns and sizes, bubble frequency and amplitude of pressure drop fluctuations.

SCOPE

Many derivations of the momentum-conservation equations for fluidized beds have appeared in the literature. When particle-particle interaction terms are neglected, a set of four dimensionless groups controlling the dynamics of fluid beds result. These dimensionless groups would be extremely valuable for fluidized bed physical modeling and process development if confidence in them could be generated.

The bubble dynamics and pressure drop fluctuations in a pair of geometrically-similar fluid beds having a scale-factor of two have been compared to test the validity of the scaling rules. The beds were dynamically scaled using the scaling relationships. The larger vessel (36 cm by 15 cm in cross-section) operated at atmospheric temperature and pressure with 200 micron glass beads, while the half-scale unit (18 cm by 7.5 cm) was operated at room temperature and a pressure of 3 atmospheres using 100 micron steel powder. Bubble size, pattern and frequency at the surface of the beds were compared using photographic records. Pressure drop fluctuations across the beds were also recorded and compared.

CONCLUSIONS AND SIGNIFICANCE

The test results show excellent agreement between the bubble dynamics and pressure fluctuations in the two scaled beds. The dimensionless spatial bubble patterns, the dimensionless bubble sizes (data not shown) and the dimensionless bubble frequencies at the top bed surface are nearly identical for the two units as required by the scaling rules. The amplitudes of the dimensionless pressure fluctuations across the beds also compare well for the two units. The results extrapolate to the predictions for bubble frequency and pressure fluctuation amplitude provided by slugging bed theory.

The tests conducted represent only a limited validation of the scaling rules for a small scaling-factor. Further testing with larger scale factors is required, and evaluation of the interpretation of the rules and test results for nonisothermal simulation situations or reactive cases is needed. Even if general validity is shown for the scaling rules, the rules are not practical for all situations and considerable development of the physical modeling technique for fluidized beds will be required. The selection of appropriate particles, gases and operating conditions to simulate large-scale behavior will limit the use of the technique in some cases. Further evaluation of limiting, simplified dimensionless groups and practical experimental scaleup techniques, will be required in order to develop this approach.

MOMENTUM CONSERVATION EQUATIONS IN FLUIDIZATION

Many attempts have been reported in the literature to describe the momentum-conservation equations for fluidized beds. Notable examples of momentum equations developed for fluidized beds have been presented by Anderson and Jackson (1), and many others. Incomplete knowledge of the terms in these equations that represent particle-particle interactions (or solid-phase pressure and stress) is evident. Some attempts to numerically solve simplified versions of these momentum-conservation equations for simple, symmetric flow situations have been made with some degree of success. For example, Blake et al (2) among others.

The mathematical modeling of the hydrodynamics of fluidized beds has not progressed to where it can be used as a practical tool for commercial design. The momentum equations and boundary conditions formulated do identify the dimensionless groups that control the solution forms. Dimensionless groups have been proposed by Broadhurst and Becker (3), Molerus (4), Horio, et al (5), Morgan and Fitzgerald (6), and Glicksman (7), among others. The generally agreed upon groups are a Reynolds number (ρ D U/μ), a Froude number (U^2/gD), the ratio of the gas to particle density, the requirement for geometric similarity and terms related to the particle-particle interaction forces. The interaction force terms can be ignored if the forces are small compared to other forces in the system or if they are functions only of the other dimensionless groups.

PHYSICAL MODELING IN FLUIDIZATION

The objective of physical modeling is to design and operate a subscale system under easily obtained laboratory conditions that has hydrodynamics similar to the full-scale system. Physical modeling of complex hydrodynamic systems has long been practiced in the process industries. Physical modeling, in the classical sense, has not been applied to the development of fluidized bed systems. The closest type of activity to physical modeling has been "cold modeling", which has meant the operation of small-scale equipment at atmospheric pressure and temperature with particles selected to have some similar characteristics with the actual bed material and air generally used as fluidizing gas. The development of fluidized beds has depended on empirical recipes leading to the evolution of systems through many stages of scaleup.

Physical modeling of fluidized bed systems will not provide all of the information necessary for the development of a fluid bed process in most cases, especially where there are high energy reactions or phase changes that result in large temperature gradients, large changes in densities, or drastic changes in particle characteristics, as with agglomeration. Physical modeling will provide in most cases an approximate picture of critical aspects such as the gas phase flow distribution, the fluidization regime, the mixing of solids and gases, the freeboard characteristics and the distributor performance. Such information, even though approximate, is far beyond the current capabilities of mathematical modeling of fluidized beds and should provide accelerated process development.

Only a few researchers have actually tested the fluidized bed scaling relationships for validity and only one application to a large-scale unit set of data has been reported. Fitzgerald et al (8), and Nicastro and Glicksman(9) have both performed small-scale testing of the validity of the scaling relationships. Results have been good in this testing. Fitzgerald et al (10) did a comparison between pressure fluctuation measurements in 2m by 2m fluidized bed combustor development unit and a small-scale test unit. He concluded that the agreement was not sufficient to show validity and that other terms neglected in the scaling relationships may be important. On the other hand, these small-scale tests were not based on a geometrically similar situation so the tests cannot be considered conclusive. Jones and Glicksman (11) have applied the scaling rules as a means to evaluate the hydrodynamics of large-scale fluidized bed combustion.

The basic scaling strategy for the scaling relationships is to start with at least a rough conceptual design of the commercial system, defining the probable range of operating conditions, the particle characteristics and the main features of the vessel and its internals. A practical sub-scale unit is then configured by selecting the test unit characteristic dimension, and then selecting the particle size, particle material, and fluidizing gas. Some iterations may be required in order to find a suitable gas and particle combination, and the gas pressure in the sub-scale unit may need to be a parameter. The sub-scale unit is then designed to be geometrically similar to the full-scale unit in all aspects. The unit will theoretically operate with dimensionless particle and gas velocity fields (V/U and V_p/U) and pressure field ($(P-P_o)/\rho U^2$) identical to the commercial unit with respect to dimensionless time ($t U/ D$) and dimensionless position (R/ D). If the ratio of the full-scale unit characteristic length to the sub-scale unit characteristic length is N ($N \equiv D_f/D_s$) then the scaling relationships indicate that $U_s = U_f\ 1/(\sqrt{N})$ and the ratio $(\rho/\mu)_s = (\rho/\mu)_f\ 1/(N\sqrt{N})$.

TEST APPARATUS

The principles applied to select the equipment for the testing of the validity of the scaling criteria in this study were:

1) An existing fluid bed unit of dimensions 36 cm by 15 cm was used for the full-scale unit (Newby and Keairns, (12)).

2) The sub-scale unit was a 1/2-scale of the full-scale unit and all of its features were exactly 1/2-scale(vessel dimensions and distributor).

3) Particles of average diameter 200 μm were used in the full-scale unit and 100 μm in the sub-scale unit, both having the same shape.

4) Atmospheric pressure, ambient temperature air was used in the full-scale unit, and the operating pressure of the sub-scale unit was adjusted to provide the proper gas density using air or carbon dioxide.

5) According to the dimensionless groups:

$$\mu_f/\rho_f = 2.82\ \mu_s/\rho_s$$

$$\rho_f/\rho_{pf} = \rho_s/\rho_{ps}$$

$$U_f = 1.41\ U_s$$

$$d_{pf} = 2\ d_{ps}$$

$$h_f = 2\ h_s$$

$$\phi_f = \phi_s$$

6) from the above, if air is used in both units then

$$P_s/P_f = 2.85 \text{ and } \rho_{ps}/\rho_{pf} = 2.85$$

and if CO_2 is used in the sub-scale unit

$$P_s/P_f = 1.45 \text{ and } \rho_{ps}/\rho_{pf} = 2.27$$

The equipment for the test is illustrated in Figure 1. The sub-scale unit is constructed from Plexiglas and is placed in a pressure shell. The beds of both units can be observed from the top and pressure taps are placed at several locations in the units.

The particles tested are described in Table 1. Steel balls were used in the 1/2-scale unit and two types of glass balls in the full-scale unit. Test results reported here are for fluidization of both units with air.

Table 1

Particle Material

Particle Type	Unit	Density (kg/m^3)	Average Diameter (μm)	Shape
steel	1/2-scale	7625	100	spherical
glass, Ballotini	full-scale	2467	200	spherical
glass, Flex-o-lite	full-scale	2931	200	spherical

Tests were conducted over a range of velocities under conditions that follow the scaling relationships. The bed depths were 60 cm and 30 cm in the full- and half-scale units, respectively. The dimensionless groups are compared in Table 2. High-speed movies of the bed surfaces were taken to determine the bubble frequency, shape and distribution. The pressure fluctuations across the beds were recorded using a pressure transducer and single-pen recorder.

Table 2

Comparison of Dimensionless Groups

Unit	$\rho\, D^{3/2} g^{1/2}/\mu$	$U_{mf}/(D\, g)^{1/2}$	ρ_p/ρ	D/d_p	Pressure (kPa)
full-scale (flex-o-lite)	1.19×10^4	0.041	2488	762	101
full-scale (Ballotini)	1.19×10^4	0.036	2088	762	101
sub-scale (steel powder)	1.16×10^4	0.036	2337	762	280

TEST RESULTS AND ASSESSMENT

The bubble frequencies as measured from high-speed movies where compared in nondimensional form and found to correspond well. This comparison for the units with both types of glass balls is shown in Figure 2. The theoretical limits for slugging are also shown in the figure and appear to correspond to the extrapolation of the curves (Baeyens and Geldart (13)). One to three bubble streets, depending on the velocity, were observed at the bed surface. These patterns, bubble shapes and spatial distributions, were identical in both beds at the same dimensionless conditions. Space is not sufficient to illustrate this visual comparison.

The amplitudes of the pressure fluctuations also agree reasonably well with the scaling relationships. Figure 3 illustrates this comparison, showing increasing pressure amplitudes with increasing velocity. The slugging limit, showing decreasing pressure amplitude with increasing velocity also appears to extrapolate to the data well.

While the test results show that the scaling criteria seem to be valid for the test conditions investigated, there is still not sufficient data to establish the criteria as being generally valid even over limited ranges of the parameters. There are some major difficulties in applying the criteria even if they are valid. For practical commercial sizes of equipment the physical model must be large enough so that the required particle size does not enter the region where particle interaction forces dominate the fluidization behavior(Geldart's Class C particles). This will require very large models in many cases. Finding the proper particles and fluid may be very difficult in some cases, where extremely large particle densities may be required. Finally, making detailed observations of the model unit behavior may be a challenge. Hopefully, the scaling criteria may be simplified into more easily applied rules over various operating regimes as suggested by Glicksman. Further critical testing of the rules, like the efforts made by Fitzgerald, are required to validate the criteria and to generate confidence in this more rational fluid bed scaling approach.

NOMENCLATURE

A = amplitude of pressure fluctuation
D = characteristic length
d_p = particle diameter
F_b = bubble frequency
g = gravitational constant
h = bed depth
N = ratio of characteristic lengths of full- to sub-scale units
P = absolute pressure
R = position vector
t = time
U = superficial velocity
U_{mf} = minimum fluidization velocity
V = gas actual velocity field
Vp = particle actual velocity field
μ = gas viscosity
ϕ = particle sphericity
ρ = gas density
ρ_p = particle density

subscripts

s = sub-scale
f = full-scale

LITERATURE CITED

1. Anderson, T. B. and R. Jackson, I & EC Fund., 6 (4), 527 (1967).

2. Blake, T. R., D. H. Brownell, Jr., S. K. Garg, W. D. Henline, J. W. Pritchett and G. P. Schneyer, Computer Modeling of Coal Gasification Reactors, Year 2, Annual Report to DOE, October 1977 (NTIS FE-1770-32).

3. Broadhurst, T E. and H. A. Becker, The Application of the Theory of Dimensions to Fluidized Beds, in Fluidization and Its Applications, Toulouse, France, October 1973.

4. Molerus, O., Physical Chemical Hydrodynamics, 4, 337, (1983).

5. Horio, M., A. Nonaka, Y. Sawa and I. Muchi, A New Similarity Rule for Fluidized Bed Scale-up, paper presented at the 1984 Annual AIChE Meeting, San Francisco, CA, November 1984.

6. Morgan, D., and T. Fitzgerald, Scaling Hydrodynamics of Fluidized Beds, paper presented at the Free Forum Session, AIChE 71st Annual Meeting, Miami Beach, FL, November 1978.

7. Glicksman, L. R., Chem. Eng. Sci., 39 (9), 1373 (1984).

8. Fitzgerald, T. J., D. Bushnell, S. Crane and Y-C. Shieh," Testing of Cold Scaled Bed Modeling for Fluidized Bed Combustors", in the Proceeding of the 7th Int. Conf. on Fluidized Bed Combustion, 766, Philadelphia, PA, October 1982.

9. Nicastro, M. T. and L. R. Glicksman, Chem. Eng. Sci., 39 (9), 1381 (1984).

10. Fitzgerald, T., D. Bushnell, S. Crane and Y-C. Shieh, Powder Technology, 38, 107 (1984).

11. Jones, T. L. and L. R. Glicksman, An Experimental Investigation of Gas Flow in a Scale Model of a Fluidized Bed Combustor, paper presented at at the 1984 Annual AIChE Meeting, San Francisco, CA, November 1984.

12. Newby, R. A. and D. L. Keairns, "Fluidized Bed Heat Transport Between Parallel Horizontal Tube Bundles", in Fluidization, 320, Cambridge University Press, Cambridge, 1978.

13. Baeyens, J. and D. Geldart, Chem. Eng. Sci., 29, 255 (1974).

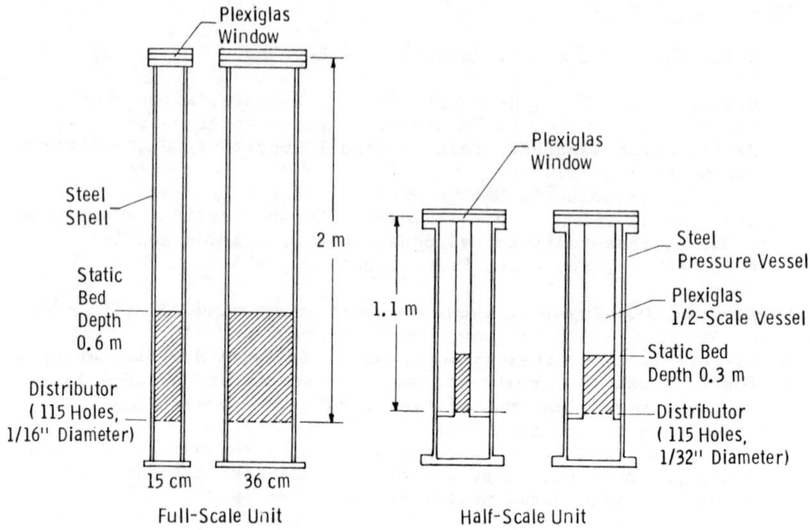

Fig. 1 - Test Units

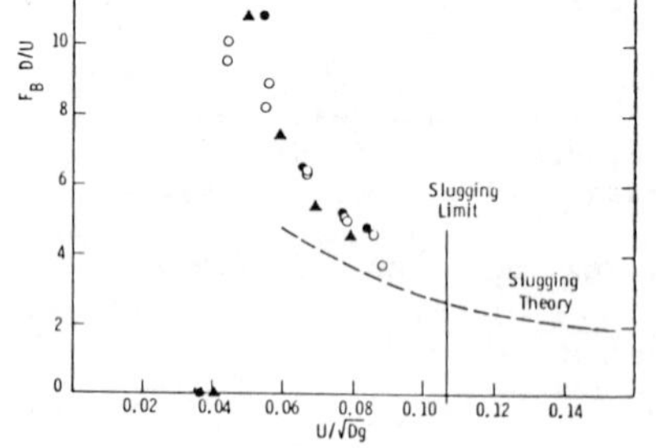

Fig. 2-Comparison of Bubble Frequencies

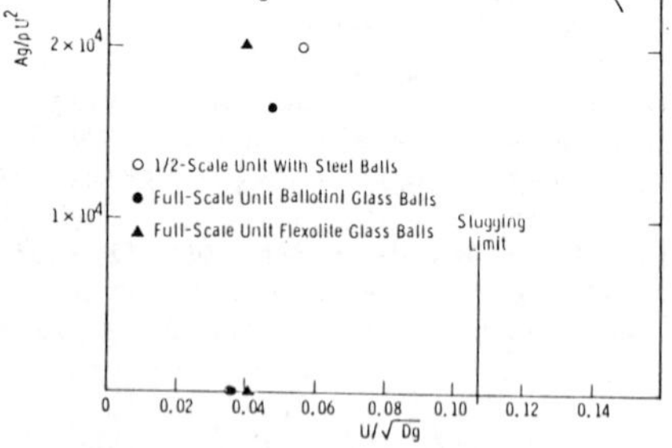

Fig. 3-Comparison of Pressure Amplitudes

INTERPHASE MASS TRANSFER DURING BUBBLE FORMATION IN FLUIDIZED BEDS

by S.-P. Sit, Alberta Energy Company, Calgary, Canada T2P 0M9

& J.R. Grace, University of British Columbia, Vancouver, Canada V6T 1W5

ABTRACT: Time-averaged concentrations were measured in the entry region of beds of 120 to 310 μm particles contained in a 152 mm column with a central orifice of diameter 6.4 mm and auxiliary tracer-free gas. Interphase transfer occurs by net outflow of gas to the dense phase followed by capture by the rising voids and by diffusion. The former mechanism is much more important in the present case.

SCOPE: The distributor region plays an important role in determining the contacting efficiency of fluidized bed reactors. Experimental studies (1 to 3) show that changing from one distributor to another, all other conditions remaining fixed, can cause major changes in conversion. Models which explicitly account for the grid region are sensitive to the interphase transfer which is adopted there (4). Mass transfer data in the grid region are scarce. When gas enters the bottom of a fluidized bed through discrete orifices, bubbles may form directly at the distributor plate or at the top of permanent spouts or jets (5 to 11). In this paper we consider only the former case, reporting our results on transfer from permanent jets in a separate paper. Experiments were carried out with four types of non-adsorbing Group B solids in a flat-bottomed cylindrical column equipped with a central orifice. To assure bubble formation right at the distributor plate and aid in interpretation of data, flow visualization was carried out in a half column of the same size, and sufficient auxiliary air was supplied to the surrounding grid to fluidize the dense phase. Methane was employed as a tracer. Gas samples were taken at different heights and radial positions in the column.

CONCLUSIONS AND SIGNIFICANCE: When bubble formation occurs directly at a orifice at the base of a fluidized bed of non-adsorbing particles, interphase transfer occurs by two mechanisms. Some gas streams through into the dense phase, later returning to the bubble phase. This represents a net convective transfer. Further exchange occurs due to diffusion from the periodically forming bubbles. The latter is less than predicted by the penetration theory and at least an order of magnitude less than that due to the convective mechanism.

The key role of the distributor plate geometry in conducting catalytic and gas-solid reactions in fluidized beds is now widely recognized. To devise models which provide a reasonable representation of the physical characteristics of that region, it is essential to understand the hydrodynamics and transfer processes which occur as gas enters beds of particulate solids. While there have been a number of recent studies of the hydrodynamics of the entry region (e.g. 7 to 9, 12), the mass transfer problem has received little attention. Investigators have tended to avoid the entry region when studying bubble-to-dense-phase exchange. For example, Stephan et al. (13) and Chavarie and Grace (14) discounted data in the lowest regions of their columns in order to avoid entry effects and obtain steady state mass transfer coefficients.

Equations for diffusional mass transfer from gas bubbles forming in liquids are of the form of the penetration theory (15). Whereas bubbles forming in liquids are impermeable, those in fluidized beds have permeable boundaries. As a consequence, additional transfer can take place due to convective flow of gas through the interface of the forming bubble. Gas flowing through the forming bubbles into the dense phase and then regaining the bubble phase higher in the bed represents a net exchange between the two phases.

EXPERIMENTAL APPARATUS AND PROCEDURE

Most of the experiments were carried out in a 152 mm i.d. plexiglass column with an expanded top section of diameter 254 mm and an overall height of 562 mm. The multi-orifice distributor plate was equipped with a central orifice of diameter 6.4 mm, fed from a separate metred air supply. There were 120 smaller (0.8 mm dia.) holes in the surrounding multiorifice plate, evenly distributed over the area and covered with nylon to prevent weeping. These provided auxiliary air from a plenum chamber below. The column is shown schematically in Figure 1. A platform was attached to the top of the column to allow careful alignment and positioning of a 3 mm o.d. steel tube for gas sampling. Methane from a gas cylinder was metered and mixed in two packed columns with the air fed to the central orifice. Gas samples were withdrawn by a vacuum pump equipped with a dust filter into a sampling valve of volume 2 cm^3. The sampling rate, regulated by a needle valve and indicated on a soap-film flowmeter, was set at 0.25 cm^3/s. Methane concentrations were measured using a Carle model 211 gas chromatograph equipped with a flame ionization detector.

Four types of particles were used in the experiments. Their key physical properties are listed in Table 1. All particles were spherical, non-porous, non-adsorbing, and belong to group B of Geldart's classification. The static height of solid particles was maintained at 0.26 to 0.30 m. Enough auxiliary air was provided to incipiently fluidize the particles under investigation. The central flow of air containing the methane tracer was then established. After steady state had been achieved, samples were taken at four different radial positions and at six vertical levels. Three or four replicate samples were analysed at the highest and lowest levels, where the variance was generally highest. Further details of the experimental equipment and procedure are provided elsewhere (16).

HALF-COLUMN FLOW VISUALIZATION

A separate "half-column" was constructed to allow flow visualization in the distributor region. Half-columns are common in spouted bed studies where they generally give reliable indications of behaviour in fully three-dimensional columns (17, 18). They have also been found useful (e.g. 8,19) in studying the entry region of fluidized beds. Observations were recorded for analysis on video tape.

A typical sequence for cases in which enough auxiliary gas was supplied to fluidize the solids (i.e. $U_{Do}=U_{mf}$) is shown in Figure 2. A bubble is seen to grow on the central orifice, with an elongated mushroom-like shape after about the first 0.06 s. When the bubble breaks away, the "stem" disappears, but a small residual bubble is left at the orifice, forming the nucleus of the next bubble. Values of bubble frequency, volume-equivalent diameter of newly formed bubbles and the nose position at severance and of the residual bubbles appear in Table 2. The frequency of bubble formation was 6 to 7 s^{-1} for all four particle types under the conditions studied (u_{or} = 10 to 50 m/s). Measured bubbles diameters were generally close to predictions from available correlations (5,6,20,22) for low u_{or}= 10 m/s, but fell well below the predicted values for the higher values of u_{or}, especially for u_{or}=50 m/s (16).

The "visible bubble flow" after severance from the orifice

$$Q_{BS} = f_{Bf} \frac{\pi}{6} D_{BS}^3 \tag{1}$$

can be compared to the volumetric flow rate through the orifice,

$$Q_{or} = u_{or} \frac{\pi}{4} D_{or}^2 \tag{2}$$

by the ratio $Y_S = Q_{BS}/Q_{or}$. Values of Y_S are given in Table 2. If the well-known "two-phase theory of fluidization" applied, we would expect Y=1 (since the auxiliary flow accounts for $U_{mf}A$). From Table 2, only 22 to 54% of Q_{or} is accounted for by Q_{BS}, the ratio decreasing with increasing u_{or}. This range is consistent with values reported by other investigators (12,20). For u_{or}= 30 m/s, the average value of Y_S from the four types of particles is 0.33, close to the ratio of 0.36 determined by Yates et al. (12) using X-rays. Since Y well above the distributor is typically 0.67 for group B solids (23), approximately half of the gas which eventually flows as visible bubbles must be accounted for by dense phase interstitial flow and throughflow near the bottom of the bed. As already noted, this represents an effective interphase transfer.

FULL-COLUMN CONCENTRATION MEASUREMENTS IN BUBBLE FORMATION ZONE

Tracer concentrations were measured in the fully cylindrical column for the hydrodynamic conditions investigated in the semi-cylindrical column described in the previous section. Profiles of time-average concentration are presented in Figure 3 for four radial positions, extending just above the bubble formation zone for 310 μm glass beads and u_{or}= 50 m/s. The bars indicate 90% confidence bands. The concentration along the axis changes very little over the first 10 orifice diameters. In the dense phase, the tracer concentration is essentially zero for the two outermost radial positions over the entire formation zone. At r=26 mm, the measured concentration increases rapidly at z ≈ 10d_{or} and the scatter in the data is much larger, probably because r was similar to the bubble radius so that samples were occasionally withdrawn from the edge of released bubbles. These profiles are similar to those shown by Stephen et al. (13). Data for the other combinations of d_p and u_{or} are tabulated elsewhere (17). Approximately 8 s is required to fill the sampling volume. Since the bubble formation frequency was typically 6 to 7 s^{-1} (see Table 2), the sampling tube when located along the axis had to encounter 40 to 50 bubbles during sampling. As newly formed bubbles lifted off from the orifice, the sampling tube, if located above the nose position of the residual bubble, would sample from the dense phase for a portion of the sampling period. Hence care is required in treating the data.

A mass balance on a bubble during the formation stage yields

$$\frac{d}{dt}(C_B V_B) = Q_{or}C_{or} - Q_{ex}C_B + k_{BD1}S_B(C_D - C_B) \tag{3}$$

where the bubble contents are assumed well-mixed. Since $dV_B/dt = Q_{or} - Q_{ex}$, Equation (3) can be rewritten as

$$V_B \frac{dC_B}{dt} = Q_{or}C_{or} + k_{BD1}S_B C_D - (Q_{or} + k_{BD1}S_B)C_B \tag{4}$$

To find the time-mean point concentration at r=0 as a function of k_{BD1}, we introduce several simplifications. V_B and S_B are written in terms of a bubble diameter, D_B, using standard relations for spheres. The dense phase concentration, C_D, is taken as zero, consistent with the outermost readings in Figure 3, and dt is replaced by dz/u_B with the velocity of the bubble nose, u_B, assumed constant over the bubble formation stage and determined from the half-column study. With these substitutions, Equation (4) can be rewritten

$$\frac{dC_B}{dz} = \frac{6Q_{or}}{\pi u_B D_B^3}(C_{or} - C_B) - \frac{6k_{BD1}C_B}{u_B D_B} \tag{5}$$

where D_B is a function of the nose position z:

$$D_B = \{D_{BR}^3 + \frac{6(Q_{or} - Q_{ex})(z - z_R)}{\pi u_B}\}^{1/3} \tag{6}$$

The boundary condition needed to solve Equation (5) is $C_B = C_{or}$ at $z = z_R$. It is further assumed that bubble formation is under constant flow conditions so that Q_{or} is constant.

Once the bubble nose reaches z_S, severance is assumed to occur, with a fully formed spherical bubble of diameter D_{BS} rising steadily and the new bubble beginning to grow with its nose at z_R. Any additional transfer during detachment is ignored. The sampling tube will, if located at $z > z_S - D_{BS}$, sample also from the fully formed bubble, whose variation in concentration must be accounted for (Equation (10) below). Because of the massive outflow from the forming bubble, the concentration of gas sampled when the tube is along the axis and between bubbles is assumed to have the same concentration as the top of the bubble below. The time-mean concentrations must be calculated separately for three different zones (see Figure 4) as follows:

<u>Zone A</u> where $0 < x \leqslant z_S - D_{BS}$: Here the sampling probe sees gas having the composition of bubbles growing on the orifice so that

$$\bar{C} = \frac{1}{t_f}\int_0^{t_f} C_B dt \tag{7}$$

<u>Zone B</u> where $z_S - D_{BS} < x \leqslant z_S + t_f u_{BS} - D_{BS}$: Here the sampling tube is exposed alternatively to gas from forming bubbles and recently

severed fully formed bubbles. The time-mean local concentration is

$$\bar{C} = \frac{1}{t_f} \{\int_{t^*}^{t_f} C_B dt + \int_0^{t^*} C_{B2} dt\} \tag{8}$$

where t* depends on the sampling tube height, x, according to

$$t^* = (x - z_S + D_{BS})/u_{BS} \tag{9}$$

C_{B2} is the concentration of the severed bubble, which is assumed to rise steadily with velocity u_{BS} so that

$$C_{B2} = C_{BS} \exp\{-6\ k_{BD2} t/D_{BS}\} \tag{10}$$

with the steady state interphase mass transfer coefficient (24):

$$k_{BD2} = U_{mf}/4 + \sqrt{4D_m\ \varepsilon_{mf}\ u_{BS}/\pi D_{BS}} \tag{11}$$

and

$$u_{BS} = 0.711\sqrt{g\ D_{BS}} \tag{12}$$

Zone C where $x > z_S + t_f u_{BS} - D_{BS}$: Now the sampling probe encounters only fully formed bubbles. The time mean concentration is

$$\bar{C} = \frac{1}{t_f} \int_{t^* - t_f}^{t^*} C_{B2} dt \tag{13}$$

The above equations can be applied to find $\bar{C}$ along the column axis for any value of k_{BD1}. The values which minimize the sum of the squares of the deviations from the corresponding experimental measurements are given in Table 3. In view of the number of assumptions, they are given to only one significant figure. These values are consistently less than those predicted by the penetration theory expression

$$k_{BD1} = \sqrt{4D_m\ \varepsilon_{mf}/\pi t_f} \tag{14}$$

The small magnitude of the fitted diffusional mass transfer coefficients probably results from interference by the massive convective outflow of gas represented by Q_{ex}. Table 3 also shows that k_{BD1} is consistently less than the corresponding single bubble steady state transfer rate, k_{BD2}, for fully formed bubbles. This is not surprising since the latter incorporates throughflow as well as diffusion.

DISCUSSION

Two effective components of mass transfer for forming bubbles have been identified and their orders of magnitudes inferred. The total diffusional transfer can be estimated for the present system by evaluating the time average of $k_{BD1}S_B$ over one cycle of bubble formation. Even for the largest derived k_{BD1} value, this results in an average of only 27 cm^3/s, an order of magnitude less than the estimated net exchange resulting from convective outflow followed by subsequent recapture. Hence the present study indicates that the diffusive mechanism is much less important than the convective mechanism. Expressed as a mass transfer coefficient based on the area of the fully formed bubbles, the latter has an order of magnitude of about 6 cm/s, larger than usual interphase mass transfer coefficients for non-adsorbing group A and B powders (25). Since bubbles are also smaller near the grid, favourable gas-solid contacting occurs in this region primarily due to convective outflow followed by recapture.

ACKNOWLEDGEMENT

The experimental work described in this paper was carried out at McGill University in Montreal. Funding was provided by the Natural Sciences and Engineering Research Council of Canada.

NOTATION

A column cross-sectional area
$\bar{C}$ time-mean local concentration
C_B bubble concentration
C_D dense phase concentration
C_{or} concentration of entering gas
D_B bubble diameter
D molecular diffusivity
D_{or} orifice diameter
d_p mean particle diameter
f_{Bf} frequency of bubble formation
k_{BD} bubble to dense phase mass transfer coefficient
Q_B visible bubble flow rate
Q_{or} gas flow through orifice
Q_{ex} flow through orifice not forming bubbles, $Q_{or} - Q_{BS}$
r radial coordinate
S_B bubble surface area
t time
t^* time defined by Equation (9)
t_f bubble formation time = $1/f_{Bf}$
U_{Do} auxiliary gas superficial velocity
U_{mf} minimum fluidization velocity
u_B velocity of rise of bubble nose
u_{or} velocity of gas through orifice
V_B bubble volume
x vertical height coordinate
Y fraction of orifice gas forming bubbles, Q_B/Q_{or}
z vertical coordinate of bubble nose
ε_{mf} bed voidage at minimum fluidization
ρ_p particle density

Subscripts:
R residual bubble
S at severance from orifice
1 formation period
2 after severance

LITERATURE CITED

1. Cooke, M.J., W. Harris, J. Highley and D.F. Williams, Instn. Chem. Engrs. Symp. Ser., No. 30, 21 (1968).
2. Walker, B.V., "Gas-solid contacting in bubbling fluidized bed," PhD thesis, Cambridge Univ., Cambridge, England (1970).
3. Bauer, W. and J. Werther, Proc. 2nd World Congress Chem. Engng., 3, 69 (1981).
4. Grace, J.R. and H.I. de Lasa, A.I.Ch.E. J., 24, 364 (1978).
5. Markevka, V.I., V.A. Basov, T.K. Melik-Skhnazarov and D.I. Orochko, Teor. Osn. Khim. Tekl., 5, 95 (1971).
6. Chiba, T., K. Terashima and H. Kobayashi, Chem. Eng. Sci., 27, 965 (1972).
7. Rowe, P.N., H.J. MacGillivray and D.J. Cheesman, Trans. Instn. Chem. Engrs., 57, 194 (1979).
8. Knowlton, T.M. and I. Hirsan, in Fluidization, J.R. Grace and J.M. Matsen (Eds.), Plenum Publishing Co., New York (1980).
9. Wen, C.Y., N.R. Deole and L.H. Chen, Powder Tech., 31, 175 (1982).
10. Yates, J.G. and D.J. Cheesman, "Distributor processes in fluidized-bed reactors", at CHISA Conference, Prague (1984).
11. Chandnani, P.P. and N. Epstein, "Spoutability and spouting characteristics of fine particles with a gas", This Conf.(1986).
12. Yates, J.G., P.N. Rowe and D.J. Cheesman, A.I.Ch.E. J., 30, 890 (1984).
13. Stephen, G.K., R.J. Sinclair and O.E. Potter, Powder Technol., 1, 157 (1967).

14. Chavarie, C. and J.R. Grace, Chem. Eng. Sci., 31, 741 (1976).
15. Clift, R., J.R. Grace and M.E. Weber, Bubbles, Drops and Particles, Academic Press, New York (1978).
16. Sit, S.-P., "Grid region and coalescence zone gas exchange in fluidized beds", PhD thesis, McGill Univ., Montreal (1981).
17. Mathur, K.B. and N. Epstein, Spouted Beds, Academic Press, New York (1974).
18. Geldart, D., A. Hemsworth, R. Sundavadra and K.J. Whiting, Can. J. Chem. Eng., 59, 638 (1981).
19. Yang, W.C. and D.L. Keairns, in Fluidization, J.F. Davidson and D.L. Keairns (Eds.), Cambridge Univ. Press, England (1978).
20. Nguyen, X.T. and L.S. Leung, Chem. Eng. Sci., 27, 1748 (1972).
21. Basov, V.A., V.I. Markevka, T.K. Melik-Akhnazarov and R.I. Orochko, Intern. Chem. Eng., 9, 263 (1969).
22. Mori, S. and C.Y. Wen, A.I.Ch.E. J., 21, 109 (1975).
23. Hilligardt, K. and J. Werther, Chem. Ing. Techn., tbp (1985).
24. Sit, S.P. and J.R. Grace, Chem. Eng. Sci., 36, 327 (1981).
25. Sit, S.P. and J.R. Grace, Chem. Eng. Sci., 33, 1115 (1978).

Table 1: Properties of solid particles

Type	d_p (μm)	ρ_p (kg/m^3)	ε_{mf}	U_{mf} (cm/s)
Glass beads	120	2500	0.52	1.2
" "	230	"	0.50	5.3
" "	310	"	0.47	8.2
Copolymer	290	1100	0.46	2.4

Table 2: Measured bubble formation properties

d_p (μm)	u_{or} (m/s)	z_R/D_{or}	z_S/D_{or}	D_{BS} (mm)	f_{Bf} (s^{-1})	$Y_S=Q_{BS}/Q_{or}$
120	30	4	15	49	6.9	0.44
230	10	3	10	37	6.5	0.54
"	30	4	14	40	6.3	0.22
"	50	5	16	50	6.9	0.28
310	30	3	12	37	7.0	0.28
"	50	4	14	48	6.0	0.22
290	30	4	18	48	6.0	0.36

Table 3: Experimental and penetration theory values of mass transfer coefficient for forming bubbles

Particle diameter, d_p (μm)	Orifice velocity, u_{or} (m/s)	Experimental k_{BD1} (cm/s)	Predicted k_{BD1} (cm/s)	Steady state k_{BD2} (cm/s)
120	30	0.5	1.0	1.5
230	10	0.0	0.95	2.6
"	30	0.7	0.94	2.6
"	50	0.0	0.98	2.5
290	30	0.2	0.88	1.7
310	30	0.7	1.2	3.3
"	50	0.8	0.89	3.2

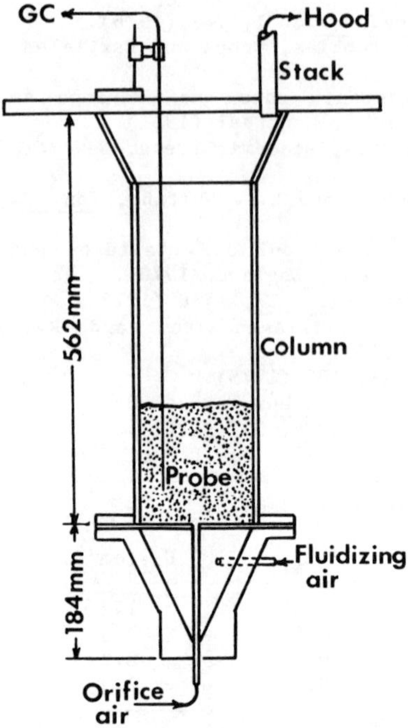

Figure 1: Schematic diagram of the fluidization column

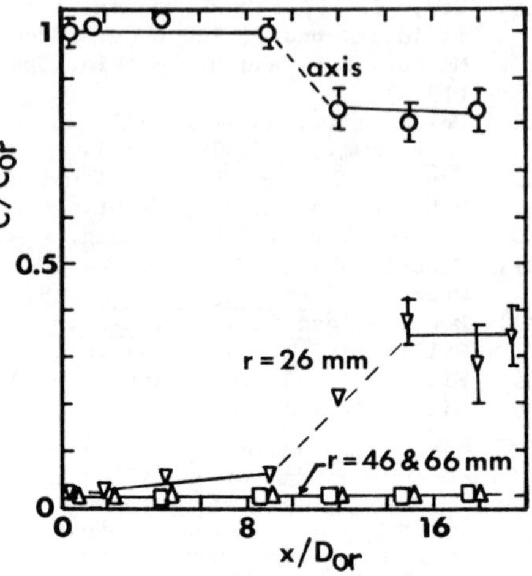

Figure 3: Measured time-mean tracer concentrations in the bubble formation zone for 310 μm glass beads and u_{or}= 50 m/s.

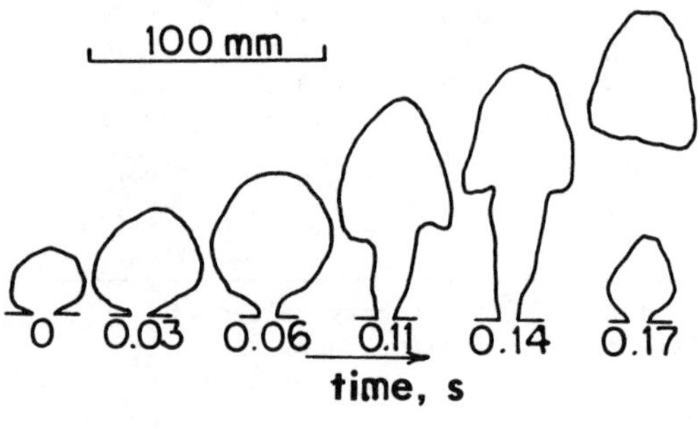

Figure 2: Bubble formation for 290 μm copolymer beads, u_{or}= 40 m/s and U_{do}= U_{mf}= 2.4 cm/s.

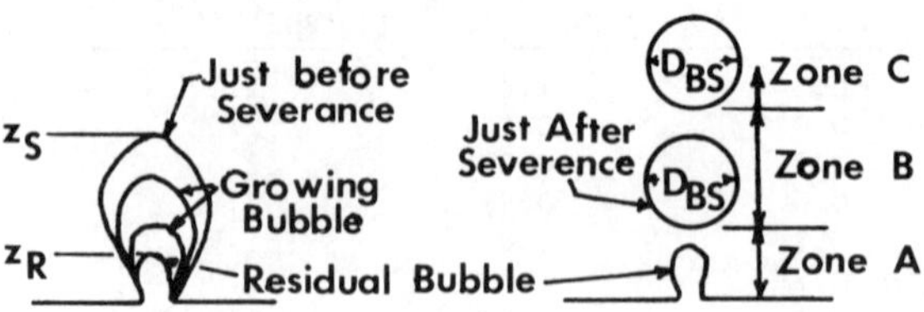

Figure 4: Zones for calculation of time-mean concentrations.

SPLIT FEED FLUIDISED BED REACTORS: MASS TRANSFER IN THE SPARGER REGION

R. J. Dry* and M. R. Judd**

*Sasol Technology (Pty) Ltd.
P.O. Box 1, Sasolburg 9570, South Africa***

**Department of Chemical Engineering
University of Natal, Durban 4000, South Africa

ABSTRACT

Interphase mass transfer between the dense phase and bubbles introduced via a secondary gas sparger has been quantified. The results suggest that sparger bubbles tend to retain their identity to an appreciable extent, and that gas-solids contact is much the same as would have been the case if all the gas had been fed via the grid.

SCOPE

Split feed systems are not uncommon in industrial catalytic fluidised bed reactors, and in such systems one gas stream usually fluidises the bed via a conventional grid while a second is introduced higher up by means of a suitable sparger. For processes involving fast reactions it is possible that bubble to dense phase mass transfer could be limiting. Under these conditions it would be helpful to know to what extent feed configuration affected gas-solids contacting, since this could dictate the incentive for considering the installation of such a system.

In this study mass transfer between the dense phase and bubbles introduced via a sparger is evaluated. The investigation is based on cold models and forms part of a broader initiative aimed at gaining a clearer understanding of how various gas feed devices might affect the performance of a certain catalytic industrial reactor. Inert gas residence time distributions were measured on a 0.64m ID mock-up of this reactor, and were interpreted on the basis of each of two models alternatively. The degree of closure between the model and the experimental gas residence time characteristics was taken to reflect the soundness of the assumptions on which the models were based, and the optimised parameters for the better-fitting model were considered indirect measurements.

CONCLUSIONS AND SIGNIFICANCE

Of the models used for data interpretation, the first describes total lateral integration between the grid and sparger bubbles and the second total segregation. The model screening results suggest that total

*** Present address: Dept. of Chemical Engineering, Monash University, Clayton, Victoria, 3168, Australia.

segregation is the better basis on which to describe this system, and this in turn suggests that bubbles from the sparger tend to retain their identify in the bed to a significant degree.

Interphase mass transfer between sparger bubbles and the dense phase, presented as the number of times the average bubble is purged on its way through the bed (i.e. crossflow ratio), is given in figure 4. A cross-plot of these data at constant gas velocity is given as figure 5, and similar data for a conventional feed system and a small slugging bed are included for reference. It appears that, on an equivalent throughput basis, there is no clear incentive to either employ or avoid a sparger from a mass transfer point of view.

APPROACH AND MODEL FORMULATION

In a previous paper (1) a technique based on gas residence-time testing and model fitting was presented for estimating the efficiency of gas-solids contact for single-feed systems. The same approach, with appropriate modifications to account for the step-change in gas flow at the level of the sparger, is employed here.

The models used previously (1) did not recognise the existence of individual bubbles - it would not have been appropriate for them to do so since the gas residence time technique has lateral averaging properties and bubble sizes were not measured. This approach quite deliberately avoids assuming the presence of "classical" bubbles in the system, since it was not at all clear whether the gas voids could be modelled in the normal manner. This uncertainty arises partly from the presence of coils in the bed, and partly from the fact that the powder used appeared to display some unusual fluidisation characteristics (6).

In the split feed case gas voids enter the region above the sparger as a mixture of (presumably) large bubbles arising from the grid and smaller ones generated at the sparger nozzles. It is not clear whether bubbles from the two sources should be regarded separate lean phases or not: if rapid coalescence between grid and sparger bubbles occurs, a model based on a single lean phase (in the region above the sparger) would probably be appropriate. If, on the other hand, coalescence does not achieve significant lateral integration of the grid and sparger bubbles, a segregated lean phase model might be more attractive. A model describing some intermediate lateral bubble mixing condition was felt unjustified, since there did not appear to be any simple way of estimating the degree to which the lean phases from the two sources merged. Application of the multiple bubble-track concept (1) was also felt unjustified in view of the increase in mathematical complexity that would have been required. It is also noted that a two-phase model was found to be more appropriate than a three-phase one for this system (1), and models describing "cloud" or "cloud-wake" phases are accordingly not considered.

The above arguments led to the formulation of two models, the first of which is the laterally mixed bubble phase or LMBP model. As shown in Figure 1, it describes a dense phase and a single bubble phase. The properties of the latter change at the height of the sparger: bubble velocity and the interphase mass transfer coefficient change from lumped values in the region below the sparger to lumped values above. The boundary conditions are similar to those of the May-Van Deemter model (2,3).

The second model describes the other extreme in terms of lateral bubble mixing, and is termed the laterally segregated bubble phase or LSBP model. As shown in figure 2, its main feature is the fact that bubbles from the sparger form a separate lean phase. The dense phase and boundary conditions are once again similar to those of the May-Van Deemter (2,3) model, and the interphase mass transfer coefficient from the grid bubbles to the dense phase changes abruptly from one value to another at the level of the sparger.

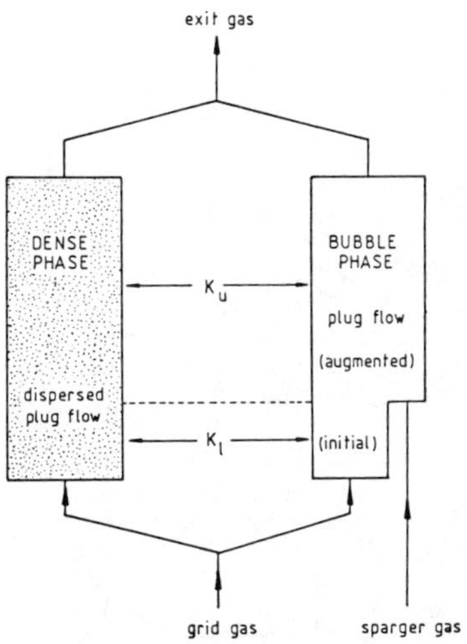

Figure 1 LMBP Model

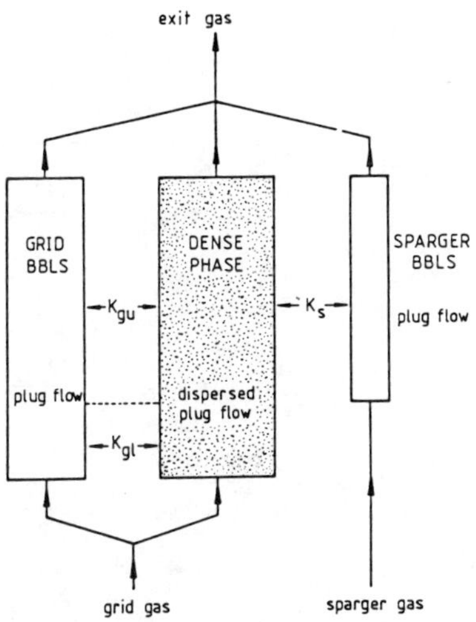

Figure 2 LSBP Model

The models were formulated in the dynamic mode, and the resulting differential equations and associated boundary conditions are given in table 1 - the full derivation is described elsewhere (4). The equations were solved in the time domain by a Crank-Nicholson finite difference procedure (5) and subsequently transformed numerically into the form of a frequency response. For full details the reader is referred to the appropriate thesis (4).

EXPERIMENTAL

A 0.64m ID cold model with dummy vertical coils and a gas sparger insert was employed - this rig is shown in figure 3. The grid and sparger

LMBP

Bubble Phase:

$0 \leq z < L_s$ $$\frac{\partial C_b}{\partial t} + U_{b\ell}\frac{\partial C_b}{\partial z} + K_\ell [C_b - C_e] = 0$$

$L_s \leq z \leq L_f$ $$\frac{\partial C_b}{\partial t} + U_{bu}\frac{\partial C_b}{\partial z} + K_u [C_b - C_e] = 0$$

Dense Phase:

$0 \leq z < L_s$ $$\frac{\partial C_e}{\partial t} + (\frac{U_d}{\varepsilon_d})\frac{\partial C_e}{\partial z} - \frac{\delta K_\ell}{(1-\delta)\varepsilon_d}[C_b - C_e] - D_d \frac{\partial^2 C_e}{\partial z^2} = 0$$

$L_s \leq z \leq L_f$ $$\frac{\partial C_e}{\partial t} + (\frac{U_d}{\varepsilon_d})\frac{\partial C_e}{\partial z} - \frac{\delta K_u}{(1-\delta)\varepsilon_d}[C_b - C_e] - D_d \frac{\partial^2 C_e}{\partial z^2} = 0$$

I.C.: $C_b(z, 0) = C_e(z, 0) = 0$

B.C.'s: (i) $C_b(0, t) = 0$

(ii) $C_b(L_s, t)$ = Dirac delta function

(iii) $\frac{\partial C_e}{\partial z} = 0$ at $z = 0$

(iv) $\frac{\partial C_e}{\partial z} = 0$ at $z = L_f$

LSBP

Grid Bubbles:

$0 \leq z < L_s$ $$\frac{\partial C_{bg}}{\partial t} + U_{bg}\frac{\partial C_{bg}}{\partial z} + K_{g\ell}[C_{bg} - C_e] = 0$$

$L_s \leq z \leq L_f$ $$\frac{\partial C_{bg}}{\partial t} + U_{bg}\frac{\partial C_{bg}}{\partial z} + K_{gu}[C_{bg} - C_e] = 0$$

Sparger Bubbles:

$L_s \leq z \leq L_f$ $$\frac{\partial C_{bs}}{\partial t} + U_{bs}\frac{\partial C_{bs}}{\partial z} + K_s[C_{bs} - C_e] = 0$$

Dense Phase:

$0 \leq z < L_s$ $$\frac{\partial C_e}{\partial t} + (\frac{U_d}{\varepsilon_d})\frac{\partial C_e}{\partial z} - D_d\frac{\partial^2 C_e}{\partial z^2} - \frac{\delta F_{g\ell} K_{g\ell}}{(1-\delta)\varepsilon_d}[C_{bg} - C_e] = 0$$

$L_s \leq z \leq L_f$ $$\frac{\partial C_e}{\partial t} + (\frac{U_d}{\varepsilon_d})\frac{\partial C_e}{\partial z} - D_d\frac{\partial^2 C_e}{\partial z^2} + \frac{\delta(f_s K_s + f_{gu}K_{gu})}{(1-\delta)\varepsilon_d} C_e$$

$$= \frac{\delta}{(1-\delta)\varepsilon_d}(f_s K_s C_{bs} + f_{gu} K_{gu} C_{bg})$$

I.C.: $C_{bg}(z, 0) = C_{bs}(z, 0) = C_e(z, 0) = 0$

B.C.'s: (i) $C_{bg}(0, t) = 0$

(ii) $C_{bs}(L_s, t)$ = Dirac delta function

(iii) $\frac{\partial C_e}{\partial z} = 0$ at $z = 0$

(iv) $\frac{\partial C_e}{\partial z} = 0$ at $z = L_f$

TABLE 1: MODEL EXPRESSIONS

nozzles were of similar geometry, and the system was operated with two parts of gas fed via the grid for every part fed via the sparger. The fluidising medium was air, and the powder milled iron oxide with an average size of 40 μ.

Methane tracer was fed into the sparger header in the form of a pseudo-random binary sequence, and continuous sampling was performed at a sparger nozzle and at some point a little below minimum expanded bed height. A ten-minute time record was captured on tape, and later processed via correlation analysis to the form of a process frequency response.

Parameter optimisation was based on a simplex technique and as before (1), two parameters per model were estimated by minimisation of an objective function. These parameters are identified in table 2. All outstanding parameters were determined in some suitable and independent manner. In particular, mass transfer coefficients below the sparger and between grid bubbles and the dense phase (in the case of the LSBP model) were estimated from results obtained previously (1). Bubble velocity was measured by the local tracer peak method (6), and dense phase voidage and velocity were assumed unaffected by the presence of the sparger. For a more comprehensive account, the reader is referred elsewhere (4).

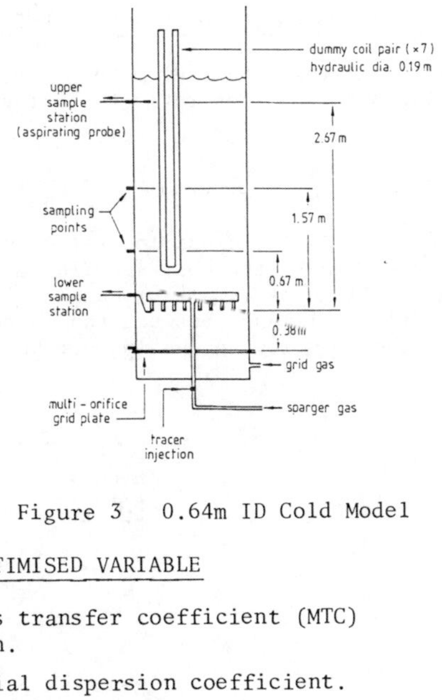

Figure 3 0.64m ID Cold Model

MODEL		OPTIMISED VARIABLE
LMBP	K_u	Interphase mass transfer coefficient (MTC) in upper region.
	D_d	Dense phase axial dispersion coefficient.
LSBP	K_s	Sparger bubble to dense phase MTC.
	D_d	Dense phase axial dispersion coefficient.

TABLE 2: OPTIMISED PARAMETER IDENTIFICATION

MODEL SCREENING RESULTS

The model screening results, presented in terms of averaged error indices (1) reflecting the relative discrepancy between the model and the experimental data, are given in table 3. The laterally segregated bubble-phase (LSBP) model fitted the experimental data with superior precision (i.e. a smaller error index) in every instance. This suggests that the LSBP model is the more appropriate for this particular system, and this in turn suggests that physically, bubbles from the sparger tend to retain their identity as they pass through the bed. It should be pointed out, though, that this implied retention of identity is based on interphase mass transfer and therefore probably applies more to the region immediately above the sparger than anywhere else in the bed.

MODEL	AVERAGE ERROR INDEX
LSBP	1.0
LMBP	6.69
No. of runs	19

TABLE 3: ERROR INDICES

MASS TRANSFER IN THE SPARGER REGION

The fact that the LSBP model emerges as the better process description allows a crossflow ratio to be logically defined as:

$$X_s = \frac{K_s(L_f - L_s)}{U_{bs}} \qquad (1)$$

where X_s is the crossflow ratio between sparger bubbles and the dense phase, and K_s the corresponding mass transfer coefficient referred to unit volume of (sparger) bubbles. The quantity $(L_f - L_s)$ represents vertical distance between the sparger and the top of the bed, and U_{bs} is the velocity in the sparger-bubble lean phase. The crossflow ratio

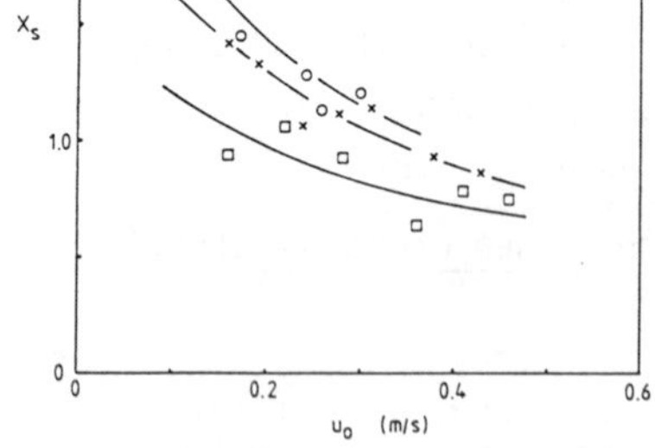

Figure 4: Crossflow Ratio vs. Total Superficial Velocity

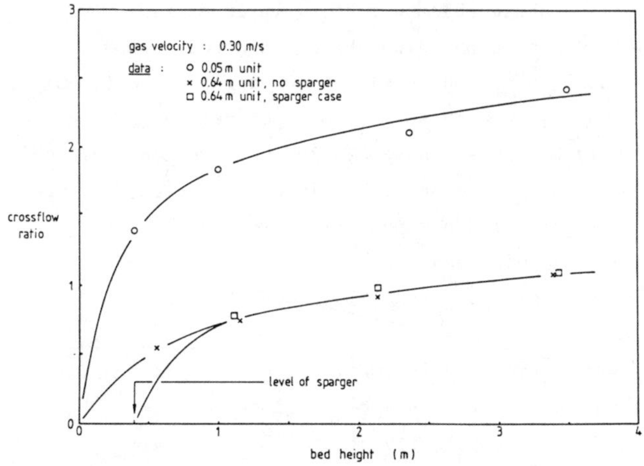

Figure 5: Axial Crossflow Profiles

X_s effectively represents the number of times the average sparger-generated bubble is purged on its way through the bed.

A plot of X_s against total superficial gas velocity is given in figure 4. The similarity between this plot and the corresponding one for a single (grid) feed (1) is quite striking, and this immediately suggests that the mechanism of interphase mass transfer in the sparger region is largely the same as that in the grid region.

A cross-plot of the data in figure 4 at constant superficial velocity is given as figure 5, and corresponding profiles for the single feed system on the same unit and for a 0.05m ID unit (1) are included for reference. From this figure it appears that gas introduced via the sparger experiences essentially the same degree of gas-solids contact as would have been the case if all the gas had been fed via the grid. There does not, therefore, appear to be any clear incentive for employing or avoiding a sparger from a mass transfer point of view.

LIST OF SYMBOLS

C_b	bubble concentration
C_{bg}	grid bubble concentration
C_{bs}	sparger bubble concentration
C_e	dense phase concentration
D_d	dense phase axial dispersion coefficient

f_s	sparger bubble phase weighting fraction
$f_{g\ell}$	grid bubble phase weighting in lower region
f_{gu}	grid bubble phase weighting in upper region
K_ℓ	bubble-dense phase MTC in lower region, erf. bbl. vol.
K_u	bubble-dense phase MTC in upper region, ref. bbl. vol.
$K_{g\ell}$	grid bbl-dense phase MTC in lower region, ref. bbl. vol.
K_{gu}	grid bbl-dense phase MTC in upper region, ref. bbl. vol.
K_s	sparger bbl-dense phase MTC, ref. bbl. vol.
L_f	minimum expanded bed height
L_s	height of sparger nozzles above the grid
t	time
$U_{b\ell}$	bubble velocity in lower region
U_{bu}	bubble velocity in upper region
U_{bg}	grid-bubble velocity
U_{bs}	sparger-bubble velocity
U_d	dense phase velocity
U_o	superficial gas velocity
X_s	sparger bubble to dense phase crossflow ratio
z	bed height co-ordinate

Greek Symbols

σ	fraction bubble hold-up
ε_d	dense phase voidage

LITERATURE CITED

1. Dry, R.J. and M.R. Judd, Pow. Tech. 43, 41 (1985).

2. May, W.G., Ind. Eng. Prog., 55 (12), 49 (1959).

3. Van Deemter, J.J., Chem. Eng. Sci., 13, 143 (1961).

4. Dry, R.J. "Gas residence time testing and model fitting : a study of gas-solids contacting in fluidised beds", Ph.D thesis, University of Natal, Durban (1984).

5. Carnahan, B., H.A. Luther and J.O. Wilkes, Applied Numerical Methods. John Wiley, New York (1969).

6. Dry, R.J., M.R. Judd and T. Shingles, Pow. Tech. 39, 69 (1984).

UNIFORMITY OF FLUIDIZATION ON A MULTI-ORIFICE GAS DISTRIBUTOR

A.P. BASKAKOV, V.G. TUPONOGOV, N.F. PHILIPPOVSKY
Ural Polytechnic Institute, Sverdlovsk, USSR

ABSTRACT

As the gas flow rate is increased through a bed lying on a multi-orifice gas distributor, the gas pressure at the distributor orifices may considerably exceed the quantity $\rho_s g(1-\varepsilon^o)H^o$. A formula has been derived for the complete -fluidization index at which all caps come into play. The experimental results obtained are in good agreement with theoretical predictions.

SCOPE

The conditions required to eliminate stagnant zones in a fluidized bed on a multi-nozzle gas distributor were investigated both theoretically and experimentally.

Whitehead (1) has found experimentally that the gas velocity which is necessary to destroy the stagnant zones (non-fluidized areas) above some caps must exceed some definite value W_G . Stagnant-zone formation with further decrease in velocity takes place at $W_{min} < W_G$. Whitehead (1) has proposed an empirical formula for calculating the value of W_{min} . Hiby (2), Yue (3) and Fakhimi (4) evaluated the possibility of stagnant zone formation theoretically, but they did not take into account a specific character of gas flow near the distributor orifices.

CONCLUSIONS AND SIGNIFICANCE

A formula is derived for calculation of the velocity of complete fluidization at which stagnant zones above all the caps are destroyed. A method is proposed to eliminate the stagnant zones in the process of putting an apparatus into operation. The usefulness of sectioning of a plenum chamber is demonstrated. The results obtained will help to design the distributor and to select the blower characteristics.

THEORY

Irrespective of whether the gas is distributed through orifices or caps, when the bed becomes fluidized the

$P = f(w)$ curve always exhibits a peculiar pressure peak, the maximum pressure drop across the bed at this instant exceeding $\Delta P_{fb} = \rho_s g(1-\varepsilon^\circ)H^\circ$ by some amount ΔP_e°. To estimate the magnitude of ΔP_e° during incipient fluidization, we consider the gas flow in the zone serviced by one orifice (Figure 1) under the following assumptions:

(1) The zone has the form of a cylinder of radius R.
(2) The gas is supplied through an orifice of radius r located at the center of the zone.
(3) The height of the cylindrical portion $H' = H^\circ - 2R/3$, as follows from the assumption (arbitrary to some extent) that the sum of the volumes of the cylinder of height H' and the hemisphere of radius R is equal to the volume of the entire material in the cylinder of height H°.
(4) The gas percolation regime is laminar: $dP/dH = w/\lambda$.

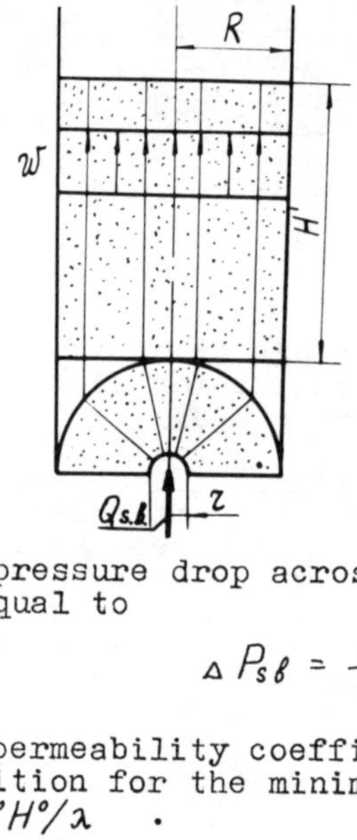

Figure 1. Diagram for calculating the pressure drop across the non-fluidized (stagnant) bed

The pressure drop across the entire fixed (stagnant) bed is equal to

$$\Delta P_{sb} = \frac{w}{\lambda}\left[\frac{R}{2}\left(\frac{R}{r}-1\right)+\left(H^\circ-\frac{2}{3}R\right)\right] \qquad (1)$$

The permeability coefficient λ will be found from the condition for the minimum fluidization $\rho_s(1-\varepsilon^\circ)gH^\circ = w^\circ H^\circ/\lambda$.

By definition, ΔP_e° is the magnitude of the excess pressure drop across the bed at the time of fluidization ($w = w^\circ$)

$$\Delta P_e^\circ = \Delta P_{sb}^\circ - \rho_s g(1-\varepsilon^\circ)H^\circ \qquad (2)$$

Substitution of Equation (1) and $r/R = \sqrt{\varphi}$ into Equation (2) yields

$$\Delta P_e^\circ = \frac{\rho_s g(1-\varepsilon^\circ)r}{2\varphi}\left(1-\frac{7}{3}\sqrt{\varphi}\right) \qquad (3)$$

If all the caps and the packing of particles in the bed were absolutely the same, the incipient fluidization would occur above all N caps at the same time on reaching a plenum pressure $P_G = \Delta P_{fb} + \Delta P_e^o + \Delta P_d^o$, i.e., at point A (Figure 2). After the onset of fluidization the redistribution of the flow above the orifices is largely facilitated, i.e., $\Delta P_e \to 0$, and the pressure in the plenum chamber decreases to the value P_D.

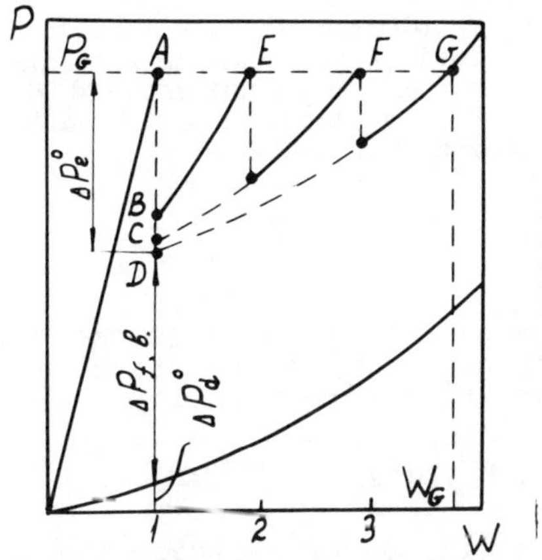

Figure 2. Theoretical dependence of plenum-chamber pressure on dimensionless fluidization velocity for 3 distributor caps

It is, however, evident that because of random doviations that arise when the gas flow rate is increased the material will first be fluidized above one or a few ($N-n$) caps. It may be assumed that the flow rates through any one of the operative caps are the same and equal to Q_{fb}, those through the inoperative ones are equal to Q_{sb}, and the ratio of the total gas flow rate for the apparatus to the total number of caps is $\bar{Q}$. Then

$$N\bar{Q} = (N-n)Q_{fb} + nQ_{sb}; \; NW = (N-n)W_{fb} + nW_{sb} \tag{4}$$

The plenum-chamber pressure P is equal to the sum of the pressure drop across the distributor and that across the bed. The distributor pressure drop is proportional to the square of the gas velocity value, i.e., $\Delta P_d = \Delta P_d^o W^2$. Taking account of the fact that the operative and inoperative caps are coupled to one plenum chamber, we have

$$P = P_{fb} = P_{sb} = \Delta P_d^o W_{fb}^2 + \Delta P_{fb} = \Delta P_d^o W_{sb}^2 + \Delta P_{sb} \tag{5}$$

Figure 2 presents a plot, constructed on the basis of Equations (4) and (5). For simplicity, the total number of caps is taken to be small ($N = 3$). If, during the initial increase of the velocity, the bed breaks down only above one cap the pressure in the plenum chamber will drop to the value P_B, if above two caps (which is less probable) then to P_c , and, finally, if above all three caps then to P_D . As the velocity is increased further, the plenum chamber pressure P rises (curves BE, CF and DG respectively). As soon as the pressure in the plenum chamber reaches again a value equal to P_a the caps in the zones that are still stagnant (points E and F) come into play for the velocity of gas percolation through the stagnant zones reaches the value of w^o.

Equations (4) and (5) allow us to obtain an expression for estimating the dimensionless velocity W_n , at which the number of inoperative caps in the bed will not exceed n . Allowing for the equalities $W_{sb} = 1$, $\Delta P_{sb} = \Delta P_e^o + \Delta P_{fb}$ which hold when the bed is fluidized above one of the caps, Equations (4) and (5) yield

$$W_n = \frac{n}{N} + \frac{N-n}{N}\sqrt{1 + \Delta P_e^o/\Delta P_d^o} \tag{6}$$

If $W_n > W_1$ no stagnant zones will be present in the bed. With increase in the number of caps N on the distributor the quantity W_1 will increase up to the velocity W_G (Figure 2)

$$W_G = \sqrt{1 + \Delta P_e^o/\Delta P_d^o} \tag{7}$$

at which complete fluidization will occur whatever the number of caps in the distributor.

Experimental Observations

The considerations outlined above were verified experimentally on an apparatus outfitted with a 0.6 x 1.2 m^2 distributor which had 18 gas distribution caps spaced a distance of 0.2 m apart. Every cap had 6 horizontal drillings with diameter of 14 mm (φ = 0.024). Corundum particles were air-fluidized. The mean particle diameter was 0.4 mm, the bulk density was ρ_s = 3960 kg/m^3. The distributor pressure drop at minimum fluidization was as high as ΔP_d^o = 390 Pa. A calculation according to Equation (3) gives ΔP_e = 1.86 kPa for these conditions.

The air flow rate was measured for each of the 18 caps with the gauges being interrogated at a frequency of 130 Hz, the readings being immediately displayed on a minicomputer. This permitted a continuous supervision of the number of operative and inoperative caps. The actuation of each orifice in the same cap was not checked. The quantity ΔP_e^o that was obtained in different experiments performed under practically the same conditions was at variance (Figure 3), since when the gas flows into the fixed bed through the orifice a cavity is produced above the orifice. The cavity size depends on many fortuitous factors. It is determined by the possibility of additional compaction of

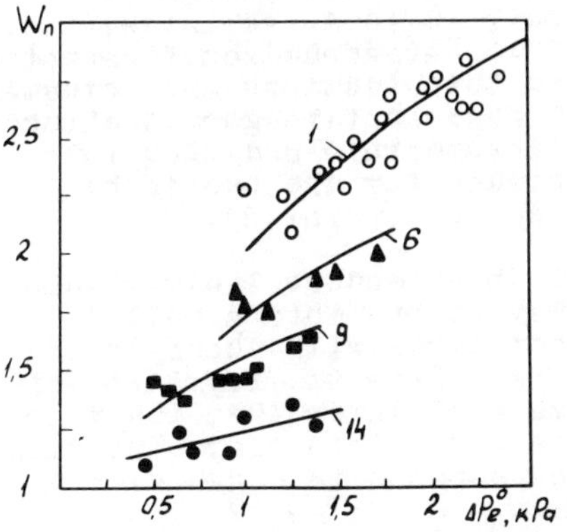

Figure 3. Experimental (points) and theoretical (curves) dependences of the dimensionless fluidization velocity W_n, at which one of n (numerals near the curves) inoperative caps comes into play, versus excess pressure drop across a stagnant bed zone ΔP_e^o for $H^o = 0.25$ m

particles near the orifice, i.e., by the initial void fraction of the bed. In Equation (3) the quantity z actually denotes the radius of the cavity during the transition of the bed to the fluidized state. The smallest values of ΔP_e^o and, accordingly, W_n were determined experimentally when refluidizing the material that had not had enough time to become firmly packed and compacted. The relation between W_n and ΔP_e^o corresponded to Equation (6).

Application of Results

To eliminate stagnant zones in large commercial fluidized-bed apparatus, it is common practice to divide the plenum chamber into a number of sections. Consider the case when a centrifugal blower operates for an apparatus whose plenum chamber is divided into two sections (Figure 4). Curve 1 presents the characteristic of the blower, when curve 2 – the air flow rate dependence of plenum-chamber pressure with only one section at work, i.e., a simplified version of the dependence exposed in Figure 2. The dashed line in Figure 4 corresponds to the region in which not all of the

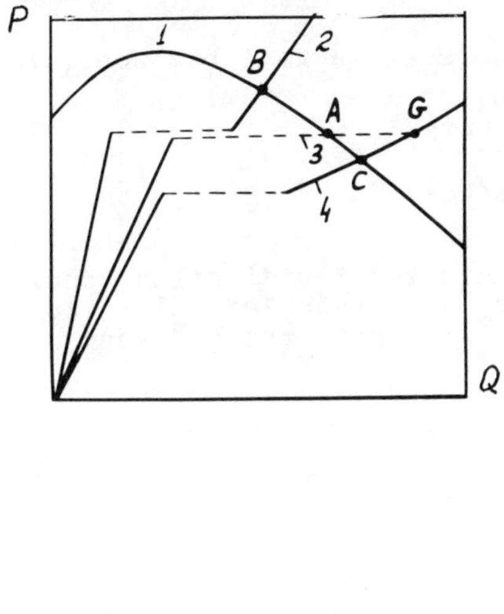

Figure 4. Effect of plenum-chamber sectioning

caps are operative. When both sections are engaged in parallel (curve 3) the gas flow rate is twice large, i.e., $Q_3 = 2Q_2$ when $P_3 = P_2$. All the stagnant zones cannot be destroyed in the regime when both sections are actuated simultaneously (point A). But if the entire gas is directed into one of the sections, the pressure and flow rate produced by the blower will suffice for the bed to be fluidized completely in this section (point B).

After complete fluidization has been successively reached in both sections, the latter may be put into parallel operation (point C). No stagnant zones will occur in that case since the velocity W_{min} at which the formation of stagnant zones begins is always much lower than the velocity W_G at which the last cap comes into play. The flow rate dependence of pressure corresponds to curve 4.

It is unreasonable to make many sections since, firstly, this complicates the design and, secondly, stalling and surging may arise in centrifugal blowers when the flow rate is highly decreased compared to the rated value.

NOTATION

g - free fall acceleration, m^2/s;
H - bed height, m;
N - number of gas distributing caps;
n - number of inoperative caps;
P - plenum-chamber pressure, Pa;
r - radius of distributor orifice, m;
R - radius of the bed cross-section serviced by a particular orifice, m;
Q - gas flow rate, m^3/s;
w - fluidization velocity, m/s;
$W = w / w^o$ - fluidization index;
ΔP - pressure drop, Pa;
ΔP_e - excess pressure drop across a stagnant bed zone, Pa;
ε - void fraction of the bed;
λ - permeability coefficient, $m^2/s \cdot Pa$;
ρ_s - density of particles, kg/m^3;
φ - free area of distributor

Super- and subscripts: (^{o}) - minimum fluidization condition; fb - fluidized bed zone; G - complete fluidization; d - cap (distributor); sb - stagnant bed zone.

LITERATURE CITED

1. Whitehead, A.B., In: Fluidization, J.F. Davidson and D. Harrison (Eds), Academic Press, London and New York, pp 781-814, (1971)

2. Hiby, J.W., Chem. Ing. Techn., 39 (10), 1125 (1967)

3. Yue, P.L., Kolaczowski, J.A., Trans. I. Chem. Eng., 60, 164 (1982)

4. Fakhimi, S., Sohrabi, M., Harrison, D., Can. J. Chem. Eng., 61, 364 (1983)

DESIGN OF DISTRIBUTORS FOR AIR GASIFICATION OF CHAR IN FLUIDIZED BEDS

R.R. Rathbone and J.F. Davidson,
Department of Chemical Engineering, University of Cambridge,
Pembroke Street, CAMBRIDGE CB2 3RA, England.

ABSTRACT

The carbon monoxide yield of an air-fluidized gasifier depends on the distributor, which influences (1) crossflow and (2) oxygen concentration in the particulate phase near the base of the reactor. Sintering may occur in dead zones above an inactive orifice, but can be avoided by suitable orifice spacing.

SCOPE

This study was prompted by the National Coal Board programme to develop an air-fluidized bed gasifier in which a combustible gas is produced from a bed of air-fluidized coal particles. The influence of the distributor on (i) the yield of carbon monoxide and (ii) the sintering of bed particles is described herein.

Shaw and Paterson (1) and Kossakowski (2) found that significant amounts of carbon monoxide could be formed as an intermediate in the combustion of char in an air-fluidized gasifier. The combustion reactions occur near the distributor, whose design may affect the quantity of carbon monoxide formed as a combustion intermediate. This study aims to examine the effect of distributor design on carbon monoxide yield and to identify a design to maximize this yield.

An air-fluidized gasifier is operated at a high temperature to maximize the rate of the gasification reactions. If the temperature is too high, the ash in the bed particles will soften, causing them to stick together. Formation of a hot spot will result in the sintering of the entire bed.

Circulation of the bed particles gives rapid heat transfer within a fluidized bed. If the bed remains fluidized everywhere, a hot spot will not develop. Dead zones of defluidized particles may occur on the distributor. If reaction takes place within a dead zone, a hot spot can develop and the bed may sinter. Part of this study was to determine the conditions under which the formation of a dead zone on the distributor leads to sintering.

CONCLUSIONS AND SIGNIFICANCE

The yield of carbon monoxide from an air-fluidized gasifier is enhanced by a distributor that promotes a rapid exchange of gas between the bubble and particulate phases. Thus a distributor that produces many small bubbles is preferable to a distributor that produces a few large bubbles. If the circulation of the bed is to be improved by an additional supply of gas through a central spout and if gases of differing oxygen concentrations are being used, e.g. air and combustor offgas, then the yield of carbon monoxide is improved if the spout gas has a low oxygen content.

Sintering does not occur in dead zones to the side of an active orifice but sintering may occur in a dead zone above an inactive orifice. Sintering will occur if the dead zone is larger than a critical size calculated from Equation (3).

These conclusions are also valid for other fluidized bed reactors as well as the air fluidized gasifier. The yield of any intermediate species from a fluidized bed reactor is governed by the same principles as the yield of carbon monoxide from the air gasifier. By using these principles an appropriate distributor can be designed to minimise undesirable intermediates or maximise desirable intermediates.

Fluidized bed reactors are often used for highly exothermic reactions when an isothermal bed is desired. The formation of a dead zone above an inactive orifice in such a reactor will produce a hot spot. Equation (3) can be used to design the distributor so that the formation of a single hot spot does not result in the total failure of the reactor.

CARBON MONOXIDE YIELD

Carbon monoxide is formed via two reaction paths in the air gasifier: the reaction between carbon and carbon dioxide, and as an intermediate in the oxidation of carbon to carbon dioxide. The combustion reactions occur close to the distributor whose design may influence the amount of carbon monoxide produced as a combustion intermediate.

Experimental

The experiments were carried out in a 0.1 m diameter gasifier described in detail by Rathbone (3). The distributor is shown in Figure 1. Fluidizing gas could be supplied to the bed through either (1) a central orifice or (2) perforations in the cone; a combination of (1) and (2) could be used. A mixture of air and nitrogen was used for fuildization. The bed material was Pharnacite char with a particle size range 0.3 to 1.0 mm. The offgas from the bed was sampled and the concentration of carbon monoxide measured by an infra red gas analyser. A plot of carbon monoxide concentration as a function of temperature was made for each of three methods of supplying gas to the bed, as follows.

(A) All the gas (60% air, 40% N_2) was supplied to the central orifice, Figure 1.
(B) Air was supplied to the central orifice and nitrogen to the perforated cone.
(C) Nitrogen was supplied to the central orifice and air to the cone.

Results and Discussion

Figure 2 shows the carbon monoxide yield as a function of temperature for each of the three distributor configurations. The carbon monoxide yield is highest when nitrogen is supplied to the central orifice and air through the cone. The yield is lowest when all the fluidizing gas is supplied to the central orifice. This behaviour may be explained by considering the hydrodynamics of the bed near the distributor and their interaction with the reaction kinetics.

Oxygen in the particulate phase reacts both with char particles to produce carbon monoxide and with the carbon monoxide to produce carbon dioxide. The amount of carbon monoxide formed by combustion is determined by the balance between these rates. The combustion of char to carbon monoxide is first order with respect to oxygen concentration but the oxidation of carbon monoxide is of half order. Thus a high oxygen concentration will favour the production of carbon monoxide since that is the higher order reaction but a low oxygen concentration will favour the oxidation of carbon monoxide. The oxygen concentration in the particulate phase is determined by the balance between the rate at which it is consumed by combustion reactions and the rate it is replenished at by crossflow from bubbles. A rapid crossflow will lead to a high concentration of oxygen in the particulate phase which will result in a higher carbon monoxide yield. The crossflow depends on bubble size. The crossflow from a few large bubbles is less than the crossflow from many small bubbles of the same total bubble volume. Thus the distributor which produces smaller bubbles will provide a greater crossflow near the base of the reactor and a higher yield of carbon monoxide. This is demonstrated by the improvement in carbon monoxide yield when air is supplied through many small holes in the distributor cone, arrangement C, Figure 2, over the case where all the fluidizing gas is supplied to the central orifice, arrangement A, Figure 2.

The highest oxygen concentrations in the particulate phase are where air is entering the particulate phase at the base of the bed. This is where the greatest proportion of carbon monoxide will be produced by combustion. A distributor which supplies a higher concentration of oxygen to the particulate phase than to the bubble phase will produce a higher yield of carbon monoxide. A larger proportion of the gas supplied through the cone will go to the particulate phase as it is supplied over a wider area of the base of the bed than gas supplied through the central orifice. This is demonstrated by the improved carbon monoxide yield, arrangement C, Figure 2, when air is supplied through the cone and nitrogen through the central orifice as opposed to arrangement B, Figure 2, when nitrogen is supplied through the cone and air through the central orifice.

SINTERING

Sintering may occur in the air-fluidized bed gasifier when there are dead zones of defluidized particles within the bed. Two different experiments were conducted to observe the appearance of a hot spot visually and to measure the temperature rise within a dead zone.

Circulation Near a Single Air Inlet

A fluidized bed of char contained in a 76 mm diameter transparent quartz tube was heated to 850°C by burning a mixture of air and propane in the bed. The fluidizing air/propane mixture was then turned off and air alone was supplied through a tube close to the quartz wall of the containing tube. The circulation of the char particles was observed through a gap in the bed insulation. The flowrate of air was varied and the appearance of hot spots and the circulation pattern of the solids were noted.

Two different regimes of solids circulation were seen, as shown in Figure 3. At high velocities a narrow jet was formed which extended to the bed surface. Solids circulation occurred within a narrow cone, with solids moving up in the centre and down on the outside, Figure 3(a). There was no hot spot anywhere within the dead zone. Visual observation showed that the circulating solids were always hotter than the dead zone.

When the flowrate of air was decreased sufficiently, the circulation pattern collapsed. A void appeared above the air entry point, Figure 3(b). There was no circulation of solids. A hot spot developed at the top of the void and slowly extended upwards. The temperature in the hot spot was higher than the temperature of the circulating particles in Figure 3(a). In some cases the temperature was high enough for aggregates to form. Once the low velocity pattern was established, a large increase in the air flowrate was necessary to break it up and return to the high velocity pattern.

It is concluded from these observations that sinters may form in a dead zone above an air inlet, but not in a dead zone to the side of an air inlet. To determine whether the defluidization of part of the bed above an orifice will lead to the formation of a sinter, it is necessary to determine the temperature rise in the hot spot.

Temperature Rise Within a Hot Spot: Experimental

The temperature rise in a defluidized char bed was measured using apparatus shown in Figure 4. A bed of char was contained within a steel tube in the centre of a 76 mm diameter bed of sand. Steel tubes of 12, 24 and 52 mm were used, thus varying the bed diameter. The steel tube was removed to obtain a 76 mm diameter char bed. The sand diameter was 0.35 to 0.5 mm; the two chars used were 0.7 to 1.0 mm Illinois No. 5 and 1.4 to 1.8 mm coke. The sand and char were heated by burning a mixture of air and propane in the bed. The propane was turned off and the flowrate of air reduced until the velocity was slightly below the value for minimum fluidization of the char; the sand was not fluidized.

The temperature was measured by a 1 mm diameter thermocouple in the centre of the bed, 20 mm above the distributor, the temperature being recorded on a chart recorder. Maximum values, T_m, are given in Table 1 together with $\Delta T = T_m - T_w$ where T_w was the temperature of the bed just before it was defluidised. It is presumed that this was also the temperature of the tube, radius r_a. Table 1 shows the superficial velocity, U, of the air at bed temperature.

TABLE 1

Temperature Rise in a Partly Fluidised Char Bed

r_a mm	U m/s	U/U_{mf} approx	T_m ^oC	ΔT K	Char
6	0.09	0.9	860	50	Illinois 5
12	0.08-0.09	0.9	860- 930	70-100	Illinois 5
12	0.18	0.9	1030	160	Coke
26	0.10-0.14	0.9	1135-1140	295-300	Illinois 5
38	0.16	0.9	1170	280	Illinois 5

Table 1 shows that the observed temperature rise in a partially fluidized bed increases with its diameter and with the velocity of the air blown through it. Presumably this arises from the balance between heat generated by combustion and lost by radial conduction. Increasing the rate of heat generation or the bed size gives a higher central temperature. This balance will be quantified in the following section.

Temperature Profile Near a Hot Spot: Theory

It is assumed that (1) the temperature reaches a steady state; (2) there is cylindrical symmetry; (3) all the heat of reaction is released at the base of the bed; (4) the molar flowrate of gas is constant across the bed and (5) the thermal conductivity is uniform throughout the hot spot. A heat balance at radius r, height x above the distributor gives

$$2\pi k \frac{\partial}{\partial r}\left(r\frac{\partial T}{\partial r}\right) + 2\pi rk\frac{\partial^2 T}{\partial x^2} - 2\pi rU\rho_g C_p\frac{\partial T}{\partial x} = 0 \qquad (1)$$

The boundary conditions are: (1) no vertical conduction through the top of the bed; (2) the heat of combustion, ΔH per mole of entering gas, is released at the base of the bed, $x = 0$; (3) $\partial T/\partial r = 0$, $r = 0$ and (4) $h(T - T_w) = -k\,\partial T/\partial r$, $r = r_a$. The solution is the series

$$T = T_w + \sum_n C_n J_o\left(\alpha_n \frac{r}{r_a}\right) \exp\left\{\frac{-\rho_g U C_p x}{2k} - \left[\left(\frac{\rho_g U C_p}{2k}\right)^2 + \left(\frac{\alpha_n}{r_a}\right)^2\right]^{\frac{1}{2}} x\right\} \qquad (2)$$

where $J_o(\alpha_n) + (k\alpha_n/hr_a)\,J_1(\alpha_n) = 0$ and

$$C_n = \frac{4\left[\Delta H/C_p - (T_w - T_i)\right] J_1(\alpha_n)}{\{1 + \left[1 + (2k\alpha_n/\rho_g U C_p r_a)^2\right]^{\frac{1}{2}}\}\,\alpha_n\left[J_1{}^2(\alpha_n) + J_o{}^2(\alpha_n)\right]}$$

h is the bed to wall heat transfer coefficient and T_i the gas inlet temperature.

From Equation 2 the maximum temperature occurs at the centre of the base of the bed, r = 0, x = 0. It is a function of the bed properties and the radius of the dead zone, r_a. As r_a increases, the maximum temperature in the hot spot increases: above a critical value sintering occurs. If T_s is the sintering temperature then the critical value of r_a can be calculated from

$$T_s = T_w + \sum_n C_n \qquad (3)$$

TABLE 2

Maximum Temperatures, T_m, from Equation (2).

C_p = 25.0 J/molK, T_i = 17°C

k W/m/K	h W/m²/K	T_w °C	ΔH kJ/mol	U m/s	r_a mm	T_m °C
0.7	100	800	44	0.09	6	974
0.7	100	800	44	0.09	12	1097
0.7	100	800	44	0.10	24	1314
0.7	100	800	44	0.10	38	1467
1.0	100	800	44	0.09	12	1036
0.7	200	800	44	0.09	12	1062
0.7	100	700	44	0.08	12	1025
0.7	100	800	79	0.09	12	1430

Table 2 gives maximum temperatures T_m calculated from Equation (2) for comparison with the experimental values in Table 1. ΔH is 44 for carbon burning to carbon monoxide and 79 for carbon burning to carbon dioxide. k is estimated as 0.7 W/m/K using the method of Yagi and Kunii (4). The theoretical maximum temperatures are larger than the experimental values. There are several possible reasons for this, as follows. The apparatus cooled during the course of an experiment so T_w may have been lower. The thermal conductivity of the bed might be larger than the estimated value of 0.7 W/m/K: if the bed sinters, the thermal conductivity of the sinter will be higher than the thermal conductivity of the bed. The theoretical maximum temperature is calculated at a point but the thermocouple has a finite size: the temperature gradient near the maximum may be as high as 50K/mm, so the temperature recorded by the thermocouple will be much lower than the maximum.

TABLE 3

Critical Radius, r_a, from Equation (3)

$T_w = 950^oC$, $T_s = 1100^oC$, $\Delta H = 44kJ/mol$

k	h	U	r_a
0.7	300	0.09	8.4mm
0.7	200	0.09	7.6mm
2.0	300	0.09	24 mm
0.7	300	0.30	1.7mm

Table 3 shows values of the critical radius for sintering, calculated for various possible conditions in an air-fluidized gasifier. The heat transfer coefficient used is for heat transfer from a wall to a fluidized bed. The small values predicted for r_a show that sintering is likely if an orifice becomes inactive.

NOTATION

- C_n coefficients of Equation (2)
- C_p gas specific heat capacity
- h bed to wall heat transfer coefficient
- J_n n^{th} order Bessel function
- k effective thermal conductivity of a packed bed
- r radial distance from bed axis
- r_a bed radius
- T temperature
- T_i gas inlet temperature
- T_m maximum temperature in a partly fluidized bed
- T_s temperature at which the bed particles sinter
- T_w wall temperature
- U superficial gas velocity at bed temperature
- x height above distributor
- α_n eigenvalue in Equation (2)
- ΔH heat release per mole of gas entering the bed
- ΔT $T_m - T_w$
- ρ_g gas molar density

LITERATURE CITED

1. Shaw, J.T. and N.B. Paterson, "Studies of the gasification of solid fuels in a fluidised bed at atmospheric pressure", in Fluidization, Davidson J.F. and D.L. Keairns (Eds). Cambridge University Press. (1978).

2. Kossakowski, E.R., "Coal Gasification in a Fluidized Bed", Ph.D. Dissertation, University of Cambridge (1981).

3. Rathbone, R.R., "Gasification and Combustion of Carbon in Fluidized Beds", Ph.D. Dissertation, University of Cambridge (1985).

4. Yagi, Sakae and Daizo Kunii, AIChEJ, 3(3), 373 (1957).

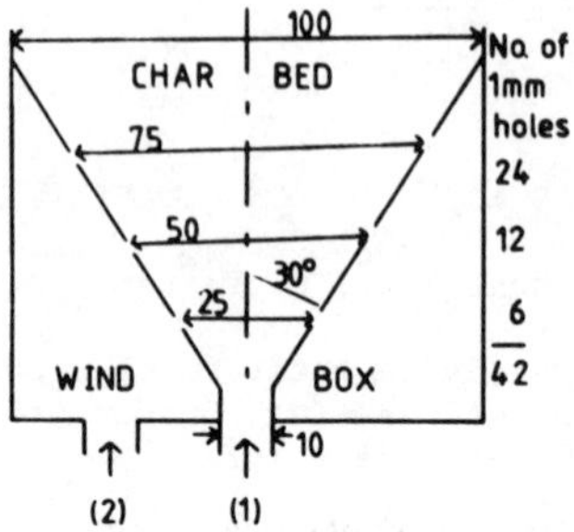

Figure 1: Distributor with Gas Inputs (1) & (2) Dimensions in mm

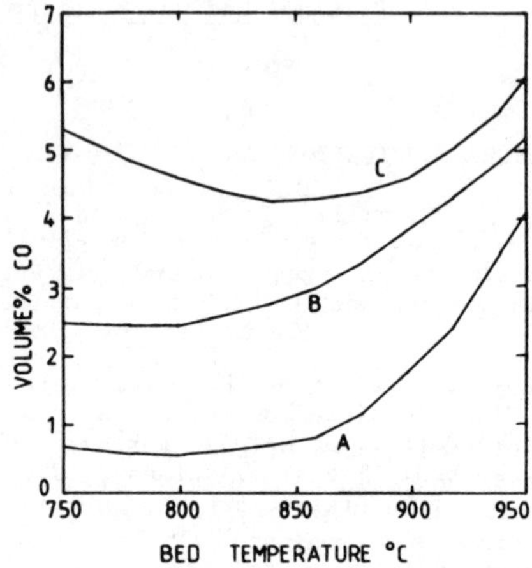

Figure 2 Effect of Distributor on CO Yield

A) Gas through (1)

B) Air through (1) and N_2 through (2)

C) N_2 through (1) and Air through (2)

Fluidizing gas: 60% Air and 40% N_2

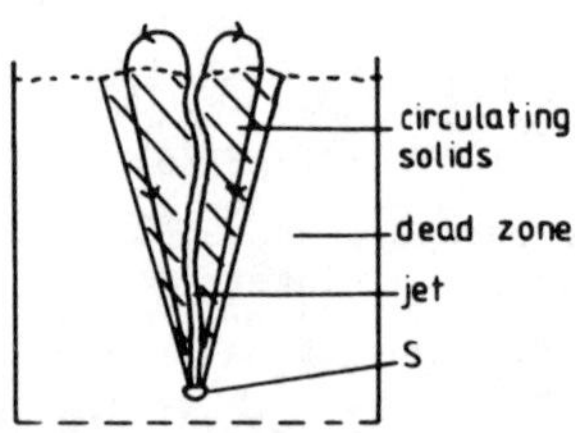

a) high velocity

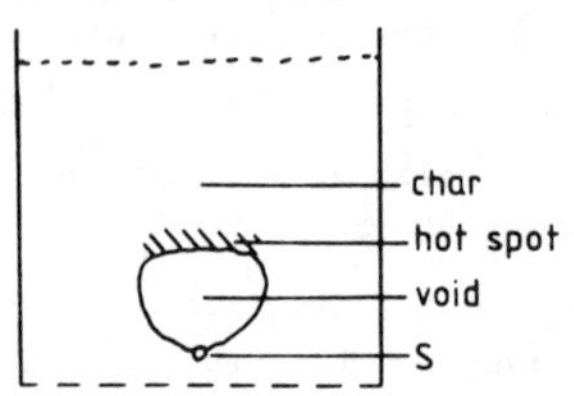

b) low velocity

Figure 3: Observed Circulation Patterns near a point air supply, S, to a bed of 1mm char at 850°C

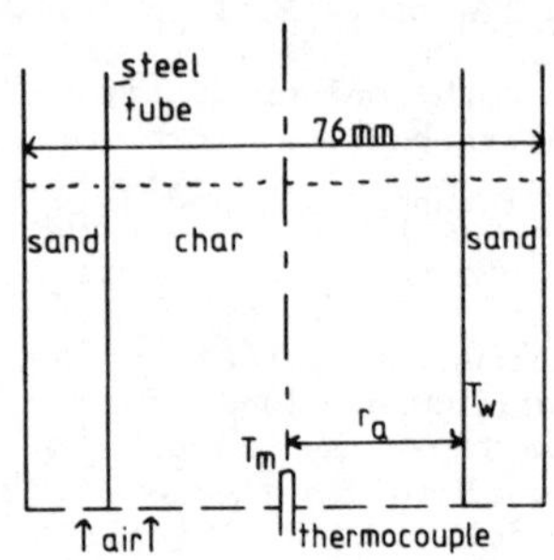

Figure 4: Temperature rise in a partly fluidized bed of char.

PRESSURE FLUCTUATIONS ASSOCIATED WITH GAS INJECTION IN FLUIDIZED BEDS

M. Filla*, L. Massimilla*, D. Musmarra*, S. Vaccaro**

* Dipartimento di Ingegneria Chimica, Università - Istituto di Ricerche sulla Combustione, C.N.R.; ** Dipartimento di Chimica, Università
P.le V. Tecchio - 80125 NAPOLI (Italy)

ABSTRACT

Different modes of gas injection in two-dimensional fluidized beds of Group B and of Group D particles are detected by pressure fluctuations at the orifice and in the bed. Frequencies of bubble generation and coalescence from photographic observation are in agreement with those from the analysis of pressure signals. Pressure perturbations spread from the jetting region at much higher velocities than predicted by available theories.

SCOPE

Gas injection in fluidized beds may have a significant role in the behavior of a number of processes, including coal combustion and gasification. The objective of this work is that of supplementing photographic information on modes of upward gas injection in a fluidized bed with instantaneous pressures in and around the jetting region. Pressure fluctuation analysis has been used previously to evaluate fluidization quality as a function of bed height, excess gas flow rate, distributor type, solids properties and bed temperature. Here the attention is focused on pressure disturbances associated with gas injection through an orifice in a bed otherwise uniformly fluidized with a plate distributor. Beds are relatively shallow to isolate injection effects from those due to bubbling in the bulk of fluidized solids. A two-dimensional column is used as a step towards an investigation in a cylindrical column with axial gas injection at the base.

CONCLUSIONS AND SIGNIFICANCE

Useful information regarding the mode of gas entry in a fluidized bed is obtained by measuring pressures at the orifice and in the bed. Depending on whether the bed is made of 200-400 μm sand Group B or 2800-3000 μm plastics Group D particles (1), pressures at the orifice are periodically higher and lower, or permanently lower than those far away in the bed. With plastics the jetting region is, therefore, a low resistance path for the flow of injected gas. This is consistent with the formation of a permanent jet and with the steady flow of solids from the bed towards such a jet, which are typical of these solids as shown in this and previous photographic studies with two-dimensional and semicylindrical columns (2,3). Power spectral density diagrams of pressure signals provide frequencies of bubble formation at the orifice (with sand) and at the tip of the jet (with plastics), as well as the numbers of such bubbles involved in coalescence immediately downstream. The frequencies are in broad agreement with those obtained with Group B particles by means of X-ray motion picture of beds in three-dimensional columns (4,5), and with Group D particles by motion pictures through transparent walls of two-dimensional and semicylindrical columns (2,3). Pressure disturbances spread from the jetting region throughout the bed within the range of bed heights tested. Time delays from phase angles of cross spectra between pairs of records indicate that pressure perturbations move at much higher speed than predicted by available theories on dynamic wave propagation in fluidized beds (6,7,8). Pressure frequential analysis is a powerful tool in the diagnostics of gas injection in fluidized beds, which can be extended from two-dimensional to full scale three-dimensional beds provided difficulties of intrusive pressure measurements in the jetting region are conveniently solved.

PREVIOUS DEVELOPMENTS

Modes of gas injection in a gas fluidized bed. Phenomena related to gas injection in a fluidized bed have been subject to different interpretations. The jetting region, in particular, has been considered as formed either of a plume of bubbles (Figure 1a) periodically detaching at the orifice (4,5,9), or of a permanent jet (Figure 1b), periodically developing to generate bubbles somewhere downstream of the orifice (10,11,12,13). X-ray examination led Rowe et al. (4) to the conclusion that jets are not typical of gas entry in a fluidized bed, except when the bed is inadequately fluidized or surfaces are present to hinder the flow of particles towards the orifice. High speed motion pictures of gas injection by Donsì et al. (2) and by Yang and Keairns (3) with two-dimensional and with semicylindrical columns with transparent walls suggest, however, that both modes may occur, depending on the type of solids. Apparently, Group A/B particles lead to the formation of a bubble plume, and Group B/D particles to permanent jets. Fluid-spouted beds, with jet crossing the whole bed, are an extreme case of the latter behavior with shallow beds.

Pressure fluctuations in a gas fluidized bed. Amplitude and frequency of pressure fluctuations in fluidized beds have been for a long time recognized as useful parameters in the diagnostics of fluidization quality. Since the early work of Shuster and Kisliak (14) their evaluation has been improved progressively by the application of statistical methods in the decoding of signals. The origin of pressure fluctuations has drawn the attention of the majority of authors interested in the subject. It appears to be by no means a simple problem because pressure depends on local bed properties, as well as on events occurring away from the point of measurement. In very shallow beds fluctuations are regular, and have been related to coherent particle oscillations by Hiby (15) and Verloop and Heertjes (16). Above a certain bed height they become irregular, and have been related to bubble formation (16,17), bubble rising motion (18,19,20), and bubble eruption at the bed surface (21,22). Each such phenomenon, together with bubble coalescence, may become dominant at different heights of a fluidized bed according to Fan et al. (23).

EXPERIMENTAL

A 390 mm x 16 mm cross section transparent column described in detail elsewhere (24) has been used. Fluidizing air is supplied through a high pressure drop perforated plate distributor, while air is discharged by a 6 mm orifice flush to the distributor on the vertical symmetry plane. Feedback of bed pressure fluctuations on injected air flowrate is prevented by a high pressure drop perforated plate inserted on the air supply line immediately upstream from the orifice. The front plate of the column is made of glass for visual and high speed cinefilm (1000 to 5000 fps) observation. The back plate is made of perspex with two series of 0.5 mm tap holes on the bed centerline and at 20 mm from one of the side walls to record pressures at up to eight locations simultaneously.

Silica sand (d_p=200 to 400 μm, ρ_p = 2500 kg/m^3; U_{mf} = 0.045 ms^{-1}; Group B) and plastic spheres (d_p = 2800 to 3000 μm, ρ_p = 1200 kg/m^3; U_{mf} = 1.0 ms^{-1}; Group D) have been used. Velocity of gas discharge through the orifice is about 30 m/s with sand and 60 m/s with plastics. The fluidizing velocity is kept at 0.35 m/s and 1.5 m/s, respectively. Static bed heights H_b are 110 and 310 mm for both materials. A shallow sand bed (H_b=15 mm) has also been tested. This range of experimental variables covers the following modes of gas discharge: fluid-spouted bed (sand, H_b=15 mm; plastics, H_b=110 mm), bubble plume (sand, H_b=110 to 310 mm), and permanent jet (plastics, H_b=310 mm).

Significant wall effects may have been expected with plastics due to the low particle to bed thickness ratio; however, high speed cinefilms of the flow pattern in this apparatus and in a semicilyndrical column 160 mm I.D. (2) gave very similar results, suggesting that this is not a major effect.

The data acquisition system consists of Schaevitz P502 strain gauge pressure transducers connected to the tap holes on the back wall, whose signals are passed through signal conditioners and admitted to a HP 3437A digital voltmeter in a sequence set by a HP 3495 scanner. Two "simultaneous" records are shifted of 1 ms. Pressure signals at a given location and at pairs of locations have been treated statistically to obtain dominant frequencies and time delays of the pressure fluctuations according to Bendat and Piersol (25). To this end a HP 9817 minicomputer stores the pressure-time records and yields the power spectral and cross-spectral density functions by Fast Fourier Transform. Each record consists of 31744 point values spaced by 20 ms.

RESULTS AND DISCUSSION

Jetting regions in beds of sand and plastics somewhat recall those in Figures 1a) and 1b) respectively when H_b=310 mm. Void contours, in particular, are very irregular compared to those sketched in these figures. Fluid-spouted bed operation is fully established with sand for H_b=15 mm and approached with plastics for H_b=110 mm. For the bed heights and velocities of section above the fractions of gas flow visible as bubbles at the tip of the nozzle (with sand) or of the permanent jet (with plastics) are about 0.6 and 0.25 respectively, which are in partial agreement with previous findings (5,26). Solids volume concentration is below 0.05 in the freshly formed bubbles with sand and about 0.2 in the jet with plastics. At the bottom of the bed particles are periodically directed back and forth with respect to the orifice with sand, and steadily towards the orifice with plastics, due to the permanent jet with this material. Particles enter the root of the jet at velocities in the range of 0.1 to 0.2 ms^{-1}.

Gas enters beds of sand as chain of bubbles as described by Yates et al. (5) for beds of alumina with comparable particle size and inlet gas velocity. Bubbles developing at the orifice at frequencies in the range 2 to 8 Hz coalesce to produce a stream of bubbles at much lower frequency. The first bubble of a sequence is usually larger than the followers. With plastics the swelling of the tip of the jet at its minimum penetration length (30 to 50 mm) opens the sequence of events by forming the roof of the leading bubble; particles conveyed upwards along the jet and entering the jet laterally build up in a more or less loose form the rear. This bubble, however, does not detach but is displaced upwards by bubbles forming next. A series of these immature bubbles can form leading to the jet maximum penetration length which under the conditions tested may approach H_b. They eventually coalesce and detach, suddenly reducing the jet to its minimum penetration length (11).

Experimental conditions under which pressure measurements have been carried out are reported in Table 1 and 2 for runs with two and four probes, respectively. Pressure tap location is defined by the distance from the distributor in millimeters, and letter (c) or (w) depending on whether the tap is on the bed centerline or near the side wall. Bed material is indicated as S (sand) or P (plastics). Results of pressure signal analysis are also reported in the tables. Together with inequalities expressing a shorthand comparison between pressure profiles at the orifice p_c and in the bed p_w, there are: relative pressure deviations, expressed as ratios of the differences between local maximum and minimum to average pressures $(p_{max}-p_{min})/p_{av}$; dominant frequencies f, as obtained from the power spectral density function; time delays t_d at frequency f, as evaluated from the phase angle of the cross spectrum of the pressure records at a pair of taps. Pressure records and power spectral density functions for some of the experimental conditions tested are given in Figures 2, 3 and 4.

Pressure profiles cross each other for sand with H_b=310 and 110 mm, whereas for plastics pressures at the orifice are always below those in the bed (Figure 2). These trends are consistent with photographic observation of an elongated region of relatively low solid concentration along the axis of the orifice in the case of plastics. Pressures at the orifice are always below those in the bed also with sand when fluid-spouted operation is observed with H_b=15 mm $(p_{max}-p_{min})/p_{av}$ values at the orifice are always larger than in the bed (Tables 1 and 2). Differences are larger the lower H_b (Table 1).

Neither free interface oscillation nor the passage of bubbles can account for $(p_{max}-p_{min})/p_{av}$ values as large as those given in Tables 1 and 2. This suggests that jetting region phenomena are responsible for the fluctuations observed throughout the bed under the experimental conditions tested. Relative positions of p_{max} and p_{min} recorded simultaneously at the orifice and in the bed (Figure 2) and at different bed heights in the bed (Figure 3) suggest that disturbances spread from the jetting region. As a whole, time delays in Tables 1 and 2 confirm this indication for sand. Time delays for plastics are not reported, being in the same order of the scanner sampling interval.

Power spectral density diagrams in Figure 4a) and 4b) for both sand

and plastics with H_b=110 mm display frequencies that are characteristic of individual bubbles forming at the orifice or at the tip of the jet, as well as of bubbles due to coalescence. For H_b=310 mm the frequency associated to the formation of coalescence bubbles prevails sharply. With these beds the same dominant frequency is shown by the diagrams from pressure data at the orifice and in the bed. On the other extreme, for H_b=15 mm with sand and H_b=110 mm with plastics, remarkably different frequencies are displayed at the orifice and in the bed. In both cases jets spouted through the bed.

Gas injection may change bed particle assemblies leading to a decrease and an increase of voidage during the intermittent development and detachment of bubbles. This in turn results in an increase and decrease of pressure gradients associated with the flow of fluidizing gas. Whereas the pressure disturbance due to a bubble rising in a fluidized bed decays rapidly with distance, that due to gas injection extends all around with significant amplitude. Time delays indicate propagation velocities of the order of 10 ms^{-1}, in agreement with those evaluated on the basis of data of Lirag and Littman (22) and Fan et al. (23). These velocities are one to two orders of magnitude higher than those calculated from equations for dynamic wave propagation velocities in fluidized beds (6,7,8). With the observed f and t_d, wavelengths should be from 1 to 10 m. This suggests that in the proximity of the orifice, the bed responds to local void disturbance as a whole as regards the permeability to gas flow. It may be worth noting that according to the Ergun equation changes of voidage in the range of 5 to 10% result in pressure deviations $(p_{max}-p_{min})/p_{av}$ like those reported in Tables 1 and 2.

ACKNOWLEDGMENTS

This work has been performed with the financial contributions of MPI and of PFE-CNR and ENEA, Rome. One of the authors (D.M.) is indebted to ENEL-CRTN, Pisa for a postgraduate scholarship. Useful discussions with Professor R. Clift during his stay in Naples are also acknowledged.

LITERATURE CITED

1. Geldart, D., Powder Tech., 7, 285 (1973)
2. Donsì, G., A. Iscaro, L. Massimilla, and C. Zucchini, "Gas Jetting in Fluidized Beds", Motion Picture, presented at the Fluidization III Conference, Henniker, N.H. (1980)
3. Yang, W.C., and D.L. Keairns, "Momentum dissipation of and gas entrainment into a gas-solid two-phase jet in a fluidized bed" in "Fluidization", Grace J.R. and Matsen J.M. (Eds.), Plenum Press, (1980)
4. Rowe, P.N., H.J. MacGillivray, and D.J. Cheesman, Trans. Instn. Chem. Engrs, 57, 194 (1979)
5. Yates J.G., P.N. Rowe and D.J. Cheesman, AIChE J., 30, 890 (1984)
6. Wallis G.B., One dimensional Two-Phase Flow, McGraw-Hill, (1969)
7. Fanucci, J.B., N. Ness, R.H. Yen, Phys. Fluids, 24, 1944 (1981)
8. Foscolo P.U. and L.G. Gibilaro, Chem. Eng. Sci., 39, 1667 (1984)
9. Davidson, J.F., "Fluidised Bed Combustion: Research and Scientific design", Fluidized Combustion Conference, ERI, CapeTown (1981)
10. Markhevka, V.I., V.A. Basov, T.Kh. Melik-Akhnazarov and D.I. Orochko, Theor. Found. Chem. Eng., 5, 80 (1971)
11. Knowlton, T.M. and I. Hirsan in Fluidization, Grace J.R. and Matsen J.M. (Eds.), Plenum Press, New York (1980)

12. Kececioglu, I., W.C. Yang and D.L. Keairns, AIChE J., 30, 99 (1984)
13. Massimilla, L., G. Donsì and N. Migliaccio, AIChE Symp. Ser., 77(205), 17 (1981)
14. Shuster, W.W. and P. Kisliak, Chem. Eng. Prog., 48, 455 (1952)
15. Hiby, J.W., "Periodic phenomena connected with gas-solid fluidization", Drinkenburg A.A.H. (Ed.), Amsterdam (1967)
16. Verloop, J. and P.M. Heertjes, Chem. Eng. Sci., 29, 1035 (1974)
17. Baird, M.H.I. and A.J. Klein, Chem. Eng. Sci., 28, 1039 (1973)
18. Kang, W.K., J.P. Sutherland and G.L. Osberg, Ind. Eng. Chem. Fundam., 6(4), 501 (1967)
19. Sitnai, O., Chem. Eng. Sci., 37, 1059 (1982)
20. Zhang, M.O., P.M. Walsh and J.M. Beér, "Determination of bubble size distributions from pressure fluctuations in fluidized bed combustor" presented at the 7th Int. Conf. on Fluidized Bed Combustion, DOE, Philadelphia (1982)
21. Tamarin, A.I., Int. Chem. Eng., 4(1), 50 (1964)
22. Lirag, R.C. and H. Littman, AIChE Symp. Ser., 67 (116), 11 (1971)
23. Fan L.T., Tho-Ching Ho, S. Hiraoka and W.P. Walawender, AIChE J., 27, 388 (1981)
24. Filla, M., L. Massimilla and S. Vaccaro, Int. J. Multiphase Flow, 9, 259 (1983)
25. Bendat, J.S. and A.G. Piersol, Random data: Analysis and Measurement Procedures, Wiley-Interscience, New York (1971)
26. Nguyen X.T. and L.S. Leung, Chem. Eng. Sci., 27, 1748 (1972)

TABLE 1 - Data and results of runs with two probes

Bed material	H_b (mm)	Probe location		Pressure intersection	$\frac{p_{max}-p_{min}}{p_{av}}$ (c)	(w)	Dominant frequencies f (Hz) (c)	(w)	Time delay t_d (ms)	Figure
S	15	3c	3w	$p_w > p_c$	2.0	0.3	34,26	13,26	--	--
S	110	15c	15w	$p_w \gtrless p_c$	2.0	0.8	3.5	3.5	10	2a)
S	310	15c	15w	$p_w \gtrless p_c$	1.5	1.2	1.6	1.6	15	2c)
P	110	15c	15w	$p_w > p_c$	0.4	0.25	6.0	2.5	--	2b)
P	310	15c	15w	$p_w > p_c$	1.0	0.9	3.2	3.2	--	2d)

TABLE 2 - Data and results of runs with four probes (H_b = 310 mm)

Probe location	Sand: $\frac{p_{max}-p_{min}}{p_{av}}$	Sand: Dominant frequencies f (Hz)	Sand: Time delay t_d (ms)	Plastics: $\frac{p_{max}-p_{min}}{p_{av}}$	Plastics: Dominant frequencies f (Hz)	Plastics: Time delay t_d (ms)
30c	1.5	1.6	--	1.0	3.2	--
30w	1.2	1.6	4.0	0.9	3.2	--
90w	1.2	1.6	11.0	0.9	3.2	--
180w	1.2	1.6	40.0	0.9	3.2	--
30w	1.2	1.6	--	0.9	3.2	--
90w	1.2	1.6	10.0	0.9	3.2	--
180w	1.2	1.6	15.0	0.9	3.2	--
240w	1.2	1.6	25.0	0.9	3.2	--

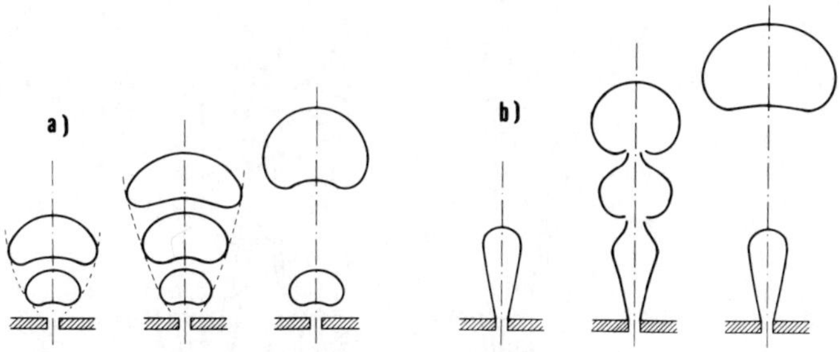

Fig. 1 - Modes of gas injection in fluidized bed.
a) bubble plume; b) permanent jet

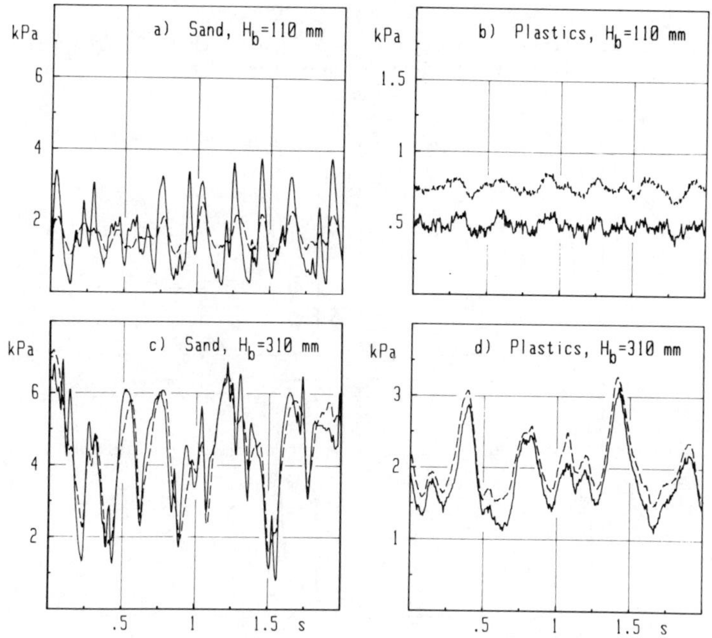

Fig. 2 - Simultaneous pressure records 15 mm above the distributor
——— at the orifice; ----- near the bed side wall

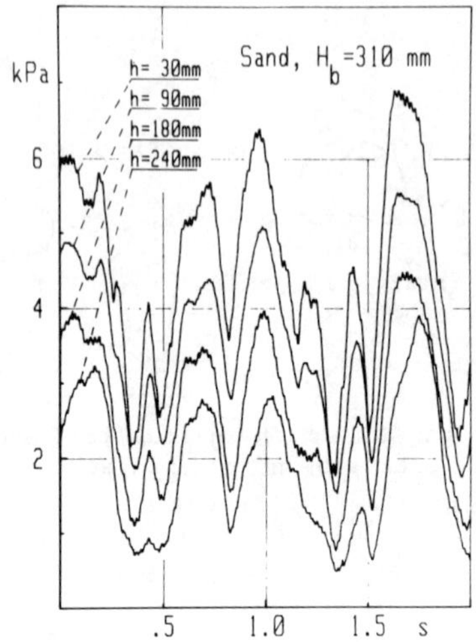

Fig. 3 - Simultaneous pressure records near the bed side wall at different heights h above the distributor

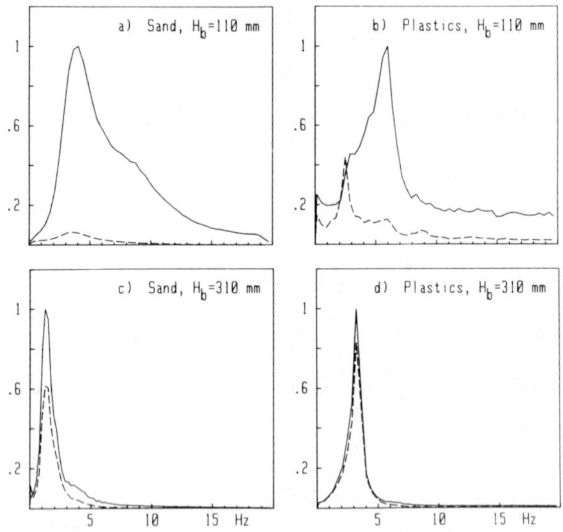

Fig. 4 - Power spectral density diagrams of the pressure records in Figure 2 (based on 635 s)
—— at the orifice; ----- near the bed side wall

JET PENETRATION INTO FLUIDIZED BEDS AT ELEVATED PRESSURES

J.G. Yates, V. Bejcek and D.J. Cheesman

Department of Chemical and Biochemical Engineering
University College London

ABSTRACT

The objective was to investigate the effect of system variables, particularly pressure, on the depth of penetration of gas jets from orifices into an incipiently fluidized bed. From this a new correlation for jet-penetration length has been developed and the conditions under which jets or bubbles form have been delineated.

SCOPE

Processes taking place at the point of entry of a gas into a fluidized bed are of importance in determining the performance of the bed as, for example, a chemical reactor. Recent experimental work has verified a mechanism whereby gas entering a bed from an orifice as a stream of bubbles causes enhanced rates of chemical reaction to occur in the region close to the distributor; such observations are in agreement with much reported industrial experience. However in a situation where gas enters not as a stream of bubbles but as a continuous jet the effects observed are likely to be quite different both qualitatively and quantitatively, and the object of the work to be described here was to investigate by means of direct observation with X-rays the conditions under which jets form at an orifice and the effect of operating parameters, particularly system pressure, on the depth of penetration of the jet into the bed.

CONCLUSIONS

(1) In the majority of situations studied here, jets that developed into a stream of bubbles were observed above an orifice in a fluidized bed.

(2) The jet penetration ratio is correlated with system properties by:

$$\frac{L_j}{d_o} = 21.2\, Fr^{0.37}\, Re^{0.05} \left[\frac{\rho_f}{\rho_p}\right]^{0.68} \left[\frac{d_p}{d_o}\right]^{0.24}$$

(3) The frequency of formation of bubbles at the extremity of the jet was approximately constant at 8 Hz.

(4) Based on this latter observation a criterion for the formation of jets as opposed to bubbles may be established.

EXPERIMENTAL METHODS

Fluidized-bed equipment

The equipment consisted of a cylindrical duralumin vessel fitted with a porous bronze distributor plate through which fluidizing gas (nitrogen) was passed into the bed of particles. This will be referred to below as the "background" flow. The distributor had an opening at its centre into which nozzles of different sizes could be fixed and through which nitrogen could be passed from a cylinder. A vareameter and a turbine meter were used to measure gas flows to the nozzle and bed respectively and a rotameter located downstream of the bed enabled the total volumetric flow of gas from the vessel to be measured. A differential manometer was used to measure the pressure drop through the nozzle-bed-cyclone and through the distributor-bed-cyclone. The complete assembly was set up inside a lead-lined enclosure which also housed the X-ray source. During operation X-rays were passed through the bed and the resulting image, registered by an image intensifier, was photographed with a ciné camera operated at 45 frames per second.

Powders used

Two powders were studied. One was a granular alumina with a surface mean particle size, d_p (determined by sieve analysis), of 260 μm and a particle density, ρ_p, of 1417 kg m^{-3}, while the other was a high shape factor coke with d_p = 320 μm and ρ_p = 1550 kg m^{-3}. Minimum fluidization velocities for the two materials were measured at four pressures in the range 1 to 20 bar absolute and the values obtained were 0.046 to 0.031 m/s for the alumina and 0.038 to 0.029 m/s for the coke.

Experimental procedures

Using X-rays film was taken of the interior of the bed operating at various values of background flow from 0 (i.e. a static bed) to 3 Q_{mf}. Two nozzles with diameters of 2 mm and 3.4 mm were used with superficial gas velocities through them of 30, 60, 120 and 180 m/s measured at bed pressure. Total system pressures of 1, 5, 10 and 20 bar absolute were employed and in all some 200 runs were carried out.

RESULTS AND DISCUSSION

Before presenting the quantitative results of the investigation it is necessary to clarify the meaning of the terms "jet", "bubble" and "jet penetration length" in the context of gas entering a fluidized bed from an orifice. The first two terms will be defined according to the proposal of Rowe et al. (1) that if a permanent flame-like cavity stands above the orifice this will be called a jet; when volumes of relatively clear gas form and detach

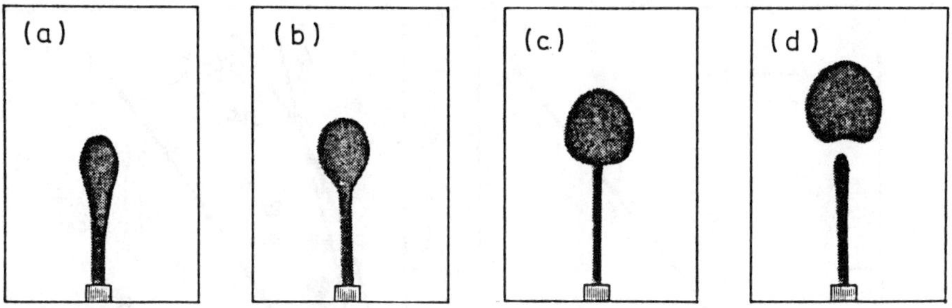

Figure 1. Consecutive frames from cine film at $^1/_{45}$ s intervals showing pattern of gas entry at 120 ms^{-1} from a 2mm diameter nozzle into an incipiently fluidized bed of alumina.

periodically from above the orifice these will be called bubbles. The jet-penetration length is defined on the basis of the observed gas entry mechanism illustrated in Figure 1 and which represents a fairly typical sequence of ciné film exposures taken at intervals of $^1/_{45}$ s. A permanent jet is clearly visible in each frame of the sequence and it is apparent that as the jet grows between frames (a) and (d) so a bubble develops at its upper end until it reaches a size where its rise velocity exceeds a critical value whereupon it detaches from and rises independently of the jet structure. The sequence of jet elongation followed by bubble growth and detachment then begins again. The jet penetration length, L_j, is defined as being the distance between the tip of the nozzle and the centre of the bubble at its point of detachment from the jet i.e. at the point shown in frame (d). This definition would appear to be consistent with that of L_{max} given by Knowlton and Hirsan (2), and used subsequently by Filla and Massimilla (3). Values of L_j were obtained by projecting the image of the film onto a horizontal screen, measuring the appropriate distance using the cursor of a Hewlett Packard HP 9874A digitiser and processing the data by means of a Commodore CBM4032 microcomputer.

The results obtained with an incipiently fluidized bed of the alumina powder and the 2 mm diameter nozzle are shown in Figures 2 to 4 where the dimensionless ratio of jet penetration length to orifice diameter (L_j/d_o) is plotted against orifice gas velocity. In Figures 2 and 3 the predictions of jet penetration ratio obtained from literature correlations are shown for comparison. It is clear that as system pressure increases the jet penetration ratio also increases, the overall trend being reflected, albeit not very precisely, by the two correlations of Wen et al. (5). The correlations of Zenz (6), Yang and Keairns (4) and (7), Wen et al. (8) and Merry (9) all show poor agreement with the measured values particularly at pressures above 1 bar.

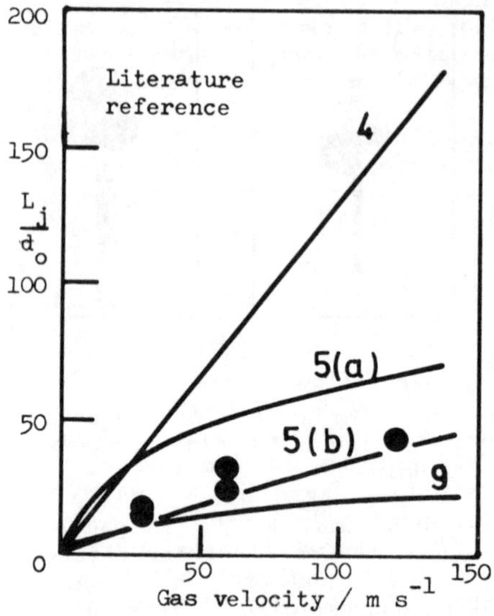

Figure 2. Jet penetration ratio. 2mm nozzle, incipiently fluidized bed of alumina; pressure = 1 bar.

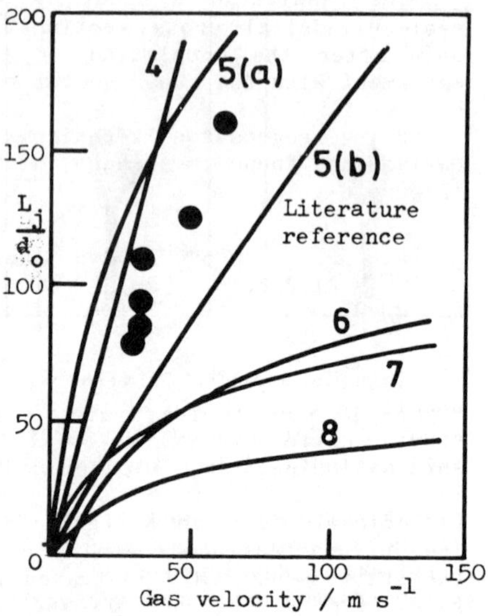

Figure 3. Jet penetration ratio. 2mm nozzle, incipiently fluidized bed of alumina; pressure = 15 bar.

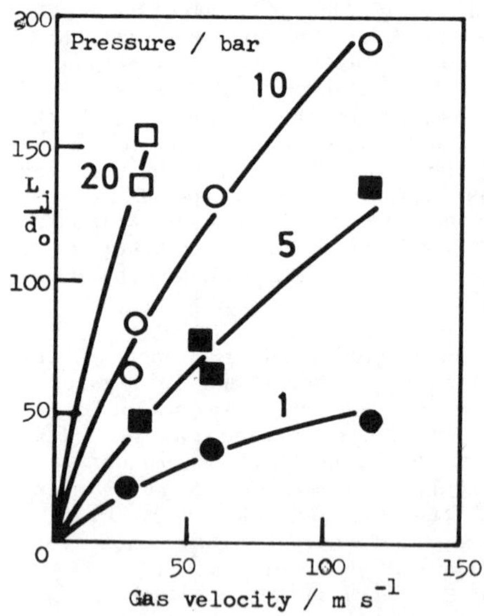

Figure 4. Jet penetration ratio. 2mm nozzle, incipiently fluidized bed of coke at various pressures.

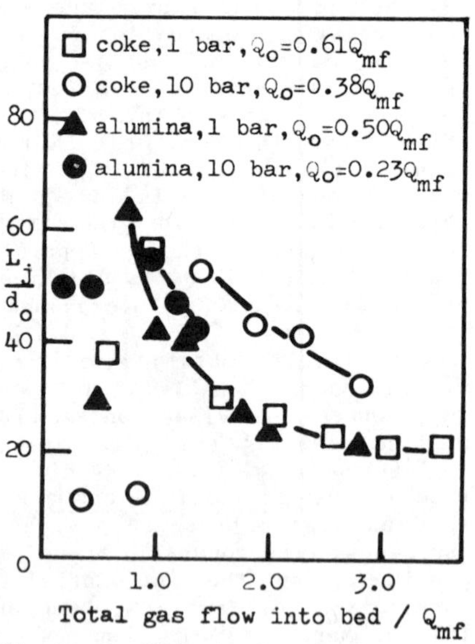

Figure 5. Effect of changing background flow.

The results of a study by Hirsan et al. (10) using a bed of semi-cylindrical cross section, (and which came to out attention only after the completion of the present work), show excellent agreement with the experimental results shown in Figures 2-4.

Using regression techniques the following relationship was derived and found to account, within ± 25%, for all the observations (Figure 6):

$$\frac{L_j}{d_o} = 21.2\ Fr^{0.37}\ Re^{0.05} \left[\frac{\rho_f}{\rho_p}\right]^{0.68} \left[\frac{d_p}{d_o}\right]^{0.24} \quad (1)$$

This is somewhat similar to the correlation of Wen et al. (5b) and to that of Hirsan et al. (10). The latter was developed from a study of jet penetration made, as was mentioned above, using a semi-cylindrical bed in which the jet orifice was located adjacent to the flat face of the vessel. This face was made of transparent plexiglass and allowed visual observations of the jet to be made. It is noteworthy and significant that the correlations for jet penetration developed from this semi-cylindrical bed and from the truly three-dimensional bed of the present study are very similar.

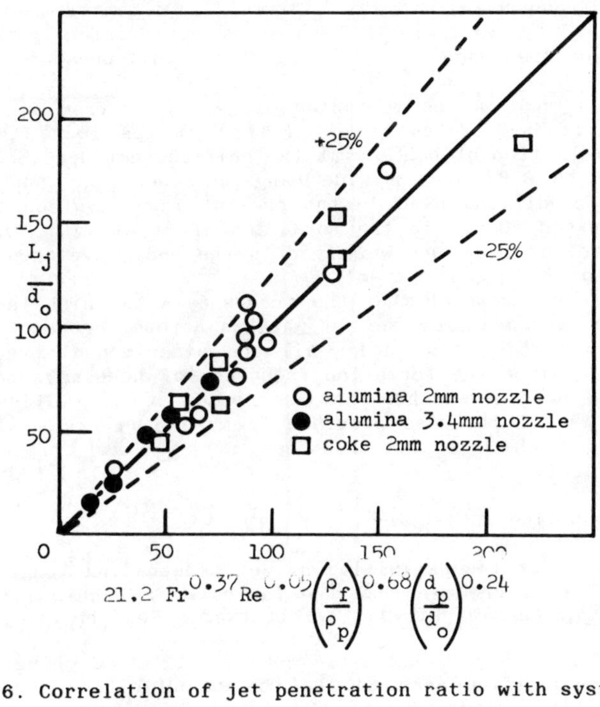

Figure 6. Correlation of jet penetration ratio with system variables.

Effect of changing background flow

Despite the fact that many individual frames of film were analysed for each set of conditions investigated, considerable scatter was found in the measurements of jet penetration length and it was felt that this may have been due to slight and uncontrolled changes in background flow. To examine this possibility a series of experiments was carried out in which the gas flow through the 2 mm nozzle was held constant at 60 m/s and the background flow was varied in the range 0 to 3 Q_{mf} at a total system pressure of 1 bar. Runs at 10 bar were conducted with a nozzle gas velocity of 30 m/s and background flows in the range 0 to 2.5 Q_{mf}; both alumina powder and coke were used in these tests. The results are shown in Figure 5 where the $^{L_j}/_{d_o}$ ratio is plotted against the total volumetric gas flow into the bed. It is clear that background flow exerts a strong effect on the jet penetration ratio, the maximum jet height being reached in all cases with a flow equal to Q_{mf} entering the bed from the nozzle and distributor. With higher values of flow through the distributor the jet height decreased and appeared to reach a constant value at total flow rates in excess of about 2.5 Q_{mf}. Similar observations were made by Hirsan et al. (10) who were able to incorporate the dependency on background flow into their correlation for jet penetration length. Work on this question is continuing using the present X-ray system and results will be reported in due course.

Bubble formation frequency

In previous work on the mechanism of gas entry from an orifice into a fluidized bed (Yates et al. (11)) it was found that the frequency of formation of bubbles at the orifice remained remarkably constant at 7 to 8 Hz over a wide range of operating conditions. The operating conditions used in the present investigation led, as has been discussed above, to the formation of a jet, which, in all cases where the jet remained submerged in the bed, developed into a bubble stream at its upper extremity.

It was observed however that the frequency with which the fully formed bubbles detached from the jet was, as before, around 8 Hz and that it retained this value under all the experimental conditions investigated. Since the formation frequency of bubbles appears to be essentially constant whether they are formed at the orifice or at the top of a jet, it becomes possible to make a prediction of the conditions under which either bubbles or jets are formed at a submerged orifice.

Criteria for bubbling or jetting

It is clear from the definition of jet penetration length given above that if L_j is greater than the radius of a bubble formed at the orifice, r_{bo}, then a jet will be observed. Now the orifice gas flow Q_o is:

$$Q_o = A_oU_o \text{ m}^3 \text{ s}^{-1} \qquad (2)$$

where A_o is the orifice area and U_o the superficial gas velocity through the orifice. Thus:

$$Q_o = \frac{\pi d_o^2 U_o}{4} \tag{3}$$

The volume of a bubble forming at the orifice is:

$$V_b = \frac{Q_o}{n_b} \quad m^3 \tag{4}$$

where n_b is the bubble formation frequency.

Assuming n_b = 8 Hz and that all the gas entering the bed from the orifice forms a visible bubble:

$$V_b = \frac{\pi d_o^2}{4} \left[\frac{U_o}{8} \right] \tag{5}$$

Now $$V_b = \frac{4}{3} \pi r_{bo}^3 \tag{6}$$

and combining (4) and (5):

$$\frac{4}{3} \pi r_{bo}^3 = \frac{\pi d_o^2}{4} \left[\frac{U_o}{8} \right] \tag{7}$$

from which:

$$r_{bo} = \left[\frac{3}{16} d_o^2 \frac{U_o}{8} \right]^{1/3} \tag{8}$$

From the empirical correlation for jet penetration ratio presented earlier:

$$L_j = 21.2\, d_o \left[\frac{U_o^2}{g d_p} \right]^{0.37} \left[\frac{\rho_f d_p U_o}{\mu} \right]^{0.05} \left[\frac{\rho_f}{\rho_p} \right]^{0.68} \left[\frac{d_p}{d_o} \right]^{0.24} \tag{9}$$

If it is assumed that for jets to form:

$L_j > r_{bo}$

then from (8) and (9) the criterion for jet formation is:

$$\left[\frac{8 d_o}{U_o} \right]^{0.33} \left[\frac{U_o^2}{g d_p} \right]^{0.37} \left[\frac{\rho_f d_p U_o}{\mu} \right]^{0.05} \left[\frac{\rho_f}{\rho_p} \right]^{0.68} \left[\frac{d_p}{d_o} \right]^{0.24} > 0.0270$$

ACKNOWLEDGEMENT
The authors wish to thank the British Gas Corporation for financial support throughout the course of this work.

NOTATION

A_o orifice area, m^2
U_o orifice gas velocity, $m\ s^{-1}$
d_o orifice diameter, m
Fr Froude number $\frac{U_o^2}{gd_p}$
g acceleration due to gravity, $m\ s^{-2}$
L_j jet penetration length, m
n_b bubble formation frequency, Hz
Q_{mf} minimum fluidization flowrate, $m^3\ s^{-1}$
Q_o orifice gas flowrate, $m^3\ s^{-1}$
Re Reynolds number $\frac{\rho_f d_p U_o}{\mu}$
r_{bo} initial bubble radius, m
V_b bubble volume, m^3
d_p particle diameter, m

Greek symbols

μ fluid viscosity, $N\ s\ m^{-2}$
ρ_f fluid density, $kg\ m^{-3}$
ρ_p particle density, $kg\ m^{-3}$

LITERATURE CITED

1. Rowe, P.N., H.J. MacGillivray, and D.J. Cheesman, Gas discharge from an orifice into a gas fluidized bed. Trans I. Chem. E., 57, 194 (1979).
2. Knowlton, T.M., and I. Hirsan, The effect of pressure on jet penetration in semi-cylindrical gas-fluidized beds. "Fluidization", J.R. Grace and J.M. Matsen (Eds), Plenum Press, New York, 315 (1980).
3. Filla, M., and L. Massimilla, Analysis of the variables controlling gas jet penetration in fluidized beds. Ind. Eng. Chem. Fundam., 23, 131 (1984).
4. Yang, W.C., and D.L. Keairns, Design and operating parameters for a fluidized-bed agglomerating combustor/gasifier, "Fluidization", J.F. Davidson and D.L. Keairns (Eds), Cambridge University Press, 208 (1978).
5. Wen, C.Y., N.R. Deole, and L.H. Chen, A study of jets in a three-dimensional gas fluidized bed, Powder Tech., 31, 175 (1982 a and b).
6. Zenz, F.A., Bubble formation and grid design. I. Chem. E. Symp. Ser., 30, 136 (1968).
7. Yang, W.C., and D.L. Keairns, Estimating the jet penetration depth of multiple vertical grid jets. Ind. Eng. Chem. Fundam., 18, 317 (1979).
8. Wen, C.Y., M. Horio, R. Krishnan, R. Khosravi, and P. Rengarajan, Jetting phenomena and dead zone formation on fluidized bed distributors, Proc. 2nd Pacific Chem. Eng. Con., AIChE, New York (1977).
9. Merry, J.M.D., Penetration of vertical jets into fluidized beds, AIChEJ, 21, 507 (1975).
10. Hirsan, I., C. Shishtla and T.M. Knowlton, paper presented at 73rd Annual AIChE Meeting, Chicago, Illinois, 16-20 November, 1980.
11. Yates, J.G., P.N. Rowe, and D.J. Cheesman, Gas entry effects in fluidized-bed reactors, AIChEJ, 30, 890 (1984).

TOMOGRAPHIC DETERMINATION OF THE VOIDAGE STRUCTURE OF GAS FLUIDISED BEDS IN THE JET REGION

J.P.K. Seville, J.E.P. Morgan and R. Clift
Department of Chemical and Process Engineering
University of Surrey, Guildford, GU2 5XH, U.K.

ABSTRACT

Gamma-ray tomography has been used to obtain maps of time-averaged voidage illustrating the transition between the jet region and the bubbling bed above a multi-orifice distributor plate. The results are consistent with the jet height correlation of Merry (18).

SCOPE

The technique of radiation tomography generates an image of a plane section through an object of interest; it is of potential value in a wide range of chemical and process engineering applications. This study demonstrates application of the method to fluidised beds by investigating the region immediately above a multi-orifice distributor in a 14.6 cm diameter bed of sand. Gamma-ray tomography is shown to yield quantitative information on the local time-averaged voidage within the bed, which has not previously been determined with accuracy. The technique may, in principle, be applied to process plant.

CONCLUSIONS AND SIGNIFICANCE

Detailed voidage maps of plane sections through the jet region of a gas fluidised bed have been obtained, at three heights above a multi-orifice distributor plate and at three gas velocities. The transition between jetting and bubbling behaviour was clearly detected. Its position is compared with predictions from the available jet height correlations (3, 16-20), and with the expression of Fakhimi et al (4) for the height of the unfluidised region between the orifices. The correlation of Merry (18), which is consistent at larger velocities with the prediction of Fakhimi et al (4), fits the data, most of the other correlations leading to overestimates of jet height. At greater distances above the distributor, tomography revealed annular regions of increased bubble activity, due to bubble coalescence, in agreement with the measurements of Werther and Molerus (23). Although the time required to obtain a tomographic image is long at present, technical improvements such as the use of increased source energies and multiple detectors will reduce this significantly. For industrial diagnostic work, where the required resolution may be low, a useful image might be obtained from comparatively few measurements obtained with high energy sources such as those currently employed on chemical process plant (24).

*   *   *

In a fluidised bed with a distributor of a type other than a porous plate, the gas and particle flow in the region immediately adjacent to the distributor differs from that in the bulk of the bed (1). For distributors of the perforated plate type considered here, the fluidising gas enters the bed at relatively high velocity at an array of discrete orifices. The gas and particle velocities and the void fractions are much higher above each orifice than in the region between the orifices. These high velocity, high voidage regions are referred to here as "jets" whether or not they take the form of permanent cavities, noting that this use of the term "jet" is consistent with its application to conventional turbulent jets (2). In normal industrial practice, and in the experiments reported here, no fluidising gas is introduced between the orifices. Unfluidised "dead zones" may then form between the orifices (3,4), and the lower parts of the jets form essentially stationary voids (5). The height of the jet region cannot be defined unambiguously, and alternative definitions will be appropriate according to which of the features of gas discharge is of concern (1). However, because the jets represent regions in which gas/particle contacting and velocities differ from the rest of the bed, it is often most appropriate to take the limit of the jet regions as the level at which the gas streams rising from the distributor break down into bubbles whose rise velocity and coalescence behaviour are the same as for bubbles in the bulk of the bed.

Measurement of the height of the jet region is not straightforward. Some workers have observed jets in "two-dimensional" beds or adjacent to transparent walls in three-dimensional beds (see 1); however, it is possible that the walls influence the structures of such jets (1,5). Rowe *et al* (5) used X-ray cinephotography to view a single jet. This technique is non-invasive, but is only applicable in the laboratory. Moreover it generates, at best, a succession of instantaneous two-dimensional projections of unsteady three-dimensional phenomena, projections which may be affected by overlap and "shadowing" of one void by another. Interpretation of the films is essentially qualitative and may therefore involve some judgement, while time-averaged conditions, required for turbulent-jet modelling of the entry region (1), cannot readily be obtained. We describe here the use of a different non-invasive technique, tomographic imaging, which is applicable to three-dimensional beds and which gives values for local time-averaged voidage. At least in principle, tomographic imaging could be applied to process plant.

EXPERIMENTAL TECHNIQUE

Radiation imaging using computerised tomography is probably best known for its role in medical diagnostics ("body scanning") (6). The technique consists of stepping a collimated beam of radiation (X-rays, in the medical application) across the object in equal distance increments, measuring the attenuation of the beam at each position, rotating the source/detector system (or the object) through a prescribed angle about an axis perpendicular to the scanning plane and repeating the linear translation. The sequence is repeated until the source/detector axis or the object has rotated through 180°. Various techniques are available (e.g. 7,8) for reconstructing a cross-sectional image of the scanned object from the set of attenuation measurements. The image conventionally consists of a matrix of picture elements ("pixels") whose grey-level (or colour) varies according to the linear attenuation coefficient of the object at the corresponding point of the cross-section. The linear attenuation coefficient itself depends on the density and the atomic number of the material at that point and on the energy of the incident radiation.

X-rays have disadvantages which make them unsuitable for non-medical applications: they are too "soft" for the majority of high density, high atomic number materials encountered, and their use is limited to qualitative work because they are not mono-energetic. Tomographic systems using mono-energetic gamma radiation have therefore been developed for quantitative non-medical use. A crude form of tomography, employing multiple gamma attenuation measurements, has previously been used to obtain voidage contours in industrial hydraulic conveying lines (9,10) and in the risers of catalytic crackers (11,12). More recently Weimer et al (13) have used rapid single attenuation measurements to obtain dense phase voidage, bubble size and bubble frequency in a fluidised bed. The tomographic scanner developed at the University of Surrey, by the Radiation Physics Group of the Department of Physics (14), is shown schematically in Figure 1. It uses a single collimated source and a single NaI detector. In this work the source was 100 mCi Am-241 with a prominent 59.6 keV peak, collimated to give a cylindrical beam of 5 mm diameter. Source and detector were aligned on an optical bench. The fluidised bed was mounted on a rotating table so that it could be translated through the beam by means of a screw connected to a stepping motor, and also rotated about a vertical axis using a second stepping motor. The whole apparatus was controlled by a microprocessor, which also served for data logging. The bed was cylindrical, 14.6 cm in diameter, and was scanned on a 5mm grid in 30 directions at 6° intervals, with 1000 photons collected at each position. The resulting tomographic image gave the time averaged attenuation in pixels 5 mm square covering the section scanned. Each scan took 6 to 7½ hours to complete. Calibration between attenuation and voidage was achieved by scanning a settled bed of material and obtaining a single attenuation coefficient averaged over the whole cross section. Without disturbing the bed, the voids were then slowly filled with water and a second average attenuation coefficient obtained. Knowing the attenuation coefficient of water, the voidage of the settled bed and hence the attenuation coefficient for the sand were obtained. Further details of the experimental technique are given by Seville et al (15). The bed particles used were 300-355 μm washed quartz sand, with density 2660 kg/m^3, fluidised by air at 25°C.

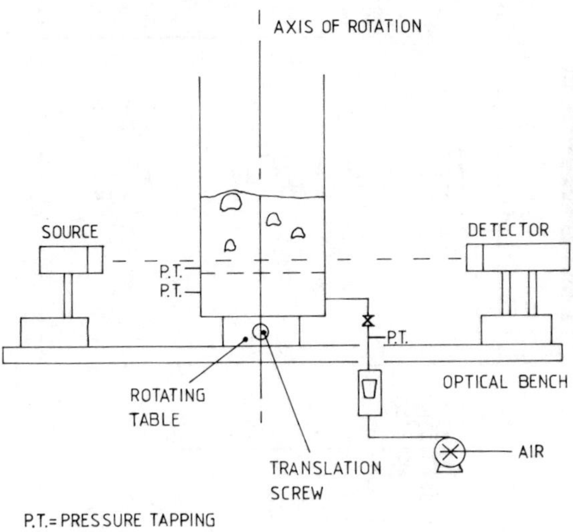

Figure 1 Experimental Apparatus

The minimum fluidisation velocity was 9.2 cm/s, and the settled bed depth was 15 cm. The distributor plate was flat with 37 orifices of 1.5 mm diameter on 24 mm triangular pitch, giving 0.39% free area. The pressure drop across the distributor in all runs was well above that required to activate all orifices (4).

RESULTS

Figure 2a is a grey-level reconstructed bed cross-section obtained 2 cm above the distributor at $3.5U_{mf}$. The scale on the left corresponds to a range of linear attenuation coefficient from 0.005 mm^{-1} (ε=92.1%; represented as black) to 0.05 mm^{-1} (ε=20.8%; represented as white). High voidage regions can be seen clearly above each orifice. Figure 2b shows the variation of attenuation coefficient along the line drawn across Figure 2a. The voidage profiles around each jet are seen to be similar at this resolution. The perspex wall of the bed, which has a lower attenuation coefficient than the sand, is not clearly distinguished.

By choosing an attenuation threshold and setting all pixels having an attenuation below this value to black and the remainder to white, a binary image can be obtained which may facilitate comparison between images. Figure 3 shows binary images with a threshold of 0.03 mm^{-1} (ε=0.525) for each of the nine combinations of experimental conditions; the lowest right-hand image is the binary form of Figure 2a. Regions of high average voidage are apparent above the distributor orifices at 2 cm height with U=3.5 U_{mf}, and are also discernible at 2 cm with 2.5 U_{mf} and 3 cm with 3.5 U_{mf}. The binary images clearly show a decrease in order of the structure with increase in scan height above the distributor, and an increase in structural detail at all heights as the velocity is increased.

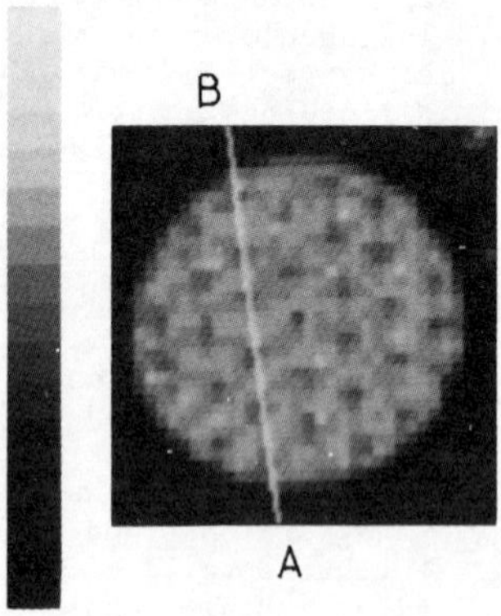

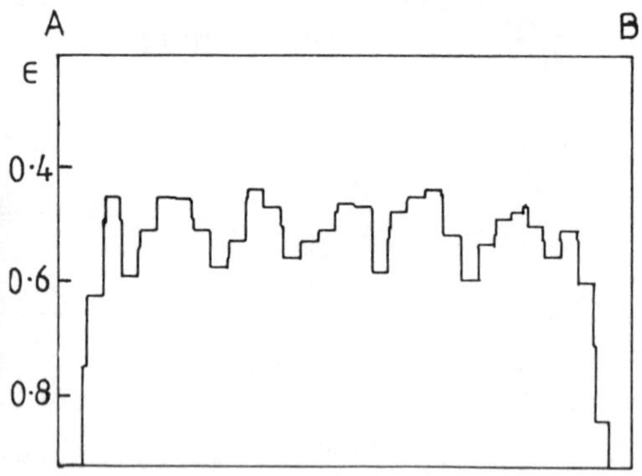

Figure 2a Reconstructed Cross Section

Figure 2b Line Section through Reconstructed Bed Cross Section

Scan Height = 2 cm; U/U_{mf} = 3.5

Fig. 3 Reconstructed Binary Bed Cross-Section (Threshold ε = 0.525)

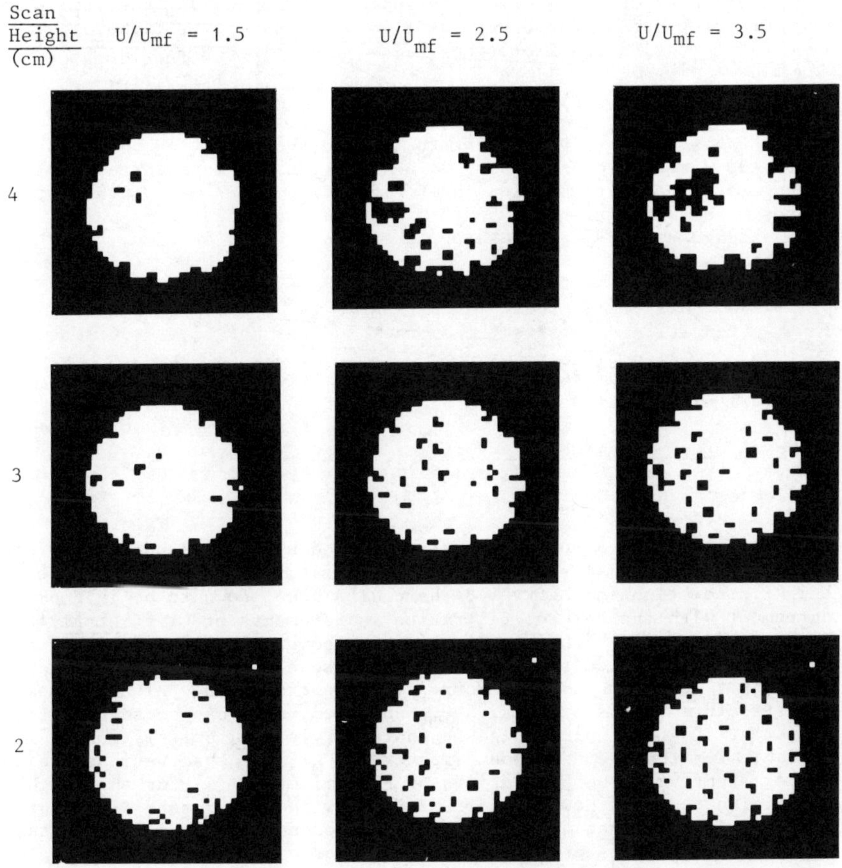

DISCUSSION

A large body of data is available on the penetration of jets into gas fluidised beds, and a correspondingly large number of correlations has been proposed to enable jet length to be predicted. Some of those relating to the case of no secondary fluidising gas (3,16-20) are presented in graphical form in Figure 4, together with a prediction due to Fakhimi et al (4) for the "height of the entrance effect" (the height above which all of the bed can be considered to be fluidised, i.e. no "dead zones" exist). Aside from the question of how best to define what is meant by jet height, it is clear that all the available correlations except those of Basov (17) and Merry (18) considerably overestimate the height of the high voidage and ordered regions observed in the lower

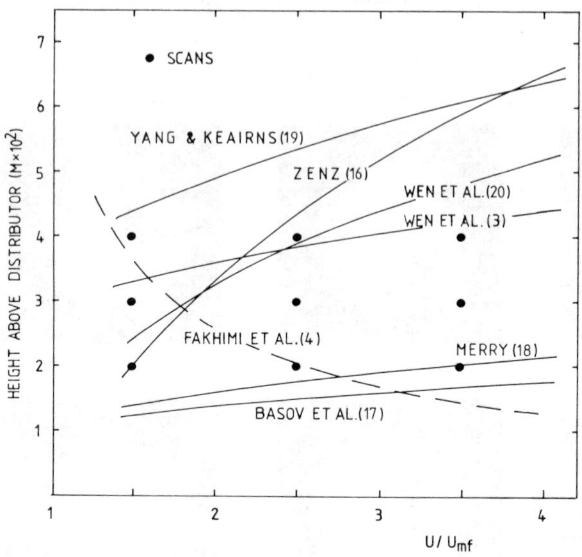

Figure 4

Jet Height and Height of Entrance Effect as functions of U/U_{mf}, for conditions of Figure 3

tomographic scans, a conclusion also reached by Ghadiri and Clift (21) for a system of similar dimensions but operating at 800°C. The jet height predictions of Merry (18) have also been shown to be in good agreement with the aerosol filtration measurements of Clift *et al* (22). The only image in which jets are visible above all the orifices is at a superficial gas velocity of 3.5 U_{mf} and a scan height of 2 cm, which falls just below the line of Merry's (18) correlation. Although some jet structure is visible at 2.5 U_{mf} and 2 cm, none can be seen at this height at 1.5 U_{mf}, which suggests that Merry's (18) correlation is reliable for these conditions. For U=3.5 U_{mf}, the observations are not inconsistent with the predictions of Fakhimi *et al* (4) for the height of unfluidised zones. However for U<2.5 U_{mf} with the present distributor, the unfluidised zones are predicted to be deeper than the jet length, so that the analysis of Fakhimi *et al* is not valid.

The scans 4 cm above the distributor indicate the results of bubble coalescence and the development of an annular region of high bubble activity near the wall, in qualitative agreement with the results of Werther and Molerus (23) on local bubble gas flow close to the distributor. As expected, development of the annular region is much less marked at the lowest flowrate considered.

This work demonstrates the potential of radiation tomography as an investigative tool in laboratory experiments. However, the technique clearly has potential in industrial applications, particularly diagnostics. For this potential to be realised it will be necessary to use more penetrating radiation sources, such as those currently employed for diagnostic work on process plant (24). It will also be necessary to reduce the total time required per scan, which is affected not only by the source strength but also by the source/detector geometry and the required resolution. Using a fan geometry with a single high-energy source and multiple detectors, the scanning period could be reduced by an order of magnitude at the present resolution. Furthermore, the time required to scan a given object is inversely proportional to pixel size to the power 4. In the work reported here the ratio of pixel size to bed diameter was approximately 1:30. In industrial diagnostic work it is possible to envisage situations in which a ratio of 1:10 would give quite adequate resolution, implying a reduction in scan time of almost two orders of magnitude. It might then be preferable to use iterative image-reconstruction techniques (7,8) which are less accurate than the analytical techniques most commonly used in the laboratory but are more suitable when the data set is limited or incomplete.

ACKNOWLEDGEMENTS

The financial assistance of the Research Committee of the University of Surrey, and the co-operation of Mr N MacCuaig and Dr W Gilboy of the Physics Department are gratefully acknowledged.

NOTATION

U Superficial gas velocity
U_{mf} Superficial gas velocity at minimum fluidisation
ε Bed voidage

LITERATURE CITED

1. Massimilla, L., "Gas Jets in Fluidised Beds", p 133 in "Fluidization", 2nd Ed., J.F. Davidson, R. Clift and D. Harrison (Eds), Academic Press, London (1985).

2. Clift, R., Chem. Engr., no 388,29, Jan. (1983).

3. Wen, C.Y., M. Horio, R. Krishnan, R. Khosravi and P. Rengarajan, Proc. 2nd Pacific Chem. Eng. Congr., Denver, Colorado 2, 1182 (1977).

4. Fakhimi, S., S. Sohrabi and D. Harrison, Can. J. Chem. Eng., 61, 364 (1983).

5. Rowe, P.N., H.J. MacGillivray and D.J. Cheesman, Trans. I. Chem. E., 57, 194 (1979).

6. Hounsfield, G.N., Brit. Prov. Patent No. 1283915 (1968).

7. Brooks, R.A. and G. di Chiro, Phys. Med. Biol., 21, 687 (1976).

8. Gilboy, W.B. and J. Foster, in "Research Techniques in Non-Destructive Testing" Vol.6, R.S. Sharpe (Ed.) Academic Press (1983).

9. Michalik, A., La Houille Blanche, 28 (1), 53 (1973)

10. Przewlocki, K. and P. Nizegorodcew, La Houille Blanche, 28 (1), 59 (1973).

11. Bartholomew, R.N. and R. M. Casagrande, Ind. Eng. Chem., 49 (3), 428 (1957).

12. Saxton, A.L. and A.C. Worley, Oil and Gas J., 68 (20), 84 (1970).

13. Weimer, A.W., D.C. Gyure and D.E. Clough, Powder Tech., 44, 179 (1985).

14. Gilboy, W.B., J. Foster and M. Folkard, Nucl. Inst. and Meth., 193, 209 (1982).

15. Seville, J.P.K., N. MacCuaig, V. Rojo, M.P. Wood, R. Clift and W.B. Gilboy, to be published.

16. Zenz, F.A., Inst. Chem. Eng. Symp. Ser., 30, 136 (1968).

17. Basov, V.A., V.I. Markhevka, T.Kh. Melik-Akhnazarov and D.I. Orochko, Int. Chem. Eng., 9, 263 (1969).

18. Merry, J.M.D., A.I.Ch.E.J., 21, 507 (1975).

19. Yang, W.C. and D.L. Keairns, Ind. Eng. Chem. Fundam., 18, 317 (1979).

20. Wen, C.Y., N.R. Deole and L.H. Chen, Powder Tech., 31, 175 (1982).

21. Ghadiri, M. and R. Clift, Ind. Eng. Chem. Fundam., 19, 440 (1980).

22. Clift, R., M. Ghadiri, J.L. Monteiro, B.K.C. Tan and K.V. Thambimuthu, p.85 in "Fluidization", D. Kunii and R. Toei (Eds.), Engineering Foundation, New York (1984).

23. Werther, J. and O. Molerus, Int. J. Multiphase Flow, 1, 123 (1973).

24. "Radiosotopes in the Chemical and Offshore Oil and Gas Industries", Imperial Chemical Industries Physics & Radiosotope Services, Billingham.

FLUIDIZATION PHENOMENA IN A LARGE JETTING FLUIDIZED BED

W. C. Yang*, B. Ettehadieh, T. C. Anestis,
R. E. Gizzie, and G. B. Haldipur

KRW Energy Systems Inc.
Madison, Pennsylvania
*Research and Development Center
Westinghouse Electric Corporation
Pittsburgh, Pennsylvania

ABSTRACT

Jet penetration, bubble dynamics, gas mixing, particle mixing, solid circulation, and particle separation were studied in a 3-meter diameter by 10-meter high cold flow model of semicircular cross-section. Two jet nozzle assemblies of 25cm and 41cm in diameter were employed with jet velocities ranging from 15 to 60 m/s and solid feed rates, from 0 to 4550 kg/hr.

SCOPE

It is generally known that fluidization phenomena depend on the size of fluidized beds. The performance in large commercial fluidized beds was frequently found to be substantially different from expectations derived from pilot plant experience. Unfortunately, the studies on fluidization phenomena in large commercial size fluidized beds are not abundant. This paper summarizes the results on cold flow experiments carried out in a large fluidized bed, 3-meter in diameter and 10-meter in height, where a large jet dominates the fluidization phenomena in the bed.

CONCLUSIONS AND SIGNIFICANCE

In a fluidized bed where a large jet is present, the fluidization is dominated by the jetting phenomena in the bed. The behavior of the jet and the bubbles emitting from it can be reasonably described by existing correlations in the literature, though extreme caution is required in selecting the applicable ones. The solids circulation pattern in a large jetting fluidized bed can be divided into three separate regions. In the central region where the jet is present, the solids move predominantly upward induced by the jetting action and the bubbles. In the outer region close to the wall, the solids move primarily downward. In between these two regions, the solids move alternately upward and downward depending on the approach and departure of the bubbles. Based on this solids circulation pattern, a model was developed to describe the radial mixing flux obtained experimentally. The experimental evidence indicated that the axial solids mixing in the central region is very fast. The gas mixing in the bed is also dominated by the jetting phenomena, especially close to the jetting zone. A dividing stream line apparently exists where a gas injected at one region will not be mixed with the gas in the other region. Continuous particle separation was also performed successfully in this large jetting fluidized bed. It is significant that several effective experimental techniques have been developed during the course of this study for collecting various fluidization data in a large commercial size fluidized bed.

EXPERIMENTAL APPARATUS AND EXPERIMENTAL CONDITIONS

The experimental apparatus was described in an earlier paper (1). A detailed description of the test facility can be found elsewhere (2). The experimental program consists of four bed configurations - a 25cm or a 41cm jet assembly with a shallow or a deep bed. Two of the bed configurations are shown in Figures 1 and 2. Both the shallow and deep bed configurations have the same bed height of 5.5 m if measured from the zero reference on the bed structure (see Figures 1 and 2). The shallow bed configuration was always with a 1.5 m high annular region around the nozzle and the deep bed configuration, without it. Thus the distance between the top of the jet nozzle to the bed surface is 1.5 m shorter for the shallow bed configuration. The conical grid, having 70° from the horizontal, has five different aeration sections. The flow to each section can be individually controlled. A total of 15 individual test plans were conducted with jet velocities ranging from 15 m/s to 60 m/s and solid feed rates of 0 to 4550 kg/hr. Crushed acrylic particles, -6 mesh, and 1400 μm in weight-mean size and 1100 kg/m^3 in density, were used as the bed material. Air was used as the fluidizing medium.

JET PENETRATION AND BUBBLE DYNAMICS

Jet penetration and bubble dynamics were studied by two different techniques - the high speed motion pictures and the force probe. They are discussed below.

Jet Penetration

Jet penetration data were obtained from frame-by-frame analysis of the high speed motion pictures taken at 64 frames per second and also via a force probe developed in-house. The additional data obtained during this study support the conclusions drawn earlier (1). Details are reported in (2).

Initial Bubble Diameter

When the bubble first detached from the jet, the largest horizontal dimension of the bubble was measured to be the initial bubble diameter. The initial bubble diameter can be correlated empirically with the total volumetric flow through the jet nozzle. Changes in solids feed rate do not seem to affect the size of the initial bubbles pinched off at the tip of the jet in all tests.

A bubble formation model was developed earlier (1), assuming continuous gas leakage from the bubble to the emulsion phase with a gas velocity of minimum fluidization velocity at the bubble boundary during the bubble formation period. The initial bubble diameters predicted using this model are compared with that obtained experimentally in Figure 3. The TP-M005-3 test series employed the 25 cm jet nozzle assembly with the shallow bed configuration at nominal jet velocities ranging from 15 to 61 m/s and solids feed rates, from 0 to 4100 kg/hr. The test series TP-M010-1 also employed the 25 cm jet nozzle assembly but with the deep bed configuration at similar operating conditions. The deep bed configuration and the 41 cm jet nozzle assembly were used in TP-M016-1 series with two nominal jet velocities of 15 and 30 m/s and with the solids feed rate ranging from 0 to 4100 kg/hr. It can be seen that the model predictions are within 20%.

Bubble Velocity

The bubbles travelled substantially faster than that expected for the fully developed bubbles with similar bubble diameters immediately after detachment from the jet. Bubble velocity decreased to the level similar to that expected for the fully developed bubbles at higher bed heights. With the larger 41cm jet nozzle, substantially larger initial bubbles were generated at the tip of the jet. Under these conditions, the bubble velocity can be interpreted more closely by the bubble velocity equation applicable for a slugging bed.

Bubble Frequency

Bubble frequency was studied by two different techniques - motion pictures and a force probe. Movies were taken at different bed elevations at a speed of 64 frames per second and the movies were analyzed frame by frame to extract the bubble frequency data. A force probe was also installed via 12 probe ports at eight different elevations as shown in Figure 2. The probe penetration into the bed was 20.3 cm. Measured oscillations were continuously processed by a spectrum analyzer (Nicolet Model 446). The power spectrum of the oscillations was calculated and the frequency of the most pronounced peak in the power spectrum was identified as the characteristic frequency of the major periodic component of density fluctuations. These characteristic frequencies were then compared with the bubble frequencies observed from analysis of the motion pictures.

Substantial bubble coalescence was found to occur in the bubble developing region up to about 12 times the jet nozzle diameter. The size of the jet nozzle does not affect the bubble frequency distribution along the bed height. The bubble coalescence pattern along the bed height depends only on the operating bed height. The frequency profiles obtained from movie analysis were confirmed by independent measurements using a force probe as shown in Figure 4. The results from motion pictures tend to have slightly higher frequencies than that from the force probe. This is because every bubble was counted in the movie analysis while only the major frequency was selected after the frequency analysis of the force probe signals. Set Points 3 to 6 in Figure 4 have a nominal jet velocity of 15 m/s and a solids feed rate ranging from 0 to 4323 kg/hr. A nominal jet velocity of 30 m/s with similar solids feed rates were employed for Set Points 7 through 10.

The depth of the jet influence zone shown in Figure 4 compares favorably with the jet penetration length observed in the motion pictures. Above the jet influence zone the frequency profiles level off into a narrow frequency band pertaining to a slugging bed, regardless of the incoming jet flow rates and solids loading. Close to the jet nozzle, the high frequencies reflect the train of disturbance due to periodical entrainment of bed particles into the jet described by Massimilla et al. (3). Thus it is a measurement of voidage wave rather than bubble frequency. This basic frequency may be approximated by the theory of bubble formation at an orifice.

GAS MIXING

The gas mixing phenomenon was investigated by injecting carbon dioxide as the tracer gas selectively into various flow streams and by taking gas samples across different cross-sections of the bed. Typical contour profiles of equal tracer gas concentrations are reported in (2).

Two important observations were made. Regardless of the incoming flow rates, the gases that are injected through the concentric jets essentially remain in the core of the bed and do not fully mix with the gases in the dense solid downflow region of the bed. The gases injected through the conical grid sections are not entrained by the incoming jets. Partial entrainment of these gases occurs at locations where bubble formation and bubble coalescence take place.

A hydrodynamic gas mixing model employing the principles of conservation of mass, momentum, and energy is being developed (4). Another simpler approach, which allows immediate extraction of both useful qualitative and quantitative information, is to utilize the experimentally measured gas velocity profiles in the jet coupled with the solution of the tracer gas conservation equation. This approach was applied successfully to the data obtained in a 30cm diameter laboratory unit with a single jet nozzle (5).

SOLIDS CIRCULATION AND MIXING

The force probe data allow the identification of three major solids flow regions in the jetting fluidized bed. At the central portion of the bed, the solids flow is induced upward primarily by jetting action at the lower bed height and by large bubbles at the higher bed height. At the outer region next to the vessel wall, the solids flow is all downward. Between these two regions the solids flow is alternately upward and downward, depending on the approach and departure of large bubbles. The central region is divided further into three separate regions axially - the jetting region, the bubble-developing region, and the developed-bubble region. Figure 5 was constructed on the basis of the force probe readings.

The solids mixing rate was studied by pulse injection of tracer particles with characteristics similar to that in the bed but with larger sizes. By taking solids samples continuously at different bed locations and analyzing by sieving, the rate of particle mixing can be calculated. The experiments were conducted only in deep bed configuration with a bed height of 5.5 meters. The experimental details and the development of a simple mechanistic model are to be published elsewhere (6).

PARTICLE SEPARATION

Two types of experiments were performed - the static and the dynamic tests. No solids were fed or withdrawn from the bed during the static tests. In the dynamic tests, material was withdrawn and fed intermittently at predetermined time duration to simulate continuous operation. Sand particles of 1600 μm were used to simulate the ash agglomerates and crushed acrylic particles of 1400 μm, char. There is some overlap in particle size distribution of two materials.

The static tests started with a bed of 20 to 40 weight percent of sand to simulate the concentration of ash in the bed. Air velocity in the annulus spearator was then adjusted to obtain a high sand concentration in the annulus. Material in the annulus was then withdrawn and recycled, and duplicate runs were carried out. During the static tests, solid samples were obtained from six different locations along the bed height to determine sand concentration profile in the bed. The bed above the jet nozzle was found to be mixed fairly well during the static tests. A typical sand concentration profile is shown in Figure 6. Substantial enrichment of sand in the annulus separator region was observed in all tests. This indicates that the jetting action indeed provides an effective way of mixing the bed material above the jet nozzle and the annular separator operated under suitable conditions can fulfill an effective particle separation function.

The objective of the dynamic tests was to determine the maximum particle separation rate obtainable under each specific set of operating conditions. The concentration profiles obtained during the static tests indicated that the interface between the acrylic-rich and the sand-rich regions of the bed was located at the top of the annulus separator. If the solid withdrawing rate from the annulus separator is larger than the particle separation rate, the interface will move downward until it reaches the solid sampling port located at the lower part of the annulus separator. A sudden drop in sand concentration in the solid sample will then be noted. From the time it takes for the interface to travel from the top of the annulus to the sampling location, the approximate particle separation rate can be estimated.

During the dynamic tests, solid samples were obtained from only one sampling port located 20 cm from the bottom of the annulus. Typical chronological data are summarized in Figures 7 and 8. When the bed was withdrawn and fed continuously at a rate less than the particle separation rate, the high sand concentration was maintained in the annulus over more than 3 hours of operation (see Figure 7). When the feeding and withdrawal rates were higher than the particle separation rate, the sand concentration in the annulus was depleted eventually as shown in Figure 8. Thus, it is important to be able to predict the intrinsic particle separation rate for each system so that the separator can be properly designed. The particle separation rates were estimated to be between 466 to 511 kg/hr under the operating conditions employed.

ACKNOWLEDGMENT

The work was performed under U. S. Department of Energy Contract DE-AC21-82MC19122.

LITERATURE CITED

1. Yang, W. C., D. Revay, R. G. Anderson, E. J. Chelen, D. L. Keairns, and D. C. Cicero, "Fluidization Phenomena in a Large-Scale Cold Flow Model," in Fluidization, Kunii and Toei (Eds.), Engineering Foundation, New York (1983).

2. Yang, W. C., B. Ettehadieh, T. C. Anestis, R. E. Gizzie, and G. B. Haldipur, "Cold Flow Scaleup Facility Experimental Results and Comparison of Performance at Different Bed Configurations," prepared by KRW Energy Systems Inc. for U.S. Department of Energy; NTIS No. FE-19122-46 (1984).

3. Massimilla, L., G. Donsi, and N. Migliaccio, AIChE Symp. Series, 77 (205), 17 (1981).

4. Ettehadieh, B., D. Gidaspow, and R. W. Lyczkowski, AIChE J., 30 (4), 529 (1984).

5. Yang, W. C., D. L. Keairns, and D. K. McLain, AIChE Symp. Series, 80 (234), 32 (1984).

6. Yang, W. C., B. Ettehadieh, and G. B. Haldipur, "Solids Circulation Pattern and Particle Mixing in a Large Jetting Fluidized Bed", accepted for publication to AIChE J., (1986).

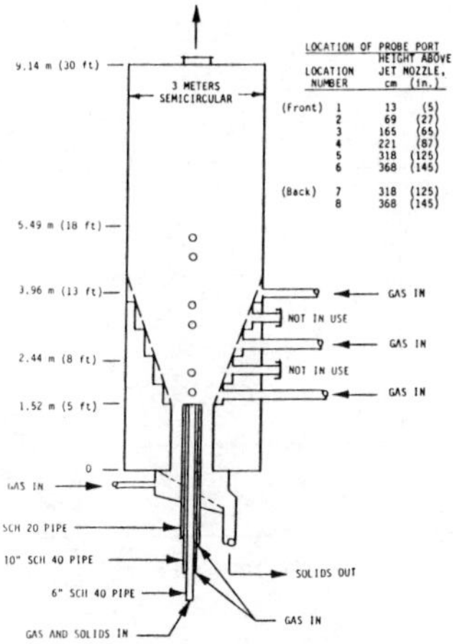

Fig. 1 Bed Configuration of 25 cm Jet Nozzle Assembly and Shallow Bed with Annulus

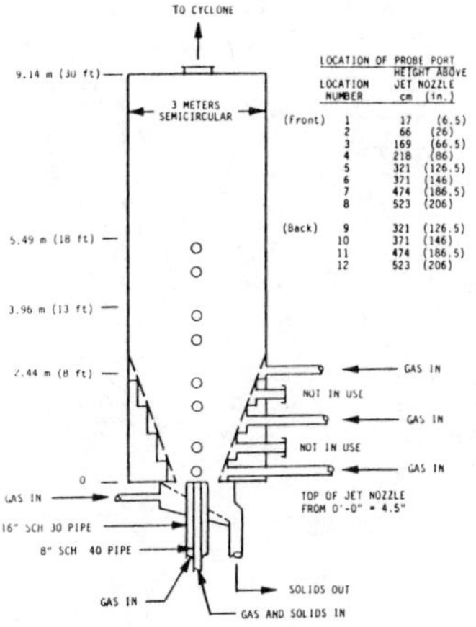

Fig. 2. Bed Configuration of 41 cm Jet Nozzle Assembly and Deep Bed Without Annulus

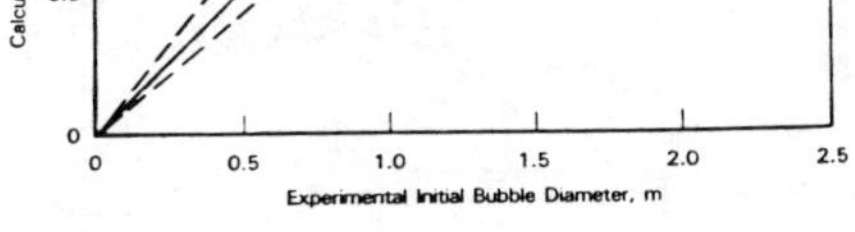

Fig. 3. Comparison of Experimental and Predicted Bubble Diameter

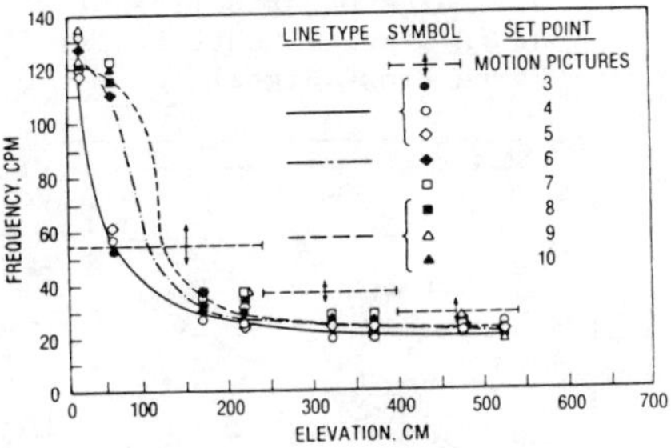

Fig. 4. Profiles of the Characteristic Major Frequencies and the Bubble Passage Frequencies Along the Bed Height - 41 cm Jet Nozzle Assembly

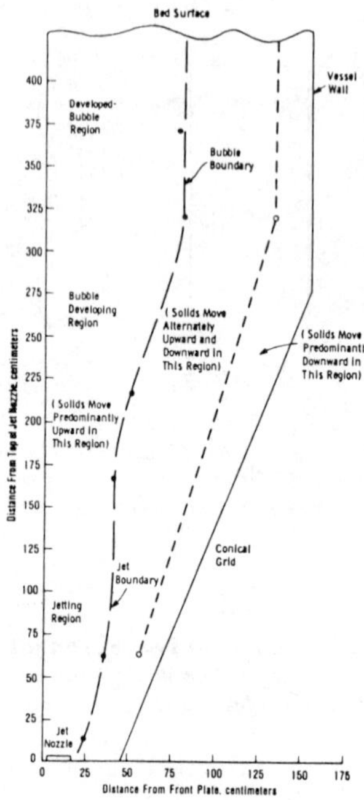

Fig. 5. Three Major Solids Flow Regions in 3-Meter Model Derived from Force Probe Signals

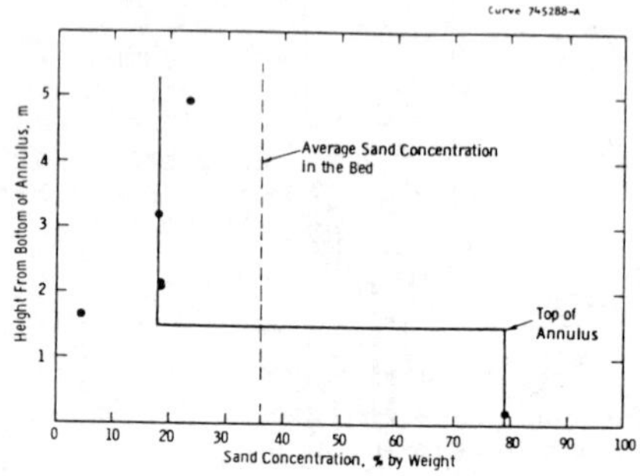

Fig. 6. Solids Concentration Profile in the Bed - Set Point 4

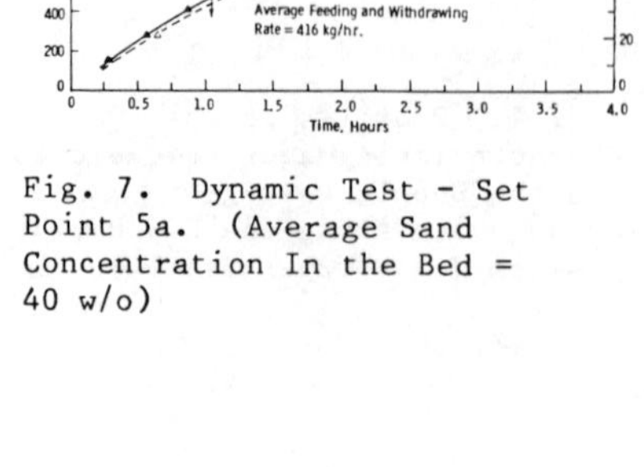

Fig. 7. Dynamic Test - Set Point 5a. (Average Sand Concentration In the Bed = 40 w/o)

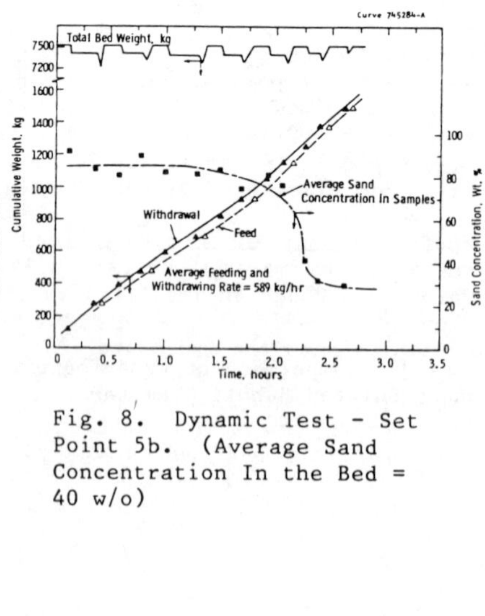

Fig. 8. Dynamic Test - Set Point 5b. (Average Sand Concentration In the Bed = 40 w/o)

GAS THROUGHFLOW IN A BUBBLING FLUIDIZED BED

Leon R. Glicksman
Thomas Yule
Department of Mechanical Engineering
Massachusetts Institute of Technology
77 Massachusetts Avenue
Cambridge, MA 02139

ABSTRACT

The gas flow through a bubble as it approaches and touches the bed surface has been numerically solved. When the throughflow results are combined with a random model for bubble coalescence a prediction for the overall dense phase gas flow results. The predicted flow agrees with recent experimental results for large-particle beds.

SCOPE

The gas flow distribution between bubbles and dense phase in bubbling fluidizeds bed has been the object of much controversy. The simple two-phase model, even including throughflow through isolated bubbles, substantially overpredicts the visible bubble flow for most cases reported. The goals of the present work are a quantitative examination of possible mechanisms of gas throughflow in a bubbling fluidized bed and a predictive model for gas throughflow in the bed as well as peak gas velocity through bubbles at the bed surface.

CONCLUSIONS AND SIGNIFICANCE

Numerical simulations were carried out to find the gas flow through bubbles or voids near a bed surface. Elongated cylindrical voids extending downward from the surface of the bed produce high gas flow through the dense phase from the distributor to the bed surface. The flow per unit cross-sectional area of the void increases with the void height-to-diameter ratio and decreased bed depth to void diameter.

When a bubble or a cylindrical void closely approaches the bed surface, the maximum gas velocity is substantially greater than the velocity through an identical void deep within the bed. The increased flow, especially for large-particle beds, is responsible for ejecting particles into the freeboard.

The distribution of deep cavities at the surface, caused by coalescence between bubbles, is predicted for bubbles in large-particle beds which are randomly spaced. The predicted dense phase gas flow is substantially greater than the flow found by the two-phase model. The new model agrees with recent data for a large-particle bed. When the bubble diameter and velocity are known, the bed expansion can be found from an overall flow balance.

BUBBLE SHAPE

When the Reynolds number for interstitial gas flow is low the flow is described by Darcy's Law. Assuming the gas and dense phase to be incompressible and the pressure at the surface of the bubbles as well as at the bed boundaries is independent of particle motion, then the solution for the pressure field and the gas bypass through the bubbles is identical to the flow through a bed with the bubbles and particles held stationary. Note that this is strictly valid only when the bubbles are not changing their volume.

Taking the bubbles as uniform in volume and the dense phase at minimum fluidization conditions the gas flow continuity expression can be written as

$$U = (1 - \delta)\, U_{mf} + \delta U_{tf} + Q_{b/A} \tag{1}$$

where U_{tf} is the average upward velocity of gas passing through the bubble voids. Alternatively (1) the gas continuity can be written as

$$U = Q_{b/A} + K\, U_{mf} \tag{2}$$

where K represents the increase of throughflow in the dense phase over the minimum fluidization velocity. The effect of bubble shape on U_{tf} for bubbles deep in the bed can be found (1) from a solution for a mathematically analogous situation: the electrical conductivity of a medium containing a dilute suspension of very conductive spheroids. An elongated spheroidal bubble with vertical axis twice the horizontal axes has a throughflow velocity of 5.8 U_{mf}, compared to 3 U_{mf} for a spherical bubble. Vertically elongated shapes approximate the conditions during the final stages of coalescence between two bubbles. However, the higher gas velocity flowing radially out of the upper surface of two submerged, coalescing bubbles should make the resulting bubble resume a more spherical shape. Thus, the increased throughflow will only persist for a short duration of the bubble coalescence process.

BUBBLE COLUMNS EXTENDING TO THE BED SURFACE

Elongated voids have been observed at the surface of a fluidized bed formed by the coalescence of bubbles at the surface. Gas from the surrounding dense phase will flow through the cavity to the freeboard. Gas velocities well in excess of U_{mf} are observed above the mouth of the cavity (2). Since the mouth of the cavity is open, the gas velocity vector is approximately vertical and there is no tendancy for the cavity to widen as it would deep within the bed. Thus, high throughflow velocities persist until the cavity leaves the bed or collapses. Cavities at the surface appear to be a likely source of excess gas flow through the bed. Very few solutions appear for this case, e.g. (3), and they correspond to semi-ellipsoids or hemispheres at the surface of deep beds.

In the present work, a numerical solution of the pressure and flow field was undertaken for several bubble or void geometries approaching and touching the bed surface including cases in which the bubble depth approaches the bed depth. Darcy's law is assumed to hold. The pressure at the bed surface is constant and the pressure below the distributor is constant. The flow through the distributor is assumed to vary linearly with the pressure difference at any local point. The bed is assumed cyclindrical with the center at the center of the void. The diameter of the bed is increased until any further increases do not influence the local throughflow through the void. Typically this occurs at a bed diameter of less than four times the bed depth. The first case considered is a bubble void in the shape of a right circular cylinder of length H, Figure 1. The pressure in the void is equal to that of the freeboard and the pressure beneath the distributor is uniform. Two parameters varied in the solutions were the bed depth, L, and β , the ratio of the pressure difference across the distributor to the total pressure difference across the bed and distributor at minimum fluidization. The pressure distribution and flow field were found from a numerical solution of Laplace's equation. The ratio of the gas velocity leaving the cavity to U_{mf} is shown on Figure 2. At a distributor pressure drop one tenth of the overall pressure drop, the velocity is a function of the bed depth only when H, the bubble depth, is close to L, the bed depth, and H/D_B is less than two. When the distributor pressure drop is one half of the total pressure drop, the results depend almost exclusively on H/D_B, irrespective of the bed depth or the bubble size.

Figure 3 illustrates the streamlines and isobars for the case of a cavity at the bed surface with H/D of 2 and L/D of 5. Increased gas flow is drawn from the distributor to the bed surface through the dense phase which remains at minimum fluidization voidage. Large pressure gradients are set up in the dense phase since the freeboard pressure is felt over the cavity surface deep within the bed. The resistance of the distributor directly below the cavity cross section moderates the velocity in this area. Note that the void draws flow away from the surrounding dense phase. The gas flow through the void, as given by its mean velocity, does not represent the net increase in gas flow through the bed. The net flow increase will, in general, be less than the cavity flow.

BUBBLES APPROACHING THE BED SURFACE

It is useful to consider how the flow increases as a sphere or elongated void approaches the bed surface. A solution for the flow must include a determination of the bubble pressure so that the net flow into the base of the bubble balances the flow from the bubble to the bed surface. A finite difference numerical solution was performed for spherical and cylindrical bubbles a distance T below the bed surface. The bed surface was assumed to remain planar. When the bubble nose is greater than a distance $D_B/4$ from the interface, the gas flow is close to that for a deep bubble. As the bubble gets closer to the surface the velocity increases dramatically. The average flow, measured from the point when the bubble nose approaches the surface to the point when the bubble wake has left the surface, is about equal to the flow for the same bubble shape deep within the bed. Details of the calculation procedure will be presented in a subsequent paper.

NET GAS FLOW WITH CAVITIES AT THE SURFACE

The net flow increase through a bed due to a cavity at the surface has been calculated for cylindrical cavities, Figure 4. To generalize the gas flow results it is assumed that the pressure drop in the bed is the sum of the pressure difference in the dense phase beneath the cavity, Δp_2, plus the pressure difference in the upper bed, Δp_1. For all cases where the distributor pressure difference is one half the total pressure difference, the pressure difference across the bed is very close to the value at minimum fluidization. The pressure difference in the lower bed is found assuming the flow in excess of Q_b/A is proportional to the product of the pressure drop in the lower bed and the lower bed permeability, thus,

$$\frac{(U - Q_b/A)}{U_{mf}} = \frac{\Delta p_2 \, (1 + 2\delta)}{\Delta p_{Bed} \, (1 - H/L)} \tag{3}$$

where the factor $(1 + 2\delta)$ accounts for the increased permeability of the lower bed when widely spaced spherical bubbles are present (1). U is the net flow rate per unit cross-section for the bed containing the cavity.

Similarly, using K_1 from Equation (2) to represent the increased permeability of the upper bed section containing the cavity, the pressure difference over the upper bed is, $(\Delta p_{Bed} - \Delta p_2)$, which yields a flow,

$$\frac{U - Q_b/A}{U_{mf}} = \frac{K_1 \, (\Delta p_{Bed} - \Delta p_2)}{\Delta p_{Bed} \, (H/L)} = K_1 \left[\left(\frac{L}{H} - \frac{\frac{L}{H} - 1}{(1 + 2\delta)} \right) \left(\frac{U - Q_b/A}{U_{mf}} \right) \right] \tag{4}$$

rearranging,

$$\frac{U - Q_b/A}{U_{mf}} = \frac{K_1 \, (L/H)}{1 + K_1 \, (L/H - 1) \, / \, (1 + 2\delta)} \tag{5}$$

where

$$K_1 = (1 - \delta) + \delta \, (U_{tf}/U_{mf})_1 \tag{6}$$

Figure 4 gives the resulting values of U_{tf}/U_{mf} for the upper bed section, when it contains a cavity. The results, similiar to those of Figure 2, are primarily a function of H/D_B and are insensitive to distributor pressure difference, β, and to column depth-to-bed depth except when the column extends from the bed surface to the distributor.

COALESCENCE MODEL FOR COARSE PARTICLES

For coarse particle beds a model for the bubble coalescence rate is available (4). This will allow a prediction of gas throughflow to be carried out. It has been shown from observations of bubble eruptions at the surface of coarse-particle beds that the bubbles are randomly distributed and their spatial distribution follows a Poisson distribution (4), (5).

The bubble coalescence rate is a function of the bubble void fraction which can be closely fit by the relation,

$$\frac{d\,(f/A)}{(f/A)\;d(z/D_B)} = -12\delta^{5/4} \tag{7}$$

where f/A is the number of bubbles passing through a unit bed cross section per unit time. The change of bubble frequency with coalescences can also be written as,

$$\frac{d\,(f/A)}{dz} = -\frac{1}{2}\,\frac{f/A}{U_b\;t_c} \tag{8}$$

where t_c is the mean time between coalescences. Combining Equations (7) and (8),

$$t_c = \frac{D_B}{24\;U_b\delta^{5/4}} \tag{9}$$

Since the bubbles are randomly distributed, as a bubble rises a distance z', in a time interval z'/U_b, the probability of it undergoing n coalescences is given by the Poisson distribution, in terms of n and $z'/U_b t_c$:

$$P_n(z') = [\exp - 24z'\,\delta^{5/4}/D_B]\quad [24z'\,\delta^{5/4}/D_B]^n/n! \tag{10}$$

To calculate the throughflow due to bubbles coalescing near the bed surface consider a spherical bubble as its nose just touches the bed surface, z' will be taken to be zero. As the bubble moves up a distance z' the probability that it has coalesced with n bubbles is given by Equation (10). When a coalescence takes place the bubble is considered to form a cylindrical cavity of diameter D_B with a depth $(2D_B - z')$. The throughflow through this cavity is given by Equations (4) and (5) together with Figure 4. At a given value of z' the average throughflow is the sum of the throughflow for a single bubble, or the throughflow for a cavity formed by one or more coalescences, each multiplied by their respective probability. The calculation is continued for step changes in z' until the bubble has left the bed.

When the bubbles are widespread and δ, the bubble voidage, is small, the probability of a coalescence near the bed surface is small. As the bubble void fraction increases, coalescences occur near the surface creating some deeper voids which permit more throughflow. If the bubble distribution follows a true Poisson probability, there should be no upper limit to the number of coalescences allowed.

However, in practice, one or two coalescences may deplete the local regions of the bed of any further voids.

Figure 5 presents the results of the random coalescence and throughflow analysis giving the average bed throughflow versus bubble voidage and bed height to bubble diameter. Bubbles which extend deeper into the bed have a greater influence on the bed throughflow. At higher bubble densities, a larger fraction of bubbles coalesce at the surface forming deeper voids. Results for isolated bubbles and uniformly spaced spherical bubbles (1) are also shown.

Figure 6 compares the predicted throughflow with the experimental results of Lord (6). Each data point represents the value of the throughflow and bubble voidage averaged over the bed volume. The predicted throughflows follow the operating line for this experiment; each operating condition has a different value of L/D_B. Results are shown for three different values for the maximum number of coalescences a bubble at the surface can undergo as it moves up a distance D_B.

A similar type of analysis can explain the throughflow found for small-particle beds. In this case, the preferred bubble tracks give a higher local bubble voidage and a higher throughflow than that based on the average bubble voidage over the entire bed cross-section.

ACKNOWLEDGEMENT

The research was sponsored by the Tennessee Valley Authority; their support is gratefully acknowledged. Computer facilities supplied by Digital Equipment Corp.

NOTATION

A	Cross Sectional Area
D_B	Bubble Diameter
f/A	Frequency of Bubble Per Unit Area
H	Bubble Depth
K	Throughflow Coefficient, Eqn. 2
L	Bed Depth
n	Number of Coalescences
Δp	Pressure Difference
P_n	Probability of n coalescences
Q	Average gas flow through bubble
Q_b/A	Visible Bubble Flow Rate
t_c	Time Between Coalescences
U	Superficial Gas Velocity
U_b	Bubble Rise Velocity
U_{mf}	Minimum Fluidization Velocity
U_{tf}	Average Throughflow Velocity, Eqn. 1
V	Gas Velocity Leaving Bubble at Bed Surface
z'	Vertical Coordinate
β	Ratio of Distributor Pressure Drop to Overall Pressure Drop
δ	Volume Fraction of Bubbles

SUBSCRIPTS

$(\ )_1$ Upper Bed, see Fig. 1

$(\ )_2$ Lower Bed

$(\ )_{Bed}$ Overall for Bed, Excluding Distributor

LITERATURE CITED

1. Valenzuela, J.V. and Glicksman, L.R., "Gas Flow Distribution in a Bubbling Fluidized Bed", to be published in Powder Technology, 1985.

2. Levy, E. and Lockwood, F. "Laser Doppler Measurements of Flow in Freeboard of a Fluidized Bed", AIChE J, Vol. 29, No. 6, p. 889, 1983.

3. Baskakov, A.P., Shuvalov, V.T. and Berg, B.V., "Investigating the Eruptions of Fine Grain Material from a Fluidized Bed," Chem. Eng. Sci., Vol. 39, 3, pp. 407-412, 1984.

4. Glicksman, L.R., Lord, W.K. and Sakagami, M., "Bubble Properties in Large-Particle Fluidized Beds", to be published in Chem. Eng. Sci., 1984.

5. Glicksman, L.R., McAndrews, G., "The Effect of Bed Width on the Hydrodynamics of Large Particle Fluidized Beds", Powder Technology, Vol. 42, 159-167, 1985.

6. Lord, W.K., "Bubbly Flow in Fluidized Beds of Large Particles", Ph.D Thesis, Dept of ME, MIT, 1983.

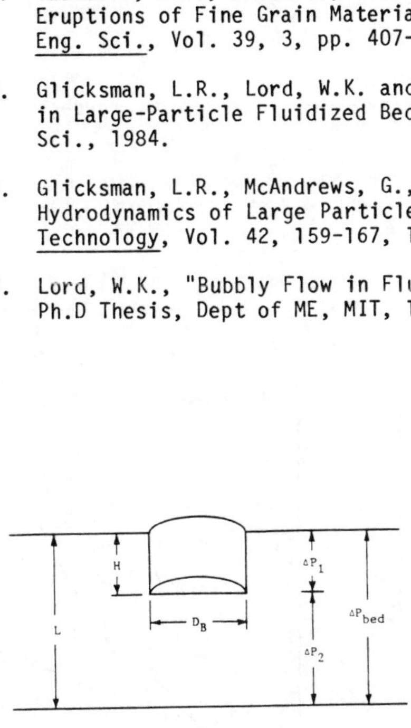

FIGURE 1 Cylindrical bubble at bed surface

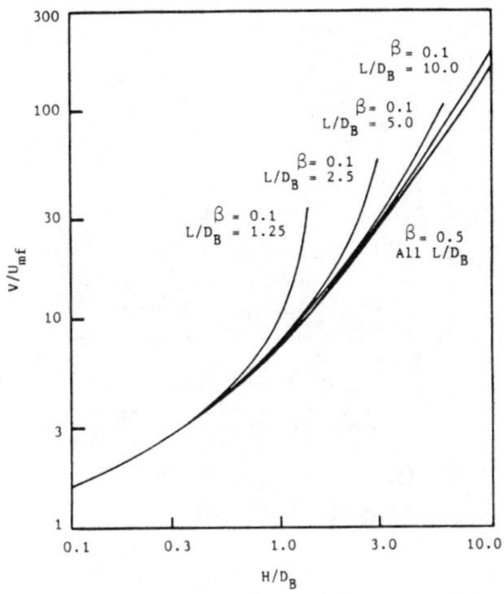

FIGURE 2 Gas velocity leaving cylindrical bubbles at bed surface

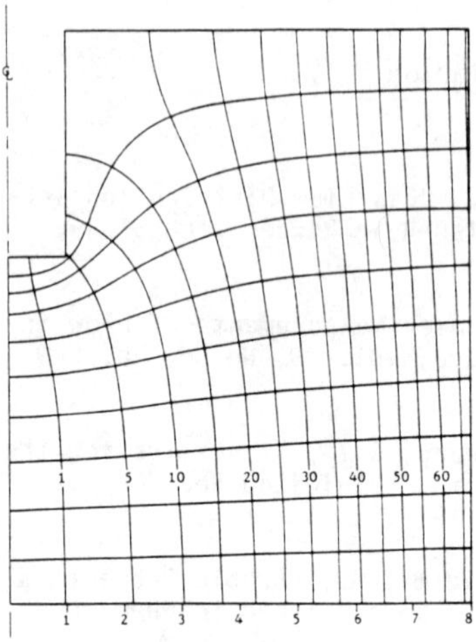

FIGURE 3 Streamlines and isobars for a cylindrical cavity at the bed surface. Given streamfunction values are multiples of $\pi D_B^2 U_{mf}/4$. The difference between any two streamfunction values gives the flow between the streamlines. In the absence of the cavity the streamfunction values would increase along the distributor face as the radial distance squared. $H/D_B = 2$, $L/D_B = 5$, $\beta = 0.1$

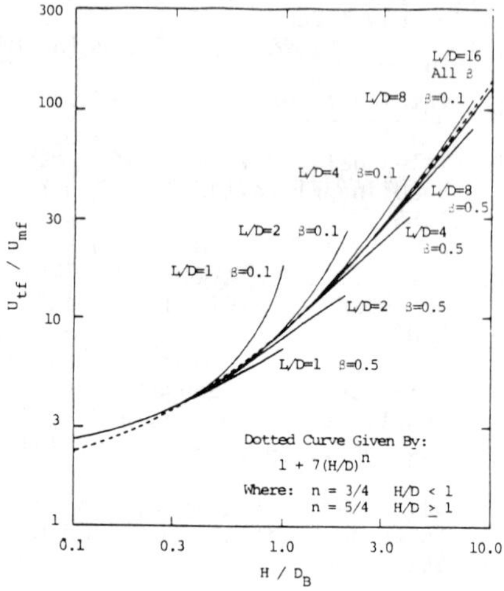

FIGURE 4 Average throughflow velocity for the upper bed section containing a cavity

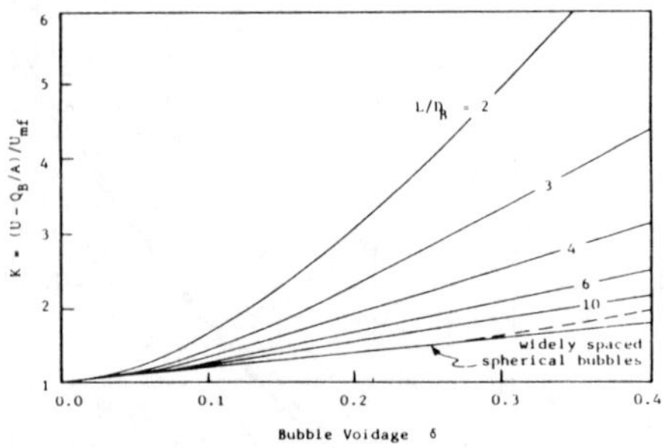

FIGURE 5 Average bed throughflow coefficient from random coalescence model. Dashed line represents uniformly spaced spheres

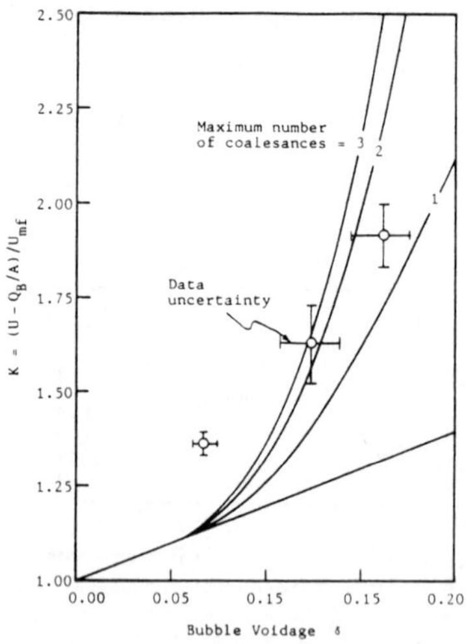

FIGURE 6 Average bed throughflow coefficient from random coalescence model. Experimental data from Lord(13). U/U_{mf}: 1.5, 2.0, 2.5, D_B: 15, 26, 37 cm U_{mf}: 58 cm/sec, bed depth = 89 cm

SPATIAL DISTRIBUTION OF BUBBLES IN THE FLUIDIZED-BED

S.Mori, M.Yamada, O.Hashimoto* and I.Yamada

Nagoya Institute of Technology
Gokiso, Showa, Nagoya, 466, Japan

*Mitsubishi Gas Chemical Corp.Inc.
Matsuhama, Niigata, 950-31, Japan

ABSTRACT

A new steady-state model was developed to simulate the time-averaged spatial distribution of bubbles in various size and types of fluidized beds. It was demonstrated that reported experimental data in both cylindrical and square types of freely bubbling fluidized-beds with different distributors could be described by the calculated results from the model proposed here.

SCOPE

Recently , many industrial processes of fluidized-bed , in which particles are classified into B and D types by Geldart's criteria , have been developed , for examples fluidized-bed coal gasifier , fluidized bed municipal waste incinerator with heat recovery , vapor polymerization reactor , vapor crystallization reactor.

It is well known that the bubble growth and the horizontal distribution of bubble frequency in these fluidized-beds are affected significantly by the size and height of the bed. Especially , the global circulation pattern of solid particles in the bed is controlled by the distribution of bubbles. Therefore , the adequate estimation of the distribution of bubbles is essentially significant for the reasonable design and operation of the fluidization process.

Previously , Miwa and Mori et al (1) , Nguyen et al (2) , Farrokhalaee and Clift (3) and Horio et al (4) proposed unsteady-state models based on procedures in which the motion of uniformly generated bubbles from distributor were followed one by one to the top of the bed within very short time intervals. Then , the computer time requirement for the numerical calculation of these model were drastically increased with the bed size and height. Recentry , Mori et al (5) proposed a new steady-state model to predict the time average spatial distribution of bubble frequency in the cylindrical fluidized-bed with axial symmetry. In this paper , this two-dimensional model was applied to develop a general three-dimensional fluidized-bed model in consideration of the horizontal distribution of bubble size.

CONCLUSION AND SIGNIFICANCE

A general steady-state model was developed to predict the spatial distribution of bubbles in freely bubbling fluidized-beds. Except one parameter , K_{bc} , which should be determined from observed data of bubble growth , the other four parameters were determined as constant values to simulate observed bubble distribution in three different types of large size fluidized-bed with different structure of distributors.

MODELING

Based on the similar assumptions of the cylindrical model proposed by Mori et al (5) , rectangular components of bulk bubble flow rates in a bed , u_x , u_y , u_z , are related as Eq.(1) from material balance.

$$\frac{\partial u_x}{\partial x} + \frac{\partial u_y}{\partial y} + \frac{\partial u_z}{\partial z} = 0 \tag{1}$$

From the local population balance of bubbles , Eq.(2) is given by the simple treatment of bubble coalescence(Mori et al (5)).

$$\frac{\partial F_x}{\partial x} + \frac{\partial F_y}{\partial y} + \frac{\partial F_z}{\partial z} = -K_{bc}u_z^{\alpha}n_b^{\beta} \tag{2}$$

The following relationship are given similar to the cylindrical model.

$$F_j = u_j/V_b \quad : (j = x,y,z) \tag{3}$$

$$u_{bz} = u_z + 0.71\sqrt{gD_b} = u_z + u_{b0} \tag{4}$$

$$D_b = (6V_b/\pi)^{1/3} \tag{5}$$

$$n_b = F_z/u_{bz} \tag{6}$$

Based on the calculated results of the horizontal interactive velocity of the bubbles-pair by the model presented by Nguyen et al(2) , it is simply assumed that the horizontal velocity of a bubble located at (x,y) which is interacted by another bubble located at (x',y') is proportional to its vertical rising velocity and the volume ratio of the bubbles. The velocity is also assumed to be described by a simple function of the horizonatl distance of bubbles as shown in Eq.(7).

$$u_{bxy}^{*} = K_{xy}\{\frac{D_b(x',y')}{D_b(x,y)}\}^3\left[1 - \frac{\{(x-x')^2+(y-y')^2\}^{1/2}}{1.25D_b(x,y)}\right]^{\gamma} u_{b0}(x',y') \tag{7}$$

This equation can be used within the following interaction area of two bubbles.

$$0 \leq \{(x-x')^2+(y-y')^2\}^{1/2} \leq 1.25D_b(x,y) \tag{8}$$

Based on the similar assumptions adopted in the cylindrical model , the net horizontal velocity components of the bubble at (x,y) in the bed are given by Eq.(9).

$$u_{bj} = -\int_{-L_x/2}^{L_x/2}\int_{-L_y/2}^{L_y/2}\left[n_b(x',y')D_b(x',y')u_{bxy}^{*}\frac{(j-j')}{\{(x-x')^2+(y-y')^2\}^{1/2}}\right]dx'dy' \quad : (j = x,y) \tag{9}$$

Then , horizontal components of bulk flow rate of bubbles are given by Eq.(10).

$$u_j = n_bV_bu_{bj} \quad : (j = x,y) \tag{10}$$

CALCULATED RESULTS

In this paper , the observed data reported by Werther (6,7) , Whitehead and Young (8) and GOLFERs (9) are compared with calculated results obtained from the model proposed here. Constant values of model paremeters of K_{xy}=0.70 , α=1.0 , β=1.5 , γ=2.0 were estimated for these calculations , however different values of K_{bc} were used to correlate observed data of bubble growth , that is , K_{bc}=0.7 for Werther's data , K_{bc}=2.0 for GOLFERs and K_{bc}=5.0 for Whitehead and Young.

At first , simulated results for Werther's (6,7) data obtained in 0.45 and 1mϕ cylindrical beds with porous plate distributors are shown in Figs. 1 , 2 and 3. Original data of bubble diameter and radial distribution of bubble flow rate at z=0.05m were used for the initial conditions of the calculation. The values of calculated bubble diameter are compared with observed data as shown in Fig.1 and calculated results of apparent bubble flow rate ($u_{Zob}/\bar{u}_Z$) at various height in beds are compared with experimental data measured by use of the capacitance probes in 1mϕ and 0.45mϕ beds in Figs. 2 and 3 , respectively. In these figures , calculated distribution of bubble flow rate along the x-axis shown by solid line are slightly different from those along the diagonal line shown by dotted line. This difference is due to four weak peaks appeared at diagonal positions in heigher bed levels as shown in Fig.4.

In Figs. 5 , 6 and 7 , simulated results for GOLFERs'(9) data are shown. GOLFERs' data were obtained in a 1 × 1m square bed with a perforated plate distributor (8mmϕ × 120holes) and then initial bubbles are

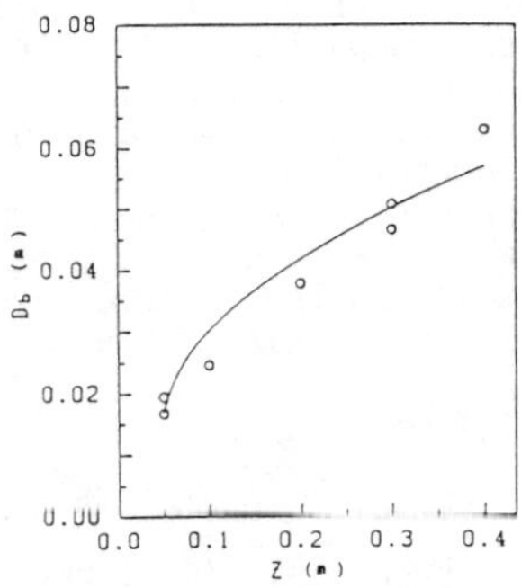

Fig.1 Bubble growth in 1mϕ bed (Werther).

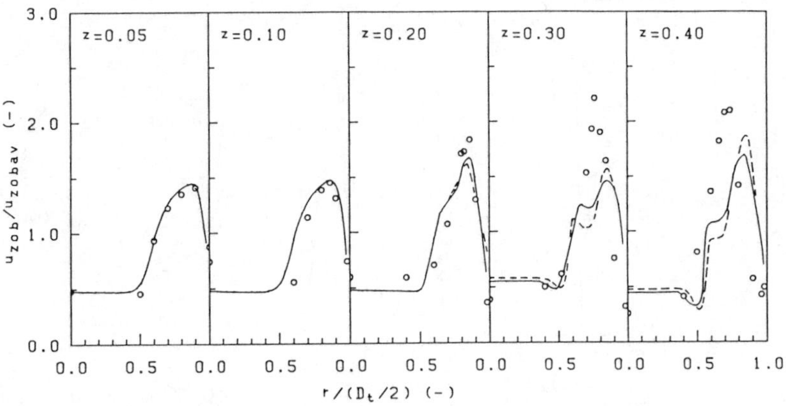

Fig.2 Apparent bubble flow rate in 1mϕ bed (Werther).

assumed to be generated homogeneously just above the distributor. In Fig.5 , it is shown that observed bubble diameter under different superficial velocities , u_0 , can be well described by calculated results for K_{bc}=2.0. In Fig.6 , bubble frequency detected by the optical method are compared with calculated results for u_0=0.084 , 0.144 and 0.200 m/s. Although experimental bed height were relatively small (L_z/L_x<1) and horizontal distributions of bubble frequency were not so sharp , it is found that observed results can be well described by calculated results and four peaks of bubble frequency neighboring the bed corners are represented at higher level in the bed as shown in Fig.7.

Whitehead and Young (8) observed bubble eruptions at the top of 1.2 × 1.2m square bed with 16 tuyeres as the distributor. Here , two cases of flow pattern of gas just above the distributor are assumed as shown in Fig.8 , that is a) almost homogeneous gas flow pattern , b) the flow pattern obtained by the superimposition of estimated gas flow pattern around each tuyeres. K_{bc}=5.0 is determined for both cases from comparison of calculated bubble diameter with observed data as shown in Fig.9. Calculated horizontal distributions of bubble frequency at each bed height for cases a) and b) are described by equi-frequency lines in Figs. 10 and 11 , respectively. In these figures , open circles show the observed position of bubble eruptions within same time intervals. Altough similar patterns of bubble frequency were estimated from both casaes of calculation , relatively sharp peaks and more reasonable bubble distribution can be obtained in case b).

NOTATION

D_b,D_t = bubble and bed diameter [m]

F_x,F_y,F_z = local bubble frequency [$m^{-2}\cdot s^{-1}$]

K_{bc} = bubble coalescence coefficient [$m^{3\beta-3-\alpha}\cdot s^{\alpha-1}$]

K_{xy} = horizontal bubble interaction coefficient [-]

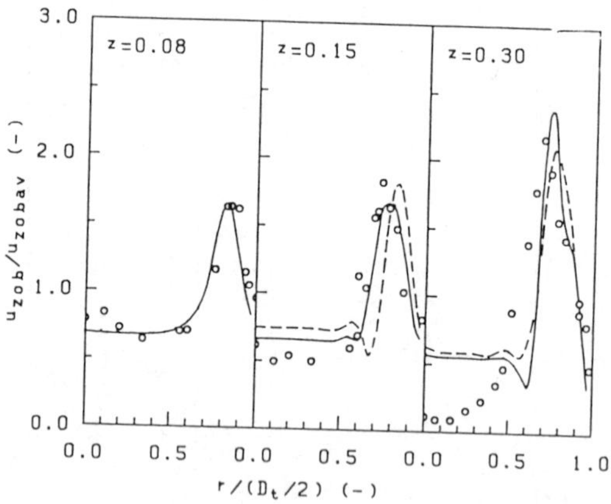

Fig.3 Apparent bubble flow rate in 0.45mϕ bed (Werther).

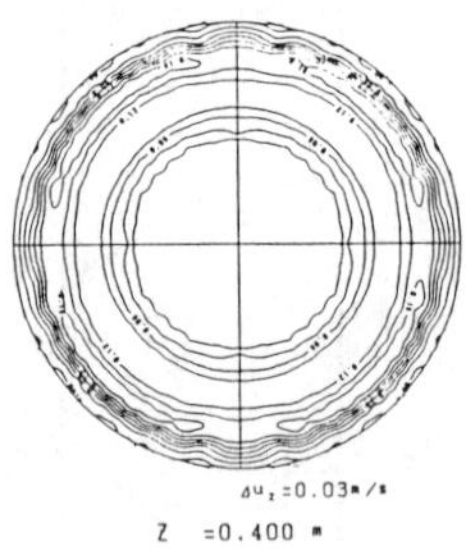

Fig.4 Apparent bubble flow pattern in 1mϕ bed at z=0.4m.

L_x, L_y = width of the square bed [m]

L_z = bed height [m]

n_b = number of bubbles in the bed [m^{-3}]

u_{bx}, u_{by} = horizontal bubble velocity [$m \cdot s^{-1}$]

u_{bz} = bubble rising velocity [$m \cdot s^{-1}$]

u^*_{bxy} = horizontal velocity of a bubble interacted by another bubble [$m \cdot s^{-1}$]

u_x, u_y, u_z = local components of bubble flow rates [$m \cdot s^{-1}$]

$\bar{u}_z$ = average bubble flow rate in the bed [$m \cdot s^{-1}$]

u_{zob} = apparent bubble flow rate measured by detector [$m \cdot s^{-1}$]

u_0 = superficial gas velocity [$m \cdot s^{-1}$]

V_b = bubble volume [m^3]

x,y,z = rectangular coodinates [m]

α, β, γ = model parameters defined in Eqs. (2) and (7) [-]

LITERATURE CITED

1. Miwa,K., S.Mori, T.Kato and I.Muchi, Intn.Chem.Eng., 12,187(1972)

2. Nguyen,T.H., J.E.Johnsson, R.Clift and J.R.Grace, "Fluidization Technology", ed. D.L.Keairns, Hemisphere Pub, vol.1,p.205(1975)

3. Farrokhalaee,T. and R.Clift, "Fluidization", ed. J.R.Grace and J.M.Matsen, Plenum, p.135(1980)

4. Horio,M., J.Liu and I.Muchi, Kagaku Kogaku Ronbunshu, 9,176(1983)

5. Mori,S., S.Miho, M.Yamada, S.Hiraoka and I.Yamada, AIChE Symp Ser., 80,No.241,10,(1985)

6. Werther,J. and O.Molerus, Int.J.Multiphase Flow, Hemisphere Pub., vol.1,p.123(1973)

7. Werther,J., "Fluidization Technology" ed. D.L.Keairns, Hemisphere Pub., vol.1,p.215(1975)

8. Whitehead,A.B. and A.D.Young, Proc.Intn Symp. on Fluidization, Eindhoven, Netherland Univ.Press. p.284(1976)

9. GOLFERs, Kagaku Kogaku Ronbunshu, 8,464(1982)

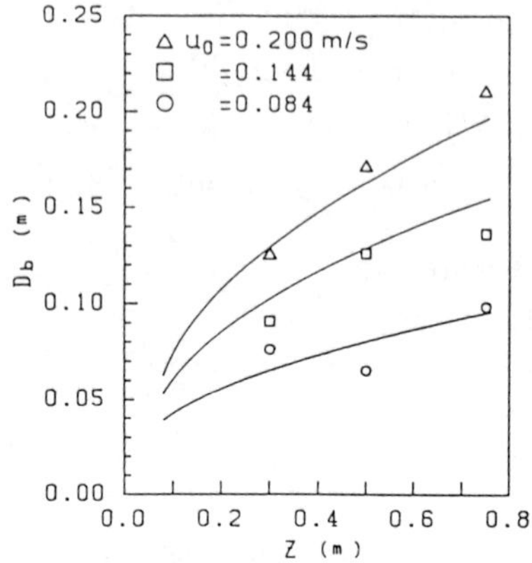

Fig.5 Bubble growth in 1×1m bed (GOLFERs).

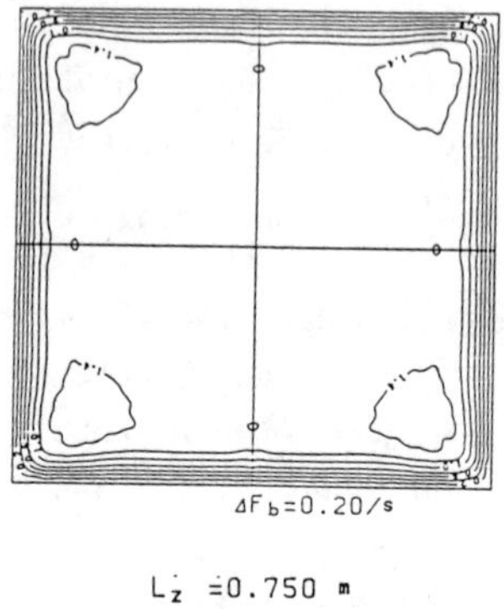

Fig.7 Bubble frequency in 1×1m bed at z=0.75m.

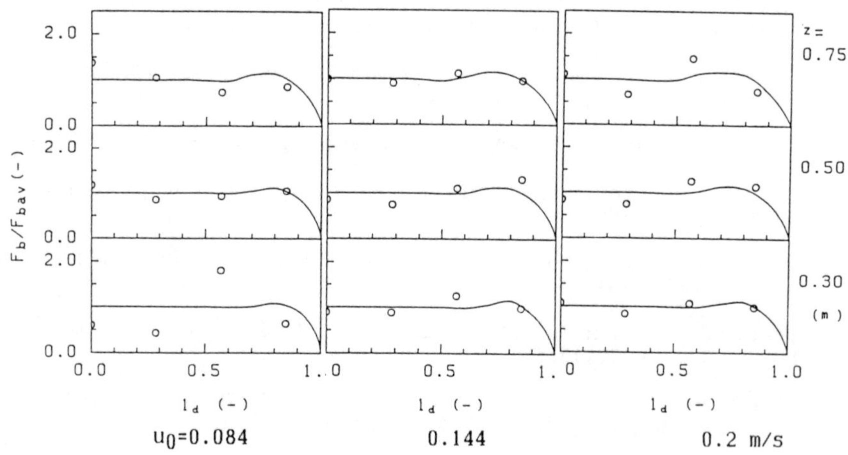

Fig.6 Bubble frequency in 1×1m bed along the diagonal line.

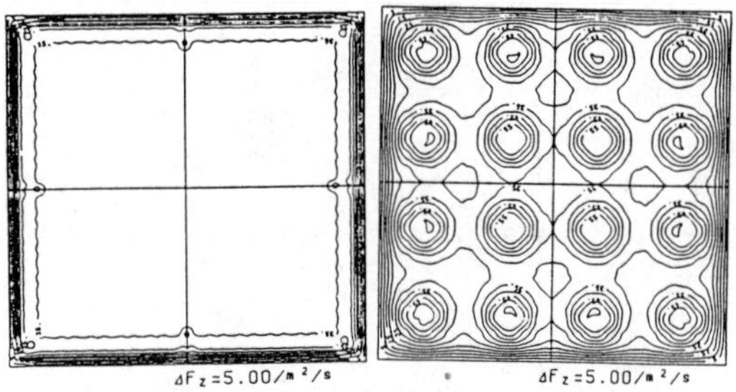

case a case b

Fig.8 Assumed initial flow pattern of gas in 1.2×1.2m bed.

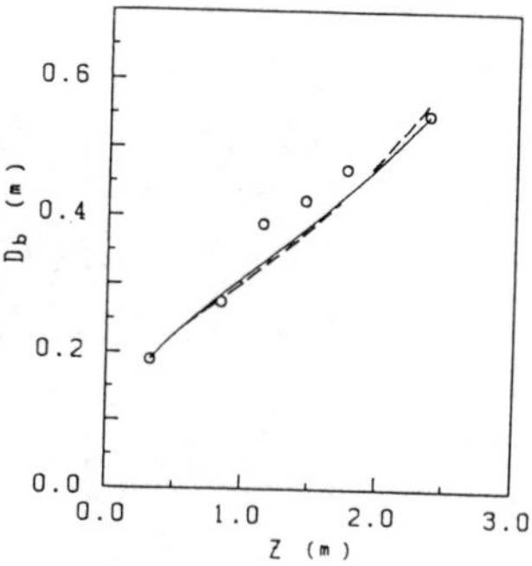

Fig.9 Bubble growth in 1.2×1.2m bed(Whitehead)

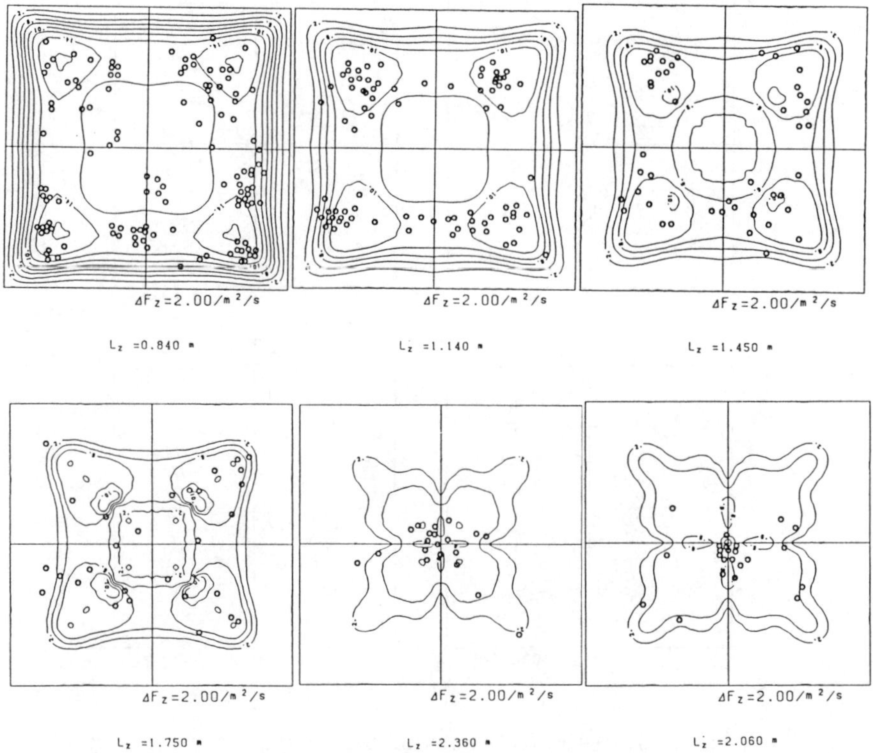

Fig.10 Calculated distribution of bubble frequency for case a) in 1.2×1.2m bed (Whitehead).

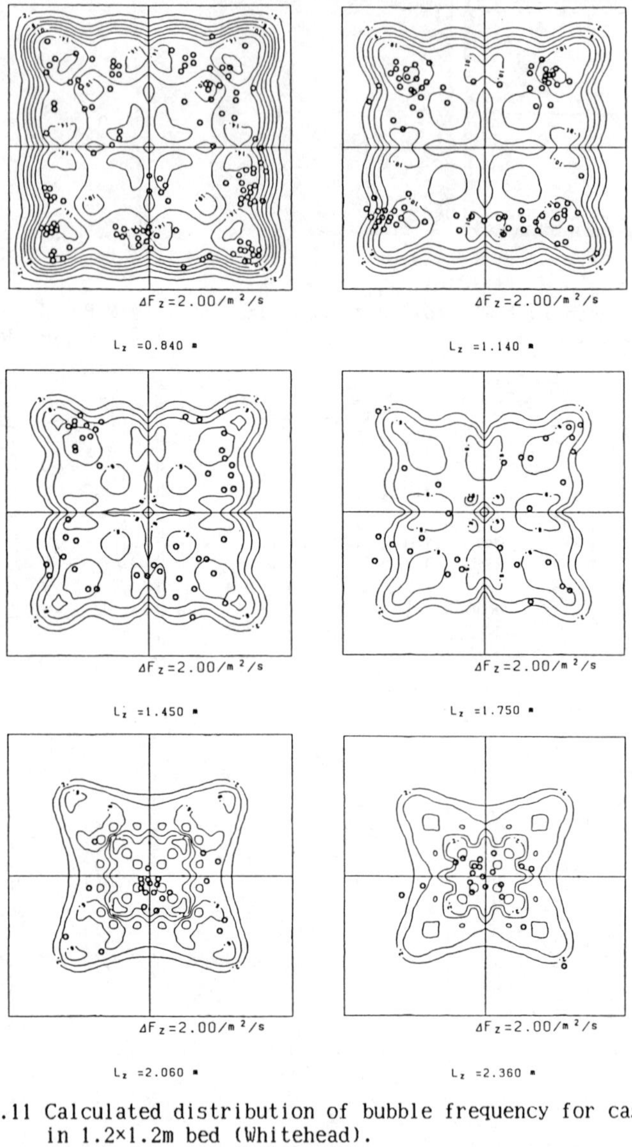

Fig.11 Calculated distribution of bubble frequency for case b) in 1.2×1.2m bed (Whitehead).

INTERPHASE MASS TRANSFER IN A THREE DIMENSIONAL FLUIDIZED BED

H. HATANO and M. ISHIDA

Research Laboratory of Resources Utilization
Tokyo Institute of Technology
Nagatsuta, Midoriku, Yokohama 227, Japan

ABSTRACT

Paired sensors consisting of a penetrative optical probe and a reflective probe were set at three heights in a three-dimensional fluidized bed. In the middle height, they were also arranged at six points. Tracer gas concentration in and around rising bubbles was measured continuously by the penetrative ones, while the bubble boundary and the zone with the prominent particle movement were detected by the reflective ones. From such profiles, both tracer gas effluence from a single bubble and throughflow between successive bubbles were clearly observed. And the entrance of the surrounding gas into bubbles accompanied by the curtain of solid particls was also detected. Based on these data, interphase mass transfer coefficient was calculated for single and paired bubbles.

SCOPE

Interphase mass transfer between bubbles and emulsion phases is a significant parameter for fluidized bed reactions. Although many theoretical and experimental researches have been conducted for the interphase mass transfer in incipiently fluidized beds, its mechanism has not been elucidated. This is because there were no measuring methods to simultaneously detect both gas concentration profiles and bubble behavior.

The object of this study is to obtain contour line map for tracer gas concentration around single and paired bubbles and their interphase mass transfer coefficients.

CONCLUSIONS AND SIGNIFICANCE

Tracer gas concentration profiles around single and paired bubbles in a three-dimensional fluidized bed were represented by contour line maps. When superficial gas velocity U_o was greater than minimum fluidization velocity U_{mf}, the gas effluence from bubble to emulsion phase occurred at the edge of the bubble bottom. Also, the entrance of the surrounding gas in the upper emulsion phase into the bubble took place accompanied by solid particles falling in it. When U_o was smaller than U_{mf}, on the other hand, steady effluence was observed at the center of bubble bottom.

In the case of paired bubbles, contour lines covered both bubbles, indicating that throughflow of tracer gas from following bubble to leading bubble occurred just before the bubble pair coalesced.

Interphase mass transfer coefficients determined in this study agreed with the empirical correlation proposed in the previous study. They showed that the effect of interaction of bubbles on them was insignificant.

INTRODUCTION

Since the first hydrodynamic model for flow of gas and solid particles around a bubble was presented, Davidson, (1), experimental investigations by direct observation of gas behavior around a bubble have been performed in a two-dimensional bed. Cloud and its effluence were observed by colored nitrogen oxide, Rowe et al., (2) and theoretical analysis was performed, Collins, (3). Tracer gas throughflow between paired bubbles was calculated together with gas circulation between paired bubbles, Shichi et al., (4). Axial gas concentration profiles around a bubble was observed in two- or three-dimensional beds by continuous gas sampling, Calderbank et al., (5), Lignola et al., (6) or optical fiber technique, Hatano and Ishida, (7).

Interphase mass transfer coefficient was also obtained in two- or three-dimensional beds by measuring tracer gas concentration in a single bubble, Chiba and Kobayashi, (8), Sit and Grace, (9). Longitudinal change in the tracer gas concentration in ozone-free and ozone-rich paired bubbles was measured, Sit and Grace, (9). Since, only one or two detectors were used in these studies, they gave only mean values, making detailed discussion for mass transfer mechanism difficult.

Optical fiber array probes have quite high potentiality to detect jet and bubble shape, Oki et al., (10) and motion of solid particles, Ishida et al., (11) in free bubbling beds. On the other hand, gas detecting system by penetrative optical fiber probe has been developed based on absorptiometry. It has quite high response speed. Furthermore, each probe can be made so thin that several probes can be set in a fluidized bed and simultaneous measurement becomes possible, Ishida and Hatano, (12).

In this study, penetrative probes were set at three levels of height and at six radii in the middle height. And two-dimensional tracer gas concentration profile around single and paired bubbles was obtained. At some points, the reflective probes were also set to simultaneously get the shapes of both bubble and wake. Furthermore, interphase mass transfer coefficients were calculated based on the longitudinal change in the gas concentration in a bubble, bubble rising velocity U_B, height D_B' and diameter D_B.

EXPERIMENTAL

Experimental apparatus and particles

The outline of the experimental apparatus is shown in Fig. 1. The column, made of transparent PVC, was 0.146 m dia. and 0.85 m high. A perforated plate with 122 holes of 1.5 mm diameter in triangular arrangement was used as gas distributor. The injector of a single bubble or a pair of bubbles was made of stainless steel tube with a bore of 4.0 mm and it was placed at the center of the column and at 0.02 m above the distributor. Optical fiber probes were set at 0.15 m, 0.25 m, and 0.35 m above the distributor.

Non-adsorptive glass beads were applied in this study (d_p=0.18 mm, U_{mf} = 0.028 m/s, ε_{mf}=0.42).

Measuring system

Six penetrative probes and eight reflective probes were used. At the lowest and the highest levels, respectively, a paired probe consisting of a reflective one and a penetrative one was installed. In the middle level, four paired probes and two reflective probes were set as shown in Fig. 1. Penetrative probes were used to draw contour line map for tracer gas concentration and to calculate the interphase mass transfer coefficient. Detailed structure of the probe was shown in the previous paper, Ishida and Hatano, (12).

Ozone which strongly absorbs 254 nm ultra-violet light was used as tracer gas in the bubble phase so that a low voltage mercury lamp was used as light source for penetrative probes. For reflective probes, a halogen lamp was used to obtain high light intensity.

Typical signals

Figure 2 (a) shows example signals detected by reflective probes for a single bubble rising in a bed at U_o=1.22 U_{mf}. When a gas bubble passed at the tip of probes, the reflected light from solid particles became weak, and thus the bubble phase can easily be recognized, as shown by the solid line in Fig. 2 (a). To decrease fluctuation of the reflected light signals from emulsion phase, three fibers in each reflective probe were used as light projector and the rest four fibers were used as light receiver. Thus, fluctuation of signal was reduced, but the motion of particles was obviously recognized in the region above and below the bubble, as shown in Fig. 2 (a). Then the bubble diameter was obtained as about 0.05 m. Also, by comparing the signals at I and II or at II and III, the bubble rising velocity and bubble height were calculated as 0.46 m/s and 0.033 m, respectively.

Figure 2 (b) shows signals detected at the same time by the penetrative probes. Ordinates show transmittance of ultra-violet light so the local tracer gas concentration can be obtained from them. Arabic numerals for the middle level II indicate the radii [cm] of the probe. Then horizontal profiles for gas concentration were obtained. For II: 0, only small amount of tracer gas was observed below the bubble. For II: -2 and -3, on the other hand, tailing of tracer gas became significant. For II: -4, the reflective probe did not detect bubble signal, while the penetrative probe detected high concentration of tracer gas. Based on bubble rising velocity, bubble height and change in tracer gas concentration in a bubble, the interphase mass transfer coefficient was calculated.

RESULTS AND DISCUSSIONS

Bubble shape and particle motion

To clarify the effect of the superficial gas velocity on the rising behavior of a single bubble, bubble shape and length of wake were measured. Figure 3 shows the relation between horizontal diameter and height for single bubbles. To obtain the bubble diameter, twelve reflective probes were set 1 cm apart on the plane II. Solid line shows the relation for spherical cap bubbles with $D_B'=D_B/1.5$, while broken line shows the relation $D_B'=D_B$. It can be said that the spherical cap bubble was obtained in this study except for the case of the low superficial gas velocity.

Length of the wake below bubbles was affected by the superficial gas velocity, as shown in Fig. 4. In this study, the wake area was hypothetically assumed as the region where motion of the solid particles was observed. When the superficial gas velocity was greater than the minimum fluidization velocity, the ratio of the height of this region to the bubble height became independent of the superficial gas velocity and it was about 1.4. This ratio was reduced significantly, as the superficial gas velocity became less than U_{mf}.

Contour line map for a single bubble

From penetrative and reflective signals, bubble locus and contour line map for tracer gas concentration were obtained, as shown in Figs. 5 (a), (b), and (c). In these figures, the position for the central axis of column was shown by a dashed line, and positions of the probes were shown by the distance from it. Then the contour lines for C/C_B=1.0, 0.5, 0.25 and 0 were shown by solid lines. The shape of the bubble detected at the middle height was shown by a broken line.

The tracer gas concentration in a bubble phase was found to be uniform and the contour line at C/C_B=1 was almost identical to the outline of the bubble, as shown in Fig. 5 (a). It is noticed that intervals of contour lines were narrow at the upper surface of bubble, but they became broader below the bubble bottom. Explicit effluence of high concentration of tracer gas was not observed in this case.

In Fig. 5 (b), another profile is shown where the tracer gas flowed significantly from the edge of bubble bottom. The concentration of the tracer gas in the effluence became higher than 0.5 C_B. Also, it is noticed in this figure that the solid particles fell into the bubble and contour lines for gas concentration entered into the bubble, i.e., fresh air in the emulsion phase entered into the bubble with the fall of solid particles. In this case, intervals of contour lines at the side of the bubble became broader than those at the upper surface of the bubble. These phenomena were observed when U_o was greater than U_{mf}.

When U_o was smaller than U_{mf}, on the other hand, tracer gas effluence was observed even at the center of the bubble bottom as well as at the edge of the bubble, as shown in Fig. 5 (c). Effluence of tracer gas at the center of the bubble bottom increased with the decrease in the superficial gas velocity. When it became 0.79 times of

U_{mf}, such steady effluence was observed almost for each bubble.

Contour line map for interacting bubbles

In order to examine throughflow or circulating gas flow between bubbles in the emulsion phase, ozone-rich bubbles were injected into the three-dimensional bed shown in Fig. 1 to elucidate such flows.

When U_o became greater than U_{mf}, tracer gas concentration profiles around a bubble pair were obtained, as shown in Figs. 6 (a) and (b). When the distance between leading and following bubbles was more than one bubble height, contour line map was similar to the one obtained for the single bubble, as shown in Fig. 6 (a).

Two bubbles approached closely in Fig 6 (b). At the highest level III, these two bubbles coalesced. It is found that tracer gas concentration in the emulsion phase between the two bubbles increased and the tracer gas of following bubble was flowing into the leading bubble. It is also found that tracer gas flowed out of the edge of the leading bubble and formed the cloud for the following bubble. It is noticed that the gas effluence from the following bubble was insignificant.

Figure 6 (c) shows the contour line map, when U_o was smaller than U_{mf}. Since the tailing for the following bubble was quite significant, the gas in the leading bubble flowed down through the following bubble.

Interphase mass transfer coefficient

Figure 7 shows the interphase mass transfer coefficient for single bubbles. The solid line shows the results calculated by the following empirical formula, Hatano and Ishida, (13).

$$k_{ob}=0.127\mathcal{D}^{1/3}\, g^{1/3}\, \alpha^{-2/9} \qquad (1)$$

The broken line covers the observed data in the above-mentiond study.

The data obtained for U_o=0.94 U_{mf} and U_o=1.06 U_{mf} agreed Equation (1) within the deviation they proposed. For U_o=0.79 U_{mf}, the obtained values became small, while they became great for U_o=1.22 U_{mf}. For the latter case, there were also bubbles consisting of fresh air and the coalescence of the ozone bubble and those bubbles might have given such high values.

Interphase mass transfer coefficient was also measured for the cases when two bubbles were coalesced. However, significant effect of such coalescence on the rate of mass transfer was not observed, in both leading and following bubbles. Similar results were obtained when ozone-rich and -free bubbles were injected into a two dimensional bed, Sit and Grace, (9).

NOMENCLATURE

C tracer gas concentration
C_B tracer gas concentration in bubble
D_B bubble diameter

D_B'	bubble height
g	gravitational acceralation
k_{ob}	single bubble interphase mass transfer coefficient
L_{wake}	length of wake
U_B	bubble rising velocity
U_e	interstitial gas velocity (U_o/ε_{mf})
U_{mf}	mimimium fluidization velocity
U_o	superficial gas velocity
α	ratio of bubble velocity to interstitial gas velocity ($=U_B/U_e$)
$\mathcal{D}$	diffusion coefficient of tracer gas
ε_{mf}	void fraction at incipient fluidization

LITERATURE CITED

1. Davidson, J.F., Trans. Inst. Chem. Eng. (London) 39,230 (1961)

2. Rowe, P.N., B.A. Partridge, and E. Lyall, Chem. Eng. Sci. (New York) 19, 973 (1964)

3. Collins, R., "The effect of some unsteady motions on gas flow patterns around a fluidization bubble, " in Fluidization, Grace, Matsen (Ed.), Plenum Press , New York and London (1980)

4. Shichi, R., S. Mori, and I. Muchi, Kagaku Kogaku, 32, 343 (1968)

5. Calderbank, P.H., J. Pereira, and J.M. Burgess, "The Physical and Mass Transfer Coefficient in the Vicinity of an Artificial Bubble Rising in a Three-Dimensional Fluidized Bed," in Fluidization Technology, Pacific Grove, Calif. (1976)

6. Lignola, P.G., G. Donsi, and L. Massimilla, "MASS SPECTROMETRIC MEASUREMENTS OF GAS COMPOSITION PROFILES ASSOCIATED WITH BUBBLES IN A TWO DIMENSIONAL BED", presented in AIChE. 74 th Annual Meeting, New Orleans (1981)

7. Hatano, H., and M. Ishida, Kagaku Kogaku Ronbunshu, 8, 219 (1982)

8. Chiba, T. and H. Kobayashi, Chem. Eng. Sci. (New York) 25, 1375 (1970)

9. Sit, S.P. and J.R. Grace, Chem. Eng. Sci. (New York) 33, 1115 (1978)

10. Oki, K., M. Ishida, and T. Shirai, "The behavior of jets and particles near the gas distributor grid in a three-dimensional fluidized bed," in Fluidization, Grace, Matsen (Ed.), Plenum Press, New York and London (1980)

11. Ishida, M., A. Nishiwaki, and T. Shirai, "Movement of solid particles around bubbles in a three-dimensional fluidized bed at high temperatures," in Fluidization, Grace, Matsen (Ed.), Plenum Press, New York and London (1980)

12. Ishida, M. and H. Hatano, 1984, "THE BEHAVIOR OF GAS AND SOLID PARTICLES IN A FLUIDIZED BED DETECTED BY OPTICAL FIBER PROBES," in Fluidization, Kunii, Toei (Ed.), Engineering Found., New York (1984)

13. Hatano, H., and M. Ishida, Kagaku Kogaku Ronbunshu, 10, 184 (1984)

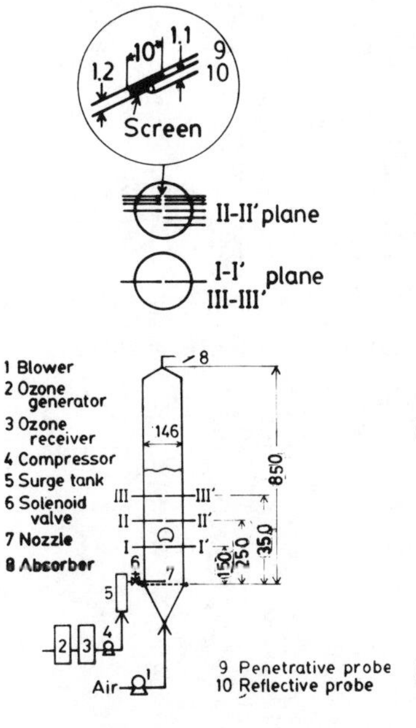

Fig. 1 Outline of apparatus and arrangement of optical probes (unit: mm)

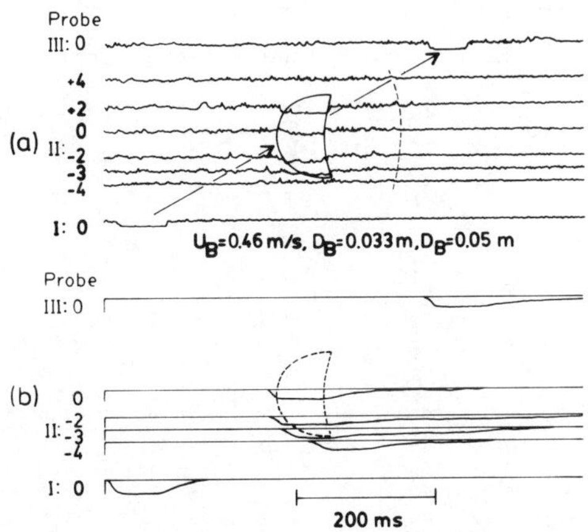

Fig. 2 Example signals for (a) reflective probes and (b) penetrative probes

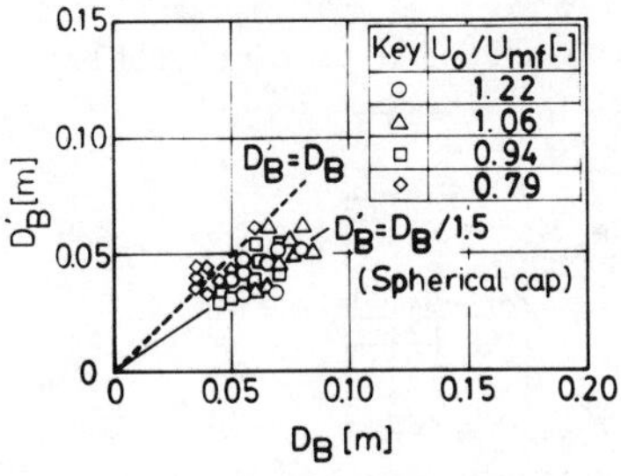

Fig. 3 Relation between bubble diameter D_B and bubble height D_B'

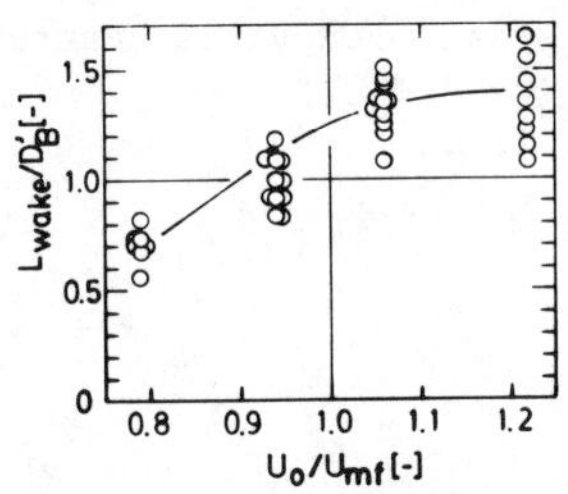

Fig. 4 Wake region under a single bubble

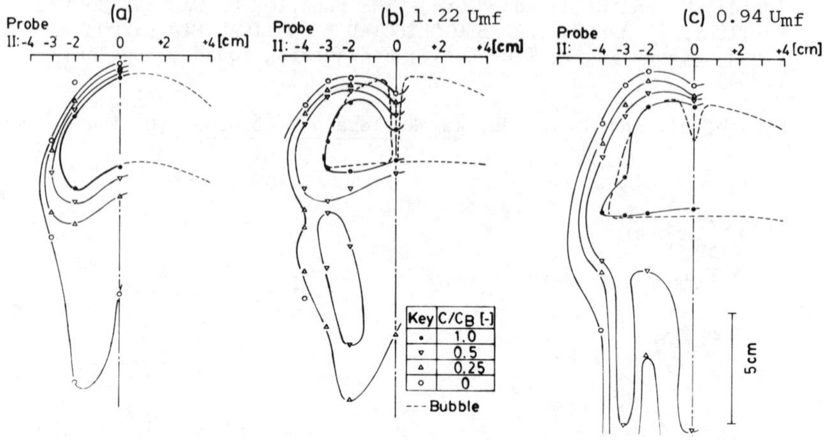

Fig. 5 Ozone concentration profiles around a single bubble

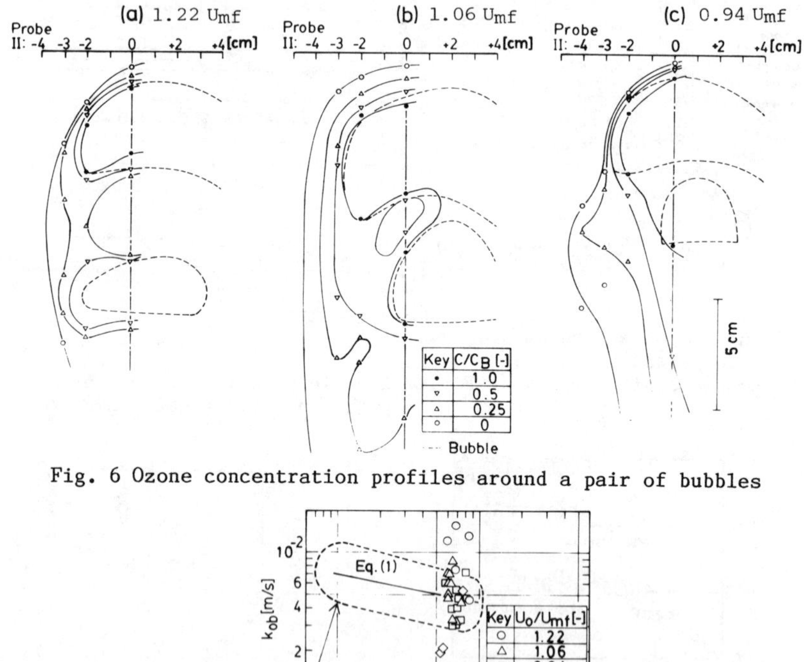

Fig. 6 Ozone concentration profiles around a pair of bubbles

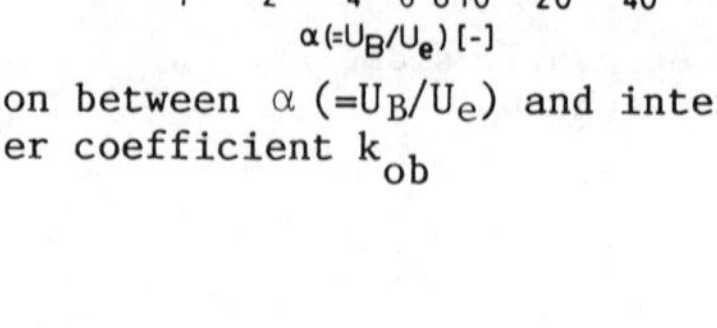

Fig. 7 Relation between α (=U_B/U_e) and interphase mass transfer coefficient k_{ob}

THE EFFECT OF TEMPERATURE ON BUBBLE PARAMETERS IN GAS-FLUIDIZED BEDS

by

C. Sishtla, I. Chan and T. Knowlton

Institute of Gas Technology
Chicago, Illinois 60632
U.S.A.

ABSTRACT

The effect of system temperature on bubble frequency, bubble velocity and bubble size in a 16.3-cm- (6.4-in) diameter fluidized bed was determined over the temperature range of 21°C (70°F) to 980°C (1800°F). It was found that bubble size, bubble frequency, and bubble velocity were essentially independent of temperature over the range investigated.

SCOPE

Bubbles dominate the behavior of gas-solids fluidized beds. Bubbles cause the intimate gas-solids mixing which results in good heat transfer and the essentially uniform bed temperatures associated with fluidized beds. However, bubbles are a mixed blessing because they also cause gas bypassing. Gas bypassing is detrimental in a fluidized-bed reactor because it allows the gas to flow through the bed without contacting the solids.

In order to understand the behavior of a bubbling, gas-solid fluidized bed, it is necessary to know the properties (size, frequency and velocity) of the bubbles. Since most industrial bubbling fluidized beds operate at elevated temperatures, knowledge of how temperature affects bubble properties is especially needed.

There are several articles in the literature which have attempted to look at the effect of temperature on bubble properties (1-9). Most of these workers have either determined the effect of temperature on either bubble size or bubble frequency. Yoshida et al. (1), Geldart and Kapoor (2), Wittman et al. (3), Stubington et al. (4), and Sitthiphong (7) investigated the effect of temperature on bubble size. Wittman et al. (3) reported no effect of temperature on bubble size over the range of 53° to 200°C. Yoshida et al. (1), Geldart and Kapoor (2), and Stubington et al. (4) reported that bubble size decreased with temperature from ambient to about 300°C. Above 300°C, no substantial change of bubble size with temperature was observed. Sitthiphong et al. (7) found that bubble size increased with temperature over the range of 427° to 757°C.

Otake et al. (5), Mii et al. (8), and Yoshida et al. (9) all reported that bubble frequency increased with temperature. Otake et al. (5) assumed that bubble size was inversely proportional to bubble frequency and, therefore, concluded that bubble size would decrease with temperature.

Because of the conflicting nature of the results as described above, an investigation was conducted to determine the effect of temperature on bubble properties over the temperature range of 20° to 980°C.

CONCLUSIONS AND SIGNIFICANCE

System temperature was found to have no effect on bubble size, frequency, or velocity over a measured temperature range of 21° to 980°C. This was found to be the case for a variety of solids ranging in size from 100 to 400 μm and in particle density from 1250 to 2563 kg/m^3. The results were in agreement with those reported by Wittman et al. (3), but differed from those reported by others (1, 2, 4, 8, 9) for "small" particles (i.e., those less than 500 μm).

EQUIPMENT AND MATERIALS

A flow sheet of the equipment used to determine the effect of temperature on bubble properties in a fluidized bed is shown in Figure 1. The fluidization column was constructed of a 16.3-cm- (6.4-in.) diameter, Type 316 stainless steel tube, 1.83 meters (6 ft) in length. The entire length of the column was enclosed by an electric resistance furnace.

The gas distributor was a multiorifice type. It contained six, 3.2-mm- (1/8-in.) diameter holes. The centers of five of these orifices were located 6.4 cm (2.5 in.) from the center of the distributor. The other orifice was located at the center of the distributor. To prevent solids weeping through the distributor, a 200-mesh screen was placed over the top of the distributor.

Solids were added to the unit from a feed hopper located immediately above the fluidization column. Nitrogen was used as the fluidizing gas in all runs. The fluidizing gas was preheated before it entered the fluidization column to reduce the load on the electric furnace. The exit gas was routed through a cyclone before being vented to the atmosphere.

The fluidization column was equipped with fifteen, 3.2-mm- (1/8-in.) diameter pressure probes (5 sets of 3) to measure bubble properties. Purge gas was used to prevent solids from plugging the probes. The location of the probes in the bed for all of the tests is shown in Figure 2. The signal from the pressure probe/transducer/demodulator system was sent to an A/D converter, which sampled the signal at a rate of 400 scans/s. An IBM PC computer was used to log the data from the A/D converter.

Five different solids were used in this study. Particle size was varied using three different sizes of Ottawa sand: -35+50 mesh, -50+150 mesh, and -120+170 mesh. The average particle sizes of these materials were approximately 400, 200 and 100 μm, respectively. The 400 and 200 μm sands were Geldart Group B solids, while the 100 μm sand was a Geldart Group A solid.

Particle density was varied by using Ottawa sand, coke breeze and lignite char. The particle densities of these materials were 2563, 1505, and 1250 kg/m^3 (160, 94 and 78 lb/ft^3), respectively. These materials had an averge particle size of 400 μm and all were Geldart Group B solids. The size distribution of these solids are shown in Figure 3.

PROCEDURE

In a typical test, solids were first loaded into the feed hopper. The high-temperature unit was then brought to temperature by flowing hot nitrogen through the column while it was empty of solids. The purge gas flow to the pressure probes and pressure taps was then set to prevent backflow of solids into the lines.

Solids were then charged to the fluidization column to a fluidized-bed height of 9.14 cm (36 in.) above the gas distributor. The minimum fluidization velocity (U_{mf}) or the minimum bubbling velocity (U_{mb}) was experimentally determined for each solid at each temperature. After the U_{mf} value was obtained, the gas velocity was set at the desired excess fluidizing velocity ($U-U_{mf}$). A range of gas velocities was used in each test. For each gas velocity used, bubble properties were then obtained at each probe location.

RESULTS AND DISCUSSION

In this paper, the variation of bubble frequency, average bubble pierced length, and bubble-rise velocity will be reported as a function of system temperature and bed height. The following ranges of these were investigated.

- System temperature, 21° to 980°C (70° to 1800°F)
- Bed height, 22.9 to 68.6 cm (9 to 27 in.)

The effects of each of these parameters on the bubble properties are discussed below.

Effect of Temperature

Tests to determine the effect of bed temperature on bubble properties were conducted with all five of the test materials over the temperature range of 21° to 980°C (70° to 1800°F). It was found that bubble size, reported here as the average bubble pierced length (ABPL) remained constant as the system temperature was increased to 980°C (1800°F). A typical plot showing this behavior is shown in Figure 4 for 400 μm sand. The ABPL in this plot was measured 22.9 cm (9 in.) above the distributor at an excess fluidization velocity of 14.7 cm/s (0.484 ft/s).

Wittman et al. (3) also reported no change in bubble size with temperature. Yoshida et al. (1) found that bubble size decreased with temperature. Their measurements were made visually in a two-dimensional bed at temperatures up to 750°C. Geldart and Kapoor (2) also reported a small decrease in bubble size with temperatures between 20° and 200°C. Above 200°C, they observed no further change in bubble size with temperature. Stubington et al. (4) reported that bubble size decreased by 5% to 15% when temperature was increased from ambient up to 300°C. Above 300°C they found no measurable change in bubble size with temperature up to 1000°C. Stubington et al. (4) postulated that the decrease in bubble size with temperature may have been due to the physical configuration of their gas distributor. It was of the bubble-cap-type with gas jetting horizontally from the holes in the bubble caps. The horizontal jet penetration will decrease as the system temperature is increased because of decreasing gas density and because U_{mf} decreases with temperature. The latter phenomenon results in the overall gas velocity being decreased to maintain a constant value of $U-U_{mf}$. Therefore, the decrease in the overall gas velocity results in a decrease in the orifice velocity which decreases the jet penetration. As a result of a decrease in jet penetration, the bubble streams from the distributor could not interact, i.e., coalesce. Therefore, a larger number of bubble streams were produced at high temperatures, which resulted in smaller bubble diameters.

Sitthiphong et al. (7) found that bubble size increased with temperature over the range of about 470° to 760°C. However, the particles used in their study were 3200 μm in size. The particle sizes used in the other reported investigations and in this study were all less than 400 μm.

Bubble frequency was also found to remain constant with temperature. A typical plot is shown in Figure 5 where the bubble frequency measured at Probe Location E for 400 μm coke breeze is plotted as a function of temperature at an excess fluidization velocity of 12.2 cm/s (0.4 ft/s). This finding contrasts with the results of others. Mii et al. (8), Otake et al. (5), and Yoshida et al. (9) all found that bubble frequency increased with temperature. Mii et al. (8) did not report, however, whether their findings were reported at constant superficial gas velocity or constant $U-U_{mf}$. Yoshida et al. (9) reported their results at a constant U/U_{mf} instead of the more accepted $U-U_{mf}$.

Since the average bubble pierced length was found to not vary with temperature, it would be expected that the average bubble-rise velocity would not change with temperature also. This is because it has been generally accepted that the bubble-rise velocity in a fluidized bed, not influenced by a wall, can be described by a Davies and Taylor (10) type relationship first proposed for gas bubbles rising through liquids —

$$U_{br} = 0.711 \sqrt{gD_B}. \tag{1}$$

Thus, if D_B is unaffected by temperature, U_{br} would also be expected to be constant with temperature. This was found to be the case and is shown in Figure 6. The data in this figure were obtained using 400 μm sand at Probe Location A at an excess fluidization velocity of 12.3 cm/s (0.404 ft/s).

Effect of Bed Height

As shown in Figure 2, the pressure probes were distributed axially along the fluidized bed, making it possible to determine the bubble characteristics as a function of bed height. The gap between the pressure probes was 1.9 cm (3/4 in.) for the probes at Locations D and E, and it was 1.3 cm (1/2 in.) for the probes at Locations A, B and C.

A typical variation of bubble frequency with height is shown in Figure 7 for 400 μm sand at an excess fluidization velocity of 10.4 cm/s (0.34 ft/s). The variation of bubble frequency with height is shown for several different temperatures for comparison.

Bubble frequency is highest near the bottom of the bed and decreases with bed height. The decrease in bubble frequency is due to bubble growth by coalescence. As two or more small bubbles coalesce into one large bubble, the number of bubbles or frequency decreases with bed height. No apparent trend of bubble frequency with temperature was observed.

The variation of ABPL with bed height is shown in Figure 8 for the 200 μm sand material at an excess fluidization velocity of 3.8 cm/s (0.125 ft/s). This curve was typical of the results obtained with the other materials, also. Data obtained at four different temperatures are shown in this figure. The ABPL increased continuously with height for all of the solids tested. This was expected since the bubble frequency was found to decrease continuously with height, indicating that coalescence was occurring faster than bubble splitting, and that the bubbles were growing.

The variation of bubble-rise velocity with bed height is shown in Figure 9 for 100 μm sand at an excess fluidization velocity of 5.5 cm/s (0.18 ft/s). Data for both the ambient temperature run and the run conducted at 650°C (1200°F) are presented. The variation of the average bubble-rise velocity with height shown in this figure is typical of that observed with all of the materials. In Figure 9, the average bubble-rise velocity first increased slightly with height, reached a maximum, and then decreased. This behavior is typical of the variation of average bubble-rise velocity in small-diameter columns. The bubble-rise velocity shown in Figure 9 is the averge bubble-rise velocity relative to the column wall, U_b. The average bubble-rise velocity relative to the emulsion is assumed to be related to the bubble-rise velocity relative to the wall by the following equation

$$U_b = (U-U_{mf}) + U_{br} \tag{2}$$

The bubble-rise velocity is proportional to the square root of the bubble diameter (Equation 1). Therefore, as the ABPL increases with height, the bubble-rise velocity would also be expected to increase continually over the height of the bed. However, in small-diameter columns, after a bubble attains a certain size its rise through the bed is hindered by the solids flowing downward near the wall of the column. Allahwala and Potter (11) have reported that the bubble-rise velocity (U_{br}) is so affected by the solids that Equation (1) cannot be used to accurately predict U_{br} when the diameter of the bubble is approximately 0.36 times the bed diameter. After the bubble reaches this diameter, U_{br} decreases or remains constant with increasing bed height until it attains the rise velocity of a slug, i.e.,

$$U_{bs} = 0.35\sqrt{gD_t}. \tag{3}$$

The bubble-rise velocity results obtained in this investigation are basically in agreement with the Allahwala and Potter findings. From Figure 9, the maximum bubble-rise velocity relative to the wall (U_b) is approximately 85.3 cm/s (2.8 ft/s). The theoretical bubble-rise velocity where the downflowing solids would have been expected to hinder the bubble would occur at a bubble diameter of 0.36 (16.3 cm). This corresponds to a U_b of about 94 cm/s (3.1 ft/s).

ACKNOWLEDLGMENT

The authors would like to thank the U.S. Department of Energy which funded the work. The work was performed under Contract DE-AC21-82MC19301.

NOTATION

D_B = bubble diameter, cm (ft)

D_T = tube diameter, cm (in.)

g = gravitational constant

U = superficial gas velocity, cm/s (ft/s)

U_b = bubble-rise velocity relative to the wall, cm/s (ft/s)

U_{br} = bubble-rise velocity relative to the emulsion, cm/s (ft/s)

U_{bs} = rise velocity of a slug, cm/s (ft/s)

U_{mb} = minimum bubbling velocity, cm/s (ft/s)

U_{mf} = minimum fluidization velocity, cm/s (ft/s)

LITERATURE CITED

1. Yoshida, K., Fujii, S. and Kunii, D., "Characteristics of Fluidized Beds at High Temperatures," Fluidization Technology 1, Keairns, D. (Ed.), 43 (1976).
2. Geldart, D. and Kapoor, D.S., "Bubble Sizes in a Fluidized Bed at Elevated Temperatures," Chem. Eng. Sci., 31, 842 (1976).
3. Wittman, K., Helmrich, H. and Schugerl, K., "Measurements of Bubble Properties in Continuously Operated Fluidized Bed Reactors at Elevated Temperature," Chem. Eng. Sci., 36 (10), 1673 (1981).
4. Stubington, J.F., Barrett, D. and Lowry, G., "Bubble Size Measurements and Correlation in A Fluidised Bed at High Temperatures," Chem. Eng. Res. Des., 62, 173 (1984).
5. Otake, T., Tore, S., Kawashima, M. and Skibata, T., J. Chem. Eng. Japan, 8 (5), 388 (1975).
6. Tone, S., Seko, H., Maruyama, H. and Otake, T., J. Chem. Eng. Japan, 7, 44 (1974).
7. Sitthiphong, N., George, A. H. and Bushnell, D., "Bubble Eruption Diameter in a Fluidized Bed of Large Particles at Elevated Temepratures," Chem. Eng. Sci., 36 1259 (1981).
8. Mii, T., Yoshida, K. and Kunii, D., "Temperature Effects on the Characteristics of Fluidized Beds," J. Chem. Eng. Japan, 6, 100 (1973).
9. Yoshida, K., Ueno, T. and Kunii, D., "Mechanisms of Bed-Wall Heat Transfer in a Fluidized Bed at High Temperatures," Chem. Eng. Sci., 29, 77 (1974).
10. Davies, R. M. and Taylor, G. I., Proc. Royal Soc. Serv. A2, 375 (1950).
11. Allahwala, S. A. and Potter, O. E., "Rise Velocity Equation for Isolated Bubbles and for Isolated Slugs in Fluidized Beds," Ind. Eng. Chem. Fund., 18 (2), 112 (1979).

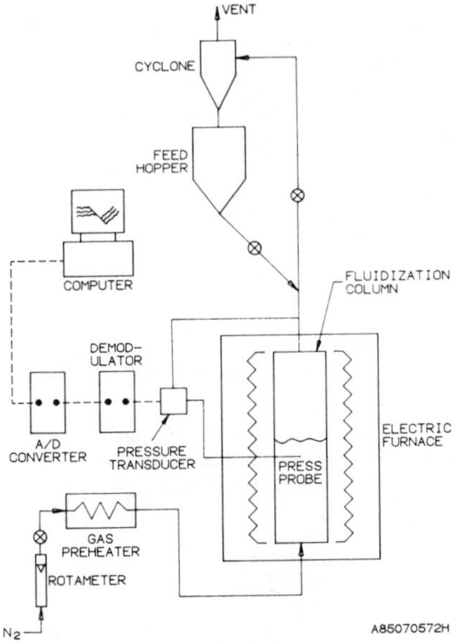

Figure 1. High-Temperature Unit Flow Sheet

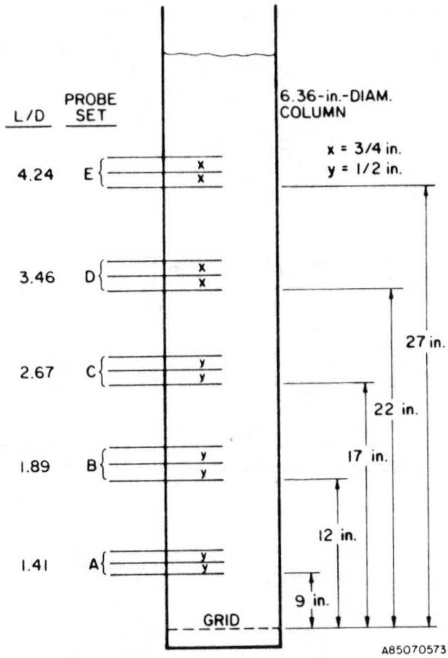

Figure 2. Pressure Probe Configuration

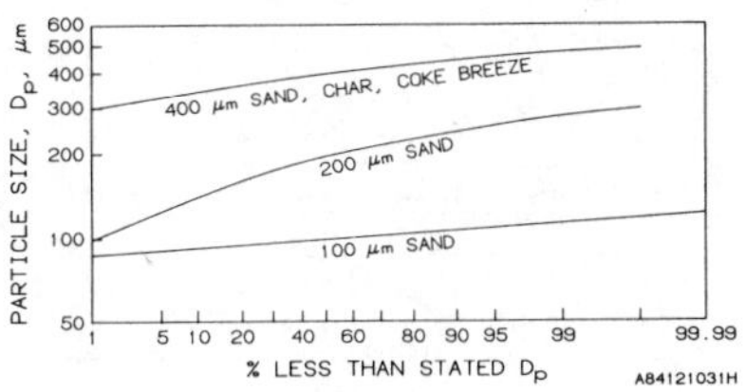

Figure 3. Particle Size Distributions of the Test Materials

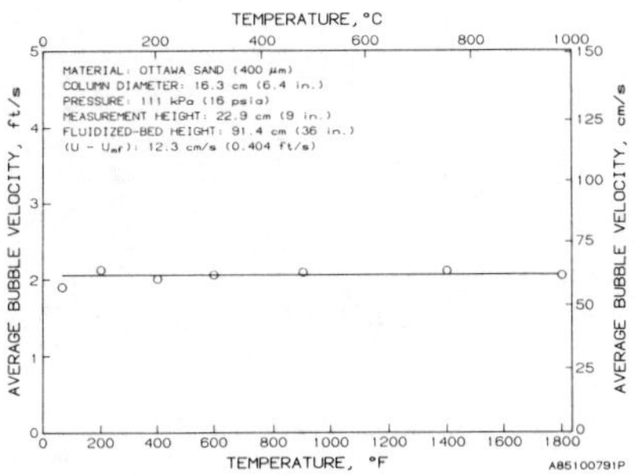

Figure 4. Average Bubble Pierced Length vs. Temperature (400 μm sand)

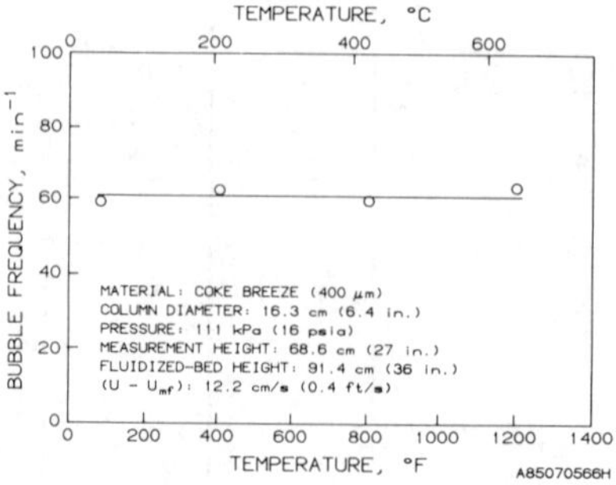

Figure 5. Bubble Frequency vs. Temperature (400 μm coke breeze)

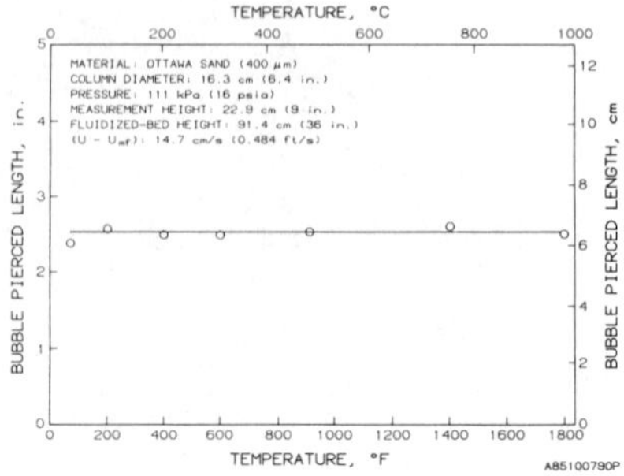

Figure 6. Average Bubble Velocity vs. Temperature (400 μm sand)

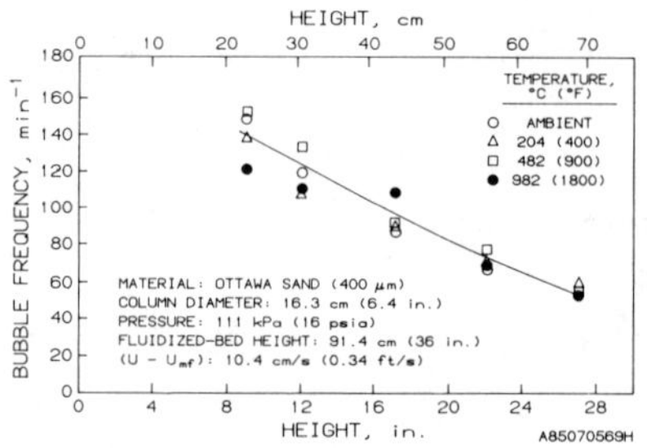

Figure 7. Bubble Frequency vs. Bed Height (400 μm sand)

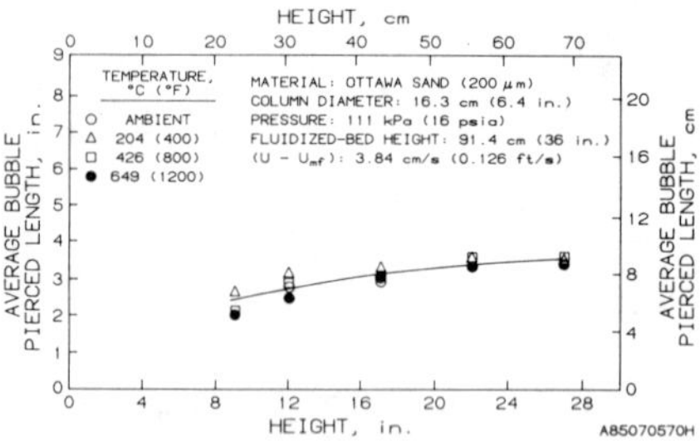

Figure 8. Average Bubble Pierced Length vs. Bed Height (200 μm sand)

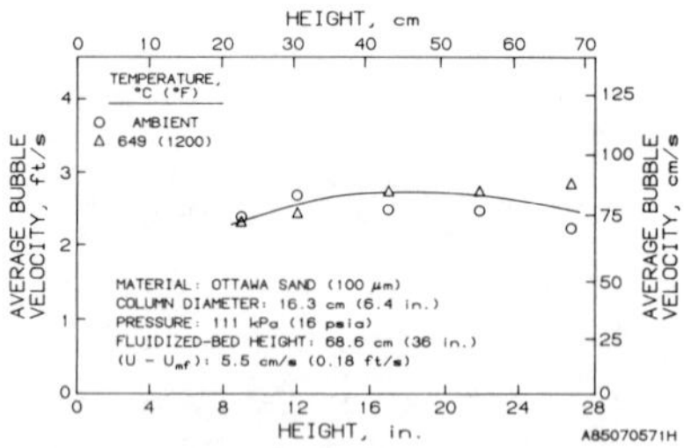

Figure 9. Average Bubble Velocity vs. Bed Height (100 μm sand)

EFFECT OF HORIZONTAL TUBES ON BUBBLE ERUPTION BEHAVIOR IN A GAS FLUIDIZED BED

E. K. Levy
R. A. Freeman
H. S. Caram
Lehigh Unversity
Bethlehem, PA 18015

ABSTRACT

A flow visualization study was made of the effects of a row of horizontal tubes on the behavior of erupting bubbles in a gas fluidized bed. The tubes lead to an increased rate of bubble coalescence and reduce bubble size. In addition, they alter the eruption mechanism for single bubbles, causing a significant fraction of the single bubbles to generate large wake eruptions.

SCOPE

Do, Grace and Clift (1), George and Grace (2), Hatano and Ishida (3,4), Levy, Caram, Dille, Edelstein (5), Levy, Dille, Caram (6) and Caram, Edelstein, Levy (7) have all reported on experiments dealing with the mechanisms by which solids are thrown upward by erupting bubbles in a gas fluidized bed. The results from these studies show that as a single bubble erupts in a three-dimensional fluidized bed, the material in the bulge or roof of the bubble is projected upward to a significantly greater distance than the material in the bubble wake. In the case of bubbles which are in the process of coalescing as they erupt from the bed free surface, the bulge material is thrown upward as was described for individual bubbles. But, in addition, the material in the wake of the trailing bubble is ejected upward at a much higher velocity and to greater heights then the bulge material. This type of wake eruption has been referred to in the literature as a wake spike and wake jet (5,7). The previous studies show that in a three-dimensional bed, wake spikes do not occur with single bubble eruptions, but require a vertical coalescence process (a double bubble) for their development. The data obtained by (5) show that almost all double bubbles result in wake spike eruptions. For conditions tested so far, anywhere from 5 to 25 percent of the bubble eruptions are of the coalescing type, the rest being bubbles which erupt individually (5,8).

In developing a mechanistic model of the particle elutriation process, it is important to know the rate at which solids are thrown above the bed by the erupting bubbles. Among other things, this requires a knowledge of the eruption mechanisms and the frequency with which each type of bubble eruption process occurs. While the previous results show that the wake spike phenomenon occurs much less frequently then the bubble bulge eruption, each wake spike event appears to be responsible for a larger fraction of projected solids then each single bubble eruption.

All of the previous bubble eruption studies were for the case of beds without internals; and yet there are a number of important applications for fluidized beds where tubes are immersed in the bed and where elutriation plays a critical role in the process. In addition, some investigators have proposed the use of internals and baffles to decrease the rate of solids elutriation. The purpose of this study was to determine the effects of horizontal tubes located within the bed on the bubble eruption process. Specifically it was desired to find how the presence of the tubes affects bubble growth, the bulge and wake ejection heights, the double bubble fraction and the frequencies of the various wake eruption types. A quantitative understanding of the effects of tubes on bubble eruptions might lead to techniques for designing the tube bundle to minimize solids elutriation. In addition, it might lead to the development of techniques for the use of internals to reduce elutriation.

CONCLUSIONS AND SIGNIFICANCE

Horizontal tubes influence coalescence and eruption behavior in a number of important ways. First, the presence of the tubes leads to an increased rate of bubble coalescence. Over the range of gas flow rates studied, without tubes present, the fraction of double bubbles reached a maximum of about 20 percent. With tubes immersed in the bed, the fraction of double bubbles increased to 30 to 40 percent. The tubes also tend to cause bubble splitting, yielding smaller bubbles in the region just above the tubes.

The tubes drastically alter the mechanism by which a single bubble erupts from the free surface of the bed. Without tubes in the bed, when single bubbles erupt, the bulge material always reaches a greater dimensionless height above the bed free surface than the wake material. With tubes present, up to 60 percent of the single bubble eruptions are characterized by the upward projection of wake material, reaching distances above the free surface which far exceed those attained by the bulge material. The wake dominated eruption of the single bubble appears to be due to the formation of elongated vertical cavities in the region between the tubes and the free surface. Comparisons with previous work by Caram et al (7) on wake eruptions indicates a strong similarity between the wake behavior with single bubbles in a bed with tubes and double bubbles in beds without tubes.

EXPERIMENTAL APPARATUS AND PROCEDURE

The study was performed in a 0.76m square by 2.44m high fluidized bed. The front and back walls of the bed were made of plexiglass plates to permit visualization of the bubbling behavior. The distributor was a 6.35mm thick steel plate drilled with 480 holes of 4.0mm diameter uniformly spaced over its surface in a square array. To prevent bed material from seeping into the plenum box, a fine mesh screen was glued to the top of the distributor plate.

Experimental data were gathered with and without tubes in the bed. Additional data without tubes were taken from (9). For the experiments with immersed tubes, a single row of 4.76cm diameter horizontal tubes was used. These were spaced 7.6cm apart from centerline to centerline and were positioned at 18 and 40cm above the distributor. Settled bed depths ranging up to 40cm above the tubes were used. The tubes were placed within the bed so that their ends were flush with the front transparent end wall, making it possible to observe the effects of the tubes on the bubbles.

All of the experiments were carried out with 264 μm monodisperse glass beads and room temperature air at atmospheric pressure. A Video-logic INSTAR IV high speed video system was used to gather data on the behavior of the sectioned bubbles immediately adjacent to the wall. These were assumed to be typical of the bubbles at the interior of the bed. The experiments were carried out with superficial gas velocities ranging from 1.1 to 3 times U_{mf} (6.2 to 17 cm/s). Higher gas velocities resulted in such chaotic motion at the free surface of the bed that it was impossible to follow the individual bubbles.

Most flow rates and bed depths were recorded for 60 seconds (7200 frames) to provide an adequate number of bubbles for analysis at each test condition. In most cases, anywhere from 40 to 160 bubbles were analyzed at each test condition. By reviewing the tapes frame by frame, quantities such as the frequencies of occurrence of the single and double bubble eruptions, bubble diameter, and bulge and wake eruption heights were measured.

Various definitions of bubble diameter appear in the literature. In this study the bubble size was characterized by measuring the frontal diameter as the bubble reached the free surface (see Figure 1). When the free surface was located somewhere within the region extending from just below to just above the tubes, the diameter of the bubble just before it entered the region of the tubes was used to characterize bubble size.

Bubbles were classified as "single" bubbles if they appeared to be unaffected by other bubbles as they erupted from the free surface. The term "double" bubble describes two bubbles which coalesce vertically as the leading bubble erupts from the free surface. The bulge height H_B is defined as the maximum height above the free surface of the bed reached by the bulge material. The wake height H_w describes the maximum height reached by material from the wake during an eruption.

RESULTS

It was found from this study that tubes influence bubble coalescense and eruption behavior in several important ways. The tubes restrict the flow area, which forces the bubbles to move sideways toward each other and leads to enhanced coalescence in the gaps between the tubes. In addition, the tubes retard some of the bubbles flowing upward from below, allowing the trailing bubbles to catch up to the retarded bubbles and coalesce in the region of the tube gaps. Figure (2) compares the fraction of the bubbles which are double bubbles for beds with and without tubes present. For the immersed tube data shown in Figure (2), the free surface of the bed was located within a few cm of the tops of the tubes. Without tubes, over this range of gas flow rates, the fraction of bubbles which were double bubbles reached a maximum of 20 percent. With tubes, the fraction of double bubbles increased to the 30 to 40 percent range.

Qualitative observations of the recorded video sequences showed that the tubes also tend to cause bubble splitting, resulting in smaller bubbles in the region just above the tubes. More data are needed to quantify the bubble size effect.

In addition to the role played by the tubes in promoting coalescence and reducing bubble size, they also dramatically alter the mechanism by which a single bubble erupts from the free surface of the bed. In previous studies without tubes (6 and 8), it was found that when a single bubble erupts, the bulge material reaches a dimensionless height above the bed surface in the range of $H_B/D_i \approx 0.7$. The bubble wake remains intact, gradually rising to a height about one half that attained by the bulge, before settling back onto the free surface of the bed. Rather than splash particles upwards as occurs with the bulge, the wake remains attached to the emulsion phase like a ripple on the surface of a pool of water. The type of bulge and wake behavior described here was found to occur with all single bubble eruptions in beds without tubes.

In contrast to this, the tubes cause a significant fraction of the single bubbles to erupt with the ejection of a large wake spike. These appear to be due to the formation of elongated vertical cavities in the region between the tubes and the free surface and lead to wake behavior similar to that previously observed with double bubbles (5 and 7).

Figure 3 illustrates the formation of such a cavity. As the top of the cavity reaches the free surface of the bed, the base of the cavity is projected upward forming a jet of wake material. In some cases, this is carried to distances far above those reached by the bulge material. Caram et al (7) developed a theoretical analysis of the wake eruption which occurs during coalescence by modeling it as the transient motion of a deep cavity at the free surface of an inviscid liquid (see Figure 4). They derived equations from which they were able to compute the initial velocity of the base of the cavity V_o as it is projected upward. The resulting solutions were found to be in good agreement with their measurements of the initial upward velocity of the wake material during a double bubble eruption.

In the present study with immersed tubes, graphs of the trajectories of the bulge, the center of the wake and the bottom edges of the wake were made for single bubbles rising above the tubes towards the free surface. From these, data were obtained on the initial upward velocities of the wake. These are shown plotted in Figure 5 along with the theoretical solution and data for the initial wake velocity obtained for double bubbles by Caram et al (7). The dashed line is a least squares curve fit of the data from the present study. There are too few data and there is too much scatter to accurately determine the slope of the curve. Nevertheless, the data show that the initial dimensionless wake velocities for the double bubble eruption process without tubes and the single bubble eruption with immersed tubes are of the same order of magnitude.

It was also found from the video sequences that in addition to the wake spike eruption, some of the single bubble eruptions resulted in an upward "spray" of wake material. Others caused an eruption somewhat resembling an upward "explosion" of wake material. Because of strong qualitative similarities between the bubble trajectories leading to the wake spike, spray and explosion phenomena, it is quite possible that the latter two are variations of the wake spike eruption. Nevertheless, more work is needed to establish the causes of the different types of wake eruptions.

An analysis of the data obtained with and without tubes in the bed showed that the tubes have little effect on the dimensionless height reached by the bulge material. However, because of the role which the tubes play in promoting large wake eruptions with single bubbles, on the average, the dimensionless height reached by the wake material is significantly greater when there are tubes in the bed. This is illustrated by comparing the ratios of the mean bulge eruption height to the mean wake eruption height for single bubble eruptions (see Figure 6 and 7). Levy et al (8) found this ratio to be approximately 1.5 for single bubbles without tubes and Levy et al (6) reported values of the order of 2, also without tubes. With tubes in the bed, the ratio ranges from 0.7 to 1.7 with most values in the 0.9 to 1.3 range.

In previous studies, it was found that at least 75 percent of the bubbles in a three-dimensional bed without tubes are single bubbles. Virtually all of these exhibit a bulge burst type of eruption where the dimensionless height reached by the bulge greatly exceeds that reached by the wake. In the present study, the following criterion was used for determining which of the eruptions to classify as wake dominated. All bubbles in which the wake eruption height was greater than the height reached by the bulge material are referred to here as having a significant wake eruption. Figures 8 and 9 show that in situations where the free surface was above the tops of the tubes, the frequency of wake dominated eruptions (spikes, sprays and explosions), ranged from 20 to 60 percent of all the single bubble eruptions. Even with the free surface located 40 cm above the tubes, the fraction of the single bubble eruptions resulting in a large wake eruption was in the 20 to 50 percent range. This indicates an important effect of the tubes which persists even when they are located relatively far below the free surface of the bed.

LITERATURE CITED

1. Do, H.T., Jr., J. R. Grace and R. Clift, "Particle ejection and entrainment from fluidized beds", Powder Technology, 6 pp. 195 to 220 (1972).

2. George, S. and J. R. Grace, "Entrainment of particles from aggregative fluidized beds" AIChE Symp. Series No. 176, Vol. 74, pp. 67 to 74.

3. Hatano, H. and M. Ishida, "Study on the entrainment of FCC particles from a fluidized bed", Powder Technology, 35 pp. 201 to 209 (1983).

4. Hatano, H. and M. Ishida, "The entrainment of solid particles from a gas solid fluidized bed", J. Chem. Eng. of Japan, Vol. 14, No. 4 p. 306 to 311 (1981).

5. Levy, E., H. Caram, J. Dille and S. Edelstein, "Mechanisms for solids ejection from gas fluidized beds", AIChE J., Vol. 29, No. 3, p. 383 May 1983.

6. Levy, E., J. Dille and H. Caram, "Single bubble eruptions in gas fluidized beds", Powder Technology, 32 pp. 173 to 178 (1982).

7. Caram, H., S. Edelstein and E. Levy, "Particle ejection due to bubble coalescense near the surface of gas fluidized beds", Proceedings Fourth Int. Conf. on Fluidization, Kashikojima, Japan, May 1983. Published by Engineering Foundation, ISBN 0-8169-0264-X.

8. Levy, E., I. Alkan, and H. Caram, "Characteristics of Erupting Bubbles in a Three Dimensional Fluidized Bed", American Institute of Chemical Engineers Symposium Series Fluidization and Fluid Particle Systems: Recent Advances, ed. by G. E. Klinzing, Vol. 80 p. 241, (1984).

9. Alkan, I., MS Thesis, Lehigh University (1984).

ACKNOWLEDGEMENTS

The authors are grateful to Dr. Morris Ojalvo of the National Science Foundation for his support of this investigation under grant CPE 8203918 and to Dr. Charles Smith of Lehigh's Department of Mechanical Engineering and Mechanics and the Air Force Office of Scientific Research for the use of the video system.

NOTATION

a - radius of idealized cylindrical cavity
D_i - bubble diameter as defined in Figure 1
g - acceleration of gravity
H - height of void as defined in Figures 1 and 4
H_B - distance above bed surface reached by bulge material
H_W - distance above bed surface reached by wake material
L_{FS} - distance above distributor of undisturbed bed free surface at U_o
L_T - distance of tubes above distributor
U^* - U_o/U_{mf}
U_o - superficial gas velocity
U_{mf} - minimum fluidization velocity
V_o - initial velocity of bubble wake as it is projected upward

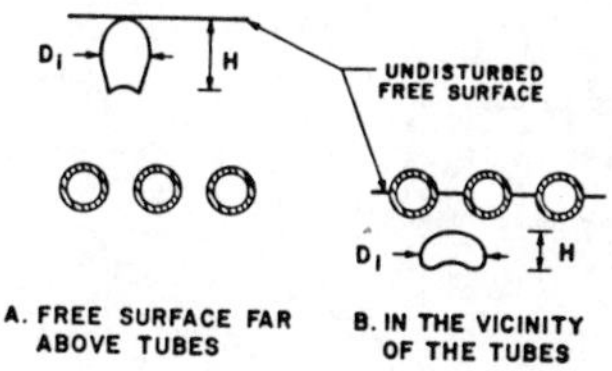

Figure 1. Sketch showing definitions of bubble diameter. D_i, and void height, H.

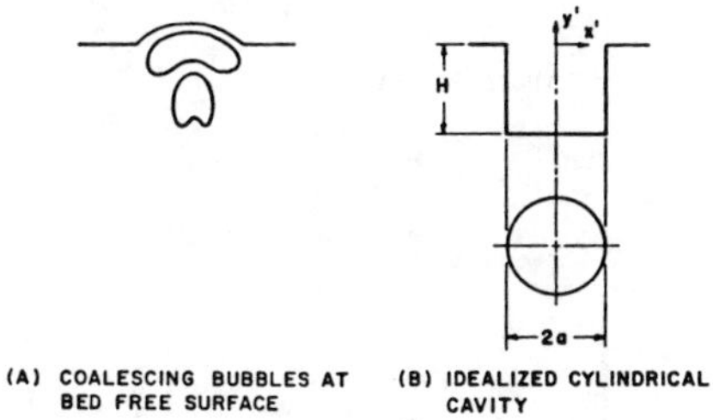

Figure 4 (A) Coalescing bubbles at bed free surface
(B) Idealized cylindrical cavity

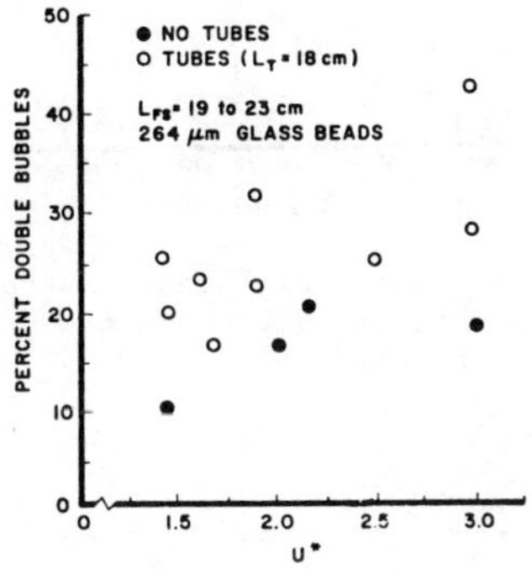

Figure 2. Double bubble fraction vs. gas flow rate.

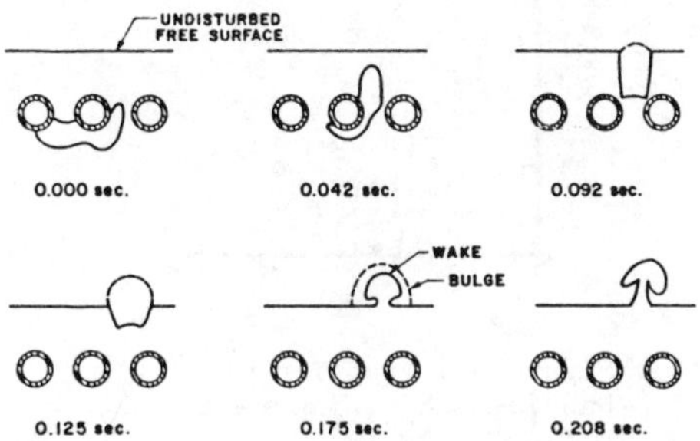

Figure 3. Wake eruption sequence ending in a wake spike. The bulge collapsed by 0.208 sec.

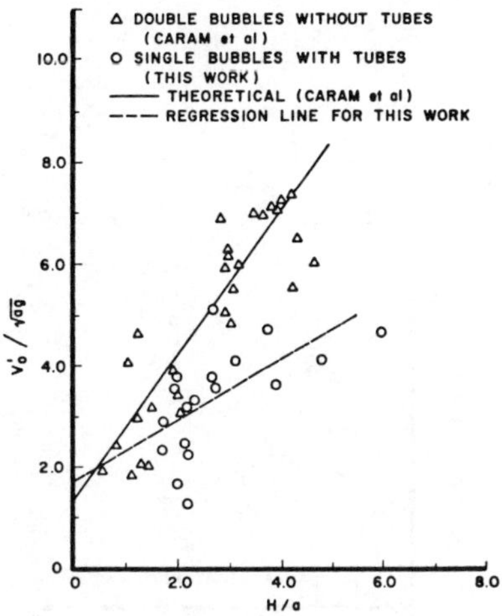

Figure 5. Non-dimensional wake velocity vs. aspect ratio for double and single bubbles.

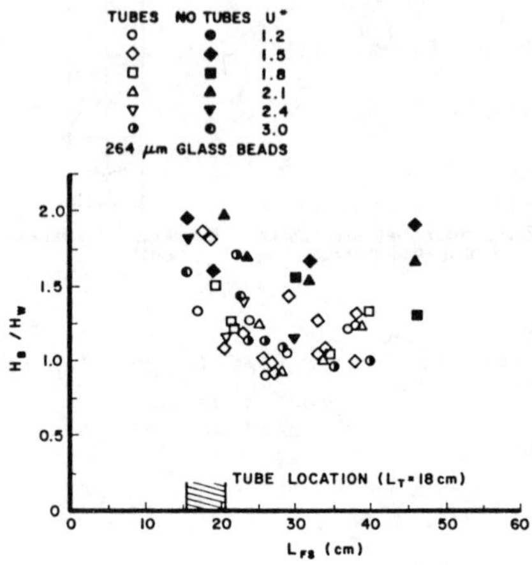

Figure 6. Ratio of bulge to wake eruption heights vs. bed depth for single bubbles.

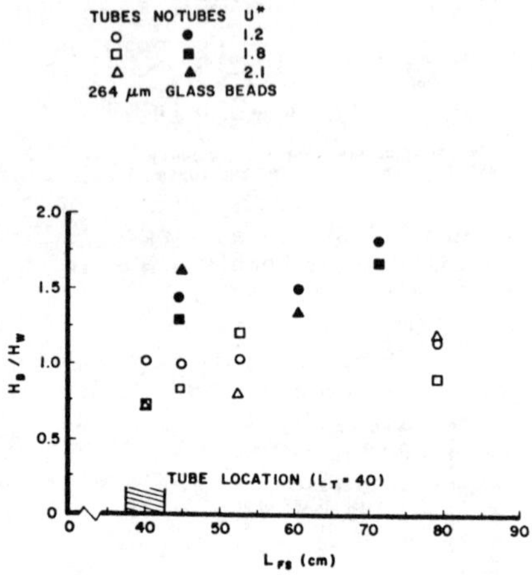

Figure 7. Ratio of bulge to wake eruption heights vs. bed depth for single bubbles.

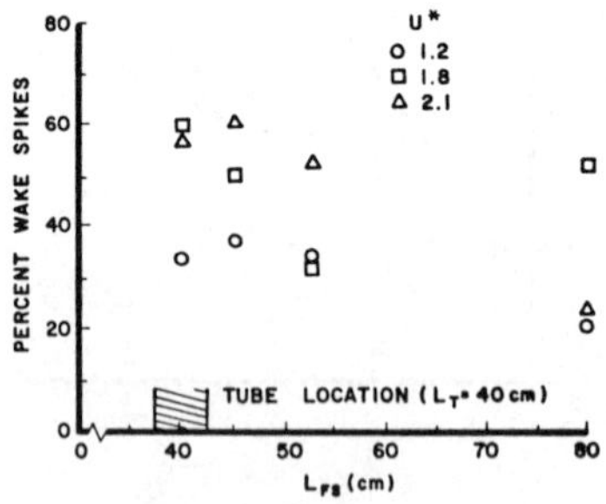

Figure 8. Fraction of single bubbles exhibiting large wake eruptions, L_t = 40 cm.

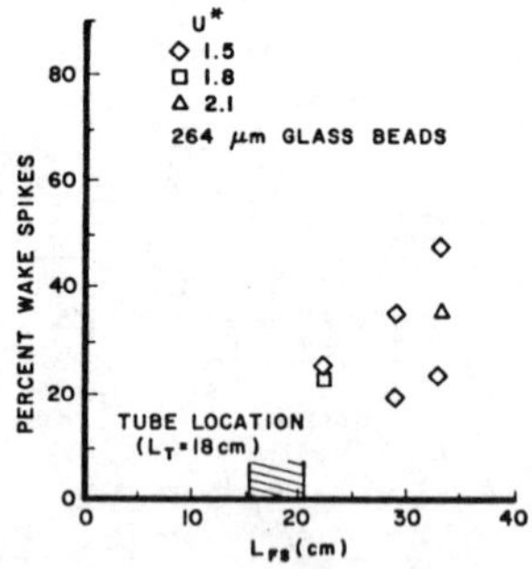

Figure 9. Fraction of single bubbles exhibiting large wake spikes, L_t = 18 cm.

THE EFFECT OF EMULSION PHASE CHARACTERISTICS ON FLUIDIZATION
-SELECTION OF POWDER PROPERTIES-

H.O. Kono, S. Chiba, T. Ells, P. Daniell, M. Suzuki

Fluidization Center
Department of Chemical Engineering
West Virginia University
Morgantown, WV 26506-6101

ABSTRACT

The effect of emulsion phase characteristics on fluidization in fine powder fluidized beds was investigated by changing powder properties. The quality of fluidization was significantly modified by the selection of powder. The experimental results were interpreted by the structural change of the emulsion phase using characterization criteria of Young's modulus of the emulsion phase and tensile strength of static powders.

SCOPE

Ideally smooth and homogeneous fluidization is achieved by keeping minimum bubble size and maximum bed expansion. These ideal characteristics are vitally important in realizing good gas-solid contacting and giving potential ability for "scale-up". There are two ways to attain these desirable characteristics; by hydrodynamically improving the characteristics or by adjusting them by selection of powders. To improve the hydrodynamics, much work has been done on the design and fundamentals of internal baffles and gas distributors as reviewed by Harrison et al. (1) and Miyauchi et al. (2). The effect of intensive circulation of the emulsion phase on fluidization was investigated in specially configured fluidized beds developed by Kono, et al. (3,4) to suppress the bubble development and simultaneously to control gas backmixing, which generally conflict each other in fluidized beds. We considered whether we could improve the particle properties beyond the ideal properties of FCC recommended by Geldart (5) in view of the reported effects of adding a greater proportion of fines (6,7) and of the assumed effect of interparticle adhesion forces by Baerns, Rietema, Massimilla, et al. (8-12).

In this study Young's modulus of the emulsion phase and tensile strength of static packed bed powder were measured and used to characterize the emulsion phase. The quality of the emulsion phase can be controlled by mixing FCC with other fine particles (d = 13μm, size range 5~100 μm) to investigate its effect on fluidization. Because the quality of the emulsion phase is not only affected by the particle size and its distribution but also by the surface characteristics of materials, two different fine powders (aluminum hydrate and limestone) were used.

CONCLUSION AND SIGNIFICANCE

By mixing FCC catalyst (d = 70μm) with aluminum hydrate $Al(OH)_3$ (d = 13 μm) and limestone $CaCO_3$ (d = 14μm) by 5-30 wt%, powders with relatively wide ranges of characteristics in Young's modulus and tensile strength were prepared. Their fluidization characteristics such as fludization velocities (U_{mf}, U_{mb}, U_{mc}), bed expansion ratio, bubble characteristics, and gas backmixing coefficients (E) were obtained. Although fine particles were added, almost no elutriation took place during fluidization within the experimental range, which was probably due to cluster formation. The minimum complete fluidization velocity (U_{mc}) of FCC containing fine particles is almost twice that of pure FCC particles. Bed expansion ratios were approximately 20% larger. Bubble shapes differ considerably from those seen in fluidized beds of FCC. The bubble shapes look irregular and more or less like a "cat's paw prints". Gas backmixing coefficients are affected by the selection of powders. Based upon the results of this investigation, fluidization states have significantly been modified by the selection of powders. The experimental results were interpreted by the structural change of the emulsion phase using characterization criteria of Young's modulus of the emulsion phase and tensile strength of static powders.

EXPERIMENTAL

FCC catalyst particles (ρ_p = 1.4 x 10^3 kg/m^3), aluminum hydrate $Al(OH)_3$ particles (ρ_p = 2.4 x 10^3 kg/m^3), and limestone $CaCO_3$ particles (ρ_p = 2.7 x 10^3 kg/m^3) were used as standard sample powders. The size distribution of each sample is shown in Figure 1. The loosely packed bed density and corresponding voidage of the bed with different weight mixing ratios (3-40%) of FCC and aluminum hydrate or limestone powders are shown in Figure 2. Microscopic pictures of pure FCC and powder prepared by mixing FCC with $Al(OH)_3$ fine powders 10 wt % are shown in Photographs 1A and 1B. The tensile strength of powder samples in static voidage range of 42-52% was measured by using apparatus developed by Hartley et al. (13). Voidage of packed samples was adjusted by changing the pressure of sample preparation. Because the adhesion forces are sensitive to humidity, each sample powder was stabilized by air with 70% relative humidity at 28°C before each measurement. The characteristic fluidization gas velocities (U_{mf}, U_{mb}, U_{mc}), bed expansion ratios, and expanded bed collapse rates were measured in a conventional fluidized bed (0.11 mID) with a sintered porous metal plate as the gas distributor. Air with relative humidity of 70% at 28°C was used as the fluidizing gas. Bed expansion ratios and expanded bed collapse rates were measured in the way proposed by Reitema (9). The calculation method for Young's modulus will be discussed in the following chapter. Gas backmixing coeffients were measured by steadily feeding tracer gas to the fluidized bed in accordance with the method reported by Kono et al. (3). The experimental apparatus (3,4) used is shown in Figures 3 and 4. Bubbling phenomena were observed in a two dimensional fluidized bed (0.01 x 0.25 x 1.15 m) with a porous metallic gas distributor.

EXPERIMENTAL RESULTS AND DISCUSSION

The tensile strengths (σ_t) of powder packed beds of FCC-aluminum hydrate and FCC-limestone with voidages of ε = 40 - 60% are shown in Figures 5A and 5B. σ_t in the emulsion phase can be extrapolated by Equation (1) modifying Rumpf's Equation (14).

$$\sigma_{t,\varepsilon_e} = \frac{1-\varepsilon_e}{1-\varepsilon} \frac{k_{\varepsilon_e}}{k_\varepsilon} \sigma_{t,\varepsilon} \tag{1}$$

where the coordination number k can be estimated by using voidage data based on the model proposed by Suzuki, et al. (15). From the collapse data shown in Figure 6, the voidages of the emulsion phase (ε_e) and loosely packed bed ($\varepsilon_{\ell p}$) were obtained as shown in Figures 2 and 7 respectively. Massimilla, et al. (11) and Rietema, et al. (12) proposed that network structures of loosely packed powder exist in the emulsion phase. As shown already in Photographs 1A and 1B, network structures of loosely packed particle were clearly observed in a static state. This type of structure probably can also be formed in emulsion phase. For the characterization of the emulsion phase, σ_t together with Young's modulus, will be useful as criteria. For the calculation of Young's modulus using the experimental data shown in Figures 2 and 7, the following assumptions were made.

1. Voidage of emulsion phase is equal to that at U_{mb}.
2. The drag force by fluid against powder in the emulsion phase can be calculated by Ergun's Equation.
3. The tensile strength of the emulsion phase prevailing at $U_o \leq U_{mb}$ is proportional to the prevailing drag force against powder in the emulsion phase.
4. The difference of tensile strength at $\varepsilon_{\ell p}$ and at ε_e is equal to net acting bed expansion pressure excluding the drag force. (Necessary pressure to change the structure of the emulsion phase.)

From assumptions 3 and 4,

$$\sigma_{t,\varepsilon_{\ell p}} - \sigma_{t,\varepsilon_e} = C(P_{\ell p} - P_e) \tag{2}$$

Rewriting Equation (2) into non-dimensional form by dividing $\sigma_{t,\varepsilon_{\ell p}} = CP_{\ell p}$.

$$1 - \frac{\sigma_{t,\varepsilon_p}}{\sigma_{t,\varepsilon_{\ell p}}} = 1 - \frac{P_e}{P_{\ell p}} \tag{3}$$

Left hand side of the Equation is non-dimensional difference of tensile strength between loosely packed bed and emulsion phase. Right hand side is non-dimensional net acting bed expansion pressure excluding the drag force. Equation (4) can be derived from Equation (1).

$$\frac{\sigma_{t,\varepsilon_e}}{\sigma_{t,\varepsilon_{\ell p}}} = \frac{(1-\varepsilon_e)\, k_{\varepsilon_p}}{(1-\varepsilon_{\ell p})\, k_{\varepsilon_{\ell p}}} \tag{4}$$

Equation (5) can be derived from Ergun's Equation.

$$\frac{P_e}{P_{\ell p}} = \frac{1-\varepsilon_e}{1-\varepsilon_{\ell p}} \left(\frac{\varepsilon_{\ell p}}{\varepsilon_e}\right)^3 \tag{5}$$

Equation (3) together with Equations (4) and (5) was plotted using experimental data on voidage obtained from Figures 2 and 7, as shown in Figure 8. As is seen in Figure 8, the assumptions 3 and 4 are thus verified to be reasonable. Therefore the emulsion phase can be assumed as an elastic body as indicated by Ritema (9,12). Young's modulus is written by Equation (6),

$$\sigma_{t,\varepsilon_{\ell p}} - \sigma_{t,\varepsilon_e} = Y\left[\frac{H_e - H_{\ell p}}{H_{\ell p}}\right] \tag{6}$$

Combining Equations (3), (5) and (6), Young's modulus can be calculated by Equation (7).

$$Y = \frac{\sigma_{t,\varepsilon_{\ell p}}(1-\varepsilon_e)}{\varepsilon_e - \varepsilon_{\ell p}}\left\{1 - \frac{(1-\varepsilon_e)}{(1-\varepsilon_{\ell p})}\left[\frac{\varepsilon_{\ell p}}{\varepsilon_e}\right]^3\right\} \tag{7}$$

Thus, Young's modulus was derived by using σ_t as internal stress in the elastic body of the emulsion phase. The effects of adding fine powder to FCC powder on Young's modulus and tensile strength of the emulsion pahse are illustrated in Figure 9. By adding fine powders to FCC powder, both Young's modulus and tensile strength of the emulsion phase become smaller. In other words, the structure of the emulsion phase becomes softer. Furthermore the tendency of tensile strength and Young's modulus of the emulsion phase looks similar in both cases of FCC-aluminum hydrate and FCC-limestone systems, as shown in Figure 9.

This result can be confirmed in Figure 10. This shows that Young's modulus and tensile strength of the emulsion phase have a close relation with each other. Those two criteria seem to be useful in characterizing the quality of emulsion phase.

Characteristics fluidization velocities (U_{mf}, U_{mb}, U_{mc}) of FCC-aluminum hydrate system are shown in Figure 11. As aluminum hydrate wt% becomes larger than 20%, channelling takes place. The upper limit of the mixing ratio to attain stable fluidization is approximately 15%. In the case of FCC-limestone systems, the upper limit is 7%. Bed expansion ratios at complete fluidization velocity of FCC with fine powder are approximately 20% higher than that of FCC alone, as shown in Figure 12. This fact is probably due to the structural change of emulsion phase. The effect of powder characteristics on gas backmixing coefficients in axial direction of beds at freely bubbling fluidization is illustrated in Figure 13. The tendency of the effect of fine powder on gas backmixing coefficients is different from each other in case of FCC-aluminum hydrate and FCC-limestone systems.

The effect of gas velocity on gas backmixing coefficients in fluidized bed with and without baffle plate is shown in Figure 14 using various types of powders. Typical examples of bubble shape in freely bubbling regime in fluidized beds of FCC and FCC-aluminum hydrate systems are shown in Photographs 2A and 2B. By selecting the fluidized powder, bubbles with "cat's paw prints" as shown in Figure 2B were observed, which are significantly different from those of FCC alone shown in Figure 2A.

NOMENCLATURE

C constant in Equation (2) (-)

d mean particle diameter (μm)

D_T bed diameter (m)
E axial dispersion coefficient of gas in dense phase (m^2/s)
H height of fluidized bed in Figure 6 (m)
H_e height of emulsion phase defined by Figure 6 (m)
H_f height of fluidized bed (m)
$H_{\ell p}$ height of loosely packed bed (m)
$k_{\varepsilon_e}, k_{\varepsilon_{\ell p}}$ coordination number in emulsion phase or loosely packed bed (-)
Δp pressure drop of fluidized bed (mH_2O)
p_e pressure drop of emulsion phase at $U_o \leq U_{mb}$ (Pa)
$p_{\ell p}$ pressure drop of loosely packed bed condition at $U_o \leq U_{mb}$ (Pa)
R cumulative undersize (%)
t time (s)
U_o superficial air velocity (m/s)
U_{mb} minimum bubbling velocity (m/s)
U_{mc} minimum complete fluidization velocity (m/s)
U_{mf} minimum fluidization velocity (m/s)
X weight fraction of fine powders (-)
Y Young's modulus (Pa)
ε voidage (-)
ε_e average voidage of the emulsion phase (-)
ε_{lp} loosely packed voidage (-)
ρ_{lp} loosely packed density (kg/m^3)
ρ_p particle density (kg/m^3)
σ_t tensile strength of powder (Pa)

LITERATURE CITED

1. Harrison, D. and Grace, J.R., "Fluidized Bed with Internal Baffles" in Fluidization, Davidson, J.F. (Ed), Academic Press, London & New York (1971).
2. Miyauchi, T. et al., "Transport Phenomena and Reaction in Fluidized Catalyst Bed," Advance in Chem. Eng., Vol. 11, Vermeulen et al. (Ed), Academic Press, London & New York (1981).
3. Kono, H.O. and Jinnai, M., AIChE Symp. Series 241, 80, 169 (1985).
4. Kono, H.O. and Ninomiya, K., "Heat Supply System for Fluidized Bed," presented at 20th National Heat Transfer Conf., AIChE, Milwaukee, WI (1981).
5. Geldart, D., Powder Technology, 7, 285 (1973).
6. Rowe, P.N. and Yacono, C.X.R., Chem. Eng. Sci., 31, 1179 (1976).
7. Geldart, D., et al., Powder Technology, 37, 25 (1984).
8. Baerns, M., IEC Fundamentals, 5, 508 (1966).
9. Rietema, K., "Application of Mechanical Stress Theory to Fluidization," in Proc. Int. Symp. Fluidization, Drinkenburg, A. (Ed), Netherlands University Press, Amsterdam (1967).
10. Donsi, G. and Massimilla, L., AIChE J., 19, 1104 (1973).
11. Donsi, G., Moser, S., and Massimilla, L., Chem. Engr. Sci., 30, 1533 (1975).
12. Rietema, K., Powder Technology, 35, 5 (1984).
13. Hartley, P.A. and Parfitt, G.D., J. Phys (E) Sci. Instrum., 17, 374 (1984).
14. Rumpf, H., Chem. Ing. Tech., 42, 538 (1970).
15. Suzuki, M., Makino, K., Yamada, M., and Iinoya, K., Int. Chem. Eng., 21, 482 (1981).

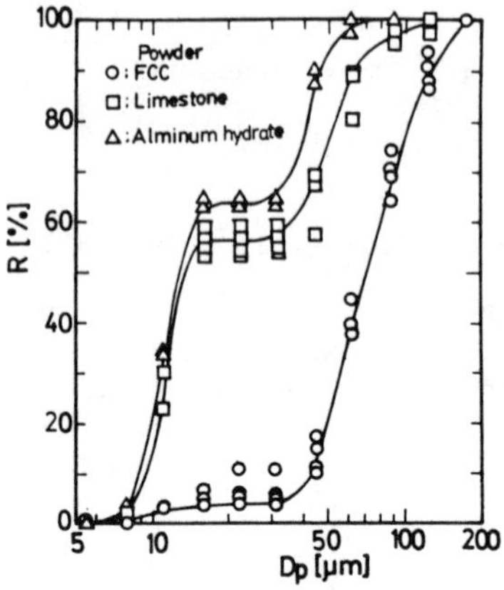

Figure 1. Particle Size Distribution of Sample Powders.

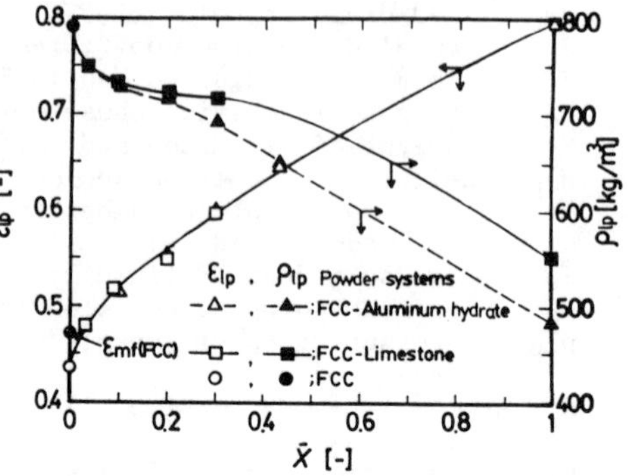

Figure 2. Loosely Packing Density and Voidage in Packed Bed Powder.

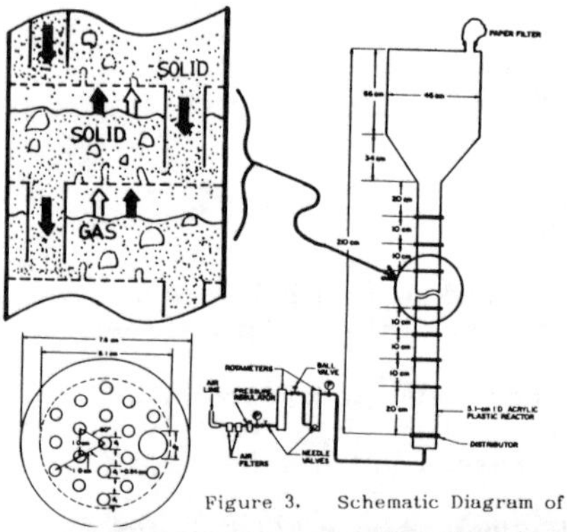

Figure 3. Schematic Diagram of Multi-Stage Fluidized Bed.

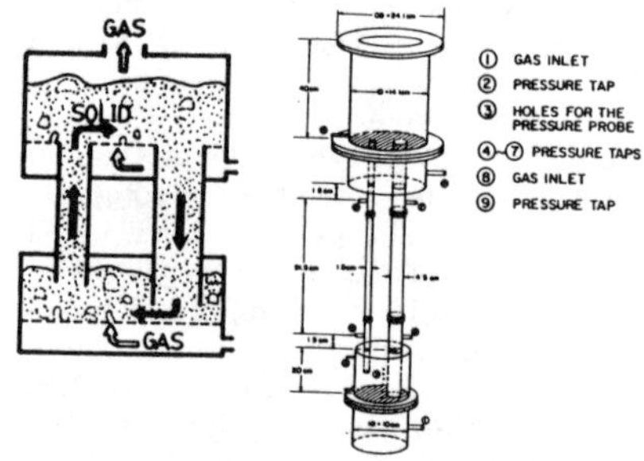

Figure 4. Schematic Diagram of Circulating Fluidized Bed.

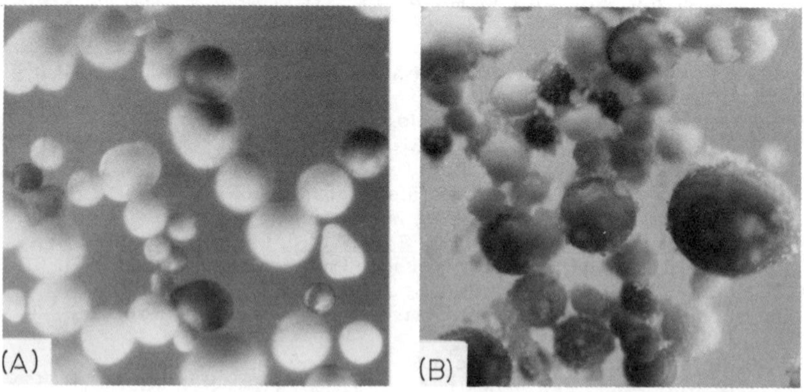

Photograph 1. Microscopic Pictures of Pure FCC and FCC-Aluminum Hydrate Mixture (10%).

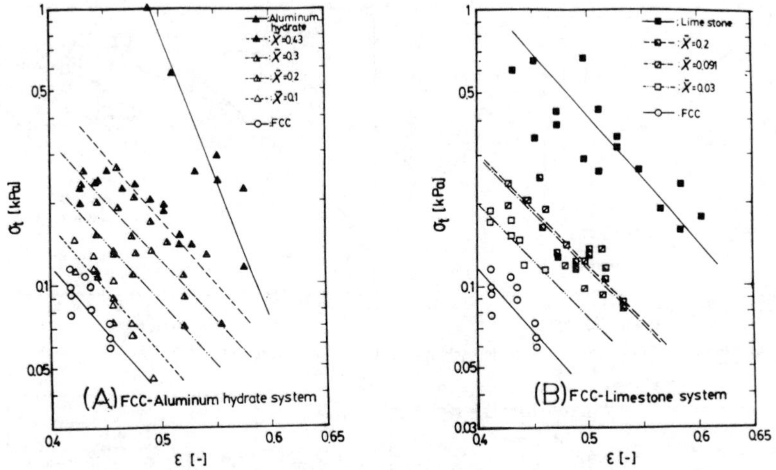

Figure 5. Tensile Strength Versus Voidage in Packed Bed Powder.

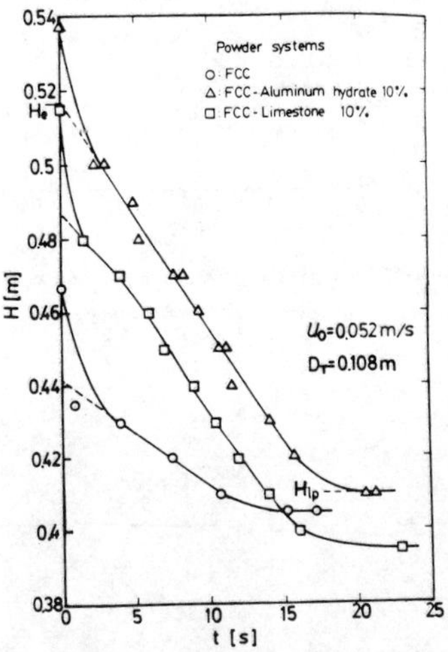

Figure 6. Examples of Collapse Test Data.

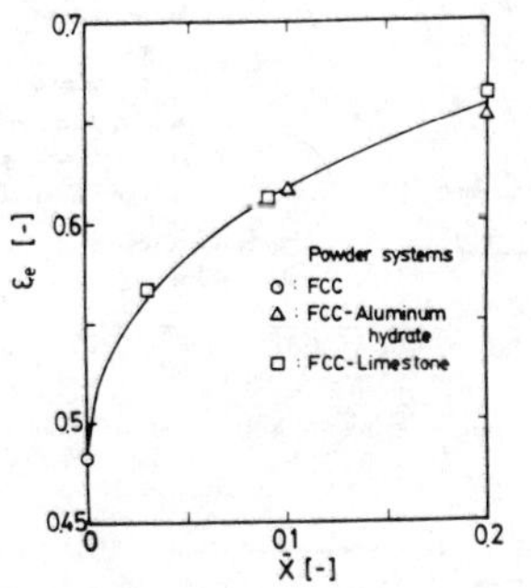

Figure 7. Voidage of the Emulsion Phase Versus Wt. Fraction of Fine Powder.

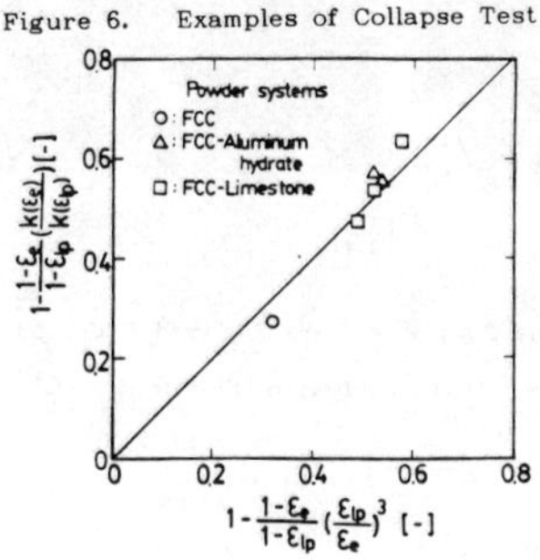

Figure 8. Confirmation of Equation (3).

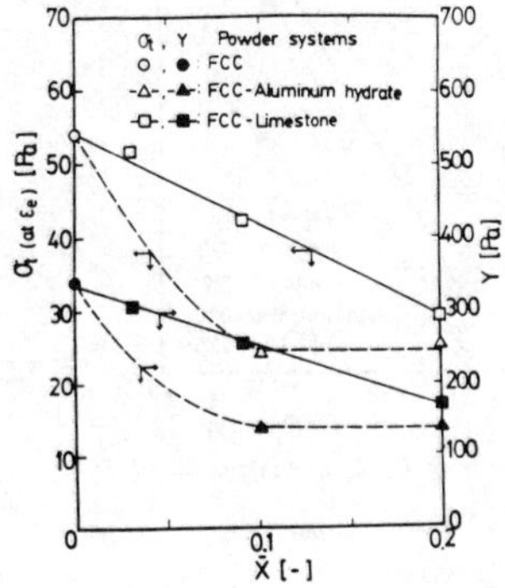

Figure 9. Young's Modulus and Tensile Strength of Emulsion Phase Versus Wt. Fraction of Fine Powder.

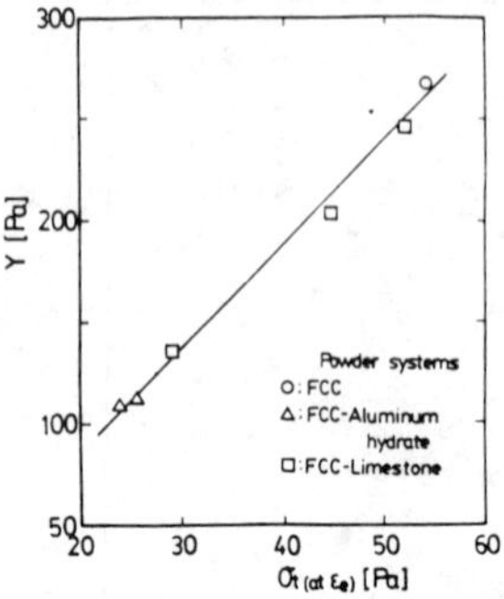

Figure 10. Young's Modulus Versus Tensile Strength of the Emulsion Phase.

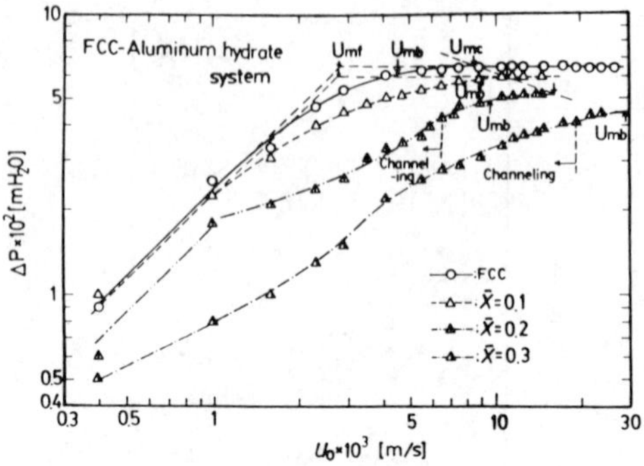

Figure 11. Characteristic Fluidization Velocities.

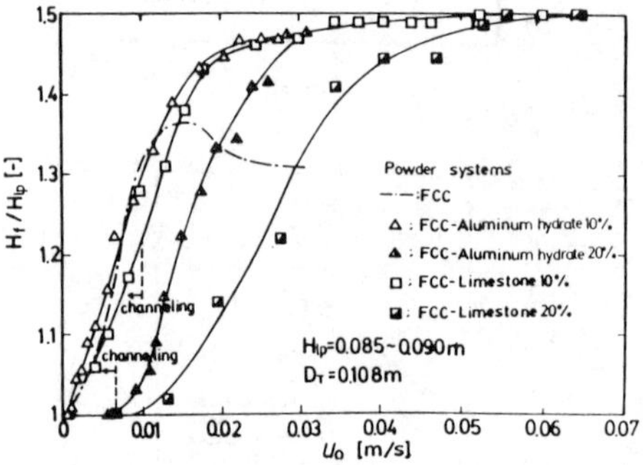

Figure 12. Bed Expansion Ratio Versus Superficial Air Velocity.

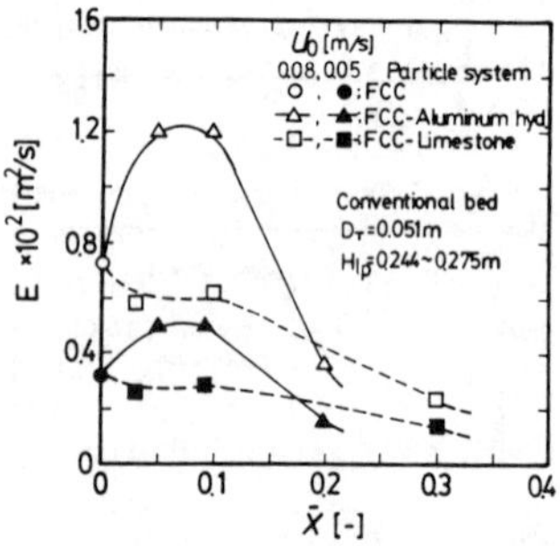

Figure 13. Dispersion Coefficient of Gas Versus Weight Fraction of Fine Powder.

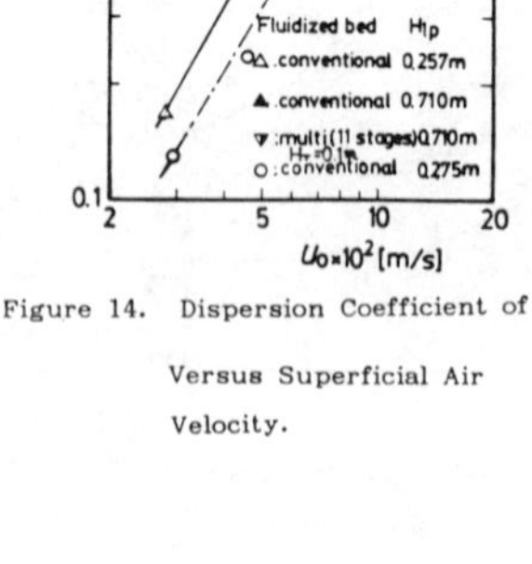

Figure 14. Dispersion Coefficient of Gas Versus Superficial Air Velocity.

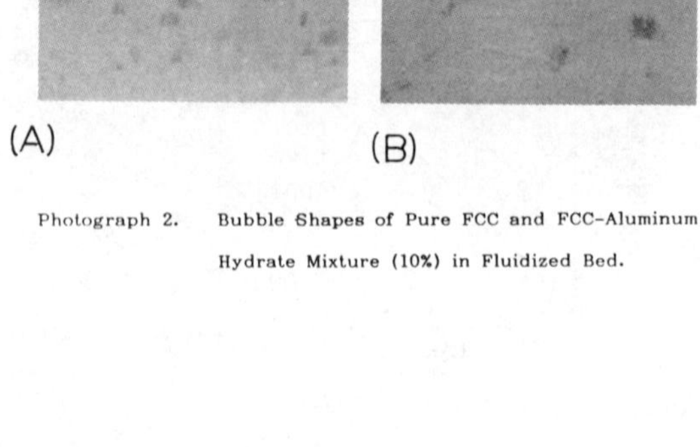

Photograph 2. Bubble Shapes of Pure FCC and FCC-Aluminum Hydrate Mixture (10%) in Fluidized Bed.

THE SIMILARITY RULE OF FLUIDIZATION AND ITS APPLICATION TO SOLID MIXING AND CIRCULATION CONTROL

Masayuki Horio, Maki Takada*, Mamoru Ishida and Norichika Tanaka*

Department of Applied Chemistry for Resources
Tokyo University of Agriculture and Technology
Koganei, Tokyo, 184

*Tsukishima Kikai Co. Ltd., 2 Tsukuda, Tokyo, 104

ABSTRACT

The similarity rule of fluidization by Horio et al. (1) was extended to the solid mixing phenomena based on the transient dispersion experiment of ferrite coated sand tracers and that of large floating particles in beds of different diameters (50, 150, 300 and 600 mm; tapered and non-tapered).

SCOPE

Establishing good mixing and uniform circulation of fluidized solids is very significant in many fluidized bed processes, since good solid mixing is one of the most essential advantages of fluidization. Solid mixing is strongly controlled by bubble motion (Rowe and Partridge (2)) and by radial bubble distribution (Merry and Davidson (3), Werther (4)). It is also controlled by bed wall tapering, structures of internals, and distributor design. So far, the scale up, i.e. the bed diameter increase, of fluidized beds has been conducted based on the principle that the bed height, bed materials and gas velocity should be maintained as identical as possible. However, since most of the above mentioned factors are global ones, it is not possible to examine them by small scale models while applying the previous principle. Horio et al. (1) presented a new similarity rule of fluidization and its validation regarding axial and radial distributions of bubbles. Although solid mixing is governed by bubbles which satisfy the similarity rule, it may not perfectly satisfy the rule because of the non-similar effects of particle-to-particle and bed-to-wall frictions. The objective of the present paper is to test if the similarity rule is applicable to the solid mixing, circulation and segregation.

CONCLUSIONS AND SIGNIFICANCE

Mixing of bed particles and that of floating large bodies in both straight and tapered columns were found to obey the Horio et al.'s similarity rule. However, the error of the transient tracer dispersion experiment tends to increase with the decrease in column scale due to the change of the time scale. With the present similarity rule, it is possible to rationally test by small scale models the design problems of practical significance, such as the global designs of the column, the distributor, internals, particle feeders and dischargers.

* * *

Fluidization technology has been increasing in popularity in the modern chemical and energy processes and a number of engineers are devoting their attention to the development and scale up of various fluidized bed processes. Nevertheless, our methodology for scale up has hardly been changed from its primitive empiricism, although scientific understandings of many aspects of fluidization have progressed. During the last decade, fluidized bed reactor models have been greatly improved by the inclusion of grid zone and freeboard models. However, most of the available models are one dimensional and are able to account for only longitudinal distributions of process variables. The direct simulation models such as the one by Pritchett et al. (5), or the bubble coalescence simulators such as those by Horio et al. (6) and Mori et al. (7) can handle the radial distribution of bubbles. However, they require too much time for computation, and the reliability of the extrapolative predictions from these computer codes has not yet been sufficiently established. If fluidization phenomena are geometrically similar like other fluid dynamic problems, the scale up methodology would be drastically rationalized. Horio et al. (1) theoretically derived a new similarity rule of fluidization phenomena based on the previous model for direct simulation (Horio et al. (6)) and experimentally validated the theory regarding radial and axial bubble distributions. The requirements from the similarity rule to estabish a geometrical similarity are that $u_0 - u_{mf}$ and u_{mf} for a fluidized bed m-times (in length; $m = D_t/D_t^o = L_{mf}/L_{mf}^o$) larger than the reference bed (denoted by superscript o) must be

$$u_0 - u_{mf} = \sqrt{m}\,(u_0 - u_{mf})^o \quad \text{(for bubble similarity)} \qquad (1)$$

$$u_{mf} = \sqrt{m}\,u_{mf}^o. \quad \text{(for interstitial flow similarity)} \qquad (2)$$

When the above conditions are satisfied, both radial and axial bubble distributions are made geometrically similar to those of the reference condition. However, the time scale of the phenomena becomes $\sqrt{m}$-times slower and the dimensionless time $t^* \equiv t\sqrt{g/D_t}$ remains as an invariant parameter. The present paper aims at extending the similarity rule to solid mixing, with particular interest in lateral dispersion which is much slower than axial mixing.

Two previous correlations for lateral dispersion of fluidized solids can be expressed in the following form:

$$(u_0 - u_{mf})L_{mf}/E_r = \alpha[(u_0 - u_{mf})/\sqrt{gL_{mf}}]^n \qquad (3)$$

In the Hirama et al. (8) correlation $\alpha = (6.2/f_w)(L_{mf}/D_t)^{0.55}$ and n=0.1. In the correlation of Shi and Fan (9), α = 22.5 and n = 0.25. Substituting Equations (1) and (2) into the above correlation, it can be confirmed that the Peclet number (the left hand side of Equation (3)) remains constant throughout the geometrically similar scale change. Therefore, as far as these correlations are concerned, the solid mixing process seems to obey our rule for geometrical similarity.

Dispersion and segregation of large floating bodies are of practical importance. Nienow et al. (10) explored the aspect of axial segregation, but little has been known on the lateral behaviors. Taking into account the gravity force, the buoyant force and the drag force due to the relative motion between a floating body and the emulsion phase, the following equation of motion for a floating body can be derived:

$$d\mathbf{v}^*/dt^* = 1 - (\rho_{bmf}/\rho_B) - (3/4)(|\mathbf{v}_r^*|\mathbf{v}_r^*\rho_{bmf}C_D/d_B^*\rho_B) \qquad (4)$$

where $d_B^* \equiv d_B/D_t$, $\mathbf{v}_r^* \equiv \mathbf{u}_r/\sqrt{gD_t}$, and $\mathbf{v}^* \equiv \mathbf{v}/\sqrt{gD_t}$. Therefore, for floating bodies with the same density, their dimensionless motions become identical, if the dimensions of the bodies are similar, and if the drag coefficient remains constant. Since the friction between the fluidized bed and a floating body may be negligible, there is a possibility that C_D remains almost constant.

EXPERIMENTAL APPARATUS AND PROCEDURE

The experiment consisted of two parts. In the first part the similarity in the mixing of bed material was examined. In the second part the similarity in distribution of floating bodies was studied.

Experimental Apparatus and Bed Materials

As summerized in Table 1 the diameter of the fluidized beds used in the experiment ranged from 50mm to 600mm. Silica sands of 2600 kg/m^3 in density were used as bed materials. The surface-to-volume mean particle size for each model was determined to satisfy Equation (2), the second condition for geometric similarity. The properties of the particles used for each model are also listed in Table 1. Although the mean particle size for each model was adjusted to satisfy Equation (2), slight deviations from it were unavoidable.

Tracer Particles

Ferrite ($CuO \cdot ZnO \cdot Fe_2O_3$) coated sands were used as the tracer for the experiment of bed material mixing. The particle diameter of the ferrite was less than 3μm. The saturated magnetization, remanence and coercive force of the ferrite were 67x10^4 C/kg, 4.6x10^4 C/kg and 10^5/4π A/m, respectively. No magnetic sticking of tracer particles was observed when there was no magnetic field. The apparent density of the ferrite coated sands was 2745 kg/m^3, which was approximately equal to that of the original sands. Since the attrition of ferrite particles became negligible after 20 minutes fluidization at u_0 = 0.337 m/s, all the tracer particles were pretreated by 20 minutes of fluidization.

The sizes of disk shaped poly-propylene (PP, density : 900 kg/m^3) tracer particles for the part 2 experiment are listed in Table 2.

Table 1. Fluidized beds and bed materials.

system **	m	$D_{t,mf}$	$D_{t,0}$	L_{mf}	plenum depth	grid orifices* diam.	grid orifices* pitch	grid orifices* number	bed materials d_{psv}	bed materials u_{mf}	bed materials $(u_{mf}/u^{o}_{mf})^2$
	-	m	m	m	m	mm	mm	-	μm	m/s	-
A	1.0	0.60	0.60	0.195	0.35	3.6	40	199	182	0.027	1
B	0.50	0.30	0.30	0.098	0.25	1.8	20	199	150	0.018	0.44
C	0.25	0.15	0.15	0.050	0.125	0.9	10	199	129	0.011	0.17
D	0.083	0.05	0.05	0.017	0.070	0.3	3.3	199	106	0.0072	0.071
A_2	1.0	0.60	0.60	0.195	0.35	3.6	80	48	182	0.027	1
B_2	0.50	0.30	0.30	0.098	0.25	1.8	40	48	150	0.018	0.44
C_2	0.25	0.15	0.15	0.050	0.125	0.9	20	48	129	0.011	0.17
B'	0.50	0.26	0.194	0.10	0.25	1.8	20	85	150	0.018	0.44
B'_2	0.50	0.26	0.194	0.10	0.25	1.8	40	19	150	0.018	0.44
C'	0.25	0.13	0.097	0.050	0.125	0.9	10	85	129	0.011	0.17
C'_2	0.25	0.13	0.097	0.050	0.125	0.9	20	19	129	0.011	0.17

*) Perforated plate; Triangular arrangement.
**) Systems with "'" are tapered (taper angle :  = 19°).
Subscript "2" denotes that orifice pitch was doubled.

Table 2. Float tracers.

system	diameter	thickness
	mm	mm
A, A_2	16	8
B, B_2, B', B'_2	8	4
C, C_2, C', C'_2	4	2

Table 3. Operating conditions.

system	m	u_0*	$u_0 - u_{mf}$	$\frac{(u_0 - u_{mf})^2}{(u^o_0 - u^o_{mf})^2}$
	-	m/s	m/s	
A, A_2	1.0	0.12	0.093	1
B, B_2, B', B'_2	0.5	0.082	0.064	0.474
C, C_2, C', C'_2	0.25	0.058	0.047	0.255
D	0.083	0.034	0.0268	0.083

*) based on the freeboard cross section.

Procedure for Bed Material Mixing Experiment

The tracer particles were placed at the center of the bed from the distributor level to the bed surface. This configuration was adopted to measure the radial dispersion coefficient with the assistance of an axisymmetric diffusion model. When the tracer particles were placed in the bed, the following procedure was very carefully taken to avoid a non-uniform packing: 1) First, sand particles were charged into the bed to the fixed height and the bed was fluidized lightly. 2) Then, a thin tube with a diameter 1/6 of the diameter of the bed was inserted at the center of the bed until the tube reached the distributor plate. 3) The sand inside the tube was then removed by a vacuum cleaner. 4) Tracer particles were poured into the tube up to the same height as the bed particles. 5) Again the bed was fluidized lightly to eliminate the compaction of the bed due to the insertion of the tube. 6) The tube was removed upward slowly. 7) Then, the bed was fluidized at the gas velocity listed in Table 3 for the specified period. 8) At the end of the fluidizing period a sectionizer (shown in Figure 1) was dropped into the bed and immediatly after this the air line was shut. 9) Sampling was done by vacuum suction and the tracer particles were separated by a magnetic separator.

In the case of system C and D the fluidizing period was too short to accurately control it manually. Therefore, a programmable sequencer (Hitachi, Hizac D28) was used to adjust the time to an accuracy of 0.1 second and to drop the sectionizer automatically by using a solenoid.

In order to obtain the dispersion coefficient, the following diffusion equation was solved numerically by SOR method subject to the initial and boundary conditions (6) and (7):

$$\partial X/\partial \tau = (1/r)(\partial/\partial r)(r\partial X/\partial r) \tag{5}$$

$$X = 1 \text{ for } 0 \le r \le r_0 \quad \text{and} \quad X = 0 \text{ for } r_0 < r \le 1, \text{ when } \tau = 0 \tag{6}$$

$$\partial X/\partial r = 0 \quad \text{at } r = 0 \text{ and } r = 1,\ \tau > 0 \tag{7}$$

where X is the weight fraction of tracer particles, r is the dimensionless radius and $\tau \equiv tE_r/R_t^2$ is the dimensionless time.

To compare the calculated results with experimental data the calculated concentration was area averaged for each sampling section. The experimentally obtained concentration was corrected by subtracting the amount of tracer in the dead zone in the range $r \le r_0$, and for $r > r_0$ by deducing the amount of sand particles in the dead zone. The dispersion coefficient was, then, determined by finding out the dimensionless time at which the difference between calculated and experimental profiles was minimized.

Procedure for Float Experiment

The PP float tracer was placed at the center but only in the region near the bed surface. This arrangement was adopted to simulate the dispersion of sludge or municipal waste incineration where the sinking of floating bodies into the bed is important. The depth of tracer zone was $L_{mf}/2.5$ for straight columns and was $L_{mf}/5$ for tapered columns. The latter part of the experiment was done in a similar manner to that of the bed material mixing experiment.

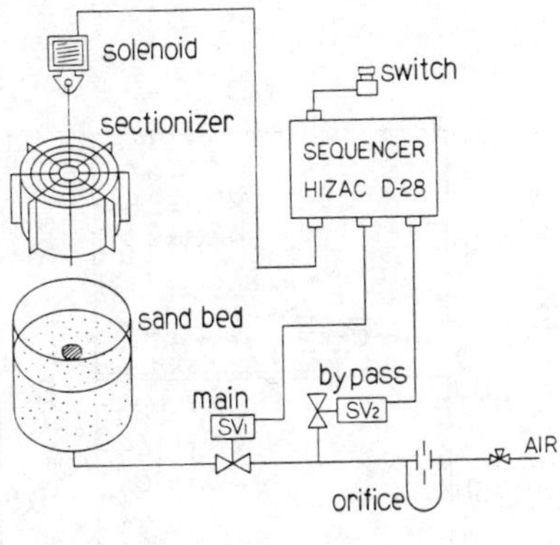

Figure 1. Bed sectionizer system.

RESULTS

The transient change of radial distribution of tracer concentration is shown in Figure 2 in comparison with the calculated curves. The Peclet number determined is shown in Figure 3 with previous experimental data and correlations. The data correspond to the conditions listed in Table 3 agree with each other within ±30% accuracy. Therefore, it can be concluded that bed material mixing also obeys the similarity rule. However, the scatter of experimetal data increases with decreases of the bed diameter. This is mainly due to the present transient response method where the experimental errors tend to increase with the decrease in the sampling period.

The distributions of float tracers in the straight columns are shown in Figures 4 and 5. The tapered model results are shown in Figure 6. It is very clear that similarity exists in the behavior of floating bodies both in the straight and tapered models.

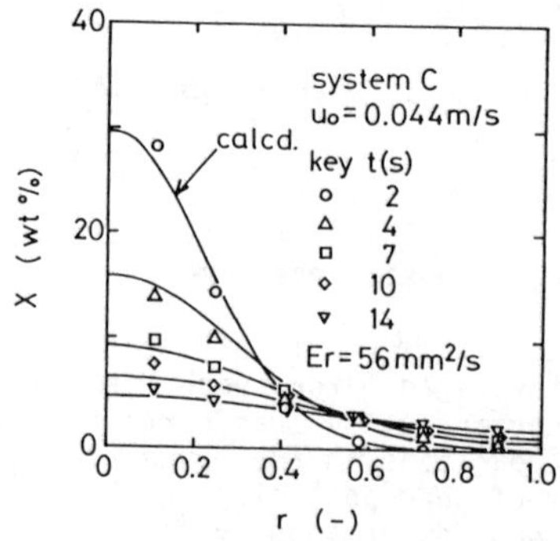

Figure 2. Computed and experimental tracer profiles.

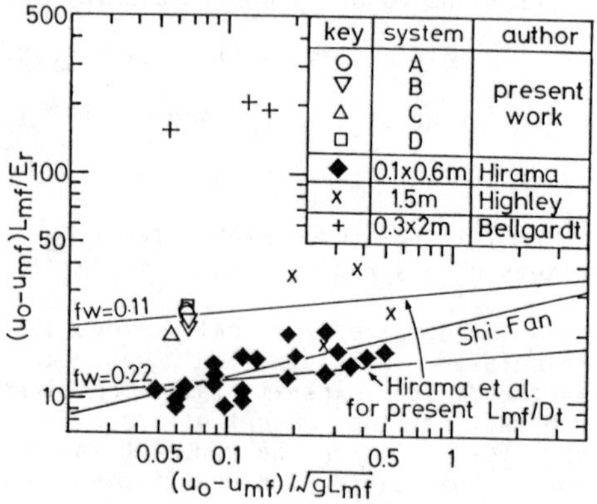

Figure 3. Peclet number vs. $(u_0 - u_{mf})/\sqrt{gL_{mf}}$.

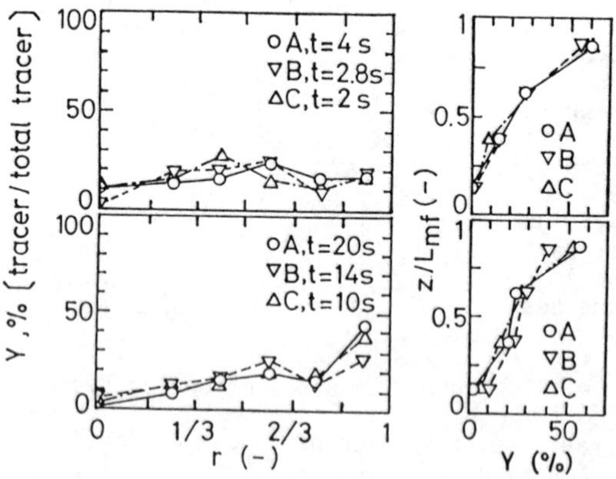

Figure 4. Float tracer distributions in straight columns.

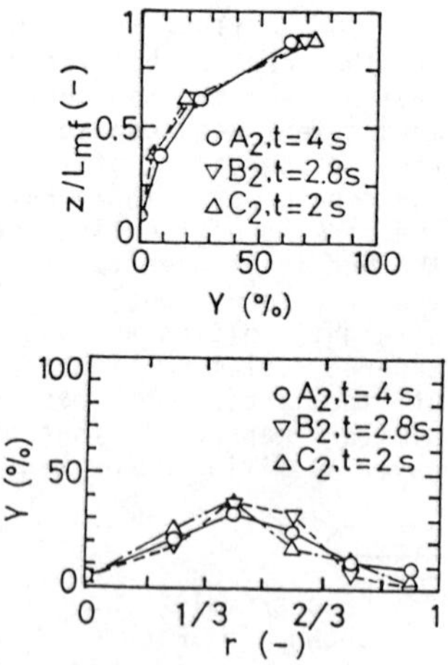

Figure 5. Effect of increased orifice pitch.

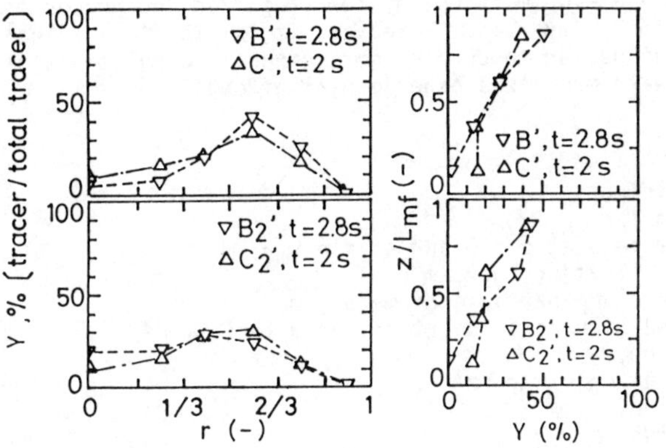

(In Figures 4, 5 and 6 axial tracer profiles are cross sectional averages and radial profiles are longitudinal averages.)

Figure 6. Float tracer distributions in tapered columns.

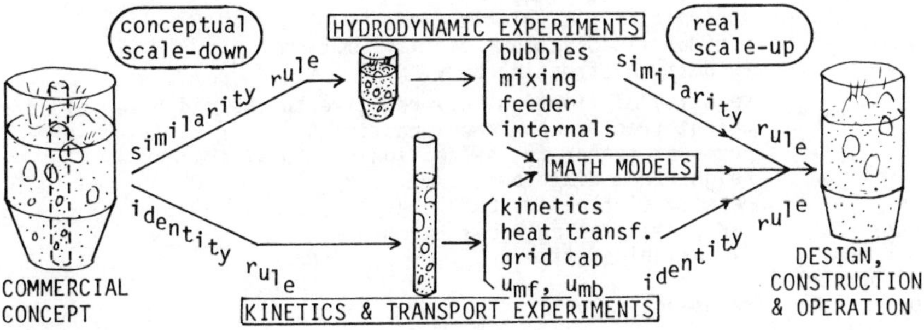

Figure 7. A rationalized development and scale-up methodology.

DISCUSSION

As shown in Figure 3 the present data as well as the data of Highley and Merrick (11), where L_{mf}/D_t was 0.2 to 0.4 similar to ours, can be consistently fitted by Hirama et al. (8) correlation with $f_w \simeq 0.11$. Hirama et al. data show lower Peclet number probably due to their semi-two dimensional bed shape (L_{mf} = 0.065 to 0.12m). However, the data by Bellgardt et al. (12), where L_{mf} was 0.82m and dry ice particles were used as float tracers, show much higher Peclet number.

The wide scatter of the data for dispersion coefficient tells us that more work should be done to develop a general correlation for lateral solid mixing. Nevertheless, it has become clear that the similarity rule is valid for solid mixing processes. Therefore, even though a generally applicable correlation is not available yet, we can predict the mixing performance in a commercial scale plant by small scale laboratory experiment. By the float tracer experiment it has been also proven that the global design items such as the wall tapering and the distributor structure, which can not be easily expressed in any correlation, can be easily simulated by small scale models. Figure 7 shows the rationalized scale-up methodology for fluidized bed processes based on the present similarity rule. Moreover, it is significant that the similarity rule helps us to develop more general correlations of fluidization phenomena with much less experimental efforts.

NOTATION

C_D	drag coefficient, -
D_t	column diameter, m
$D_{t,0}$, $D_{t,mf}$	column diameters at z = 0 and at z = L_{mf}, m
d_B	diameter of floating body, m
d_{psv}	surface-to-volume particle diameter, m
E_r	radial dispersion coefficient of bed material, m^2/s
f_w	wake fraction, -
L_{mf}	bed height at $u_0 = u_{mf}$, m
m	magnification, -
$R_t \equiv D_t/2$	column radius, m
r,r'	dimensionless radius, -, and radius, m, respectively
t	time, s
$t^* \equiv t\sqrt{d/D_t}$	dimensionless time, -
u_0	superficial gas velocity, m/s
u_{mf}	minimum fluidization velocity, m/s
v	velocity of floating body. m/s
$\mathbf{v}_r$	velocity of floating body relative to the bed material, m/s
X	weight fraction of tracer particles, -
Y	per cent tracer in the section per total tracer, -
z	height from distributor, m
ρ_B	density of floating body, kg/m^3
ρ_{bmf}	bed density at $u_0 = u_{mf}$, kg/m^3
$\tau \equiv tE_r/R_t^2$	dimensionless time, -
(Superscript)	
o	reference value

LITERATURE CITED

1. Horio, M., T. Sawa, A. Nonaka and I. Muchi, "A New Similarity Rule for Fluidized Bed Scale Up", AIChE J. in print (1986).
2. Rowe, P.N., and B.A. Partridge, Proc. Symp. on Interaction between Fluids and Particles, Inst. Chem. Eng., London, p. 135 (1962).
3. Merry, J.M.D., and J.F. Davidson, Trans. Inst. Chem. Eng., 51, 361 (1973).
4. Werther, J.,Powder Technology, 15, 155 (1976).
5. Pritchett, J.W., T.R. Blake and S.K. Garg,, AIChE Symp. Ser., 74 (176), 134 (1978).
6. Horio, M., J. Liu and I. Muchi, Kagaku-Kogaku-Ronbunshu, 9, 176 (1983).
7. Mori, S., S. Miho, M. Yamada, S. Hiraoka and I. Yamada, AIChE Symp. Ser., 80 (241), 10 (1985).
8. Hirama, T., M. Ishida and T. Shirai, ibid., 1, 272 (1975).
9. Shi, Y.F. and L.T. Fan, Powder Technol., 41, 23 (1985).
10. Nienow, A.N., P.N. Rowe and T. Chiba, AIChE Symp. Ser. 74 (176), 45 (1978).
11. Highley, J., and D. Merrick, ibid., 67 (116), 219 (1971).
12. Bellgardt, D., M. Schoessler and J. Werther, Proc. of 8th Intl. Conf. on Fluidized-Bed Combustion, vol. I, p. 115 (1985).

ON THE PACKET MECHANISM OF MIXING IN A FLUIDISED BED

N.N.Bobkov and Yuri P.Gupalo

Institute for Problems in Mechanics, USSR Academy of Sciences, 101 Vernadsky Ave., Moscow 117526, USSR

ABSTRACT: We study the steady motion of local inhomogeneities of the porosity in a developed fluidized bed, simulating them by spherical "packets" of solid particles, and we also consider the perturbed motion of the solid and fluid phases near the packets. On the basis of a model of two interpenetrating interacting continua - ideal fluids that simulate the solid and fluid phases - we determine the velocity fields of the phases outside and inside a packet, the rate of its steady ascent or descent, and the flow rate of the fluidizing medium through the packet. We investigate the existence, shape, and size of regions of circulation of the fluidizing medium generated by moving packets.

SCOPE: The mixing of the solid and fluid phases in a fluidized bed has a dramatic effect on fluid-bed industrial processes. The mixing of the fluid phase is usually associated with the motion of bubbles, i.e., cavities free of solid particles and the mixing of the solid phase with the transport of heavy particles in the wake of a bubble and also directed macrocirculation motions in the bed. At the same time it is well known that, especially in the case of fluidization by gases, there are appreciable fluctuations in the porosity of the fluid bed in a wide range. Their theoretical description can be based on the notion of packets of particles carrying mass, momentum, and thermal energy. The concept of a packet of particles was apparently first put forward by Mickley and Fairbanks (1). The porosity of a packet can vary from a value near unity (for bubbles) to the porosity of the unfluidized bed (for bunches of particles in a state near close packing). The sizes of packets, or "clusters" were estimated by Yerushalmi et al. (2) in the case of fast fluidized bed.

CONCLUSIONS AND SIGNIFICANCE: It is shown that the motion of the fluidizing fluid and solid particles is governed by the physical parameters of the fluid and solid phases, and also by the ratio of the porosities inside and outside a packet. As in the case of a cavity free of solid particles (bubble), two flow regimes exist: a circulation regime associated with the formation in the neighborhood of the packet of closed streamlines of the fluidizing fluid and a through-flow regime. A region with circulation ("cloud") entirely contains a packet (Figure 1a) if it is buoyant (i.e., of the type of a local rarefaction), as in the case of a bubble in a fluidized bed. However, for a dense packet (i.e. a local increase in the concentration of the solid phase) the fluid flow pattern is changed qualitatively, and the cloud is contained within the packet (Figure 2a). We determine the conditions of transition to the through-flow regime of motion of the fluid phase through a packet (Figures 1b and 2b) whose porosity is close to that of the ambient fluidized bed. The cloud size is calculated and the parameters characterizing the flow regimes for a packet and a bubble are compared (Figure 3).

A proper understanding of the hydrodynamics of fluid motion inside and outside a packet is important from the point of view of chemical reactor design.

FORMULATION OF PROBLEM OF THE MOTION OF FLUID AND SOLIDS NEAR AND INSIDE A PACKET IN A FLUIDIZED BED

We consider a fluidized bed as a double continuum consisting of interpenetrating interacting continua (3) that simulate the phases of the solid particles (the disperse phase) and the fluidizing fluid (the dispersion medium). We identify both phases with ideal incompressible fluids, i.e., in the momentum conservation equations we ignore the deviators of the stress tensors, retaining however the viscosity-dependent terms in the equation for the interphase interaction force. Under these assumptions, the complete system of equations of the steady two-phase flow has the form

$$d_f\varepsilon(\vec{v}\nabla)\vec{v}=-\nabla p_f+d_f\varepsilon\vec{g}-\vec{f},\quad d_s\rho(\vec{w}\nabla)\vec{w}=-\nabla p_s+d_s\rho\vec{g}+\vec{f}$$
$$\nabla(\varepsilon\vec{v})=0,\quad \nabla(\rho\vec{w})=0,\quad \rho=1-\varepsilon,\quad \Phi(p_s,\varepsilon,\ldots)=0 \qquad (1)$$

The last relation in (1) is the equation of state of the disperse phase and serves to close the system. In what follows, we use its simplest form assuming that the concentrations of the particles outside and inside a packet are constant. A similar assumption was made by Davidson (4) in analyzing the motion of a bubble in a fluidized bed. Therefore, in what follows the pressure p_s occurs in the equations only as a formal term whose structure does not directly influence the final results.

For the interphase interaction force we take the relation (5)

$$\vec{f}=-\rho\nabla p_f-\rho(\vec{w}-\vec{v})F(\varepsilon,a_p,\mu_f) \qquad (2)$$

where a_p is particle radius.

The surface of a packet, which is a macroscopic region with porosity different from that of the ambient two-phase medium, is, essentially, a discontinuity surface in the fluidized bed. The system of boundary conditions on that surface in the case of axial symmetry has the form (6)

$$\varepsilon v_n = (\varepsilon v_n)', \quad \frac{1}{2} d_f (v_n^2 - v_n'^2) = p_f - p_f', \quad \rho w_n = (\rho w_n)', \quad w_\xi = w_\xi' \qquad (3)$$

Here and in what follows, the prime denotes the flow parameters within the packet; n and ξ are, respectively, the normal and tangential natural coordinates associated with the surface of the packet.

As a simplest model of a packet, we consider a fludized sphere that moves uniformly with velocity U_d and constant porosity ε' in the unbounded fluidized bed, whose porosity ε is also assumed to be constant. We assume that the packet surface is impermeable for solid particles, so that $w_n = w_n' = 0$. Further, for the practically important case of fluidization by gases the density of the fluidizing fluid can be assumed to be small compared with the density of the solid particles.

We introduce a spherical coordinate system r , θ , φ with origin at the center of the packet and polar axis directed downwards. We divide the complete space into two regions: the exterior ($r > a$) and the interior ($r < a$) of the packet, and in each of these regions we formulate the basic equations and the boundary conditions on the surface $r = a$ in the framework of the above described model.

The concentration of solid particles outside the packet is fairly high, and therefore in the momentum conservation equation of the fluidizing fluid we ignore the terms proportional to the density d_f compared with the gradient (proportional to the weight of unit volume of the bed) of the pressure p_f (for this, the condition $1 > \rho \gg d_f/(d_f + d_s)$) must obviously be satisfied). Taking into account (2), we obtain the following system of equations governing the flow fields of fluid and solid phases outside the packet:

$$\vec{v} - \vec{w} = -k \nabla p_f, \quad d_s \rho (\vec{w} \nabla) \vec{w} = -\nabla (p_s + p_f) + d_s \rho \vec{g}, \quad \nabla \vec{v} = 0, \quad \nabla \vec{w} = 0 \qquad (4)$$

Here, the first Equation (1) has been rewritten by means of (2) in a more compact form and corresponds to the well-known Darcy law, the coefficient of proportionality being

$$k = \frac{\varepsilon}{(1-\varepsilon) F(\varepsilon, d_p, \mu_f)} = \frac{4 d_p^2 \varepsilon^2}{c_1 \mu_f (1-\varepsilon)^2} \qquad (c_1 \approx 150)$$

As the second equation in (4) we have taken the sum of the momentum conservation equations of both phases, ignoring the negligibly small terms. This equation has the same form as the Euler equation for the flow of an ideal incompressible fluid with density $d_s \rho$ and pressure $p_f + p_s$.

In the internal flow region, the form of the basic system of equations depends essentially on the packet porosity ε'. However, in the considered case, in which the concentration of particles in a packet is high, the system (4) also remains valid for the flow within the packet (the velocities, densities, pressures, and concentrations must be replaced by the corresponding primed quantities for the internal flow).

The formulation of the problem admits a natural generalization to the case of arbitrary particle concentration both inside and outside the packet. One can consider, for example, the problem of the motion of a cluster of particles suspended in a gas.

In the framework of our model of a packet, we write the system of boundary conditions (3) on the discontinuity surface $r = a$ in spherical coordinates in the form

$$\varepsilon v_r = (\varepsilon v_r)', \; p_f = p_f', \; w_r = w_r' = 0, \; w_\theta = w_\theta' \tag{5}$$

The following conditions must also be added. First, the condition of constancy of the pressure gradient in the fluid phase far from the packet; second, the pressure must be a bounded function everywhere within the packet:

$$(\partial p_f/\partial y)_{r \to \infty} = -v_0/k; \quad |p_f(r,\theta)| < \infty, \; 0 \leq r \leq a \tag{6}$$

Further, the velocity vector of the solid phase far from the packet has the form

$$\vec{w}_{r \to \infty} = \{\pm U_d \cos\theta, \pm U_d \sin\theta, 0\} \tag{7}$$

where the plus and minus signs correspond to buoyant and sinking packets, respectively. Finally, the velocity components of the solid phase within the packet must be finite, i.e.,

$$|\vec{w}(r,\theta)| < \infty, \; 0 \leq r \leq a \tag{8}$$

VELOCITY AND PRESSURE FIELDS OF THE FLUID AND SOLID PHASES

It follows from the first and last two of Equations (4) that p_f and p_f' are harmonic functions. Because of the axial symmetry of the pressure fields and the conditions (6), we shall seek the functions p_f and p_f' in the form

$$p_f = \frac{v_0}{k} r\cos\theta + \sum B_n P_{n-1}(\cos\theta) a^{n+1} r^{-n}, \quad p_f' = \sum A_n P_n(\cos\theta) a^{1-n} r^n \tag{9}$$

where $P_n(\cos\theta)$ is a Legendre polynomial of degree n; A_n and B_n are coefficients to be determined.

Substituting Equations (9) in the system of boundary conditions (5) for the radial components of the velocities and the pressures, using the first Equation (4), and equating the coefficients of the same P_n, we obtain an infinite algebraic system for determining A_n and B_n. Solving it, we find the pressure distribution of the fluid outside and inside the packet:

$$p_f=\frac{v_0}{k}r\cos\theta\left(1+\gamma_1\frac{a^3}{r^3}\right),\quad p_f'=\frac{3v_0}{k'}\gamma_2 r\cos\theta;\quad \gamma_1=\frac{\varepsilon k-\varepsilon' k'}{2\varepsilon k+\varepsilon' k'},\quad \gamma_2=\frac{\varepsilon k'}{2\varepsilon k+\varepsilon' k'} \tag{10}$$

To construct the flow fields of the phases, we consider the second and fourth Equations (4) and introduce a stream function ψ_s (respectively, ψ_s' in the interior region) for the flow of the solid phase:

$$w_r=(r^2\sin\theta)^{-1}\partial\psi_s/\partial\theta,\quad w_\theta=(r\sin\theta)^{-1}\partial\psi_s/\partial r$$

As a result, the equation of motion of the solid phase can be reduced to a nonlinear partial differential equation for the stream function that admits solution of the form (7) (A, B, C are constants)

$$\psi_s=(Ar^{-1}+Br^2+Cr^4)\sin^2\theta \tag{11}$$

In the flow outside the packet the condition (7) and the third condition in (5) are satisfied by the function

$$\psi_s=\pm\frac{U_d r^2\sin^2\theta}{2}\left(1-\frac{a^3}{r^3}\right) \tag{12}$$

which corresponds to potential flow of an incompressible ideal fluid past a sphere. In (12) and what follows, the upper and lower signs correspond, respectively, to motion of the packet along and in the opposite direction to the flow of the fluidizing fluid at infinity.

In the flow region within the packet, we obtain $A=0$ from (11), satisfying the condition (8). Using now the third and fourth of the boundary conditions (5), and also Equation (12), we obtain an expression for the stream function ψ_s' of the solid phase corresponding to a spherical Hill vortex:

$$\psi_s'=\pm\frac{3U_d}{4a^2}r^2(r^2-a^2)\sin^2\theta \tag{13}$$

The exterior velocity field of the fluidizing fluid can be found **from** the first Equation (10), the first Equation (4), and the expression (12), in which the plus and minus signs correspond to a rarefied packet ($\varepsilon'>\varepsilon$) and a dense packet ($\varepsilon'<\varepsilon$), respectively. The interior velocity field can be sought similarly by means of the second Equation (10) and (13). We write the stream functions ψ_f and ψ_f' of the fluid outside and inside the packet, respectively, in the dimensionless form

$$\psi_f=\left[\pm\delta-1+\frac{1}{r^2}(2\gamma_1\mp\delta)\right]\frac{r^2\sin^2\theta}{2},\ \psi_f'=\left[\pm\frac{3}{2}\delta(r^2-1)-3\gamma_2\right]\frac{r^2\sin^2\theta}{2} \tag{14}$$

Here $\bar{\psi}_f=\psi_f/v_0a^2$, $\bar{r}=r/a$, $\delta=U_d/v_0$ and the bars above the dimensionless parameters have been omitted.

REGIMES OF MOTION OF THE FLUIDIZING FLUID NEAR A PACKET: REGIONS OF CIRCULATION

Fluid Flow Rate Through the Packet. Using the first Equation (14) to determine the radial component of the fluid velocity on the boundary of the packet and then integrating it over the hemisphere $r=a$, we arrive at the following expression for the flow rate:

$$J=3J_0\frac{\alpha}{\alpha+2}\ ;\ J_0=\pi a^2v_0\varepsilon,\ \ \alpha=\frac{\varepsilon'k'}{\varepsilon k}=\frac{\varepsilon'^3(1-\varepsilon)^2}{\varepsilon^3(1-\varepsilon')^2} \tag{15}$$

Setting $\varepsilon'=1$ in (15), we obtain Davidson's well-known result (4): the fluid flow rate through a bubble is equal to $3J_0$.

Through-Flow and Circulation Flow Regimes. The condition for the existence of a circulation region is obviously the vanishing of the radial velocity component of the fluid on some surface, which will be the boundary of the region. For an ascending packet ($\varepsilon'>\varepsilon$) this condition is satisfied on a concentric sphere of radius a_c outside the packet and determined by the expression

$$\left(\frac{a_c}{a}\right)^3=\left(\delta+2\frac{\alpha-1}{\alpha+2}\right)(\delta-1)^{-1} \tag{16}$$

In the flow perturbed by a descending packet ($\varepsilon'<\varepsilon$), the radial component of the velocity of the fluid also vanishes on a spherical surface, but this is now frozen within the packet, so that the radius of the cloud is less than the radius of the packet and is determined by

$$\left(\frac{a_c}{a}\right)^2=1-\frac{2\alpha\varepsilon}{\varepsilon'(2+\alpha)\delta} \tag{17}$$

Since a_c is essentially positive, it follows from (16) and (17) that in the case of an ascending packet a cloud exists only under the condition $\delta>1$, i.e., when the steady velocity U_d of the packet exceeds the interstitial fluid velocity v_0. In the case of a descending packet, the existence of a circulation region requires the velocity to satisfy the inequality $U_d>2\alpha\varepsilon v_0/\varepsilon'(2+\alpha)$.

It can be shown that in the first of the two cases described above a cloud is formed under the condition that ratio α of the permeabilities of the bed inside and outside the packet exceeds a certain critical value α_*, while in the second case this ratio must be less than some different critical value α_{**}. If $\alpha_{**}<\alpha<\alpha_{*}$ a packet of either type moves in the through-flow regime, i.e., closed streamlines of the fluid are not formed. The critical parameters $\alpha_{*}=\alpha_{*}(\varepsilon,N)$ and $\alpha_{**}=\alpha_{**}(\varepsilon,N)$ are known functions of the porosity of the bed outside the packet and of the number $N=2/3\, C_1 \mu_f d_s^{-1}(2a_p)^{-2} a^{1/2} g^{-1/2}$, which is related to $\delta = U_d/v_0$ by the equation

$$\delta(\varepsilon,\varepsilon',N) = N(1-\varepsilon)\varepsilon^{-2}\left(\frac{|\alpha-1|}{\alpha+2}\right)^{1/2}$$

We consider as an example a fluidized system in which particles of an alumosilicate catalyst of diameter $2a_p=5\cdot10^{-5}$m and density $d_s = 2.2\cdot10^{3}$ kg/m^3 are suspended in an airflow with viscosity $\mu_f = 2\cdot10^{-5}$ kg/ms . We take the size of a packet moving in a bed of porosity $\varepsilon = 0.4$ equal in order of magnitude to the mean size of the bubbles in the system, i.e., $a = 10^{-2}$ m. In this case, $N \cong 11.6$. For such a packet, we construct the streamlines of the fluid for the various flow regimes decribed above.

It follows from the analysis of the Equations (16), (17) that already when there is a small deviation of the concentration of particles in the packet from the critical values the cloud surface effectively coincides with that of a packet. For a sinking packet, the cloud rapidly expands with increasing concentration and fills the complete packet from inside, while for an ascending packet the cloud rapidly contracts, and its boundary approaches the boundary of the packet from outside. To show clearly the circulation mechanism of the fluid within the cloud, the deviation of the internal porosity from the critical values in the numerical examples considered below has been chosen to be small.

The flow pattern in the neighborhood of a packet in the form of a local rarefaction for $\varepsilon'-\varepsilon = 1.6\cdot10^{-4}$ ($\delta\cong1.04$) is shown in Figure 1a and for $\varepsilon'-\varepsilon = 9.3\cdot10^{-5}$ ($\delta \cong 0.79$) in Figure 1b. Curves 1-9 in Figure 1a correspond to the following values of the dimensionless stream functions Ψ_f and Ψ_f' : $\Psi_f' = -0.52, -0.5$; $\Psi_f'=\Psi_f = -0.31, -0.19, -0.098$; $\Psi_f = 0, 0.1, 0.2, 0.33$. In Figure 1b, these values are as follows: $\Psi_f = \Psi_f' = 0, -0.073, -0.25, -0.43, -0.48, -0.481, -0.492$; $\Psi_f' = -0.5, -0.504, -0.53, -0.61$. In the first case, the packet ascends in the circulation regime, in the second in the through-flow regime.

Figure 2 shows the disturbed flow pattern in the case when the packet is a local density enhancement. In Figure 2a, for which $\varepsilon'-\varepsilon = -10^{-4}$ ($\delta \cong 0.83$), we show a packet sinking in the circulation regime, the cloud being contained entirely within the packet. On streamlines 1-11, the functions Ψ_f and Ψ_f' take the following values: $\Psi_f' = 0.0057, 0.005, 0.004, 0.001$; $\Psi_f'=\Psi_f = 0, -0.073, -0.25, -0.43, -0.5, -1.78, -3.45$. For the packet shown in Figure 2b, $\varepsilon'-\varepsilon =$

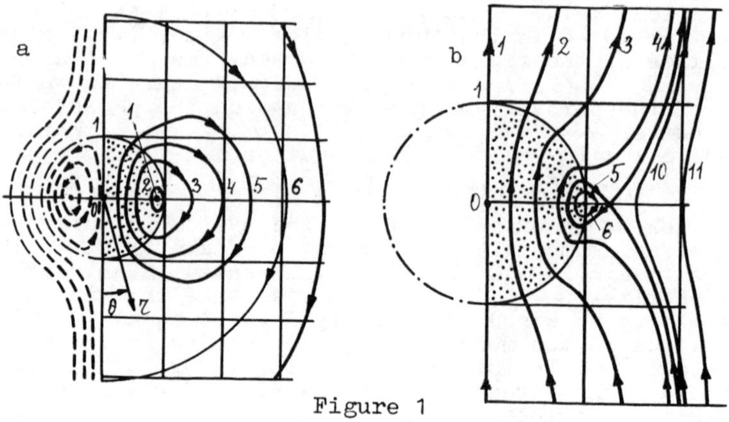

Figure 1

$= -6.5\ 10^{-5}$ ($\delta \cong 0.15$), and it sinks in the through-flow regime, distorting very weakly the uniform flow. In this case, the stream functions take the following values with increasing number of the curve: $\Psi_f' = \Psi_f = 0$, -0.073, -0.25, -0.43; $\Psi_f = -0.5, -1.24, -2.52$.

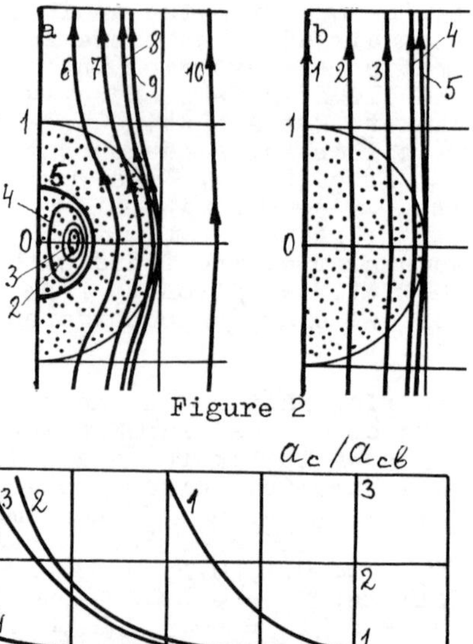

Figure 2

Size of the Region of Closed Circulation. In Figure 3a, we have plotted in accordance with Equation (16) the ratio of the cloud radius a_c for an ascending packet when $\alpha > \alpha_*$ to the cloud radius a_{cb} for a bubble of equal size vs. α. Curves 1-4 in Figure 3 correspond to $N = 1, 10, 11.6, 100$. It can be seen that compared with a bubble of equal size a packet that generates circulation disturbs the normal regime of mass transfer in a larger region. At the same time, however, the cloud contracts very rapidly with increasing ε. At α only slightly exceeding α_*, $a_c \approx a_{cb}$. In Figure 3b a_c / a vs. α is shown for a sinking packet ($\alpha < \alpha_{**}$).

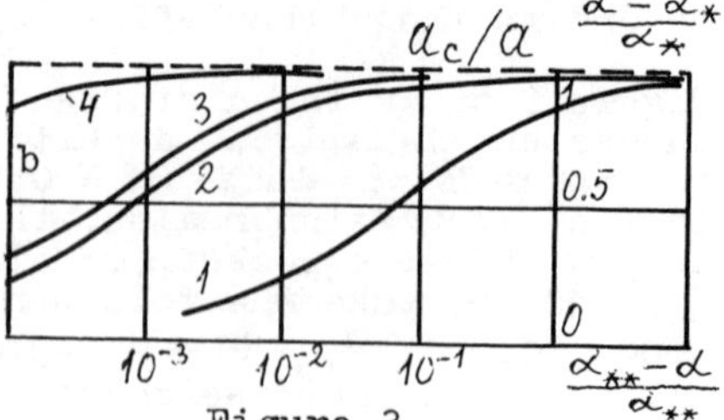

Figure 3

NOTATION

- a packet radius
- a_c cloud radius for a packet
- a_{cb} cloud radius for a bubble
- c_1 Karman constant
- d_f fluid density
- d_s solids density
- $\vec{f}$ interphase interaction force
- $\vec{g}$ gravitational acceleration
- J flow rate through the packet
- k constant in Darcy law
- p_f fluid phase pressure
- p_s solid phase pressure
- r radial coordinate
- U_d packet velocity
- $\vec{v}$ interstitial fluid velocity
- v_0 undisturbed interstitial fluid velocity
- $\vec{w}$ solid phase velocity
- $y = -r\cos\theta$ vertical coordinate
- $\alpha = \varepsilon' k'/\varepsilon k$
- $\delta = U_d/v_0$
- ε bed porosity
- θ angle in spherical coordinates
- μ_f fluid viscosity
- $\rho = 1-\varepsilon$ concentration of solid particles
- ψ_f stream function for fluid phase
- ψ_s stream function for solid phase

Superscript $'$ denotes the quantities within the packet.

LITERATURE CITED

1. Mickley, H.S. and Fairbanks, D.F., AIChE J., 1(3), 374 (1955).

2. Yerushalmi, J., Cancurt, N.T., Geldart, D. and Liss, B., AIChE Symp. Ser., 74 (176), 1 (1978).

3. Rakhmatulin, Kh.A., Prikl. Mat. Mech. (Moscow) 20 (2), 184 (1956).

4. Davidson, J.F. and Harrison, D., Fluidized Particles, Cambridge University Press, Cambridge (1963).

5. Jackson, R., Chapter 3, in Fluidization, Davidson, J.F. and Harrison, D. (Ed.), Academic Press, London (1971).

6. Borodulya, V.A. and Gupalo, Yu.P., Mathematical Models of Fluid-Bed Chemical Reactors [in Russian]. Nauka i Teknika, Minsk (1976).

7. Milne-Thompson, L.M., Theoretical Hydrodynamics, 4th ed., Macmillan, London (1960).

HYDRODYNAMICS OF A RECIRCULATING FLUIDISED BED

O. Ta'eed, M. Mohamed and B.M. Gibbs
Dept. of Fuel & Energy, Leeds University, UK.

ABSTRACT

A novel tracking technique was employed to determine both particle vectors and stream bandwidth in a cold 2-D fluidised bed. Controlled recirculation was induced by 2 techniques, draft tubes and jets. Optimum conditions were found to be sensitive to bed and jet velocities, and the static bed height.

SCOPE

A large number of fluidised bed applications require the bulk transfer of fluidised solids. The fluidised bed offers a number of advantages over most methods of contacting, such as high rates of heat and mass transfer and solids mobility. A recirculating fluid bed (incorporating a centralised jet and draft tube) has additional advantages by controlling the recirculation of fluidised solids and improving the solids handling. In these systems, given process conditions, it is important to be able to predict the recirculation rates of the bed material and to ensure circulation is obtained throughout the bed. Tests carried out in two-dimensional cold fluidised beds have been found useful in identifying the critical design parameters.

The application of a recirculating fluidised bed with a draft tube was probably first described by Taskaev and Kozina (1956) (1). They utilised the bed for low temperature carbonisation of coals in a 15cm diameter column with a 2.5cm diameter draft tube. Later a 'seeded coal process' was developed by Curran et al. (1973) (2) employing the same concept to smear the 'liquid' raw coal undergoing the plastic transition onto the seed char and the recirculating char during low temperature pyrolysis. Recent work done by Mohamed and Gibbs (1985) (3) on gasification of biomass in a fluidised bed has shown that further improvements to the process can be obtained by using recirculation of the bed material to dry and partially burn the wet biomass in the downcomer prior to pyrolysis/gasification in the draft tube section. The main advantage being the reduction of capital cost, since previously this has been achieved using twin fluidised beds.

However, one of the largest concerns when using recirculating fluidised beds to commercialise many chemical processes is the scale-up. This is believed to be due to the absence of an experimentally verified hydrodynamic theory. Although Yang and Keairns (1978) (4) and LaNauze (1976) (5) have studied the hydrodynamics involved and presented models for the recirculation rates, their work has been restricted, by their tracking technique, to the measurement of bulk flow rather than individual particles. They only considered moving solids (of unknown bandwidth) in the downcomer section. The present study utilised a novel method by Ta'eed et. al. (6) to investigate the individual particle movement within the whole bed as well as the bulk flow to enable overall recirculation rates and patterns to be evaluated effectively.

In the present study three sets of data have been collected regarding the tracking of particles to determine circulation rates. The first set of data concerns the circulation rates and flow patterns in an empty bed (i.e. no obstacles) of sand. The second set of data concerns the circulation rates and flow patterns incorporating a draft tube and a centralised jet, (Figure 1(a)). Some of this data has been discussed elsewhere (7). This paper compares the draft tube data with the third set of data collected on the circulation rates and flow patterns employing a novel method of inducing recirculation by using two jets, pushing air into the bed horizontally and vertically, as shown in Figure 1(b). Measurements were taken to explain the effect of the following parameters on the hydrodynamics of the recirculating fluidised bed :

(i) Twin jet mass flow rate of 66, 80 and 97 litres/min, (11, 13.33 and 16.17 x 10^{-4} m^3/s).
(ii) Draft jet flow rate of 50 l/min (8.33 x 10^{-4} m^3/s).
(iii) Fluidising velocity through the bed of 16.7, 20.82 and 27.5 cm/s.
(iv) Bed height for draft tube of 30, 35, and 40 cm.

Only particle vector measurements are presented in this paper, but other results were also gathered (bubble vectors, dispersion, etc.). Since the two dimensional bed determines the relative velocities (not absolute) in the bed, the results are comparative.

CONCLUSIONS & SIGNIFICANCE

It is clear that the results of previous modelling on recirculating fluidised beds have only been as good as the particle tracking techniques which they employed. It has been shown that:

1. Induced circulation rates improved towards a limit with increasing bed velocity whilst using draft tubes. Similar trends were observed with jets, but random and narrow bandwidth circulation were obtained for high and low bed velocities respectively.
2. The particle bulk velocity (utilised in previous studies) in itself was found to be a poor indication by which to judge the quality of circulation, since it fails to adequately portray the extent (bandwidth) of circulation.
3. Circulation patterns were seen to be very sensitive to the bed velocity, a variation of 0.67 to 1.1 U_B/U_{mf} produced significant circulation differences.
4. Optimum static bed height for good circulation was found to be at the height of the draft tube.
5. There was an optimum jet flow rate for which there was sufficient particle entrainment whilst limiting the distortions caused to the circulation patterns by jet penetration.

A two-dimensional glass bed of size 0.5m x 1.0m x 0.008m with a high density polythene distributor plate was employed (Figure 2). For one set of data the draft tubes were placed on either side of the centralised jet at equal distances of 0.04m, as shown in Figure 3(a).

Since the bed anatomy is axially symmetrical, only 2 jets were employed to simulate the second circulation method (Figure 3(b)). The dimensions of the draft tubes/jets were designed to allow for circulation across the tubes where the particle motion was induced whilst supporting the stationary parts of the bed. This enabled the optimum position and size of the tube to be located with respect to bed velocity. Bed material used was sand (average diameter 0.414mm, sphericity 0.68, density 2.452 Kg/m^3, U_{mf} = 25 cm/s) with a static bed height of 30cm. The bed was surrounded by four 60cm fluorescent strip light fittings with reflectors, using 365.5nm blacklight blue tubes. A small proportion of the sand was dyed with a fluorescent sulphur based dye which adheres to the sand well without altering the sphericity, density and size significantly. Under the UV light, the particles provide a sharp contrast between the 'black' sand and the fluorescing particles. A camera with a 1 second open shutter and UV filter is placed behind a rotating chopper. The chopper speed is measured accurately with a photo-electric diode connected to a timer, this enables the velocity of the particles to be evaluated. The developed negatives are projected down from an enlarger onto a 50cm bit pad, and by digitizing the control points, the tracks may be recorded and processed both numerically and in graphical 'picture' format.

This technique eliminates the errors encountered using large tracer particles, as in the case of 'radio pills'. It also eliminates the errors arising from any intrusive obstacles which impede the flow, such as aerials which would need to be inserted if the radio pill method was adopted. Using this technique, Mohamed & Ta'eed (1985) (6) have shown previously that a large tracer does not necessarily follow the path of the bed material accurately. Accuracy measurements have shown that the technique is susceptible to the usual photographic constraints such as parallax and misalignment of the camera and enlarger to the bed and digitizer respectively. These produce recording errors of particle location of 1·2% maximum, depending on whether the position of the particle is at the extremes of the camera optical field or whether it lies central to the optical axis of the lens. The chopper motor had a uniform rotation speed within 1·1.4% error margin, but the speed was always time-averaged. Overall, the velocity data is accurate to within a 4·5% maximum limit.

RESULTS

The importance of a recirculating fluidised bed is the controlled induced recirculation within the bed. It is therefore essential to optimise the recirculation rates in the required directions, while eliminating them in the other directions. The increasing bandwidth of recirculation is also of importance, since this enables the maximum recirculation rates, by the positioning of jets and draft tubes, to be evaluated.

Draft Tube

Figures 4 to 6 represent the effect of increasing the bed velocity on particle velocity vector patterns, Table 1 shows the numerical data gathered. At low U_B/U_{mf} (0.668), the downward velocity is high but the stream bandwidth is narrow. Coupled with this is the fact that the circulation is contained within the top half of the bed, leaving most of the bed static. Increasing U_B/U_{mf} (0.833), lower bulk velocity is obtained but a much wider bandwidth is gained with an associated circulation throughout the entirety of the bed. At high U_B/U_{mf} (1.1), further improvements are observed. It should be noted, however, that further work in this paper (Figure 13) shows that a high U_B/U_{mf} can lead to random circulation and distort the required flow patterns. Figures 5, 7 and 8 can be compared for the effect of bed height at a fixed U_B. All possess similar bandwidths, but the mean downward velocity is significantly higher when the bed height is approximately equal to the height of the draft tubes (35 cm). In addition, a greater percentage of particles were travelling downwards at this bed height (96.7% of all recorded particles).

Jets

Figures 9 to 11 and Table 2 show the effect of decreasing jet flow rate whilst keeping U_B constant (with associated three dimensional velocity profiles). Downward (left) velocity increases with increasing V, but the percentage of downward-left moving particles is significantly higher (60% as opposed to 40%) at medium jet flow rate (80 l/min). Coupled with this is the decrease in percentage of up-right flowing particles at that flow rate. The reason for the poor performance of the extremes of jet velocities is that at high V, the particles can be seen to be repelled from the mouth of the jets due to the jet penetration, whilst at low flow rate there is insufficient flow rate to entrain the particles. Figures 11, 12 and 13 represent increasing U_B with a constant V. Figure 12 shows that whilst bulk particle down-left velocity is relatively high, the bandwidth is severely limited. Figure 13 shows that the higher particle velocities can be obtained with high bed velocity, but that this leads to a distortion of the recirculation pattern (there is only 23.5 % of particles travelling in the required direction). In contrast, Figure 11 reveals low particle velocities, but the entire bed maintains the desired circulating pattern. In conclusion, it can be seen from Figure 14 that the highest down velocity is not a true indication of good circulation since the conclusions drawn from that figure would not have taken adequate account of the essential stream bandwidth of downward flow and the extent to which the entire bed was being utilised.

NOTATION

H	Static Bed Height.
U	Superficial Gas Velocity.
V	Jet Volumetric Flow Rate.

Subscripts

B	Bed.
H	Horizontal Jet.
J	Draft Tube Jet.
mf	Minimum Fluidisation.
V	Vertical Jet.

LITERATURE CITED

1. Taskaev, N.D. and Kozina, M.I., Trudy Akad, Nauk Kirgiz S.S.R., 7, p.105, (1965).

2. Curran, G.P., Pasek, B., Pell, M. and Gorin, E., "Pretreatment of bituminous coals for pressure gasification", presented at the Fluidised Bed Combustion Symposium, ACS, Chicago, (1973).

3. Mohamed, M. and Gibbs, B.M., Proc. Annnual Research Conf., IChemE, London, (April 1985).

4. Yang, W.C. and Keairns, D.L., AIChE Symposium Series, 176(74), p.218, (1978).

5. LaNauze, R.D., Powder Technology, 15, p.117, (1976).

6. Ta´eed, O., Mann, N. and Gibbs, B.M., Proc. 11th Annual Research Conference, IChemE, Bath, pp. 217-222a, (April 1984).

7. Mohamed, M., Ta´eed, O. and Gibbs, B.M., Proc. 1st Conference on Circulating Fluidised Beds, Canada, (November 1985).

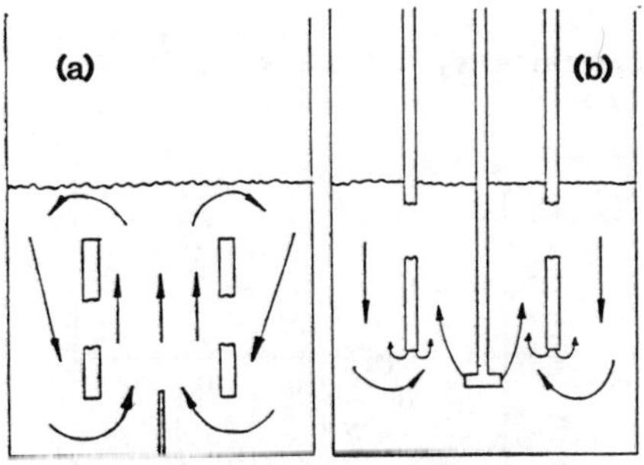

Figure 1 : Circulation Patterns

(a) Draft Tubes (b) Jets

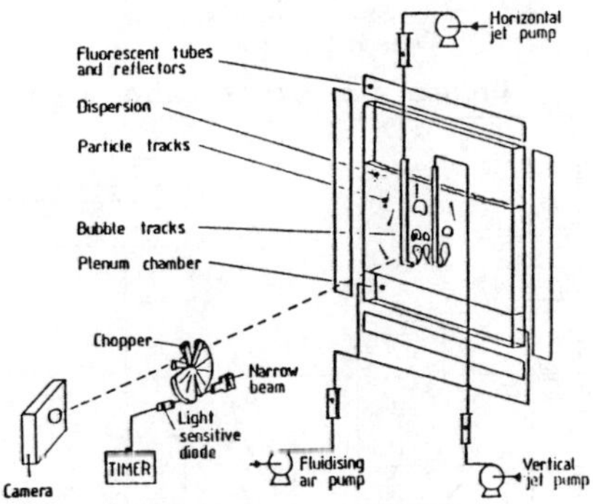

Figure 2 : Tracking Technique

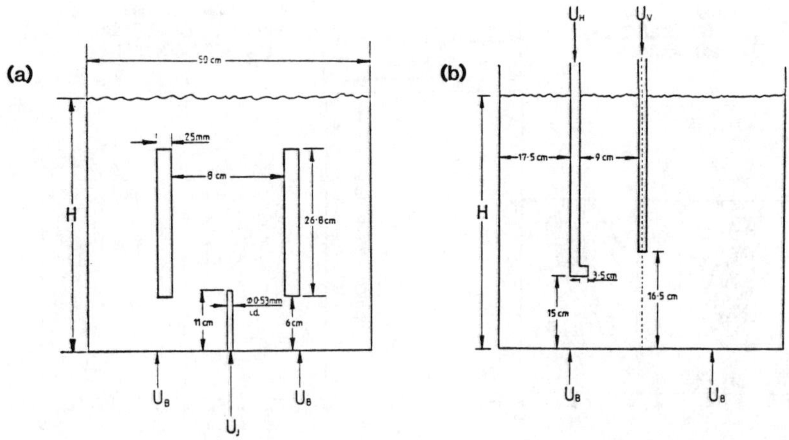

Figure 3 : Experimental Dimensions

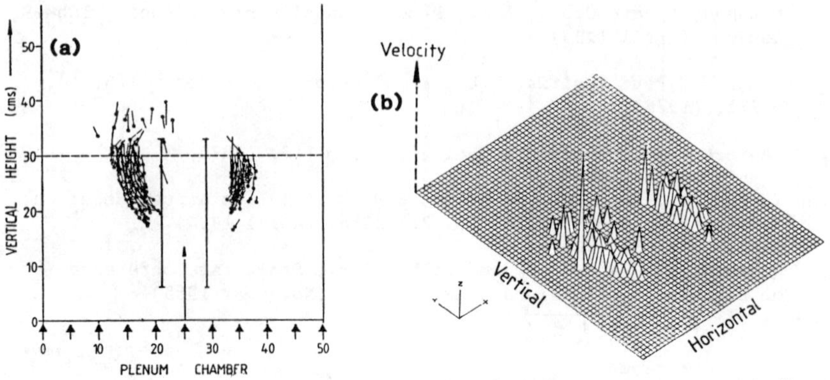

Figure 4 : Draft Tube U_B – 16.70cm/s, V_J – 8.33 x $10^{-4} m^3/s$

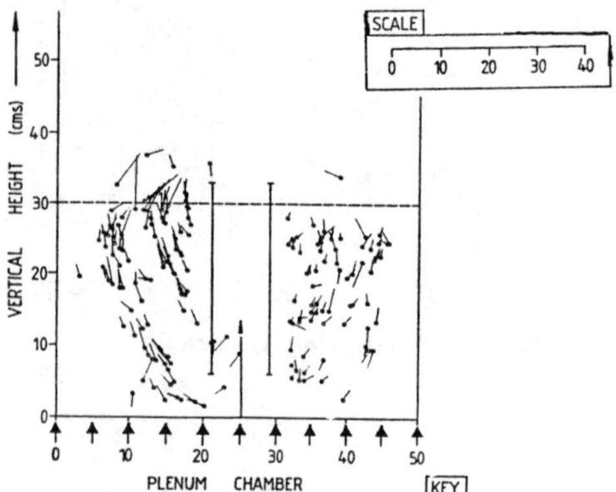

Figure 5 : Draft Tube
U_B – 20.83cm/s

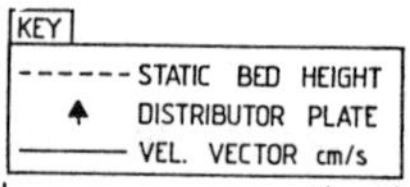

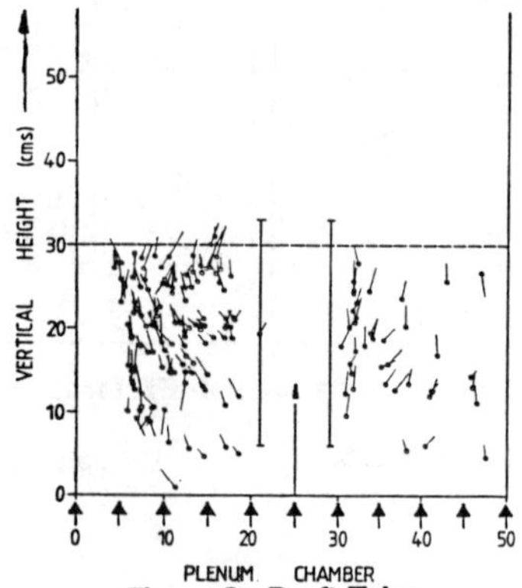

Figure 6 : Draft Tube
U_B — 27.50cm/s

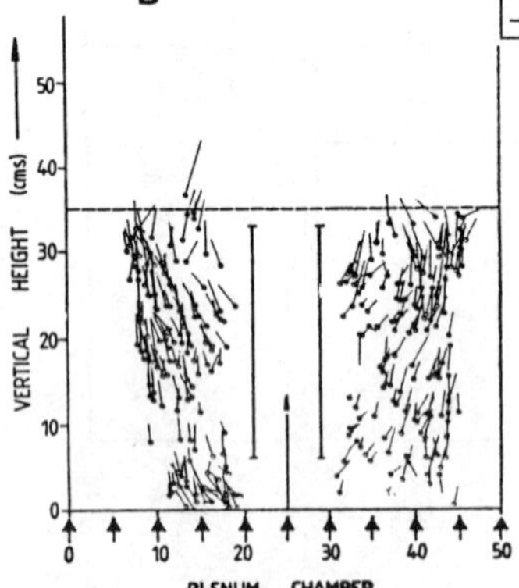

Figure 7 : Draft Tube, H – 35cm
U_B – 20.83cm/s

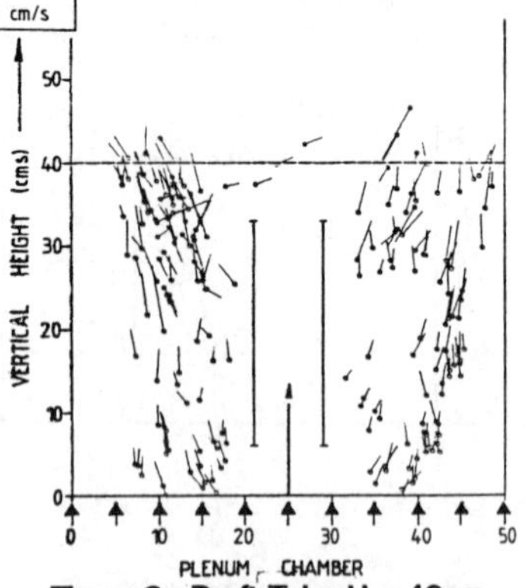

Figure 8 : Draft Tube H – 40cm ;
U_B – 20.83cm/s

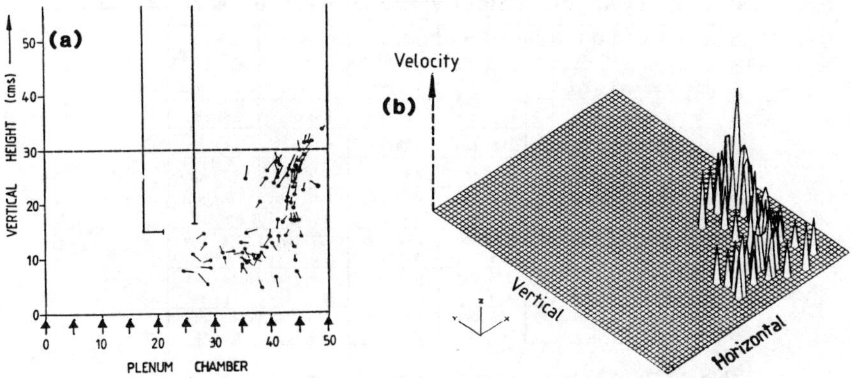

Figure 9 : Jets U_B - 20.83cm/s, V - 16.19 x $10^{-4} m^3/s$

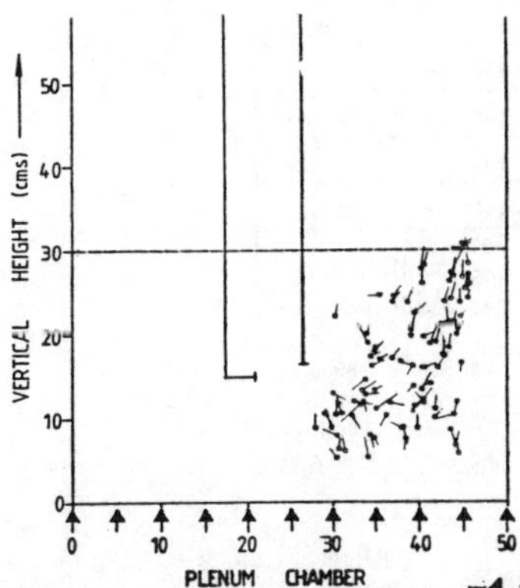

Figure 10 : Jets V - 13.33 x $10^{-4} m^3/s$
U_B- 20.83cm/s

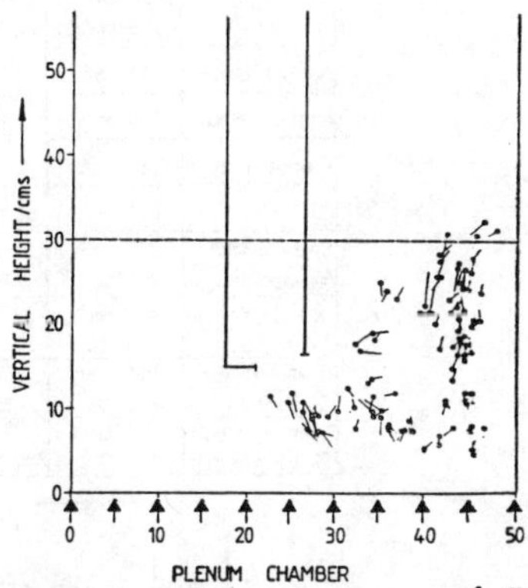

Figure 11 : Jets V - 11.0 x $10^{-4} m^3/s$
U_B- 20.83cm/s

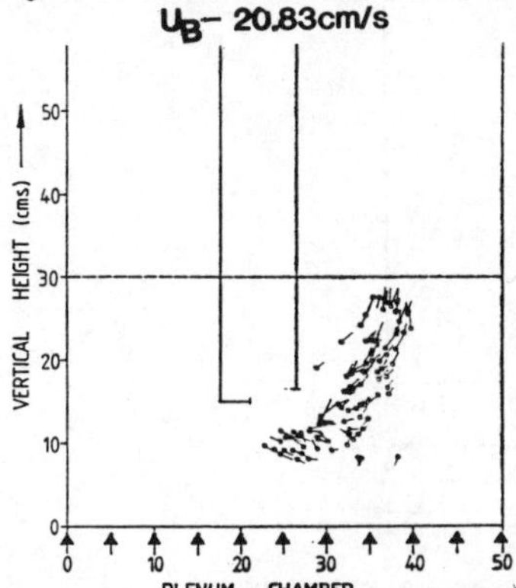

Figure 12 : Jets V - 11.0 x $10^{-4} m^3/s$,
U_B- 16.70cm/s

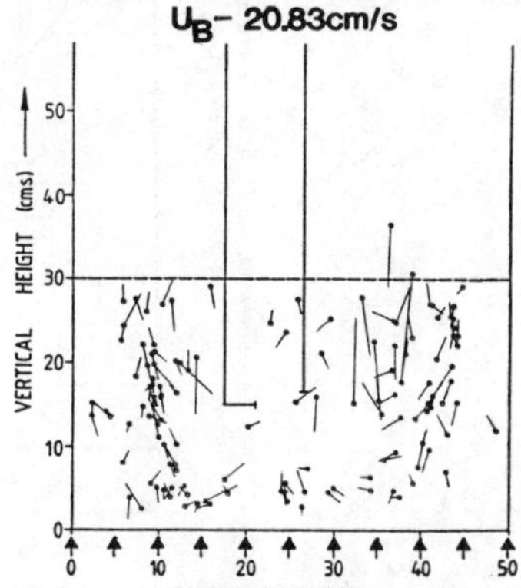

Figure 13 : Jets V - 11.0 x $10^{-4} m^3/s$
U_B- 27.50cm/s

Table 1 : Particle Velocity For Draft Tube Circulation

	H = 30 cm U_B = 16·70 cm/s	H = 30 cm U_B = 20·83 cm/s	H = 35 cm U_B = 20·83 cm/s	H = 40 cm U_B = 20·83 cm/s	H = 30 cm U_B = 27·50 cm/s
Max.(av.) velocity	4·94 cm/s	4·07 cm/s	5·33 cm/s	4·97 cm/s	4·35 cm/s
Min. (av.) velocity	4·70 cm/s	3·55 cm/s	4·62 cm/s	4·46 cm/s	3·84 cm/s
Average velocity	4·82 cm/s	3·81 cm/s	4·98 cm/s	4·73 cm/s	4·09 cm/s
S.D. (av.)	± 0·12 cm/s	± 0·26 cm/s	± 0·35 cm/s	± 0·24 cm/s	± 0·25 cm/s
Up-right (av.) % particles	15·28 cm/s 1·2 %	4·27 cm/s 7·2 %	5·40 cm/s 2·3 %	6·35 cm/s 3·5 %	4·00 cm/s 6·2 %
Up-left (av.) % particles	4·33 cm/s 2·0 %	4·08 cm/s 2·8 %	2·69 cm/s 1·0 %	4·27 cm/s 9·0 %	4·83 cm/s 2·5 %
Down-left (av.) % particles	4·74 cm/s 42·1 %	3·76 cm/s 43·9 %	4·70 cm/s 46·7 %	4·58 cm/s 38·5 %	4·53 cm/s 33·8 %
Down-right (av.) % particles	4·67 cm/s 54·7 %	3·76 cm/s 46·1 %	5·26 cm/s 50·0 %	4·83 cm/s 49·0 %	3·81 cm/s 57·5 %

Table 2 : Particle Velocity For Jet System Circulation

	U_B = 16·70 cm/s V_H = 5·07×10^{-4} m^3/s V_V = 5·93×10^{-4} m^3/s	U_B = 20·83 cm/s V_H = 5·07×10^{-4} m^3/s V_V = 5·93×10^{-4} m^3/s	U_B = 20·83 cm/s V_H = 6·14×10^{-4} m^3/s V_V = 7·19×10^{-4} m^3/s	U_B = 20·83 cm/s V_H = 7·45×10^{-4} m^3/s V_V = 8·72×10^{-4} m^3/s	U_B = 27·50 cm/s V_H = 5·07×10^{-4} m^3/s V_V = 5·93×10^{-4} m^3/s
Max.(av.) velocity	2·76 cm/s	2·56 cm/s	2·75 cm/s	3·39 cm/s	4·69 cm/s
Min. (av.) velocity	2·30 cm/s	2·17 cm/s	2·25 cm/s	2·49 cm/s	4·29 cm/s
Average velocity	2·53 cm/s	2·36 cm/s	2·50 cm/s	2·91 cm/s	4·47 cm/s
S.D. (av.)	± 0·21 cm/s	± 0·19 cm/s	± 0·24 cm/s	± 0·42 cm/s	± 0·22 cm/s
Up-right (av.) % particles	2·25 cm/s 9·0 %	2·33 cm/s 22·0 %	2·54 cm/s 9·0 %	2·97 cm/s 22·5 %	4·70 cm/s 24·2 %
Up-left (av.) % particles	2·32 cm/s 19·0 %	2·34 cm/s 21·0 %	2·54 cm/s 13·0 %	2·52 cm/s 13·8 %	4·49 cm/s 27·3 %
Down-left (av.) % particles	2·65 cm/s 65·0 %	2·37 cm/s 40·0 %	2·46 cm/s 60·0 %	3·29 cm/s 40·0 %	4·85 cm/s 23·5 %
Down-right (av.) % particles	2·35 cm/s 7·0 %	2·38 cm/s 17·0 %	2·57 cm/s 18·0 %	2·46 cm/s 23·7 %	3·86 cm/s 25·0 %

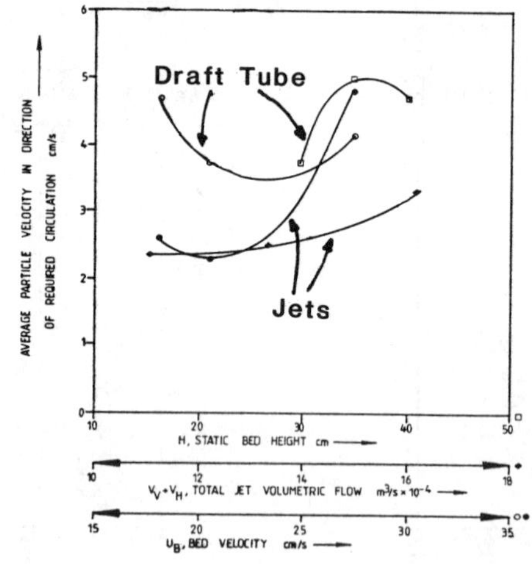

Figure 14 : A Comparison Of Draft Tube & Jet Circulation Methods

EFFECT OF A DRAUGHT TUBE ON FUEL GRAIN CIRCULATION IN A FLUID BED COMBUSTOR

H.A. MASSON

INIEX
Rue du Chéra 200
B - 4000 Liège (Belgium)

ABSTRACT

The effect of a vertical draught tube on the fuel grains behaviour has been investigated by a radioactive tracing technique. It is shown that the profiles of up- and downward velocity, the ones of presence probability and of frequency of passage become smoother than in a normal fluid bed.

SCOPE

In a fluid bed combustor, the solid consists mainly of ash and a few percent of fuel grains. It has been shown previously (1) that segregation (or non-mixing) may happen if the densities and the sizes of the fuel grains and the ash are too different. Gluing a tiny Co^{60} source on a fuel grain enables the operator to track the displacements of this tracer with a γ detector and appropriate electronics (2). Shallow fluidized beds ($H \leqslant 20$ cm) may generally be regarded as perfect mixers for the solid phase. Some applications require deeper beds to achieve a longer residence time of the fuel in the bed (e.g. gasification).
To organise the circulation patterns of the fuel grains, it has been suggested to put a vertical cylindrical tube (draught tube) in the deep beds (3,4) in order to smooth circulation properties over the bed height. The search for an optimum geometry and position of this draught tube in the bed is an important step in the design of such systems.
The objective of the present study is to use a radioactive tracing technique to investigate the effect of the draught tube position and diameter on :

- the fraction of time spent by the tracer at different levels in the bed (presence probability) ;
- the number of passages of the tracer at a given level (frequency of passage) ;
- the up- and downward velocities of the marked grain ;
- the probability for the tracer to change from an up- to a downward velocity ;
- the random diffusion like character of its trajectories.

CONCLUSIONS

It has been shown that the presence of a draught tube in a deep fluid bed equipped with a pyramidal distributor makes the presence probability,

the frequency and the velocity profiles smoother, while the probability for a direction change remains unaffected.
The optimum geometry of the draught tube is the one where its bottom reaches the upper extremity of the pyramidal distributor while its top is at the same level than the static bed height.
Changing the diameter of the draught tube acts only significantly on the downward velocity, which increases with the diameter of the draught tube.

EXPERIMENTAL FACILITIES

The cold model consists of a 180 x 180 mm perspex column, one metre in height, equipped with a pyramidal distributor having a slope of 45°. The bed is filled with solid up to 500 mm. Air is supplied by a Roots compressor and metered with a rotameter.
To follow solid circulation, a very small Co^{60} source (γ ray emitter of 300 μcuries and 300 μ in size) is glued on a fuel element (900 μ). The volume and the weight contribution of the tracer are neglectable. The emitted γ rays are detected by two photomultipliers equipped with NaI crystals and connected to fast ratemeters (τ = 10 ms). Two overlapping horizontal measurement windows defined by lead block collimators divide the bed in five zones and allow to identify eight different transitions corresponding to the passage of the tracer from one zone to another. They are labelled from A to H. A digital code is associated to each transition, as shown on Fig. 1. The general scheme of the installation and its electronic chain were described previously (2).
The identification code is generated by combining the state signal of one channel and the signal of a slope detector relative to the other channel. This slope detector is just a differentiator of the corresponding state signal. The transition code is sent to an one-line Apple IIe microcomputer with video display and printing facility.
For each experimental condition, 6000 transitions are recorded as well as the time interval between successive transitions (time resolution : 0.1 ms).
Subsequent retrieval of the data permits the operator to build up a frequency table for the individual transitions and also for some remarkable chains of transitions (doublets and triplets).
For a given pair of transitions, the time intervals between them may be printed in a chronological order. These times may be later condensed in a histogram. From there it is possible to compute the :

1) *Time averaged vertical velocities*
The upward velocity is given by

$$U_u = \frac{L}{\overline{(B.\ C)}}$$

where L is the width of the measurement window and where $\overline{(B.\ C)}$ represents the mean of the time interval distribution between a B transition followed directly by a C one. Symmetrically :

$$U_D = \frac{L}{\overline{(E.\ F)}}$$

represents the downward velocity.

2) In parallel, it is also possible to count the cumulative presence time of the tracer in a measurement window. This presence time divided

by the total time for a run gives the *presence probability* of the particle at the measurement level.
In absence of any kind of segregation, this presence probability equals the ratio of the window volume over the total bed volume. The presence probability of the tracer divided by the volume fraction occupied by the window may be used as a local mixing index.

3) *Mass flow of fuel through an horizontal plane*
If the tracer does not segregate, it passes as often as any other grains through a given horizontal plane. Necessarily, an ascending passage must be followed by a descending one.
The mass flow of fuel M' is thus given by :

$$M' = \frac{N}{2} \cdot M_{fuel}$$

In this expression M_{fuel} is the charge of fuel in the bed and N is the number of passages per time unit of the tracer at the measurement level.

4) Another parameter needed to describe the circulation of a large particle in a fluid bed is the *probability for direction change* once this particle is engaged in an upward or a downward motion. Especially, it is important to know if the tracer "has a memory" of its previous history or if its behaviour may be described by its present state and a transition probability associated to this state. If the second proposition is correct, the process is a markovian one, and may be described following a purely diffusive mechanism (2).
For a large number of observed transitions AB/A is an unbiased estimator of the conditional probability to continue upwards once the tracer is engaged in an ascending displacement. If the system has no memory, AB/A must be equal to ABC/AB or to ABCD/ABC.
A symmetrical reasoning applied to downward displacements leads to :

$$\frac{GH}{H} = \frac{FGH}{GH} = \frac{EFGH}{FGH}$$

It is noticeable that A equals H and AB equals GH, from continuity consideration, so that, in any case :

$$\frac{AB}{A} = \frac{GH}{H}$$

The "memory effect" may be tested by computing BC/B x BCD/ABCD and EF/E x EFG/EFGH , which equals 1 if the process is markovian.

CHARACTERISTICS OF THE SOLIDS

Sand and anthracite fines were used in the present study.
The density of sand was found, by pycnometry to be equal to 2.65 10^{+3} kg/m^3 ; for anthracite, we have found 1.4 kg/m^3.
Pressure drop - gas velocity measurements were performed for a bed of sand and for the same bed where 10 % of fines were added.
As was expected from the literature (1), the addition of 10 % of anthracite affects only slightly the value of U_{mf}. However in this case, a part of the bed remains unfluidized up to more than 2 U_{mf}.

EXPERIMENTAL CONDITIONS

All the experiments were conducted with a mixture of 90 % sand and 10 %

anthracite.
The sand had been sieved and the size fraction 50 μm - 200 μm retained ($\overline{dp}$ = 100 μm). Anthracite grains are all under 2200 μm and their average diameter is of about 1000 μm.
The minimum fluidization velocity of the mixture is 30 mm/s. However, the whole charge is fluidized only if the gas velocity becomes higher than 70 mm/s.
The geometric parameters characterizing the draught tube are defined on Fig. 2. Table 1 summarizes the experimental conditions used. The static bed height Ho has been kept constant and equal to 500 mm.

Table 1. Identification of the experimental runs.

Draught tube			Superficial gas velocity u = 110	180	260	430 mm/s
ø (mm)	H (mm)	Δ (mm)		Basic design		
100	500	0	1	2	3	4
100	400	0		6		
100	400	100		11		
100	500	100		15		
0	0	0		23 (reference case)		
50	500	0		26		
160	500	0		29		

EXPERIMENTAL RESULTS

1) Effect of the draught tube in standard configuration (case 2, 23 ; Fig. 3 to 6).

To achieve good mixing in a fluid bed it is generally accepted that the bed ratio (height/diameter) must be equal to one. For deeper beds, several superposed 'mixing cells' may appear. As a result the vertical profiles of physical quantities describing solid circulation would present local sharp fluctuations. This may be seen for the up- and downward velocity, as well as for the solid mass flow on Fig. 3 and 5. These profiles become much smoother if a draught tube is immerged in the bed, without, on the average, any significant velocity modification but a reduction of the mass flow by a factor of about 1.5. Mass flows are measured over a single line ; velocity measurements are integrated over the height of a window. This result may thus indicate a reduction in the frequency of small scale loops, while long range systematic displacements remain unaffected.
On the other hand, the presence probability profiles are flat in both the cases and show no effect of the presence of the draught tube.

The probability for direction change remains also flat, without any effect of the presence of the draught tube.
From these figures it was decided to further focuse only on velocity and mass flow profiles to study eventual effects of the operating conditions or of the geometry of the draught tube on solid circulation.

2) Effect of draught tube geometry (Fig. 7, 8, 9 ; case 2, 6, 11, 15, 26, 29).

Most of the geometries lead to the same flat profile of solid mass flow ; a slight reduction in mass flow may be identified in case 6, which corresponds to the situation where the draught tube upper extremity is deeply submerged below the surface of the layer. The same situation leads to a reduction in up- and downward velocities.
On the other hand, when the draught tube is emerging from the bed, the upward velocity (in the central part of the system) is slightly increased against the other cases.
Finally, if the draught tube diameter increases, the downward velocity (in the reduced annular section) increases but shows some reduction in the lower part of the bed, where the more difficult passage must be realized between up- and downward part of the system.
There is thus no net advantage to choose another geometry than the basic one. Any deviation from it leads to increased friction at the bottom of the tube or to increased solid velocity ; both situations lead to erosion and eventually to erratic behaviour in the solid flux.

3) Effect of gas velocity in the case of the standard configuration (Fig. 10, 11, 12 ; case 1, 2, 3, 4).

Although the general shape of the profiles remains the same, there is a general increase in the mass flow and the velocities when the gas velocity increases.
The upward velocities range between .25 and .9 m/s, while the downward ones range between .05 and .7 m/s.
This result may be qualitatively explained by the fact that increasing gas velocity increases bubble flow, according to the '2-phase theory of fluidisation'. At the same time an increased bubble flow promotes increased long range solid circulation at the scale of the whole bed by wake and cloud contribution to solid vertical transport.

LITERATURE CITED

(1) ROWE, P.N. et al., *Trans. Inst. Ch. Engrs*, 50, 324 (1972).

(2) MASSON, H.A. et al., *Proc. 4th Eng. Found. Conf. on Fluidization*, Kunii Daizo and Toei Ryozo Ed., Academic Press, N.Y. (1984), p. 185.

(3) YANG, W.C. and KEAIRNS, D.L., *AIChE Symp. Series*, 176 (74), 218 (1978).

(4) ESCUDERO, J.B. et al., *Proc. 4th Journées européennes sur la Fluidisation*, 336, Toulouse (1985).

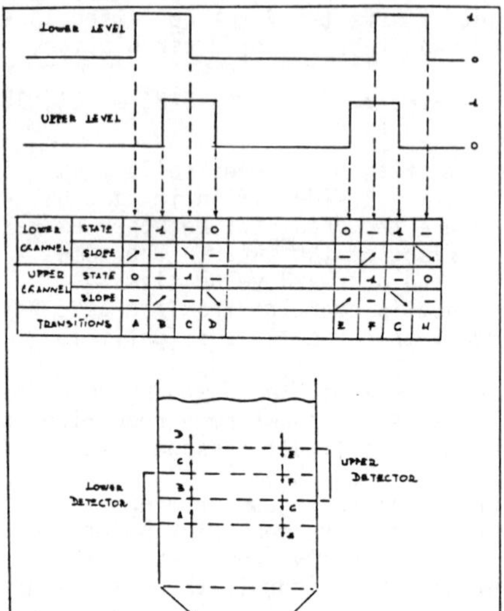

Figure 1 :
Transition identificator logic.

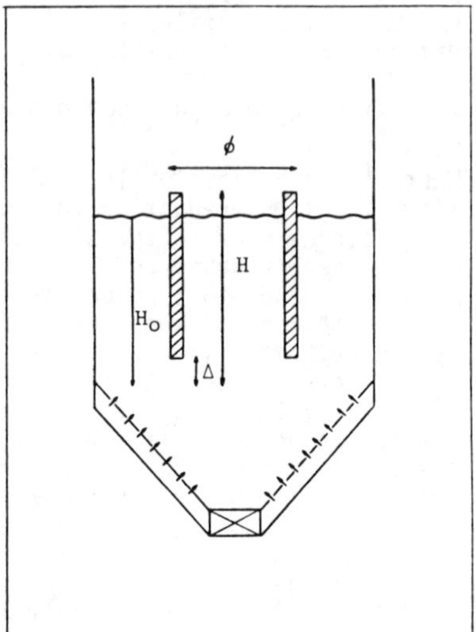

Figure 2 : Parameters identification.

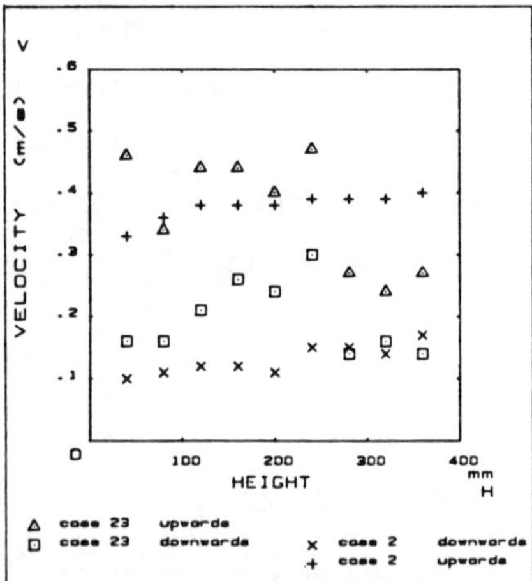

figure 3 : Effect of drought tube on solid velocity (upwards and downwards)

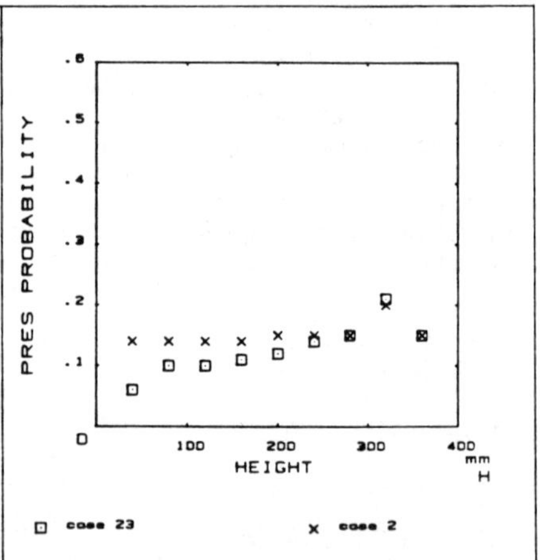

figure 4 : Effect of drought tube on presence probability

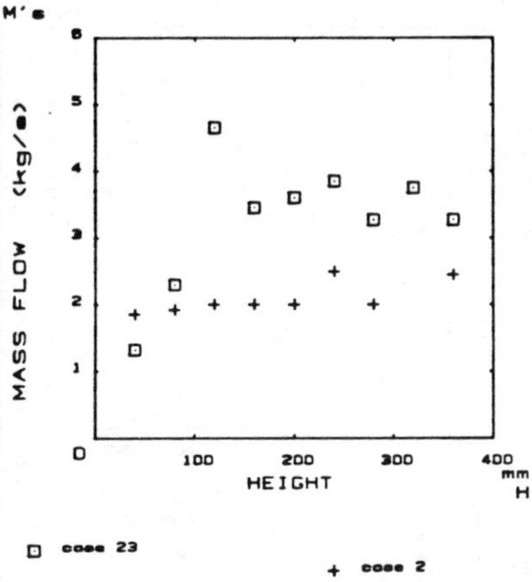

figure 5 : Effect of drought tube on solid mass flow

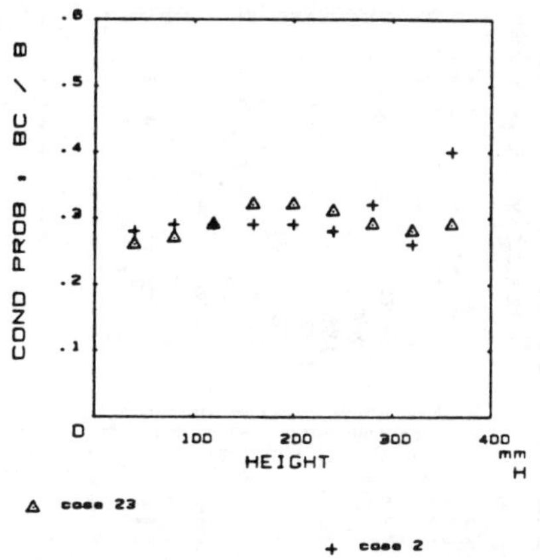

figure 6 : Effect of drought tube on conditional probability

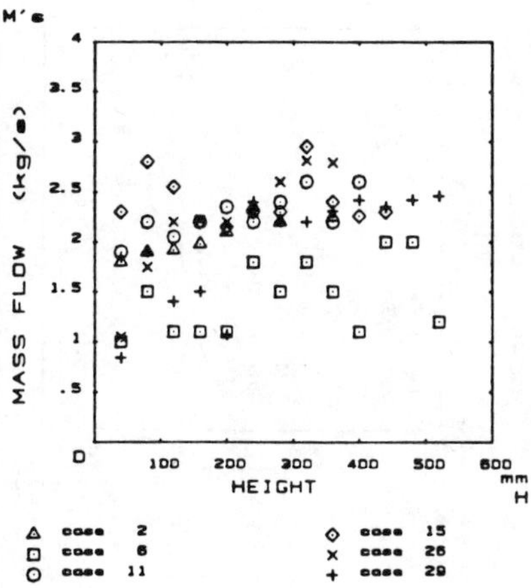

figure 7 : Sensibility of solid mass flow on drought tube geometry

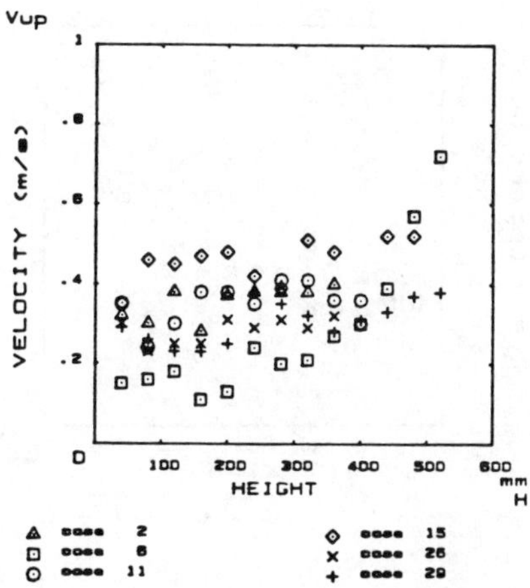

figure 8 : Sensibility of upward velocity on drought tube geometry

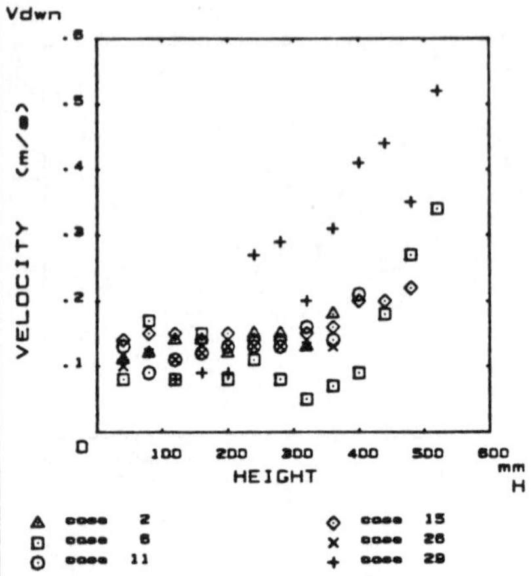

figure 9 : Sensibility of downward velocity on drought tube geometry

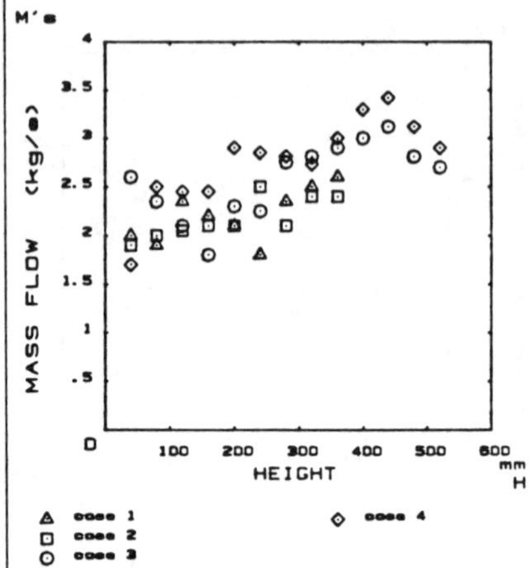

figure 10 : Effect of gas velocity on solid mass flow

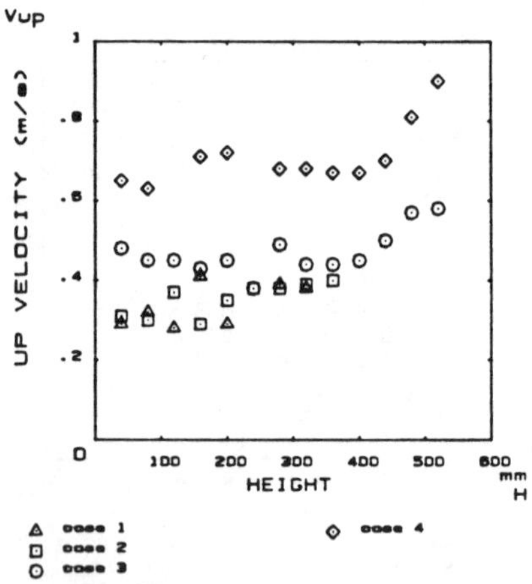

figure 11 : Effect of gas velocity on upward velocity

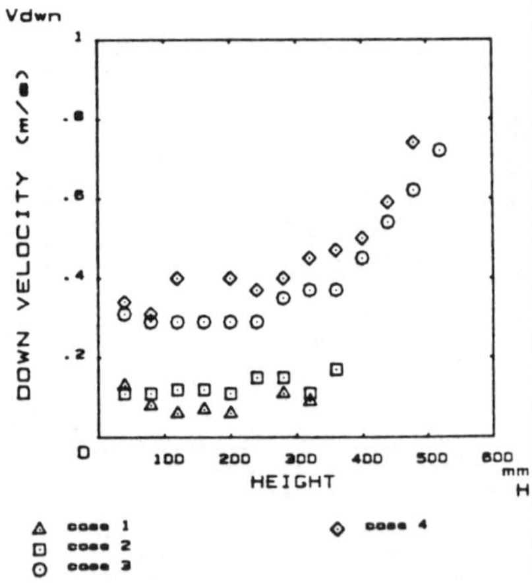

figure 12 : Effect of gas velocity on downward velocity

PREDICTION OF THE STEADY STATE SEGREGATION PATTERN IN GAS FLUIDISED BEDS WITH PARTICLES IN THROUGHFLOW

T. Chiba*, S. Chiba** and A.W. Nienow***

* Coal Research Institute, Hokkaido University, Japan.
** Government Industrial Laboratory, Hokkaido, MITI, Japan.
*** Department of Chemical Engineering, University of Birmingham, United Kingdom.

ABSTRACT

A solid mixing and segregation model for a batch gas-fluidised bed has been developed for the prediction of the steady state segregation pattern in beds of binary systems of solid particles in throughflow. The model predicts the unsteady state discharge composition and the segregation pattern reasonably well except at high jetsam concentrations.

SCOPE

There are already many industrial processes where gas fluidised beds of binary or higher-order systems are operated with solids flowing through them continuously. However, in spite of considerable work on batch mixing and segregation of multi-component particle system, little analytical work has been done on the steady state characteristics of segregating gas fluidised beds in throughflow. Hence, the present work aims to predict the change of discharge solid composition with time (the "break-through" curve) and the steady state solid concentration profile in the bed (the "segregation pattern"). This has been done by a simple model developed by combining our previous model for predicting the batch segregation pattern with typical experimental findings of continuous mixing and segregation.

CONCLUSIONS AND SIGNIFICANCE

When a gas fluidised bed of size- and density-different binary particles is operated with solids flowing through it continuously, the segregation pattern of the bed and the discharge composition vary with time depending on the initial bed composition and the height of the solid discharge. However, regardless of the initial bed composition, a steady state is reached when the composition at the discharge height becomes equal to the feed composition.

Based upon the above fact, the breakthrough curves can be predicted by a pseudo-steady state model which assumes an invariable segregation pattern for a short time and employs the Naimer-Chiba-Nienow model for predicting the batch segregation pattern. The model predictions agree well with the observed break-through curves, particularly for lower discharge heights.

INTRODUCTION

As shown in Figure 1, particle movement in a gas fluidised bed is solely caused by the bubbles. Each bubble carries particles in its wake and lifts them towards the bed surface. The rising bubble also causes drift of the surrounding particles. These upward motions are counterbalanced with downward motion in the rest of the bed, resulting in the circulation over the whole bed. While rising with the bubble, the wake exchanges a part of its particles with the bulk through continuous gathering and intermittent shedding. When the bed consists of two component particles, one, the flotsam, tending to rise and another, the jetsam, tending to sink, segregation is caused by the preferential descent of jetsam particles through the temporarily disturbed region behind each bubble. These mechanisms give rise to the typical segregation pattern also shown in Figure 1.

Gibilaro and Rowe (1) first formulated a mathematical model reflecting conceptually the above mixing and segregation mechanisms. Then, the model parameters for individual rate processes such as wake exchange and jetsam descent were examined and expressed as a function of bubble characteristics and particle properties (Chiba, et al. (2) and Tanimoto, et al. (3)). Based on these studies and correlations in the literature between bubbling characteristics and operating variables, a model for predicting the segregation pattern in a batch bed without any adjustable parameters was developed which showed a good agreement between predicted and observed patterns in low jetsam systems (Naimer, et al. (4)). The relevant equations are shown in Table 1.

Nienow and Naimer (5) conducted an experimental study to relate batch experiments to the behaviour of beds with solids in throughflow. Figure 2 shows their essential findings for continuous operation. During the unsteady state period, the outlet composition is a function of the initial bulk bed composition. However, a steady state is reached

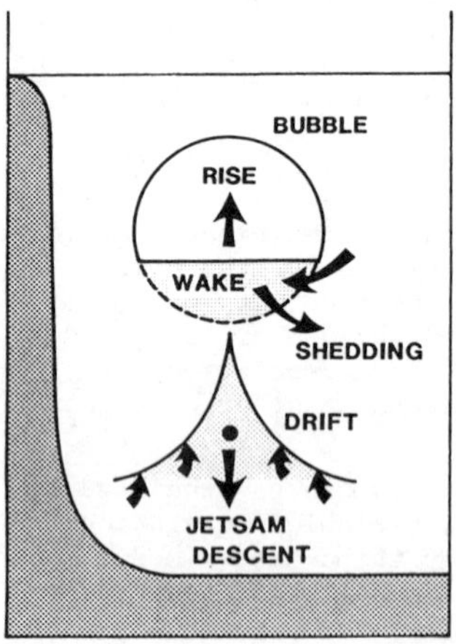

Figure 1. Mechanism of solid mixing and segregation and the typical segregation pattern

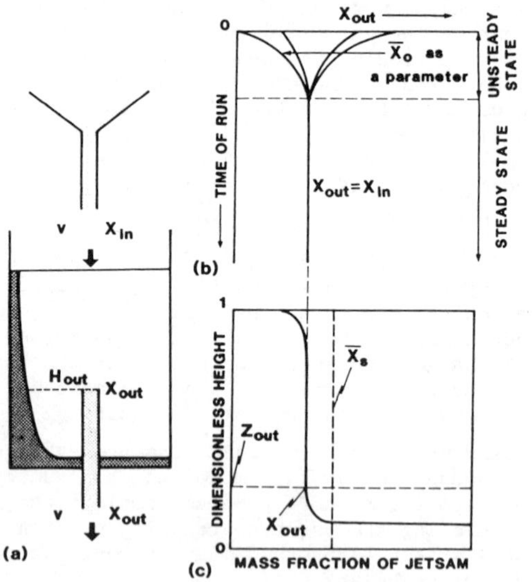

Figure 2. Break-through curve (b) and steady state segregation pattern (c) in continuous operation (a).

independently of the initial composition when the outlet composition becomes equal to the inlet composition. All these results are described in detail in a recent review article, (Nienow and Chiba, (6)).

In the present work, these theoretical and experimental results are utilised to develop a model for the prediction of the behaviour of beds with solids in throughflow.

MATHEMATICAL MODELLING

The segregation pattern in a batch bed depends not only on the gas velocity but also on the bulk bed composition. If, in a continuous operation, the outlet composition differs from the inlet, the bulk bed composition varies with time resulting in the corresponding change of the segregation pattern. Here, such changes are described by a model which assumes a pseudo-steady state for the segregation pattern.

If change of the bulk bed composition is assumed to be negligibly small for a short time, Δt, prediction of the segregation pattern can be done by models for batch mixing and segregation. In the present case, the Naimer-Chiba-Nienow model (4) is employed because of its ability to predict the segregation pattern directly from given operating variables and particle properties. As shown briefly in Table 1, the model is based on the Gibilaro and Rowe model and is applicable to cases where

Table 1. Basic equations for the prediction of a batch segregation pattern

Case 2 of the Gibilaro and Rowe model :

$$(C_B/C_{Bo})^{1+\lambda}\{(1-C_B)/(1-C_{Bo})\}^{1-\lambda} = e^{-\gamma Z} \quad ;\ \text{for the bulk phase}$$

$$C_W = C_B + C_B(1-C_B)/\lambda \quad ;\ \text{for the wake phase}$$

where

$$Z = h/H\ , \quad \lambda = w/k \quad \text{and} \quad \gamma = qH/w$$

Correlations for the model parameters :

$$w = u_B f_B A_B \quad ;\ \text{circulation rate}$$

$$k = 0.45(\rho_J/\rho_F)(d_J/d_F)^{1/3}\, u_B A_B \quad ;\ \text{segregation rate}$$

$$q = 1.5 f_W u_{mf} A_B/(d_B \varepsilon_{mf}) \quad ;\ \text{exchange rate}$$

where

$$A_B = a_B'/(1-a_B') \quad \text{and} \quad a_B' = d_B f_B/(1.5u_B)$$

Bubbling Characteristics :

$$d_{Bo} = (6G/\pi)^{2/5}/g^{1/5} \quad ;\ \text{initial bubble diameter}$$

$$d_B = d_{Bo} + 0.00014(\overline{d\rho})(u/u_{mf})h \quad ;\ \text{bubble diameter}$$

$$u_B = u - u_{mf} + 0.711\sqrt{g d_B} \quad ;\ \text{bubble rise velocity}$$

$$f_B = 1.5(u - u_{mf})/d_B \quad ;\ \text{bubble frequency}$$

$$u_{mf} = u_F(u_J/u_F)^{\overline{x}^2} \quad ;\ \text{minimum fluidisation velocity}$$

the bulk bed concentration of jetsam is not so high. From the above assumptions, the segregation pattern at time $\bar{t}$ can be predicted on the basis of the bulk bed concentration at t - Δt, $\bar{C}(t-\Delta t)$, and the outlet concentration $\bar{C}_{out}$ (t) is obtained from the segregation pattern, i.e., C at the discharge height. The bulk bed composition for the next period from t to t + Δt is then assumed as

$$\bar{C}(t) = \bar{C}(t - \Delta t) + \{C_{in} - C_{out}(t)\}\Delta t / \bar{t} \tag{1}$$

where C_{in} and $\bar{t}$ represent the feed concentration of jetsam and the mean residence time of particles, respectively. Figure 3 describes the computation flow chart in which the concentrations of jetsam are expressed as the weight fractions converted by

$$X = 1/\{1 + \rho_F(1 - C)/(\rho_J C)\} \tag{2}$$

The calculations were simply iterated for $0 < t < 5\bar{t}$ with Δt = 10 s which was verified to be an appropriate interval by preliminary calculation. Furthermore, for simplicity, the bubble diameter was assumed to be invariable over the whole bed and was taken as the value at half of the fluidised bed height.

EXPERIMENTAL

The continuous fluidised bed apparatus is shown diagramatically in Figure 4. The fluidised bed {2} was 0.099 m i.d. and 0.50 m in height with a perforated plastic distributor through the centre of which passed

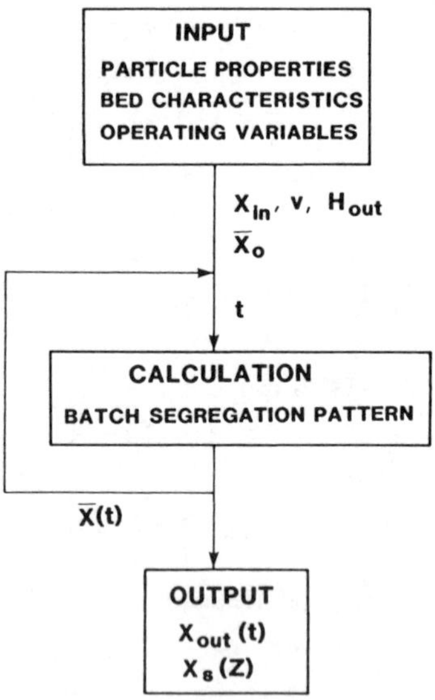

Figure 3. Computation flow chart

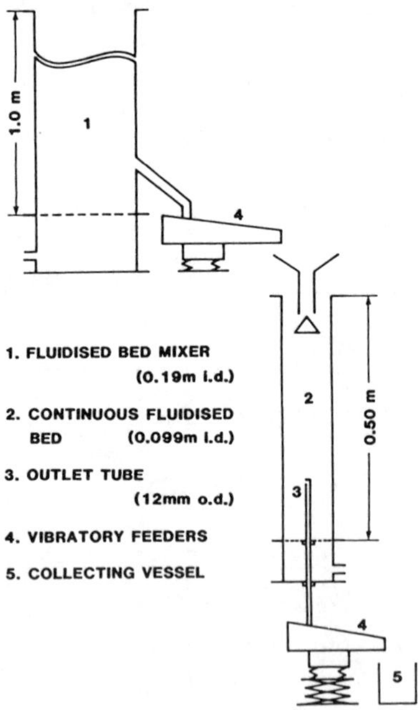

Figure 4. Schematic diagram of experimental apparatus

a 12 mm o.d. brass pipe {3} for the discharge of the bed material. Particles of known composition were fed to the bed from the 0.19 m i.d. fluidised bed mixer {1} (a very vigorously fluidised bed such that the feed to the test bed was well-mixed). The flow rates of particles at the inlet and the outlet were adjusted to keep a steady fluidised bed height of 0.105 m by the vibrators {4}. All particles from the outlet were collected in a beaker {5} every two to five minutes and the composition of each collected sample was analysed by sieving and weighing.

The bed material used was 0.090 mm copper shot (the jetsam) and 0.214 mm glass beads (the flotsam). Some of their relevant properties are shown in Table 2. The experimental procedure was almost the same as that described by Nienow and Naimer (5).

RESULTS AND DISCUSSIONS

It has been reported that when jetsam and flotsam particles are loaded into the lower and the upper layers respectively, batch fluidisation of a segregating bed gives a steady-state segregation pattern quite rapidly (7). This was confirmed from observed segregation patterns which were obtained for different intervals of fluidisation. Figure 5 shows the observed variations in the mixing index M, the ratio of the average weight fraction of jetsam particles in the region where $X < \bar{X}_o$ to the bulk bed weight fractions $\bar{X}_o$, with time for different initial bed loadings. As can be seen, the mixing index levels off within about 15 s for $\bar{X}_o = 0.10$ and about 30 s for $\bar{X}_o = 0.20$. Hence, in the present experiments all continuous runs were started after fluidising the initial bed for a few minutes. Measurement of the break-through curve was conducted for fixed values of the feed concentration of jetsam X_{in} and the mean residence time $\bar{t}$. Their relative variations during each run were held within 8% of the average value.

Figure 6 shows effect of the initial bulk bed concentration $\bar{X}_o$ on the break-through curve and the steady state segregation pattern for cases of the discharge location near the bottom of the bed ($H_{out} = 0.015$ m). The change of break-through curves with $\bar{X}_o$ seems very similar to that reported previously (5). With $\bar{X}_o$ greater than the feed concentration X_{in}, the outlet concentration X_{out} decreases monotonously with time while it increases with $\bar{X}_o$ less than X_{in}. In despite of these different break-through curves, a unique steady state is attained when X_{out} becomes equal to X_{in}, giving the same segregation pattern within experimental error.

Predictions of the break-through curve by the present model were made by assuming the bed voidage ε_{mf} and the wake volume fraction f_W respectively

Table 2. Properties of particles used

Particles	Particle Density [kg/m³]	Bulk Bed Density [kg/m³]	Mean Diameter [mm]	u_{mf} [m/s]
Copper Shot (CS90)	8900	5280	0.090	0.034
Glass Beads (GB214)	2520	1450	0.214	0.051

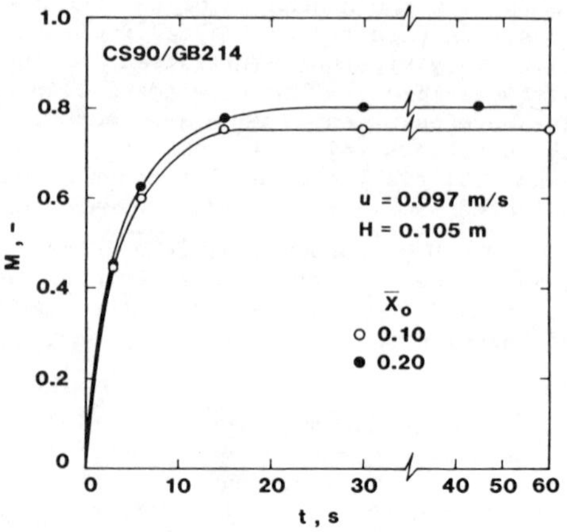

Figure 5. Variation of mixing index with time in a batch operation

as 0.50 and 0.25. Also, since preliminary calculations showed no essential difference in the predicted break-through curves and the steady state segregation patterns with Δt shorter than 30 s, a fixed value of Δt = 10 s was used. As shown by the solid lines in Figure 6, the model explains fairly well the observed effect of $\bar{X}_o$ on the break-through curves. It is less successful in predicting the steady-state segregation pattern. The steady state model suffers from the same type of weakness (4).

Figure 7 shows predicted (based on X_{in}=0.1) and experimental (based on X_{in}=0.094 to 0.097) break—through curves and segregation patterns as a function of discharge height. The experimental data show that discharge height has very little effect which is in good agreement with previous experimental work. However, since a steady state is obtained when $X_{out}=X_{in}$, this suggests that for a given feed concentration the holdup of jetsam particles in the bed $\bar{X}_s$ at steady state would increase with increasing discharge height H_{out} depending on the slope of the concentration profile. However, as shown in Figure 7, the observed effect of H_{out} on the steady state segregation pattern is much less than predicted by the model; the observed segregation patterns are much steeper than predicted and hence, the change of $\bar{X}_s$ with H_{out} is smaller, particularly for the higher discharge. This lack of agreement arises because the fundamental Gibilaro-Rowe model inherently predicts segregation patterns which show a change in concentration with bed height when the mechanism of wake exchange is superimposed on those of bulk circulation and preferential jetsam segregation. In practice, except at high concentrations, actual segregation patterns always have a constant composition over most of their height (6,7). The fact that experimentally at steady state X_{in} and X_{out} are slightly greater than X_s at the upper take off points implies additional local segregation at withdrawal. This segregation is not allowed for in the model and for solids in throughflow it enhances the fundamental divergence between theory and experiments.

Although a little improvement in the prediction could be expected by allowing for the axial distribution of bubble diameter and the change of wake volume fraction with bubble diameters, further work is needed on developing a model for both batch operation and solids in throughflow for segregating beds of binary systems of particles, particularly at high jetsam concentrations.

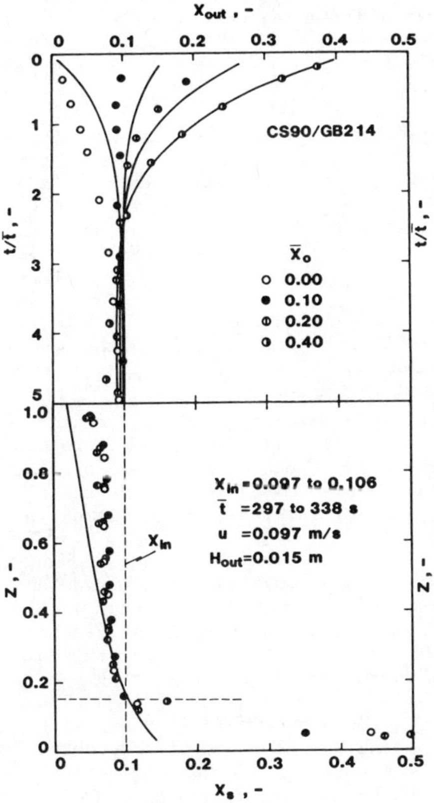

Figure 6. Effect of initial bed composition on break-through curve and steady-state segregation pattern

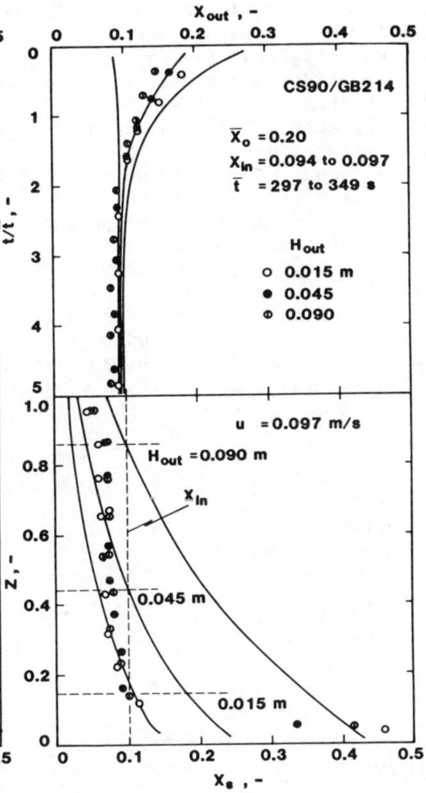

Figure 7. Effect of discharge height on break-through curve and steady state segregation pattern

NOTATION

a'_B volume fraction of bubbles and wakes in bed (—)

A_B $a'_B/(1-a'_B)$ (—)

$\underset{\sim}{C}$	volume fraction of jetsam particles	(—)
$\bar{C}$	bulk bed volume fraction of jetsam particles	(—)
$\bar{d}$	mean particle diameter of mixture	(mm)
d_B	mean bubble diameter in bed	(m)
d_{Bo}	bubble diameter just formed	(m)
f_B	point bubble frequency	(s^{-1})
f_W	volume fraction of wake in bubble and wake region	(—)
g	gravitational acceleration	(m/s^2)
G	effective gas flowrate for bubble formation	(m^3/s)
h	height from gas distributor in bed	(m)
H	height of fluidised bed	(m)
k	segregation rate based on cross-sectional area of bulk phase	(m/s)
M	mixing index	(—)
q	exchange rate based on cross-sectional area of bulk phase	(s^{-1})
$\underset{\sim}{t}$	time of continuous run	(s)
$\bar{t}$	mean residence time of particles in bed	(s)
Δt	short time interval	(s)
u	superficial gas velocity	(m/s)
u_B	bubble rise velocity	(m/s)
v	the volumetric flow rate of solids through the bed	(m^3/s)
w	circulation rate based on cross-sectional area of bulk phase	(m/s)
$\underset{\sim}{X}$	weight fraction of jetsam particles	(—)
$\bar{X}$	bulk bed weight fraction of jetsam particles	(—)
Z	dimensionless height (h/H)	(—)

Greek Symbols

ε	bed voidage	(—)
λ	w/k	(—)
γ	qH/w	(—)
$\underset{\sim}{\rho}$	particle density	(kg/m^3)
ρ	mean particle density of mixture	(kg/m^3)

Subscripts

B	bulk phase	in	at inlet
W	wake phase	out	at outlet
F	flotsam particles	o	initial
J	jetsam particles	s	at steady state
mf	at minimum fluidisation		

LITERATURE CITED

1. Gibilaro, L.G. and P.N. Rowe, Chem.Eng.Sci., 29, 1403 (1974).
2. Chiba, S., H. Tanimoto, T. Chiba and H. Kobayashi, J.Chem.Eng. Japan, 12, 43 (1979).
3. Tanimoto, H., S. Chiba, T. Chiba and H. Kobayashi, J.Chem.Eng. Japan, 14, 273 (1981).
4. Naimer, N.S., T. Chiba and A.W. Nienow, Chem.Eng.Sci., 37, 1047 (1982).
5. Nienow, A.W. and N.S. Naimer, Trans.Instn.Chem.Engrs., 58, 181 (1980).
6. Nienow, A.W. and T. Chiba, "Fluidisation of Dissimilar Materials" in "Fluidisation", Ch.10, (J.F. Davidson, R. Clift, D. Harrison, Eds.), Academic Press, 2nd Ed. (1985), pp.357-382.
7. Rowe, P.N., A.W. Nienow and A.J. Agbim, Trans.Instn.Chem.Engrs., 50, 324 (1972).

THE NON-STEADY STATE SEGREGATION OF PARTICLES IN GAS FLUIDIZED BEDS

P.J.M. Valkenburg, J.C. Schouten and C.M. van den Bleek,
Laboratory of Chemical Technology, Delft University of Technology,
Julianalaan 136, 2628 BL Delft, The Netherlands

ABSTRACT

Non-steady state segregation experiments were carried out to investigate the influence of increase of density by SO_2-take-up, on the axial particle distribution of sorbent material during fluidized bed combustion of coal. In a 0.1 m ID gas fluidized bed, representing the "tube-like" space between vertically mounted cooling tubes, coarse sorbent material (d_p = 2.7 mm) and equally shaped, but slightly higher density tracer particles were used. It is shown that a pulse-like input of tracer particles results in a significant maximum in the particle distribution curve. The commonly used one-dimensional Taylor dispersion model is not able to match the measured data. Hence two models were derived which in principal are able to describe a with time shifting maximum in the axial distribution curve of the segregating component.

SCOPE

From a particle point of view a gas fluidized bed is usually assumed to be well mixed. However, segregation may occur when the bed contains particles which differ in density and/or in size, where segregation is most influenced by density differences. Much has been published on the measuring and modelling of the steady state segregation in fluidized beds. Recently Beeckmans et al. (1) have given a review on the research on segregation. In the fluidized bed combustion of coal the rate and extent of segregation of sorbent material, coal and ash have to be well understood in order to select the proper operating conditions.
The objective of the study presented is the modelling of the non-steady state segregation of sorbent particles which differ in density from fresh material. This phenomenon is occurring because of the intake of sulfur dioxide by the sorbent material.

CONCLUSIONS AND SIGNIFICANCE

From non-steady state segregation experiments using coarse particles it has become clear that a pulse-like input of slightly higher density tracer particles at the top of a fluidized bed results in a maximum in the axial particle distribution. This maximum shifts downwards with increasing mixing time. A model describing the non-steady state segregation by means of a dispersion coefficient $\mathbb{D}$ and a dimensionless, Peclet-like parameter λ is introduced to predict the measured tracer distribution. The agreement between the experiments and the model is in general good. The estimated dispersion coefficient is linear to the superficial gas velocity in the investigated range of 1.05 m/s to 1.25 m/s. The dimensionless number λ decreases with the gas velocity. Since λ

governs the extent of segregation at steady state, and a large absolute value of λ will result in a high degree of segregation, this model can also predict the superficial fluidizing velocity at which segregation will disappear. When λ equals zero the net convective upward flow of particles equals the net convective downward flow of particles and the rate of mixing is only dependent on the dispersion coefficient $\mathbb{D}$. So, this model can be considered as an extension to the one dimensional Taylor dispersion model and is suitable for modelling the non-steady state segregation. The presented compartments in series model can also predict the measured maximum in the tracer particle distribution. It can, however, only be used at relatively short mixing times, since the model can not describe steady state segregation.

EXPERIMENTAL

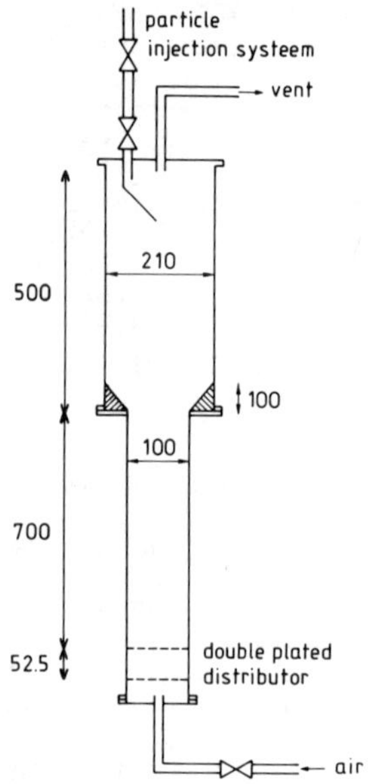

Figure 1.
The 0.1 m ID fluid bed

The mixing or segregation of solids in a fluid bed can be measured by several experimental techniques (Oki et al. (2)). For practical reasons abrupt defluidization of the bed and layer-by-layer analysis was chosen to investigate the non-stationary segregation of fresh and partly spent sorbent particles. The experiments were carried out in a 0.1 m ID fluid bed made from Plexiglass (Figure 1) at ambient conditions. This diameter is chosen to represent the "tube-like" space between vertically mounted cooling tubes in a fluidized combustor. At the top of the freeboard section a particle injection system, a pipe with two valves to create an airlock, is used to inject tracer particles. The fluidized material in these experiments is a synthetic sorbent material, alumina (Al_2O_3) impregnated with calcium oxide, also in use in other research projects on the sulfur dioxide capture in this Laboratory. Details on the preparation are given by Noordergraaf (3). Some physical properties of these particles are listed in Table 1.

Table 1. Physical properties of the alumina impregnated with CaO.

shape	cylinders
length	2.8 ± 0.1 mm
diameter	2.6 ± 0.1 mm
density	1250 kg/m^3
U_{mf}	0.9 m/s

The non-steady segregation experiments were carried out with coloured alumina particles. Due to the colouring there was an increase in particle density of 15%, comparable with a sulfur capture of 70%.

Overall three different coloured kinds of particles are used during one experiment. Each is dropped onto the bed with a certain interval in time using the injection system. After abruptly stopping the air flow, the settled bed material is collected in five layers and the number of tracer particles in each layer is counted or weighed. In this way the axial tracer distribution is known for three mixing times at a fixed value of U_o. In all experiments the ratio of the settled bed height and bed diameter was kept at a constant value of 1.5.

RESULTS

The experiments point out that a pulse-shaped input of the tracer at the top of the bed moves downwards into the bed as the mixing time increases.
In Figures 2 and 3 this experimental result is illustrated. Since the

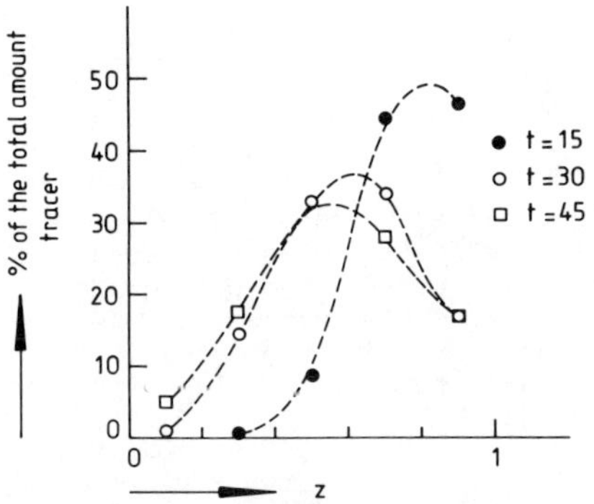

Figure 2. The measured axial tracer distribution at U_o = 1.20 m/s for mixing times t = 15, 30 and 45 s.

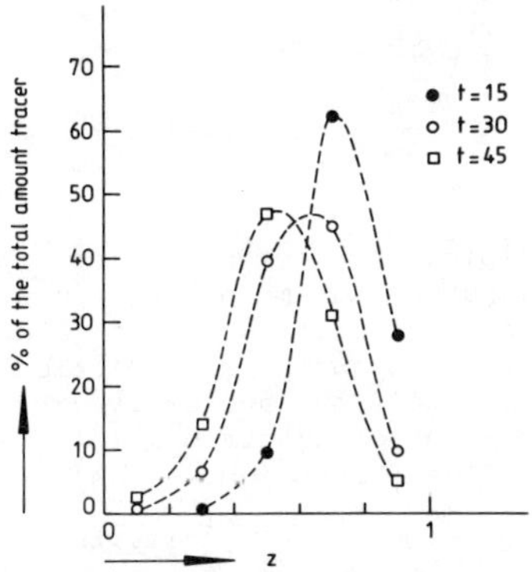

Figure 3. The measured axial tracer distribution at U_o = 1.05 m/s for mixing times t = 15, 30 and 45 s.

settled bed material is divided into five layers, five points of the curves at mixing times t = 15, 30 and 45 s are measured.
To describe the observed particle distributions two models are used: A model completely suited for modelling the non-steady state segregation and a model originally derived to use as a non-steady state mixing model.

The λ-model

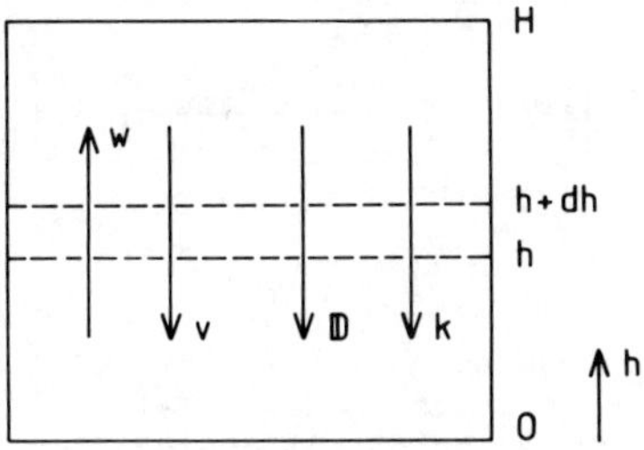

Figure 4. Schematical representation of the λ-model

The first model is based on the assumption that the flow of particles in a fluidized bed can be characterized by three mechanisms: convection, dispersion and segregation, see e.g. Gibilaro and Rowe (4). This model is schematically showed in Figure 4. It is assumed that particles are moving downwards and upwards with a velocity v and w.
In a slugging fluidized bed the velocity w is related to the slug rise

velocity while the velocity v is related to the descending velocity of particles in a gas slug. Dispersive flow in the fluid bed is characterized by a dispersion coefficient $\mathbb{D}$. Due to density and/or size differences between particles segregation may occur. This segregative flow of particles is represented by a segregation velocity k and is assumed to be proportional to the concentration of the tracer at every height in the bed. Hence the flow of particles due to segregation is treated in the same manner as the convective flows. From a mass balance over a slice of the fluid bed with thickness Δh the following partial differential equation can be derived:

$$\frac{dc}{dt} = (v-w+k)\frac{dc}{dh} + \mathbb{D}\frac{d^2c}{dh^2} \qquad (1)$$

The following boundary conditions are applied for $0 \leqslant t \leqslant \infty$

$$\text{at } h = H\text{: } \mathbb{D}\frac{dc}{dh}F + (v-w+k)\,c\,F = N\delta(t)$$

$$\text{at } h = 0\text{: } \mathbb{D}\frac{dc}{dh}F + (v-w+k)\,c\,F = 0$$

So, dropping N tracer particles onto the bed is represented by a pulse-input at the mean bed height H. Since no particles can leave the fluid bed the net flow of particles at the bottom of the bed equals zero. From the analytical solution of Equation (1), using the above boundary conditions and derived by Laplace transformation, the concentration of tracer particles as a function of the mixing time t and dimensionless bed height z can be calculated;

$$\begin{aligned} c(z,t) = c_o \cdot \exp\left\{\frac{\lambda}{2}(1-z)\right\} \cdot \Bigg\{ & 2\sum_{n=1}^{\infty}(-1)^n \exp\left(-\mathbb{D}\left(\frac{\lambda^2+(2n\pi)^2}{4H^2}\right)t\right)\cos(n\pi z) + 1 \\ & - 2\sum_{n=1}^{\infty}(-1)^n \cdot \frac{\lambda^2}{\lambda^2+(2n\pi)^2}\cos(n\pi z)\left[\exp\left(-\mathbb{D}\left(\frac{\lambda^2+(2n\pi)^2}{4H^2}\right)t\right) - 1\right] \\ & - \frac{\lambda z}{2} - \frac{\lambda}{\pi}\sum_{n=1}^{\infty}\frac{(-1)^n}{n}\exp\left(-\mathbb{D}\left(\frac{\lambda^2+(2n\pi)^2}{4H^2}\right)t\right)\sin(n\pi z) \\ & + \frac{\lambda}{\pi}\sum_{n=1}^{\infty}\frac{(-1)^n}{n}\frac{\lambda^2}{\lambda^2+(2n\pi)^2}\sin(n\pi z)\left[\exp\left(-\mathbb{D}\left(\frac{\lambda^2+(2n\pi)^2}{4H^2}\right)t\right) - 1\right]\Bigg\} \end{aligned} \qquad (2)$$

The parameter λ is a "Peclet-like" dimensionless number, the ratio of the convective and dispersive flow of particles in the bed. To calculate the number of tracer particles in a layer from z_1 to z_2 the expression for the concentration in Equation (2) is integrated.

$$N(t) = \int_{z_1}^{z_2} c(z,t)\ HF\ dz \tag{3}$$

Since from the experiments the number of particles in five layers is obtained, Equation (3) is used to compare the results of the model calculations and the experiments. A Simplex computer routine is used to estimate the parameters λ and $\mathbb{D}$. Typical results are showed in the Figures 5 and 6. The model seems to describe the experiments fairly

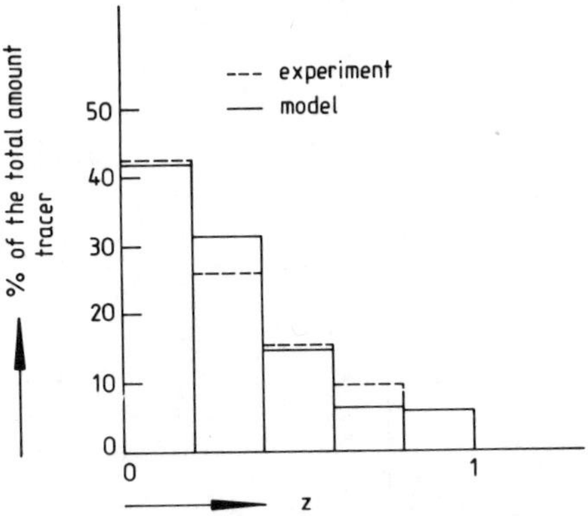

Figure 5. Comparison of the measured and the calculated particle distribution.
U_o = 1.1 m/s, t = 60 s, Δ = 10.5%

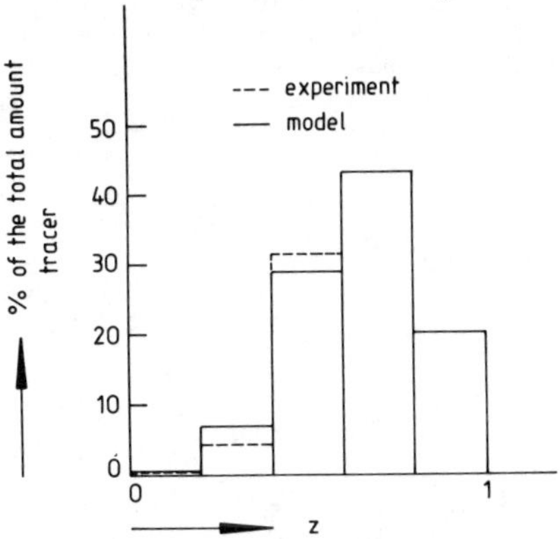

Figure 6. Comparison of the measured and the calculated particle distribution.
U_o = 1.1 m/s, t = 45 s, Δ = 5.3%

well. The disagreement between a measured particle distribution and the calculated one is defined as:

$$\Delta = \sum_{i=1}^{5} \text{abs}\ (N_{i,EXP} - N_{i,MOD}) / \sum_{i=1}^{5} N_{i,EXP} \tag{4}$$

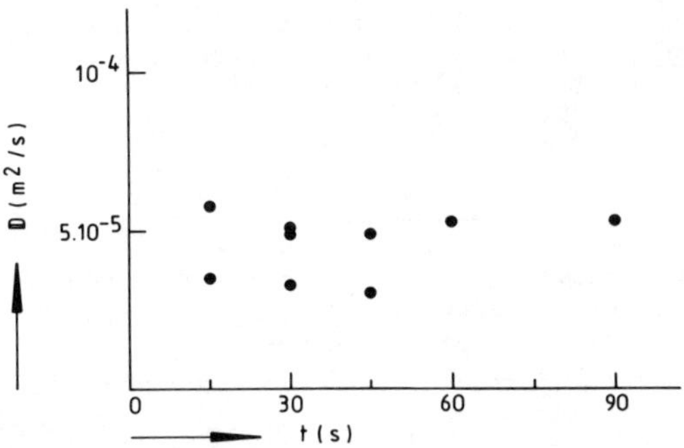

Figure 7. The estimated value of the dispersion coefficient $\mathbb{D}$ versus the mixing time t for U_o = 1.1 m/s

A typical value for the disagreement calculated by Equation (4) is 5.3%. The average disagreement is 10.6%. This figure is based on all the 38 experiments performed, the superficial fluidizing velocity ranging from 1.05 m/s to 1.25 m/s and for mixing times up to 90 seconds.

$\mathbb{D}$ showed little or no dependency on time, as it should be (Figure 7). In this figure duplo and triplo measurements are plotted to illustrate the variance in the calculated dispersion coefficient. λ showed also little dependency on time, however with a higher variance. Using the time averaged value of $\mathbb{D}$ and λ in the model calculations the average disagreement between model and experiment therefore increases from 10.5 to 30 %. From Figure 8 it can be seen that in the investigated velocity range the estimated dispersion coefficient is linear to the superficial fluidizing velocity. It is also clear that $\mathbb{D}$ equals zero at a gas velocity near the minimum fluidization velocity U_{mf}. In Figure 9 the relation between the dimensionless number λ and the

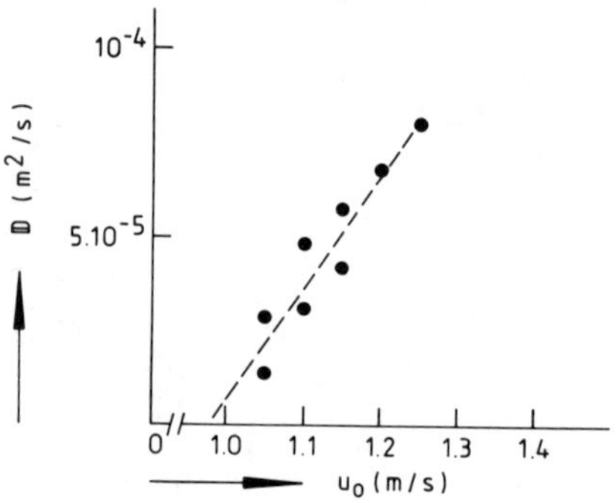

Figure 8. The dispersion coefficient $\mathbb{D}$ versus the superficial fluidizing velocity U_o

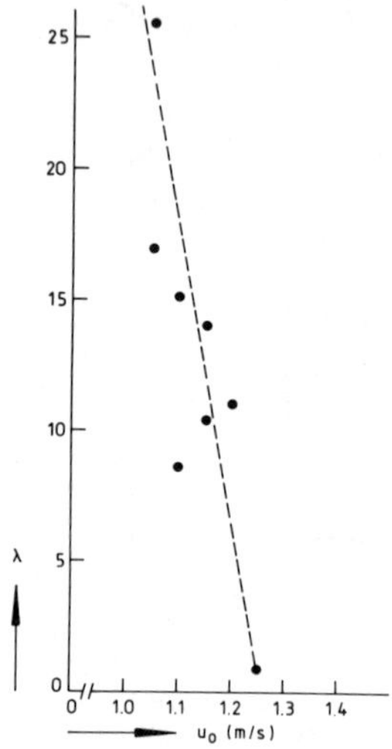

Figure 9. The "Peclet-like" number λ versus the superficial fluidizing velocity

fluidizing velocity U_o is given. As expected the dimensionless number λ is decreasing with increasing gas velocity. If λ equals zero no steady state segregation will occur (see Schouten et al. (5)).

The Chained Compartments Model

This model consists of a chain of equally sized, ideally mixed compartments and assumes the existence of a circulating flow of particles ϕ. The total number of ideally mixed compartments is M, 1/2 M compartments in series representing the upward moving phase and the same number representing the downward moving phase. M is chosen to be a multiple of 10 since experimentally the amount of tracer particles is determined in exactly five layers.

In contradiction with the bubble assemblage model described by Ishida & Wen (6) the presence of crossflow of particles between two adjacent compartments is omitted, since model calculations pointed out that a maximum in the tracer distribution could only be realized when the

exchange particle flow was chosen relatively small to the circulating particle flow.
From a mass balance over one ideally mixed compartment i and using the Dirac function $\delta(t)$ to represent the introduction of the tracer particles, the concentration of the tracer in compartment i can be obtained:

$$n_i = \frac{N}{\phi}\frac{1}{\tau}\exp(-t/\tau)\sum_{m=1}^{\infty}\frac{(t/\tau)^{M(m-1)+i-1}}{(M(m-1)+i-1)!} \tag{5}$$

The mean residence time τ in compartment i can be written as:

$$\tau = V/M\,\phi) = H/(M\,u) \tag{6}$$

The particle velocity u and the number of compartments in series M are calculated by matching the theoretical solution to the measured tracer particle distribution. For M = 10, 20, 30 and 40 the particle circulating velocity u is calculated to give the best fit. The average disagreement between model and experiment, calculated with Equation (4), was the smallest for M = 20 namely 18.2%. It must be stated, however, that the average disagreement for M = 30 and M = 40 was not significantly greater, 19.7% and 20.0% respectively. Overall can be concluded that the tracer particle distribution measured at low superficial fluidizing velocities and short mixing times can be best described using a high number of compartments in series (M = 40). A small number of compartments in series (M = 10) gives the best results for measurements at high gas velocities and long mixing times. This can be explained because the number of compartments in series governs the extent of dispersion of the pulse-like input of tracer particles in the bed; the higher M the more ideal plug flow of the particles is approached. The circulating particle velocity u in the fluid bed tends to increase with the superficial fluidizing velocity U_o for a fixed number of compartments in series M and a fixed mixing time t.
The relation between u and U_o, however, is sofar not very clear.

ACKNOWLEDGMENT

The authors wish to thank all students at the Delft University who contributed to this research. Akzo Chemie Ned. BV is gratefully acknowledged for the supply of the alumina particles. So is Ms. C. Monna for typing out the manuscript.

NOTATION

c	tracer particle concentration	(number/m^3)
c_o	overall tracer particle concentration in the bed (=N/HF)	(number/m^3)
$\mathbb{D}$	dispersion coefficient	(m^2/s)
F	cross-sectional bed area	(m^2)
H	mean bed height	(m)
h	height in the fluid bed	(m)
i	number of compartment	
k	segregative velocity of particles	(m/s)

M	total number of compartments	
N	total number of tracer particles	
N(t)	number of tracer particles in a layer from z_1 to z_2	
$N_{i,EXP}$	experimental number of tracer particles in layer i	
$N_{i,MOD}$	calculated number of tracer particles in layer i	
n_i	tracer particle concentration in compartment i	(number/m^3)
t	mixing time	(s)
U_o	superficial fluidizing velocity	(m/s)
U_{mf}	minimum fluidization velocity	(m/s)
u	superficial circulating velocity of particles	(m/s)
V	volume of the fluid bed	(m^3)
v	convective downward velocity of particles	(m/s)
w	convective upward velocity of particles	(m/s)
z	dimensionless height in the bed	

Greek symbols

Δ	disagreement between model and experiment	(%)
λ	dimensionless number (= $(v-w+k)H/\mathbb{D}$)	
ϕ	circulating particle flow	(m^3/s)
τ	mean residence time in compartment i	(s)

LITERATURE CITED

1. Beeckmans, J.M., L. Bergström and J.F. Large, Chem. Eng. J., 28, 1 (1984).

2. Oki, K., W.P. Walawender and L.T. Fan, Powder Tech., 18, 171 (1977).

3. Noordergraaf, I.W., PhD Thesis, Delft University of Technology, Delft, The Netherlands (1985).

4. Gibilaro, L.G. and P.N. Rowe, Chem. Eng. Sci., 29, 1403 (1974).

5. Schouten, J.C., P.J.M. Valkenburg and C.M. van den Bleek, to be published.

6. Ishida, M. and C.Y. Wen, AIChE Symp. Ser. 128 (69), 1 (1973).

LARGE PARTICLE MOTION ON THE DISTRIBUTOR OF A FLUIDIZED BED OF SAND

C.R. Bemrose, J.S.M. Botterill, J. Bridgwater and A.W. Nienow

Department of Chemical Engineering, University of Birmingham, P.O. Box 363, Edgbaston, Birmingham, B15 2TT, U.K.

ABSTRACT

Large particles which have settled through a uniformly sized bed of smaller sand onto a drilled plate distributor are not entrained by the bubble wakes. Observation of their movement shows that they remain on the distributor and are carried by the "gulf streaming" currents generated within the bed.

SCOPE

A fluidized bed combustor design capable of burning dirty coal is desirable as it would obviate the need for the initial coal cleaning stage. To be successful, the system would need to be able to control the degree of particle segregation and to remove selectively the large shale and ash particles from the bed. Thus, the use of a spouted bed in conjunction with a fluidized bed has been proposed as a means of handling a mixed size distribution and maintaining a mixed bed of particles which otherwise would have segregated in the combustion zone (Rogers and Cooke (1)). Elsewhere in the bed, segregation will occur. Nienow and Chiba (2) have set out some simple rules to determine which component of a binary mixture becomes jetsam and sinks through the bed and Cooke et al (3) have suggested the use of a sloping distributor across which the jetsam would flow to some low offtake point from which the unwanted components would be removed.

Whitehead et al. (4) have investigated the mobility of large (6.7 and 12.7 mm) magnetite particles in a bed of magnetite having an incipient fluidization velocity of 0.5 m/s. They found that the large particles placed at the bottom of the bed were dispersed into an irregular pattern but the behaviour of large particles in either a fluidized or a spouted bed of smaller mean particle size is not yet sufficiently understood or defined to permit reliable design. The work presented here is part of a study initiated to increase understanding and quantify the motion and behaviour of such particles in a fluidized bed. It involves direct observation of the movement of large particles across the surface of a drilled plate distributor made of transparent Perspex.

CONCLUSIONS AND SIGNIFICANCE

Large particles sank through the fluidized bed to reside on the drilled plate distributor. Gulf streaming of bed particles was observed and the large particles were seen to move with the stream once a certain excess gas velocity was exceeded. They could have large lateral mobility across the distributor, their velocity depending upon the design of the distributor. Circulation cells developed which depended on the overall bubbling behaviour. These can be modified by distributor design. The large particles tend to migrate towards regions containing more bubbles and this can cause them to move up a sloping distributor. By suitable distributor design, controlled particle translation across the distributor can be achieved leading, for example, to improved provision for the removal of non-combustible material from coal combustors.

EXPERIMENTAL

The work reported has been done in a 0.4m x 0.15m Perspex bed with a drilled plate distributor also made of transparent Perspex (Table 1). A spout section could be added at the end of the bed if desired. The bed could be tilted so that the distributor sloped at up to 10° from the horizontal. All the tests have been at ambient temperature with a shallow bed of 590 μm sand (97% between 500 μm and 850 μm. The nominal density of the sand was 2600 kg/m^3 (bulk density 1480 kg/m^3), and its minimum fluidizing velocity 0.35 m/s.

Large tracer particles were made more visible by coating them with fluorescent paint and their motion has been observed visually from beneath the transparent distributor. They would begin to move once the gas velocity exceeded the minimum fluidizing velocity of the bulk bed material, U_{mf}, by about 0.2 m/s but the value was dependent on the size, shape and orientation of the tracer particle. A single tracer particle was then tracked visually across a grid marked on the distributor to give locations and times which were stored in a computer file. From the stored values, velocity and direction of movement could be deduced, and various mean values calculated.

A track was begun either by physically locating the particle at the desired starting point (sometimes with the excess velocity reduced below the threshold value until ready to commence the tracking) or by injecting the particle at the desired location by dropping it down a vertical injection tube at the start of tracking.

RESULTS

Preliminary tests established that the large particles rapidly sank through the fluidized bed and thereafter remained on the distributor. There was no evidence of regular entrainment in bubble wakes followed by redeposition. The shale was occasionally temporarily lifted off the distributor at the highest superficial velocities of 0.69 m/s so far used.

The implications of this persistent settling will be discussed later, but it should be noted that particle observation through the distributor is a practical and useful technique as removal of large particles from the bed must be effected at distributor level. Although the particles were not lifted from the distributor, rapid movement of the particles was also sometimes observed.

One Dimensional Tests

It was anticipated that large particles could be induced to move down a sloping distributor because they sink to the bottom of a fluidized bed. They could then be removed or alternatively entrained in a spout at the base of the slope. Attempts were made to characterize and quantify the movement by tilting the fluidized bed laterally. A series of 5 different particles (Table 1) was first placed at the higher side of the distributor and then at the lower and their lateral progress monitored versus time for different distributor angles (including horizontal).

TABLE I Experimental Details

Distributor type

	Distributor A	Distributor B
	2 mm holes on 10 mm square pitch.	0.8 mm holes on 5 mm square pitch.
Bed height	0.11 to 0.125 m	0.11 to 0.125 m
Superficial velocity U	0.61 to 0.69 m/s	0.62 to 0.64 m/s
Excess velocity $U-U_{mf}$	0.26 to 0.34 m/s	0.27 to 0.29 m/s
Bed pressure drop	1.3 to 1.7 kPa	1.3 to 1.7 kPa
Distributor pressure drop	0.45 to 0.6 kPa	1.45 to 1.65 kPa

Large Tracer Particles

	Marble	Pebble			Shale
	16.5 mm dia.	Large	Medium	Small	
Mass g	6.2	10.5	5.6	3.2	0.55
Density kg/m^3	2580	2560	2670	2460	1600

Figure 1 shows the arithmetic mean velocity for a 16.5 mm glass marble in a bed of sand as a function of distributor angle and starting position. With the marble starting from the position labelled 'Right' the particle moved up not down the slope with a velocity which increased with distributor angle. Starting from the 'Left' side, there was some motion down the distributor, but at angles up to -8°, there was again a tendency for the particle to move up the slope. These observations are consistent with the motion being a combination of movement up a slope and towards some particular location.

For the other particles (large pebble, medium pebble, small pebble and piece of shale) motion from the 'Right' side followed the same pattern. Motion from the 'Left' side showed features similar to the marble but was somewhat more erratic. This is thought to be due to the starting positions being distributed along the length of the bed, and, as described in the next section, the influence of two-dimensional effects and non-uniformity of behaviour.

Two Dimensional Tests

The one dimensional tests indicated that the motion of large particles across the distributor was affected by other factors besides the distributor angle.

Two dimensional observations of particle tracks versus time were then made for a large particle from a 5 x 7 array of starting positions on the distributor. One such track is shown in Figure 2, where the observations of x and y position were made every 15 seconds. It was clear that, in spite of a random element in the movement, there were several preferred zones to which the particle tended to move. Once near a preferred zone, the particle appeared to move randomly in that region. Data from these experiments were analysed in several ways. Firstly each track was treated individually. The arithmetic mean of the x and y coordinates for each measurement was computed and the point furthest from the mean point discarded. This computation was then repeated, each time discarding the most distant point, until all the remaining points were within some arbitrary distance (10 mm) of the mean position of those points. Figure 3 shows the starting and calculated "mean" finishing positions for each of the 35 tracks in one experiment and confirms the visual observation of preferred zones or collection points. In each experiment the collection points were grouped together and the x and y coordinates of a mean value were obtained.

The series of tests included measurements with different particles and the two different distributors. For the various particles, groups of collection points showed similar patterns, but these were distinctly different for the two distributors, as shown between Figures 4(a) and (b). The number of points which have been grouped together is indicated beside the mean valued plotted.

In addition to location information, these experiments produced velocity information. In general each individual track could be considered to have two separate components:

(a) movement from the starting point to the collection zone (the translational component); (b) random motion around the collection point (the random component). For each track a translational velocity was defined and calculated as:

$$\frac{\text{distance from start to first point within 10 mm of collection}}{\text{time taken to traverse this distance.}}$$

The mean value of translational velocity, $\overline{V}$, for each group of collection points is also shown in Figure 4 and demonstrates that higher velocities were obtained with distributor B than with A. This is more clearly shown in Figure 5, where mean distance travelled is plotted against time taken for that movement. Individual data points have considerable scatter, but the mean values exhibit significant trends.

The values of translational velocity are

	By calculation (weighted mean)	From Figure 5
Distributor A	1.85 mm/s	1.1 mm/s
Distributor B	2.97 mm/s	1.7 mm/s

The discrepancy between the two assessments is due to the weighting of the mean value in the calculation procedures since there are many more points at the shorter time intervals.

Figure 5 exhibits an apparent change of slope for distributor A at distances of the order of 100 mm. Comparison with Figure 4(a) suggests that no starting point was more than about 100 mm from a collection point, and few tracks were longer than this. For distributor B however there were some much longer tracks as shown in Figure 3 which apparently passed close to one collection point en route for a more distant one.

Spouting Fluidized Bed

The technique of spouting can be used to lift large particles and make them circulate to some degree in a fluidized bed. This idea has been examined in a qualitative manner by appending a spout at the end of the normal fluidized bed. Airflow to the fluidizing and spouting regions could be separately controlled. The observed characteristics are listed below:

(a) With the bed fluidized and the spout in operation, a marble landing on the bed sank to the distributor and remained there. Motion of the particle on the distributor (even when the distributor was tilted) did not bring the particle into the spouting region, except on very rare occasions. In general the particle tended to shun the spouting region.

(b) With the spout alone in operation, the bed surface settled to the angle of repose (∿ 40°). A marble inserted into the spout could be spouted and was normally seen to land on the sloping bed surface from whence it rolled or slid back into the spout and was re-entrained.

(c) As the air supply to the bed was gradually increased without changing the airflow in the spout, the bed's angle of repose reduced. The marble then continued to circulate by spouting and rolling/sliding down the reduced slope of the surface of the packed bed until the slope of the bed surface reduced to about 20°.

(d) Immediately the fluidizing airflow was increased to U_{mf}, the bed surface became horizontal and any large particle landing on the bed sank to the distributor and ceased to circulate [see (a) above].

In this work spouting velocities up to 70 m/s (at the air inlet) were used.

DISCUSSION

Movement of material in the wake of bubbles formed in a fluidized bed is a well-known phenomenon (see for example Rowe (5)), but the forces acting on a given particle to cause its motion are very complex. Considering the presence of large, spherical particles in a bed of sand, there are at least four ways of approaching the problem.

1. The U_{mf} of the sand used was 0.35 m/s. For 16.5 mm marbles, the U_{mf} can be calculated by the Ergun equation (6) to be ∿ 3.7 m/s. This would suggest the need for an excess gas velocity > 3.4 m/s to lift the marbles.
2. Chiba and Nienow (7) have examined the defluidization of large particles due to segregation. They calculated the maximum diameter of a particle which could be transported upwards in a bubble wake, d_j as:

$$d_j = \left(\frac{3}{4} f_w\right)^{1/3} D_{Bo}$$

where D_{Bo} = estimated initital bubble diameter $= \left(\frac{6(U-U_{mf})A}{\pi g^{\frac{1}{2}} n_d}\right)^{2/5}$

For $U-U_{mf}$ = 0.35 m/s, d_j = 8.5 mm (distributor A) or 4.9 mm (distributor B). Conversely an excess velocity of 2.1 m/s (distributor A) or 7.4 m/s (distributor B) would be needed to lift 16.5 mm marbles.
3. Kinetic energy in the air emerging from distributor jets could be equated with the energy required to lift a particle. This is an extremely complex problem which involves hole spacing on the distributor, size, shape and position of particle (since the air will not impinge normally on the particle surface). This would however, give rise to lateral motion through an air jet impinging on a non-horizontal surface of a large particle.
4. The density of glass marbles was approximately 2600 kg/m^3, which is comparable with the density of sand. However, overall, a fluidized sand bed has an apparent density of about 1500 kg/m^3 because of its voidage and thus an isolated glass marble would be expected to sink rapidly as noted by Masson et al (8) in conformity with the simple flotsam-jetsam rules of Nienow and Chiba (2).

The above suggests that large particles are unlikely to be lifted and mixed in the bulk of a fluidized bed of sand as defined in this paper.

Most of the work on large particle jetsam segregation in gas fluidized beds (see, for example, Chiba and Nienow (7)) has been conducted in small cylindrical beds (∿ 120 mm. diameter). In such beds, gulf streaming is eliminated by a well-designed distributor. However, in this work with larger beds, gulf streaming was always visible both through the distributor and at the wall although the distributors were carefully made by NC drilling and the experiments were conducted with an operating pressure drop in the range ∿ 0.25 to 1 times the bed pressure drop. This tendency for gulf streaming and the formation of bulk flow cells fits in well with the work of Whitehead et al (4) on larger beds. When the bed distributor is inclined, the reduced pressure drop through the shallower bed at the upper end leads to preferential gas flow in this region. Thus, there is a greater excess gas velocity there resulting in a higher rate of bubble flow and hence an increased upward flow of bulk bed material in bubble wakes. This enhanced flow at one end has to be fed from the rest of the bed by bulk flow up the slope, i.e., gulf streaming up the slope.

Large particles were seen to be transported along the distributor with the general bed movement and tended to congregate at those locations where bubbling was predominant. With sloping beds, this location was at the top of the slope. With other beds, location of preferential bubbling and the associated size and location of gulf stream circulation cells would be dependent on the detailed distributor design.

It is suggested that the mechanism of movement is linked directly to gulf streaming. However, the parameters which control the rate of particle movement are still being investigated. For distributor A, Figure 5 suggests that it might be possible to equate cell size (if not location) with a change of slope of the distance/time graph.

ACKNOWLEDGEMENTS

We gratefully acknowledge receipt of an SERC grant for this work and helpful discussions with NCB research staff.

NOTATION

A	Bed area	m^2
D_{Bo}	Initial bubble diameter	m
d_j	Minimum diameter of jetsam particle lifted in bubble wake	m
f_w	Wake volume fraction	-
g	Acceleration due to gravity	m/s^2
n_d	Number of holes in distributor	-
U	Superficial velocity	m/s
U_{mf}	Minimum fluidization velocity	m/s
$\overline{V}$	Mean value of translational velocity	m/s

LITERATURE CITED

1. Rogers, E.A., and Cooke, M.J. "The recovery of energy from coal mining waste by fluidized bed combustion", Paper 37 presented at Symposium on the Reclamation, Treatment and Utilization of Coal Mining Wastes. National Coal Board, Durham (September 1984).

2. Nienow, A.W., and Chiba, T., "Fluidization of dissimilar materials", in Fluidization (2nd ed.), Davidson, J.F., Clift, R., and Harrison, D. (Ed). Academic Press, London (1985), pp. 357-382.

3. Cooke, M.J., Rogers, E.A., Dando, R.L. and Gauld, D.W., "Ash management in the fluidized bed combustion of coal", DISC/27/232, 3rd Int. Fluidized Combustion Conference, Institute of Energy, London (Oct. 1984).

4. Whitehead, A.B., Dent, D.C., and Close, R. "Some effects of distributor type on size segregation in gas-solid fluidized beds", in Fluidization, Kunii, D., and Toei, R., (Ed), Engineering Foundation (1984), pp. 515-522.

5. Rowe, P.N., "Experimental properties of bubbles", in Fluidization (1st ed.). Davidson, J.F., and Harrison, D. (Ed), Academic Press, London (1971), pp. 122-191.

6. Ergun, S., Chem. Eng. Prog., 48, 89 (1952).

7. Chiba, T., and Nienow, A.W., "Defluidization of large particles due to segregation", in Fluidization, Kunii, D., and Toei, R, (Ed), Engineering Foundation (1984), pp. 195-202.

8. Masson, H.A., Rios, G.M., Dang Tran, K., and Bourtembourg, "Shape and density effects on the behaviour of a large isolated body moving in a gas solid fluid bed", ibid., pp. 185-193.

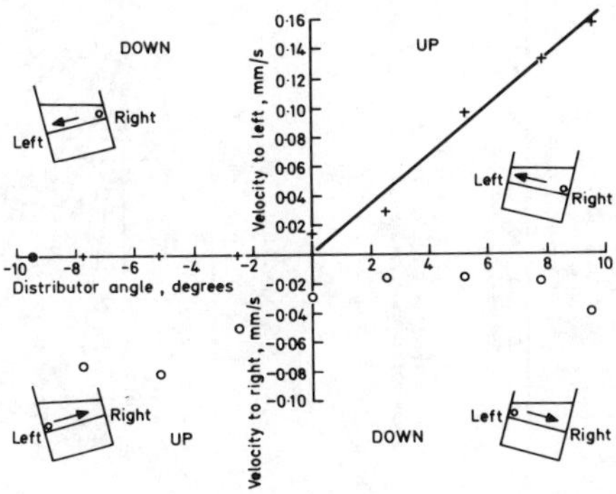

Figure 1. Mean velocity of marble across angled distributor.
+ marble starting at 'Right' side
0 marble starting at 'Left' side

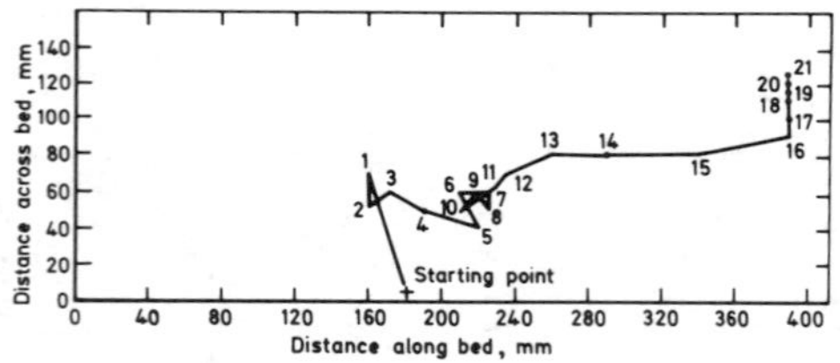

Figure 2. Typical track of large pebble - Distributor B. Steps represent distances traversed in 15 seconds.

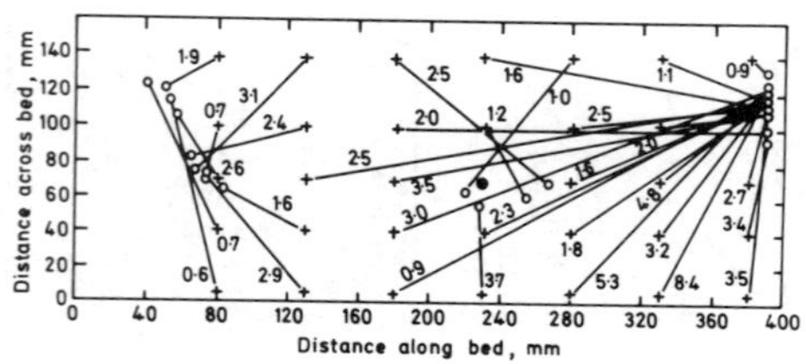

Figure 3. Starting and finishing points for large pebble tracks - Distributor B.
+ starting point. 0 collection point.
Mean velocity along each track labelled in mm/s.

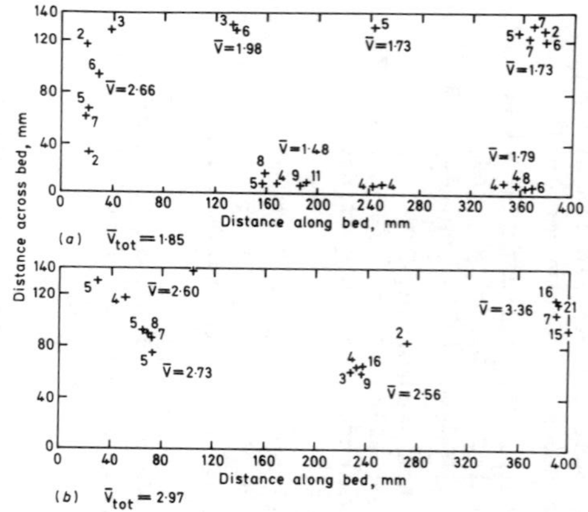

Figure 4. Mean collection points for two distributors, showing mean values of translational velocities.
(a) distributor A (2 mm holes on 10 mm square pitch)
(b) distributor B (0.8 mm holes on 5 mm square pitch)

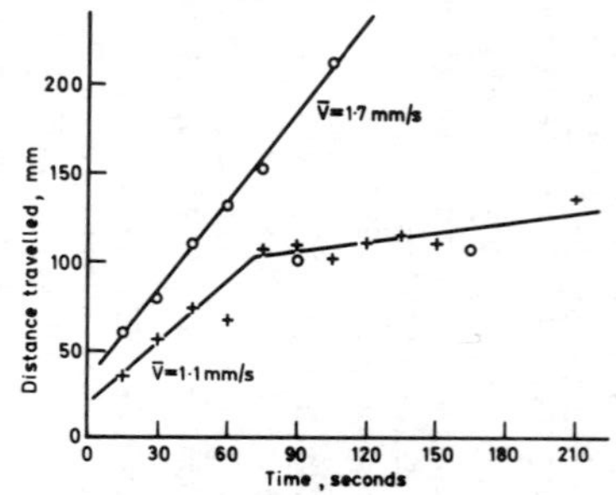

Figure 5. Mean values of distance travelled versus time for two distributors.
+ distributor A
0 distributor B

The Effect of Temperature and Gas Velocity Distribution on the Fluidization of the Sticky Particles

H. Arastoopour, A. Z. Gu, and S. A. Weil
Department of Gas Engineering
Illinois Institute of Technology/Institute of Gas Technology
3424 South State Street
Chicago, Illinois 60616 USA

ABSTRACT

We have constructed a 15 cm ID visually observable fluidized bed to obtain information about the fluidized bed under agglomerating conditions and study the effect of gas distribution and temperature variation on the rate and extent of agglomeration. The gas distributors consist of porous cones and a central jet. The temperature distribution and extent of agglomeration were measured at different operating conditions. Jet height, fluidizing air temperatures, and proper gas distribution can enhance the agglomeration of the sticky particles in the fluidized bed without causing detrimental sintering.

SCOPE

Agglomeration in fluidized bed reactors is sometimes required for successful operation and sometimes for the first sign of sinters or defluidization. This work reports on the role of distributor design and its effect on agglomeration rates and on temperature distributions as part of a continuing study to enhance our understanding and ability to design for thermal agglomeration.

CONCLUSION AND SIGNIFICANCE

The rate and extent of agglomeration of the sticky particles in the fluidized bed may be controlled by the shape of the gas distributor and the jet, fluidizing air temperatures, and flow rates. There is an optimum cone angle for fluidizing the gas distributor on which the mixing between particles and gas streams will result in a larger hot jet zone and a maximum amount of generated agglomerates without sintering. Based on our results this optimum half cone angle is about 45°.

The temperature distribution is one of the most important and dominant parameters in controlling the extent of agglomeration in the fluidized bed.

The rate of formation of agglomerates increased with jet and fluidizing air temperatures, and decreased with jet air velocity. The rate of agglomeration is more sensitive to jet air temperature than to the auxiliary air temperature.

As the fluidized bed approaches the softening temperature of the bed material, the particles become sticky and tend to agglomerate. Operating conditions should be properly set to prevent excessive agglomeration, which can lead to unacceptable sinters and defluidization, Sigell (1), Gluckman, et al.(2) and Huang, et al.(3).

A fundamental study leading to a successful correlation of particle characteristics, gas rates, and agglomeration rates would allow the rational design of systems that would achieve the desirable, and avoid the undesirable types of agglomeration phenomenon.

Experimental data is needed to obtain fundamental information on the rate of agglomeration and flow behavior of beds containing agglomerating material. In this part of a continuing study, Arastoopour, et al. (4) and Huang (5), we focus on how the bed behaves under agglomerating conditions and how the agglomeration rate of sticky particles depends on different operating variables with primary emphasis on distributor design.

Apparatus

The experimental setup is similar to Arastoopour et al. (6). Figure 1 shows the schematic diagram of the experimental setup. The fluidized bed column is made of a 15.2 cm ID glass pipe with 183 cm length which permits visual observation of the gas and solids flow patterns. The gas distributors consist of a porous section which has a two layer laminate 0.106 cm thick and a jet nozzle placed in the center of the porous distributor. A porous flat plate and three different porous cones with half cone angles of 30°, 45°, and 60° were used.

The jet air and auxiliary air streams were heated and regulated, as their flow rates were measured independently. The auxiliary air stream, which flowed through the conical porous plate, was the primary fluidizing gas. The bed temperature was measured using a thermocouple inserted from the top of the bed. To measure the central jet air temperature, a thermocouple was axially located inside at the tip of the jet nozzle. A thermocouple was also installed below the porous section of the gas distributor to measure the auxiliary air temperature. The outlet bed air passed through a cyclone to collect the elutriated particles.

Materials

To simulate systems of interest, the solid particles temperature should be dependent on stickiness. This provides a temperature range for particles to agglomerate in the bed. The bed material we chose was polyolefin with a softening temperature of 122°C and a melting temperature of 130°C. This material is generally safe and convenient. Table 1 shows the size distribution of our polyolefin feed particles.

Table 1. SIZE DISTRIBUTION OF FEED MATERIAL

Particle Size (μm)	Weight Percentage (%)
+1180	0
-1180+850	0.49
-850+710	3.75
-710+600	7.23
-600+500	11.30
-500+425	14.84
-425+355	15.44
-355+300	17.19
-300+250	12.14
-250+212	7.51
-212+180	5.30
-180+150	3.37
-150	1.44

Operating Procedure

The fluidized bed was initially charged with 800 grams of the polyolefin powder. The auxiliary and jet air flow rates and temperatures were adjusted to the desired values. The auxiliary flow rate was adjusted to have complete fluidization in the bed. The temperature of the auxiliary air was slightly lower than the softening temperature of the bed material to avoid sintering on the porous section of the gas distributors, and have a bed temperature appropriate for agglomeration when the jet temperatures were above the softening temperature. The bed was sampled from inside the column with buckets that were lowered down inside the fluidized bed, from the top of the columns at 2 hour intervals after jet temperature became steady. The particle size distributions were determined by sieve analysis. In this report, agglomerates are defined as all particles greater than 850 μm.

Result and Discussion

Effect of Gas Redistributors. An important factor in the extent of agglomeration in the fluidized bed is the rate of particle entrainment into the high temperature jet zone. The entrainment rate depends on the stress distribution among the particles in the vicinity of the jet zone. The direction, magnitude, and component of the stress toward the jet zone depends on the cone angle. To study the effect of the cone angle on the extent of agglomeration and temperature distribution in the vicinity of the high temperature jet zone, four distributors with half cone angles (α) with respect to vertical axes of 30°, 45°, 60°, and 90° (flat), were constructed and used. A central jet of 1.27 cm diameter was placed at the apex of each of the inverted cone distributors. The jet and auxiliary air temperatures were 136.7°C and 112.8°C, respectively. The jet and auxiliary air velocities were 14.9 m/s and 0.26 m/s, respectively. Figure 2 compares the weight percentage of the agglomerates obtained with the four gas distributors after 2, 4, 6 and 8 hours of steady temperature operation. Using the distributor with a cone angle of α = 90° (flat plate distributor), caused negligible agglomeration to occur. With a cone angle of 60°, the amount of agglomerates is significant. The extent of agglomeration increased radically as we decreased the cone angle to 45°. After 6 hours the amount of agglomerates generated with the 45° cone is six times that obtained with the 60° cone. However, in the runs with a half cone angle of α =30° gas distributor, the agglomeration was less than that with α = 45°. This implies that although the steeper inverted cone directed the stress on the particles in the cone section more effectively toward the jet zone, the amount of particles in the cone section decreased enough to result in lower entrainment of the particles. Because of this competitive effect there exists an optimum cone angle for maximum rate of agglomeration. This half cone angle is probably around α = 45°. We expect this angle could be the function of particle size distribution, internal friction angle of the particles, and jet and auxiliary air velocities. For the runs with α = 45°, sinter free operation was obtained over 7.5 hours. By then, however, more than 15% of the bed materials were large agglomerates which segregate and initiate a sinter in the conical section of the gas distributor.

Figure 3 shows the axial temperature measurement at the center line of the fluidized bed with the different gas distributor geometries. In Figure 3, $\Delta T = T - T_B$ and $\Delta T_o = T_j - T_B$, where T_j and T_B are jet and bed temperatures respectively. The jet temperature decreased and approached bed temperature as the jet penetrated into the bed. With the flat distributor, the jet is surrounded by materials essentially at bed temperature, and the heat transfer between the two is most efficient. With the conical distributors, where the auxiliary air temperature is higher than the bed temperature and there is a large ratio of distributor area to bed materials within the conical section, bed materials tend to be at a higher temperature than bed temperature. In these circumstances cooling of the jet would be delayed. Among the conical distributors, the steeper cones should have warmer materials in the conical section, and therefore less heat loss from the jet. However, direct mixing of auxiliary air with the jet air to get steeper cones will result in more heat loss from the jet and a significant decrease in jet temperature due to the mixing of two the two air streams. These opposing effects apparently result in a minimum temperature decrease rate; that is, a maximum hot jet zone which in turn results in a maximum rate of agglomerate generation. According to our experiments this optimum angle is about $\alpha = 45°$.

The Effect of Jet Air Temperature

In our thermally agglomerating fluidized-bed system, the jet air was supplied at temperatures higher than the melting temperature of the bed material. Agglomerates were generated in the fluidized bed presumably from partially melted particles in the hot jet zone. Figure 4 shows the effect of the jet air temperature on the amount of agglomerates generated during 8 hours of steady-state operation, with our 60° half cone angle distributor and 1.27 cm jet at the center. The auxiliary air temperature was 112.8°C and the auxiliary and jet air velocities were 0.26 and 14.9 m/s, respectively. The higher jet temperature resulted in more agglomerates. For all jet temperatures, the amount of agglomerates in the bed increased linearly with the residence time. No segregation of large particles or sinters were observed after 8 hours of steady operation. The rate of generation of agglomerates increased, exponentially with jet temperature. Figure 5 shows the rate of generation of particles larger than 850 μm versus jet temperature. The sensitivity of the rate to the jet air temperature is shown by an increase of five times for less than 3°C increase in jet air temperature.

The Effect of Fluidizing Air Temperature

The effect of auxiliary air and bed temperatures on the extent of agglomeration is shown in Figure 6, at a jet air temperature of 139.4°C. Three runs were conducted at auxiliary air temperatures of 107.2°, 110°, and 112.8°C, corresponding to bed temperatures of 100.7°, 101.3°, and 104.7°C, respectively. All are lower than the softening temperature of the bed material. In runs with auxiliary air temperatures equal to or higher than the softening temperature, sinters were formed on the porous cone immediately after the temperature of the auxiliary air next to the porous cone achieved the softening temperature. As the bed temperature increases with our increase in either auxiliary or jet air temperatures, the agglomeration is more sensitive to the auxiliary air temperature because of the much larger auxiliary air flow rate. With all other conditions the same, in a higher bed temperature more particles will reach and remain at the melting point because it is hotter before it is entrained into the jet. With a better chance to melt and become sticky, it generates agglomerates faster. This increase in the rate of agglomeration is also exponential with auxiliary air or bed temperatures, but it is not as sensitive as with jet air temperature. The conical angle of the auxiliary air distributor plays an important role in determination of limiting jet and auxiliary air temperatures for sinter free operation under agglomerating conditions in the fluidized beds.

ACKNOWLEDGEMENT

Thanks to the Institute of Gas Technology for financial support of this project and the Gas Research Institute Basic Research Division for initial support of this project.

LITERATURE CITED

1. Sigell, J. H., "Defluidization Phenomena in Fluidized Beds of Sticky Particles at High Temperatures," Ph.D. Thesis, City University of New York, 1976.

2. Gluckman, M. T., Yerushalmi, J., and Squires, A. M., "Defluidization Characteristics of Sticky Agglomerating Beds," in Keairns, D. L. (Ed.), Fluidization Technology, Vol. 2, 395-422, Washington, DC, Hemisphere Publishing Corp. (1976).

3. Huang, C. S., Arastoopour, H., and Weil, S. A., "Fluidization Behavior of Sticky Particles," paper presented at 1983 Fine Particle Society Meeting, Honolulu, Hawaii, August 1-5, 1983.

4. Arastoopour, H., Gu, A. Z., and Weil, S. A., "The Effect of Gas Distributors on the Agglomeration Process in Fluidized Beds," proceeding of 4th International Symposium on Agglomeration, C. E. Capes (Ed.), 443-450, Bookcrafters Inc., MI (1985).

5. Huang, C. S., "Fundamentals of Fluidization of Agglomerating Particles," Ph.D. Thesis, Illinois Institue of Technology (1985).

6. Arastoopour, H., Weil, S. A., Gu, A. Z., and Huang, C. S., "A Fundamental Study in Support of the Understanding of the Agglomeration of Coal in Coal Gasifiers, paper presented at the 20th Intersociety Energy Conversion Engineering Conference, Miami Beach, Florida, August 18-23, 1985.

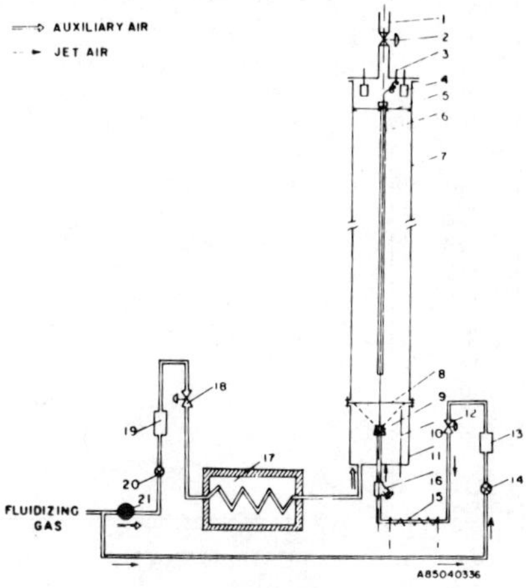

Figure 1. SCHEMATIC DIAGRAM OF EXPERIMENTAL SETUP

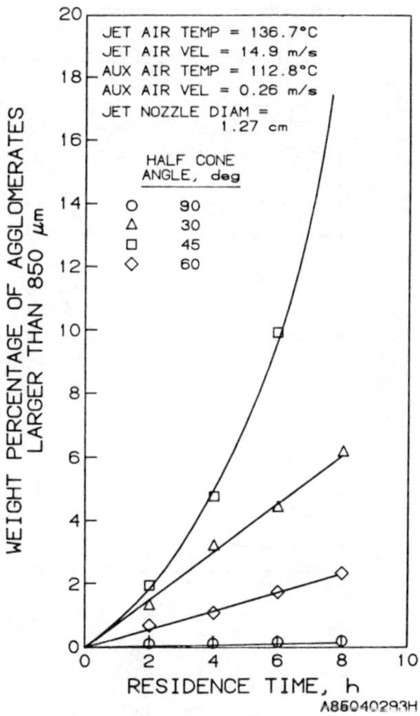

Figure 2. WEIGHT PERCENTAGE OF GENERATED AGGLOMERATES LARGER THAN 850 μm WITH DIFFERENT GAS DISTRIBUTORS

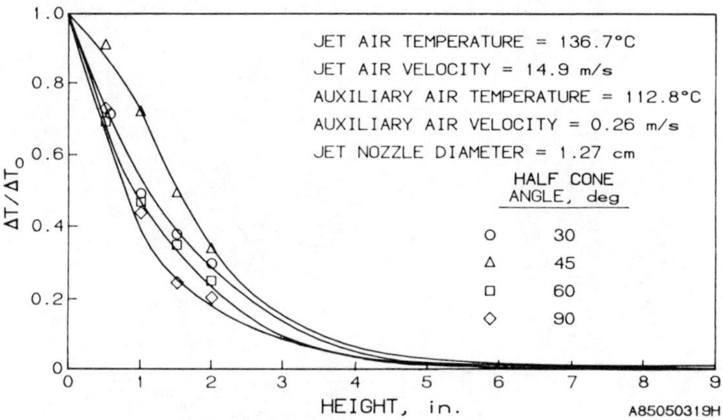

Figure 3. TEMPERATURE DISTRIBUTION ALONG THE CENTRAL AXIS OF THE FLUIDIZED BED WITH DIFFERENT GAS DISTRIBUTORS

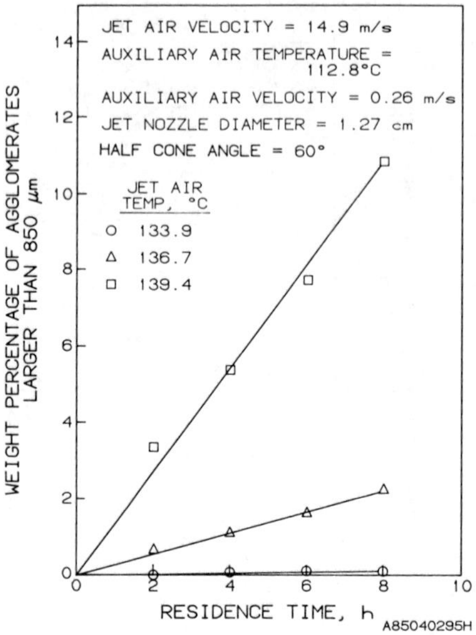

Figure 4. WEIGHT PERCENTAGE OF AGGLOMERATES LARGER THAN 850 μm AT DIFFERENT JET TEMPERATURES

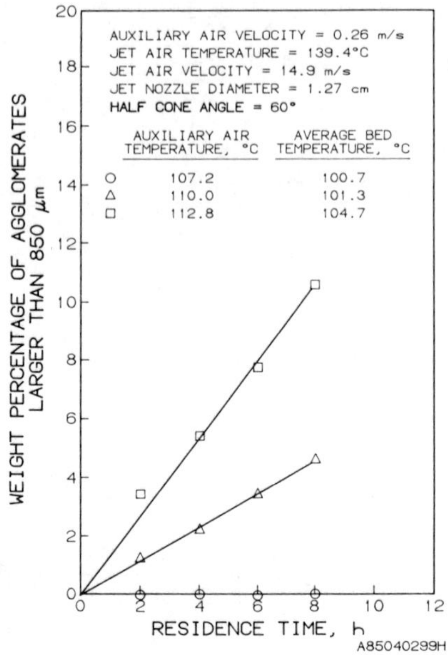

Figure 6. WEIGHT PERCENTAGES OF AGGLOMERATES LARGER THAN 850 μm AT DIFFERENT AUXILIARY OR BED TEMPERATURES

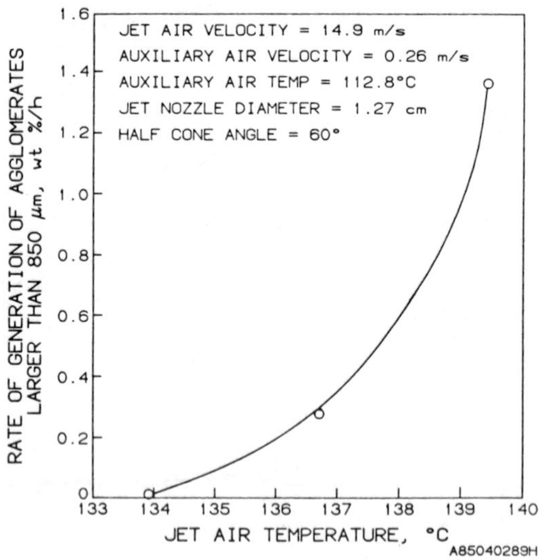

Figure 5. RATE OF GENERATION OF THE AGGLOMERATES VERSUS JET AIR TEMPERATURE

FLUIDIZATION OF NOVEL TENDRILLAR CARBONACEOUS MATERIALS

Edward F. Brooks
and
Thomas J. Fitzgerald

Chemical Technology Operations
TRW Inc.
One Space Park
Redondo Beach, California 90278

ABSTRACT

A family of tendrillar (fibrous) carbonaceous materials having unusual and desirable fluidization properties has been developed. When fluidized, the material agglomerates into a low density Geldart Group A, exhibiting high bed expansion and rapid mixing without bubbles. Similar results are obtained when the material is mixed with other fine powders.

SCOPE

This paper describes bed expansion and collapse, gas mixing, and heat transfer experiments performed in fluidized beds of Tendrillar Carbonaceous Material (TCM). This material when fluidized by itself has several unusual fluidization properties, the most obvious of which are large bed expansions, typically in the range of 100% to 500%, an absence of traditional bubbling, and low entrainment rates even at high bed expansion. In mixtures, the TCM can improve the fluidization of a wide variety of other materials, including very cohesive powders which will not fluidize at all by themselves. Depending on the nature of the other material, these improvements may include reduction of bubble size or elimination of bubbles, reduction of elutriation, and improved heat transfer. This applicability in mixtures greatly broadens the potential usefullness of the TCM.

CONCLUSIONS AND SIGNIFICANCE

This is the second paper on fluidization of TCM. In the first paper, Brooks and Fitzgerald (1) dealt with fluidized particle size, bed expansion, and entrainment. A brief synopsis of those results is relevant here. It was shown that TCM when fluidized forms agglomerates several hundred microns in size, as confirmed by optical analysis and inferred from entrainment and bed expansion data. It was also shown that bed expansion and entrainment increased with increasing bed temperature, at constant pressure and superficial velocity. The low entrainment rates reported, typically 10^{-3} kg/m^2s, put the fluidized TCM well outside the range noted by Yerushalmi and Avidan (2) for turbulent fluidization, where the solids rate was typically 10 kg/m^2s or more. The primary conclusion in the first paper was that TCM has a fundamental particle size which should make it a very cohesive Group C material, but that when fluidized, it forms stable agglomerates in the Group A range and fluidizes in a non-bubbling mode. A similar type of behavior has been reported for a copper/alumina gel material by Chaouki, et al. (3).

The present paper continues in the same vein of comparing and contrasting the behavior of TCM with that of commonly used fluidization materials. Tracer gas tests performed with TCM and with microballoons of similar size and density showed tracer gas bypassing and backmixing characteristic of a bubbling bed for the microballoons, but a tracer gas mixing pattern for the TCM which is consistent with a one dimensional convection plus diffusion gas flow. Experiments on rates of de-aeration of the TCM qualitatively agree with the results for Group A powders reported by Geldart and Wong (4), although the TCM bed expansion was substantially greater. In heat transfer tests where the heat flow was from an immersed surface to the bed, TCM behavior was consistent with what would be expected for a non-bubbling material. The heat transfer coefficient for the TCM fluidized alone was quite high in view of the low fluidized density of the material. The addition of TCM to an alumina bed substantially improved its heat transfer even though the fluidized density of the bed decreased.

TCM is of interest and significance because of its disinclination to bubble, which makes it a potentially useful tool to investigate the physics of bubbling. In an industrial sense, the ability of the TCM to improve the fluidization characteristics of other materials, which can include improved gas/solid contact, more vigorous solids mixing, and lower entrainment, may make it possible to improve various existing fluid bed processes, and to invent new ones using materials which could not previously be fluidized effectively.

MATERIAL CHARACTERIZATION

A photomicrograph of a typical TCM sample is shown in Figure 1. The material consists primarily of carbonaceous fibers with diameters in the range of 0.1 to 0.3 microns. The bulk density range for the TCM is from about 20 to about 400 kg/m^3. The fibers agglomerate into clusters of various sizes depending on how the material is handled. The TCM can be dispersed in an ultrasonic bath to a mass mean size of about 5 microns or less. At rest, the TCM behaves as a cohesive powder - it typically has a negative angle of repose and will not flow readily due only to gravity.

EXPERIMENTS

Bed Expansion and Rate of De-Aeration

Two samples of TCM having bulk densities of 35 and 220 kg/m^3 were fluidized in a two dimensional cold flow unit, the former at superficial velocities of 0.07 and 0.09 m/s, and the latter at superficial velocities of 0.12 and 0.20 m/s. The test unit had inside dimensions of 0.016 m deep, 0.5 m wide, and 1.84 m tall. For deaeration experiments, a valve closely coupled to the inlet plenum (volume equivalent to about 3% of the test unit volume) was closed, and settling bed height as a function of time was plotted from a video tape of the event. Test results are shown for each TCM sample at one fluidization velocity in Figure 2. The other runs were omitted for clarity. Results are quite similar to those shown by Geldart and Wong (4) for Group A powders. The high density material had a diffuse bed top, the initial collapse of which accounts for the steep slope for the first two seconds. The bed top for the low density sample was more uniform, so no initial sharp drop was seen. Each material then exhibited a linear drop in bed height (Geldart and Wong's "hindered settling stage") followed by a slower final compaction (Geldart and Wong's "solid consolidation stage"). Collapse times were similar to those shown by Geldart and Wong, but the bed expansions of about 200% for the high density TCM and about 550% for the low density TCM are much higher than any reported by those authors. Since their hindered settling stage and solid consolidation stage take place without bubbles in the bed, and since the TCM fluidizes in an essentially bubble free mode, it is reasonable to say that the observed TCM deaeration characteristics are exactly what would be expected, based on their work.

Gas Mixing

A transient response gas tracer experiment was performed in a 0.10 m diameter cold flow test unit to determine gas mixing characteristics in a bed of TCM with a bulk density of 180 kg/m^3 and results were compared to a similar experiment in a bed of alumina microspheres with a bulk density of 400 kg/m^3 and a particle size range of 60 to 300 microns. Sampling probes were installed on the centerline of the test unit at various heights above the perforated plate distributor. The fluidizing gas was nitrogen. Helium was used as the tracer gas and a helium leak detector was used as the diagnostic instrument. Each material was fluidized by itself at a superficial velocity of 0.20 m/s, and helium was injected through one of the probes for a period of about 30 seconds and detected through the others. Results are presented in Figure 3 for both materials. The helium injector probe was located on the test unit centerline about 0.51 m above the perforated plate distributor, and the sampling probes were located at heights of 0.21, 0.31, 0.41 and 1.3 m above the distributor plate. The left vertical scale indicates the relative maximum concentration of tracer gas sensed by the detector (open symbols) and the right vertical scale indicates the time from helium injection until a first response was recorded at the detector probe (blackened symbols). In the microballoon test, the tracer gas was first detected at the downstream probe, and later at the three upstream probes. The pattern of arrival times at the upstream probes is consistent with the well characterized bubbling mode in which bubbles travel upward near the test unit centerline, with solids and some gas backmixing downward along the wall to the distributor, then back up along the centerline with the bubbles. In the case of the TCM, the helium was detected at the downstream probe and at the two closest upstream probes at virtually the same time, but was not detected at the furthest upstream probe until much later. Also note the very steep concentration gradient upstream of the injector probe. The data for the TCM are consistent with a vigorous, random local mixing pattern superimposed on a one dimensional convection pattern, as opposed to the top to bottom mixing pattern produced by bubbling in the microballoon bed. Also note that the concentration gradient upstream of the injector probe was much steeper for the TCM than for the microballoons, indicating the more limited backmixing for the TCM. From this test data, we concluded that the gas and, by extension, the solids mixing patterns were quite different for the two materials, and were consistent with bubbling fluidization of the microballoons and non-bubbling fluidization of the TCM.

Heat Transfer

A series of experiments was performed in a 0.16 m diameter test unit to look at heat transfer characteristics of various materials. In this particular series, a resistance heater with a diameter of 0.019 m and a length of 0.15 m was mounted vertically in the test unit. Heat transfer from this heater to the bed was measured, based on the readings of six thermocouples attached to the heater and eleven thermocouples immersed in the bed. The fluidizing gas was nitrogen and the operating pressure was slightly above atmospheric. Results are reported in Table 1 for experiments with an angular alumina with a bulk density of 1700 kg/m^3 and a particle size range of 60 to 200 microns, and for TCM with a bulk density of 200 kg/m^3.

The TCM fluidized alone had a surprisingly high heat transfer coefficient in view of its very low fluidized density. The heat transfer coefficient for the alumina is similar to that for sand reported by Botterill, et al (5). When 5% TCM by weight was added to an alumina bed, the fluidized density decreased significantly, while the heat transfer coefficient remained about the same. When the TCM concentration was increased to 12%, the heat transfer coefficient increased substantially, clearly showing a synergistic effect between the two materials.

The alumina was on the B side of the Geldart Group A/B border. In a transparent 2-D cold flow unit, a bed of the alumina was found to have large bubbles, with somewhat slow (especially compared to the TCM) solids mixing except in the vicinity of bubbles. When TCM was mixed with the alumina, bubbling was reduced as the TCM concentration increased, and bulk solids mixing was observed to be more rapid. Based on the heat transfer results for each material alone and on the observation that the fluidized density of a mixture of TCM and alumina is less than for the alumina alone, we would tend to expect that the heat transfer coefficient of a mixture would be less than that of the alumina alone. These effects, however, were counteracted by the increase in solids mixing, especially at the 12% TCM concentration. The scope of the study did not permit us to vary the alumina/TCM ratio over a wide enough range to identify the maximum achievable heat transfer coefficient.

In a separate series of experiments, heat transfer to and from a copper tube with water flowing through it was measured in beds of various materials, and a curious result was obtained in a bed of TCM. With the bed at room temperature, heat transfer coefficients similar to those in Table 1 were measured for both hot and cold water. However, when the bed temperature was increased to about 680°K, so that the temperature differential between the bed and the water was about 400°K, it was found that the TCM clung to the tube, forming an insulating layer and resulting in very low heat transfer. Similar but less severe behavior was reported by Hoelen and Stemerding (6) for Group A powders which have substantial bed expansion, and is attributed at least partially to a local decrease in bed expansion in the vicinity of the cold heat transfer surface. Further investigation of this phenomenon needs to be performed in the more commercially applicable temperature differential range of about 20°K to 100°K, rather than the more extreme 400°K at which our testing was performed.

OBSERVATIONS ON TCM BEHAVIOR

The issue of why fluidized TCM should not bubble has been addressed informally, and there appear to be several contributing factors. As noted by Geldart (7), graphite coating of particles has been used to reduce problems associated with static electricity, so the fact that TCM is carbonaceous and hence conductive is likely an important feature. Secondly, it is broadly known that bubbles tend to be less stable in low density solids, so that the very low density of the TCM is rather certainly desirable for bubble free fluidization. A third feature of the TCM felt to be critical is the highly irregular particle shape of the TCM, which fortuitously seems to allow the material to agglomerate to sizes desirable for fluidization. The combination of density and shape was investigated when a TCM sample was compacted into smooth spheroidal particles of about 100 to 200 microns in size, with a bulk density of about 650 kg/m^3. When fluidized, this granular material bubbled freely much like any other material of similar size and density. As the granules broke up and the smooth surfaces disappeared, the material gradually returned to its normal non-bubbling state.

The effect of TCM in mixtures with another material depends on the properties of the other material as well as the mixture ratio. Energy and momentum transfer tend to be the key parameters governing the interaction. The alumina used in the heat transfer experiments is at about the upper limit of particle mass which will stay well mixed with the TCM. With larger, more dense Group B materials, the TCM forms a dual layer fluid bed where the TCM sits on top of the heavier material and there is little interaction between the two layers. It has been shown that bubbles emerging from the lower layer in this configuration are rapidly dissipated in the TCM layer. As the density and size of the second material decrease into the lighter and smaller Group A powders, the TCM will mix readily with the other powder and tend to stay well mixed. When the TCM is combined with another material having a size of about 70 microns and a bulk density of about 780 kg/m^3, which is typical of a cracking catalyst for example, a 5% concentration by mass of TCM is sufficient to effectively eliminate bubbles and substantially reduce entrainment of the other material. As we go from Group A powders to cohesive Group C powders, the amount of TCM needed to produce good quality fluidization in a mixture begins to increase, compared to that required for a mixture with an easily fluidized Group A material. We have achieved good results in mixtures with materials near the C/A border using about a 10% by mass concentration of the TCM. Mixtures with highly cohesive powders which will not fluidize at all by themselves may require about a 50% by mass concentration of TCM to achieve good fluidization. A 50/50 mixture of TCM and an extremely cohesive kieselguhr of the same bulk density was found to have fluidization properties, such as bed expansion, very similar to those for the TCM fluidized alone. As a general rule, for mixtures with up to 50% TCM by mass and a second material with a bulk density of 700 kg/m^3 or greater, we have observed bed expansions in the range of 50% to 150% at a superficial velocity of about 0.3 m/s.

LITERATURE CITED

1. Brooks, E. F., and T. J. Fitzgerald, "Aggregation and Fluidization Characteristics of a Fibrous Carbon," presented at the 77th Annual Meeting, AIChE, Chicago (1985).

2. Yerushalmi, J. and A. Avidan, "High-Velocity Fluidization," in Fluidization, Second Edition, Davidson, Clift, and Harrison (Ed.), Academic Press Inc. (London) Ltd. (1985).

3. Chaouki, J., C. Chavarie, D. Klvana, and G. Pajonk, "Effect of Interparticle Forces on the Hydrodynamic Behavior of Fluidized Aerogels," Powder Technology 43, 117 (1985).

4. Geldart, D., and A. C. Y. Wong, "Fluidization of Powders Showing Degrees of Cohesiveness-II. Experiments on Rates of De-aeration," Chem. Eng. Science, 40 (4), 653 (1985).

5. Botterill, J. S. M., Y. Teoman, and K. R. Yuregir, "Temperature Effects on the Heat Transfer Behavior of Gas Fluidized Beds." AIChE Symp. Ser. 77 (208), 330 (1981).

6. Hoelen, Q. E. J. J. M. and S. Stemerding, "Heat Transfer in a Fluidized Bed. Part 1. The Effect of Thermal Driving Force on the Heat Transfer Coefficient." Powder Technology 30, 161 (1981).

7. Geldart, D. "Types of Gas Fluidization," Powder Technology 7, 285 (1973).

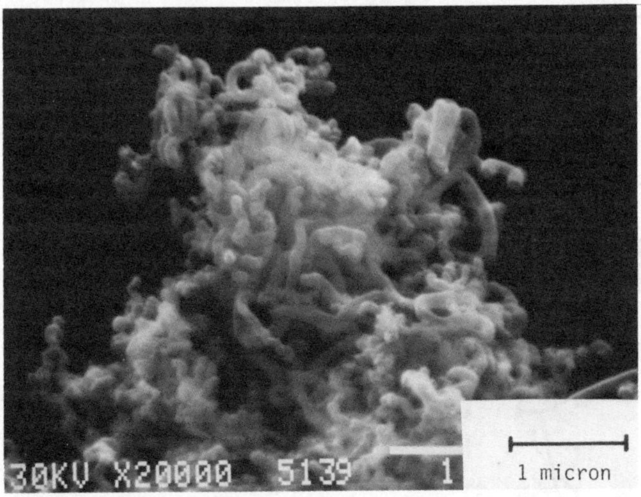

Figure 1. Photomicrograph of TCM Sample

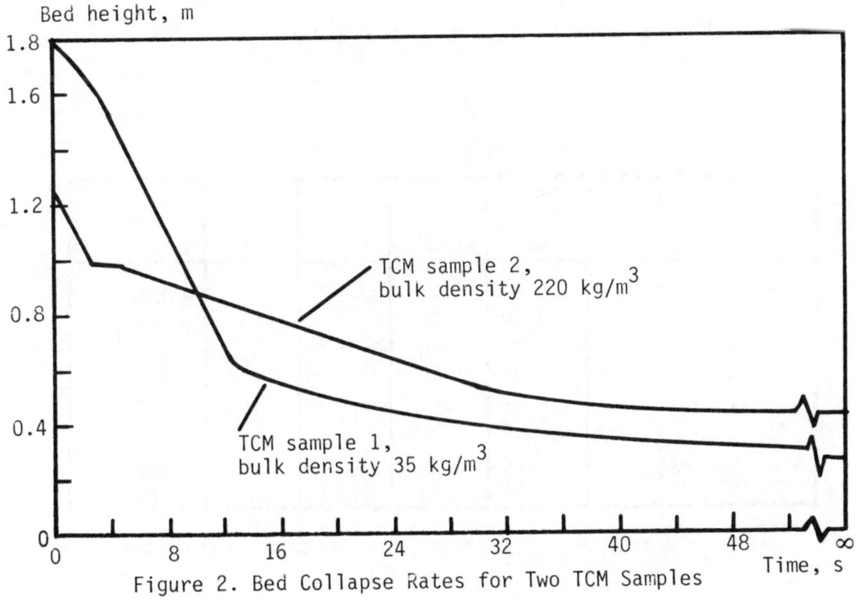

Figure 2. Bed Collapse Rates for Two TCM Samples

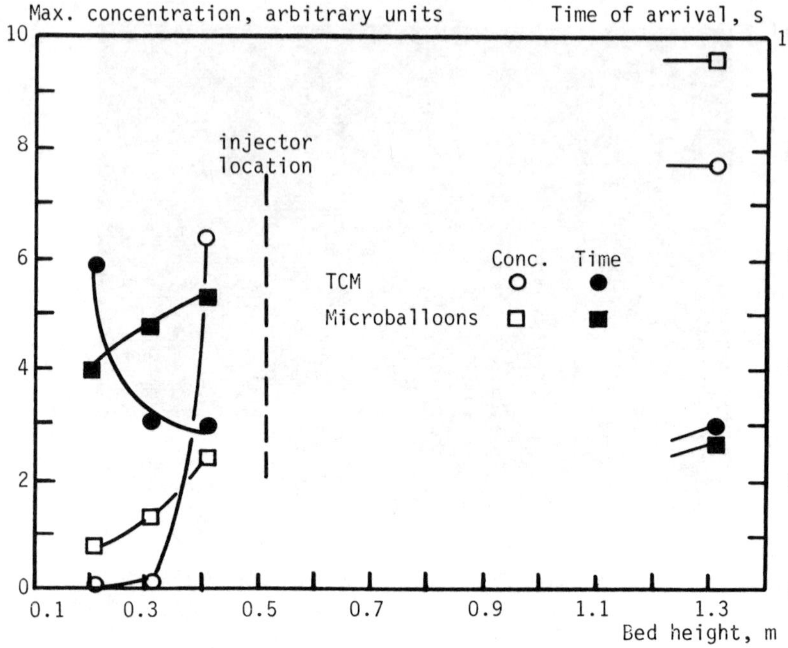

Figure 3. Arrival Time and Concentration Data for TCM and Microballoon Beds

Material	u_s m/s	ρ_f kg/m^3	T_{bed} °K	ΔT °K	h_o W/m^2°K
TCM	0.20	170	773	91	270
	0.50	80	769	83	300
Alumina	0.20	1260	785	111	610
	0.42	1310	777	107	690
95% Alumina, 5% TCM	0.44	820	781	117	710
88% Alumina, 12% TCM	0.18	950	794	100	970
	0.44	780	771	78	1200

Table 1. Heat Transfer Data for TCM, Alumina, and Mixtures

A MOBILE BED PROCESS FOR FIBROUS MATERIALS

R Legros, C A Millington and R Clift
Department of Chemical and Process Engineering
University of Surrey, Guildford, Surrey GU2 5XH, England

ABSTRACT

A novel type of spout fluid bed has been developed, for processing fibrous materials. The bed uses gas jets, introduced at relatively high velocity through the sloping sides of the distributor, to entrain gas from the freeboard. The drag of the entrained gas puts the bed into tension in its upper layers near the bed axis, thereby disentangling the fibres and maintaining controlled circulation of the particles.

SCOPE

Industrial processing techniques are needed to mobilise masses of fibrous material in a controlled fashion, and to contact them with a gas. Some specific operations are blending and drying of tobacco, but there are many potential applications to lignocellulosic and other materials. While fluidised beds are commonly used for treating granular materials, and spouted beds are employed for larger or non-spherical particles, neither of these techniques is suitable for processing materials in which the individual particles take the form of long fibres. The present work was undertaken to develop a novel technique for handling and processing fibrous materials. The device relies on using a number of gas jets to impose forces on the material to support it and to put it into tension so that the fibres are disentangled. The material can then be maintained as a toroidally-circulating controlled inventory of separated fibres, enabling the required processing operations to be carried out.

CONCLUSIONS AND SIGNIFICANCE

It is possible to maintain controlled mobilisation of a mass of long fibrous material by a technique which combines spouting and fluidisation. The technique differs from more common spout-fluid beds in using a distributor with inclined walls through which gas enters as jets at relatively high velocity. These wall jets entrain gas downwards from the freeboard. The resulting distribution of forces puts the bed into tension in its upper layers. When sufficient tensile stress is achieved, the fibres are disentangled and controlled continuous circulation of the material within the bed can be maintained. Analysis of the forces on the bed leads to a quantitative condition for the limit of controlled mobilisation. The technique has been developed initially for such operations as blending and drying of tobacco, but it should be applicable to many fibrous materials.

* * *

Spouted beds were originally developed for processing agricultural materials, such as wheat, which cannot readily be handled in conventional fluidised beds (1). The range of agricultural products which can be professed in fluidised or spouted beds has steadily been extended. For example, Drona et al (2), have described the drying of paddy in spouted beds, while Akao and Okamoto (3) have shown that wheat bran powder can be fluidised controllably and used as the substrate for micro-organism culture. However, vegetable materials which take the form of long fibres cannot be processed readily in fluidised or spouted beds, because they form "mats" in which the fibres are so entangled that free movement is prevented. Therefore a gas/particle contacting device for processing fibrous materials must disentangle the fibres, as well as supporting and mixing them. Use of internal moving parts or of a granular diluent to facilitate fluidisation generally causes breakage or other degradation of the material being processed. Therefore it is desirable to use solely the action of the gas flows, and to keep the disentangling action as gentle as possible while maintaining controlled mobilisation of the fibrous material. A device will now be described which meets these requirements for cut tobacco, and which should be applicable to a wide range of fibrous materials.

DESCRIPTION OF MOBILE BED PROCESS

Figure 1 shows the contacting device. Further details are given by Legros *et al* (4), while Legros (5) describes less successful configurations tested in the course of reaching the design in Figure 1. The distributor (a) is an inverted truncated square pyramid. The inclination of the walls depends upon the material to be mobilised; an angle of 30° to the vertical is appropriate for cut lamina tobacco. The size of the lower square section is probably more critical than that of the upper; stable operation has been achieved with lower sections 3 or 5 cm square in columns in which the freeboard was 30 or 15 cm square. The inclined surfaces contain a number of orifices (b) positioned in horizontal rows, the lowest row being at the junction of the base and the inclined wall. The number of rows is dependent upon the size of the device; three rows are necessary for a 15 cm square unit and five for a 30 cm unit. The number, pitch and diameter of the orifices and the row spacing depend upon both the solid medium and the size of the device; but a minimum of 12 x 3.5 mm diameter holes (3 per side) was found necessary in the lowest row to mobilise tobacco. The 15 cm square unit had in total 9 holes per side while the 30 cm square unit had 14 holes per side. The lower square section of the distributor contains a single orifice (c) positioned on the central vertical axis.

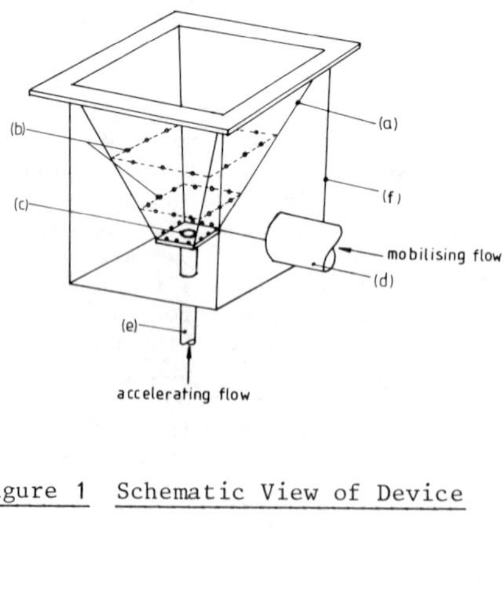

Figure 1 Schematic View of Device

Air at 15°C and 50% relative humidity was introduced as two independently controlled and metered supplies from a common low pressure source. Flow through the orifices in the inclined walls (the mobilising flow) was from a common plenum (f), no control being exercised on the flow through individual orifices. The second supply (the accelerating flow) was connected to the single orifice in the base. A square section perspex column, 1 m high, was mounted above the distributor and the exhaust air passed through a cyclone before discharge to atmosphere.

Cut Lamina Tobacco

British-American Tobacco Company blend 72 cut lamina tobacco was used. It is prepared by removing the leaf flesh or lamina from the stem to produce large irregular pieces of lamina, typically up to 10 cm square but mixed with many smaller pieces. The pieces, which are often folded over, are laid in layers with material from different sources, compressed by bands and cut normal to the lamina surface. The pitch of the cutting process is variable and for 'blend 72' is specified as 35 cuts/inch. The resulting fibrous particles are approximately 0.1 mm thick (the leaf thickness), 0.72 mm wide (the cutting pitch) but vary between 0.5 and 10 cm in length. Because of folding during processing the particles are twisted along their long dimension and have a number of "kinks" or "knees". The resulting product is entangled so that the separation of individual particles is difficult.

The tobacco used here had an initial moisture content of 18%, determined by a standard oven test. Moist tobacco particles are flexible and resilient, but become brittle as the moisture content is reduced; the transition depends on both temperature and moisture-content. Under the experimental conditions, particles containing 18% moisture were flexible and resisted breakage while those containing 8% moisture were brittle and easily fractured. Care was taken to ensure that the tobacco was always in the flexible condition, although no measures were taken to disentangle the fibres before charging to the column.

Qualitative Observations

In the experimental program, the 15 cm square device was loaded with either 50 g or 100 g: a 100 g charge resulted in a full distributor. The corresponding charge for the 30 cm device was 800 g. Increasing the charge to utilise part of the column above the distributor gave unstable operation. By independent control of the mobilising and accelerating flows, conditions for stable mobilisation of the tobacco were investigated.

No mobilising flow. Operation without mobilising flow, equivalent to conventional spouting, proved impossible for this fibrous material. Increasing the accelerating flow from zero initially causes the tobacco mass either to expand or to be displaced slightly upwards from the inclined walls of the distributor. Further increase forms a void above the central orifice, with individual particles or small entangled clumps or particles expelled from the bed at high velocity, the bulk of the solid remaining immobile. The displaced solid was either transported out of the column or settled onto the immobile mass. Unlike operation with a granular material, particles were not entrained from the annulus into the spout. High accelerating flows transported all the charge out of the column.

No acclerating flow. When mobilising flow is started, the bed moves away from the inclined walls of the distributor. As the flow is increased the upper bed surface becomes domed and eventually motion is apparent in the surface from the central vertical axis towards the walls. At this point, the entire bed becomes mobile with a "rolling" circulating pattern. Gas and solid move upwards near the bed axis, while the particles move outwards near the surface and downwards near the inclined walls of the distributor. With increasing gas flow, the motion becomes more violent until eventually the rolling mobilisation is disrupted and the charge is transported out of the column. Once mobilisation is established, it is possible to reduce the gas flow below the level required to initiate motion without losing circulation. The motion of the solid charge suggested that the fibres had been largely disentangled.

Both mobilising and accelerating flows. Stable mobilisation is possible over a range of flows when mobilising and accelerating flows are used together. The behaviour of the bed is the same as that observed with no accelerating flow, but the mobilising flow needed to initiate motion is less than that necessary with no accelerating flow. Figure 2 shows schematically the range of stable mobilisation. At lower flows the charge acts as a high resistance packed bed, while at higher flows the solid approaches a state of lean phase fluidisation and/or pneumatic transport. The upper boundary of the stable region is somewhat subjective and hence more difficult to define than the lower boundary. The lower boundary defines a state of incipient mobilisation in which the motion of the solid is just maintained and any reduction in either gas flow rate causes motion to stop and the bed to settle.

Mobilisation of tobacco in the 30 cm square unit was more stable than in the 15 cm square unit. No mobilisation occurred when a two-dimensional model, 2 cm thick, of the smaller unit was tested. It seems that the ratio of particle length to maximum distributor dimension may be important in a two-dimensional unit. Sieve analysis of the tobacco before and after 10 minutes of mobilisation showed that breakage of particles was slight: the analyses showed no significant difference. Some dust and small sections of lamina were collected in the cyclone, but this material amounted to less than 2% of the initial bed mass.

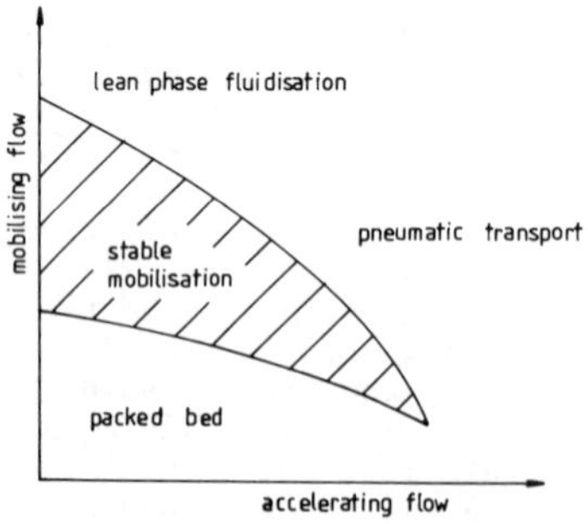

Figure 2 Regime of Stable Mobilisation (schematic)

MECHANICS OF MOBILISATION OF A FIBROUS MASS

A tobacco plus presents a high resistance to gas flow, with pressure gradient proportional to the square of the superficial gas velocity for the range of velocities used in this device.

Velocity profiles were taken with

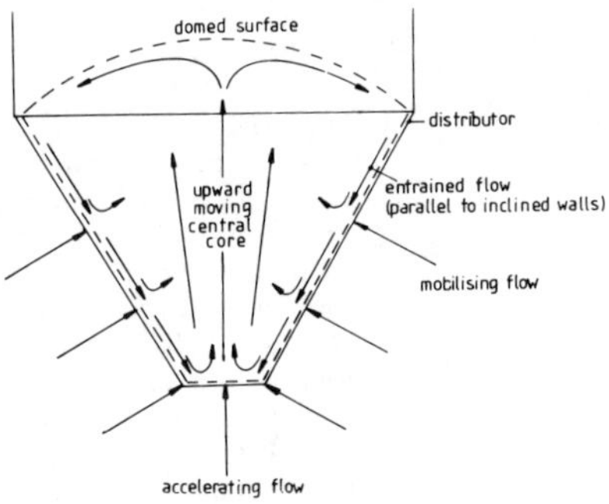

Figure 3 Motion of Mobilised Tobacco (schematic)

a pitot tube across the top of an empty distributor through which both mobilising and accelerating flows were passed. A central core of upward moving gas was detected, with downward flow close to the inclined walls. Integration of the profiles indicated that the volumetric flow rate in the central core was up to twice the sum of the mobilising and accelerating flows. Thus the injected gas flows entrain a comparable flow of gas downward from the column above the distributor region. Above an immobile charge, the velocity profile was flat, as expected with the gas flow distributed uniformly by the high resistance of the tobacco plug.

Rapid fluctuations in velocity and direction of local gas flow prevented a profile from being determined above a mobile bed. However, other evidence suggests that the flow pattern is similar, at least qualitatively, to that in empty column. A perspex distributor was constructed for the 15 cm column, so that the tobacco could be observed directly. Figure 3 shows schematically the observed motion of the mobilised solid. A central expanding core of upward moving solid was visible, starting at the base of the device, with a layer approximately 1 cm deep of particles moving rapidly downwards parallel to the inclined walls of the device. The outward motion of the particles toward the walls in the upper surface of the bed was also confirmed.

In a separate experiment it was found that mobilisation was disrupted when a mask, which just fitted the square column and contained a square central hole, was lowered towards the solid. The distance of the mask above the upper surface of the bed at which disruption occurred was reproducible, and decreased as the hole in the mask was made larger. The mask was found to inhibit entrainment from the freeboard down the inclined walls, so that this flow appears to be required for stable mobilisation. The disentanglement of the fibres which takes place in the upper zone of the bed probably results from the action of tensile forces, and this is consistent with the "rolling" motion of the bed.

Analysis of the Forces on a Plug of Tobacco

Using these observations, the forces acting within a bed of tobacco at the point of incipient mobilisation may be analysed. The analysis is reported in more detail by Legros (5). Because the distributor pressure drop is high, equal flow is assumed through each mobilising orifice. For a high-resistance medium, it may be shown (5) that a gas jet exerts a force on the solid approximately equal to its momentum flux. Therefore, for the accelerating (A) and each mobilising (M) orifice, the force exerted on the solid by the air flow is given by:

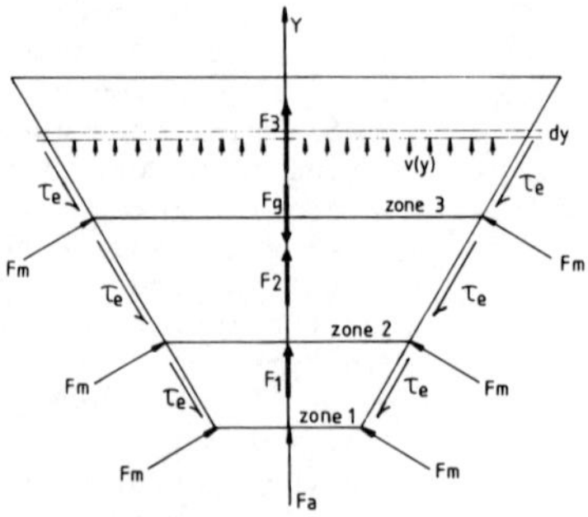

Figure 4 Forces on Tobacco Bed

$$F_M = 4\rho M^2/n^2\pi d_m^2 \qquad (1)$$

$$F_A = 4\rho A^2/\pi d_A^2 \qquad (2)$$

The upward flow rate of gas in the bed varies with depth but is assumed constant within each of the zones defined in Figure 4. The upward flow in any zone is the summation of all flows entering below its upper boundary and one third of the flow entrained downwards at the wall. Thus, for a distributor with the same number of holes in each row, the flows in the three zones are given by:

$$Q_1 = M/3 + A + E/3 \quad (3) \qquad Q_2 = 2M/3 + A + 2E/3 \quad (4) \qquad Q_3 = M + A + E \quad (5)$$

The entrained flow, E, is unknown but a trial and error use of the force balance (see below) indicates that a first estimate of 10% of the mobilising flow is realistic. The velocity profile of the upward moving gas is assumed to be flat because of the high resistance of the tobacco bed. The vertical force exerted by the gas flow on the bed within the element from y_{i-1} to y_i above the bottom of the distributor is then

$$F_i = \int_{y_{i-1}}^{y_i} \beta\rho V^2 S dy \qquad (6)$$

where the constant β may be evaluated from the relationship between velocity and pressure drop for a packed bed. This integral may be evaluated to give the drag force F_i in each zone of the bed as

$$F_i = \phi\beta\rho Q_i^2/2\tan\alpha \qquad (7)$$

where the term ϕ describes the geometry of the distributor and is a function of the positions of the rows of mobilising orifices and the dimension of the distributor base. The gas flow entrained from the freeboard of the bed is assumed to exert a distributed shear stress, τ_e, along the inclined boundary of the solid. The gravitational force on the bed, F_g, is known from the bed mass, assuming uniform bed density. Summing these forces in the vertical direction

$$\tau_e = (F_1 + F_2 + F_3 - F_g + nF_M \sin\alpha + F_A)/S_e \cos\alpha \qquad (8)$$

From Equation (8), τ_e is evaluated and used to obtain a new estimate for E so that the flow rates in the three zones can be refined. This trial-and-error process is repeated until the vertical force balance is satisfied.

Observation of mobilised tobacco suggests that the upper layers of the bed are in tension and the lower section in compression. Consider the forces on a half-bed section, shown in Figure 5, with a distributed normal stress acting along the central plane. The local normal force per unit height acting on the central plane is denoted by Σ, taken to be

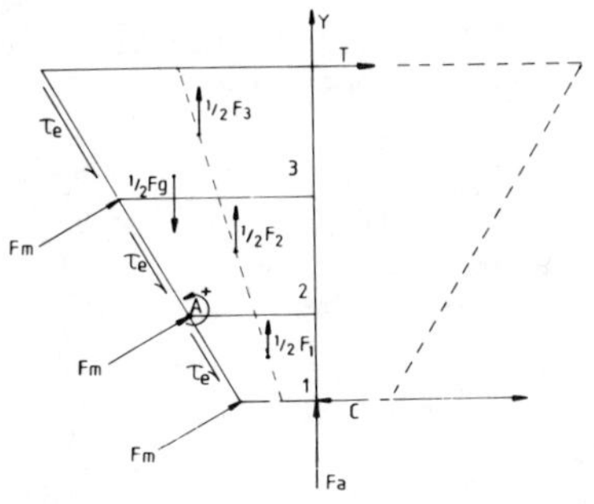

Figure 5 Distribution of Forces on Half-Bed

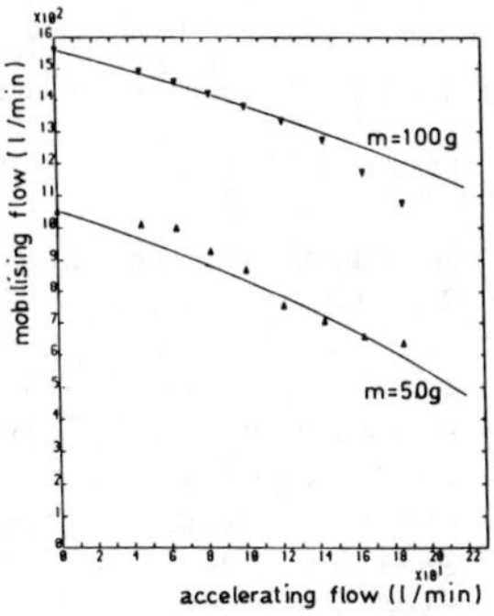

Figure 6 Conditions for Incipient Mobilisation

positive in tension. Assuming Σ to vary linearly from -C at the bottom to T in the surface:

$$\Sigma = (T + C) \frac{y}{b \cos \alpha} - C \qquad (9)$$

Note that C, T and Σ are effectively normal stresses integrated across the width of the bed at the level in question. Resolving forces in the horizontal direction gives

$$(\tau_e \, Se \sin \alpha)/4 + (n \, F_m \cos \alpha)/4 + (T - C)(b \cos \alpha)/2 = 0 \qquad (10)$$

and taking moments about point A yields

$$M_{PS} + M_{DF} + M_{F1} + M_{F2} + M_{F3} + M_{Fg} + M_{FA} = 0 \qquad (11)$$

where M_{DF} is a function of C. Eliminating C between Equations (10) and (11) and substituting the appropriated geometric data for the 15 cm square distributor yields

$$\left.\begin{aligned} &0.1\text{kg charge: } T = 1.95 \times 10^5 M^2 + 6.07 \times 10^5 MA + 1.58 \times 10^6 A^2 - 3.9 \\ &0.05\text{kg charge: } T = 1.77 \times 10^5 M^2 + 9.79 \times 10^5 MA + 2.95 \times 10^6 A^2 - 5.3 \end{aligned}\right\} \qquad (12)$$

Evaluating the tensile force on the surface, using experimental values for A=0, gives T as 128 N/m for a 0.1 kg charge and as 80 N/m for a 0.05 kg charge; the corresponding mean tensile stress across the bed is also higher for the larger charge. Probably the apparent dependence of the value of T at incipient mobilisation on charge results from the simplifying assumptions made in the analysis, in particular the implicit assumption of constant stress along a bed section. However, inserting these values into Equation (12) gives a quantitative relationship between the mobilising and accelerating flows at incipient mobilisation. The relationship is compared in Figure 6 with experimental values for charges of 0.05 and 0.1 kg. The close agreement suggests that the analysis summarised here is valid. Data are now being obtained from the 30 cm square unit, to investigate the applicability of the analysis to scale-up.

ACKNOWLEDGEMENTS

The work described here has been supported by the Group Research and Development Centre of British-American Tobacco Company Ltd.

NOTATION

A	- accelerating gas flow	(m^3/s)
b	- length of inclined side of distributor	(m)
C	- compressive force per unit height at base of bed	(N/m)
d_A	- diameter of accelerating orifice	(m)
d_m	- diameter of mobilising orifice	(m)
E	- total gas flow entrained down the walls of the distributor	(m^3/s)
F_A	- force exerted by accelerating flow	(N)
F_g	- gravitational force on bed	(N)
F_i	- drag force due to upwards gas flow in zone i	(N)
F_M	- force exerted by gas flow through each mobilising orifice	(N)
M	- mobilising gas flow rate	(m^3/s)
M_{DF}	- moment about A due to distributed force	(N.m)
M_{FA}	- moment about A due to force of accelerating flow	(N.m)
M_{Fi}	- moment about A due to drag force in zone i	(N.m)
M_{Fg}	- moment about A due to gravitational force on plug of tobacco	(N.m)
M_{PS}	- moment about A due to force at mobilising orifices	(N.m)
n	- number of mobilising orifices	(-)
Q_i	- upwards gas flow in zone i	(m^3/s)
S	- local cross sectional area	(m^2)
S_e	- area of inclined walls	(m^2)
T	- tensile force per unit height at surface of bed	(N/m)
V	- local gas velocity	(m/s)
y	- vertical distance measured from the lower square section	(m/s)
α	- angle of inclination of the distribution wall from the vertical	(o)
β	- composite proportionality constant	(m^{-1})
ρ	- gas density	$(kg.m^3)$
Σ	- tensile force per unit height within bed	(N/m)
$\emptyset$	- factor dependent on bed geometry	(-)

LITERATURE CITED

1. Mathur, K.B. and N. Epstein, "Spouted Beds", Academic Press, New York (1974).
2. Drona, A.K., B.P.N. Singh and S.C. Bhattacharya, p. 405 in "Fluidization", D. Kunii and R. Toei (Eds), Engineering Foundation, New York (1984).
3. Akao, T. and Y. Okamoto, ibid. p. 31.
4. Legros. R., C.A. Millington and R. Clift, U.K. Patent Appl. no. 8501959 (1985).
5. Legros, R., Ph.D. Thesis, University of Surrey (1986).

SPOUTABILITY AND SPOUT DESTABILIZATION OF FINE PARTICLES WITH A GAS

Pratap P. Chandnani and Norman Epstein
Department of Chemical Engineering
The University of British Columbia
Vancouver, B.C., Canada V6T 1W5

ABSTRACT

A criterion for the cold gas spoutability of fine particles is developed and flow regimes are mapped. The criterion applies also to coarse particles. Termination of stable gas spouting for fine particles is shown to occur due to choking of the spout.

SCOPE

Spouting is a well established gas-solids contacting technique for coarse particles ($d_p > 1$ mm) which are not readily fluidized. Spouting of fine particles ($d_p < 1$ mm), however, is a sparsely touched subject. The objectives of the present work were to develop a reliable gas spoutability criterion, to map the flow regimes bracketing gas spouting of fine particles and to investigate the gas spouting termination mechanism which applies to fine particles.

CONCLUSIONS AND SIGNIFICANCE

The criterion for gas spoutability of particles, both fine and coarse, is that the gas inlet orifice diameter must be less than 25 times the particle diameter. Particle density appears to play no role in spoutability. The regimes which follow the introduction of a gas jet into a fixed packed bed of fine particles were identified as steady spouting, incoherent spouting, bubbling and slugging.

Termination of steady spouting of fine particles, whether by excessive gas flow or by excessive bed height, is shown to occur due to choking of the spout. Equations for choking in risers are successfully applied to predict the end of stable spouting.

The results contribute to a greater knowledge and better understanding of gas spoutability and spout termination mechanisms.

RELEVANT PRIOR WORK

Based on some experiments with particles as small as 165 μm glass beads, Ghosh (1) postulated that particles can be made to spout only if the gas inlet orifice diameter is less than 30 times the particle diameter. Most results of other workers (2,3) corroborate this criterion for stable spouting.

Geldart (4) proposed the criterion

$$(\rho_p-\rho_f)d_p^{\ 2} > 10^6 \ g(\mu m)^2 cm^{-3} \tag{1}$$

to characterize particles which tend to form stable spouts, and he classified such particles as his 'Group D'. Molerus (5) rederived the condition for spoutability as

$$(\rho_p-\rho_f)gd_p > 15.3 \ N/m^2 \tag{2}$$

Both of these criteria suggest that increasing the particle size and the density difference between the particles and the fluid enhance their spoutability, and that there is no effect of fluid inlet size on spoutability.

Littman et al. (6) have observed qualitatively that, in the case of air spouting of relatively fine particles, the mechanism for termination of steady spouting by increasing the bed depth appeared to be slug formation in the spout rather than fluidization of the annulus, as in coarse particle spouting (7) and water spouting of fine particles (8).

EXPERIMENTAL

The first part of the present work involved testing the generalizations of the previous investigators (1,4,5) to derive a better criterion for gas spoutability of fine particles. The results were also used to obtain regime maps, like the one shown in Figure 2a, which summarize the different 'states' of a system under various operating conditions.

Most runs were made in a 60° included cone angle, 152 mm diameter plexiglass half column with a wide variety of solids ranging in size from 170 to 1000 μm and in density from 900 to 8900 kg/m^3, with ten gas inlet orifice diameters varying from 2.8 to 28 mm. The spouting gas was always atmospheric air, pre-conditioned to 80% relative humidity in order to eliminate electrostatic effects. A few runs were made with 98 μm particles and a 36.4° half-conical base.

For obtaining the regime maps (e.g. Figure 2a), solids were poured into the column, starting with a low bed height, and their loose packed bed height measured. The air flow rate was gradually increased and the various regimes identified visually. The regime maps were constructed by noting the transition points with decreasing flow rate, and confirmed by increasing and then decreasing the flow rate once more. This procedure was then repeated for larger bed heights. Maps were constructed for various gas inlet and particle sizes.

Gas velocity in the annulus at various bed levels was measured by locating a pre-calibrated differential pressure probe, similar to one used in an earlier study (9), midway between the spout and the centre (i.e. the vertical bisector) of the curved wall. For every particle species used, the minimum fluidization velocity was determined by admitting air into a large number of evenly distributed auxiliary gas flow lines in the half-conical base of the column using bed heights in excess of 800 mm to ensure uniform fluidization.

RESULTS

Spoutability

Steady spouting was achievable only when the gas inlet diameter was reduced to less than 25 times the particle diameter. Beyond this limit, bubbling or slugging occurred. This conclusion was established for a very wide variety of solids (Figure 1). The lower limit of each vertical line represents the highest inlet size employed at which steady spouting was achievable and the upper limit represents the lowest inlet size used at which no spouting occurred. Line AA' in Figure 1, obtained by linear regression of the mid-point of each of the inlet size ranges, is used as representative of the critical inlet diameter. Hence for steady spouting to occur,

$$d_i < (25.4 \pm 9.8\%)\ d_p \qquad (3)$$

This criterion provides a thorough refinement of the earlier postulate by Ghosh (1), which was based on more limited data.

It is apparent from Figure 1 that particle density has no significant effect on spoutability, thus showing the criteria of Geldart (4) and Molerus (5) to be inadequate. This inadequacy was also manifested by the fact that steady spouting could be achieved, given a small enough gas inlet orifice, on either side of the dividing lines represented by both equations (1) and (2), and also by the fact that it could not be achieved on either side of these boundaries without the appropriate orifice.

It should also be noted that the criterion given by equation (3) is consistent with the available data on cold gas spouting of coarse solids (7), as well as with more recent data obtained from our laboratories (10), in which 3.45 mm polystyrene spheres in a 0.914 m diameter half-column showed steady air spouting at $d_i/d_p = 22.0$, incoherent, unstable or unsteady spouting at $d_i/d_p = 26.7$ and bubbling at $d_i/d_p = 29.5$. Given the criterion for coarse particle spouting that (11) $D_c \geqslant d_i(U_t/U_{mf})^{1/2}$, it follows from Equation (3) that D_c may have to exceed $25d_p(U_T/U_{mf})^{1/2}$ for stable spouting.

Regime Maps

When particles with $d_p < 300$ μm were employed, provided Equation (3) was satisfied, the regimes observed with increasing gas flow were: fixed bed, bubbling and steady spouting (12). The solids circulation was very poor and a large stagnant zone of solids existed at all times in the lower outer periphery of the annulus.

For particles with $d_p > 400$ μm, the region of circulation was much enhanced and much greater bed depths could be spouted, thus yielding a more practical system. Provided the criterion of Equation (3) was met, a typical regime map was exemplified by Figure 2a. Steady spouting was characterized by a steady spout-annulus interface, and the presence of a stable well-defined fountain. As the flow rate was increased, incoherent spouting (I.S.) set in when the parabolic fountain shape became distorted and waves started to appear along the spout-annulus interface near the bed surface. The transition from I.S. to bubbling occurred when the amplitude of the waves became equal to the spout radius, leading to a pinch point and thereby the formation of a bubble. When the diameter of the rising bubbles became equal to the column diameter, slugging set in. These transition points were virtually reproducible when the flow rate was kept constant and the bed height increased (e.g. MNOPQ on Figure 2a).

Increasing the gas inlet size reduced the region of steady spouting, which completely disappeared when $d_i > 25.4\ d_p$ (Figure 2b). Particle size increase, in contrast, led to a larger region of steady spouting.

Minimum Spouting Velocity

For the particles larger than 400 μm, the boundary between the fixed bed and steady spouting regions (c.f. Fig. 2a) represents the minimum spouting velocity, U_{ms}, as it does also for coarser particles ($d_p > 1$ mm), while for particles smaller than 300 μm, it was the transition from bubbling to spouting that was taken as U_{ms}. For all the particle, inlet orifice and bed height combinations of the present study which gave rise to steady spouting, U_{ms} could be predicted within ±16% (12) by means of the empirical Mathur-Gishler (3) equation,

$$U_{ms} = (d_p/D_c)(d_i/D_c)^{1/3}(2g(\rho_p-\rho_f)H/\rho_f)^{1/2} \qquad (4)$$

which has been shown to correlate a large range of coarse particle data to ±10% (7).

Gas Flow Distribution

As in the case of coarse particle spouting (9), the superficial gas velocity in the annulus, U_a, was found to increase with vertical distance from the gas inlet and to be independent of bed height. However, unlike the case of coarse particle spouting, where U_a at the maximum spoutable bed height is typically about 0.85-1.00 times U_{mf} (9), $U_a(H_M)$ in the present study was consistently found to be in the range 0.55-0.70 times U_{mf}. The failure to achieve annulus velocities close to U_{mf} at $Z = H = H_m$ points to a spouting termination mechanism other than fluidization of the annulus, as will be detailed below.

Maximum Steady Spouting Velocity

The phenomenon of transition from steady to incoherent spouting is visually comparable to choking of a riser due to solids overloading, the riser in this case being the spout. The hypothesis that this transition is caused by choking of the spout was tested by comparing the superficial gas velocity, U_{SH}, at the top of the spout, where the solids concentration is greatest and the gas velocity smallest, with the choking velocity, U_c, estimated after Leung (13) from (14)

$$2gD(\varepsilon_c^{-4.7}-1)/(U_c-U_T)^2 = 0.00874\ \rho_f^{0.77} \text{ (SI units)} \qquad (5)$$

in which ε_c was calculated from Smith's (15) semi-empirical equation for vertical pneumatic transport,

$$(n\varepsilon_c^{n-1}(1-\varepsilon_c)U_T)^2/gD = 0.5 \qquad (6)$$

and D was taken as the measured spout diameter. Equation (8) is considered to be particularly applicable to risers with relatively small diameters (14) similar to those of the spouts in the present study (< 30 mm). The spout velocity, U_{SH}, was obtained from a gas balance at the top of the bed:

$$U_{SH}A_{SH} + U_{aH}(A_c - A_{SH}) = UA_c \qquad (7)$$

where U, U_{aH} and A_{SH} were measured or determined at the regime transition point, and A_c was simply $\pi D_c^2/4$. Table 1 compares U_{SH} with U_c and it is seen that the agreement is better than satisfactory, as choking predictions go (13).

Maximum Spoutable Bed Height

The spout choking hypothesis was reinforced by consideration of the maximum spoutable bed height, H_M, which was found to increase with decreasing D_i and increasing d_p, in line with trends for fine particle spouting reported earlier (2). Empirical or other equations for H_M in coarse particle spouting (7,11,16) either show H_M as decreasing monotonically with increasing d_p or, even when they do show the correct trend with d_p, overpredict H_M in the present study by more than 100% (12). These equations, when they have some theoretical basis, assume fluidization of the annulus as the spouting termination mechanism. If, however, H_M is predicted by first solving for U_c by means of Equations (5) and (6), then putting $U_{SH} = U_c$ and $U_{aH} = U_a(H_M)$ into Equation (7) and solving for U, and finally equating this U to U_{ms} in Equation (4) and solving for H ($=H_M$), the resulting values of H_M are overpredicted by no more than 30% (see Table 2). At least one-half of this overprediction can be accounted for by the observation of Geldart et al. (17) that H_M in a half column is about 15% less than in a full column. The remaining discrepancy can be attributed to errors inherent in the use of inexact empirical equations such as (4), (5) and (6).

The qualitative observation by Littman et al. (6) that, for fine particles, termination of gas spouting with increasing H results from instabilities which develop within the spout, rather than from fluidization of the annulus, is thus borne out quantitatively.

NOTATION

A_c = cross-sectional area of cylindrical or semi-cylindrical column
A_{SH} = cross-sectional area of spout at top of bed
D = riser diameter
D_c = cylindrical or semi-cylindrical column diameter
d_i = gas orifice inlet diameter
d_p = mean particle diameter

g = acceleration of gravity
H = loose-packed bed height = height of annulus
H_M = maximum spoutable bed height
I.S. = incoherent spouting
n = Richardson-Zaki index
Q_{mf} = volumetric flow rate of gas at minimum fluidization
U = superficial gas velocity based on cylinder cross-section
U_a = superficial gas velocity in annulus at a given value of Z
U_{aH} = U_a at top of annulus
$U_a(H_M)$ = U_{aH} at top of bed of height H_M
U_c = superficial choking velocity in standpipe
U_{mf} = superficial minimum fluidization velocity
U_{ms} = superficial minimum spouting velocity
U_{SH} = superficial gas velocity at top of spout
U_T = terminal free settling velocity of particles
Z = vertical distance from gas inlet orifice
ε_c = voidage at choking
ρ_f = gas density
ρ_p = particle density

LITERATURE CITED

1. Ghosh, B., Indian Chem. Engr., 7(1), 16 (1965).

2. Rooney, N.M. and D. Harrison, Powder Technol. 9, 227 (1974).

3. Mathur, K.B. and P.E. Gishler, A.I.Ch.E. J., 1, 157 (1955).

4. Geldart, D., Powder Technol., 7, 285 (1973).

5. Molerus, O., Powder Technol., 33, 81 (1982).

6. Littman, H., M.H. Morgan III, D.V. Vukovic, F.K. Zdanski and Z.B. Grbavcic, Can. J. Chem. Eng., 55, 497 (1977).

7. Mathur, K.B. and N. Epstein, Spouted Beds, Academic Press, New York (1974).

8. Kim, S.J., "Model and Experimental Verification of the Flow in the Annulus of a Spouted Bed of Small Particles Spouted with Water", PhD thesis, Rensselaer Polytech. Inst., Troy, N.Y. (1982).

9. Epstein, N., C.J. Lim and K.B. Mathur, Can. J. Chem. Eng., 56, 436 (1978).

10. Lim, C.J. and J.R. Grace, private communication (1984).

11. Littman, H., M.H. Morgan III, D.V. Vukovic, F.K. Zdanski and Z.B. Grbavcic, Can. J. Chem. Eng., 57, 684 (1979).

12. Chandnani, P.P., "Gas Spouting of Fine Particles", MASc Thesis, Univ. of British Columbia, Vancouver, B.C. (1984).

13. Leung, L.S., "The Ups and Downs of Gas-Solid Flow: A Review in Fluidization, J.R. Grace and J.M. Matsen, eds., Plenum Press, N.Y. (1980).

14. Punwani, D.V. M.V. Modi, P.B. Tarman, "A Generalized Correlation for Estimating Choking Velocity in Vertical Solids Transport", Proc. Int. Powder and Bulk Solids Handling and Processing Conference, Powder Advisory Centre, Chicago (1976).

15. Smith, T.N., Chem. Eng. Sci., 33, 745 (1978).

16. McNab, G.S. and J. Bridgwater, "Spouted Beds -- Estimation of Spouting Pressure Drop and the Particle Size for Deepest Bed", Proc. European Congress on Particle Technology, H. Brauer and O. Molerus, eds., Nuremburg (1977).

17. Geldart, D., A. Hemsworth, R. Sundavadra and K.J. Whiting, Can. J. Chem. Eng., 59, 638 (1981).

Table 1. Prediction of Maximum Stable Spout Velocity ($=U_c$) by Choking Criterion Compared with Experimental Value ($=U_{SH}$)

Solids	d_i mm	H mm	ε_c Eq.(6)	U_c m/s Eq.(5)	U_{SH} m/s Expt.	Dev'n %
Sand (401 μm)	6.0	410	0.969	5.24	5.87	-10.73
Sand (401 μm)	6.0	300	0.969	5.21	6.95	-25.04
Sand (516 μm)	6.0	500	0.972	6.16	6.71	- 8.20
Sand (516 μm)	6.0	410	0.973	6.04	7.12	-15.17
Sand (516 μm)	6.0	300	0.973	5.98	7.59	-21.21
Sand (516 μm)	12.7	410	0.973	6.02	5.58	+ 7.89
Glass (710 μm)	6.0	500	0.979	9.68	8.64	+12.04
Glass (710 μm)	6.0	410	0.980	9.61	11.44	-16.00

Table 2. Prediction of Maximum Stable Bed Height by Choking Criterion Compared with Experimental Value

Solids	d_i mm	H_M Predicted	H_M Expt.	Deviation %
Sand (401 μm)	6.0	58.6	46.8	+25.2
Sand (516 μm)	6.0	71.2	58.5	+21.7
Sand (516 μm)	12.7	62.6	48.5	+29.1
Glass (710 μm)	12.7	66.1	50.8	+30.1

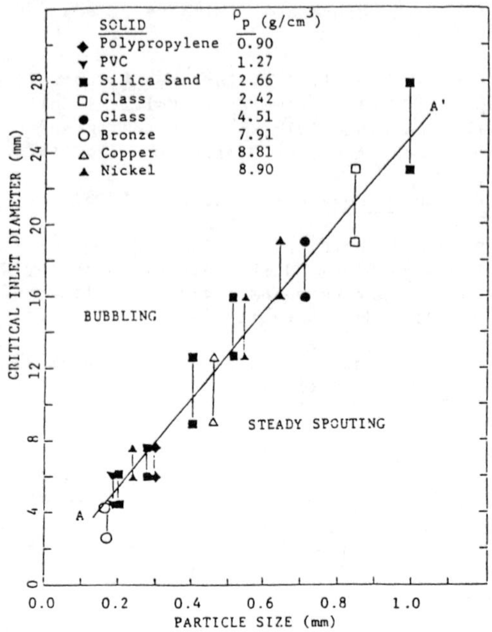

Figure 1. Critical gas inlet diameter range for steady spouting vs. particle diameter.

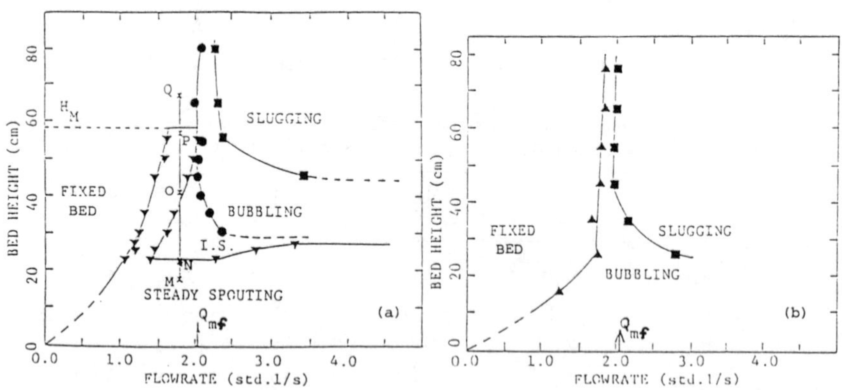

Figure 2. Regime maps for 516 μm silica sand with D_c = 152 mm and (a) d_i = 6.0 mm
(b) d_i = 19.05 mm

A Model for Calculating the Minimum Fluid Flowrate and Pressure Drop in a Conical Spouted Bed

by

Dž. E. Hadžismajlović[1], Ž. B. Grbavčić[1], D. V. Vuković[2], D. S. Povrenović[2] and H. Littman[3]

(1) Institute for Chemistry, Technology and Metallurgy, Belgrade University, 11000 Belgrade, Njegoševa 12., Yugoslavia

(2) Faculty of Technology and Metallurgy, Belgrade University, 11001 Belgrade, P.O.B. 494, Yugoslavia

(3) Department of Chemical Engineering and Environmental Engineering, Rensselaer Polytechnic Institute, Troy, N. Y. 12180-3590, USA

ABSTRACT

A new model has been developed for calculating the minimum spouting flowrate and pressure drop in a conical spouted bed which fits our experimental data for those quantities to 8.4% and 13.1% on average.

SCOPE

Conical spouted beds have very attractive characteristics which can be exploited effectively in processes such as grain drying where low pressure drop and low fluid velocities in the annulus are desirable. These low pressure drops relative to those in cylindrical and cone-base cylindrical spouted beds make it possible to avoid the use of compressors for the air supply in most cases.

The minimum spouting flowrate and pressure drop are basic parameters required for the design of these beds. The data reported here were obtained from pressure drop - flowrate curves using 5 mm dried peas in conical half columns with cone angles of 30 and 60°.

There is only one model in the literature for predicting the minimum spouting flowrate and one for predicting the pressure drop. The model developed in this paper is based on the concept of dividing the bed into a large number of equal cylindrical segments each of which, except that at the spout inlet, is treated as a spout-fluid bed. This works because the radially averaged fluid velocity in the annulus is essentially independent of bed height.

CONCLUSIONS AND SIGNIFICANCE

Using our model, we were able to obtain excellent predictions of both the minimum spouting flowrate and the bed pressure drop. The gas velocity in the annulus rises to a peak near the spout inlet and then decreases. It is always much less than the minimum fluidizing velocity which accounts for the relatively low pressure drop in conical spouted beds.

INTRODUCTION

Conical spouted beds have received less attention in the literature than ordinary spouted beds despite the fact that they have some very attractive processing characteristics. They operate stably with low bed pressure drop and low gas velocities in the annulus as compared with cylindrical and cone-base cylindrical spouted beds. In addition, they do not have a maximum spoutable height limitation and can be built in large sizes. A significant amount of information about them is given in Romankov and Rashkovskaya (1) and by Mathur and Epstein (2).

The minimum spouting flowrate and pressure drop are important design parameters and it is the purpose of this paper to provide a model for calculating those quantities. There are some correlations in the literature for predicting the minimum spouting flowrate (3-8) and a model due to Kmiec (9). Correlations for predicting the pressure drop at minimum spouting (10,11) and the peak pressure drop (12-14) have also been proposed.

EXPERIMENTAL

The minimum spouting flowrate and pressure drop were obtained from pressure drop - flowrate curves in the same manner as for cylindrical spouted beds. The data obtained were for 5 mm spherical dried peas in conical half-columns with cone angles of 30 and 60° (see Figure 1). The peas had a density of 1275 kg/m^3, a minimum fluidizing velocity of 1.40 m/s and a voidage at minimum fluidization of 0.44. Six inlet tube diameters varying in size from 25 to 100 mm were used with a 100 mm base so that d_i/d_b varied between 0.25 and 1. In two other columns d_i/d_b was equal to 1 with d_b equal to 25 and 50 mm.

The conical half-column was equipped with taps located in the flat vertical face to measure static pressures in the spout and annulus. These measurements were used to determine the axial pressure profile at the spout-annulus interface and the gas velocity distribution in the annulus by the method of Mathur and Gishler (15).

MODEL FOR CALCULATION

The minimum spouting velocity is calculated from a material balance on the bed. Thus

$$V_{mS} = (\pi/4) \left\{ \left[[d_b + 2H \tan(\alpha/2)]^2 - (d_S)^2_{mSF} \right] u_{AH} + (d_S)^2_{mSF} u_{mF} \right\} \quad (1)$$

The key to our model is based on the observation that the radially averaged gas velocity in the annulus, u_A, (or the pressure gradient, dp_A/dz), is essentially independent of bed height as seen in Figure 2. When this is the case, material added to the bed will not affect the axial gas velocity distribution in the original bed. This remarkable condition has also been observed in cylindrical spouted beds (16,17). Thus, if the bed is divided into n equal cylindrical segments of height, H/n, as shown in Figure 1, each segment acts as a spout-fluid bed except for the first segment. Using the work of Hadžismajlović et al. (18), their equation 7 can be written for a segment as:

$$(u_{AH})_i/u_{mF} = 1 - [1 - ([(u_A(0)]_i/u_{mF})][1 - (H/n)/(H_{mSF})_i]^3 \tag{2}$$

where $$[u_A(0)]_i = (u_{AH})_{(i-1)} (A_A)_{(i-1)}/(A_A)_i \tag{3}$$

$$(A_A)_i = (\pi/4)\{[(D_c)_i^2 - [(d_S)_{mSF}]_i^2] \} \tag{4}$$

$$(D_c)_i = d_b + \frac{H}{n}(2i-1) \tan(\alpha/2) \tag{5}$$

Starting at the bottom of the bed with $[u_A(0)]_{i=1}=0$, $(D_c)_{i=1}=d_b + \frac{H}{n}\tan(\alpha/2)$ and n=100, the calculation was carried out on an IBM 1130 computer segment by segment until the top segment of the bed was reached, giving u_{AH}. Equations for calculating $(H_{mSF})_i$, given in the appendix, are derived in reference 18.

Note in equation 2, that $(u_{AH})_i$ decreases with z despite the fact that $[u_A(0)]_i$ increases with z, because $(H_{mSF})_i$ also increases with z. Equation 3 adjusts the velocity between segments for the increase in area resulting from the conical shape of the bed.

The annulus pressure drop at minimum spouting is calculated from the integral of $u_A(z)$. Thus

$$\Delta P_{mS}/\Delta P_{mF} = \int_0^1 (u_A(z)/u_{mF})\, d\,(z/H) \tag{6}$$

It is important to note that our model can be used to calculate both V_{mS} and ΔP_{mS} in cylindrical and cone-based cylindrical columns because $u_A(z)$ is independent of H in those columns also.

RESULTS

The data in Figure 2 show that $u_A(z)$ falls linearly with distance from the bed inlet and is approximately independent of bed height. No data points are shown for z < 100 mm because the annulus is very thin there making the gas velocity measurement inaccurate. In addition, the particle velocities were not measured in the lower part of the annulus although they probably influence the gas velocities.

The model fits the gas velocity profile quite well and shows that the velocity passes through a maximum, as it must, in the lower part of the bed. Although u_A falls with z in the upper part of the bed, $V_A(z)$ always increases monotonically. Note that $u_A(z)$ is significantly lower than u_{mF}.

When $d_i/d_b = 1$ and $d_i = 25$ mm, the spout is bounded by the wall of the bed near the spout inlet. There is, in fact, no annulus there. In such cases, the model gives significantly higher predictions for the pressure drop than was found experimentally (see Table 1).

The data in Table 1 fit our model for V_{mS} to about 8.4% on average. In order to test Kmiec's model (9), we assumed $\varepsilon = 0.44$ and found that his model predicted V_{mS} values which averaged 67.5% below our experimental ones (Table 1). On the other hand, we were able to fit Kmiec's data (9) for agalit using his column V with our model to about 25%. The pressure drop data as seen in Table 1 fit our model on average to about 13.1% and the Mukhlenov and Gorshtein correlation (10,11) to about 28.0%. It also shows that ΔP_{mS} is much smaller than ΔP_{mF}. The peak pressure drop differed by about 29.3% and 38.6% on average from those calculated from the Mukhlenov and Gorshtein (12) and Kmiec (14) correlations, respectively. The correlation of Gelperin et al (13) was found to be unreliable.

The axially averaged spout diameters differed by about 10% from those calculated from the relations given in the appendix. However, the shape of the spout is different than what the model predicts. The actual spout widens near the inlet reaches a maximum near where the model predicts the maximum in the annulus gas velocity and then narrows. In the model, the spout widens slightly from bottom to top.

DISCUSSION

It is possible to calculate u_{AH} by solving a boundary value problem for flow in the annulus as has been done for cylindrical beds by Littman et al. (19) and for conical base cylindrical beds by Rovero et al. (20). We did not solve the boundary value problem for the conical spouted bed, however, because we have not as yet developed a general pressure boundary condition at the spout-annulus interface for conical beds similar to that given for cylindrical spouted beds by Morgan and Littman (21). In Rovero et al. (20), the velocity and pressure distributions in the annulus were calculated using the vector form of the Ergun equation in the field (22) and the Epstein and Levine (23) axial pressure distribution at the spout-annulus interface. Their theoretical and experimental annulus velocities behave in the way our model predicts in Figure 2 up to the cone-cylinder junction. Beyond that height, $u_A(z)$ rises monotonically to a value slightly less than u_{mF} at $H = H_m$. Thus the conical bed provides a way of reducing the gas velocity in the annulus continuously as the bed height is raised and for this reason does not have a maximum spoutable height. The low pressure drop in conical beds is of considerable practical importance because ordinary blowers can be used for the air supply. These features of the conical bed make it attractive for applications such as grain drying.

It is important to note finally that if one is to accurately represent u_A (z) near the spout inlet, it is probably necessary to take into account the high particle velocities and the shape of the spout there.

ACKNOWLEDGMENT

This work was supported by the Research Council of Serbia and the National Science Foundation through Grant F5F035Y provided by the Yugoslav-US Joint Board on Scientific and Technological Cooperation.

APPENDIX

Equations for calculating the spout diameter and maximum spoutable height are:

$$[(d_S)_{mSF}]_i = (D_S)_i \left\{ 1 - [1 - (H/n\ (H_m)_i)]^3 \right\}^{1/2} \qquad (1a)$$

where $(D_S)_i/d_i = \{[2.10e^{-(0.0180/A)} + 1.00]/3.10\}$

$$\{0.862 + 0.219\ [(D_c)_i/d_i] - 0.00530\ [(D_c)_i/d_i]^2\} \qquad (\underline{24}) \qquad (2a)$$

$$(H_m)_i = [0.218 + (0.005/A)][(D_c)_i^2/d_i] \qquad (\underline{25}) \qquad (3a)$$

$$\frac{(H_{mSF})_i\ (D_S)_i\ \{1 - [1 - ((H_{mSF})_i/(H_m)_i)]^3\}^{0.5}}{(D_c)_i^2 - (D_S)_i^2\ \{1 - [1 - ((H_{mSF})_i/(H_m)_i)]^3\}^{0.5}} =$$

$$\left\{ \sum_{k=0}^{5} A_k([u_A(0)]_i/u_{mF})^k \right\} + \{1 - ([u_A(0)]_i/u_{mF})\}$$

$$\{0.345\ [(D_S)_i/(D_c)_i]^{-0.384} - 0.596\} \qquad (\underline{18}) \qquad (4a)$$

where $A_0 = 0.596,\ A_1 = -0.432,\ A_2 = 0.056$

$A_3 = -0.297,\ A_4 = 0.568,\ A_5 = -0.425$

NOTATION

- A — $[\rho_f/(\rho_p - \rho_f)][u_{mF}\ u_T/gd_i]$
- A_A — annulus area
- A_c — column area
- A_S — spout area
- d_b — diameter of the base of the column (Figure 1)
- d_i — spout inlet tube diameter (Figure 1)
- d_p — particle diameter
- d_S — average spout diameter
- $(d_S)_{mSF}$ — average spout diameter in bed of height, H_{mSF}, at the minimum spout-fluid flowrate
- ΔP_{mF} — pressure drop at minimum fluidization
- ΔP_{mS} — pressure drop at minimum spouting
- u_A — average gas velocity in the annulus
- u_{AH} — superficial velocity at the top of the annulus, $u_A(z)$ at $z = H$
- $[u_A(0)]_i$ — $[V_A(0)]_i/[A_A]_i$
- u_{mF} — minimum fluidizing velocity
- u_{SH} — superficial velocity of gas at the top of the spout
- u_T — terminal fall velocity of a particle

D_C	column diameter	V_A	gas flowrate in annulus
D_S	average spout diameter in bed of height, H_m, at the minimum spouting flowrate	$V_A(0)$	annulus flowrate at bed inlet
g	gravitational acceleration	V_{mS}	minimum spouting flowrate
H	bed height	ε	bed voidage
H_m	maximum spoutable height in a spouted bed	ε_{mF}	bed voidage at minimum fluidization
H_{mSF}	maximum spoutable height in a spout-fluid bed	ρ_f	fluid density
P_A	pressure in the annulus	ρ_p	particle density
ΔP_A	pressure drop in the annulus	α	cone angle

LITERATURE CITED

1. Romankov, R. G. and N. B. Rashkovskaya, "Drying in a Suspended State", 2nd Ed., Chem. Publ. House, Leningrad Branch, (1968), in Russian.
2. Mathur, K. B. and N. Epstein, "Spouted Beds", Academic Press, New York, 1974.
3. Goltsiker, A., N. B. Rashkovskaya and P. G. Romankov, Zh. Prikl. Khim. (Leningrad) 37, 1030 (1964).
4. Gorshtein, A. E. and I. P. Mukhlenov, Zh. Prikl. Khim. (Leningrad), 37, 1887 (1964).
5. Kmiec, A., Chem. Eng. J. 13, 143 (1977).
6. Nikolaev, A. M. and L. G. Golubev, Izv. Vyssh, Ucheb Zaved. Khim. Khim. Tekhnol. 7, 855 (1964).
7. Tsvik, M. Z., M. N. Nabiev, N. V. Rizaev, K. V. Merenkov and V. S. Vyzgo, Uzb. Khim. Zh. 11 (2) 50 (1967).
8. Wan-Fyong, F., P. G. Romankov and N. B. Rashkovskaya, Zh. Prikl. Khim. (Leningrad) 42, 609 (1969).
9. Kmiec, A., Can. J. Chem. Eng., 61 (3), 274 (1983).
10. Mukhlenov, I. P., and A. E. Gorshtein, Zh. Prikl. Khim. (Leningrad), 37, 609 (1964).
11. Muklhenov, I. P., and A. E. Gorshtein, Vses. Konf. Khim. Reactrom Novosibirsk., 3, 553 (1965).
12. Mukhlenov, I. P. and A. E. Gorshtein, Khim. Prom. (Moscow) 41, 443, 1965.
13. Gelperin, N. J., V. G. Ainshtein and L. P. Timokhova, Khim. Machinostr. (Moscow), No. 4, 12, 1961.
14. Kmiec, A., "Bed expansion and mass transfer in fluidized beds", Wroclaw Technical University Monographs No. 19, Wroclaw, 1980.
15. Mathur, K. B. and P. E. Gishler, AIChE Journal, 1, 157 (1955).
16. Grbavčić, Ž. B., D. V. Vuković, F. K. Zdanski and H. Littman, Can. J. Chem. Eng., 54, 33 (1976).
17. Epstein, N., C. J. Lim and K. B. Mathur, Can. J. Chem. Eng., 56, 436 (1978).
18. Hadžismajlović, Dž. E., Ž. B. Grbavčić, D. V. Vuković and H. Littman, Can. J. Chem. Eng. 61, 343 (1983).

19. Littman, H., M. H. Morgan, III, P. V. Narayanan, S. J. Kim, J.Y. Day, G. M. Lazarek, Can. J. Chem. Eng. 63, 188 (1985).
20. Rovero, G., C. M. H. Brereton, N. Epstein, J. R. Grace, L. Casalegno and N. Piccinini, Can. J. Chem. Eng., 61, 289 (1983).
21. Morgan, M. H., III and H. Littman, "General Relationships for the Minimum Spouting Pressure Drop Ratio, $\Delta P_{mS}/\Delta P_{mF}$, and the Spout-Annular Interfacial Condition in a Spouted Bed", in "Fluidization", Eds. J. R. Grace, and J. M. Matsen, Plenum Press, New York (1980), 287.
22. Stanek, V. and J. Szekely, AIChEJ 20, 974 (1974).
23. Epstein, N. and S. Levine, "Non-Darcy Flow and Pressure Distribution in a Spouted Bed" in "Fluidization", Eds., J. F. Davidson and D. L. Keairns, Cambridge Univ. Press, England, (1979), 98.
24. Littman, H., M. H. Morgan III, D. V. Vuković, F. K. Zdanski and Ž. B. Grbavčić, Can. J. Chem. Eng. 57, 684, 1979.
25. Littman, H., M. H. Morgan III, D. V. Vuković, and F. K. Zdanski in "Fluidization", Eds. J. F. Davidson and D. L. Kearns, Cambridge Univ. Press, England (1979), 381.

Table 1. Minimum spouting flowrate and pressure drop data for dried peas

	d_b[m]		$V_{mS}[m^3/s] \times 10^3$					P_{ms}[kPa]				
α	d_i[m]	H[m]	expt	our model	%	Kmiec model	%	expt	our model	%	correl. (10,11)	%
60°	0.100	0.15	18.9	17.4	7.9	3.9	79.4	0.40	0.40	0.0	0.52	30.0
	0.025	0.20	25.1	23.8	5.2	4.5	82.0	0.50	0.55	10.0	0.60	20.0
		0.25	30.0	30.3	1.0	5.0	83.8	0.60	0.68	13.3	0.67	11.7
	0.100	0.15	21.0	21.9	4.2	8.2	61.2	0.60	0.54	10.0	0.85	41.7
	0.050	0.20	27.4	29.9	9.1	9.2	65.9	0.80	0.71	11.3	0.98	22.5
		0.25	31.2	38.5	23.4	10.3	67.7	0.90	0.88	2.2	1.11	22.2
	0.100	0.15	22.1	22.9	3.6	9.8	55.7	0.63	0.57	9.5	0.96	52.4
	0.060	0.20	30.2	31.1	2.9	11.2	62.9	0.80	0.76	5.0	1.10	37.5
		0.25	34.4	40.1	16.5	12.4	64.0	1.05	0.93	11.7	1.24	18.1
		0.30	42.6	49.6	16.4	13.5	68.0	1.20	1.09	9.2	1.34	11.7
	0.100	0.20	36.0	34.1	5.5	19.2	46.7	0.84	0.83	1.2	1.54	83.3
	0.100	0.25	43.0	43.9	1.9	21.1	50.9	0.98	1.01	3.1	1.72	75.5
		0.30	51.2	53.3	4.1	22.9	55.3	1.24	1.20	3.2	1.88	51.6
60°	0.025	0.15	12.9	11.3	12.4	3.9	69.8	0.50	0.69	37.0	0.57	14.0
	0.025	0.20	17.1	16.8	1.8	4.5	73.7	0.56	0.85	51.1	0.65	16.1
		0.25	21.8	22.8	4.5	5.0	77.1	0.67	0.99	48.0	0.72	7.5
		0.30	25.5	29.3	14.9	5.4	78.8	0.78	1.13	45.0	0.79	1.3
	0.050	0.15	15.7	16.3	3.8	8.2	97.8	0.70	0.73	3.6	0.90	28.6
	0.050	0.20	20.8	23.4	12.5	9.3	55.3	0.80	0.92	14.5	1.03	28.7
		0.25	26.0	31.4	20.8	10.3	60.4	0.90	1.09	21.1	1.15	27.8
		0.30	32.6	40.0	22.7	11.1	66.0	1.10	1.25	13.8	1.24	12.7
30°	0.100	0.15	14.3	12.1	9.8	2.7	81.1	0.60	0.63	5.0	0.48	20.0
	0.025	0.20	18.4	16.7	9.2	3.1	83.2	0.80	0.89	11.3	0.55	31.2
		0.25	22.7	20.8	8.4	3.4	85.0	1.18	1.14	3.6	0.61	48.3
	0.100	0.15	15.0	14.6	2.7	5.6	62.7	0.70	0.78	11.3	0.74	5.7
	0.050	0.20	18.9	18.8	0.5	6.4	66.1	1.00	1.07	6.8	0.90	10.0
		0.25	22.8	23.5	3.1	7.0	69.3	1.30	1.35	3.8	1.01	22.3
		0.30	26.7	28.4	6.4	7.6	71.5	1.60	1.62	1.5	1.11	30.6
			OVERALL AV.		8.4		64.5			13.1		28.0

H = 0.15m P_{mF} = 1.10 kPa H = 0.25m P_{mF} = 1.83 kPa

H = 0.20m P_{mF} = 1.47 kPa H = 0.30m P_{mF} = 2.20 kPa

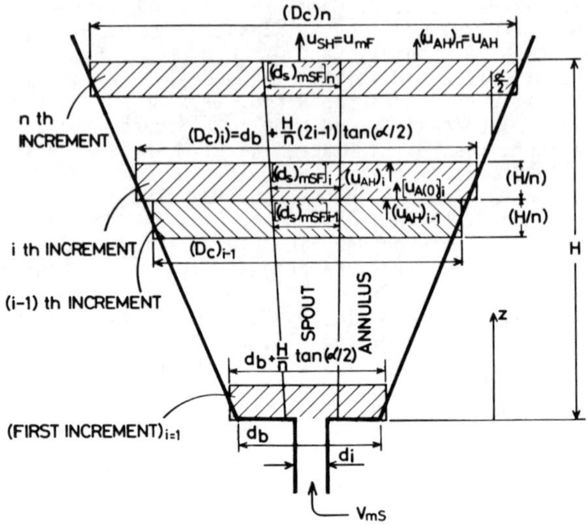

Figure 1 Model for calculation and schematic diagram of bed.

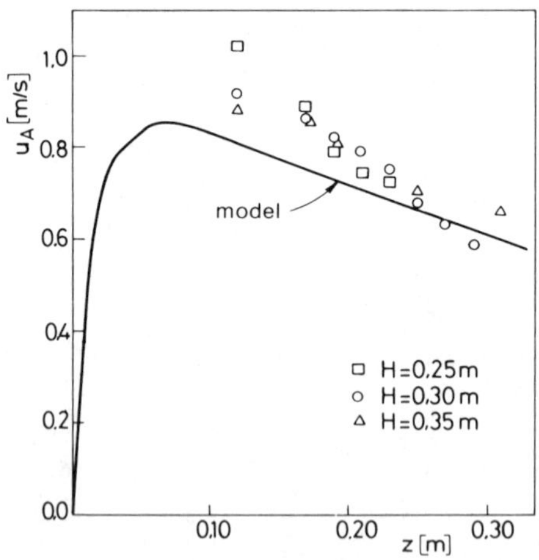

Figure 2 Gas velocity in the annulus vs distance from inlet at the minimum gas flowrate, d_i=60 mm, d_b=100 mm, α=60°.

Analytical Expressions for the Spout/Annulus Pressure Drop and Pressure Gradient in Spouted Beds of Coarse Particles

by

Morris H. Morgan, III

Department of Chemical Engineering
and Environmental Engineering
Rensselaer Polytechnic Institute
Troy, New York 12180-3590

ABSTRACT

New relationships for the spout pressure drop, the spout pressure profile and the spout pressure gradient are derived via a variational approach. A lower theoretical limit is also established for the spout pressure drop ratio at the maximum spoutable height for systems that terminate by annular fluidization.

SCOPE

To properly design spouted beds the spout pressure profile must be known explicitly. Several models (1,2) have been proposed to predict this quantity. Earlier models predicted a unique profile for all spouted bed systems. Later experiments (4) with water spouted systems indicated that a family of such profiles exist and that the fluidization profile is a lower-bound for these systems. None of the earlier models satisfied the empirically observed Grbavcic condition.

The Morgan and Littman (4) model recently confirmed by Littman et al. (3) satisfies the Grbavcic condition and generates a family of spout pressure profile curves. It's major drawback is a requirement for detailed spout pressure drop data. The model developed here overcomes this informational requirement and generates the spout pressure gradient explicitly.

CONCLUSIONS AND SIGNIFICANCE

Excellent agreement is obtained between theory and experimental data (within 8%). The spout pressure profile is not unique and a family of such curves is possible. The shape of the profiles are solely determined by the magnitude of the pressure drop ratio, $(\Delta P_{mS}/\Delta P_{mF})_{H_m}$. This latter quantity according to theory must exceed 0.50.

INTRODUCTION

The determination of the spout/annulus pressure profile in a spouted bed is crucial to the design of these systems. Without a knowledge of this profile a determination of either the annular flow field or fluid cross-flow between the spout and annulus in such systems is not possible. Both of these latter quantities are essential to the effective modeling of these systems as chemical reactors.

Various pressure profile models have been proposed for spouted bed systems. A comprehensive review of most of the literature models may be found in references (1,2). Recently, Littman et al. (3) established the generality of the Morgan and Littman (4) spout pressure profile shown below. That model was derived from the empirically established Grbavcic (1) condition. (See Figure 1).

$$\frac{P_S}{\Delta P_{mS}} = 1 - \frac{z}{H}\frac{(\Delta P_{mS}/\Delta P_{mF})_z}{(\Delta P_{mS}/\Delta P_{mF})_H} \tag{1}$$

This condition relates that $-dP_a/dz \neq f(h)$ but varies only with height, H. It follows from this condition that particle loadings above a given level, z has no effect on the pressure field below it.

A major drawback with the application of Equation 1 is the requirement that spout pressure drop data be available. Usually such detailed information is not known and empirical correlations must be used to predict the needed information.

In this paper, a theory which assumes a prior knowledge of the spout pressure drop ratio at the maximum spoutable height, $\Delta P_{mS}/\Delta P_{mF}|_{H_m}$ is used to derive analytical relationships for the bed pressure gradient, $-dP_a/dz = f(z)$. It is then shown that direct integration of the resulting expression leads to analytical relationships for both the overall bed pressure drop ratio, $\Delta P_{mS}/\Delta P_{mF}|_{H_m}$ and spout/annulus interfacial profile.

Theoretical Basis for the Variational Approach

Grbavcic et al. (1) established that the pressure gradient in spouted beds, $-dP_a/dz \neq f(h)$ and depended only upon the bed level, z. That result is highlighted in Figure 1 where data for both air and water spouted systems are summarized. A few general observations about the given data may be made. First, for beds at the maximum spoutable height, H_m it is observed that the pressure gradient at the top of the bed is essentially equal to the fluidization pressure gradient, $\Delta P_{mF}/H_m = (1 - \varepsilon_{mF})(\rho_p - \rho_f)g$. For bed heights less than, H_m, the pressure gradient at the top of the bed will be smaller than $\Delta P_{mF}/H_m$. Second, the pressure gradient at the bottom of the bed is zero. This is typically the case for flat-bottomed columns. With conical beds, $dP_a/dz|_{z=0}$ may be positive indicating the possibility of a flow reversal in the beds as noted by Rovero et al. (5). Third, the pressure gradient, $-dP_a/dz$ increases montonically with height, z.

Thus, if the bed pressure drop, $\Delta P_{mS}/\Delta P_{mF}|_{H_m}$ is also known for a spouted bed system, the problem of predicting the pressure gradient, f(z) may be posed as an Euler-Lagrange variational problem. The following system of equations may then be formulated for a spouted bed at H_m.

$$J(\zeta) = \int_0^1 \sqrt{1 + (g')^2} \; d\zeta \qquad (2)$$

$$\zeta = 0 \quad ; \quad g(0) = 0 \qquad (3)$$

$$\zeta = 1 \quad ; \quad g(1) = 1 \qquad (4)$$

and

$$\int_0^1 g(\zeta)\, d\zeta = \left.\frac{\Delta P_{mS}}{\Delta P_{mF}}\right|_{H_m} \qquad (5)$$

A closed form solution (6) exists for this system of equations and is given below

$$g(\zeta) = C_2 - \frac{1}{\lambda}\,[1 - (\lambda z + C_1)^2]^{1/2} \qquad (6)$$

The parameters C_1, C_2 and λ appearing in this equation are integration constants. The numerical values for these constants determined using Equations 3-5 are summarized in Table 1. Clearly, C_1, C_2 and λ are strongly affected by the magnitude of $\Delta P_{mS}/\Delta P_{mF}\,|\,H_m$. The relationships used to evaluate these constants are

$$0.707 \leq C_1 \leq 1.0 \qquad (7)$$

$$\lambda = (1 - C_1^{\,2})^{1/2} - C_1 \qquad (8)$$

$$C_2 = \frac{(1 - C_1^2)^{1/2}}{\lambda} \qquad (9)$$

Expressions for both the spout pressure profile and the spout pressure drop follows directly from the integration of the two equations below

$$\left.\frac{\Delta P_{mS}}{\Delta P_{mF}}\right|_h = \frac{\int_0^h g(\zeta)\, d\zeta}{\int_0^h d\zeta} \qquad (10a)$$

$$\left.\frac{\Delta P_{mS}}{\Delta P_{mF}}\right|_h = C_2 - \frac{1}{2\lambda^2 h}\,[\,(\lambda h + C_1)[1-(\lambda h + C_1)^2]^{1/2} - C_1(1-C_1^2)^{1/2}$$

$$+ \sin^{-1}(\lambda h + C_1) - \sin^{-1} C_1] \qquad (10b)$$

$$\frac{P_S(\zeta)}{\Delta P_{mS}} = 1 - \frac{\int_0^\zeta g(\zeta)\, d\zeta}{\int_0^1 g(\zeta)\, d\zeta} \tag{11a}$$

$$\frac{P_S(\zeta)}{\Delta P_{mS}} = 1 - \tag{11b}$$

$$\left[C_2\zeta + \frac{\zeta}{2\lambda^2} \frac{[(\lambda\zeta + C_1)[1-(\lambda\zeta + C_1)^2]^{1/2} - C_1(1-C_1^2)^{1/2} + \text{Sin}^{-1}(\lambda\zeta + C_1) - \text{Sin}^{-1}C_1]}{(\Delta P_{mS}/\Delta P_{mF}\Big|_{H_m}} \right]$$

Equations 10a and 11a are derived from the Grabavcic condition. The predicted spout pressure profiles are shown in Figure 2. Note that a family of spout pressure profiles are possible and that their shape is solely determined by the magnitude of $P^* = \Delta P_{mS}/\Delta P_{mF}\Big|_{H_m}$. One also observes that as P* increases, the resulting spout pressure profile is shifted toward the P* = 1.0 (diagonal) line. Such behavior indicates that the spouted bed system is becoming more fluid-bed like in character.

The fundamental difference between equations 1 and 11b revolves around the fact that $P_S(\zeta)/\Delta P_{mS}$ is an explicit function of ζ in the latter case. Given accurate $\Delta P_{mS}/\Delta P_{mF}$ (ζ) data equations 1 and 11b will produce identical results.

Theoretical limits for the maximum spoutable pressure drop ratio, P*

Morgan et al. (6) recently suggested that coarse particle spouting systems may be characterized by the dimensionless parameter, $C_o = \Delta P_{mS}/\Delta P_{mF} + (\rho_f u_{mS}^2 (D_c/d_S)^4/\Delta P_{mF})$. They concluded for stable spouting that $0.215 \leq C_o \leq 0.785$ in order to hold the boundary conditions represented by Equations 3 and 4. They maintained that the upper boundary or limit, $C_o = 0.785$ represented beds at the maximum spoutable height, H_m and that

$$\frac{\Delta P_{mS}}{\Delta P_{mF}}\Bigg|_{H_m}^{max} \approx 0.785 - (\rho_f u_{mS}^2 (\frac{D_c}{D_S})^4/\Delta P_{mF}) \tag{12}$$

The above relationship is consistent with literature observations (1,2).

A lower limit for $(\Delta P_{mS}/\Delta P_{mF})\Big|_{H_m}$ may also be established as outlined in the argument below. On physical grounds one may assume that the radial crossflow between the spout and annulus, $u_r(z) \geq 0$ and decreases montonically with vertical distance z. Thus it follows that $u_r'(z) \leq 0$. Since $u_r \propto (-d^2p_a/dz^2)$ - from continuity consideration - one may conclude that $d^3P_a/dz^3 \leq 0$. If the latter is valid then $(- dP_a/dz)$ is a concave downward function and

$$\frac{\Delta P_{mS}}{\Delta P_{mF}}\Bigg|_{H_m}^{min} \geq 0.500 \tag{13}$$

Thus the theoretical bounds for P* are for systems that terminate by annular fluidization.

$$0.500 \leq \left(\frac{\Delta P_{mS}}{\Delta P_{mF}}\right)\Bigg|_{H_m} \leq 0.785 \qquad (14)$$

RESULTS

Data fitting the format of either Equations 6, 10b or 11b may be used to test the accuracy of the proposed variational approach. Since Grbavcic (7) reported his data as reduced pressure drop versus bed height, those experimental results are compared to the predictions obtained from Equation 10b. That analysis was conducted as follows: Given a single value for $\Delta P_{mS}/\Delta P_{mF}\big|_{H_m}$ the constants C_1, C_2 and λ were determined using Equations 7, 8 and 9 and 10b. Next $(\Delta P_{mS}/\Delta P_{mF})_H$ was calculated using those predetermined constants for experimental cases were $H < H_m$. The predictions were then compared with the experimental values for bed heights less than H_m. The results are presented in Figure 3. As highlighted there the majority of experimental data points fall within the ± 10% error lines. The average overall percent deviation is less than 8%. The range of experimental condition represented by Grbavcic data are summarized in Table 2. As expected the largest deviation from prediction occurs at small bed heights were a constant measurement error will have more of a pronounced effect on accuracy.

DISCUSSION

Clearly, the reliability of the method depends on the accuracy of $(\Delta P_{mS}/\Delta P_{mF})_{H_m}$. Any error in the estimation of this quantity carries through the calculation. Figure 2 indicates however that large errors in $(\Delta P_{mS}/\Delta P_{mF})_{H_m}$ will have only a modest effect on the shape of the spout pressure profile. Thus in the absence of an experimental values for $(\Delta P_{mS}/\Delta P_{mF})_{H_m}$ this quantity may be estimated from correlation. Furthermore Equation 14 shows that this ratio is restricted to values between 0.50 and 0.785. The Grbavcic et al. (1) and Lefroy and Davidson (8) models are also depicted in this figure for comparison purposes. Interestingly, these two models are reasonable fits of their respective variational counterparts, P* = 0.75 and P* = 0.64.

With the Grbavcic data (water spouted systems) the ambiguity of whether the spout termination mechanism is slugging in the spout or annular fluidization is completely avoided because such systems always terminate by the latter mechanism.

Even though this variational approach assumes a prior knowledge of the bed's condition at H_m, this approach is not restricted to such cases. If, in addition to the overall bed pressure, the bed pressure gradient at the top of the spout is also known then an Euler-Lagrange variational description of this system is possible for any bed $H < H_m$.

NOTATION

C_o	$(\Delta P_{mS}/\Delta P_{mF}) + [\rho_f u_{mS}^2 (D_c/d_S)^4/\Delta P_{mF}]$
C_1 C_2	integration constants defined by Equations 7 and 9
d_i	inlet nozzle diameter, m
d_S	spout diameter, m
d_p	particle diameter, m
D_c	column diameter, m
D_S	spout diameter for a bed at H_m, m
$f(\zeta)$	spouted bed pressure gradient
g	gravitational constant, m/s^2
$g(\zeta)$	$f(z)/(1 - \varepsilon_{mF})(\rho_p - \rho_f)$ g
h	H/H_m
H	spout bed height, m
H_m	maximum spoutable height, m
$J(\zeta)$	defined by Equation 2
P^*	$\Delta P_{mS}/\Delta P_{mF}$
P_a	annulus pressure
$P_S(\zeta)$, $P_S(z)$	spout-annulus pressure at ζ,z, Pascal
ΔP_{mF}	minimum fluidization pressure drop, Pascal
ΔP_{mS}	minimum spouting pressure drop, Pascal
u_{mS}	minimum superficial spouting velocity, m/s
$u_r(z)$	superficial radial velocity at spout/annulus interface, m/s
$u_r'(z)$	derivative of $u_r(z)$, m/s^2
z	vertical distance from spout inlet nozzle, m
ε_{mF}	minimum fluidization voidage
λ	defined by Equation 8
ρ_f	fluid density, kg/m^3
ρ_p	particle density, kg/m^3
ζ	z/H

LITERATURE CITED

1. Grbavcic, Z. B., D. V. Vukovic, F. K. Zdanski and H. Littman, Can. J. Chem. Eng., 54, 33 (1976).

2. Epstein, N., C. J. Lim and K. B. Mathur, Can. J. Chem. Eng., 56, 436 (1978).

3. Littman, H., M. H. Morgan, III, P. V. Narayanan, S. J. Kim, J. Y. Day and G. M. Lazarek, Can. J. Chem. Eng., 63, 188 (1985).

4. Morgan, M. H., III and H. Littman, "General Relationships for the Minimum Spouting Pressure Drop Ratio, $\Delta P_{mS}/\Delta P_{mF}$, and the Spout-Annular Interfacial Condition in a Spouted Bed", in "Fluidization", Eds. J. R. Grace, and J. M. Matson, Plenum Press, New York (1980), 287.

5. Rovero, G., C. M. H. Brereton, N. Epstein, J. R. Grace, L. Casalegno and N. Piccinini, Can. J. Chem. Eng., 61, 289 (1983).

6. Morgan, M. H., III, J. Y. Day and H. Littman, Chem. Eng. Sci., 40, 1367 (1985).

7. Grbavcic, Z. B., M. S. Thesis, 1975, University of Belgrade, Belgrade, Yugoslavia.

8. Lefroy, G. A. and J. F. Davidson, Trans. Instn. Chem. Engrs., 47, T120 (1969).

TABLE 1

Integration Constants for the Euler-Lagrange Solution

$\Delta P_{mS}/\Delta P_{mF}$ | H_m	C_1	C_2	λ
0.500	0.707	∞	0.0
0.663	0.950	-0.490	-0.638
0.676	0.960	-0.412	-0.680
0.689	0.970	-0.334	-0.727
0.704	0.980	-0.255	-0.781
0.729	0.990	-0.166	-0.849
0.735	0.995	-0.111	-0.900
0.749	0.997	-0.084	-0.927
0.785	1.000	0.000	-1.000

TABLE 2

Range of Experimental Conditions (7)

dp :	0.935×10^{-3}m	-	4.11×10^{-3}m
d_i :	4.30×10^{-3}m	-	27.5×10^{-3}m
D_c :	0.04m	-	0.11m
h :	0.10	-	1.00

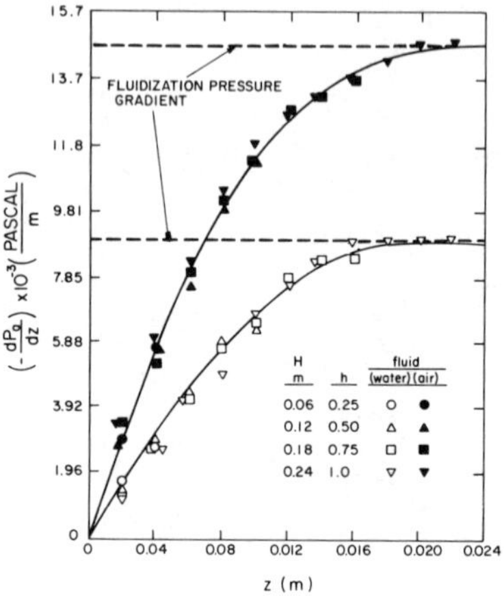

FIGURE 1: SPOUT-ANNULUS INTERFACIAL PRESSURE GRADIENT (D_c = 0.110 m, dp = 2.5 x 10^{-3}m) VERSUS BED HEIGHT

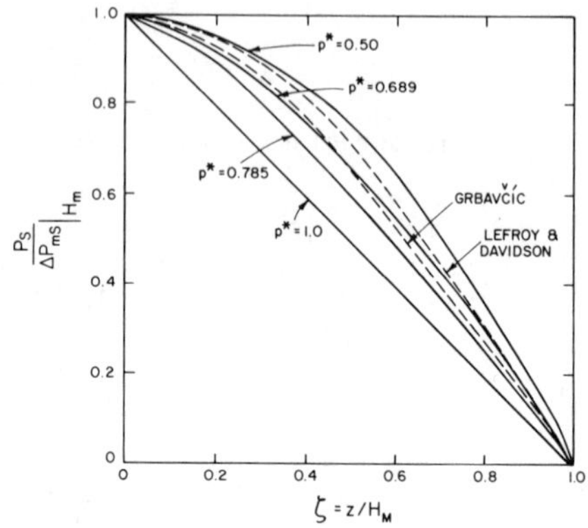

FIGURE 2: NORMALIZED PRESSURE AT SPOUT-ANNULUS INTERFACE VERSUS DIMENSIONLESS BED HEIGHT.

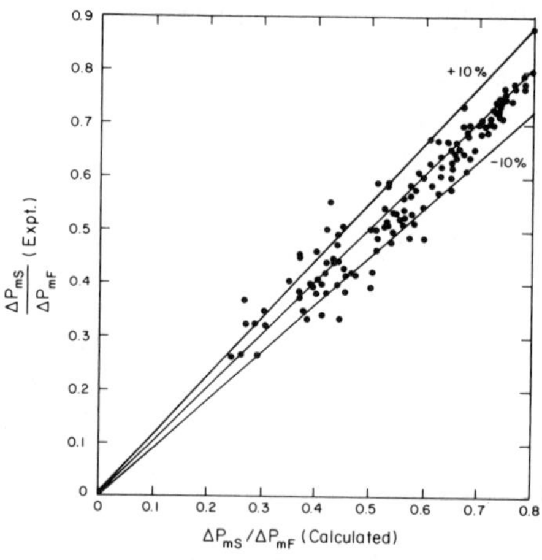

FIGURE 3: A COMPARISON BETWEEN EXPERIMENTAL AND CALCULATED SPOUTED PRESSURE DROP RATIOS (DATA FROM REFERENCE (7)).

The Minimum Spouting Characteristics of Small Glass Particles Spouted with Water

by

H. Littman and S. J. Kim*

Department of Chemical Engineering and Environmental Engineering
Rensselaer Polytechnic Institute
Troy, New York 12180-3590

*Department of Chemical Engineering
Chonnam National University
Kwangju, 500-05, KOREA

ABSTRACT

Experimental data are reported for the bed pressure drop, maximum spoutable height and spout diameter at minimum spouting which show that the spouting of small glass particles in water represents a new spouting regime. The transition from coarse to fine particle spouting occurs when particles appear in the inlet to the spout.

SCOPE

The minimum spouting characteristics of beds of glass particles 0.275 to 0.995mm in diameter spouted with water were studied in a cylindrical half-column 50.8mm in diameter with spout inlet tube diameters varying from 3.18 to 19.05mm in order to explore the spouting behavior of small particles in liquids. Since no study of the basic spouting parameters in the minimum spouting condition is available, u_{mS}, ΔP_{mS}, H_m, D_S and d_S data were obtained, correlated and compared with correlations for coarse particles. The lack of correlation between data for coarse and fine particles is explained in terms of a regime change.

CONCLUSIONS AND SIGNIFICANCE

1. The basic spouting parameters such as the bed pressure drop, maximum spoutable height and spout diameter at minimum spouting do not correlate with data for coarse particles.

2. Stability considerations show that these data are representative of a new regime of spouting which occurs when particles appear in the inlet to the spout.

3. The voidage in the spout is more uniform axially than that in coarse particle systems.

Equipment and the Measurement of the Bed Properties

The spouted bed is a cylindrical half-column, 50.8mm in diameter and 1.3m high. It is made of Plexiglas, has a flat base and was designed so that semi-circular inlet tubes of different diameters could be attached to the column. The flat surface of the half column has a non-uniform rectangular grid of 43 pressure taps all connected to piezometer tubes. The water temperature was generally kept at about 23°C and an air separtor was used to remove dissolved air from the water. A schematic diagram of the apparatus is given in Figure 1. Details can be found in Kim (1).

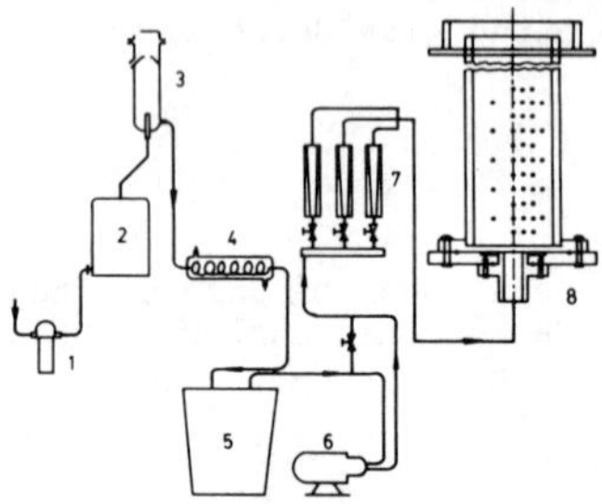

Figure 1. Schematic Diagram of Experimental Apparatus; 1. Water Filter, 2. Water Heater, 3. Air Separator, 4. Water Cooler, 5. Water Tank, 6. Pump, 7. Rotameter, 8. Spouted Bed; Arrows indicate the direction of water flow.

Determination of u_{mS}, ΔP_{mS}, H_m, D_S and d_S

The bed pressure drop, annulus height, jet penetration and spout fountain height are shown in Figure 2 as a function of flowrate for 0.995mm particles. The bed is shallow enough for the jet to penetrate through it. With increasing flowrate, the jet begins to penetrate the

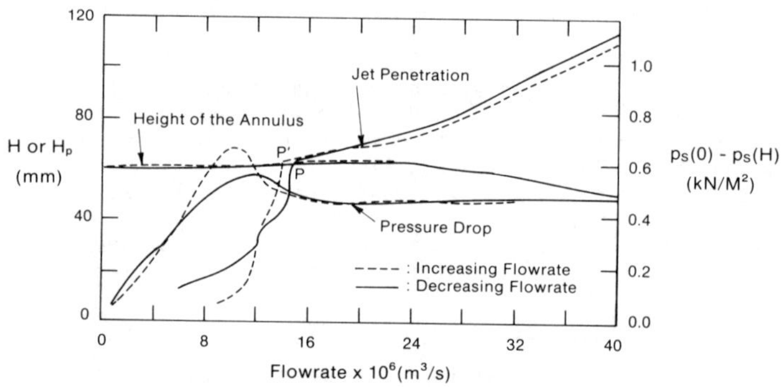

Figure 2. Pressure Drop, Annulus Bed Height, Jet Penetration and Spout Fountain Height as a Function of Flowrate; $d_p = 0.995$mm, $d_i = 6.35$mm, $\Delta P_{mF} = 0.607$ kN/m^2

bed and as the flowrate is increased, penetration continues rapidly with decreasing bed pressure drop until the jet reaches the top of the bed at point P'. On decreasing the flowrate from above the minimum spouting, the jet ceases to penetrate the bed at point P. This flowrate, as seen in Figure 2, is defined as the minimum spouting flowrate. ΔP_{mS} is the measured pressure drop, $[p_S(0) - p_S(H)]$, along the axis of symmetry in the minimum spouting condition.

Above a certain bed height, the jet does not penetrate to the top of the bed regardless of flowrate because the bed expands faster than the jet can penetrate it. The criterion for determining H_m and $(u_{mS})_{H_m}$ for these beds was formulated by observation of the change in the shape of the jet with decreasing flowrate. As the flowrate is lowered from above $(V_{mS})_{H_m}$, the jet narrows at the top finally assuming a flame tip shape. $(V_{mS})_{H_m}$ is the highest flowrate for which the flame tip shape is visible and the height of the jet to the tip of the flame at $(V_{mS})_{H_m}$ is denoted H_m. $(\Delta P_{mS})_{H_m}$ is the pressure drop at $(V_{mS})_{H_m}$ in a bed of height, H_m. The general behavior of the bed is unchanged by increasing d_i.

The spouted bed properties in a bed of height, H_m, in the condition of minimum spouting are given in Table 1. The spout diameter listed there is the mean diameter and ε_{mS} is the average void fraction based on H_m. ϕ_S was determined from photographs assuming that the particle shape could be modeled as an ellipsoid.

TABLE 1 - Experimentally Measured Spouted Bed Properties in Beds of Height, H_m, in the Condition of Minimum Spouting

d_p mm	ρ_p kg/m^3	ϕ_S	d_i mm	H_m mm	D_S mm	$(u_{mS})_{H_m}$ m/s x 10^3	$(\Delta P_S)_{H_m}$ kN/m^2	ε_{mS}
0.275	2570	0.81	3.18	165.0	9.6	1.00	1.170	0.395
			6.35	158.0	9.9		1.096	0.402
			12.70	152.6	10.3		1.062	0.405
			19.05	146.0	11.1		1.037	0.404
0.460	2480	0.98	3.18	133.4	11.5	2.93	0.905	0.408
			6.35	120.7	11.9		0.830	0.417
			12.70	104.0	13.0		0.717	0.416
			19.05	97.0	13.5		0.685	0.419
0.774	2470	1	3.18	96.8	13.4	6.47	0.630	0.414
			6.35	90.5	13.8		0.599	0.420
			12.70	84.0	14.7		0.554	0.424
			19.05	82.1	15.3		0.543	0.424
0.995	2670	1	3.18	83.7	15.0	16.00	0.677	0.433
			6.35	82.0	15.6		0.668	0.438
			12.70	77.8	16.2		0.629	0.435
			19.05	76.0	17.1		0.625	0.439

Correlation of Data and Comparison with Data for Coarse Particles

The minimum spouting velocity data are well correlated by the dimensionless form of the Mathur and Gishler relation (2) with u_{mF} replaced by $(u_{mS})_{H_m}$, that is

$$u_{mS}/(u_{mS})_{H_m} = h^{1/2} \quad ; \quad h > 1/4 \tag{1}$$

A modified Grbavcic et al. (3) equation gives predictions which are high. When $h < 1/4$ neither of the equations fit the data well predicting significantly higher values for u_{mS} than found experimentally.

The effect of bed height on $\Delta P_{mS}/\Delta P_{mF}$ at minimum spouting shows that the experimental values increase with h reaching an asymptotic value of about 0.85. This is considerably above the maxima predicted by the Mamuro and Hattori (4) and Lefroy and Davidson (5) theories (0.75 and

0.64 respectively). The data are well correlated by the equation

$$\Delta P_{mS}/\Delta P_{mF} = 0.37 + 1.32h - 1.43h^2 + 0.59h^3 \; ; \; h > 0.2 \qquad (2)$$

with standard error of 0.03. Based on experimental data for the spouting of coarse particles with air and water, Grbavcic et al. (3) proposed the following equation for $\Delta P_{mS}/\Delta P_{mF}$.

$$\Delta P_{mS}/\Delta P_{mF} = 1.5h - h^2 + 0.25h^3 \qquad (3)$$

This equation predicts considerably lower values of $\Delta P_{mS}/\Delta P_{mF}$ than found experimentally. The Morgan and Littman (6) theory predicts some of the high values of $\Delta P_{mS}/\Delta P_{mF}$ observed in experiments, but it is unsatisfactory in general.

The difference between the fine and coarse particle data is also indicated in the effect of h on the minimum spouting pressure drop ratio, $\Delta P_{mS}/(\Delta P_{mS})_{H_m}$. Our data correlate as

$$\Delta P_{mS}/(\Delta P_{mS})_{H_m} = 1.084h - 0.084 \quad ; \quad h > 0.2 \qquad (4)$$

while for coarse particle, the Grbavcic et al. theory gives

$$\Delta P_{mS}/(\Delta P_{mS})_{H_m} = 2h^2 - (4/3)h^3 + (1/3)h^4 \qquad (5)$$

Equation 5 which represents coarse particle data very well predicts values for the ratio which are generally low. This indicates very substantial differences in the spout pressure profile which in turn have a substantial effect on the flow of fluid from the spout to the annulus and thereby on the annulus flow field.

The H_m - D_S relationship derived by Littman et al. (7) for coarse particles

$$H_m D_S/(D_C^2 - D_S^2) = 0.345(D_S/D_C)^{-0.384} \qquad (6)$$

satisfies the fine particle data quite well overall.

An equation for predicting H_m in terms of D_C, d_i and the fluid-particle properties is given by Littman et al. (8) as

$$(H_m d_i)/D_C^2 = 0.218 + 0.005/A \qquad (7)$$

Equation 7, which has been very successful in predicting H_m for coarse particles, is plotted in Figure 3 along with the present data and the data of Grbavcic et al.(3). Note that mA is independent of d_i. H_m in systems of small particles, drops off rapidly as A is reduced and it is clear that a different type of equation is necessary to describe H_m data in the low A regime ($A < 0.02$). Since in our experiments ρ_f and ρ_p are essentially constant, A is a function of d_p and d_i. H_m decreases as d_p and d_i increase and the effect of d_i on H_m is small compared to that of d_p. An empirical equation of the form $H_m = a\, d_i^b d_p^c$ was used to correlate the data and a least square fitting gives

$$H_m = 98.34\ d_i^{-0.07}\ d_p^{-0.50} - 7.76 \quad ; \quad \text{mm} \tag{8}$$

with standard error of 3.3mm for H_m. Note that H_m is only a weak function of d_i as might be expected because of the low inlet jet momentum of the fluid. The effect of d_i on H_m, however, increases considerably as the inertial force of the jet increases and H_m varies as d_i^{-1} asymptotically for large particles (8). The effect of dp on H_m also changes with inlet jet momentum from $dp^{-0.5}$ to being independent of dp. The non-dimensional form of equation 8 indicates that other variables may be important.

The spout diameter in the minimum spouting condition follows the equation

$$d_S(H,\ u_{mS})/D_S(H_m,\ (u_{mS})_{H_m}) = 0.72h + 0.28 \tag{9}$$

Littman et al.'s (8) correlation of D_S for coarse particles

$$D_S/d_i = \{[2.10e^{-0.018/A} + 1.0]/3.10\}[0.862 + 0.219D_C/d_i - 0.0053(D_C/d_i)^2] \tag{10}$$

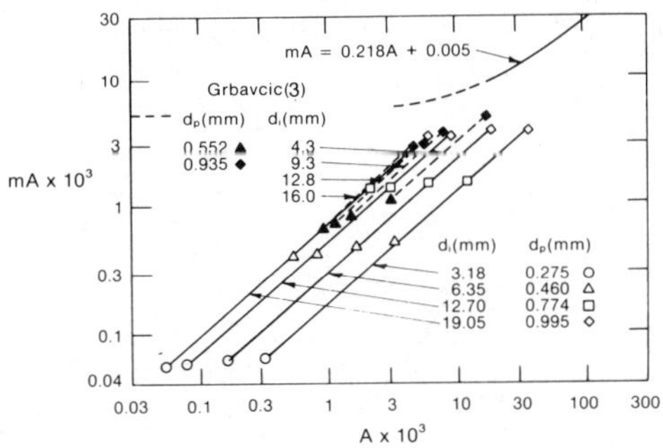

Figure 3. Dimensionless Maximum Spoutable Height Relationship-Effect of d_p and d_i

does not predict the small particle data at all. Those data show that D_S/d_i increases linearly with D_C/d_i for each particle size and that the slope of such a plot increases with particle size.

For the largest inlet size (small D_C/d_i), Equation 10 is in reasonably good agreement with experimental data. However, it underpredicts D_S/d_i as d_i decreases for a given D_C. The reason for this lies in the failure Equation 7 to predict the data in the low A regime. Note that Equation 10 is derived by combining Equations 6 and 7. In the absence of a general relationship for D_S, the best that can be done if H_m is known is to calculate D_S from Equation 6.

DISCUSSION

Equations representing the minimum spouting characteristics of small particles do not in general correlate with those for coarse particles. To understand why this is so, it is necessary to look at stability conditions for spouted beds.

Morgan et al. (9) have calculated the spout voidage distribution at minimum spouting using a variational technique and the boundary conditions $\varepsilon_S(0) = 1$ and $\varepsilon_S(H) = \varepsilon_{mF}$. Their solution shows that there is no variational solution at minimum spouting for which ε decreases monotonically that can satisfy the aforementioned boundary conditions unless a particular parameter, C_0, is bounded by 0.215 and 0.785. Their axial voidage profile with C_0 equal to 0.785 (labeled VAR) is given in Figure 4.

C_0 in the systems studied is essentially equal to $\Delta P_{mS}/\Delta P_{mF}$ and from the data in Table 1 is clearly greater than 0.785. In addition, we always observed particles in the inlet to the spout at minimum spouting showing contrary to what is observed for coarse particles, that $\varepsilon_S(0) < 1$. The transition to the new regime occurs, we believe, with the release of the boundary condition, $\varepsilon_S(0) = 1$ and the simultaneous rise in C_0 above 0.785. Support for this explanation lies in the fact that $d\varepsilon_S(0)/dz$ is infinite when C_0 equals 0.785 [see the VAR curve in Figure 4].

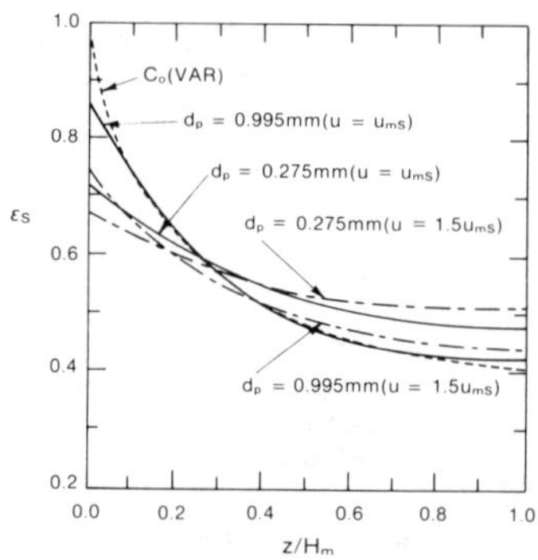

Figure 4. Voidage Profiles in the Spout

The voidage distribution in the spout in low inertia systems can be calculated very simply using only the spout pressure profile as is apparent once the momentum equation is simplified to represent this case. For the fluid and particle flow through an element of the spout assuming a constant spout diameter and radially averaged velocities, pressure and voidage

$$\rho_f \frac{d}{dz}(\varepsilon_S u_S^2) + \rho_p \frac{d}{dz}[(1-\varepsilon_S)v_S^2] = -\frac{dp_S}{dz} - \rho_f g - (1-\varepsilon_S)(\rho_p - \rho_f)g \quad (11)$$

In our work, the inertial term is small at minimum spouting. Neglecting the terms on the left hand side of Equation 11, we obtain

$$-[dp_S/dz + \rho_f g] = (1 - \varepsilon_S)(\rho_p - \rho_f)g \quad (12)$$

which shows the simple connection between p_S and ε_S.

Using the experimental axial pressure profile in the spout at minimum spouting in a bed of height H_m, ε_S was calculated from Equation 12 for the particle sizes 0.275 and 0.995 mm. The results are shown in Figure 4 along with the VAR profile. In all cases, $\varepsilon_S(0) < 1$. Note that increasing the particle size raises the value of $\varepsilon_S(0)$ toward its asymptotic limit of one. At the top of the spout, our experimental value of $\varepsilon_S(H)$ is essentially equal to ε_{mF}. Thus, we define a spouting system as being in the coarse particle regime if at minimum spouting in a bed for which $H \leq H_m$, $\varepsilon(0) = 1$, and as being in the fine particle regime if $\varepsilon_S(0) < 1$.

The voidage profiles with $u/u_{mS} = 1.5$ were calculated using Equation 12 and the experimental spout pressure profile. As seen in Figure 4, they are, as expected flatter, than those at minimum spouting.

The spout voidage distributions also tell a great deal about the particle circulation patterns. Whenever the pressure profiles are linear as they are generally in the top portion of the spout, Equation 12 shows that the spout voidage is constant. The particle continuity equation then becomes

$$(1 - \varepsilon_S)(dv_S/dz) = (dW_S/dz)/\rho_p(\pi/4)d_S^2 \quad (13)$$

We have observed that the particle velocities in the top portion of the bed do not vary significantly with z indicating that there is very little particle circulation there from the annulus into the spout. In general, we observe that particles enter the spout primarily in the lower part of the bed. The particle residence time distribution in the annulus is, therefore, much narrower than when using coarse particles. Lowering the particle diameter or increasing the inlet fluid velocity above the minimum both narrow the particles residence time distribution in the annulus in the direction of what would occur by inserting a draft tube in the bed.

ACKNOWLEDGMENT

The authors are grateful for support of this work by the National Science Foundation under Grant Number CPE-7903736. One of the authors, S. J. Kim, gratefully acknowledges the support of this work by the Korea Science and Engineering Foundation.

LITERATURE CITED

1. Kim, S. J., "Model and Experimental Verification of the Flow in the

Annulus of a Spouted Bed of Small Particles Spouted with Water," PhD thesis, Rensselaer Polytechnic Institute, Troy, N.Y. (1982).
2. Mathur, K. B. and P. E. Gishler, AIChE J., 1, 157 (1955).
3. Grbavcic, Z. B., D. V. Vukovic, F. K. Zdanski and Howard Littman, Can. J. Chem. Eng., 54, 33 (1976).
4. Mamuro, Tadashi and Hirotsugu Hattori, J. Chem. Eng. Japan, 1, 1 (1968).
5. Lefroy, G. A. and J. F. Davidson, Trans. Instn. Chem. Engrs., 47, T120 (1969).
6. Morgan, M. H., III and Howard Littman, "General Relationships for Minimum Spouting Pressure Drop Ratio, $\Delta P_{mS}/\Delta P_{mF}$, and the Spout-Annular Interfacial Condition in a Spouted Bed," Fluidization, Plenum Press, New York, pp. 287 (1980).
7. Littman, Howard, M. H., III, Morgan, D. V. Vukovic, F. K. Zdanski, Z. B. Grbavcic, Can. J. Chem. Eng., 55, 497 (1977).
8. Littman, Howard, M. H. III, Morgan, D. V. Vukovic, F. K. Zdanski, and Z. B. Grbavcic, Fluidization-Proceedings of the Second Engineering Foundation Conference, Cambridge Univ. Press, Cambridge, pp. 381-86 (1978).
9. Morgan, M. H., III, J. Y. Day and Howard Littman, Chem. Eng. Sci., 40, 1367 (1985).

NOMENCLATURE

A — $[\rho_f/(\rho_p - \rho_f)][u_{mF}\, u_T/gd_i]$

C_o — $\int_0^1 \frac{1 - \varepsilon_S(\frac{z}{H})}{1 - \varepsilon_{mF}}\, d(\frac{z}{H}) = \frac{\Delta P_{mS}}{\Delta P_{mF}} + \frac{\rho_f\{[D_C/(d_S)_{mS}]^4\, u_{mS}^2 - \varepsilon_{mF}\, u_{SH}^2\}}{[(1 - \varepsilon_{mF})(\rho_p - \rho_f)\, gH]}$

d_i spout inlet tube diameter
d_p particle diameter
d_S spout diameter
D_C column diameter
D_S spout diameter ($H=H_m$, $u=u_{mS}$)
g gravitational acceleration
h H/H_m
H bed height
H_p jet penetration or spout fountain height
H_m maximum spoutable height
m $(H_m d_i/D_C^2)$
mA $[\rho_f(\rho_p-\rho_f)][u_{mF}u_T H_m/gD_c^2]$
p_S fluid pressure in the spout
ΔP_{mF} min. fluidization pressure drop
ΔP_{mS} min. spouting pressure drop
$(\Delta P_{mS})_{H_m}$ $\Delta P_{mS}(H=H_m)$
u_{mS} min. spouting velocity
u_S interstitial fluid velocity in spout
u_{SH} u_S at $z = H$
u_T terminal velocity of particle
V_{mS} flowrate at min. spouting
v_S interstitial particle velocity in spout
W_S mass flow rate of particles in spout
z vertical coordinate measured from spout outlet
ε bed voidage
ε_{mF} voidage at min. fluidization
ε_{mS} voidage at min. spouting
ε_S spout voidage
μ fluid viscosity
ρ_f fluid density
ρ_p particle density
ϕ_S particle shape factor

PARTICLES HOLD-UP AND AXIAL PRESSURE DROP IN VERTICAL PNEUMATIC TRANSPORT REACTOR (RISER)

Kunio Kato, Yasuo Ozawa, Hiroaki Endo,
Masaru Hiroyasu and Tamotsu Hanzawa

Department of Chemical Engineering
Gunma University
Kiryu, Gunma 376 JAPAN

ABSTRACT

The particles hold-up and axial pressure drop were measured in the pneumatic transport reactor in the range of 3-10 cm tube diameter. The particles hold-up was affected by the superficial gas velocity, the terminal velocity of transported particles, particle feed rate per unit cross section and the tube diameter. The empirical equation for the particles hold-up in vertical pneumatic transport reactor was obtained.

The axial pressure drop was analyzed from the knowledge of the particles hold-up. The axial pressure drop consists of the following three individual contributions ; the pressure drop due to particle hold--up, the pressure drop due to conveying fluid, the pressure drop due to friction of particles in tube. The empirical equation for the solid friction factor was obtained.

The experimental pressure drop was well agreed with the calculated pressure drop based upon this analytical method.

SCOPE

Gas-solid pneumatic unit is widely used for the fast gas-solid reaction and drying of particles. Recently the reactor of F.C.C. process has become a pneumatic transport reactor (riser). To analyze these processes, it is very important to know the particle hold-up.

When particles are transported pneumatically, it is very important to estimate the axial pressure drop. Since the axial pressure drop is strongly affected by the particles hold-up, it must be analyzed from the knowledge of the particle hold-up. The object of this work is to establish the relationship between the particle hold-up and the operating conditions, and to establish the analytical method for the axial pressure drop based upon the particles hold-up.

CONCLUSIONS AND SIGNIFICANCE

1. An empirical equation for the particles hold-up in the pneumatic transport unit was obtained.

2. The axial pressure drop consists of three individual contributions, that is, the pressure drop due to the particle hold-up, the pressure drop due to conveying fluid and the pressure drop due to friction of

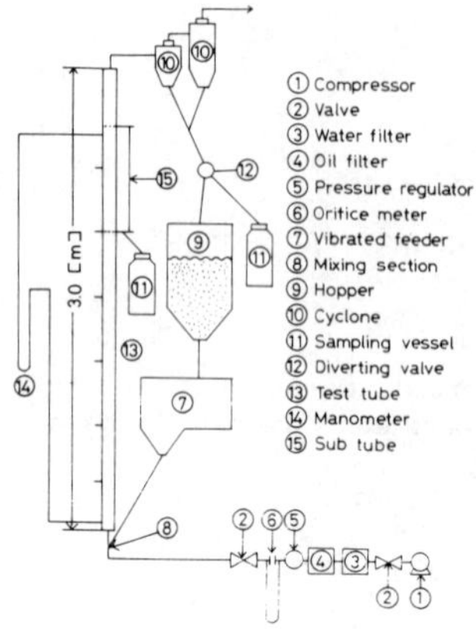

Fig.1 Experimental apparatus

	$\bar{d}_s$ [μm]	ρ_s [Kg·m^{-3}]	U_t [m·s^{-1}]
F.C.C - catalyst	43	8.40×10^2	0.15
(silica alumina)	50	1.50×10^3	0.27
Activated Alumina	120	1.30×10^3	0.61
	150		0.76
	385	9.00×10^2	1.55

Table 1 Properties of solid particles

the particles. To calculate the pressure drop due to the friction of particles, the solid friction factor must be obtained. An empirical equation for the solid friction factor was obtained.

3. The experimental axial pressure drop was well agreed with the calculated pressure drop based upon this analytical method.

INTRODUCTION

To analyze reaction performance of a gas-solid pneumatic reactor and the drying characteristics of a pneumatic conveying dryer, it is very important to know the particles hold-up in the unit. It seems that there is no literature to measure directly the particle hold-up in the unit.

The axial pressure drop of gas-solid pneumatic conveying system has been investigated by many investigators[1,2,3,4,5,6,7,8]. There are many approaches to analyze the axial pressure drop. But particle hold-up is the most important factor for the analysis of the pressure drop.

In this study the particles hold-up and the axial pressure drop were measured in the pneumatic transport unit in the range of 3-10 cm tube diameter. The effects of the superficial gas velocity, the terminal velocity of the particles, the feed rate of the particles and the tube diameter upon the particles hold-up were investigated. The axial pressure drop is analyzed from the knowledge of the particles hold-up and the average particles velocity. The solid friction factor is correlated with the operating conditions.

EXPERIMENTAL APPARATUS AND PROCEDURE

Figure 1 shows a schematic diagram of the experimental apparatus. The vinyl chloride tubes of 3 m length with 4.0, 6.6, 9.6 cm inside diameter were used as the vertical transport tube. Air from a compressor passed through oil filter to remove oil mist and was fed to the tube. The particles were fed from a vibrating feeder. The particles

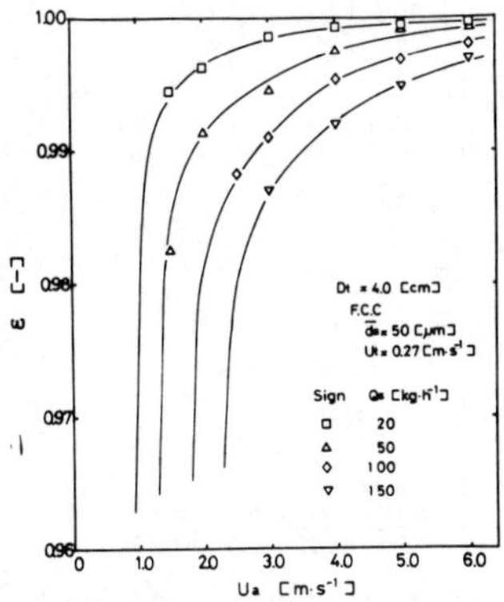

Fig.2 Relation between ε and Ua

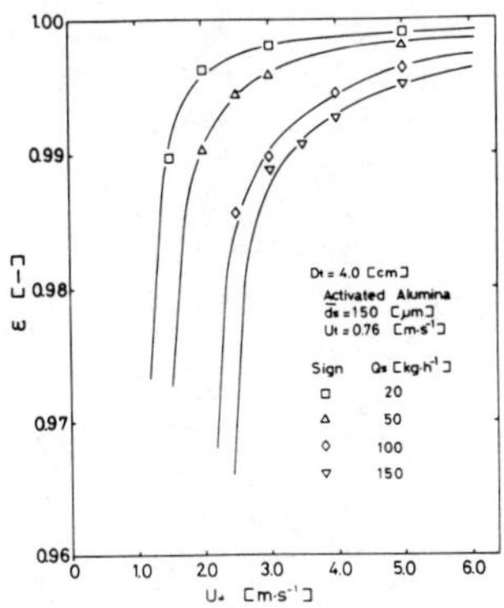

Fig.3 Relation between ε and Ua

were separated from the air by cyclones. The feed rate of the particles was measured by weighting the particles collected in the sampling vessel below the cyclones per constant time interval.

Pressure taps mounted on the tube wall were provided at 30 cm intervals along the transport tube. A water manometer or micro pressure gage was employed to measure the pressure drops.

The particles hold-up in the tube was measured as follows. The same size subtube with 60 cm length as the transport tube was equipped with 1.5 m above the bottom of the transport tube. When the pneumatic transport of the particles reached the steady state, the subtube was suddenly slided to the corresponding transport tube and the particles in the tube were collected in the sampling vessel and the particles hold-up was determined by weighting the particles. Table 1 shows the properties of F.C.C. catalyst particles and activated alumina particles used in this work.

THE PARTICLES HOLD-UP IN THE TUBE

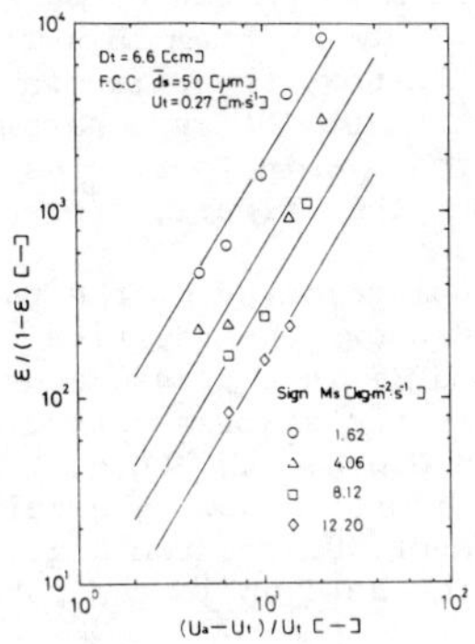

Fig.4 Relation between ε/(1-ε) and (Ua-Ut)/Ut

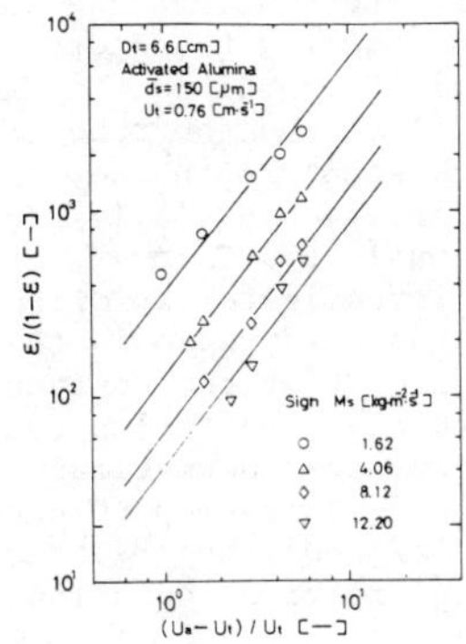

Fig.5 Relation between ε/(1-ε) and (Ua-Ut)/Ut

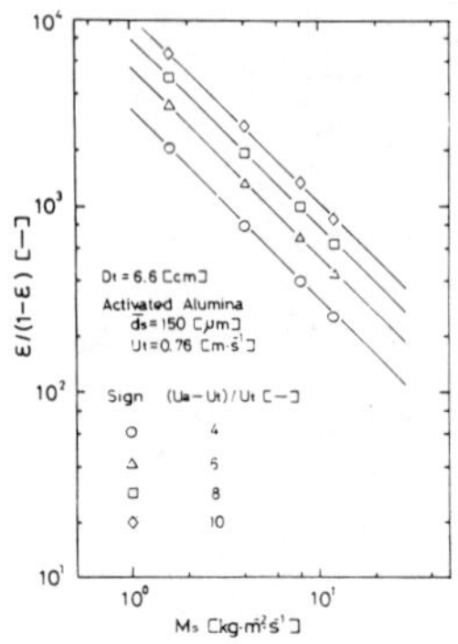

Fig.6 Relation between $\varepsilon/(1-\varepsilon)$ and Ms

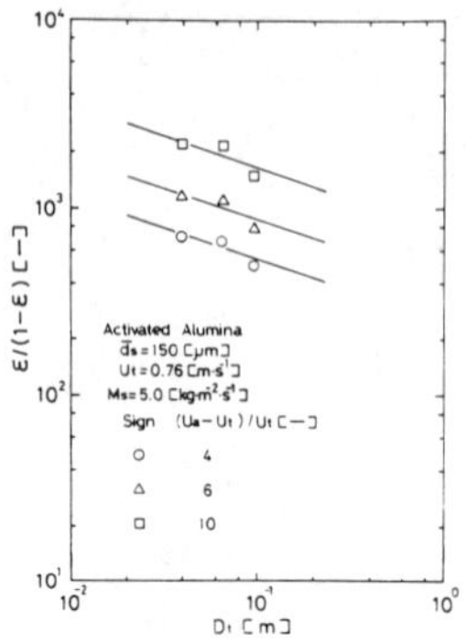

Fig.7 Relation between $\varepsilon/(1-\varepsilon)$ and Dt

Figures 2 and 3, respectively, show the relation between the void fraction in the tube and the gas velocity with the feed rate of the particles as a parameter in the case where F.C.C. catalyst particles ($\bar{d}s$=50 μm) and activated alumina particles ($\bar{d}s$=150 μm) are pneumatically transported in the vertical tube. From Figs.2 and 3, the void fraction in the tube decreases with increase of the feed rate of the particles. If the gas velocity is smaller than a certain value, the void fraction suddenly decreases with decrease of the gas velocity and the particles are not able to transport by choking.

The void fraction in the tube is affected by the superficial gas velocity, mass flux of the particles, the tube diameter and the terminal velocity of the particles. Figures 4 and 5, respectively, show the relation between the void ratio $\varepsilon/(1-\varepsilon)$ and (Ua-Ut)/Ut with mass flux of the particles Ms as a parameter in the case where F.C.C. catalyst particles ($\bar{d}s$=50 μm) and activated alumina particles ($\bar{d}s$=150 μm) are pneumatically transported in 6.6 cm I.D. tube. From these figures, $\varepsilon/(1-\varepsilon)$ is proportional to $\{(Ua-Ut)/Ut\}^{1.5}$.

Figure 6 shows the relation between $\varepsilon/(1-\varepsilon)$ and Ms with (Ua-Ut)/Ut as a parameter in the case where activated alumina particles ($\bar{d}s$=150 μm) are pneumatically transported in 6.6 cm I.D. tube. From this figure, $\varepsilon/(1-\varepsilon)$ is proportional to $Ms^{-1.0}$. Figure 7 shows the relation between $\varepsilon/(1-\varepsilon)$ and the tube diameter Dt with (Ua-Ut)/Ut as a parameter at the constant Ms in the case where activated alumina particles ($\bar{d}s$= 150 μm) are pneumatically transported. From this figure, $\varepsilon/(1-\varepsilon)$ is proportional to $Dt^{-0.4}$.

To investigate the effect of the terminal velocity of the particles Ut upon the void ratio in the tube, Figure 8 shows the relation between $\varepsilon/(1-\varepsilon)$ and Ut at the constant (Ua-Ut)/Ut and Ms in the case where F.C.C. catalyst particles ($\bar{d}s$=50 μm) or activated alumina particles ($\bar{d}s$=150 μm) are pneumatically transported in 6.6 cm I.D. tube. From Fig.8, $\varepsilon/(1-\varepsilon)$ is proportional to $Ut^{1.8}$. Figure 9 shows the relation between $\{\varepsilon/(1-\varepsilon)\}\cdot Ms^{1.0}\cdot Dt^{0.4}\cdot Ut^{-1.8}$ and (Ua-Ut)/Ut. From Fig.9, the following empirical equation for the particles hold-up in the tube is obtained.

$$\varepsilon/(1-\varepsilon) = 230\{(Ua-Ut)/Ut\}^{1.5}\cdot Ms^{-1.0}\cdot Dt^{-0.4}\cdot Ut^{1.8} \quad (1)$$

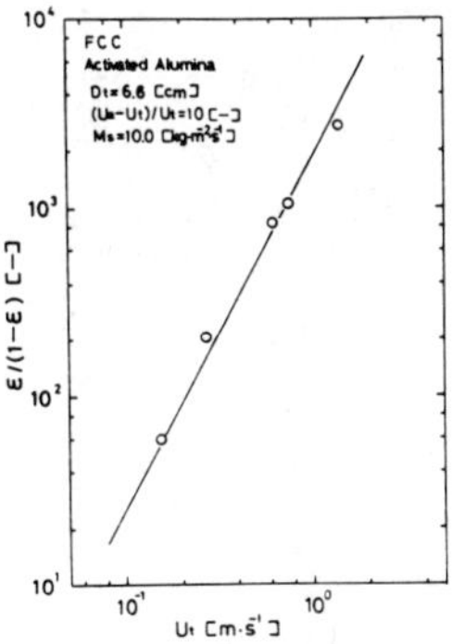

Fig.8 Relation between ε/(1-ε) and Ut

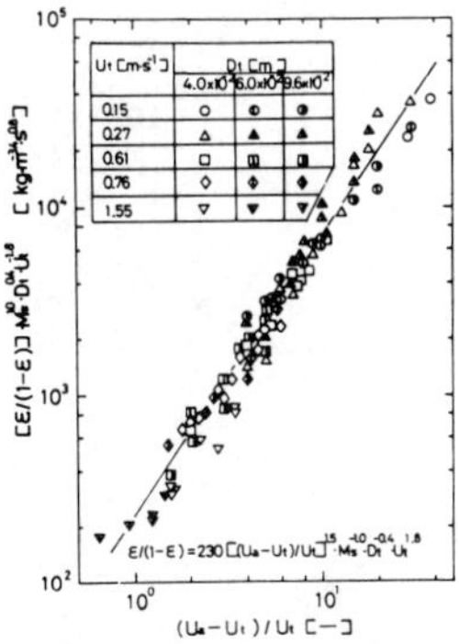

Fig.9 The correlation for the particles hold-up in the tube

The application range of Eq.(1) is as follows.

$0.7 < Ua[m \cdot s^{-1}] < 10$, $0.1 < Ut[m \cdot s^{-1}] < 2.0$, $0.5 < Ms[kg \cdot m^{-2} \cdot s^{-1}] < 40$,

$0 < m$(mass flow ratio)$[-] < 20$, $0.03 < Dt[m] < 0.10$

THE AXIAL PRESSURE DROP IN THE TUBE

Figure 10 shows the relation between the axial pressure drop per unit length of tube and the gas velocity with the feed rate of the particles as a parameter in the case where F.C.C. catalyst particles ($\bar{d}s$=50 μm) are pneumatically transported in 4.0 cm I.D. tube. From Fig.10, the axial pressure drop is affected by the feed rate of the particles and the superficial gas velocity. Figure 11 shows the relation between the axial pressure drop and the gas velocity with the terminal velocity of the particles as a parameter in the case of constant Qs. From Fig.11, when the terminal velocity of the particles is small, the pressure drop becomes small.

The axial pressure drop in the vertical tube consists of three individual contributions due to acceleration of the particles, hold-up of the particles and friction of the particles and gas.
The following equation is obtained.

$$\Delta P_T = \Delta P_A + \Delta P_S + \Delta P_F \qquad (2)$$

The acceleration term ΔP_A is almost zero except that in the lower part of the tube.
The pressure drop due to the particles hold-up is expressed as:

$$\Delta P_S = \rho_s(1-\varepsilon) \cdot g \cdot \Delta L \qquad (3)$$

The friction term ΔP_F is separated into two terms due to fluid alone and to the particles:

$$\Delta P_F = \Delta P_{FA} + \Delta P_{FS} \qquad (4)$$

The friction due to conveying fluid alone is usually defined by the Fanning equation. The axial pressure drop due to the friction of particles is expressed as[4)]

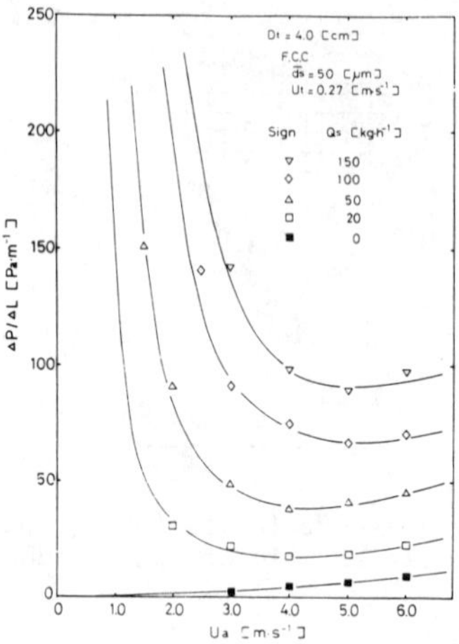

Fig.10 Relation between $\Delta P/\Delta L$ and Ua

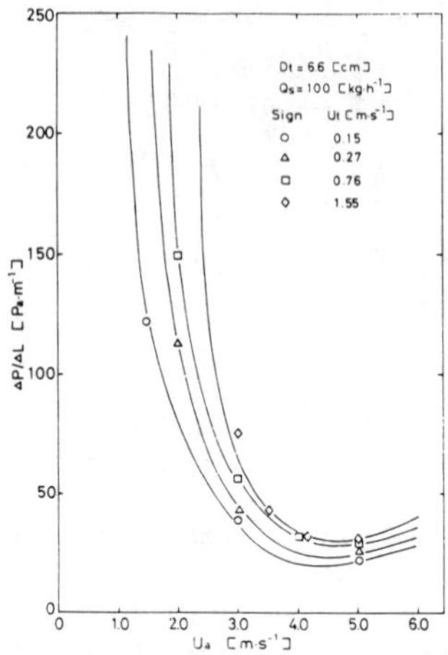

Fig.11 Relation between $\Delta P/\Delta L$ and Ua

$$\Delta P_{FS} = 2\ fs\ (1-\varepsilon)\ \rho_s\ U_s{}^2/D_t \qquad (5)$$

where

$$Us = \frac{M_s}{\rho_s\ (1-\varepsilon)}$$

From Eqs.(2), (3) and (4), ΔP_{FS} is calculated. A solid friction factor is calculated from Eq.(5). Fig.12 shows the relation between fs and Froude number, $Us/\sqrt{g\ D_t}$. From Fig.12, the following empirical equation for fs is obtained.

$$fs = 0.038\left(\frac{Us}{\sqrt{g\ D_t}}\right)^{-0.77} \qquad (6)$$

The application range of Eq.(6) is as follows

$$0.95 < \varepsilon < 0.999, \quad 10^{-1} < Fr < 50$$

From the information of particles hold-up and fs, the axial pressure drop in the tube for various operating conditions is calculated from Eq.(2).

Figure 13 compares the experimental axial pressure drop per unit length of tube $\Delta P_{T\ exp}/\Delta L$ of this experiment with the calculated one $\Delta P_{T\ cal}/\Delta L$ from Eq.(2) - (5). From Fig.13, $\Delta P_{T\ exp}/\Delta L$ is well agreed with $\Delta P_{T\ cal}/\Delta L$.

Figure 14 compares the experimental pressure drop per unit

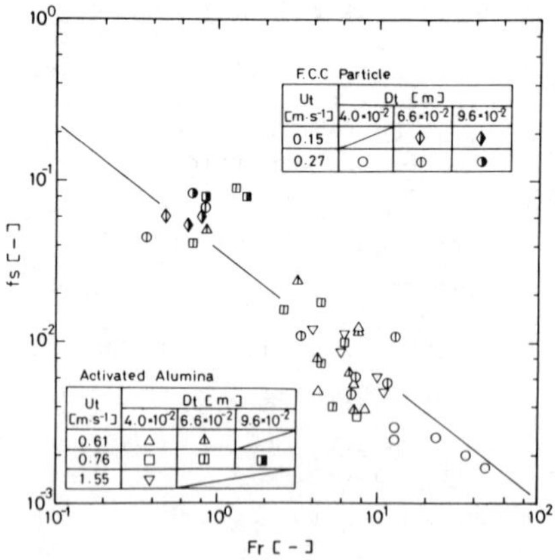

Fig.12 The correlation for solid friction factor of particles

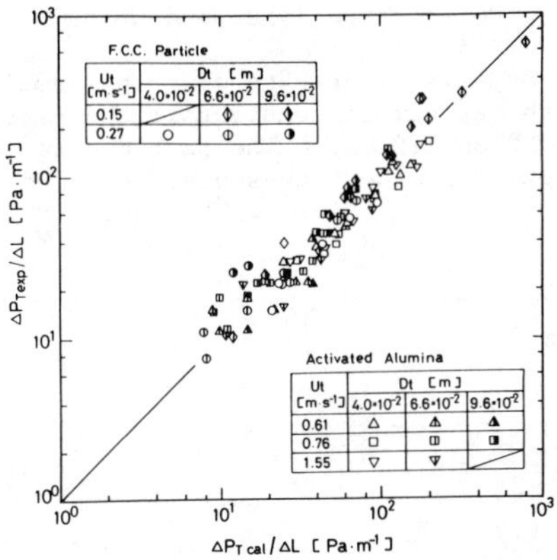

Fig.13 Relation between $\Delta P_{T\,exp}/\Delta L$ and $\Delta P_{T\,cal}/\Delta L$

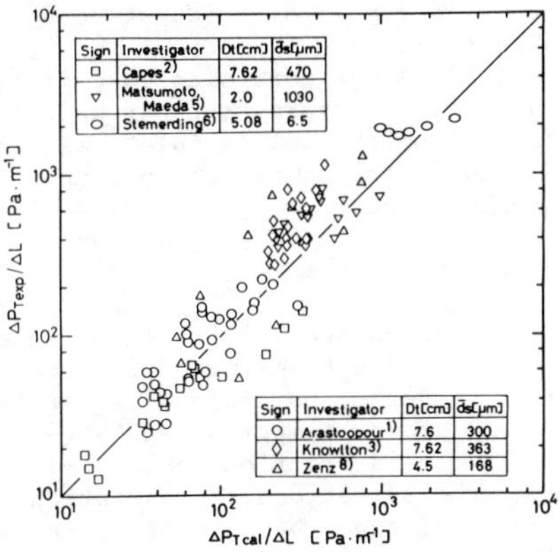

Fig.14 Relation between $\Delta P_{T\,exp}/\Delta L$ and $\Delta P_{T\,cal}/\Delta L$

length of tube $\Delta P_{T\,exp}/\Delta L$ obtained by the other investigators[1,2,3,5,6,8] with the calculated one. From Fig.14 the experimental pressure drop obtained by the other investigators is also well agreed with the calculated one.

Therefore, the axial pressure drop in the case where the particles are pneumatically transported in the vertical tube can be estimated from this method.

DISCUSSION

The particles hold-up in the tube is correlated by Eq.(1). The effect of the operating conditions upon the void ratio $\varepsilon/(1-\varepsilon)$ is considered qualitatively. The relative velocity (Ua-Ut) is the most important factor which affects the particles hold-up in the tube. Since the particles velocity increases with increase of (Ua-Ut), the void fraction ε increases. Accordingly the ratio $\varepsilon/(1-\varepsilon)$ is proportional to $(Ua-Ut)^{1.5}$. As the particles hold-up increases with increase of mass flux of the particles in the tube, the ratio $\varepsilon/(1-\varepsilon)$ is proportional to $Ms^{-1.0}$. The radial gas velocity distribution becomes large in the case of large tube diameter and the particles near the tube wall flow downward along the tube wall and the average particles velocity becomes small. Therefore, the void ratio decreases with increase of the tube diameter.

Generally, the axial pressure drop became small in the case where the void fraction was large. The void fraction decreased with increase of the tube diameter at the constant gas velocity and at the constant mass flux of the particles. Consequently, the axial pressure drop increases with increase of the tube diameter.

The axial pressure drop in the tube ΔP_T consists of three individual contributions ΔP_S, ΔP_{FA} and ΔP_{FS}. When the gas velocity is low ($Ua/Ut < 5.0[-]$), ΔP_T is strongly dependent on ΔP_S term. When the gas velocity is high ($Ua/Ut > 8.0[-]$), ΔP_T is strongly dependent on ΔP_{FS} term.

Since the particles used in this experiment were considerably small, the axial pressure drop per unit length was almost constant if the distance from the solids inlet was larger than 120 cm. The particle hold-up and the pressure drop per unit length were the representative value of the reactor.

NOTATION

A	cross-sectional area of tube (m^2)
D_t	inside diameter of tube (m)
$\overline{ds}$	average particle diameter (μm)
Fr	froude number $Us/\sqrt{g\,D_t}$ (-)
fs	solid friction factor (-)
g	gravitational acceleration ($m \cdot s^{-2}$)
ΔL	distance between pressure taps (m)
Ms	mass flux of particles ($kg \cdot m^{-2} \cdot s^{-1}$)
m	mass flow ratio (-)
$\Delta P/\Delta L$	axial pressure drop per unit length of tube ($Pa \cdot m^{-1}$)
ΔP_A	axial pressure drop due to acceleration of particles (Pa)
ΔP_F	axial pressure drop due to hydrodynamic resistance of particles and fluid (Pa)
ΔP_{FA}	axial pressure drop due to hydrodynamic resistance of fluid (Pa)
ΔP_{FS}	axial pressure drop due to hydrodynamic resistance of particles (Pa)
ΔP_S	axial pressure drop due to particles hold-up (Pa)
ΔP_T	axial pressure drop (Pa)
$\Delta P_{T\,exp}/\Delta L$	experimental axial pressure drop per unit length of tube ($Pa \cdot m^{-1}$)
$\Delta P_{T\,cal}/\Delta L$	calculated axial pressure drop per unit length of tube ($Pa \cdot m^{-1}$)
Qs	feed rate of particles ($kg \cdot h^{-1}$)
Ua	superficial gas velocity ($m \cdot s^{-1}$)
Ut	terminal velocity of particles ($m \cdot s^{-1}$)
ε	void fraction in the tube (-)
ρ_a	density of gas ($kg \cdot m^{-3}$)
ρ_s	density of particles ($kg \cdot m^{-3}$)

LITERATURE CITED

1) Arastoopour, H., C.H. Wang and S.A. Weil ; Chem. Eng. Sci., 37, No.9, 1379(1982)
2) Capes, C.E. and K. Nakamura ; Can. J. Chem. Eng., 51, 31(1973)
3) Knowlton, T.M. and D.M. Bachovchin ; Fluidization Technology, 2, 253(1976)
4) Konno, H. and S. Saito ; J. Chem. Eng. Japan, 2, No.2, 211(1969)
5) Matsumoto, S., H. Sato, M. Suzuki and S. Maeda ; J. Chem. Eng. Japan, 15, No.6, 440(1982)
6) Stemerding, S. : Chem. Eng. Sci., 17, 599(1962)
7) Wen, C.Y. ; AIChE Journal, 24, No.3, 548(1978)
8) Zenz, F.A. ; Ind. Eng. Chem., 41, 2801(1949)

MODELLING OF A CATALYTIC CRACKER STANDPIPE

J.R. Bernard

ELF FRANCE
Center de Recherche de Solaize
69360 St. Symphorien d'Ozon
B.P. 22, France

C.S. Teo and L.S. Leung
Department of Chemical Engineering
University of Queensland
St. Lucia, Queensland, Australia, 4067

ABSTRACT

The flow of gas-solid mixture down a stripper to regenerator spent catalyst standpipe in a FCC is studied experimentally by using radioactive tracers. Strong erosion of the lines in the lower part of the stripper and in the spent catalyst standpipe is caused by intense turbulent catalyst flow in those regions. A two-valve model is proposed to explain the mechanism leading to such turbulence. The steam injection at the stripper is postulated to act as a restriction to catalyst flow giving rise to coexistence of lean phase fluidized flow and dense phase fluidized flow in the standpipe system. Intense turbulence and catalyst recirculation would occur when the high velocity lean phase stream hits the surface of the dense phase catalysts.

SCOPE

In solid circulation systems, standpipes are often used to convey the solid material from a lower pressure region to a higher pressure region and to provide gas seal in one direction, Jones (1).

The objective of the present study is to investigate the flow behaviour of a stripper to regenerator spent catalyst standpipe system in a fluidized catalytic cracking plant. This system was troubled with extensive erosion of the lines at high solid circulation rate. Radioactive tracer measurements brought pertinent information on the flow behaviour of the system. A mathematical model is proposed to describe the phenomena.

CONCLUSIONS AND SIGNIFIANCE

A two-valve model has been proposed to explain the cause of erosion in a stripper-spent catalyst standpipe system. The steam injection at the stripper is postulated to act as a restriction (a valve) to catalyst flow entering the lower stripper section. As a result, coexistence of lean phase fluidized flow and dense phase fluidized flow may occur in the system. Intense turbulence and catalyst recirculation would occur when the high velocity lean phase stream hits the surface of the dense phase catalysts. Such postulate is consistent with the experimental observations using radioactive tracers.

The present work is significant in two aspects : (i) experimentally, the flow behaviour in a commercial standpipe is revealed using a radioactive tracer technique ; and (ii) theoretically, the inclusion of the top restriction (by steam injection) in addition to the bottom restriction (slide valve) in the model is important and essential.

INTRODUCTION

Figure 1 is a schematic diagram of a fluid catalytic cracking plant showing catalyst circulation along the riser reactor, fluid bed disengager, stripper, spent catalyst standpipe, regenerator and regenerated catalyst standpipe. The flow of the spent catalyst to the regenerator is controlled by a slide valve on the spent catalyst standpipe in response to the dense fluidized bed level (indicated by a pressure differential) in the disengager.

Catalyst flow behaviour in the standpipes is extremely important for the plant stability. The gas entrained down with the solids may assume different relative velocities with respect to the catalyst velocities giving rise to a variety of possible flow regimes such as non-fluidized flow, lean phase fluidized flow, dense phase fluidized flow, flooding etc..

DESCRIPTION OF STRIPPER-STANDPIPE OPERATION

Figure 2 shows the stripper to spent catalyst standpipe section. At the stripper, stripping steam is fed via an annular box. The steam rises countercurrently to the downflowing catalyst through baffles. The stripped catalyst descends in the understripper without any internals and then flows via an expansion joint into a 60 deg. inclined standpipe. Depending on the operating conditions, the catalyst fluxes are between 40 and 100 $kg/m^2/s$ in the stripper and between 400 and 1000 $kg/m^2/s$ in the standpipe, while the superficial velocity of stripping steam, if taken 100% upwardly without any baffle would be between 0.08 and 0.3 m/s.

After a change to a low coke yield catalyst, the circulation rate rose to its maximum, i.e. 800 to 1000 $kg/m^2/s$ in the standpipe and strong erosions of the lines appeared, resulting in leaks and shutdowns. During these operations, every indicator remained at normal levels. The only abnormal plant indication was a differential pressure reading between the stripper and the top of slide valve on the spent catalyst standpipe, which fell dramatically as the stripping steam rate was increased as shown in Table 1 in order to minimize the coke yield.

GAS AND CATALYST BEHAVIOUR IN STRIPPER-STANDPIPE BY THE TRACER METHOD

Stripping steam at both the low rate (0.275 m^3/s) and high rate

($0.867\ m^3/s$) was traced by injecting pulses of Argon 41 (half life 110 min) just upstream from the distribution box. Radioactive equilibrium catalyst (Lanthanum 140, half life 40h) was injected at the same place to be mixed with the circulating catalyst after leaving the stripping box. The locations of the external detectors are shown in Figure 2.

Steam Flow

The tracer studies showed that a fraction of the steam rose countercurrently to the catalyst through the stripper baffles and the rest was entrained downwardly by the spent catalyst. For $0.867\ m^3/s$ stripper steam flow, 66% flowed upwards and 34% was entrained downwards. For $0.275\ m^3/s$ stripping steam, the corresponding figures were 40% and 60% respectively.

Steam flow in the understripper. The steam flow descending to the standpipe was recorded by the three detectors around the vessel between the stripping steam distribution point and the spent catalyst standpipe/ As can be noted on Figure 3, two detectors (5,7) failed to record much tracer, while the larger part was detected by the third one (6) which was closer to the standpipe axis. These curves indicate that the steam flowed in less than half the cross section of the understripper, just above the standpipe. The Peclet number ($Pe = \frac{UL}{D}$) of this stream was 8 to 10, indicating a dispersed plug flow.

Steam flow in the standpipe. Figure 4 shows the residence time distribution of steam recorded by detectors 8 and 9 in the standpipe at a S.S. rate of $0.867\ m^3/s$. The curves do not change much in shape after moving down from detector 8 to detector 9, which was only possible when back mixing was small. Indeed Pe was greater than 20, indicating that the steam was flowing in plug flow. Steam velocities were 1.5 m/s and 0.6 m/s for $0.867\ m^3/s$ S.S. and $0.275\ m^3/s$ S.S. respectively.

Catalyst Flow

Good injection of radioactive catalyst was realised only at $0.875\ m^3/s$ S.S.

Catalyst flow in the understripper. Perfect similarity with steam flow was observed in terms of velocity and Peclet number. The same channelling occurred as with steam (Figure 5).

Catalyst flow in standpipe. In the first section covered by detector 8, Pe=1, indicating heavy back mixing. In contrast, in the second section covered by detectors 8B and 9, there was a plug flow with a time shift of the residence time distribution curves without any change in shape (see Figure 6) and Pe=1000.

These phenomena of turbulence and then plug flow of catalyst can be interpreted by using a two-valve model developed by Teo (2).

MODELLING THE STRIPPER-STANDPIPE SYSTEM

In the present study, the catalyst circulation rate is high so that the stripper working at 0.875 m3/s steam may act as a restriction to catalyst flow. The restriction due to the stripping steam injection is assumed to behave like a valve here defined as the TOP VALVE. An increase of steam injection rate is taken to correspond to a reduction of the top valve opening. By defining the slide valve as the BOTTOM VALVE, we have a two-valve system, i.e. a standpipe with a vertical section of larger diameter (understripper section) and an inclined section of smaller diameter (spent catalyst standpipe) between a top valve and a bottom valve through which catalyst descends (see Figure 7). The two-valve model developed by Teo (2) may then be applied to this two-valve system.

Depending on the operating conditions, the entire standpipe can be filled with dense phase fluidized flow (DENFLO), or part of the pipe is in DENFLO. In the latter case, we have coexistence of a lean phase flow (LEANFLO) section above a DENFLO section. The interface of coexistence is at a height z_p above the bottom valve. Above z_p solids descend at a high voidage in LEANFLO and at a very high solid velocity. The high velocity solid stream hits the surface of the dense solids at the interface causing intense turbulence and mixing. This turbulence is believed to be responsible for the erosion of the pipe lines internally.

With the coexistence of LEANFLO above DENFLO, the steam and catalyst may stream past the vertical standpipe section into the inclined pipe section as shown in Figure 8.

Model

The two-valve standpipe system can be modelled by writing equations to each of the following four sections :

(i) bottom valve (slide valve) ;
(ii) inclined standpipe ;
(iii) understripper ;
and (iv) top valve (stripping steam distribution/injection point)

The equations describing each of four sections have been described elsewhere, Teo (2).

Model Predictions

In the present study, interactions between the standpipe operation with the rest of the catalytic cracking operation have not been taken into account. The interactive effects are nevertheless important and should be taken into consideration in a further study. The effects of each independent parameter on standpipe operation will now be discussed.

Effect of valve openings. Figure 9 and Figure 10 plot z_p and W_s respectively against the top valve openings, Acons, for different bottom valve openings, Aopen. When Aopen=1, the bottom valve is fully open, but the flow area of the slide valve represents only a fraction of the inclined pipe cross section. Hence, the bottom valve is always a restriction. As for the top valve, when Acons=1, the top valve is fully open, but represents a fraction of the stripper cross-sectional area due to the geometry of the stripping steam injection point.

Without steam injection, Acons=1 is an order of magnitude larger than Aopen=1. As steam is introduced, Acons will be reduced, until at a sufficiently large steam rate, the effect of the steam injection point may interact with the bottom valve in determining the flow behaviour in the standpipe system.

For a given set of parameters, such as keeping terminal pressures at both ends of the standpipe system constant, for each bottom valve opening, an increase in the top valve opening (corresponds to a reduction of steam injection rate) leads to an increase in z_p in Figure 9 and increase in W_s in Figure 10.

If the top valve opening is constant, then for constant terminal pressures, an increase in the bottom valve opening leads to reduction of z_p in Figure 9 and a corresponding fall in W_s in Figure 10.

Notice in both Figures 9 and 10 that each curve exists within a certain range of Acons. For each curve, beyond the upper limit of the ratio of top valve to bottom valve opening, the coexistence of LEANFLO above DENFLO model is no longer valid. This is because the flow regime is now replaced by DENFLO throughout the system. Such flow regime has been modelled by Teo (2).

However, for each curve when the ratio of top valve opening to the bottom valve opening is small and below the lowest limit, then LEANFLO throughout is likely although the model has not been extended to consider such a situation.

Effect of the terminal pressure difference, $P_B - P_T$. Figure 11 shows the effect of the terminal pressure difference, $P_B - P_T$ on z_p for Aopen=0.7. z_p increases as the terminal pressure difference increases. W_s is not affected by the terminal pressure difference as long as z_p has not reached the top of the understripper. For DENFLO throughout regime, increase in terminal pressure difference leads to a dramatic decrease in W_s (see Teo (2).

Overall Comparison of Plant Data with Model Predictions

The plant data are summarized in Table 1. The data reveal the complex interrelationship of the various independent parameters and their effects on z_p. It is apparent from the table that the position of the DENFLO-LEANFLO interface (i.e. z_p) is sensitive to various operating parameters. This is consistent with the outcome of the model. The table suggests a trend that as steam injection rate increased, the pressure build-up on the standpipe decreases. This shows the predominant effect of the steam injection rate on z_p.

From the radioactive tracing studies, for gas flow at the lower part of the stripper, two detectors (5 & 7) failed to record much tracer gas, while the larger part of the tracer gas was detected by the third one (i.e.6), which was closer to the standpipe axis. The flow area of the gas was less than half the area of the cross section of the understripper. For catalyst flow in the understripper, similar observation was obtained. This is consistent with the postulate that LEANFLO occurs in the understripper with a core of gas-solid moving down at high velocity bypassing the stagnant region covered by detectors 5 and 7. This high velocity stream of gas-solid mixture is believed to be responsible for the strong erosion at the lower part of the understripper. In the inclined standpipe section, plug flow of the gas phase was observed. However, intense turbulence and recirculation of catalyst occured in the section covered by detector 8, while perfect plug flow occured in the section covered by detectors 8B and 9. This is consistent with the postulate that LEANFLO coexists above DENFLO and that the high velocity lean phase stream hits the dense phase interface at z_p (see Fig. 8).This causes intense turbulence and recirculation of catalyst in the DENFLO regime.

Although the data show the same trends than the model, there are some quantitative differences. For instance, the computed solid rate is lowe lower than the real rate. This may be due to uncertainties on different factors like the fluidization properties of catalyst in the standpipe medium, sloped instead of vertical standpipe, rough modelling of the stripper constriction.

NOTATION

Symbol	Definition	Unit
Acons	fraction of top valve opened	-
Aopen	fraction of bottom valve opened	-
D	dispersion coefficient	m^2/s
L	distance between tracer injection point and detector	m
P_B	bottom terminal pressure of standpipe	Pa
P_T	top terminal pressure of standpipe	Pa
Pe	Peclet number, UL/D	-
U	average velocity of steam	m/s
W_s	solid mass rate	
z_p	relative height of dense phase fluidized flow section above slide valve in standpipe and understripper	-

LITTERATURE CITED

1. Jones, P.J., "Downflow of gas-solid mixtures in bottom-restrained vertical standpipes", PhD Thesis, University of Queensland, Brisbane (1981).

2. Teo, C.S., "A theoretical and experimental study of gas-solid flow in standpipes", PhD Thesis, University of Queensland, Brisbane (1985)

Table 1. Variation of Plant Conditions Versus the Stripping Steam (S.S.) Rate

S.S. Rate	ΔP between stripper and top of slide valve	Spent Catalyst Standpipe		Catalyst Circulat. Rate	Terminal Pressure Difference	Regenerated Catalyst Standpipe	
		Valve opening	P across valve		$P_B - P_T$	Valve opening	Δ P across valve
m3/s	Pa	%	Pa	kg/s	bar	%	Pa
0.246	44000	60	56000	242	0.38	50	35000
0.419	21000	66	31000	239	0.33	46	35000
0.578	15000	70	28000	236	0.28	56	30000
0.867	13000	67	26000	234	0.33	45	29000

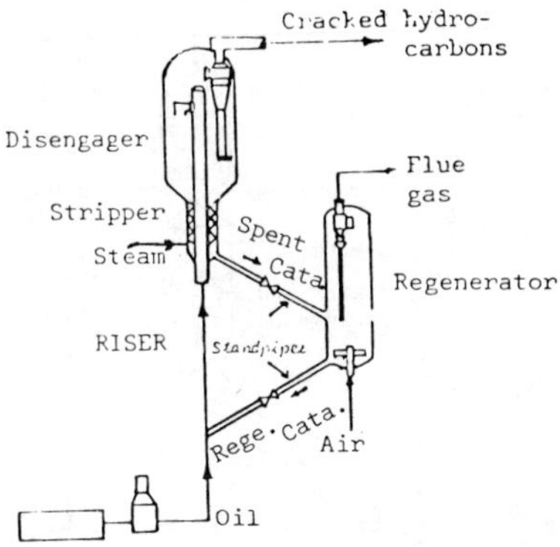

Figure 1. Typical scheme of a FCC plant

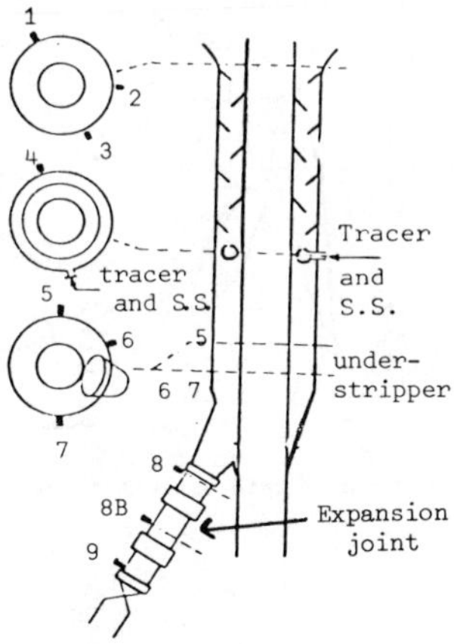

Figure 2. Stripper to spent catalyst standpipe section and positions of detector

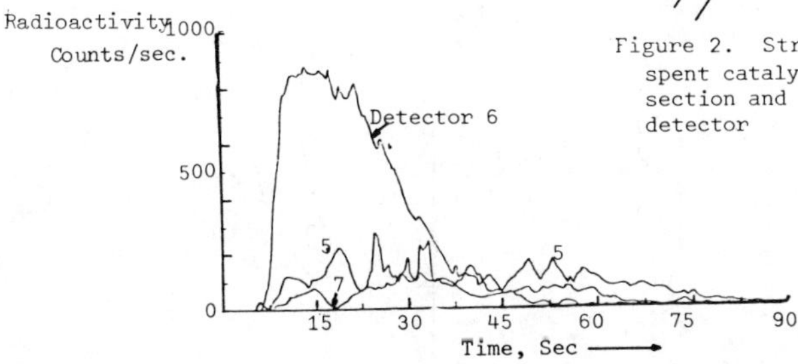

Figure 3. Tracing curves for steam flow in the understripper (0.867 m^3/s S.S.)

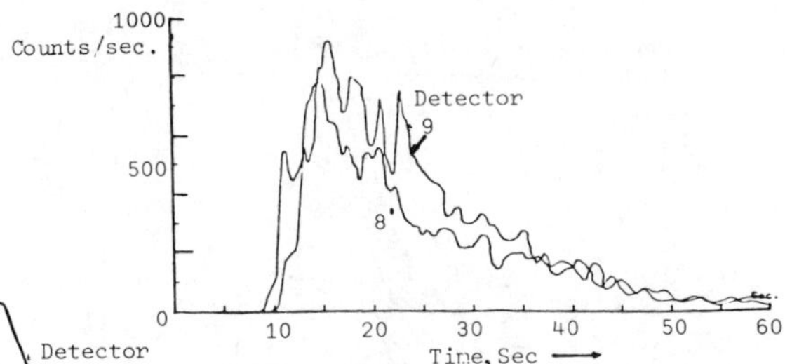

Figure 4. Tracing curves for steam flow in the spent catalyst standpipe (0.867 m^3/s S.S.)

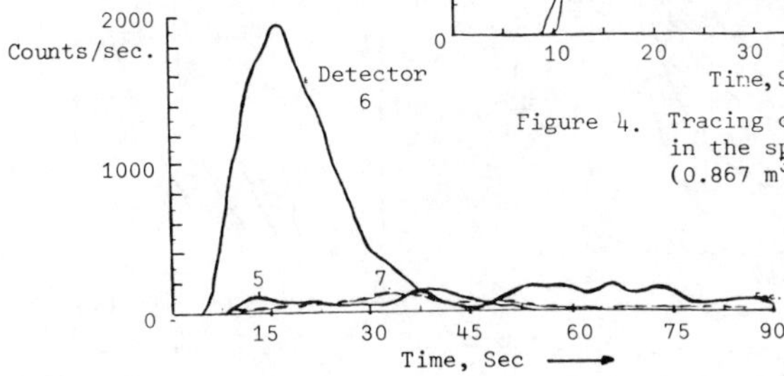

Figure 5. Tracing curves for catalyst flow in the understripper

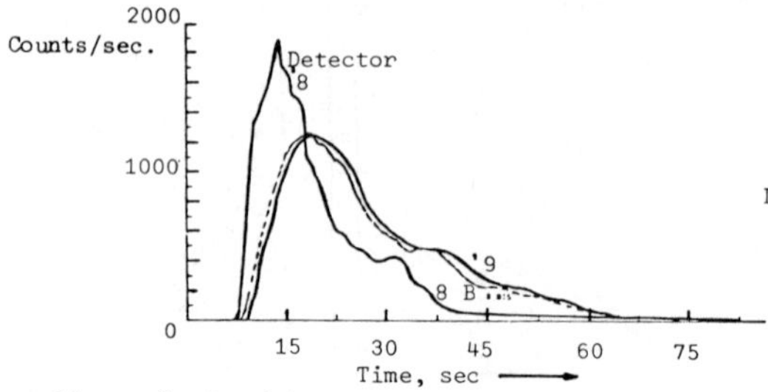

Figure 6. Tracing curves for catalyst flow in the spent catalyst standpipe

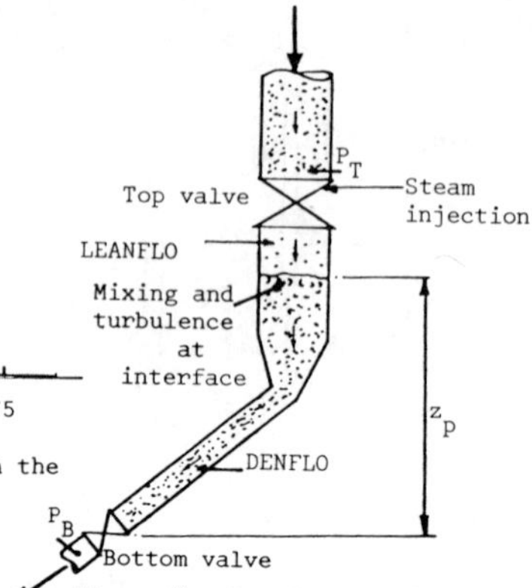

Figure 7. Two-valve system

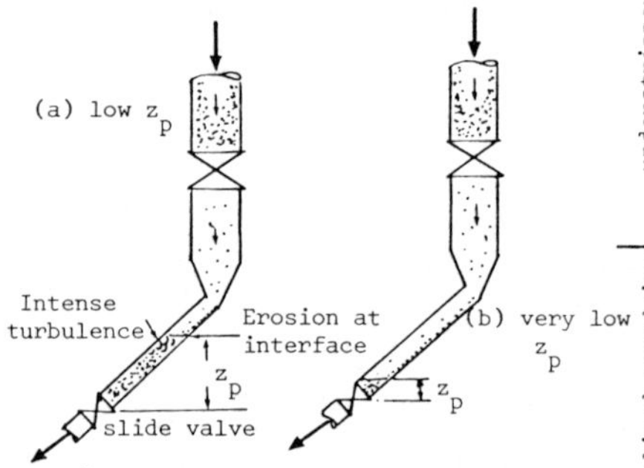

Figure 8. Variation of position of DENFLO-LEANFLO interface

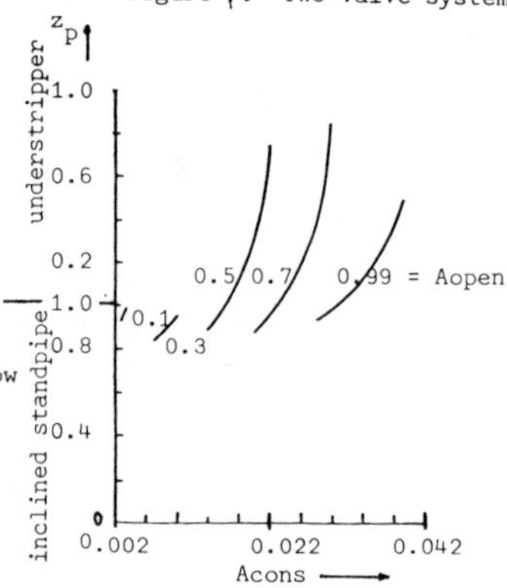

Figure 9. Effect of valve openings on z_p. P_B = 2.9 bar, P_T = 2.6 bar

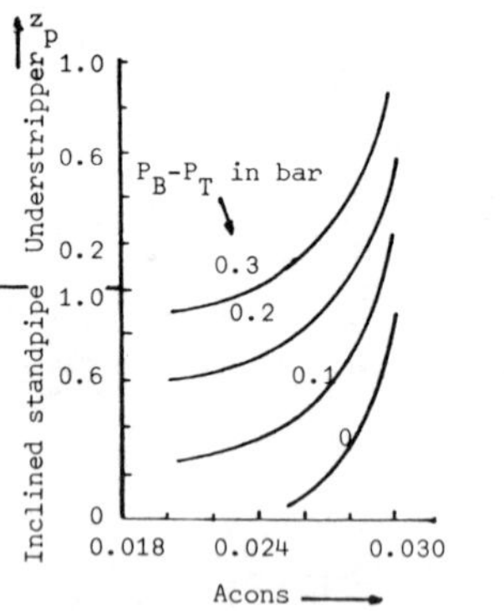

Figure 11. Effect of terminal pressure difference on z_p, Aopen = 0.7, P_T = 2.6 bar

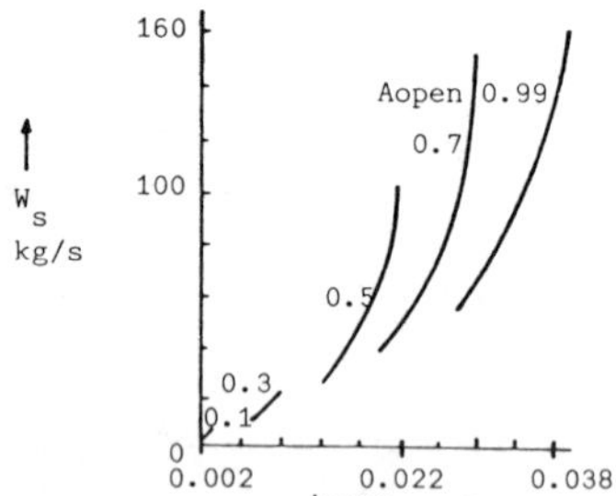

Figure 10. Effect of valve openings on W_s, P_B = 2.9 bar, P_T = 2.6 bar

TRANSITION TO TURBULENCE?

M.J. Rhodes and D. Geldart
Postgraduate School of Powder Technology
and Chemical Engineering,
University of Bradford,
BRADFORD
West Yorkshire, U.K.

ABSTRACT

A model is proposed to explain the transition velocity, U_k, in terms of the rates of increase of bed expansion and freeboard hold-up with gas velocity. Theory and experiment are compared using the results of studies on a wide variety of solids in three expanded-top columns of different scale.

SCOPE

This report summarises the results of an extensive study of the bed transition from the strongly two-phase slugging flow in small fluidized beds, with its attendant large pressure fluctuations, to a state of apparently greater homogeneity indicated by diminished variations in pressure - the so-called turbulent regime characterised by the velocity, U_k. A model is presented explaining this transition. Experimental evidence in support of the model is drawn from our own study and from the work of others. Correlations are presented which allow prediction of transition velocities using particle properties and equipment scale.

CONCLUSIONS AND SIGNIFICANCE

We believe that the transition characterised by the velocity, U_k, is actually a transfer of solids from the fluid bed to the freeboard as opposed to a fundamental change in the bed structure. The transition, which would more accurately be described as "refluxing transport", is akin to choking, and is explained in terms of entrainment and bed expansion; the velocity at which the transition occurs increases with particle size, particle density and equipment size.

INTRODUCTION

The terms "turbulent regime" and "transition to turbulent flow" first appeared in a paper by Kehoe and Davidson (1) dealing with their studies of slugging fluidized beds. They noted that, at high gas velocities, the strongly two-phase nature of the slugging bed gave way to an apparently more homogeneous condition having much smaller fluctuations in bed density as detected by capacitance probes. Lanneau (2) had earlier observed the same phenomenon in his studies of slugging. Geldart and Rhodes have recently proposed (3) that what was being observed was not a transition to a new "turbulent regime" of fluidization but rather a transition to refluxing pneumatic transport, or in the case of Massimilla (4) and of Carotenuto et al. (5), the exposure of the capacitance probes to freeboard as the bed level fell due to entrainment.

Other workers Abed, (6) Avidan and Yerushalmi (7) and Canada and Staub (8) have used the term "turbulent regime" to describe phenomena which are clearly very different from those originally observed by Lanneau (2) and by Kehoe and Davidson(1). The study reported in this paper is of a transition equivalent to that observed by Lanneau, and Kehoe and Davidson, and we present the results of an extensive experimental study together with a proposed model.

DEFINITION OF TERMS

The following terms are used in this paper

bed: the region in the equipment where fluidised solids are the continuous phase and gas bubbles or slugs are the discontinuous phase

freeboard: the region above the bed in which the gas is the continuous phase and the solids are the discontinuous phase

freeboard holdup: the quantity of solids in the freeboard

the transition: the change to small fluctuations in pressure characterised by the velocity U_k

EXPERIMENTAL

Experiments were carried out in three expanded-top units with the dimensions:

1. 25 mm x 300 mm x 1.5m high to expanded section
2. 76 mm diameter x 4 m high to expanded section
3. 152 mm diameter x 4 m high to expanded section

Each of these was fitted with external cyclones returning solids into the expanded section. Measurements of pressure drop and amplitude of fluctuations in pressure drop were made across a section of the apparatus just above the distributor (tappings were 0.1 m apart in unit 1 and 0.3 m apart in units 2 and 3) unless otherwise stated. Transition velocities, U_c and U_k were determined by plotting amplitude of pressure fluctuations against superficial gas velocity (Figure 1). This is the method used by Yerushalmi and co-workers (9). Visual observations also played an important part in corroborating the results obtained from pressure drop measurements. Details of the solids used in the investigation are given in Table 1.

THE MODEL

Consider a fluidized bed operating just above minimum fluidization velocity. As the gas velocity is gradually increased the bed expands causing the height of the bed, H, to increase. At the same time, entrainment from the bed surface increases with increasing gas velocity. This causes the solids holdup in the freeboard and in the solids separation and return system to increase.

We have, therefore, two processes working in opposition: bed expansion causing an increase in H with increasing gas velocity, and freeboard holdup causing a decrease in H with increasing gas velocity.

In a slugging bed, the rate of increase of H with gas velocity (strictly $U-U_{mf}$) due to expansion is steady (10). However, the rate of decrease of H due to increasing freeboard holdup increases dramatically with gas velocity. (It is widely accepted that entrainment rate varies as the third or fourth power of gas velocity, and it is quite probable that freeboard holdup follows a similar trend).

The gas velocity, U_c, at which the rates of the two opposing processes are equal is the start of the transition and represents the point at which the maximum bed height is achieved.

Further increase in gas velocity beyond U_c results in a net decrease in bed height as more solids are transferred to the freeboard. The gas velocity, U_k, at which the bed height is just zero represents the end of the transition. The freeboard has grown at the expense of the bed until all the solids that were originally in the bed are in the freeboard which now fills the equipment (Figure 2).

In a small diameter fluidized bed where pressure fluctuations can be monitored, U_c represents the velocity at which the fluctuations are greatest and U_k represents the velocity at which the fluctuations are levelling out at a much lower value. The reason for this is that the amplitude of pressure fluctuations, caused by the slugging which occurs, depends largely on the height of solids above the pressure tappings recording the fluctuations (Table 2). At U_c this height is a maximum; at U_k this height is zero and any pressure fluctuations are due to small variations in solids holdup in the section of equipment monitored.

DIRECT EVIDENCE IN SUPPORT OF THE MODEL

For several Geldart (11) Group A powders measurements of H_i and ΔH were made in units 1 and 3 at gas velocities below transition. In order to deduce H_i and ΔH, the pressure drop across the entire bed, Δp, and the pressure drop per unit height of the bed $\Delta p/L$ were recorded (Δp was measured using a pressure probe whose position could be adjusted to follow the position of the bed surface as determined visually). H_i and ΔH could then be calculated as follows:

bed density $$\rho_B = \frac{\Delta p/L}{g}$$

ideal bed height $$H_i = \frac{M}{\rho_B A}$$

mass of solids remaining in dense bed, $M_b = \frac{\Delta p . A}{g}$

depth of bed of density ρ_B that would be occupied by freeboard solids $\Delta H = \frac{M - M_b}{\rho_B A}$

H_i and ΔH were plotted against superficial gas velocity, the curves extrapolated where necessary, and U_c and U_k determined according to the proposed model. Results are presented in Table 3 and agreement between predicted and observed values of U_c and, particularly, U_k are, in all but two cases, good.

OTHER EVIDENCE IN SUPPORT OF THE MODEL

(i) The proposed model predicts a transition which begins at the top of the bed and moves downwards. This was indeed confirmed by our visual observations and measurements (Table 4) and also by Avidan (12).

(ii) The model predicts a decrease in U_k with decreasing bed depth, the effect becoming more pronounced for very shallow beds (Figure 2). This is also confirmed by our measurements (Table 5) and by observations of other workers (2).

(iii) Since, according to the model, the transition depends largely on the degree of entrainment from the bed surface at a particular gas velocity, one would expect the transition to occur at lower gas velocities for powders with smaller mean size or lower density. Our observations indeed confirm this and suggest the empirical relationship:

$$U_k = C \rho_p^{\ x} d_p^{\ y} \quad (U_k,\ m/s;\ \rho_p,\ Kg/m^3;\ d_p\ \mu m)$$

A similar trend was noted by other workers (1,9) (see Table 6 for values of C, x and y).

(iv) The proposed model envisages the transition as a transfer of solids from the bed to the freeboard. Our observations indicate that the freeboard holdup comprises a central rapidly-rising dilute phase core surrounded by a slowly falling dense phase in the wall region where gas velocities are lowest. There is interchange of solids between the two regions. The amount of solids which can be supported in the freeboard in this way at a particular gas velocity should depend on the ratio of surface area of wall to the flow area. This ratio decreases with increasing scale of equipment and we would therefore predict an increase in U_c and U_k with increasing bed diameter. The results of Yerushalmi and co-workers (9,12) showed the same trend although the opposite trend was claimed.

(v) According to the proposed model, the transition velocity is largely dependent on the degree of entrainment from the bed surface. May and Russell quoted in Zenz and Weil (13) claim that carryover rate is directly proportional to absolute gas pressure. Hence, at higher pressures, for a given gas velocity the freeboard would hold a greater proportion of the solids inventory. The model would therefore predict a lower transition velocity at higher pressures. Lanneau's results (2) are in agreement with this prediction (at 10 psig U_k = 6 ft/s, at 60 psig U_k = 4 ft/s).

DISCUSSION

The transition discussed here is that described by Lanneau, and by Kehoe and Davidson. We believe that what Kehoe and Davidson called the turbulent regime was actually the refluxing transport of solids in the freeboard which had grown at the expense of the bed. The voidage immediately after the transition would be, typically, 0.97. The onset of this transition is signalled by a decrease in amplitude of pressure fluctuations and an increase in voidage at the velocity U_k. In equipment which allows no net through-flux of solids it represents the transfer of solids from bed to freeboard. In equipment with an imposed solids flux, U_k will increase with solids flux and represents the velocity at which the gas can just entrain all the solids fed into the equipment. In this situation U_k is synonymous with the choking velocity U_{ch}.

Today, however, the phrase 'turbulent regime' is used to describe the condition which is found in a large bubbling bed of fine catalyst at velocities above 0.3 m/s. Under these conditions the bubbles, which reach a maximum size of just a few centimetres, are rapidly splitting and coalescing and give rise to very small pressure fluctuations. The bed has a definite surface and the voidage is typically 0.7-0.8. This "transition" is not accompanied by a decrease in amplitude of pressure fluctuations nor a sudden increase in voidage but it is claimed (12) that it can be detected by a change in slope of the Richardson-Zaki plot. It is apparent that some redefinition has occurred.

CONCLUSIONS

1. The transition, characterised by velocity, U_k represents the onset of refluxing pneumatic transport and is akin to choking.

2. A model based on the transfer of solids into the freeboard is proposed, and is supported by experimental observations.

3. The transition velocity increases with increasing particle density and size.

ACKNOWLEDGEMENT

The authors wish to thank the Science and Engineering Research Council for providing the finance for this and a much more extensive study of recirculating fluidized beds.

NOTATION

A	Bed cross sectional area
C	Proportionality constant
d_p	Volume-surface mean particle diameter

g	Acceleration due to gravity
H_i	Ideal bed height assuming no loss of solids to the freeboard, calculated from bed density
ΔH	Depth of bed that would be occupied by the solids held up in the freeboard
H	Actual bed height
M	Original mass of solids in the bed
M_b	Mass of solids remaining in the bed
Δp	Pressure drop across the entire dense bed
$\Delta p/L$	Pressure drop per unit height of the dense bed
U	Superficial gas velocity
U_c	Superficial gas velocity at maximum amplitude of pressure fluctuations
U_k	Superficial gas velocity at which pressure fluctuations level out
U_{mf}	Minimum fluidization velocity
x	Exponent
y	Exponent
ρ_B	Bed density
ρ_p	Particle density

LITERATURE CITED

1. Kehoe, P.W.K. and Davidson, J.F., "Continuously Slugging Fluidized Beds", "Chemeca 70", Chem. Eng. Conference, Australia, Butterworth & Co. (Australia) Ltd., 97-116 (1971).
2. Lanneau, K.P., Trans. Inst. Chem. Eng., 38, 125 (1960).
3. Geldart, D., and Rhodes, M.J., Plenary Lecture, First International Conference on Circulating Fluidized Beds, Halifax, Nova Scotia (1985).
4. Massimilla, L., "Behaviour of Catalytic Beds of Fine Particles at High Gas Velocities", A.I.Che.E. Symposium Series, 128, Vol., 69, 11-13 (1971).
5. Carotenuto, L.., Crescitelli, S., and Donsi, G. Quad. Ing. Chim. Ital. 10, 185 (1974).
6. Abed, R., in Fluidization IV., ed. D. Kunii and R. Toei , Eng., Fnd. p.137 (1983)
7. Avidan, A.A. and Yerushalmi, J., Powder Technology, 32, 223-232, (1982).
8. Canada, G.S., McLaughlin, M.H. and Staub, F.W. A.I.Ch.E. Symposium Series, 74, No. 176, 27 (1976).
9. Yerushalmi, J. and Cankurt, N.T., Powder Technology, 24, 187-205 (1979).
10. Matsen, J.M., Hovmand, S., and Davidson, J.F., Chem. Engng. Sci., 24, 1743
11. Geldart, D., Powder Technology, 7, 285-292 (1973)
12. Avidan, A.A., "Expansion and Solids Mixing in High Velocity Fluidized Beds", Ph.D thesis, The City College, New York (1980)
13. Zenz, F.A. and Weil, N.A., A.I.Ch.E.J., 4, 472 (1958)

Table 1

Powder	Symbol used	ρ_p kg/m^3	d_p (μm)
Vermiculite	1	327	223
Alumina 25	2	1015	42
CBZ-1	3	1308	38
E-Cat	4	1618	40
9G	5	1908	64
FRF-5	6	2335	66
Sand	7	2665	69
Sand	8	2600	270
Sand	9	2600	606
Sand	10	2660	1090
Sand	11	2648	134

Table 2

Relative height of bed surface above top pressure tapping at U = 0.4 m/s	Relative amplitude of pressure fluctuations
1	1
1.5	2.1
2.0	3.5
3.0	5.8
4.0	8.9
4.5	10.6

Table 3

Powder	Unit	Observed Values U_C	Observed Values U_K	Predicted Values U_C	Predicted Values U_K
7	1	1.50	1.86	1.42	1.97
6	1	1.10	1.61	1.20	1.62
1	1	0.75	0.97	0.59	0.97
3	3	1.04	1.44	0.47	1.25
5	3	1.00	2.12	0.80	2.16
2	3	0.92	1.38	0.80	1.20

Table 4

Distance of upper pressure tapping from distributor (m)	U_C	U_K
0.2	0.66	2.06
0.6	0.66	1.70
0.9	0.66	1.42

Table 5

Static bed depth (m)	U_C	U_K
0.9	1.20	2.33
0.8	1.20	2.30
0.6	1.20	2.52
0.4	1.20	2.3
0.3	1.10	2.05
0.2	1.10	1.70

Table 6

Unit	1		2		3
Powder Group	A	B	A	B	A and B
C	0.00114	0.0056	0.0022	0.011	0.0049
x	0.62	0.53	0.58	0.51	0.49
y	0.58	0.40	0.59	0.33	0.59

Table 7

Unit / Powder	1 U_C	1 U_K	2 U_C	2 U_K	3 U_C	3 U_K
1	0.6	0.97	1.15	1.55	1.20	1.98
2	0.27	0.68	0.9	1.11	0.92	1.38
3	0.26	0.7	0.9	1.30	1.04	1.44
4	0.28	1.24	0.65	1.32	1.0	1.45
5	0.00	1.37	1.1	2.06	1.0	2.12
6	0.63	1.57	1.15	2.39	1.37	2.52
7	1.05	1.65	1.1	2.58	1.88	2.80
8	1.6	3.05	1.8	3.84	-	-
9	-	-	2.0	5.36	-	-
10	-	-	2.4	6.68	-	-
11	1.24	2.15	0.9	2.94	-	-

Table 1 Properties of solids used in the investigation

Table 2 The effect of bed depth on the amplitude of pressure fluctuations

Table 3 Comparison of observed and predicted values of transition velocities

Table 4 The effect of pressure tapping position on the transition velocities (Solids = 5 : Unit = 2 : pressure tapping separation = 0.1m)

Table 5 The effect of bed depth on the transition velocities (Solids = 7 : Unit = 2)

Table 6 Values of constants for the correlation $U_k = C\rho_p^{\,x} d_p^{\,y}$

Table 7 The effect of equipment size on transition velocities

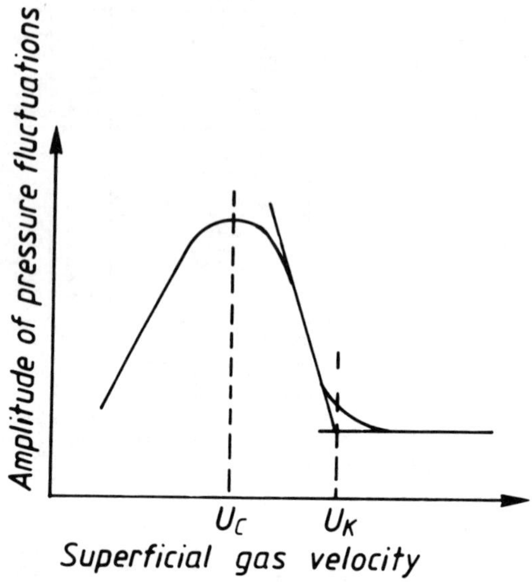

Figure 1 Definition of transition velocities from pressure fluctuations

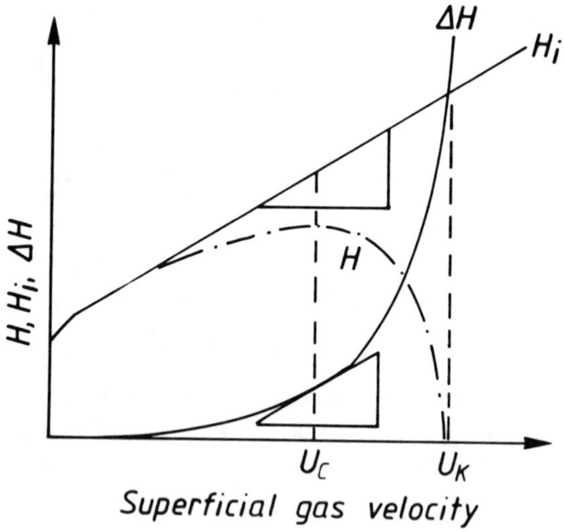

Figure 2 Proposed model for transition velocities

A CRITERION FOR TRANSITION FROM BUBBLING TO TURBULENT FLUIDIZATION

Jin Yong, Yu Zhiqing, Wang Zhanweng, Cai Ping
Department of Chemistry & Chemical Engineering,
Tsinghua University, Beijing, China

ABSTRACT

Investigations of the transition from bubbling to turbulent fluidization were conducted in a gas-solid fluidized bed based on the on-line computer analysis of pressure fluctuation. In order to study the effects of density and size of the particles as well as the geometric conditions of the bed on the transition, a series of experiments were performed.

SCOPE

Flow patterns in fluidized bed chemical reactors have a strong influences on the results of reactors. Since commercial catalytic reactors are run successfully in the turbulent fluidized region, the purpose of this investigation was to study the factors that influence the flow pattern transition in fluidized beds.

CONCLUSIONS AND SIGNIFICANCE

1. The technique of analysing pressure fluctuation in fluidized bed was applied successfully in this work and found to be very effective and reliable in fluidization studies.

2. The effect of powder properties and the fluidized bed geometry on the transition from bubbling to turbulent region was investigated. A correlation was obtained for the flow pattern transition velocity, Uc:

$$\frac{U_c}{\sqrt{g d_p}} = \left[K \left(\frac{D_f}{d_p}\right) \left(\frac{\rho_p - \rho_f}{\rho_f}\right) \right]^n$$

3. Addition of a new U_c - axis to Geldart's classification diagram is suggested. It is found that the Geldart's classification of particles also predicts flow pattern transition.

Knowledge of the behavior of gas-solid fluidization is essential in order to understand flow patterns and many investigations have been reported in this subject. Lanneau's work (1) was on the uniformity of fluidized bed of fine powder (alumina, $d_p = 61\mu m$) and indicated that bed nonuniformity index increased with increasing velocity to a maximum value and then decreased. Over the velocity range of 3 to 5 ft/sec, a condition of "almost a uniform or 'particulate' fluidization was approached." As a matter of fact it implied that different flow patterns exiet in the fluidized bed.

The concept of turbulent fluidization was first put forward by Kehoe and Davidson (2). They described the transition from bubbling to turbulent region of fluidization and proposed that turbulent fluidization would occur when $U > 3\ U_t$.

In the experiments in a cold model bed (152 mm I.D.), Yerushalmi (3) found a maximum magnitude of pressure fluctuation along with the increase of gas velocity. This maximum value correspondes to the transition from bubbling to turbulent fluidization, and over this point, the higher the gas velocity is, the lower is the pressure fluctuation magnitude. The experimental results also showed that the transition velocity is a function of density and size of solid, but the correlation was not reported.

In recent paper Avidan and Yerushalmi (4) showed a curve of bed voidage versus superficial velocity and found there was a break in the curve of the derivative of the voidage. This break point was explained as the transition point between bubbling and turbulent regions. For FCC catalyst, the gas velocity corresponding to this point was about 0.4 m/sec.

EXPERIMENTAL

The experiments were carried out in equipment (280 × 280 × 5000 mm high) made from plexiglass with a branched pipe distributor. In order to investigate the influence of particle properties on the transition of flow pattern, 8 kinds of particles were employed which belong to A and B regions in Geldart's classification diagram (5). (Figure 1)

For the purpose of observing the influence of bed's geometry on the transition of flow pattern, three kinds of beds (the free fluidized bed, the bed with vertical tubes and the bed with pagoda baffles) (Jin Y. et al) (6) were used which were different from each other in structure. An inductive transducer was fixed in the middle of expanding height of the fluidized bed to measure the pressure fluctuation. A micro-computer was used to collect data at a frequency of 128 per second and to do the on-line analysis in both the time and frequency domains. The still height of the bed is 450, 620 and 820 mm.

THEORETICAL APPROACH

Fluidized bed pressure fluctuation are stationary random signals which can be analysed in domains of time and frequency. Ordinarily, the maximum frequency of pressure fluctuation is not greater than 20 HZ and the sampling interval used in our experiments was 1/128 second. Therefore the sampling frequency was much higner than the Nyquist frequency required by sampling theory.

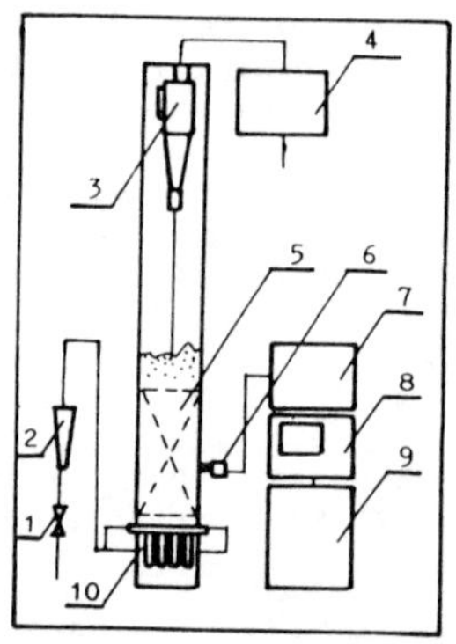

1. Valve
2. Rotometer
3. Cyclone
4. Filter
5. Baffles
6. Transducer
7. Amplifier
8. Oscillograph
9. Micro-computer
10. Distributor

Figure 1. Experimental equipment

In order to clear away high frequency

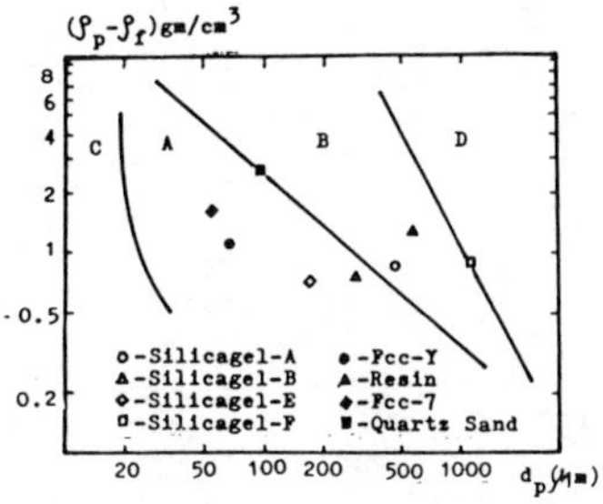

Figure 2. Powder classification, following Geldart.

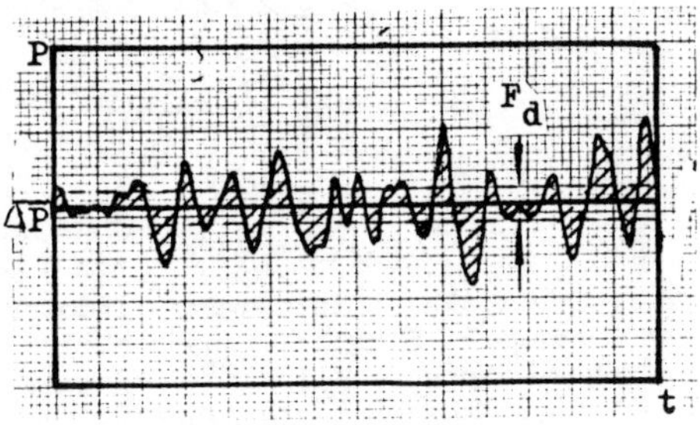

Figure 3. The mean amplitude of the pressure fluctuation.

noise, the signals were passed through a low filter. The mean pressure difference $\overline{\Delta P}$ between the measuring location and the bottom of the bed was calculated from the following equation:

$$\overline{\Delta P} = \frac{1}{T} \cdot \sum_{i=1}^{N} \Delta P_i \Delta t \tag{1}$$

Two parameters were computed as follows:

1. F_d-The mean amplitude of the pressure fluctuation. (Figure 3)

$$F_d = \frac{\text{The total shaded area in the figure at sampling time}}{\text{Sampling time}}$$

$$= \frac{\int_0^T \left| \Delta P_i - \overline{\Delta P} \right| dt}{T} \tag{2}$$

2. F_p - Nonuniformity coefficient in the time domain:

$$F_p = \frac{F_d / \overline{\Delta P}}{f} \times 100 \tag{3}$$

The pressure fluctuation in a fluidized bed are due to bubble behavior. If bubbles break up and coalesce very frequently, their average size is small. This gives a small mean amplitude F_d and a high fluctuation frequency. The higher the frequency the better the fluidization, and vice versa. Hence F_p had definite meaning as follows:

$$F_p \quad \frac{\text{bubble size}}{\text{extent of the bubbles breaking up and renewing}}$$

In analysing the signals with respect to frequency, the pressure fluctuation P(t) in time was transformed into signal P(f) in the frequency through the use of Fast Fourier Transformation (FFT) and then the signal was expressed in terms of linear combination of the individual frequency powers. In this way the power spectrum G(f) of the signal is obtained as follows (Figure 4).

$$G(f) = \lim_{T\to\infty} \frac{1}{2T} \left| \int_{-T}^{T} P(t)\, e^{-i2\pi ft}\, d_t \right|^2 \qquad (4)$$

$$G(f) = \lim_{T\to\infty} \frac{1}{2T} \left| P(f) \right|^2$$

Where

$$P(f) = \int_{-\infty}^{+\infty} P(t)\, e^{-i2\pi ft}\, d_t \qquad (5)$$

To reduce leakage the Hamming Window Function was selected to correct the signals in Fourier Transformation. With respect to B group particles and those close to B in the Geldart's classification the form of power spectrum of the pressure signal is a single peak. The major frequency, f_m and frequency bandwidth of the half peak value, Δf could be found from the power spectrum (Figure 4). Obviously f_m indicates the bubble size and the extent of bubble breaking and renewing. A high value of f_m implies good fluidization. The frequency bandwidth Δf is related to the distribution of bubble size and bed nonuniformity. On the basis of this physical concept a new parameter was defined as follows.

3. Nonuniformity coefficient F_f in the frequency domain

$$F_f = \frac{F_d \,/\, \overline{\Delta P}}{f_m \,/\, \Delta f} \times 100 \qquad (6)$$

It can be seen that F_f is a dimensionless parameter and has a definite physical meaning, so it may reflect bed nonuniformity more comprehensively. It was found that in our experiments when the gas velocity increased in the bubbling region, the bubbles coalesced violently, therefore F_d, F_p and F_f also increased rapidly. As the gas velocity exceeded U_c, because the velocity of bubbles breaking increased, the bed was transformed from bubbling to turbulent fluidization, the pressure fluctuation in the bed began decreasing gradually and F_d, F_p and F_f were also reduced. The maximum values of F_d, F_p and F_f provide a quantitative criterion for distinguishing the flow regime in the fluidized bed.

At the maximum values of F_d, F_p and F_f, the transition velocities, U_c,

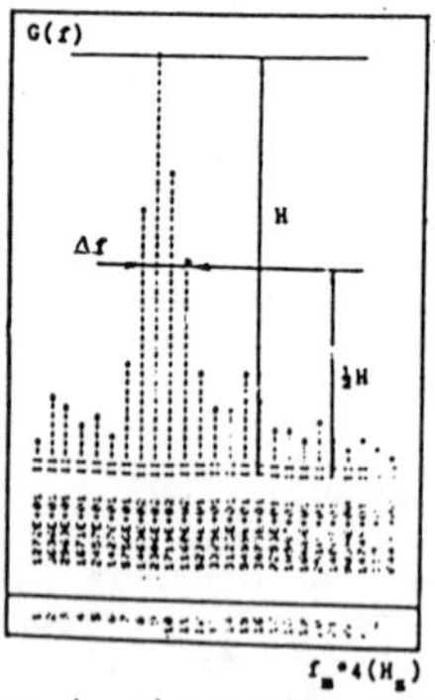

Figure 4. The power specturm diagram.

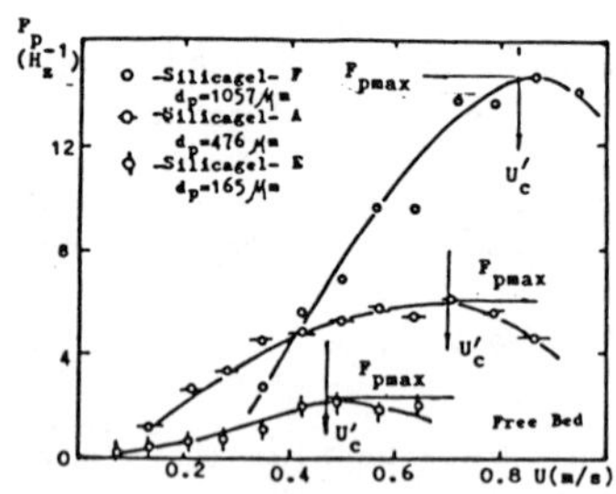

Figure 5. The particle size effect.

Uc' and Uc" were defined as seen in Figure 5, 6 and 7. At the same experimental conditions Uc, Uc' and Uc" were approximately the same. It can also be noted that parameters F_d, F_p and F_f were equivalent and each of them can be considered as an independent criterion to distinguish transition from bubbling to turbulent.

THE FACTORS INFLUENCING THE TRANSITION OF THE FLOW PATTERNS

1. The particle size effect

The particle size effect was studied by using different size of silica gel (d_p= 165, 476, 1057μm; ρ_p = 711, 834, 844 kg/m^3). The experiment indicated that larger particles have larger transition velocity and greater nonuniformity of the bed, as shown in Figure 5.

2. The particle density effect

In order to examine the effect of particle density on the transition of flow pattern, two groups of particles with different densities and roughly the same diameter were employed, as shown in Figure 6. Both the maximum nonuniformity of bed F_p and the transition velocity U_c increase with increasing the particle density.

3. The geometric structure effect

It is noted that in the design of a commercial chemical reactor, the quality of fluidization can be improved by adding fine particles which have suitable density. But a bed of fine particles would result in decrease of the operation velocity, therefore the production capacity of the reactor would be reduced. In some cases, for example, in the Fischer -Tropsch process (Ikeda) (7), due to the larger density of the ferritic catalyst, the operation quality of fluidized bed may not be satisfactory

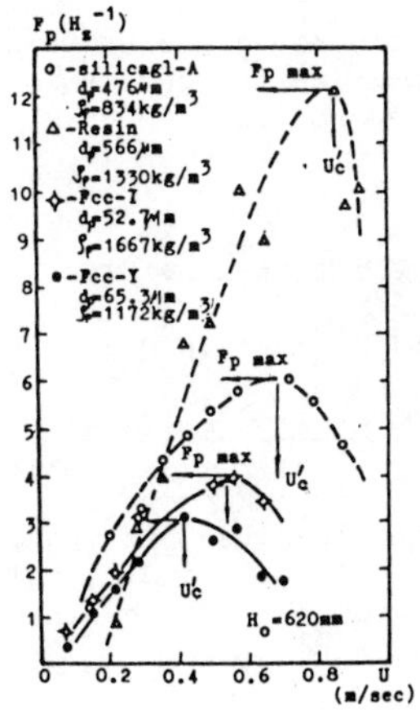

Figure 6. The particle density effect.

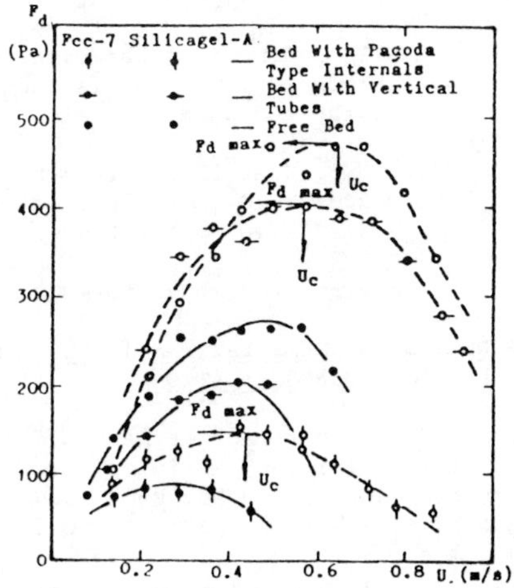

Figure 7. The geometric structure effect.

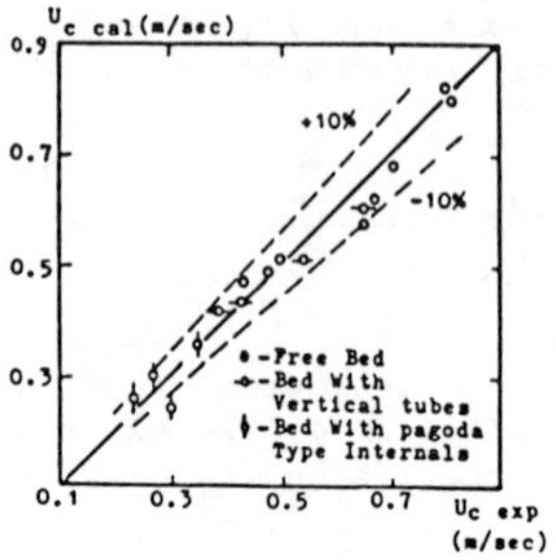

Figure 8. Comparison between experimental and predicted U_c.

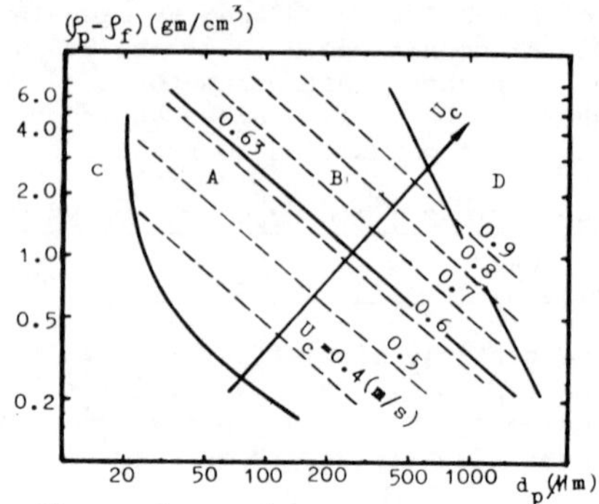

Figure 9. U_c isoplethes for free bed.

and further steps are needed to improve it.

In this paper, the effect on transition velocity Uc of internal baffles was studied and the experimental results are shown in Figure 7. It can be seen from the figures that the internal baffles improve operation quality, and the transition from bubbling to turbulent regime occurs at smaller Uc. The extent of the improvement depends on the configuration and structure of the internal baffle. For example, an internal baffle with pagoda shape has been shown to be much better than the vertical tubes.

The experimental data can be correlated as follows:

$$\frac{U_c}{\sqrt{g d_p}} = \left[K \left(\frac{D_f}{d_p}\right) \left(\frac{\rho_p - \rho_f}{\rho_f}\right) \right]^n \tag{7}$$

Where n = 0.27 and D_f is a parameter called performance diameter which has length dimension. The KD_f is found to characterize the geometric structure of the beds and its value is obtained as follows:

$KD_f = 3.67 \times 10^{-3}$ for free bed (280 X 280 mm)
$KD_f = 2.32 \times 10^{-3}$ for bed with vertical tubes
$KD_f = 3.42 \times 10^{-4}$ for bed with pagoda type internal baffles.

The average error for above formula is about 6.5% as seen in Figure 8.

DISCUSSING ON THE RELATIONSHIP BETWEEN TRANSITION OF FLOW PATTERN AND PARTICLE CLASSIFICATION

According to the effect of particle properties on fluidized behavior, Geldart (5) dividing particles into four groups -- A, B, C and D. This classification method has been accepted generally so far. The empirical dividing line between particle regions A and B is

$$(\rho_p - \rho_f)^{1.17}\, d_p = 0.906 \tag{8}$$

Equation (7) can be rearranged as follows

$$(\rho_p - \rho_f)^{1.17} d_p = \left(\frac{\rho_f}{KD_f}\right)^{1.17} \left(\frac{U_c}{\sqrt{g}}\right)^{1/0.23} \quad (9)$$

Comparing (8) and (9), the following equation can be obtained:

$$U_c = \left[0.906 \ (KD_f/\rho_f)^{1.17} \ g^{1/0.46}\right]^{0.23} \quad (10)$$

In the cold model experiments, it is found that the dividing line between particle regions A and B is just one of the lines of equal U_c values (isoplethes), that is U_c=0.63 m/s. A group of U_c isoplethes parallel to the dividing line can be drawn as shown in Figure 9. When developing to A region, the value U_c of isoplethes becomes smaller; When developing to B region, it becomes larger. Thus, a quantitative scale is added to the qualitative description, and a new axis U_c can be drawn in Geldart's classification diagram.

Geldart's experiments for distinguishing A and B region were performed near incipient fluidization. They are quite different from those encountered in the industry. By drawing a U_c-axis on Geldart's diagram, a clear physical meaning is found over a wide velocity region (from incipient to turbulent fluidization) and a new meaning can be attached to Geldart's classification. When the transition velocity isoplethes are drawn on Geldart's diagram With the same approach for the fluidized bed with vertical tubes and that with pagoda baffles, it can be seen that the isopleth U_c=0.63 m/s has moved to the right some distance (Figure 10 and 11). It means that region A becomes wider by the action of baffles. Some particles which belong to B region in free bed could have behaviour of fine powder in a bed with baffles. In other words, by the action of baffles the fluidized quality of different particles could be improved.

From a comparison among Figure 9, 10 and 11, it can be seen that the scale of U_c-axis has been enlarged in a fluidized bed with baffles. It means that in a fluidized bed with baffles, the effect of particle characters on the fluidized quality is much smaller than in a free bed. In a commercial fluidized reactor with baffles, if there are some remarkable changes of particle size owing to wearing and entrainment, the reaction could be still kept stable. The sensitivity of fluidization quality to particle parameters is reduced By the effect of baffles.

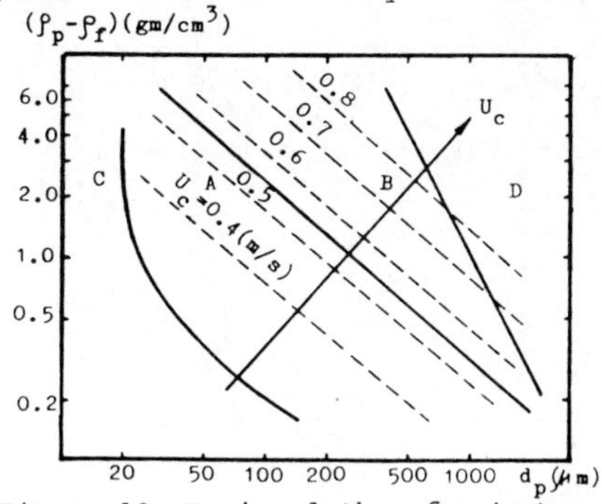

Figure 10. U_c isoplethes for bed With vertical tubes

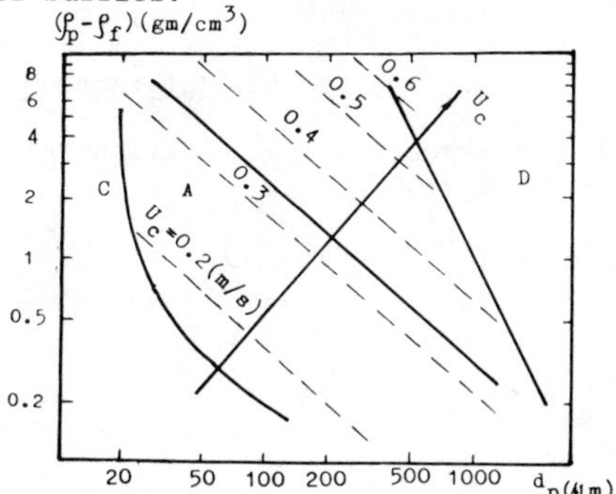

Figure 11. U_c isoplethes for bed With pagoda baffles

NOTATION

D_f — performance diameter of the fluidized bed, m

f — mean frequency of the pressure fluctuation (time domain), HZ

f_m — major frequency of the pressure fluctuation (frequency domain), HZ

N — sampling number; N = 6000

T — total sampling time, sec

U_c, U_c', U_c'' — transition velocity from bubbling to turbulent fluidization, m/sec

f — frequency bandwidth of the half peak value, HZ

p — mean pressure difference, mm H_2O

t — sampling interval, sec

G(f) — power spectrum

ACKNOWLEDGEMENT

The authors would like to thank Prof. P. N. Rowe, Prof. Mooson Kwauk and Prof. J. D. Wang for their helpful and valuable suggestions.

REFERENCES

1. Lanneau, K. P., Trans. Instn. Chem. Eng., 38, 125 (1960).

2. Kehoe, P. W. K. and Davidson, J. F., Inst. Chem. Eng., (London) symp. ser. 33, 97 (1971).

3. Yerushalmi, J. and Cankurt, N. T., Powder Technology, 24,187-205 (1979).

4. Avidan, A. A. and Yerushalmi, J., Powder Technology, 32, 223 (1982).

5. Botterill, J. S. M., "Fluid-bed heat transfer (gas-fluidized bed behavior and its influence on bed thermal properties)", Academic Press. (1975).

6. Jin Yong et al, International Chem. Eng., 22 (2), 269 (1982).

7. Ikeda Y., 'fluidization 85' second China-Japan symposium, 1 (1985).

DISTRIBUTION OF PARTICLES IN THE FREEBOARD REGION ABOVE A FLUIDIZED BED.

B. Hoggen, T. Lindstad, T.A. Engh
Norwegian Inst. of Technology.
SINTEF Division of Metallurgy.
N-7034 Trondheim-NTH

A model describing the entrainment of solid particles in the freeboard region of a fluidized bed is proposed. Experimental data from an industrial roaster has been used to calculate the ejection velocity distribution at the bed surface.

SCOPE

When the fluidized bed is used as a chemical reactor the freeboard region acts as a space for additional reactions between particles and gas. For particles with a small residence time in the bed this region can become very important.

Extensive studies on entrainment and elutriation have been made. The existing studies may be broadly classified as: empirically-based works by Leva(1951),Wen and Hashinger (1960),Merrick and Highley(1972); or theoretically-based works by: Zenz and Weil(1958),Andrews(1960),Chen and Saxena (1978),George and Grace (1978). Most of these studies gives correlations for the elutriation rate above TDH(transport disengaging height). Only the work of George and Grace (7) provides information about particle ejection velocities and the gas flow field in the freeboard. For beds operated in the steady state the above studies seem to agree that the particle flux decays exponentially with height above the bed.

The purpose of the present study is to correlate the entrainment data from an industrial roaster (10 m I.D.) and to calculate the particle vertical ejection velocities necessary to give the particle flux distribution in the freeboard.

CONCLUSIONS AND SIGNIFICANCE

In this work measurements were performed on mass fluxes in the freeboard region of an industrial fluidized bed roaster (10 m ID) and of particle velocities in a laboratory model.

A mathematical model for the ejection velocity distribution is proposed based on the assumption that the flux is decaying exponentially up through the freeboard and using the equation of motion for the ejected particles.

The model gives a near Gaussian distribution Figure(3) with a maximum at $\approx u_B$. About 10wt% of the particles has a ejection velocity greater than $2u_B$. This "tail" of the distribution is very important for the mass flux in the freeboard.

MODEL

Particle trajectories.

Particle trajectories are calculated by numerical solution of the equation of motion

$$\frac{dv_z}{dt} = -3/4\ \frac{C_D \rho_g v_r |v_r|}{\rho_s d_p} - \frac{(\rho_s - \rho_g)g}{\rho_s} \qquad (1)$$

$$v_r = v_z - u_g$$

For the drag force experienced by a particle we have used the same relationship between the drag coefficient (C_D) and the particle Reynolds number (N_{Re}) as Do et al.(8)

$$C_D = (24/N_{Re})(1 + 0.15 N_{Re}^{.687}) + 0.42/(1 + 4.25 \cdot 10^4 N_{Re}^{-1.16}) \qquad (2)$$

were $N_{Re} = \rho_g v_r d_p / \mu$

Predicted trajectories were obtained by numerical integration of Equation (1) using the 4.5-order Kutta-Merson process with the initial conditions $v = v_0$ and $h = 0$ at $t = 0$.

Particle flux in the freeboard.

If we look at a number of particles with size d_i and ejection velocity v_{0j}, then all of these particles will get the same local velocity calculated by numerical integration of Equation (1). Along their trajectories the flux will remain constant.

$$F(d_i, v_{0j}, z) = F_0(d_i, v_{0j}) \qquad (3)$$

where $F_0(d_i, v_{0j})$ is the particle flux for particles with size d_i and ejection velocity v_{0j}.
This means that it is enough to calculate the maximum height for each trajectory, $z_{max}(d_i, v_{0j})$, and then use Equation (3) below this height. For particles with terminal velocities

less than u_g, z_{max} is equal to infinity and the flux remains constant throughout the freeboard. The upward flux at a height z for particles with size d_i is given by

$$F_{z(d_i)up} = \sum_j F(d_i, v_{0j}, z) \tag{4}$$

where

$$F(d_i, v_{0j}) = \begin{cases} F_0\,(d_i, v_{0j}) & \text{when } z_{max}(d_i, v_{0j}) \geq z \\ 0 & \text{when } z_{max}(d_i, v_{0j}) < z \end{cases}$$

The total flux in height z will be

$$F_{z,up} = \sum_{ij}\sum F(d_i, v_{0j}, z)$$

The downward flux $F_{z,down}$, is 0 for particle with terminal velocity less than u_g. For the rest of the particles $F_{z,down} = F_{z,up}$.

The distribution of particle vertical ejection velocity.

From the literature (9) the particle flux in the freeboard for particles with terminal velocities greater than u_g, can be found from

$$F_z = F_0 \exp(-az) \tag{5}$$

This expression must be the result of a distribution of particle ejection velocities $f(v_0)$ for each particle size d_i. As stated above a particle with size d_i will rise to a height z_{max}. The ejection velocity that gives $z_{max} = z$ is defined as v* ie $v^* = v_0(z_{max}, d_i)$. All particles having an ejection velocity greater than v* will contribute to the flux in this height.

This gives

$$F_z = F_0 \int_{v^*}^{\infty} f(v_0)\,dv_0 \tag{6}$$

also

$$\int_0^{\infty} f(v_0)\,dv_0 = 1$$

Combining Equation (5) with (6) gives

$$F_0 \exp(-az) = F_0 \int_{v^*}^{\infty} f(v_0)\,dv_0 \tag{7}$$

This equation can be differentiated with respect to z and rearranged

$$f(v^*) = -\left(dv^*/dz\right)^{-1} \cdot a \cdot \exp(-az) \qquad (8)$$

This can best be illustrated with an example. If we ignore the drag force (ie large particles) Equation (1) becomes

$$dv^*/dz = -g/v^* \qquad (9)$$

$$z_{max} = v_0^2/(2g)$$

This can be inserted into equation (8)

$$f(v_0) = \frac{av_0}{g} \exp\left(\frac{-av_0^2}{2g}\right) \qquad (10)$$

The above distribution is plotted for a=3.0 in Figure (1)

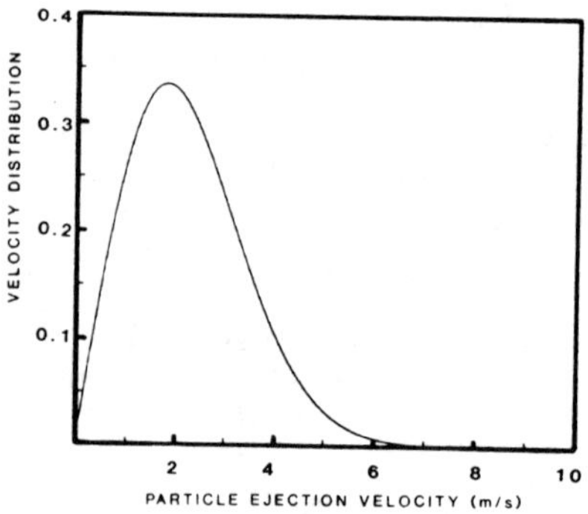

Figure 1. The distribution of the vertical ejection velocity for large particles.

The above example shows that a Gaussian distribution of the vertical starting velocity gives an exponential distribution of the flux in the freeboard.

EXPERIMENTAL.

The measurements were carried out in an industrial roaster at Borregaard A/S. The roaster is cylindrical with a diameter of 10m and the hight is 5m. The distributor consists of nozzles. Air is used as fluidizing fluid. Bed depth varied between 1.1 and 1.5 meters.

A catching device was constructed from stainless steel. The dimensions of the device was critical because it had to be introduced into the freeboard through the only possible opening which was the opening for the start up burner.

Manipulator arms were mounted on the sampler so that it could be rotated into correct position in a short time. It is possible to rotate each sampler about the two axes of the catching device to minimise the shadowing effect.

Bed material consists of porous hematite ore $\rho_S \approx 3400$. Particle size of the bed 3 % < 150 μm 30 % > 500 μm (agglomerates). Roaster gas is ambient air.

RESULTS.

The measurements were carried out using the catching device described above. Three series of measurements were taken and the available data for the fluid bed roaster is given in Table 1.

TABLE 1 Experimental conditions for the FB-reactor.

	series 1	series 2	series 3
Solid Feed.(kg/h)	17000	13100	13200
Roaster gas.(Nm^3/h)	43000	34500	34000
Bed Temp.(K)	1100	1030	1040
Distributor pressure (Pa)	19500	22900	18900
CALCULATED			
Bed height(m)	1.1	1.5	1.2
Bubble diam.(m)	0.4	0.4	0.4
Bubble vel.(m/s)	1.9	1.9	1.9

Seven samples were obtaind at a time, and each sample was sieved on standard (ASTM:II) laboratory sieve. Least square fits for each size range were calculated and Figure 2 gives an example of these results.

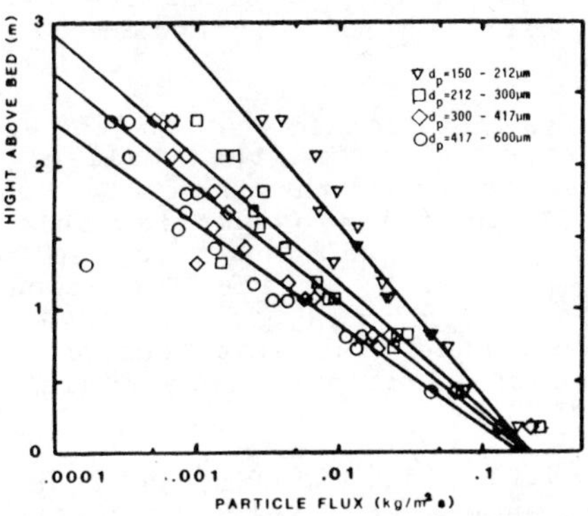

Figure 2 Fluxes calculated from measurements in an industrial FB-roaster, series 1.

The calculated value of "a" from these measurements were then used as input to the model described above and the resulting ejection velocity distributions for each size range are shown on Figure 3.

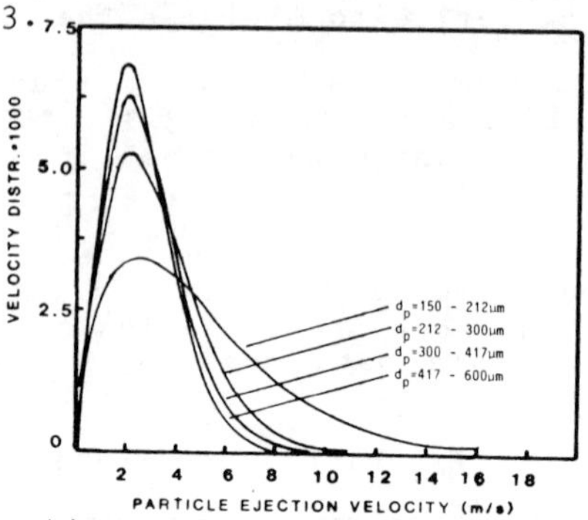

Figure 3 Calculated ejection velocity distribution from series 1.

DISCUSSION.

The results in Figure 2 from the industrial FB-roaster, giving aproximately straight lines, confirms the generally accepted theory that the elutriation rate of a particle size group decays exponentially with height above the bed.

The model for the freeboard gives ejection velocity distribution shown in Figure (3). George and Grace(7) has measured the vertical ejection velocities. These results are much the same as ours, but surprisingly(7) has not detected particles with ejection velocities less than $\approx u_B$.

Our results show that the ejection velocity distribution is not the same for each particle size. The tail of the ejection velocity distribution increases with decreasing particle size.

Uneven gas flow is not considered in the model. This can lead to errors especially for the smaller particles where the terminal velocity for the particle is close to u_g. George and Grace(7) used the assumption that particles were transported in a jet of gas, but this led to an incorrect distribution of particles in the freeboard. We have not included the effect of a gas jet, but nevertheless we seem to have a more resonable flux distribution in the freeboard and the same type of ejection velocity distribution at the bed surface.

For the coarse solids (d_p > 300μm) we found that 10wt% of the particles had an ejection velocity greater than 4.0 m/s. This is approximately $2 \cdot u_B$ (u_B is calculated from the procedure given by Geldart(10)).

By means of a laser-doppler anemometer we tried to measure this velocity distribution in a cold laboratory model Hoggen(11). As the size distribution was wide the noise level got too high, but it was possible to measure particles velocities up to $2 \cdot u_B$.

NOTATION.

C_D = Drag coefficient
F = Mass flux of particles, $kg/m^2/sec$.
F_0 = Mass flux of particles at z=0
N_{Re} = Reynolds number $\rho_g d_p v_r / \mu$
a = Constant defined in Equation(5)
d_i = Particle diameter, m.
d_p = Particle diameter, m.
u_g = Superficial gas velocity, m/sec.
u_B = Bubble velocity, m/sec.
v_0 = Particle velocity at z=0, m/sec
v_r = $v_z - u_g$
v^* = Ejection velocity that just brings a particle to z.
μ = Gas viscosity, kg/m/sec.
ρ_g = Gas density, kg/m^3.
ρ_s = Particle density, kg/m^3.

Subscripts
i refers to discrete particle size range
j refers to discrete particle ejection velocity range
z refers to freeboard hight z

LITERATURE CITED.

1. Leva,M., "Elutriation of Fines from Fluidized Systems" Chem. Eng. Prog, 47 1 1951.

2. Wen,C.Y. and Hashinger,R.F. "Elutriation of Solid Particle From a Dense Phase Fluidized Bed". AIChE J., 6 2 1960.

3 . Merrick, D. and Highley, J., "Particle Size Reduction and Elutriation in a Fluidized Bed Process." AIChE Symposium Series. 70, No 137, 366 (1974).

4. Zenz, F.A., and Weil, N.A. "A Theoretical-Empirical Approach to the Mechanism of Particle Entrainment from Fluidized Bed," AICHE J., 4, No 4, 472)1958).

5. Andrews,J.M., "Kinetic Study of Fluidized Solids Entrainment," I & EC, 52 No. 1, 85 (1960).

6. Chen, T.P., and Saxena, S.C., "A Theory of Solids Projection from a Fluidized Bed Surface as a First Step in the Analysis of Entrainment Processes." Fluidization edited by J.F. Davidson and D.L. Keairns, Cambridge University Press, 151 (1978).

7. George, S.E. and Grace, J.R., "Entrainment of Particles from Aggregative Fluidized Beds," AIChE Symposium Series, 74, No. 176, 67 (1978).

8. Do, H.T., Grace, J.R. and Clift, R., "Particle Ejection and Entrainment from Fluidized Beds," Powder Technol., 6, 195 (1972).

9. Kunii, D., and Levenspiel, O., "Fluidization Engineering," John Wiley and Sons. Inc., 302 (1969).

10. Geldart, D. "Gas Fluidization.",The Center for Professional Advancement, Amsterdam (1982). Lectures.

11. Hoggen,B. "Distribution of particles in the freeboard region above a fluidized bed." PhD thesis, Norwegian Inst. of Technol., Trondheim(1984). In Norwegian.

FREEBOARD PHENOMENA IN A TURBULENT FLUIDIZED BED

Gangqiang Chen, Guanglin Sun, Gan-Tang Chen

UNILAB Research Center of Chemical Reaction Engineering

Zhejiang University

Hangzhou, Zhejiang, People's Republic of China

ABSTRACT

In a Ø800 mm turbulent fluidized bed, the variations of entrainment rate, particle concentration and particle size distribution along the freeboard height were measured. A mathematical model involving axial and radial terms of particle flux was proposed. By this model, all the above mentioned distributions can be predicted.

SCOPE

The entrainment rate, particle concentration and particle size distribution along the freeboard height greatly affect the overall performance of commercial fluidized reactors, but there are only limited data available in the literature for large scale units. It is, therefore, seldom possible to derive reliable mathematical models which can be used for design purpose.

In the present study, the variations of the entrainment rate, particle concentration and particle size distribution along the height at superficial gas velocities varied from 0.2 to 0.55 m/s were measured in a Ø800x7000 mm fluidized bed of fine particles ($d_p = 63$ µm). Based on the phenomena observed, a mathematical model was derived on the assumption of two flow regions (the upflow and downflow regions) with axial and radial flux terms involved as parameters. The predicted values agree quite well with the experimental data.

CONCLUSIONS AND SIGNIFICANCE

It was found that for turbulent fluidization of polydispersed fine particle systems, the entrainment rate varies with bed height as hyperbolic function rather than exponential.

By taking into consideration both the upflow and downflow particle streams and by introducing a radial flux term, a mathematical model was derived and the calculated results were found to be quite satisfactory as compared with the experimental data.

Fine particle fluidized beds operating in turbulent regime received much attention in recent years because of their importance in the field of catalytic and non-catalytic processes accompanied with strong heat effects, such as the oxychlorination of ethylene, ammoxidation of propylene and coal combustion. The freeboard region is usually considered as a space for the disengagement of entrained particles and the TDH is conventionally defined as the height at which no more particles could further be disengaged by gravity. Thus only those particles of which the terminal velocities are lower than the operating gas velocity will be carried out of the freeboard. However, the true feature of the freeboard phenomena are more complicated because of the following facts:

1. The particles present in the freeboard may amount to a considerable fraction of the bed material. The finer the particle and the higher the gas velocity, the larger will be this fraction.

2. The residence time of gas in the freeboard is much longer than that in the dense bed, because the gas velocity is lower while the travel distance is longer.

3. The temperature drop in the freeboard is usually not large for commercial scale fluidized beds, because the specific surface area of the wall is relatively small.

4. The interaction between particles is evident and the effect of wall on the flow of gas and particles should not be ignored

It is, therefore, clear that the freeboard is not only a space for the disengagement of particles, but also an important reaction zone especially to high temperature turbulent fluidization processes.

The aim of the present work is to reveal the real situation in the freeboard of a large unit and develop a relevant mathematical model which can be used as a stepstone for the modelling of turbulent fluidized bed reactors.

EXPERIMENTAL

The arrangement of the experimental equipment is shown in Fig.1. The whole bed is made of Plexiglas with diameters of 0.8 m for the main section and 1.2 m for the removable enlarged section. The total height is 7 m. The average diameter of particles is 63 μm and its bulk density 960 kg/m^3. The static bed height was varied from 1.15 to 1.98 m while the superficial gas velocity was varied from 0.2 to 0.55 m/s.

In order to measure the quantities at various axial and radial positions, two sampling methods have been adopted. One is to collect all the particles out of the column section of different heights, another is to suck out the gas-solid mixture at local points through sampling tubes with its volumetric flow rate strictly equal to that at these local positions.

MATHEMATICAL MODEL

The freeboard phenomena can be schematically represented by Fig.2. The key points are

1. The particles ejected from the bursting bubbles or entrained from the bed possess the same size distribution as in the dense bed. They move in different directions, but the initial velocity distribution is regarded as the same for all sizes.

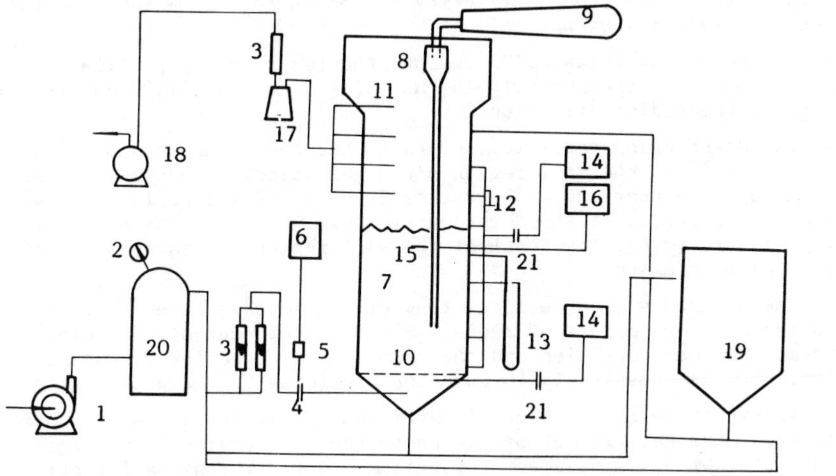

Fig. 1 Schematic diagram of the experimental facility

1. Blower 2. Pressure gauge 3. Rotameter 4. Orifice meter 5. Pressure transducer 6. Flowmeter 7. Fluidized bed 8. Cyclone 9. Filter bag 10. Gas distributor 11. Particle sampling probe 12. Pressure taps 13. U-manometer 14. Reactor 15. Capacitance probe 16. Capacitometer 17. Particle collector 18. Vacumn pump 19. Storage tank of particles 20. Buffering tank 21. Strainometer

Fig. 2 Particle flow in the freeboard
1. Dense phase 2. Bubble

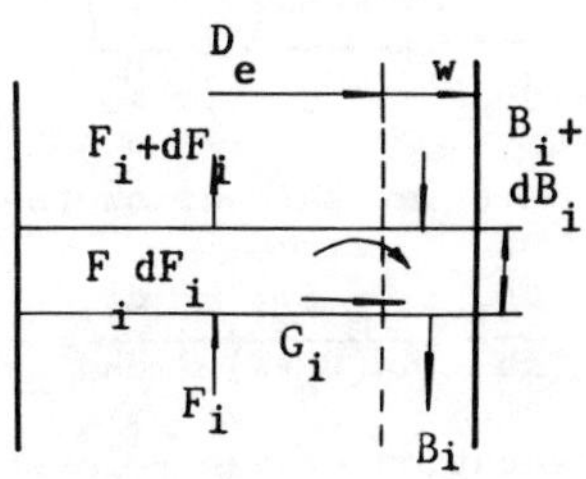

Fig. 3 Particle flow in the freeboard region

2. The upflowing particle will change its direction downward at certain height where the gas velocity decreases to become equal to the terminal velocity of particle.

3. In the upflow ("central") region, the gas velocity profile is assumed to decay exponentially in the axial direction (1,2) and is uniform in the radial direction.

4. The downflowing particles return to the bed along random paths and mostly near the wall region since gas velocity becomes zero or even negative near the wall surface. Thus, a unified wall region was assumed to represent the downflow region. Referring to the experimental data (3,4), the depth of the wall region was taken as 5% of the column diameter.

5. The departure from vertical flow of particles can be attributed to the combined effect of radial velocity component of particle, impact between particles and the turbulent fluctuation of gas streams. The net result is to push the particles to the wall region.

6. An effective radial flux of particles, G, is introduced to characterize the net flow of particles to the wall region (see Fig.3). The value of G_i is assumed proportional to (1) the overall particle concentration and the flux of that fraction at the definite cross section; (2) the lateral motion of particle or, consequently, the lateral fluctuation of gas stream and the particle diameter. Thus, G can be represented by

$$G_i = k \cdot u_g^n d_{pi}^m F_i \rho \tag{1}$$

where k, n and m are parameters to be determined from experimental data.

Based upon the above mentioned knowledges, a mathematical model for the freeboard phenomena was formulated below:

1. Material balance of the upflowing particles of the ith fraction

$$\frac{dF_i}{dh} + F_i \frac{dx_i}{dh} + \frac{4}{D_e} G_i = 0 \tag{2}$$

2. Material balance of the downflowing particles of the ith fraction

$$\frac{dB_i}{dh} + \frac{D_e^2 F_i}{4W(D_e+W)} \frac{dx_i}{dh} + \frac{D_e G_i}{W(D_e+W)} = 0 \tag{3}$$

3. Equation of gas velocity

$$u_g = U_o + \left(\frac{U_o}{\varepsilon_o}\left(\frac{D_t}{D_t-2W}\right)^2 - U_o\right) e^{-ah} \tag{4}$$

4. Equation of motion of the ith fraction particle with initial velocity u_j

$$\frac{du_{pij}}{dh} = \frac{-3}{4} \frac{C_D \rho_p (u_{pij}-u_g)|u_{pij}-u_g|}{d_{pi} \rho_p u_{pij}} - \frac{(\rho_p-\rho_g)g}{\rho_p u_{pij}} \tag{5}$$

where C_D is the drag coefficient and can be calculated by

$$\begin{cases} C_D = 24/Re_p & Re_p = d_p \rho_g (u_p - u_g)/\mu < 0.4 \\ C_D = 10/Re_p^{0.5} & 0.4 \leq Re_p \leq 500 \\ C_D = 0.43 & Re_p \geq 500 \end{cases}$$

5. Equation of particle concentration

$$\rho = [A_e(\rho_F) + A_w \rho_b]/(\pi D_t^2/4) \tag{6}$$

where $\rho_F = \sum_{i=1}^{N} \sum_{j=1}^{M} (F_i Y_{ij}/u_{pij})$

$$\rho_B = \sum_{i=1}^{N} (B_i/u_{ti})$$

Boundary conditions:

For h=0, $F = F_{io} = F_o X_i$

$$B_i = B_{io} \tag{7}$$

$$u_{pij} = u_{pijo}$$

In the above equations, F_o was calculated by

$$F_o = 0.222 \rho_p (U_o - U_{mb}) \tag{8}$$

This equation was obtained from the literature (5) with U_{mf} replaced by U_{mb}. The equation for prediting U_{mb} can be found elsewhere (6). The initial velocity distribution of particles was derived on the assumption that particles ejected from the upperhalf spherical surface of gas bubble possess the same velocity in their normal direction. Thus we have

$$u_{pijo} = u_r \cos \theta_j \tag{9}$$

$$Y_{ijo} = \cos \theta_j - \cos(\theta_j + d\theta_j) \tag{10}$$

where θ_j denotes the jth discrete angle measured from the vertical axis of bubble and u_r was taken as to the ascending velocity of bubble at the bed surface.

By using method of numerical iteration, the mathematical model was solved to obtain the results given below.

RESULTS AND DISCUSSIONS

Fig.4 shows typical data of the entrainment distribution at some gas velocities. It was found that the entrainment decreases more rapidly than in the case of exponential decay. These curves can be best fit by using equations of the hyperbolic form rather than the exponential e.g., at U_o=0.40 m/s, the curve can be represented empirically by

$$e=0.6299+0.1506h^{-3} \tag{11}$$

with an average deriation of 7.54%, while for the exponential form

$$e=0.7194\exp(-0.4621h)$$

the deviation is 23.46%. Thus, we see that the entrainment decreases most rapidly in the splashing zone which is about 1 m height. Furthermore, it is hard to define a TDH uniquely as far as the wall effect is ever present.

Fig.5 shows the variation of average diameter of particles along the freeboard height while Fig.6 gives the variation of weight fraction of the individual particle sizes.

The relation between F at outlet and U_o was found to be

$$F=13.31\cdot U_o^{4.06} \tag{12}$$

and was plotted in Fig.7. On the same plot, the dotted line was calculated by the method of Zenz and Weil (1). The slopes of these lines are just the same.

Based on our experimental data, the parameters in eq. (1) were found to be

$$k=60.04\cdot U_o^{-5.806} \tag{13}$$

$$n=1.507, \qquad m=1.028$$

Fig. 8 and Fig. 9 show the comparisons between the experimental data and the curves calculated by the proposed model. The agreements are quite satisfactory.

In Fig. 10, the calculated variation of particles size distribution along the freeboard height was plotted. The feature of these curves is just the same as what found in the literature elsewhere. By using this model, the detailed state of the freeboard can be predicted.

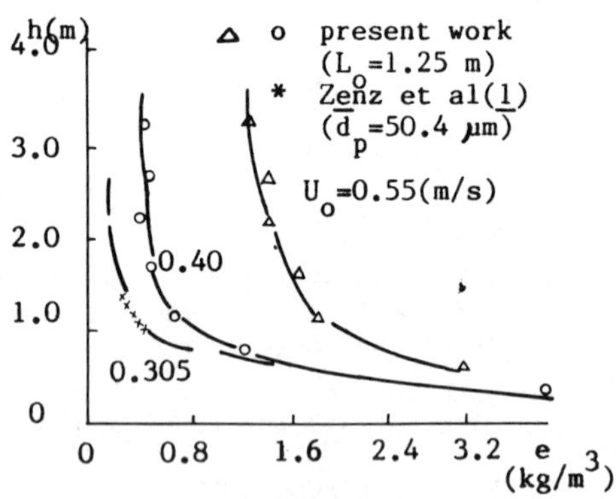

Fig.4 Entrainment distribution curves

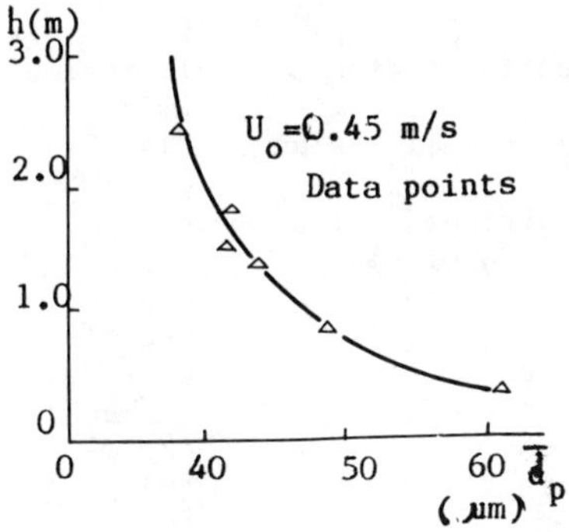

Fig.5 Variation of d_p with h

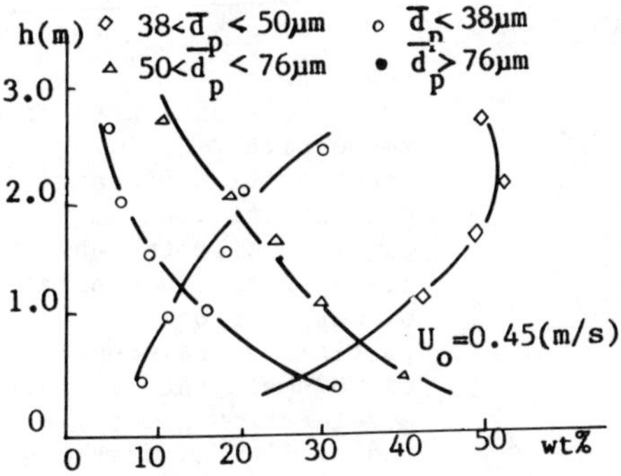

Fig.6 Variation of particle weight fraction with h

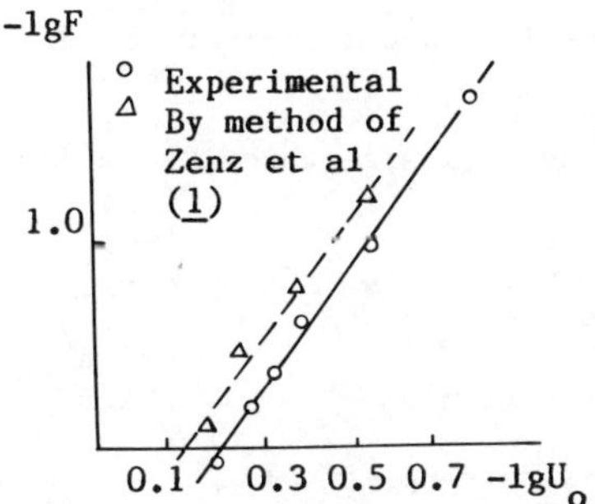

Fig.7 Correlation of F

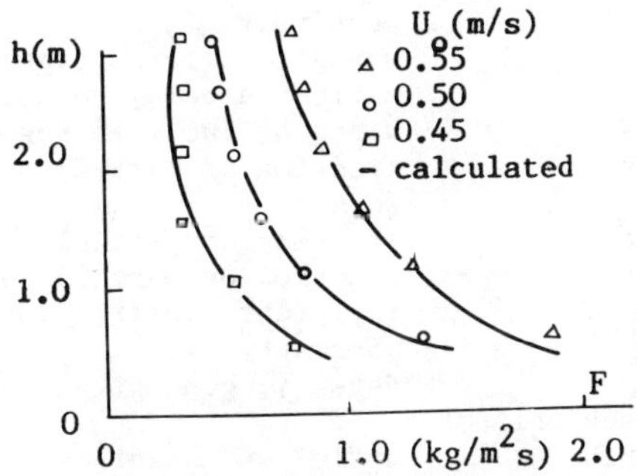

Fig.8 Comparison between calculated F and experimental values

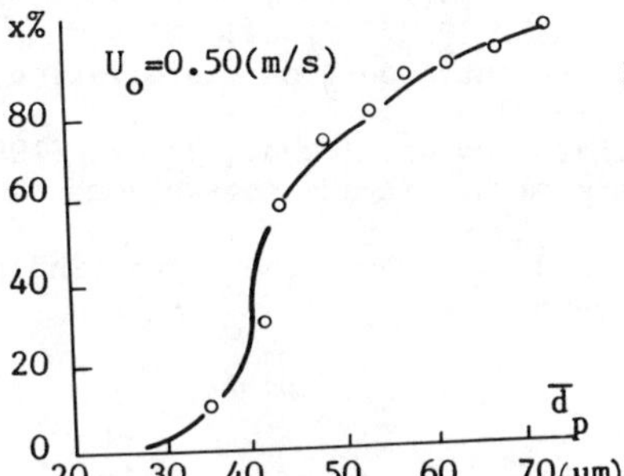

Fig9 Particle size distribution at outlet

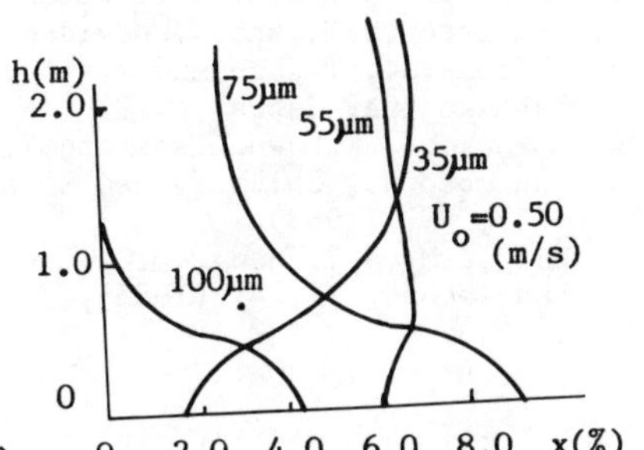

Fig.10 Variation of particle size distribution

NOTATION

a	constant	(—)
A_e, A_w	cross-sectional area of central region and wall region respectively	(m^2)
B	downflow flux of particles in wall region	(kg/m^2s)
C_D	drag coefficient	(—)
d_p, $\bar{d}_p$	particle diameter and the averaged value	(m)
D_e	diameter of the central region of bed	(m)
D_t	diameter of bed	(m)
e	particle entrainment	(kg/m^3)
F	entrainment rate of particles	(kg/m^2s)
g	acceleration gravity	(m/kgs^2)
G	radial flux of particle flow	(kg/m^2s)
h	coordinate of freeboard height	(m)
TDH	transport disengaging height	(m)
k,m,n	parameters	
M,N	discrete number of velocity distribution and size distribution of particles respectively	
u_g	gas velocity	(m/s)
u_r, u_t	normal velocity and terminal velocity of particle respectively	(m/s)
U_o, U_{mb}, U_{mf}	superficial, incipient bubbling and incipient fluidization velocity respectively	(m/s)
W	width of the wall region	(m)
x	fraction of particles of certain size which turns downward	(—)
X	fraction of particle in bed	(—)
Y	fraction of particle with velocity u_j	(—)
ε_{mf}	incipient fluidization voidage	(—)
μ	viscosity	(Ns/m^2)
ρ	density; concentration of particles	(kg/m^2)
subscripts:		
F,B	upflow and downflow particles respectively	
i,j	index for particle size and particle velocity discretes	
o	at bed surface	

LITERATURE CITED

1. Zenz,F.A. and N.A. Weil,AICHE J. 4,(4)472 (1958)
2. Pemberton,S.T. and J.Poavidson, CES, 39,829 (1984)
3. Morooka,S., T.Kago and Y.Kato, 4th.Int. Conf. on Fluidization, Kashikojima, Japan, p.291 (may,1983)
4. Morooka,S, K.Kawazuishi and Y.Kato, Powder Techn., 26, 75 (1980)
5. Tanimoto H., Chiba T. and Kobayashi H., Kagaku Kogaku Rombunshu, 9(5), 497 (1983)
6. Wang,Z.M.,N.Y.Zhang, B.C. Chen and G.T. Chen, Chem. Reaction Eng. and Techn., 1, 47 (1985),(Chinese)

ON BED VOIDAGE AND APPARENT DILUTE PHASE HOLD-UP IN HIGH PRESSURE-TURBULENT FLUIDIZED BEDS OF FINE POWDERS

Alan W. Weimer and Karl V. Jacob*

Engineering Research Laboratory - 1776 Building
The Dow Chemical Company
Midland, MI 48674 U.S.A.

*Michigan Applied Science and Technology Labs
Dow Chemical U.S.A. - 677 Building
Midland, MI 48667 U.S.A.

ABSTRACT

Relative localized bed density fluctuations and expanded bed height, L_f, are measured in a 0.097 m I.D. fluidized bed of 66 and 108 μm carbon powders fluidized by a synthesis gas mixture at pressures of P = 2070, 4140, and 6210 kPa. High bed voidages and low relative localized bed density fluctuations indicate that the 66 μm powder is fluidized turbulently at the higher pressures. Bed expansion, L_f, increases with increasing pressure, P, and is found to be the result of both increasing dense phase voidage, ε_D, and increasing apparent dilute phase hold-up, δ_B.

SCOPE

Most commercial, catalytic, fluidized bed chemical reactors utilize fine powders and operate at near atmospheric pressure with gas velocities high enough to achieve turbulent fluidization. Although it is desirable to operate high pressure fluidized beds in the turbulent regime as well, few studies have investigated the effect of pressure on the transition to turbulent flow. Furthermore, these studies have generally been limited to pressures below P = 2500 kPa.

The transition to turbulent flow is marked by improved fluidization "quality" (i.e. "smoother" fluidization characterized by the absence of large discrete voids), and increased bed voidage, ε. Of fundamental interest is whether the increased bed voidage, ε, results from increased dense phase voidage, ε_D, and/or increased dilute phase hold-up, δ_B.

One method of quantifying fluidization "smoothness" is to measure instantaneous bed density, $[\rho_{Bed}]_c$, at localized regions in the bed. A rapid response nuclear density gauge is mounted to a movable assembly mechanism which allows fluctuating bed density, $[\rho_{Bed}]_c$, to be monitored at any axial position, z, in the bed. Relative localized bed density fluctuations quantify fluidization "quality", while time averaged density measurements, $[\rho_{Bed}]_c^t$, are used to determine expanded bed height, L_f, and, hence, bed voidage, ε.

A principle objective of this work is to quantify the effect of pressure, P, particle size, d_p, and gas velocity, u_o, on localized bed density fluctuations and bed voidage, ε, in an industrial, pilot scale, fluidized bed. A second objective is to determine the effect of dense phase voidage, ε_D, and dilute phase hold-up, δ_B, on overall bed voidage, ε.

CONCLUSIONS AND SIGNIFICANCE

The following conclusions can be drawn from the present study:

1. Relative localized bed density fluctuations decrease with increasing pressure, P, and increase with increasing particle size, d_p, for systems of fine powders (Figure 2). For the two powders studied in this work, the effect of pressure is more significant for the 108 μm powder than for the 66 μm powder.

2. At all pressures, P, relative localized bed density fluctuations for the 108 μm powder increase with distance above the gas distributor, z (Figure 3), possibly indicating that voids are continuing to grow up the bed, not reaching a maximum stable bubble size at the higher pressures, P.

3. For fine powders, higher pressure, P, results in higher bed voidage, ε, and allows higher voidage, ε, to be obtained at a lower superficial gas velocity, u_o.

4. The increasing bed expansion, L_f, which accompanies increasing operating pressure, P, results from both higher dense phase voidage, ε_D, and higher apparent dilute phase hold-up, δ_B (Figures 5 and 6).

5. Bed voidage exceeding $\varepsilon = 0.75$ (Figure 4) and relative localized bed density fluctuations less than 1.5% (Figure 2), indicate that the 66 μm powder is most likely turbulently fluidized at all gas velocities for bed operation at P = 4140 and 6210 kPa.

INTRODUCTION

Most commercial, catalytic, fluidized bed chemical reactors employ fine powders and operate at near atmospheric pressure with superficial gas velocities, u_o, high enough to be within the turbulent flow regime. Turbulent fluidization is characterized by smaller gas voids which continually coalesce and split, tracing tortuous passages as they rise through the bed. This somewhat homogeneous structure is marked by strong interaction between dense and dilute phases, resulting in higher gas conversions than would be realized by operation in the bubbling regime.

At near atmospheric pressures in small diameter beds, the transition from bubbling or slugging to turbulent fluidization is characterized by a sudden decrease in pressure fluctuations (1-3). Although significant quantitative data are lacking, it has been found that the transition to turbulence takes place at lower u_o as P is increased (4,5). Since fluidized beds are now being considered for higher pressure catalytic processes (e.g. Fischer-Tropsch Synthesis), it is necessary to understand the relationship between P, u_o, particle size, dp, and the transition to turbulent flow. The ability to operate in turbulent flow at lower u_o is desirable in terms of reduced catalyst attrition and entrainment. Hence, an objective of this study is to examine the effect of increasing P on decreasing local density fluctuations and increasing ε for systems of fine powders. It is also of fundamental importance to investigate whether the increasing ε with increasing P results from increasing ε_D and/or increasing δ_B due to the presence of smaller and more slowly rising voids.

EXPERIMENTAL

Apparatus

The experiments are carried out in a 0.097 m I.D. x 2.8 m, steel, fluidized bed. Feed gas is compressed and fed to the bed through a porous metal distributor. Pressure in the bed is maintained by a back-pressure control valve, located downstream of the bed. Gas flow rates are controlled by mass flow controllers upstream of the compressor and are verified downstream of the back-pressure control valve via a calibrated gas dry test meter. Fast action on-off solenoid valves are located upstream and downstream of the pressurized bed and can isolate it instantaneously. A schematic of the experimental system is shown in Figure 1.

Physical Properties of Powder and Gas

The fluidized powders are commercial grade, granular, activated carbon powders screened to give mean particle sizes of dp = 66 and 108 μm. The fluidizing medium is a synthesis gas mixture having an average molecular weight of M_w = 17 and a viscosity of $\mu_g = 1.56 \times 10^{-5}$ Pa-s at T = 293K. Physical properties of the powders and experimentally determined minimum fluidization voidages, ε_{mf}, at different P are given elsewhere (6,7).

Nuclear Radiation Density Gauge

A rapid response (1 ms time constant) nuclear density gauge (Ohmart Densart 3400; 500 mCi Cs-137 point source) is mounted to a movable scanning mechanism constructed so that the gauge can range over the entire height of the fluidized bed. The scanning mechanism is interfaced to an on-line microprocessor which controls a motor that moves the density gauge to specified locations. The on-line microprocessor monitors (up to 100 points per second) and performs statistical calculations on the fluctuating bed density signal. The application of this type gauge is described in detail elsewhere (8).

EXPERIMENTAL MEASUREMENTS

Cross-sectional, centerline, bed density, $[\rho_{Bed}]_c$, is measured up the bed by moving the density gauge to various axial locations and analyzing the output signal. The output signal from the density gauge is an instantaneous measurement of bed density, averaged over the volume of the penetrating collimated beam of gamma radiation, $[\rho_{Bed}]_c$. This fluctuating density measurement is integrated over time to give a time averaged value, $[\rho_{Bed}]_c^t$. As the size of voids passing through the collimated beam increases, the amplitude of the density gauge output signal increases. Likewise, an increase in void frequency results in an increase in the frequency of fluctuations in the density gauge output signal.

The "quality" of fluidization is quantified by calculating the relative density fluctuations as the ratio of the standard deviation of the instantaneous density measurement to the time averaged density measurement, $\sigma[\rho_{Bed}]_c/[\rho_{Bed}]_c^t$. These calculations are made with an on-line microcomputer from measurements taken with the density gauge located at a distance z = 1.2 m above the gas distributor. Measurements are taken at operating pressures of P = 2070, 4140, and 6210 kPa for both the 66 and 108 μm powders and for u_o approaching 0.08 m/s. Expanded bed height, L_f, is consistently determined at all operating conditions through use of the movable density gauge assembly, moving the density gauge up or down and calculating $[\rho_{Bed}]_c^t$.

RESULTS AND DISCUSSION

Dense phase voidage, ε_D, and dense phase superficial gas velocity, u_{Do}, for both powders, have been determined experimentally from pressurized collapsed bed experiments and are reported elsewhere (6,7).

Effect of Pressure, P, and Superficial Gas Velocity, u_o, on Relative Localized Bed Density Fluctuations

Relative localized bed density fluctuations measured at z = 1.2 m are shown in Figure 2 for both powders at all pressures and all excess gas velocities, u_o-u_{Do}. Clearly, the "quality" of fluidization is improved at higher pressures for both powders with the pressure effect being substantial for the 108 μm powder and modest for the 66 μm powder. Fluidization at higher excess gas velocities generally increases the relative localized density fluctuations for both powders; however, the gas velocity effect is more substantial for the 108 μm powder than for the 66 μm powder. The relative localized density fluctuations are always higher for the 108 μm powder, relative to the 66 μm powder - independent of pressure. Even at P = 6210 kPa for the 108 μm powder, the relative fluctuations are higher than for the 66 μm powder at P = 2070 kPa.

Since fluctuations in bed density are indicative of gas voids, these data indicate that voids are generally smaller at higher P for both powders and are always smaller for the 66 μm powder, regardless of P for the pressures studied here. These findings are in agreement with results previously reported (7) from cross-correlations of fluctuating differential pressure measurements.

Relative localized bed density fluctuations are shown in Figure 3 at different axial distances, z, for the 108 μm powder at all P. At all P, relative density fluctuations increase with z. One explanation for this phenomenon is that voids are moving toward the axial centerline with increasing distance above the gas distributor and that the increasing relative fluctuations are the result of this movement (density gauge measurement is biased across the axial centerline (8)). Another possibility, however, is that voids are continuing to grow up the bed at high pressures - that voids are generally smaller at higher pressures, but continue to grow, not reaching a maximum stable size.

Effect of Pressure, P, and Superficial Gas Velocity, u_o, on Bed Voidage, ε

Bed voidage, ε, is shown in Figure 4 for both powders at all P and all u_o. Bed voidage, ε, is higher for the 66 μm powder relative to the 108 μm powder and is affected by P to a larger degree. Clearly, higher P increases ε and allows higher ε to be obtained at lower u_o. For the 66 μm powder, $\varepsilon = 0.8$ is obtained at $u_o = 0.072$ m/s for P = 2070 kPa, while the same ε is obtained at $u_o = 0.036$ for P = 4140 kPa and $u_o = 0.025$ m/s for P = 6210 kPa. The increasing ε with higher P is due to increasing ε_D and/or increasing dilute phase hold-up, δ_B.

Effect of Pressure, P, on Transition to Turbulence

The high ε shown in Figure 4 for the 66 μm powder are above those indicative of turbulent fluidization for fine powders fluidized at atmospheric pressure (1-3). Furthermore, the fact that higher ε can be achieved at lower u_o as P increases indicates that turbulent fluidization should be achievable at lower gas velocities relative to operation at lower pressures. Nonetheless, quantitative indications of the transition to turbulent flow, such as the sudden decrease in bed P fluctuations with increased u_o which occur during low pressure studies, are not seen in the high pressure studies here.

Effect of Pressure, P, and Superficial Gas Velocity, u_o, on Apparent Dilute Phase Hold-up, δ_B

It is well known that, for the fluidization of fine powders, higher operating P results in higher ε (as shown in Figure 4 for this study). Guedes de Carvalho and Harrison (9) postulated this increased bed expansion to be the result of an increased δ_B caused by the formation of small and slowly rising bubbles at higher P. Their experiments were limited to visual observations and a maximum P of 2200 kPa. Subzwari et al. (10) used cinephotography to observe the fluidization of 60 μm mean sized cracking catalyst at P to P = 700 kPa. From these observations, they discounted Guedes de Carvalho and Harrison (9) and suggested that an increased bed expansion must be due to an increased ε_D, accompanying the higher P. Piepers et al. (11) determined ε_D from collapsed bed experiments for a 59 μm mean sized cracking catalyst at P to 1500 kPa and concluded that an increasing total bed expansion is not only the result of an increasing dilute phase hold-up, δ_B, but is also due to higher ε_D at higher P. Barreto et al. (12) also measured higher ε_D with higher P from collapsed bed experiments of Group A powders at P to 2000 kPa and found δ_B to increase with increasing P. However, these data were contradicted by visible bubble hold-up, f_B, measurements made from analyses of x-ray pictures of the bed. They postulated that bubble wakes have higher voidages than ε_D, thus accounting for the higher bed expansion.

Dense phase voidage, ε_D, has been found to increase with P for both the 66 μm and 108 μm powders (6,7), having a substantial effect on the 66 μm powder and a modest effect on the 108 μm powder. The bed expansion above that due to the dense phase alone (i.e. L_f/L_D) is shown in Figures 5 and 6 for the 66 and 108 μm powders, respectively, at all P and all excess superficial gas velocities, u_o-u_{Do}. Clearly, δ_B increases with increasing P for both powders. For the 108 μm powder, the bed initially contracts at lower u_o-u_{Do}, but then expands at higher excess velocities. These findings are consistent with those of previous investigators (5). From these experiments, it may be concluded that the higher bed expansions which occur at higher operating pressures are the result of both higher ε_D and higher δ_B.

NOTATION

d_p	mean particle size
f_B	visible dilute phase hold-up
L_D	height of the expanded dense phase alone
L_f	expanded bed height
M_w	gas molecular weight
P	pressure
u_o	superficial gas velocity
u_{Do}	superficial dense phase gas velocity
u_{mf}	minimum fluidization velocity
δ_B	apparent dilute phase hold-up
ε	total bed voidage
ε_D	dense phase voidage
μ_g	fluidizing gas viscosity
$[\rho_{Bed}]_c$	cross-sectional centerline instantaneous bed density
$[\rho_{Bed}]_c^t$	time averaged cross-sectional centerline bed density
$\sigma[\rho_{Bed}]_c$	standard deviation of instantaneous bed density

LITERATURE CITED

1. Yerushalmi, J., N.T. Cankurt, D. Geldart, and B. Liss, AIChE Symp. Ser. No. 176, 74, 1 (1978).

2. Yerushalmi, J. and N.T. Cankurt, Chemtech, 8, 564 (1979).

3. Yerushalmi, J. and N.T. Cankurt, Powder Technol., 24, 187 (1979).

4. Canada, G.S., M.H. McLaughlin, and F.W. Staub, AIChE Symp. Ser. No. 176, 74, 19 (1978).

5. Chitester, D.C., R.M. Kornosky, L.S. Fan, and J.P. Danko, Chem. Eng. Sci., 39, 253 (1984).

6. Weimer, A.W. and G.J. Quarderer, AIChE Symp. Ser. No. 241, 80, 79 (1984).

7. Weimer, A.W. and G.J. Quarderer, AIChE J., 31, 1019 (1985).

8. Weimer, A.W., D.C. Gyure, and D.E. Clough, to be published in Powder Technol. (1985).

9. Guedes de Carvalho, J.R.F. and D. Harrison, Inst. Fuel Symp. Ser. No. 1, 1, B1-1 (1975).

10. Subzwari, M.P., R. Clift, and D.L. Pyle, in Fluidization, Davidson, J.F. and D.L. Keairns (Ed.), Cambridge Univ. Press, Cambridge, 50 (1978).

11. Piepers, H.W., E.J.E. Cottaar, A.H.M. Verkooijen, and K. Rietema, Powder Technol., 37, 55 (1984).

12. Barreto, G.F., J.G. Yates, and P.N. Rowe, Chem. Eng. Sci., 38, 1935 (1983).

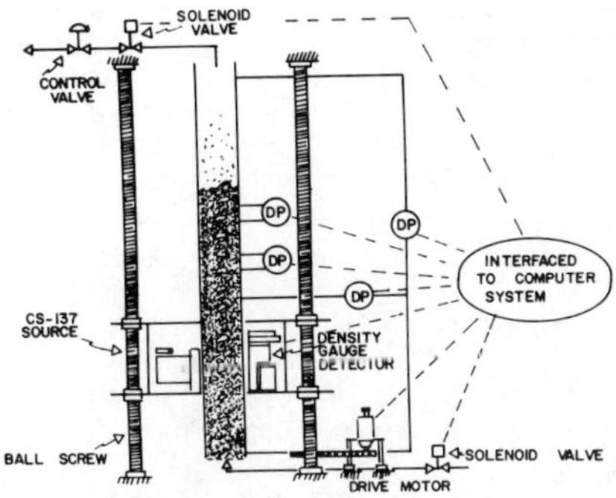

Figure 1. Schematic of Experimental Apparatus.

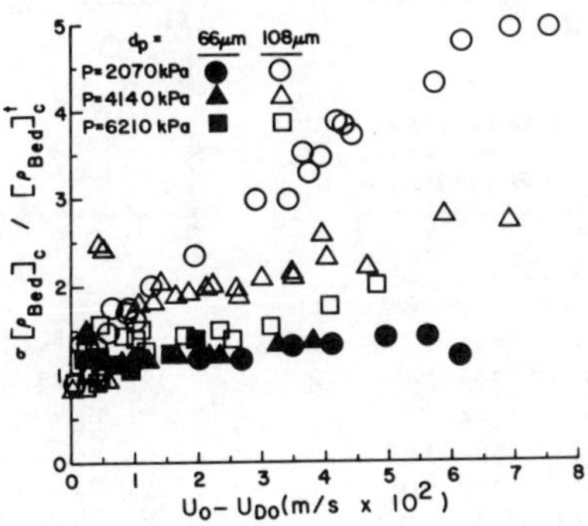

Figure 2. Relative Localized Bed Density Fluctuations. Density Gauge Fixed at z = 1.2 m.

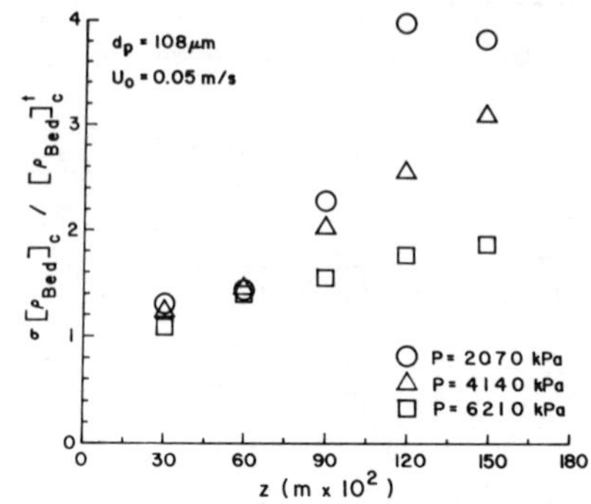

Figure 3. Relative Localized Bed Density Fluctuations. Density Gauge moved to Various Axial Locations.

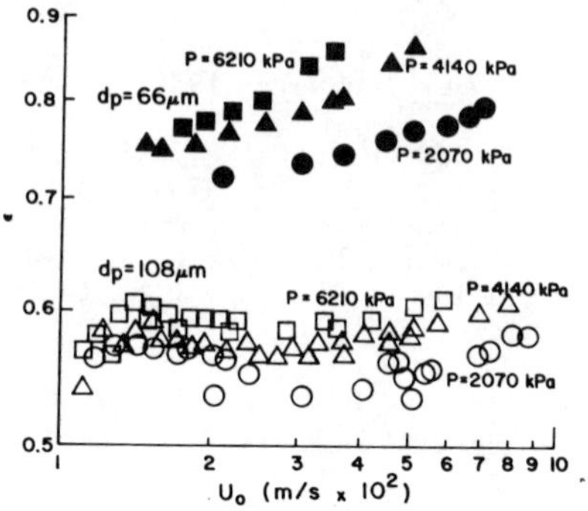

Figure 4. Effect of Superficial Gas Velocity, u_o, on Bed Voidage, ε; P = 2070, 4140, and 6210 kPa; d_p = 66 μm, 108 μm.

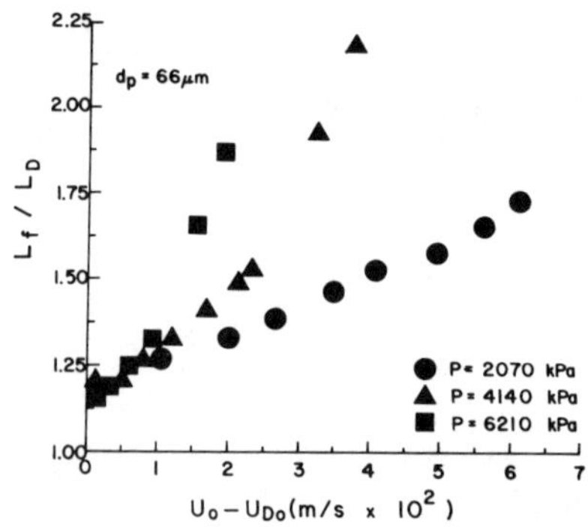

Figure 5. Effect of Excess Superficial Gas Velocity, u_o-u_{Do}, on Bed Expansion Ratio, L_f/L_D (Apparent Dilute Phase Hold-up, δ_B); d_p = 66 μm; P = 2070, 4140, and 6210 kPa.

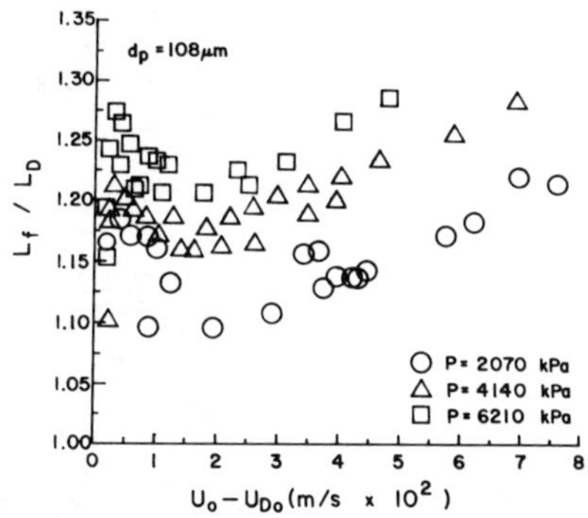

Figure 6. Effect of Excess Superficial Gas Velocity, u_o-u_{Do}, on Bed Expansion Ratio, L_f/L_D (Apparent Dilute Phase Hold-up, δ_B); d_p = 108 μm; P = 2070, 4140, and 6210 kPa.

INTERACTIVE FORCES IN CONTRAFLOW BALL/PARTICULATE FLUIDIZED BEDS

John E. Gwyn
Shell Development Company
P. O. Box 1380
Houston, Texas 77251, USA

ABSTRACT. Balls raining through a column of entraining particulates provide a convenient method for countercurrent contacting of the solid phases. The interaction between the balls and particulates retards the flow and increases the concentration of each by a factor of five over that for either solid alone.

The interaction and gas/solids drag forces are analyzed to determine the range of operating conditions under which the contraflow can exist. The operability of the system was demonstrated in a 20 cm diameter column.

SCOPE. The concept of raining balls in contraflow to entrained particulates was developed as a means of contact heat exchange between the solids phases. The concept may also be applied for mass transfer operations (1,2,3).

In order to define the operating characteristics of raining ball systems it was necessary to derive the equations for the interactive force between the three phases. The unsteady state form of the equations was used to evaluate entrance effects for balls and particulates. The acceleration terms were also used to drive the iterative calculations to steady state conditions in the column.

CONCLUSIONS AND SIGNIFICANCE. High mass fluxes of balls can rain through entraining particulates over a range of conditions that would represent turbulent fluidization to pneumatic transport for the particulate solids alone. However, the interaction between the solid phases retards both flows dramatically resulting in dense beds, typically 500 kg/m^3. This enhances transfer between phases and, hence, requires relatively small vessels for given process requirements.

One dimensional force equations account for the nature of the interaction and retardation of flows. This permits calculation of feasible operating ranges for raining ball operation and of equipment size for given applications.

BACKGROUND. The raining ball concept was developed as a key operation in the Shell Pellet Heat Exchange Retorting (SPHER) process for retorting of oil shale (1,2,3). In this process two ball flow loops are utilized, Figure 1. In one hot raining balls supply heat to turbulently fluidized shale in the retort vessel. The partially cooled balls are separated from the bottom of the retort, reheated in a lift pipe combustor, and added back into the top of the retort. In the second loop hot spent shale is entrained upward in a liftpipe against downflowing balls; the spent shale is cooled and the raining balls are heated. These heated balls are pneumatically transported to the top of a raw shale liftpipe to rain countercurrent to the raw shale. The cooled balls are transported back to the spent shale liftpipe. Hence, the shale feed is preheated by heat recovered from the spent shale and the thermal efficiency of the process is enhanced.

More detailed descriptions of the process are presented elsewhere (1,2,3). In brief, a number of factors are important to a viable process design. In approximately process sequence the more important factors are:

Shale Feed Preparation. Shale is pulverized to a fluidizable size range with a maximum particle size of about 1.5 mm and a minimal content of ultra fines (dust). The maximum size is determined by segregation tendencies and the ease of separating the shale from the balls.

Shale Preheater. The raw shale is added near the bottom of a fast fluidized (entraining) bed preheater. The shale is contacted by heat carrying raining balls in countercurrent flow. These balls are introduced at the top of the preheater vessel by a distributor in a way to achieve uniform contacting and limit their falling velocity. These hot balls are transported pneumatically from the spent shale heat recovery vessel. The smallest sized balls, i.e., those with the highest surface to volume ratio, that will fall sufficiently fast through the preheater are desirable for effective heat exchange. Choice of ball material is governed by heat capacity, density, chemical inertness, and resistance to fracturing or abrading under flow conditions. Countercurrent staging of four to six equivalent mixing stages of the raining-ball/fast-fluidized flow regime is effected by choosing height to diameter ratios greater than four. The shale must also have sufficient freeboard to disengage from their contact with the balls before exiting and separation from carrier gas. Temperature approaches of 40°C at the top and 30°C at the bottom were calculated for the design case.

Balls collecting in a boot at the bottom of the preheater are stripped of shale particles by an elutriating gas. Elutriation of shale particles from between the balls is easily achieved when the ball/shale size ratio is at least 4/1. The balls are then metered into a transport line where they are pneumatically conveyed to the spent shale heat recovery section.

Retort. Shale from the preheater is metered to the lower portion of the retort via a hopper and standpipe. Steam or gas injection is required at the bottom to initiate fluidization. Vapor evolved by retorting adds greatly to the flow of fluidizing gas as it rises through the retort. Since retorting is reaction rate limited (4) five to seven minutes shale residence time with minimal bypassing (about four mixing stages) is desired to achieve near maximum oil yields. Process shale is removed at the top of the retort through an overflow weir. Absorbed and entrained vapors are removed from the retorted shale by steam in a spent shale stripper. Stripped shale is sent to the heat recovery section.

Hydrocarbon vapor product is passed through high efficiency cyclones near the top of the retort to reject entrained shale back into the bed. The vapor is routed to a fractionator for further processing.

Hot heat carrying balls are added at different levels of the upper portion to distribute the heat transfer according to the process needs and to avoid overheating at the top of the retort. The ball diameters need to be of size (say 6 mm) to assure a reasonable velocity (say 0.1 m/s) through a turbulently fluidized bed of shale. The balls collect in a boot at the bottom of the retort and are stripped of shale and hydrocarbon by use of superheated steam. The balls are reheated in a vertical pneumatic transport combustor, disengaged form combustor gas, and fed back to the retort through a hopper system.

Heat Recovery. The heat recovery is similar to the preheat section in that fast fluidized spent retort shale is cooled by contacting it in countercurrent flow with balls from the preheat section. The cooled shale is recovered and conditioned for disposal. The heated balls are transported to the preheat section.

Two separate ball circuits are used and the optimum size and material may differ for each.

There are many other factors in a final design concerned with economics and operability (1,2). It is apparent that the raining ball flow regime and characteristics are vital in design and sizing of the process vessels. Hence, attention is now turned to defining these.

INTERACTIVE FORCES IN RAINING BALLS/FLUIDIZED PARTICULATES BEDS. One dimensional equations were defined assuming ideal, plug flow conditions. Turbulence effects were accounted for in the definition of drag coefficients. Specifically, the buoyancy forces of the fluidized bed on the balls is adapted from Doig and Roper (7); the forces on the fluidized bed are based on a modification of Ergun's equation (5,6). These forces per unit cross section and unit height are:

1. GAS IMPACT AND DRAG ON THE SHALE PARTICLES

$$F_{gs} = \frac{3C_{Ds}\rho_g\phi_s}{4\,\phi_g d_{ps}}(U_g - U_s)|U_g - U_s| + \frac{C_{DAs}\,\phi_s^2\,\mu_f}{\phi_g^2\,d_{ps}^2}(U_g - U_s)$$

2. GAS IMPACT AND DRAG ON THE BALLS

$$F_{gb} = \frac{3C_{Db}\,\rho_g\phi_b}{4\,\phi_g d_{pb}}(U_g - U_b)|U_g - U_b| + \frac{C_{DAb}\,\phi_b^2\,\mu_f}{\phi_g^2\,d_{pb}^2}(U_g - U_b)$$

3. BALL-SHALE INTERACTION (KNOCK-BACK)

$$F_{sb} = \frac{3C_{Dsb}}{4}\left(\frac{d_{ps} + d_{pb}}{d_{pb}}\right)^2 \frac{\rho_s\phi_s\phi_b}{d_{pb}}(U_s - U_b)|U_s - U_b|$$

4. GRAVITATIONAL FORCES

$$G_{rs} = g\phi_s\rho_s,\quad G_{rb} = g\phi_b\rho_b$$

5. WALL FRICTION, GENERALLY NEGLIGIBLE

These equations are solved by assuming initial velocities and calculating accelerations. These accelerating forces are used to drive the velocities until zero accelerations are achieved according to the equations below.

1. GAS DRAG ON THE SHALE EQUALS THE BALL KNOCK-BACK PLUS GRAVITY FORCES ON SHALE

$$\rho_s \phi_s U_s \frac{dU_s}{dt} = F_{gs} - F_{sb} - G_{rs}$$

2. GAS DRAG PLUS SHALE KNOCK-BACK FORCES EQUALS THE GRAVITY FORCE ON THE BALLS

$$\rho_b \phi_b U_b \frac{dU_b}{dt} = F_{gb} + F_{sb} - G_{rb}$$

3. TOTAL DRAG ON THE GAS EQUALS THE SUM OF THE GRAVITY FORCES ON THE SHALE AND ON THE BALLS

$$\Delta P/L = g\phi_s \rho_s + g\phi_b \rho_b$$

This form is also useful for calculating the effect of inlet conditions (e.g. velocities) on operability of the process.

Example. After a number of calculations and experimentations 6 mm alumina balls (3400 kg/m^3) and 914 μm shale (1600 μm maximum, 2400 kg/m^3) at mass fluxes of 98 and 73 kg/m^{2}s were chosen as design base conditions for the preheat section of SPHER.

Calculation conditions and results are summarized in Table 1. The strong interaction effects are emphasized by comparing the contraflow with the flow of each in the absence other phase. By these calculations, shale in the absence of balls would occupy only 3.3% of the riser volume and would create 79 kg/m^2 pressure drop. Raining balls in the absence of shale would occupy only 0.2% of the volume and create 7 kg/m^3 pressure drop. In joint contraflow they would occupy 8.7 and 8.8%v respectively and would cause 508 kg/m^3 pressure drop. The interactive forces between balls and shale amount to 261 kg/m^3, approximately one half that of the gas drag.

TABLE 1: EXAMPLE

CALCULATED CONTRAFLOW RAINING BALLS/ENTRAINED PARTICULATES

INPUT CONDITIONS - DESIGN CASE

	GAS	PARTICULATES	BALLS
MASS FLUX, kg/m^2/s	5.87	73.4	97.9
SUPERFICIAL VEL., m/s	4.57	0.030	0.027
DENSITY, kg/m^3	1.28	2408	3400
PARTICLE DIAM. mm.	-	0.914	6.34

DRAG COEFFICIENTS: GAS/PART. = 1.472[6]; GAS/BALLS = 0.66[7]
PART./BALLS = 1.950[7]; FINES VISCOUS = 150.0[6] BALLS = 150.0[6]
MAXIMUM PACKING FRACTION: PART. = 0.50, BALLS = 0.50, MIX = 0.75

CALCULATED RESULTS

FOR FINES FLOW ONLY (FAST FLUIDIZED BED)

FREE FALL VELOCITY	= 3.90 m/s
FAST FLUIDIZED VELOCITY	= 0.93 m/s RISING
GAS LINEAR VELOCITY	= 4.73 m/s
GAS VOLUME FRACTION (VOIDS)	= 0.9673 v/v
PRESSURE DROP (NO WALL FRIC.)*	= 79 kg/m^3

FOR RAINING BALLS ONLY

FREE FALL VELOCITY	= 18.24 m/s
BALL VELOCITY	= -13.63 m/s FALLING
GAS LINEAR VELOCITY	= 4.58 m/s
GAS VOLUME FRACTION	= 0.9979 v/v
PRESSURE DROP (NO WALL FRIC.)*	= 7.2 kg/m^3

FOR SIMULTANEOUS BALL/SHALE FLOW

	GAS	PARTICLES	BALLS
RISE VELOCITY, m/s	5.54	0.35	-0.33
PRESSURE DROP, kg/m^3*	508	209	299
BALL/PARTICLES INTERACTION FORCE kg/m^3		261	
VOLUME FRACTION OF PHASE	0.825	0.087	0.088

*kg force x 9.807 = Newtons

Sensitivity to Conditions. The calculations were repeated for a number of mass fluxes and gas velocities. The calculated shale and ball velocities are indicated in Figure 2, while pressure drops (i.e., effective bulk densities) are indicated in Figure 3.

The solids velocities are fairly stable with increasing gas velocities over a wide range. Above 4 m/s gas velocity and at high mass loadings (say 100 $kg/m^2/s$) balls are entrained upward and the system fails to function in counterflow. In general, shale entrainment velocity is enhanced by shale mass flux loading. Below 2 m/s gas superficial velocity the shale behaves more like a dense bed and balls fall at about 0.3 m/s as long as the bed is well fluidized (no defluidized zones).

The solids holdup does vary considerably with conditions, increasing with decreasing gas velocity in the lower velocities range. At low velocities the gas does not entrain the shale as rapidly; at high velocity the fall of balls is retarded.

AIR/ALUMINA BALLS/CRUSHED SHALE DEMONSTRATION UNIT

A 0.19m ID x 5.5m tall test column was built to demonstrate operability in the raining ball/elutriating particulate flow regime. The shale circuit portion is illustrated in Figure 4. The ball hopper, standpipe, liftpipe, and cyclone portions of the ball circuit are not shown.

Most efforts were concentrated on ball and shale introduction and removal and system control schemes. However, some scant pressure drop data was obtained as given in Table 2. The measured values are less than half the calculated values in the upper portions and only approach calculated values in the bottom section.

Thus, the interaction is about half the calculated values. This presumably results from localized addition of the balls in the central 11% of area so that some height is required to break up the ball stream and distribute the balls randomly over the cross section.

TABLE 2

SPHER - 0.190 METER COLUMN PRESSURE DATA

Column Configuration

A. Shale Injector - 0.101 Meter I.D. Side Arm
B. Shale Disengaging - Annulus Pellet Inlet to 0.190 Meter I.D. Pipe

Shale Rate, $kg/m^2/sec$	33.4		45.1		48.8	
Pellet Rate, $kg/m^2/sec$	51.3		64.4		73.2	
Superficial Gas, m/sec	5.3		5.3		5.3	
Column Bottom Pressure, cm Hg gauge	5.6		5.3		-	
Column ΔP, kg/m^3	Meas	Calc	Meas	Calc	Meas	Calc
Top 1.22m	62	221	99	251	125	278
Mid 1.75m	77		111		**	
Btm 1.32m	141		208		**	

** Not measured.

CONCLUDING REMARKS. The development of the raining ball concept was carried far enough to indicate it to be a viable approach. Obviously, further developments are required to achieve the ideal ball dispersions and ball/particulate interaction.

ACKNOWLEDGMENT. The author wishes to thank S. C. Roberts, D. E. Hardesty, G. L. Johnson, and G. P. Hinds, Jr. (retired), all of Shell Development Company, Houston, for the contributions to the Shell Pellet Heat Exchange Retorting process.

LITERATURE CITED

1. Oil Shale, Tar Sands, and Related Materials, H. C. Stauffer, ed., pp. 167-182, ACS Symposium Series No. 163, 1981.
2. 13th Oil Shale Symposium Proceedings, J. H. Gary, ed., Colorado School of Mines Press, pp. 35-45, 1980.
3. U.S. Patent No. 4,110,193, dated August 29, 1978.
4. Hinds, G. P. Jr., AIChE Annual Meeting, Miami, November 1978.
5. Ergun, S., CEP 48, 89 (1952).
6. Ibid., Ind. Eng. Chem. 41, 1179 (2949).
7. Doig, I. D., and Roper, G. H., IEC Fundamentals, 7 No. 3, 459-470 (August 1968).

NOMENCLATURE

C_D, C_{DA} = drag coefficients
d_p = ball or particle diameter
F = force
G_r = gravitational force
ϕ = volume fraction of phase

g = gravitation
U = velocity
μ = viscosity
ρ = density

subscripts:
b = balls or pellets
g = gas
s = particulates

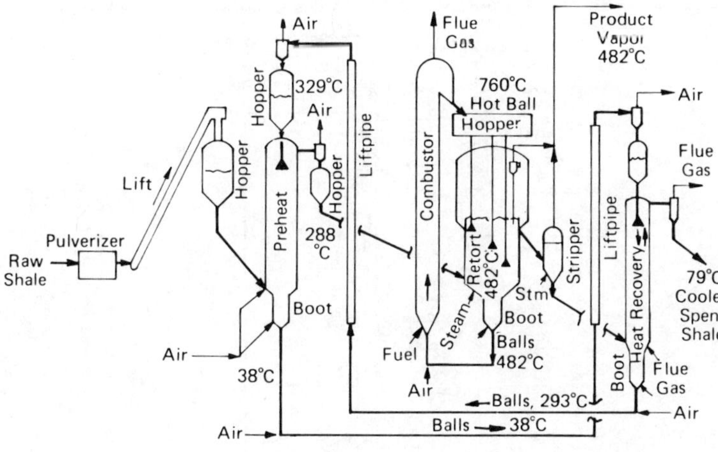

Figure 1. Raining Ball Oil Shale Pyrolysis - Schematic

08895

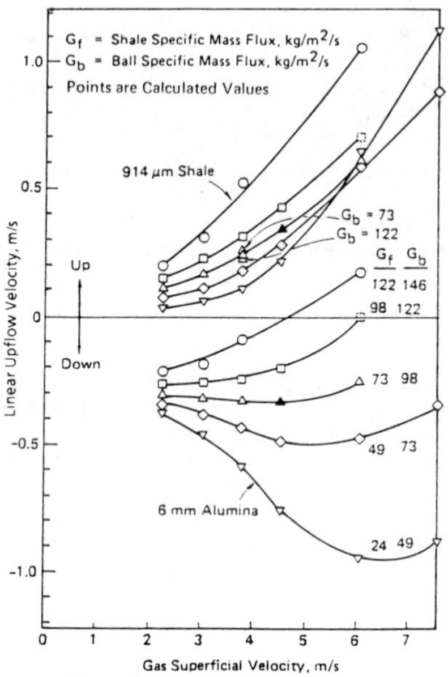

Figure 2. Raining Ball - Fast Fluidized Bed

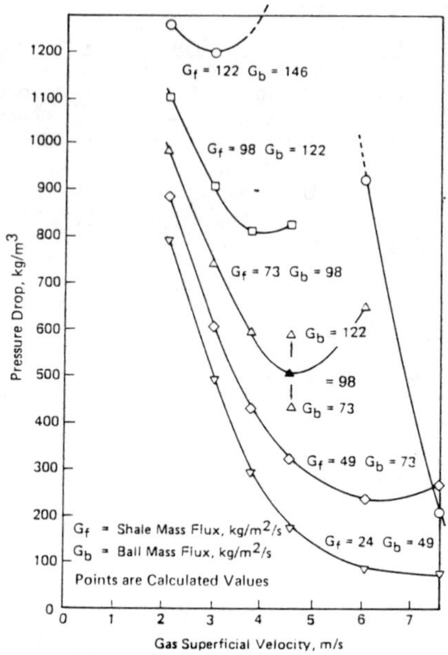

Figure 3. Raining Ball - Fluidized Shale Bed

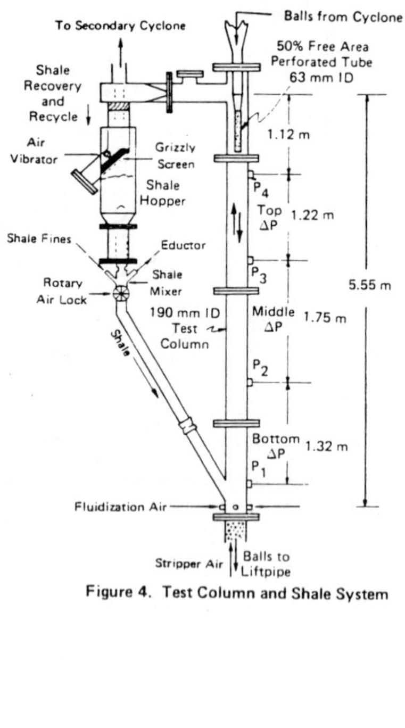

Figure 4. Test Column and Shale System

RADIAL VARIATION IN VOID FRACTION IN A FAST FLUIDIZED BED

H. Weinstein, M. Shao, M. Schnitzlein, and R.A. Graff
Department of Chemical Engineering
The City College of
The City University of New York
New York, New York 10031

ABSTRACT

The radial variation of void fraction in a transported, high velocity (fast) fluidized bed of laboratory scale is investigated experimentally. It is shown that a strong radial gradient exists in the lower dense phase region. The gradient moderates in the central transition region but persists through the upper dilute phase region of the bed.

SCOPE

The transported, high velocity (fast) fluidized bed has many desirable characteristics which make its use as a catalytic chemical reactor highly attractive. There is, however, at present no real predictive modeling capability with which to scope and/or design fast bed reactors for new processes or even for significant process alterations in operational units.

Recent laboratory investigations of fast fluidized bed behavior have concentrated on the "global" fluid mechanics of the gas-solid flow to accumulate a data base both for the inception of a modeling effort and for use in an interpolative way for scoping and pilot unit design. The initial thrust of experiments was to relate void fraction in the bed to gas velocity and solid rate. Further studies showed that strong axial void fraction variation often existed in laboratory beds and that the imposed pressure difference across the fast bed in laboratory systems with solid recirculation had a strong effect on void fraction in the fast bed. There was also evidence that strong radial gradients of void fraction exist in laboratory size fast beds. In this paper the existence of the radial gradient is demonstrated and quantified.

CONCLUSIONS

1. The radial distribution of void fraction in the lower dense phase region of a fast fluid bed shows a dilute core-dense annulus structure. The local void fraction decreases monotonically from at or near the center to the wall. The rate of decrease is small in the central half-radius and becomes much larger in the outer half-radius. Changes in gas velocity and/or solid rate do not change the nature of the variation in the lower dense phase region.

2. When the fast bed consists of a lower dense phase region and an upper dilute phase region separated by a transition zone, the dense annulus region void fraction increases in magnitude much more through the transition region than does that of the dilute core region. In the dilute phase upper region the void fraction profile shows only a slight curvature with the void fraction in the annular region approaching that of the dilute core.

3. In the lower dense region of the bed, the void fraction profile exhibits very low frequency oscillation about a relatively time invariant mean value.

INTRODUCTION

Transported, high velocity (fast) fluidized beds have been the subject of a number of recent laboratory investigations. Initially, the focus of these investigations was to relate the bed average void fraction, to the flow parameters of the system, gas velocity and solid rate(1), (2), (3).

It became clear in further studies studies that an axial variation of solid loading often existed in the bed (3), (4), (5), (6), (7) and that this axial variation was influenced by the pressure difference imposed across the length of the bed (6). It was also claimed that a fast fluidized bed consisted of a dilute core and a dense phase wall region (3). The claim was based on visual observation of plate film obtained with the use of a small, portable x-ray unit. This strong radial variation of solid loading was later clearly demonstrated for turbulent fluid bed with capacitance probe measurements (8) and for a fast bed with X-ray densitometry (9).

In this work, a large x-ray mounted on a 15 cm diameter fast fluidized bed is used to quantitatively measure the radial variation in void fraction across the fast bed. The x-ray system is mounted on an elevator which carries it along the bed so that axial changes in the radial variation can also be determined. The radial variation of void fraction along the bed is reported as a function of gas velocity and solid rate at two axial positions.

EXPERIMENTAL APPARATUS

A schematic of the fast fluidized bed circulating system (2) is shown in Figure 1. High velocity fluidization is maintained in the 15.2 cm I.D., 8.5 m tall column shown on the left. On the right is a 34.3 cm I.D. round vessel, in which the test solids are maintained in low velocity, bubbling fluidization. This so-called "slow" bed serves the double purpose of a standpipe and a storage for the solid.

Solid can be maintained in the 15.2 cm column at any desired fluidization condition through independent control of the solid inventory, the gas velocity and the solid rate from the slow bed to the bottom of the 15.2 cm column. Solids exiting the 15.2 cm column are separated from the gas in a primary cyclone, which forms an extension of the vessel that houses the slow bed. An additional smaller cyclone acts as the secondary cyclone and gas leaving it flows to a bag filter.

The X-ray system consists of a Norelco model MG 150/300 generator, an industrial unit, and a 150 kV tubehead. The maximum output of the system is 30 mA at 100kV or 20mA at 150 kV. The tubehead has focal spots of 1.5 and 4mm either of which can be considered as a point source of x-rays in this study.

The X-ray tubehead and the cassette holder for plate film used in recording images are mounted on opposite sides of the fluidized bed. Figure 2 is a scale drawing with side removed of the carriage which supports and positions these components plus additional image recording devices. The entire carriage assembly rides up and down a 3m long center section of the fluidized bed. The bottom position of the traverse is approximately 3m above the air injection nozzles at the bottom of the U-tube.

12 by 14 plate film (Kodak Industrex type M) is positioned in the plate film cassette holder. Initial test exposures at various tubehead potentials gave a clear image at voltages between 75kV and 110 kV. The resolution appeared to be optimal at about 100 kV. For plate film exposures, operating conditions are limited to a maximum of 10 mA and overall image density was controlled by adjusting exposure time.

All the runs were made with a single Engelhard cracking catalyst, HFZ-33. The characteristic properties of the solid, which fit Geldart's Class A, (10) are shown in the following table:

Size Range	dp	Density	Sphericity
20-120μm	59μm	1.46 g/cc	~1.0

The solid has almost identical fluidization properties to those of HFZ-20, a solid used in previous investigations (5), (6), (7), (9).

DATA REDUCTION

The data contained on the exposed plate film were analyzed with a chordal absorptometry technique. The optical density of the image on the film was read along a horizontal line with a Macbeth optical densitometer. The readings were corrected for x-ray scatter and for the absorption of the plexiglas column by using data from the film image of stepped blocks of aluminum and plexiglas. These stepped blocks were attached to the x-ray carriage alongside the bed and were always in the x-ray-to-film field. Extrapolations to zero thickness of the scatter-connected data from these stepped blocks were used to obtain a measure of the incident radiation intensity for each plate film exposure.

The solid density within the bed was expressed as a polynomial function in the radial distance from the center of the bed. This description presumes axial symmetry for the solid density. The coefficient of the polynomial term in r to the first power was set to zero as another axial symmetry constraint.

Polynomials of third to ninth order were integrated along the chords through the bed which corresponded to the data points read from the film. The resulting sets of equations in the unknown coefficients of the polynomials and the optical density data were inverted to yield the coefficents of each polynomial equation. The polynomial expressions differed negligibly from each other in describing the density variation. This result indicated that a third order polynomial was sufficient to fit the data. A preliminary set of measurements on the system with a packed bed of the same catalyst powder provided the basis for interpreting the polynomial coefficients for the fluidization runs in terms of void fraction or solid density.

RESULTS AND DISCUSSION

In order to test the time invariance of the radial profile of void fraction as is obtained from a 5 second time exposure a series of 5 second time exposures were taken at 5 minute intervals with both gas velocity and solid rate fixed. The set of profiles is shown in Figure 3. It is seen that the profiles change position and curvature continuously implying the presence of fluctuations with frequencies considerably smaller than 0.2 Hz. The radial average value of $1 - \varepsilon$ for this set profiles is 0.192 ± 10%. This can be considered constant within the experimental error essentially indicating time invariance of the mean value of solids and hence void fraction. However, any profile calculated from a given time exposure is not a true time average. The presence of these very low frequency fluctuations was also observed by Abed (8), who reported that a 3 minute averaging time for his capacitance probe reading was required in order to obtain "steady state" readings.

Data were obtained to illustrate the behavior of the radial void fraction, in response to three different types of flow variations. In the first variational scheme, the solid valve was kept fully open and the gas velocity was raised from 1.1 m/s (a turbulent fluid bed) to 5 m/s. The solid rate increased with increasing gas velocity and represents the maximum solids carrying capacity at a given gas velocity in this apparatus with a 2m inventory of solids in the slow bed. These data, taken at the bottom position of the x-ray traverse, are presented in Figure 4 and show that the shapes of the profiles remain essentially similar within the experimental error (~± 10% of $1 - \varepsilon$) and the oscillation bandwidth. The magnitude of the mean void fraction increases with increasing velocity as has been shown previously with differential pressure measurements (7).

In Figure 5, data are presented for the second variation in which the gas velocity is varied while the solid rate is held approximately constant. Profiles are shown at two axial positions, the bottom of the x-ray traverse and 1.6m above, for gas velocities, of 3.1, 4.1 and 5.1 m/s at a solid rate of approximately 123 kg/m^2-S. Adjacent pairs of profiles are for the two locations and a single velocity, starting at 3.1 m/s for the bottom pair. The figure shows that the inflection point in the axial void fraction profile separating the lower dense phase from the upper dilute phase was well above the upper measuring location for the 3.1 m/s case. Both profiles are essentially the same for this case. At 4.1 m/s, the inflection point which always moves downward as velocity increases is probably low enough so that the upper location is in the dense-to-dilute phase transition region and some flattening of the profile is seen. At 5.1 m/s, the inflection point has retreated to below the lowest point of the the x-ray traverse and both profiles are in the dilute phase region. However, the lower profile fo the pair indicates that the annular region is still thinning out at this location.

Data obtained at the two axial positions with the velocity held approximately constant at 3.1 m/s and with solid rate varying from 50 to 121 kg/ms are presented in Figure 6. It has been shown that at constant gas velocity, solid rate variation has a small effect on radial average solids resulting in about a 25% increase in $1 - \varepsilon$ for a 100% increase in solids rate. However, decreasing solids rate causes the inflection point location to fall to lower and lower positions and at low enough solids rate, no dense phase region is present in the fast bed. Figure 6 can then be interpreted loosely as indication the change in radial variation of void fraction at successive axial positions up the column. It shows clearly that the dilute core thins out going up the bed from a $1 - \varepsilon \sim 0.1$ to about 0.01, while the dense annulus disappears. The wall region $1 - \varepsilon$ decreases from about 0.4 to about 0.02 in this variation of solid rates.

ACKNOWLEDGEMENT

The National Science Foundation provied support for this work under an Industry-University Cooperative Research Program Grant, number CPE-8103627. Exxon Research & Engineering Company is the cooperating industrial organization.

LITERATURE CITED

I. Yerushalmi, J., Turner, D.H., and Squires, A.M., I&EC Proc. Des. & Dev., 15, 47 (1976).

2. Avidan, A.A., "Bed Expansion and Solid Mixing in High Velocity Fluidized Bed," Ph.D. Dis., City University of New York, 1980.

3. Bierl, T.W. and Gajdos, L.J., "Phenomenological Modeling of Reaction Experiments in Risers," Final Report, DOE/MC/14249-1149 (DE 820008670), Environmental Research and Technology, Inc., February, 1982.

4. Li Youchou, and Kwauk, M., in Fluidization, J.R. Grace and J.M. Matsen, Eds., Plenum Press, New York, 1980.

5. Weinstein, H., Avidan, A.A., Jean-Louis, J. and Meller, M., "Axial Variations in a Transported High Velocity Fluidized Bed," Presented at the 1981 Fall Annual Meeting of the AIChE, New Orleans, LA, November, 1981.

6. Weintein, H., Graff, R.A., Meller, M. and Shao, M.J., in Fluidization, D. Kunii and Toei, Eds., Engineering Foundation, New York, 1984.

7. Weinstein, H., Meller, M., Shao, M-J., and Parisi, R.J., in Fluidization and Fluid Particle Sytems: Theories and Applications, T.M. Knowlton., Ed., AIChE Symposium Series 234, Vol. 80, 1984.

8. Abed, R., in Fluidization, D. Kunii and R. Toei, Eds., Engineering Foundation, New York, 1984.

9. Weinstein, H., Shao, M-J., and Wasserzug, L., in AIChE Symposium Series 241, 1985.

10. Geldart, D., Powder Technology 7, 285 (1979).

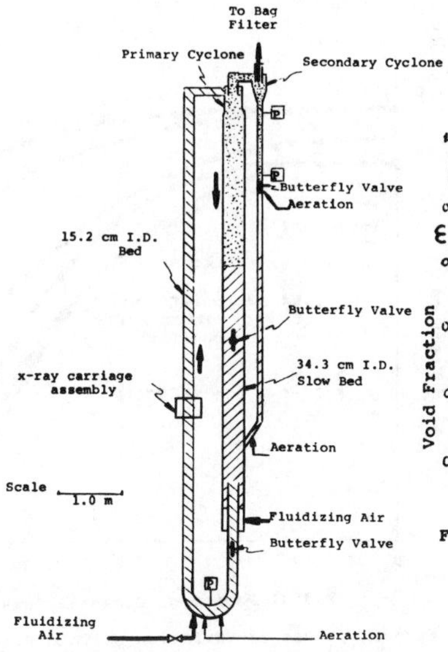

Figure 1. The 15.2 cm I.D. Circulating Fluid Bed System.

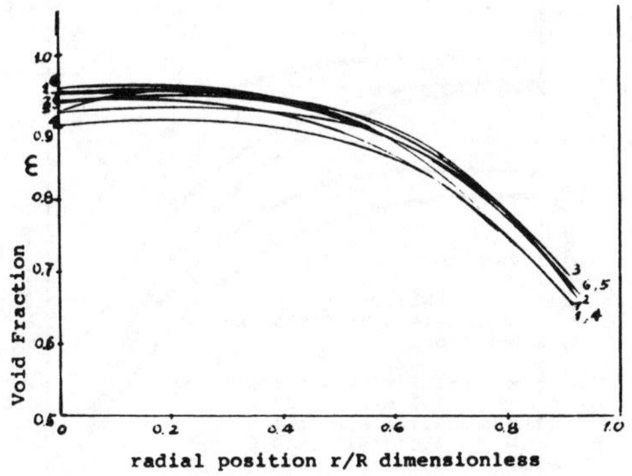

Figure 3. **Radial Variation of Void Fraction at 5 minute intervals, at U_g=3.2 m/s G_s=87 kg/m^2-s, I_s=2.5 m, h_x=0, and at 100kV, 5 mA, 5 second exposure time.**

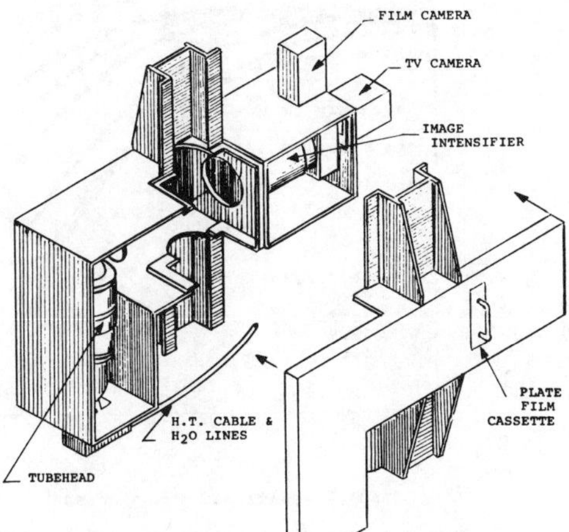

Figure 2. Scale drawing of the x-ray carriage assembly - the fluidized bed fits into the slot on the assembly.

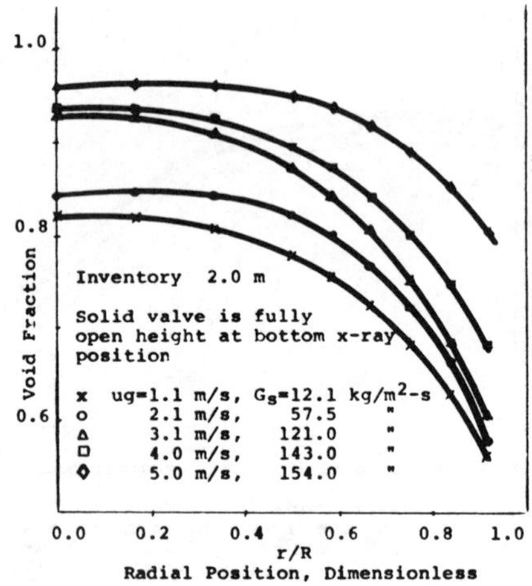

Figure 4. Radial Variation of Void Fraction at Different Velocities, Open Valve

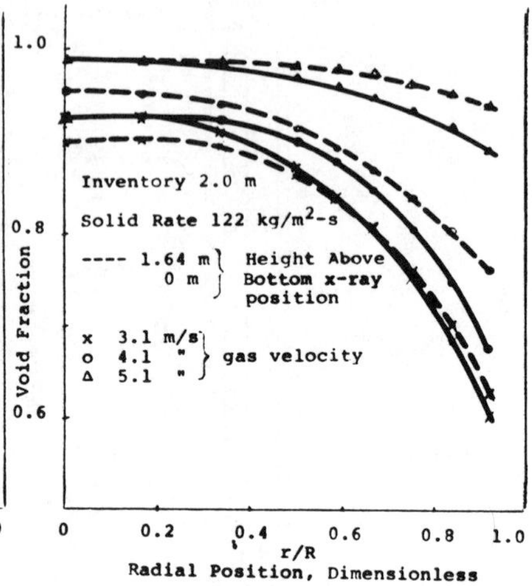

Figure 5. Radial Variation of Void Fraction, Constant Solids Rate

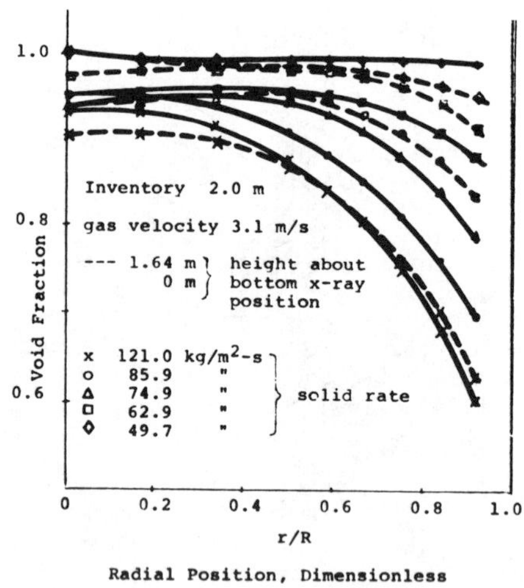

Figure 6. Radial Variation of Void Fraction, Constant Gas Velocity

PARTICLE MASS FLUX PROFILES AND FLOW REGIME CHARACTERIZATION IN A PILOT-SCALE FAST FLUIDIZED BED UNIT

L. MONCEAUX, M. AZZI, Y. MOLODTSOF and J.F. LARGE

Université de Technologie de Compiègne
B.P. 233 - 60206 Compiègne Cédex - France

Abstract

This paper reports the overall and local characterisation of the hydrodynamics of the vertical flow of 59 μm cracking catalyst particles, carried by atmospheric air in a 144 mm ID column for particle volumetric concentrations up to 15 %, and gas velocities varying between 2 m/s and 7 m/s. In the fully-developed flow region the existence of a similar profiles regime with a sharp transition, at volumetric concentrations of about 5 %, into a denser regime is confirmed experimentally.

Scope

The Circulating Fluidized Bed (CFB) stands between the dilute phase vertical pneumatic transport and conventional dense phase or turbulent fluidized bed. Most of the previous studies devoted to the hydrodynamics of a CFB (1 to 6) investigated a volumetric concentration range corresponding to the in-line transition of the suspension flow from turbulent to fast fluidization, i.e. about 20 %. On the other hand many authors examined the dilute phase vertical pneumatic transport (i.e. volumetric concentrations of about 1 %), and recent studies suggested the existence at low solids concentration of a so-called similar profiles regime from the fully-developed vertical pipe flow of a gas-solid suspension (9).

In this paper, the intermediate concentration range is carefully investigated in order to show the link between dilute phase pneumatic conveying and the denser CFB flow of the gas-solids mixture. The air velocities were in the typical CFB range (2 to 7 m/s) and the volumetric concentration of the 59 μm cracking catalyst particles varied from zero to 15 %.

The hydrodynamics of the flow is characterized by both overall and local measurements. The longitudinal pressure profile was measured along the riser by 12 pressure taps and particle concentration in the fully-developed region was deduced from the pressure gradient. Gas and solids flowrates were measured by conventional methods. A non-isokinetic sampling probe allowed the determination of the radial solids mass fluw profile.

Conclusions and Significance

In all experiments, a fully-developed flow region characterized by an essentially constant longitudinal pressure gradient was observed. Its length was reduced with increasing concentration, especially at low gas velocities, by the extension of the acceleration region.

In the fully-developed flow region, the existence of the theoretically predicted similar profiles regime has been proven by the radial particle mass flux profiles, as long as the volumetric solids concentration does not exceed 5 % ; the similarity factor is either the particles mean concentration or the mass flowrate, since the mean solids velocity depends only upon the gas velocity as in dilute phase pneumatic transport.

Finally the radial solids mass flux profiles show an important particle recirculation flow near the wall, at low gas velocities. The recirculation rate increases with concentration in the denser regime beyond the transition point.

The pressure gradient vs solids flowrate curves show a sharp transition to a denser regime where the mean solids velocity decreases with increasing concentration, and where the radial profiles are no longer similar.

INTRODUCTION

According to its conventional definition, the fast fluidization regime, or as it is often called in process industries, the circulating fluidized bed (CFB), is obtained when the particle entrainment rate above the bed reaches an important level compared to the bed inventory. When the height of the column exceeds several times its diameter, the flow of the gas-solids mixture tends towards a fully-developed pipe flow in the upper part of the column. This has been shown by the pressure drop profiles measured by Yerushalmi et al. (1) and longitudinal particle concentration profiles presented by Kwauk et al. (2, 3).

In industrial installations, the stationary operation of a fast fluidized bed is mostly obtained by the introduction of secondary air in order to enhance particle entrainment above the turbulent bed located at the bottom which is fed with solids at the desired rate.

According to the classification proposed by Yerushalmi et al. (e.g. 4) fast fluidization is linked to the flow in transport reactors (called risers in the oil industry) and the flow in dilute phase pneumatic transport in an order of decreasing concentrations, typically from about 20 % by volume down to 0.

The aim of this work was to investigate carefully the link between dilute phase pneumatic transport and the denser flow of a CFB. The object was to point out the regime transition characterized by both overall and local measurements of flow variables and to take advantage of the recent findings in the field of dilute phase flow (7, 8, 9) for a better and more detailed understanding of the hydrodynamic mechanisms involved in a CFB. The present paper is mainly concerned with the fully-developed flow region of a CFB since the results on the acceleration region have been previously published (10, 11).

EXPERIMENTAL APPARATUS

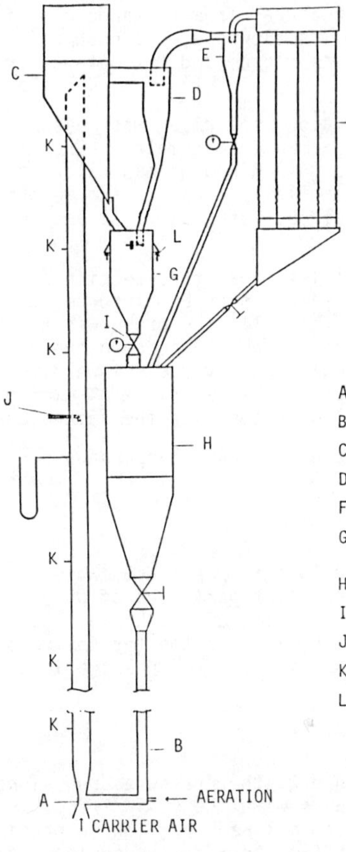

A : Venturi type solid injector
B : L-valve
C : Disengagement chamber
D, E Cyclones
F : Bag filters
G : Solids flowrate measurement hopper
H : Storage hopper
I : Pneumatic valve
J : Sampling probe
K : Pressure taps
L : Force transducers

Figure 1 : Schematic diagram of the experimental apparatus

We used a riser 0.144 m ID and 8 m high made mostly from plexiglass (Figure 1). Neither a dense phase fluidized bed located at the bottom nor secondary air injection were present, and the carrying gas (atmospheric air) was directly blown into the column. The solids were fed at a controlled rate by means of an L-valve in the vertical air flow at the throat of a venturi nozzle. At the top of the column the suspension entered a 0.6 m ID disengaging chamber and then a two-stage cyclone separator. The remaining fines were retained by bag-filters.

The gravity driven bulk flow of the particles collected by the disengaging chamber and the primary cyclone, was fed into a primary hopper supported by a special system equipped with three calibrated strain gauges. Then particles entered the larger secondary hopper located underneath through a pneumatic valve. Finally, the solids were fed into the vertical pipe of the L-valve by gravity.

The superficial gas velocity was measured from a standard orifice plate meter. In order to measure the solids flowrate, the pneumatic valve located at the primary hopper outlet was closed and the outputs from the strain gauges were recorded.

The longitudinal pressure profile along the riser was measured by means of the 12 pressure taps connected to water manometers. The mean solids concentration in the fully-developed flow region was calculated assuming that the pressure gradient was essentially equal to the weight of the solids hold-up as it had been proposed by Yerushalmi (5) and recently confirmed by Muzyka et al. (12).

The radial particle mass flux profiles in a cross-section were determined with the help of a non-isokinetic sampling probe. In order to measure particle recirculation rates near the wall, the probe was turned through 180°. The mass flux profiles given in Figure 3 concern the net flux, i.e. the algebraic difference between the upwardly and downwardly directed fluxes. The recorded data were reproducible within ± 5 % limits and the integrated profiles fitted the solids rates with the same accuracy.

A more detailed description of the experimental apparatus and operating procedure is given in references (10, 11).

EXPERIMENTAL RESULTS

The experimental results presented in this paper are those obtained using a group A fine powder : cracking catalyst particles of 59 μm mean diame- (30 % below 40 μm) and an apparent density of 900 kg/m3. The air velocities were varied from 2 m/s to 7 m/s, i.e. in the typical velocity range of CFB operating conditions. Volumetric solids concentrations up to 0.15 were investigated.

Axial Pressure Profiles

As a typical example shown in Figure 2, the pressure drop along the line, at a constant gas velocity, increases with the solids rate. In all cases, the axial pressure profile exhibited a linear portion (emphasised in Figure 2 by a solid-line) which testifies to a fully-developed flow -in which both radial concentration and velocity fields are independent of the axial coordinate (12). The lower part of the column is occupied by the acceleration zone where particle velocity increases from zero to its final value. In the upper portion of the line ($z > 6$ m) the disturbance due to the disengagement chamber is in evidence.

This behaviour is similar to that reported by Yerushalmi (1) under similar operating conditions. Axial solids concentration profiles measured e.g. by Kwauk (2, 3) show also that above the acceleration zone, the particle concentration becomes independent of the axial coordinate.

It should be noted that in our experiments at lower gas velocities (e.g. 2 m/s) the fully-developed flow region disappears when the solids rate is increased, since particle acceleration requires a greater length, and the in-line transition from the accelerated to fully-developed flow becomes difficult to observe.

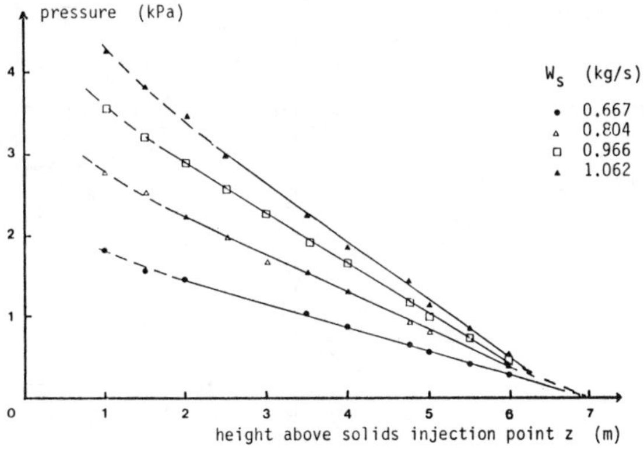

Figure 2 : Axial pressure profiles
at a constant gas velocity (U = 2,9 m/s) and various solids flowrates

Similar Profiles Regime

The slope of the straight lines in Figure 2 is essentially equal to the acceleration of gravity multiplied by the mean solids concentration :

$$-\frac{\partial p}{\partial z} \simeq \rho_s \, \overline{\alpha}_s \, g \tag{1}$$

where $\overline{\alpha}_s$ denotes the solids volumetric concentration (i.e. volume fraction) averaged over a cross-section :

$$\overline{\alpha}_s = \frac{1}{\pi R^2} \int_0^R 2 \pi r \, \alpha_s(r) \, dr \tag{2}$$

Recently Molodtsof (7) showed that in dilute flow (mathematically, when $\overline{\alpha}_s$ tends towards zero) solids concentration and velocity profiles of the form :

$$\alpha_s(r \,;\, U, \overline{\alpha}_s) = \overline{\alpha}_s . f(r \,;\, U) \quad \text{and} \quad V_s(r \,;\, \overline{\alpha}_s, U) = V_s(r \,;\, U) \tag{3}$$

(where $f(r \,;\, U)$ and $V_s(r \,;\, U)$ are independent of $\overline{\alpha}_s$) are asymptotic solutions of the general equations of multiphase mixtures in the case of a vertical fully-developed flow. Our experimental results confirm the existence of such a so-called similar profiles regime in which the local fluxes are proportional to the solids rate as can be seen in Figure 3.

Typical particle mass flux profiles measured by our sampling probe are presented in Figure 3 in a non-dimensionalised form : reduced flux (i.e. the local net flux divided by the solids rate) vs the relative distance to the centerline. It can be seen that as long as the solids rate does not exceed 0.55 kg/s, the reduced particle mass flux profiles are identical at a constant gas velocity. The reduced mass flux profile changes significantly from the unique solid-line curve for greater solids concentrations. However, the theoretical approach cannot predict the limiting value of $\overline{\alpha}_s$ beyond which the similar profiles will no longer be observed, i.e. the transition point.

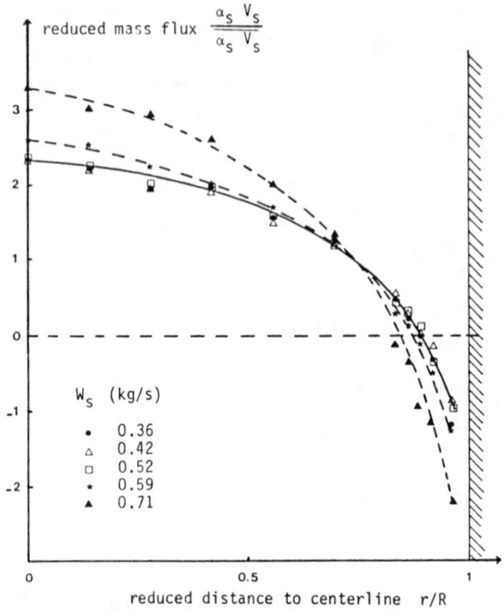

— similar profiles regime
-- profiles beyond the transition point

Figure 3 : Reduced radial profiles of particle mass flux at a constant gas velocity (U = 2.7 m/s) and various solids flowrates

Transition to Denser Regime

A typical plot of the axial pressure gradient (which is approximately proportional to $\overline{\alpha}_s$) vs the solids feed rate at a constant gas velocity is shown in Figure 4. As may be seen, for low solids rate $\overline{\alpha}_s$ is strictly proportional to W_s. In other words, the mean solids velocity $\overline{V}_s$ defined by the following equation :

$$\overline{\alpha}_s \cdot \overline{V}_s = \frac{1}{\pi R^2} \int_0^R 2 \pi r \alpha_s(r) V_s(r) dr \quad (4)$$

is independent of the mean solids volumetric concentration $\overline{\alpha}_s$ as expected in the similar profiles regime. However, a sharp transition occurs at a volumetric concentration of about 5 %, and the slope of the straight line changes significantly.

Beyond this transition point which always appeared in our experiments for $\overline{\alpha}_s \sim 0.05$ the mass flux profiles are no longer similar as was shown in Figure 3. This 5 % limit has

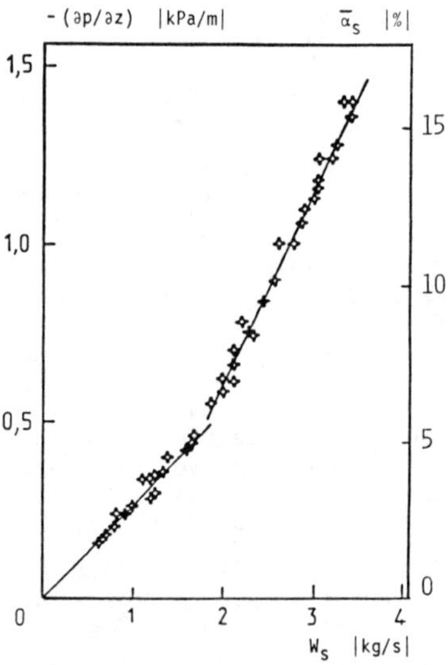

Figure 4 : Variations of the axial pressure gradient with the solids flowrate at a constant gas velocity (U = 4.6 m/s)

been previously suggested by Avidan and Yerushalmi (6).

Particle recirculation near the wall recorded in Figure 3 is present for low gas velocities ($U < 4$ m/s) and explains the high slip velocities in a CFB (always operated at low gas velocities).

The relative importance of the internal recirculation rate increases with concentration beyond the transition point as shown in Figure 3. Consequently the mean solids velocity decreases in the denser CFB regime with increasing solids flowrate (Figure 5), and the mean slip velocity increases with W_S, at a constant gas flowrate, only for concentrations greater than the transition value.

Figure 5 : Variations of the mean solids velocity with the solids flowrate at a constant gas velocity ($U = 2.9$ m/s)

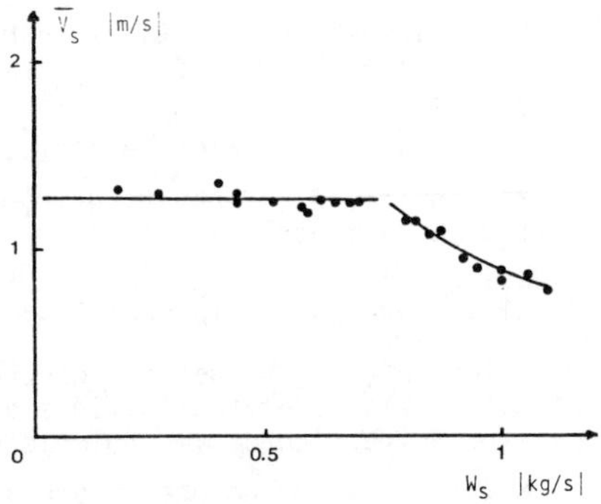

CONCLUSIONS

In the fully-developed flow region of a CFB either solids flowrate or mean concentration is, together with the superficial gas velocity, the relevant variables of the hydrodynamic behaviour of the mixture.

At low concentration a similar profiles regime exists in which solids velocity and reduced concentration profiles are determined only by the superficial gas velocity for a given gas-solids and wall material system.

A sharp transition, at a volumetric concentration of about 0.05, to a denser regime has been noticed. Beyond this transition point, the reduced particle mass flux profiles depend on the solids flowrate.

NOMENCLATURE

g	Acceleration of gravity
p	Pressure
r	Radial coordinate
R	Pipe radius
U	Superficial gas velocity
$V_S(r)$	Radial profile of solids velocity
$\overline{V}_S$	Mean solids velocity (averaged over a cross-section)
W_S	Solids mass flowrate
z	Axial coordinate (upwardly directed)

$\alpha_s(r)$ Radial solids volumetric concentration profile

$\overline{\alpha}_s$ Mean solids volumetric concentration (averaged over a cross-section)

ρ_s Solids density

LITERATURE CITED

1. Yerushalmi, J., Turner, D.H., Squires, A.M., "The fast fluidized bed", Ind. Eng. Chem., Process Des. Dev., 15 (1) (1976)

2. Li, Y., Kwauk, M., "The dynamics of fas fluidization", presented at Fluidization III, Henniker (1980)

3. Li, Y., Chen, B., Wang, F., Wang, Y., Guo, M., "Rapid fluidization", Int. Chem. Eng., 21 (4), (1981)

4. Yerushalmi, J., Cankurt, N.T., "Further studies of the regimes of fluidization", Powder Technol., 24, 187-205 (1979)

5. Yerushalmi, J., Cankurt, N.T., "High velocity fluid beds", presented at Chemtech (Sept. 1978)

6. Avidan, A.A., Yerushalmi, J., "Bed expansion in high velocity fluidization", Powder Technol., 32, 223-232 (1982)

7. Molodtsof, Y., "Equations générales probabilistes des écoulements polyphasiques et applications aux mélanges gaz-solides", Thèse d'Etat, Université de Technologie de Compiègne (1985)

8. Muzyka, D.W., "The use of probabilistic multiphase flow equations in the study of the hydrodynamics and heat transfer in gas-solids suspensions", Ph.D. Thesis, Univ. Western Ontario (1985)

9. Molodtsof, Y., Muzyka, D.W., "Solutions semblables asymptotiques des équations générales probabilistes des mélanges gaz-solides", presented to the 7e Congrès Français de Mécanique, Bordeaux (1985)

10. Monceaux, L., Azzi, M., Molodtsof, Y., Large, J.F., "Overall and local characterization of flow regimes in a circulating fluidized bed", presented to the 1st Int. Conf. on Circulating Fluidized Bed, Halifax (1985)

11. Monceaux, L., "Etude des régimes d'écoulement dans un lit circulant", Thèse de Docteur-Ingénieur, ENSPM, Rueil-Malmaison (1985)

12. Muzyka, D.W., Molodtsof, Y., Large, J.F., Bergougnou, M.A., "Application of probabilistic multiphase flow equations to dilute phase pneumatic transport", presented to the 33rd Can. Chem. Eng. Conf., Toronto (1983)

FLOW STRUCTURES IN FAST FLUIDIZED BEDS

E.-U. Hartge, Y. Li* and J. Werther

Technical University of Hamburg, D 2100 Hamburg 90, W.-Germany

*Institute of Chemical Metallurgy, Academica Sinica, Beijing, China

ABSTRACT

Local flow structures have been measured in fast fluidized beds of 0.05 m and 0.4 m diameter, respectively. Fine quartz sand and Al_2O_3-powder were used. Vertical profiles of the solids concentration were found to be similar in the small and in the large column, with a lower dense region and a dilute region in the upper part of the beds. Fiber optic probes in connection with an in-situ calibration procedure were used to locally measure solids concentrations inside the fast fluidized bed. Horizontal solids concentration profiles revealed the existence of characteristic solids distributions in the small as well as in the large diameter column. Steep concentration profiles were found in the dense region of the bed, whereas the dilute region was characterized by comparatively flat profiles.

SCOPE

Fast fluidized beds have in recent years been successfully applied in several processes on the large industrial scale. Fundamental research in the field of fast fluidization, however, has largely been carried out on a comparatively small scale. Nearly all academic investigations available in the literature have been carried out in beds with diameters less than 0.15 m (Yerushalmi and Avidan, (1)) the only exception being the work of Schmidt (2,3) who investigated the combustion of heavy oil in a pilot scale fast fluidized bed of 1 m diameter. The main objective of the present study is the investigation of changes in the flow structure if the size of the fast fluidized bed is increased from the laboratory scale to the pilot scale.

Another aim of the present work is to look deeper into the detail of the local structure of the gas-solids flow in the fast fluidized bed. The first observation of a characteristic axial solids concentration profile in the fluidized bed, with a denser bottom region and a dilute region near the top of the bed is due to Schmidt (2,3). Yerushalmi, Squires and co-workers (4,5) and particularly Weinstein and his colleagues (6) have later on studied the axial solids distribution in greater detail. Kwauk and his co-workers (7,8,9) presented a model of the formation of the solids concentration profile which finally led to a calculation procedure for the axial profile of the solids holdup.

Information about the horizontal distribution of solids in the fast fluidized bed is fairly scarce. Qin and Liu (10) investigated the lateral distribution by means of a fiberoptic probe. Traversing a 0.09 m diameter bed with their probe they found a characteristic profile with increased voidage at the bed center and increased solids concentration at the wall, respectively. This same shape of the solids concentration profile was detected by Gajdos and Bierl (11). In the present investigation a multifiber optic probe in connection with an in-situ calibration procedure has been used to measure horizontal solids concentration profiles in beds of 0.05 and 0.4 m diameter, respectively.

CONCLUSIONS AND SIGNIFICANCE

Measurements in fast fluidized beds of 0.05 m and 0.4 m diameter, respectively, yielded similar, although not quantitatively comparable profiles in the small and large diamater column. The density profiles could not be predicted by the correlations presented by Li et al. (9), a possible explanation of this disagreement being the different geometry and the different solids feed design used by these latter authors. The solids concentration in the upper dilute region turned out to be the same in the small and in the large diameter bed, if only the same gas velocities and solids mass flow rates were applied.

Characteristic horizontal solids concentration profiles were found both in the small and in the large diameter bed. The profiles were quite different in the lower dense and in the upper dilute region. In the latter one a relatively flat profile with a slight increase of the solids concentration near the wall was observed whereas the lower dense region was characterized by nearly parabolic concentration profiles of great steepness. Further insight into the nature of the gas-solids flow in the fast fluidized bed was obtained by measurements of the amplitude probability distribution of the probe signal. In the lower dense region a fairly broad distribution of solids concentrations was observed in the core of the bed which could be indicative of a mixture of clusters and voids whereas in the annular region close to the wall a more homogeneous suspension of greater density is possibly descending. The upper dilute region is characterized by a more homogeneous flow structure.

As a conclusion of the present work it may be stated that the flow structure of the fast fluidized bed is obviously more complicated than expected up to now. Not only a vertical variation in the fluid mechanics but also horizontal profiles will have to be taken into account in future fluid mechanical models of the fast fluidized bed.

EXPERIMENTAL

Two experimental installations have been used. The laboratory scale unit (Fig. 1) contains a fast fluidized bed of 0.05 m diameter and 3.3 m height. The pilot scale unit (Fig. 2) comprises a fast fluidized bed section of 0.4 m diameter with a height of 7.8 m. A more detailed description may be found in (12). In most experiments quartz sand was used with a surface mean diameter of 58 microns. Some experiments were also carried out with Al_2O_3-powder (surface mean diameter 64 microns).

Since a previous comparison of the solids holdup calculated from the differential pressure drop with solids holdup values measured by

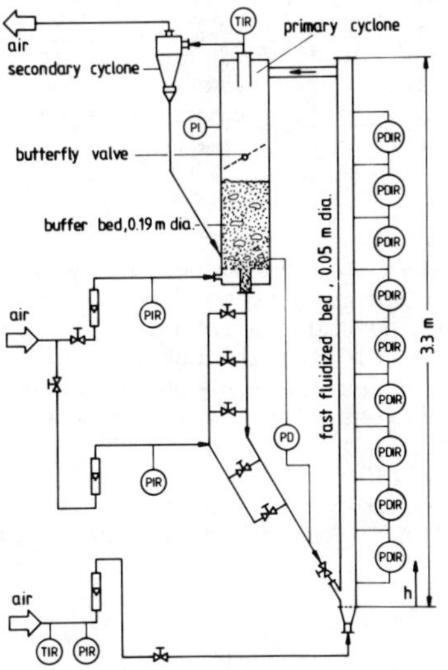

Fig.1

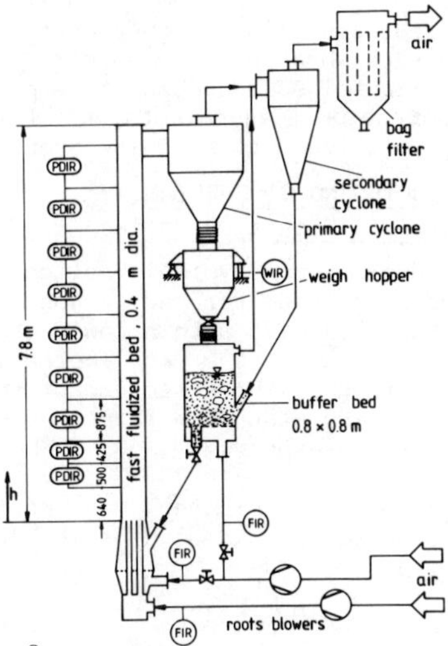

Fig.2

γ-ray absorption yielded satisfactory agreement after taking account of the different methods of averaging with both measuring techniques (12) it was decided to use the simpler differential pressure drop measurement for the determination of the average solids holdup in the respective volume element of the bed.

Local measurements of the solids concentration were carried out with capacitance and fiber optic probes, respectively. Severe troubles with electrostatic charging in the case of the capacitance probes were overcome by humidifying the fluidizing air (12). Since this latter means, however, created new difficulties further work was focused on the development of multifiber optic probes. These probes were mounted with their faces perpendicular to the direction of flow and contained both light emitting fibers and fibers for the transmittal of the reflected light. Good performance was achieved with a design where a bundle of 700 fibers was used. Each fiber had a diameter of 0.05 mm and the bundle had a diameter of 2.0 mm. Round about half of the fibers were emitting and receiving light, respectively. Light emitting and light receiving fibers were randomly mixed at the probe tip. This probe gave a linear relationship between the voltage of the probe signal U and the solids concentration c_v which was checked in a liquid fluidized bed. Since the whole system was quite sensitive to the positioning of the laser light supply as well as to break of single fibers due to bending an in-situ calibration procedure was adapted in order to obtain reliable and reproducible solids holdup values. Taking advantage of the linearity of the probe's response it was necessary to determine two points of the calibration line only, the one point being the measurement under empty tube conditions. The second point was obtained by traversing the probe over the diameter of the bed. From the integral of the probe signal over the bed's cross-sectional area an average voltage level $\overline{U}$ was obtained which was subsequently compared to the average solids hold-up $\bar{c}_v$ calculated from the differential pressure drop measurement. The calibration line is thus determined by the two points $\bar{U}(\bar{c}_v)$ and $U_o(c_v=0)$. This calibration procedure was repeated for each measurement of a new horizontal solids concentration profile. Reproducability of the concentration measurements was checked by repeated measurements of profiles and was found to be quite good.

RESULTS AND DISCUSSION

Vertical Density Profiles

Figure 3 shows typical vertical density profiles measured in the small and large diameter bed, respectively. The measurements show the typical shape (2,4,6,7) with a lower dense region and a dilute region in the upper portion of the bed. Increasing the velocity at constant solids flow rate reduces the extension of the dense region, whereas an increase in the solids mass flow rate at a constant gas velocity causes the dense region to expand.

The curves in Figure 3 have been obtained by fitting the expression for the vertical voidage profile, suggested by Kwauk and his co-workers (7,8),

$$\ln\left(\frac{\varepsilon - \varepsilon_a}{\varepsilon^* - \varepsilon}\right) = -\frac{1}{z_o}(z - z_i) \tag{1}$$

to the measured solids concentration values. Numerical values for the four fitting parameters ε_a, ε^*, z_i and z_o have been determined by a least squares method with the additional restriction that ε^* is always greater than or equal to the measured voidage at the top of the fluidizing column. Equ. (1) is seen to give a good description of the measured solids profiles.

One of the present authors (Y.L., (9)) has recently derived empirical correlations for the determination of the parameters ε_a, ε^*, z_i and z_o on the basis of measurements in a fast fluidized bed of 0.09 m diameter using a variety of bed materials. Numerical values for the parameters

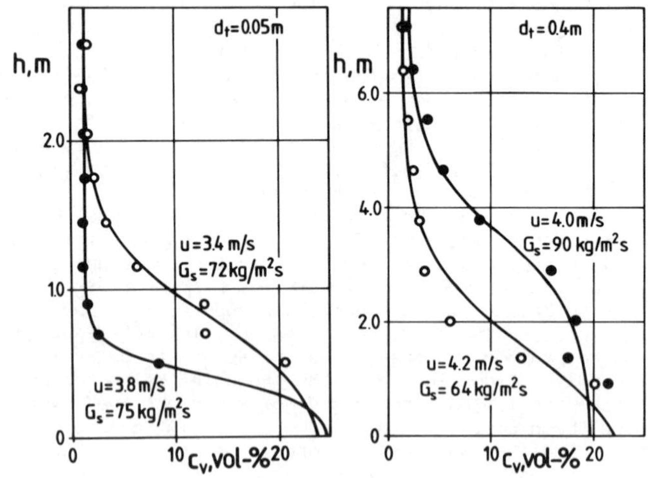

Fig. 3

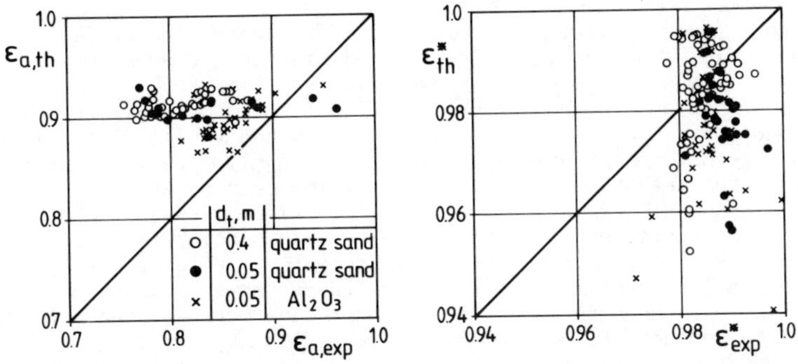

Fig. 4

ε_a and ε^* have been calculated from the correlations cited in (9) for the values of the operating parameters used in the present investigation. These values $\varepsilon_{a,th}$ and ε^*_{th} are compared in Figure 4 with experimentally determined values of ε_a and ε^*, respectively. The general tendency of the ε_a-plot is that Li's correlation tends to overestimate ε_a although the general accuracy of the prediction, particularly with the Al_2O_3-powder in the 0.05 m diameter bed, is not too bad. The scatter in the ε^* plot is fairly large and indicates that possibly physical mechanisms are influencing ε^* which are not sufficiently considered in Li's model. Values of the parameter z_i calculated from the correlation reported in (9) were found to be in disagreement with values obtained in the present investigation. Thus it is not possible at present to calculate the vertical density profile for the given designs on the basis of available correlations. As has recently been suggested by Weinstein et al. (6) the pressure distribution influenced by the solids return system and the solids feed design will have to be taken into account.

Although the density profiles in the small bed and in the large diameter rig look quite similar at first sight they are not directly comparable. What can only be stated at present is that the voidage ε^* in the upper dilute region is the same in the small and in the large diameter bed if the two rigs are operated at the same values of gas velocity and solids mass flow rate (Fig. 5). Further work on this aspect is going on at the Technical University of Hamburg.

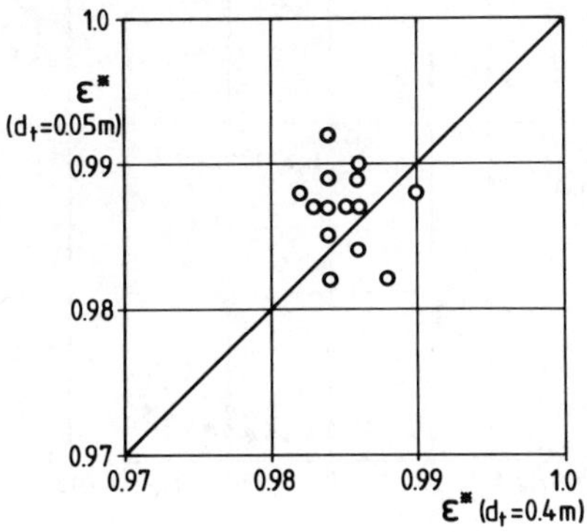

Fig. 5

Horizontal Density Profiles

Figure 6 shows typical horizontal solids concentration profiles measured in both fluidization rigs using the calibration procedure described above. It is astonishing and unexpected that the profiles in the small diameter bed and in the large diameter bed look quite similar, if plotted against the dimensionless distance from the vessel center-line. The shape of the concentration profiles is obviously determined by the location of the measuring level in the bed. The concentration profiles measured in the dilute region are relatively flat, with a slight increase of the concentration close to the wall. This distinct wall region is completely absent in the profiles measured in the lower dense region where nearly parabolic profiles of extraordinary steepness are observed with solids concentrations at the center-line between 5 and 10% compared to concentration values up to 30 or 40% at the wall. The steepness and regularity of the solids concentration profiles in the dense region can certainly not simply be explained by a distributor effect, since both installations are equipped with completely different gas distributing devices. Although a lot of measurements have been carried out - on the total some 70 solids concentration profiles have been measured under various operating conditions in the two fast bed installations - a direct quantitative comparison of the horizontal profiles in the small and large diameter bed turned out to be difficult. Obviously further and deeper insight into the local structure of this region is needed.

A first step into this direction was done by analyzing amplitude probability distributions of the probe signals obtained at different distances of the vessel center-line. Figure 7 shows some first results obtained in the 0.4 m diameter bed. The measurement in the height of 2.45 m above the distributor is typical of a measurement in the dense region of the fast bed. Near the center-line of the bed, where the average solids concentration is around 10% the distribution of solids concentrations

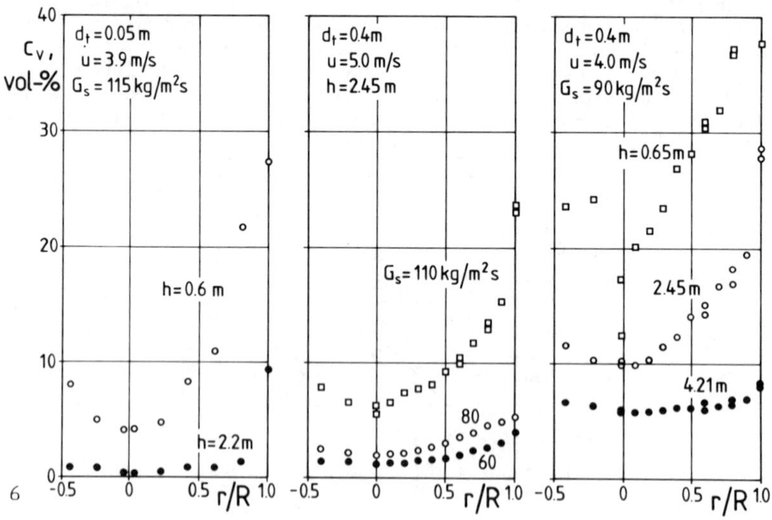

Fig. 6

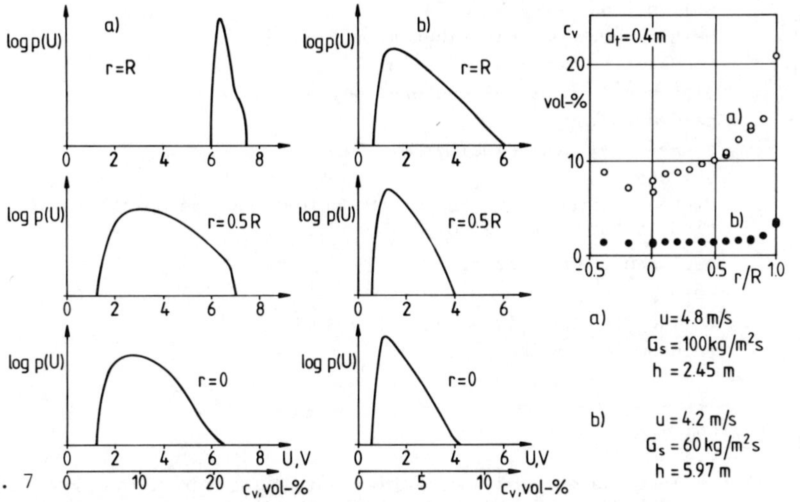

Fig. 7

ranges from about 3 to round about 20%, whereas close to the wall a fairly narrow distribution of solids concentrations between 20 and 25% is measured. It seems as if in the core of the bed a more dilute suspension with a broad distribution of densities exists which could be regarded as a mixture of clusters and voids whereas in the annular region close to the wall a more homogeneous suspension of greater density is possibly descending. The measurements in the upper dilute region reveal a completely different flow pattern. The local distribution of solids concentration is simply extended slightly to higher concentration values near the wall, i.e. the concentration range from 1 to 7% at r=0 is extended to round about 12% near the wall. It seems as if the flow structure is more homogeneous over the bed's cross-sectional area in this latter case.

It should certainly be born in mind that the solids concentration does only give limited information about the flow structure. Further insight about flow regions in the bed will require the measurement of local particle or cluster velocities, and this will be the next goal of the present investigations.

ACKNOWLEDGEMENT

One of the authors (Y.L.) gratefully acknowledges financial support of his stay at the Technical University of Hamburg by UNIDO and the Carl-Duisberg-Gesellschaft. The authors would furthermore like to thank Mr. M. Schnitzlein for valuable help in the early stage of this work.

NOTATION

c_V solids volume concentration, -

d_t bed diameter, m

G_s solids flow rate, kg $m^{-2}s^{-1}$
h height above gas distributor, m
p probability density (cf. fig. 7)
r distance from vessel center-line, m
R column radius, m
u superficial gas velocity, m s^{-1}
U voltage, V
z_i distance between the top of the column and the point of inflection in the vertical voidage profile, m
z_o parameter, equ. (1), m
ε voidage, -
ε^* voidage in the dilute region, -
ε_a voidage in the dense region, -

LITERATURE CITED

1. Yerushalmi, Joseph and Amos Avidan, in Fluidization, Davidson, J.F., Clift, Roland and David Harrison (Eds.), Academic Press, London (1985).

2. Schmidt, H.-W., Dissertation, Universität Karlsruhe, W.-Germany (1971).

3. Schmidt, H.-W., Proc. 3rd. Int. Conf. Fluidized Bed Combustion, II-1-1 (1973).

4. Yerushalmi, J., Turner, D.H. and A.M. Squires, Ind. Eng. Chem., Proc. Des. Dev., 15, 47 (1976).

5. Yerushalmi, J. and A.M. Squires, AIChE Symp. Ser., 73 (161), 44 (1977).

6. Weinstein, H., Graff, R.A., Meller, M. and M.J. Shao, in Fluidization, Kunii, D. and R. Toei (Eds.), Engineering Foundation, New York, 299 (1983).

7. Li, Y., Chen B., Wang, F., Wang Y. and M. Guo, J. Chem. Ind. Eng., (China) 143 (1979), and Int. Chem. Eng., 21, 670 (1981).

8. Li, Y. and M. Kwauk, in Fluidization, Grace, J.R. and J.M. Matsen (Eds.), Plenum Press, New York/London, 537 (1980).

9. Li, Y., Chen, B., Wang, F. and Y. Wang, in Fluidization, Science and Technology, Kwauk, M. and D. Kunii (Eds.), Science Press, Beijing, China, 124 (1982).

10. Qin, S. and G. Liu, in Fluidization, Science and Technology, Kwauk, M. and D. Kunii (Eds.), Science Press, Beijing, China, 258 (1982).

11. Gajdos, L.J. and T.W. Bierl, "Topical Report for the US DOE, Contract No. EX-C-76-01-2449", US Department of Energy, Washington D.C., (1978).

12. Hartge, E.-U. and J. Werther, "Analysis of the Local Structure of the Two Phase Flow in a Fast Fluidized Bed", 1st Int. Conf. on Circulating Fluidized Beds, Halifax (1985).

HOLDUPS AND MODELS OF THREE PHASE FLUIDIZED BED

Tsung-ting Hu, Bao-tien Yu, Y-ping Wang

Dept. of Chemical Engineering, Tianjin University
Tianjin, The People's Republic of China

ABSTRACT

This work was concerned with the measurement of the phase holdups and the models of a three-phase fluidized bed. Empirical correlations for gas and solid holdup distribution and derived expression for averaged gas holdup were obtained.

SCOPE

The object of this paper is to present the measurement and the analysis of the phase holdups in a gas-liquid-solid three-phase fluidized bed. The experiments were performed in plexglass columns of three sizes. The three phases were air, water and glass beads respectively. The gas phase holdups were measured by electric conducting probe, of which the signals were also recorded on oscillograms. An optical fibre technique was used for solid phase holdups. Empirical correlations for different phase holdups, which are important parameters for the design of reactors, were obtained. No information about the effect of column diameters on average gas phase holdups in three-phase fluidized beds has ever been reported, and only a few measured data of local solid phase holdups have been presented in the literature.

CONCLUSIONS AND SIGNIFICANCE

The radial gas holdup profile was found to be parabolic, The effects of superficial gas and liquid velocities on the mean gas holdup are shown on Figure 5 with break points which are different for columns of different diameters. Horizontal lines on the right of these break points are defined as developed fluidized region and oblique lines on the left as undeveloped region. A predicting formula for the starting velocity of a developed region was obtained. Averaged gas holdups within the developed region were related with superficial gas and liquid velocities and with column diameters by Equation (9). This work presents a specialized technique for the measurement of solid holdups, some experimental data on solid phase holdups which are scarce in literature, and an expression (10) for the solid holdup radial distribution.

EXPERIMENTAL SETUP

The measurements were performed on the system of air-water flowing up through three sizes (66, 92, 284 mm. i.d.) plexglass columns. The suspended solid phase was glassbeads of 1.125 mm. diameter with an apparent density of 2.4065 kg/L. The detail scheme of experimental apparatus may be found in Reference (1).

MEASUREMENT METHOD AND DATA PROCESSING

The local gas holdups were measured by using the electric conducting probe method (Figure 1). The construction and principle of measurement of probe were basically the same as described in the literature (2).

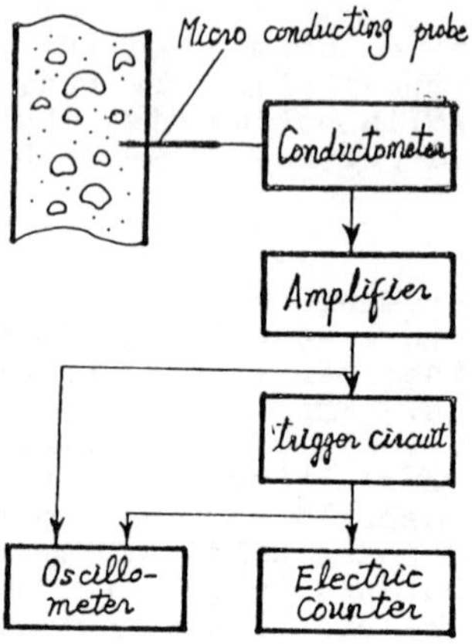

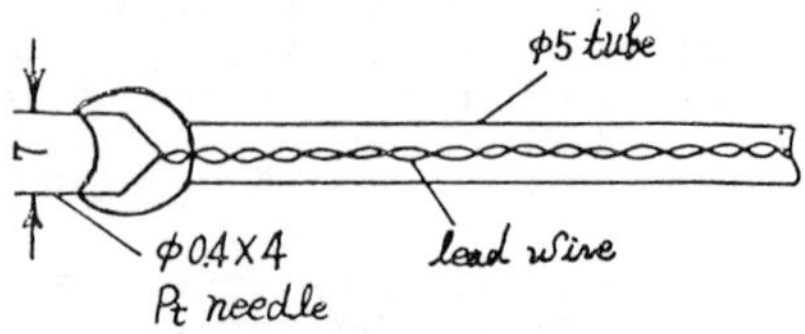

Fig.1 Measurement system for ε_g

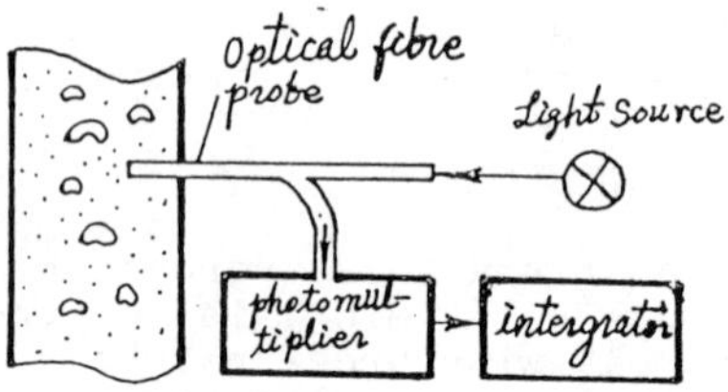

Fig.2 Instrumentation for optical fibre measurement

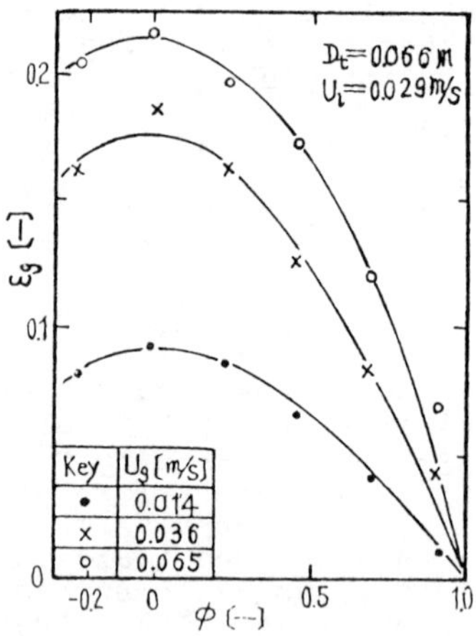

Fig.3 Radial profile of ε_g

The local solid holdups were measured by means of fibre optics technique (Figure 2). The details of this probe were described in Reference (3). Let N_s be the integrated values of reflected signals. It is a function of solid holdup of fluidized bed.

$$N_S = \int_0^T V(t)\,dt = f(\varepsilon_S) \quad (1)$$

From the experiments with a L-S bed, a linear relation between ε_S and N_S can be exnressed by Equation (2).

$$\varepsilon_S = 1.1845 \times 10^{-5}\ N_S - 0.1371 \quad (2)$$

The main obstacle to the measurements of solid holdups by fibre optics is that gas bubbles reflect light as well as solid particles. In order to get a value for solid particles alone, we run a liquid-solid fluidized bed, operating with the same bed expansion ratio of the three-phase bed. The combined values of reflected signals from g-L-s and L-s beds were measured separately and designated as N_{g-L-s} and N_{L-s} respectively. Then, evidently, the difference between the two should be the part of reflected signals caused by gas holdups. Hence, we have the integrated values for gas holdups as

$$N_g = N_{g-L-s} - N_{L-s} \quad (3)$$

It was found from experimental data, that N_g increases with gas holdups, but decreases with the increase of solid holdups, essentially in a linear dependence.

$$N_g = 68105.9\ \varepsilon_g - 30172.1\ \varepsilon_s + 8031.4 \quad (4)$$

Therefore, the solid holdup can be calculated from Equation (5), which is obtained by combining Equations (2) and (4).

$$\varepsilon_S = 1.8432\times10^{-5}\ N_{g-L-s} - 1.2554\varepsilon_g - 0.36136 \quad (5)$$

RESULTS AND DISCUSSION

LOCAL GAS HOLDUPS

Figure 3 shows the typical radial distribution of gas holdups in the three-phase fluidized bed. Points on this Figure show that the radial profile of gas holdups is very uneven, greatest holdup at the center of the bed and approaching zero at column wall. All the data of ε_g obtained from our three size columns were reasonably described by a parabolic equation

$$\varepsilon_g = \varepsilon_{gmax}\left[1 - \phi^2\right] \quad (6)$$

It is in agreement with results in the literature (2).

Figure 4 gives a comparison of the predicted values from Equation (6) and the experimental data.

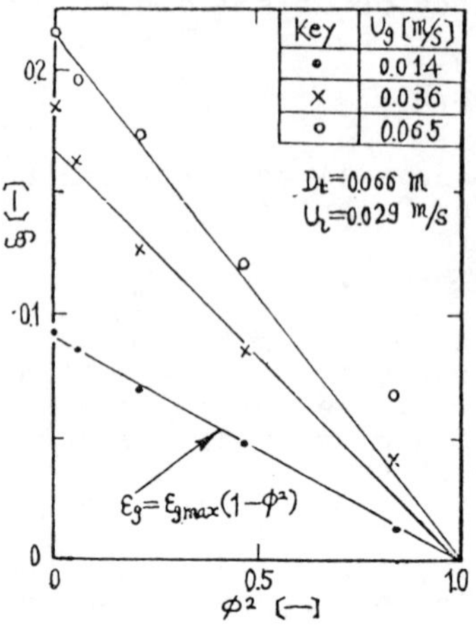

Fig.4 Comparison of experimental and calculated ε_g

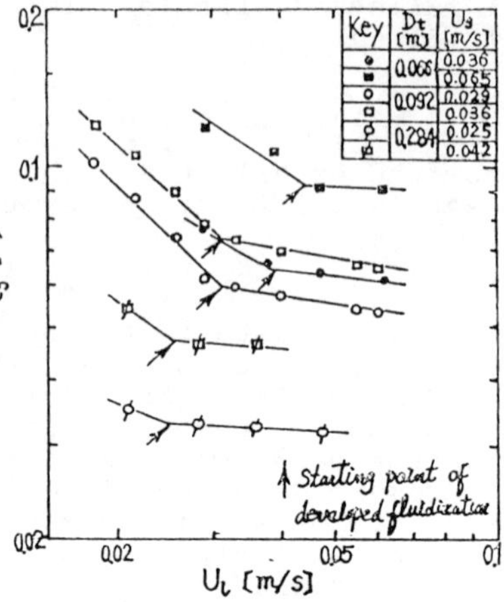

Fig.5 Effect of U_L, U_g and Dt on $\bar{\varepsilon}_g$

MEAN GAS HOLDUPS

Figure 5 shows the effects of U_g and U_L on mean gas holdups, $\bar{\varepsilon}_g$, predicted from the values of local gas holdups according to the following formula.

$$\bar{\varepsilon}_g = \frac{\int_0^1 \varepsilon_g 2\pi\phi d\phi}{\int_0^1 2\pi\phi d\phi} \tag{7}$$

From Figure 5 it can be seen that the effects of U_g and U_L on $\bar{\varepsilon}_g$ were more complicated. A turning point of liquid velocity exists. For liquid velocity beyond this point,

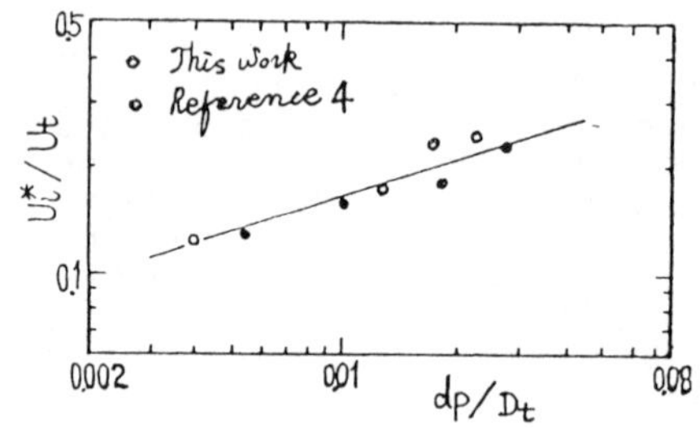

Fig.6 Effect of dp/Dt on U_L^*/U_t

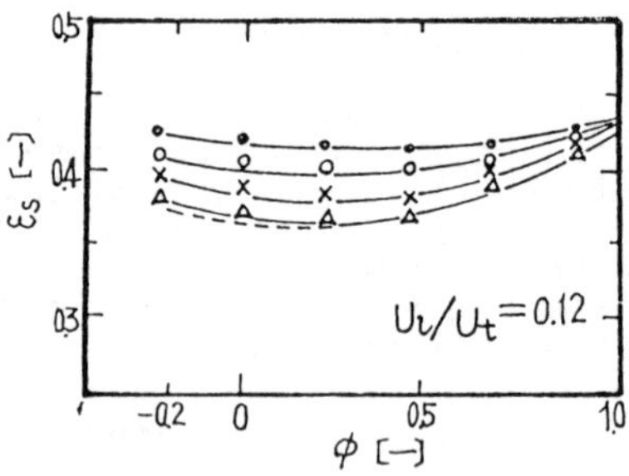

Fig.7(a) Radial profile of ε_s

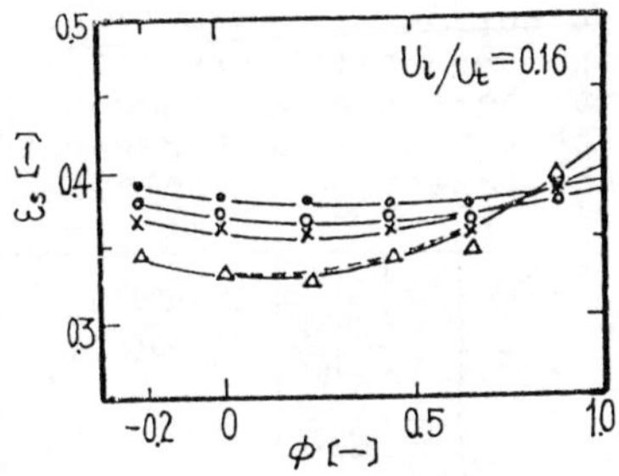

Fig.7(b) Radial profile of ε_s

Symbol	•	o	x	△
U_g cm/s	1.43	2.87	4.87	6.64

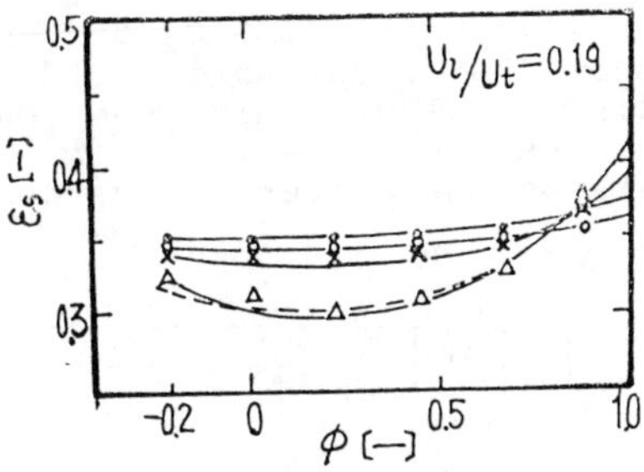

Fig.7(c) Radial profile of ε_s

the effects of U_L on $\bar{\varepsilon}_g$ were almost leveled. One can define this region as developed fluidized region. However, at U_L lower than this point, $\bar{\varepsilon}_g$ goes up as the liquid velocity is decreasing. This is defined as undeveloped fluidized region. Mooroka et. al. (2) in their study indicated that the hydrodynamic behaviour of 12 cm diameter three-phase fluidized bed at $U_L/Ut=0.2$ belongs to the developed fluidized region. The experimental results of this work show that the starting liquid velocity of the developed fluidized region decreases with increasing bed diameter and increases with increasing solid particle size. The data of this work and part of those from reference (4) constitute the points on Figure 6. A regression line for the starting liquid velocity of the developed fluidized region was obtained.

$$\frac{U_L^*}{U_t} = 0.7\left(\frac{d_p}{D_t}\right)^{0.3} \tag{8}$$

The average deviation was 10.4%.

Based on the data of $\bar{\varepsilon}_g$ measured in beds of three different diameters, a correlation for $\bar{\varepsilon}_g$ within the developed fluidized region was obtained:

$$\bar{\varepsilon}_g = 0.0645\ U_g^{0.842}\ U_L^{-0.096}\ D_t^{-0.419} \tag{9}$$

Equation (9) shows that column diameter significantly affects $\bar{\varepsilon}_g$ in our experimental conditions. The similar results were also observed in a gas-liquid bubble column (5).

LOCAL SOLID HOLDUPS

The radial profiles of solid holdups are plotted in Figure 7 for different superficial gas and liquid velocities in the three-phase fluidized bed of 28.4 cm i.d. and Figure 8 shows the radial distributions of ε_s for different column diameter at almost the same U_g and U_L. The curves on these Figures show that local solid holdups along the radial direction distribute unevenly, the lowest at the center of bed and highest near the wall. It means that the highest gas holdups at the center correspond to the lowest solid holdups.

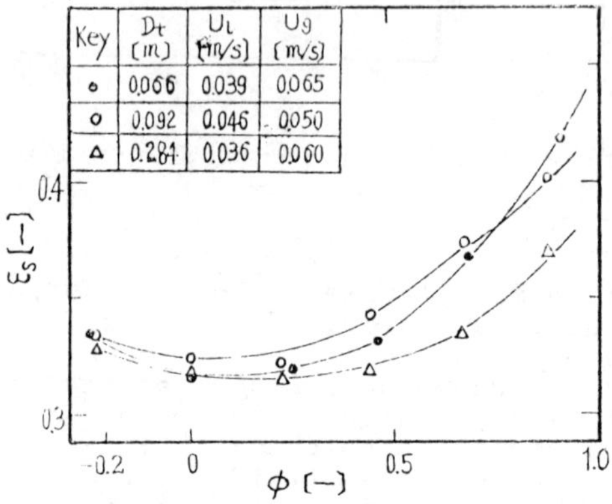

Fig.8 Effect of D_t on solid holdup distribution

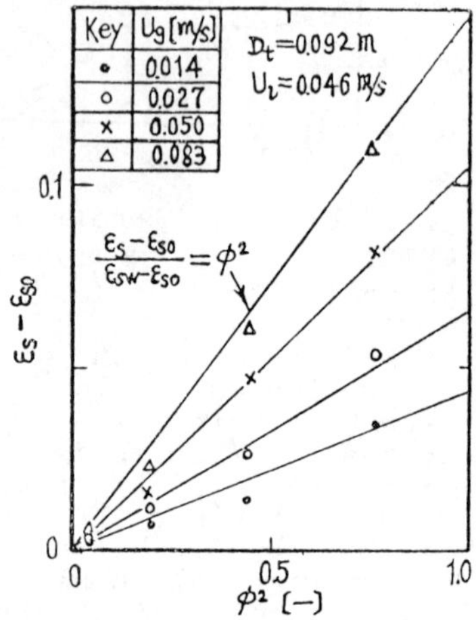

Fig.9 Comparison of experimental and calculated ε_s

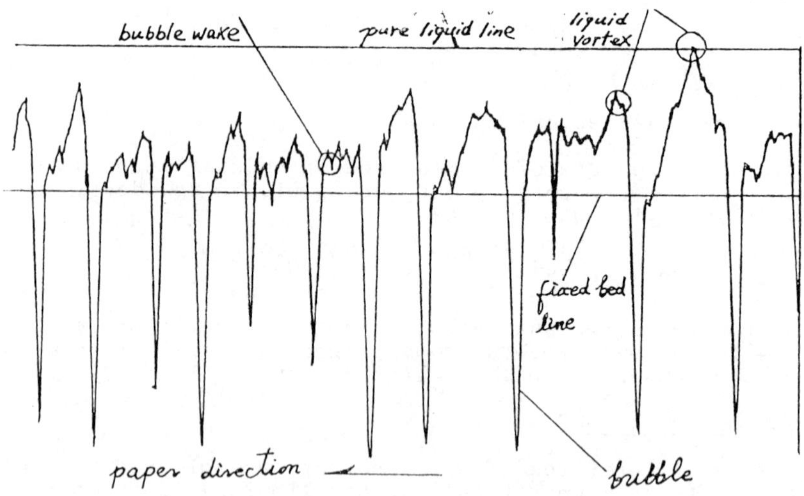

Fig.10 Oscilloscope recording

When the bed diameter is increased, the profile becomes flatter.

On the basis of experimental data of this work, an expression for the radial profile of ε_s was obtained as:

$$\frac{\varepsilon_s - \varepsilon_{so}}{\varepsilon_{sw} - \varepsilon_{so}} = \phi^2 \tag{10}$$

Figure 9 gives a comparison of the calculated values from Equation (10) and experimental data. Equation (10) is also a better fit to data of Linneweber (6).

From Equations (6) and (10), we can derive the following:

$$\frac{\varepsilon_s}{\varepsilon_s + \varepsilon_L} = \frac{\bar{\varepsilon}_s}{\bar{\varepsilon}_s + \bar{\varepsilon}_L} \tag{11}$$

$$\frac{\varepsilon_s}{\bar{\varepsilon}_s} = \frac{1 - \varepsilon_g}{1 - \bar{\varepsilon}_g} \tag{12}$$

Equation (12) means that the radial distribution of the local solid holdup is only affected by the gas radial profile.

On Figure 7 (dashed lines) were also shown the solid holdup radial profile predicted by Equation (12). It is observed from Figure 7 that the calculated profiles of ε_s were in agreement with experimental data.

MODEL OF THREE PHASE FLUIDIZED BED

As is well known, a wake model has been presented for predicting the phase holdups characteristics of a three-phase fluidized bed. This model has commonly been set up by assuming that the bed can be divided into a liquid-fluidized phase, a gas bubble phase and a bubble-wake phase, and that the bubbles and their wakes travel at the same velocities. The principal differences amongst the investigators were their assumptions of x, the ratio of solid holdup in the wake to solid holdup in the liquid-fluidized phase. Thus Stewart and Davidson (7) and most others assumed x=0, while Østergaard (8) took x as unity. And considerable vortex shedding were first observed and suggested by Rigby and Capes (9) to contribute to the bed contraction. During the measurement of local gas holdup in this work, the oscillograms (Fig.10) obtained by the electric conducting probe display the wave shape signals following gas bubbles. We find in some of the wakes that the solid particle concentration was almost the same as that of the liquid-fluidized phase.

In addition, a large number of vortices occurred between

bubbles, just as shown in Figure 10. Liquid vortices appear on the ossillogram as wave peaks. Most of the peaks show that there are only small amount of solid particles within the vortices and some of them shoot up so high to show that the concentration of solid particles is nearly zero. These vortices were not shed from wakes, we presumed. They were most possibly formed by violent agitation of gas. Therefore, we propose a four-zone model for a three-phase fluidized bed: that is, gas bubbles phase, wake phase, liquid-fluidized phase and the fourth, liquid vortex phase. This part of experiment is still going on in our laboratory. No quantitative informations could be obtained from this kind of oscillograms up to this moment.

NOTATION

DT = column diameter
dp = particle diameter
N = integrated values of reflected signals
r = radial coordinate
R = bed radius
t = time
T = sampling interval
U = superficial velocity
U_L^* = starting liquid velocity of developed fluidized region
U_t = terminal velocity of a single particle in stagnant liquid
V = e.m.f. signal of solid particle converted from reflected light
ε = holdup
ε_{so} = solid holdup at the bed center
ε_{sw} = solid holdup near the wall
ε_{gmax} = gas holdup at the bed center, i.e. maximum gas holdup
Ø = r/R, dimentionless radial coordinate

Subscripts: g=gas phase; L=liquid phase; s=solid phase

LITERATURE CITED

1. Tsung-ting, Hu and Zhang, Ying., J. Chem. Ind. and Eng. (Chinese), (1) 78 (1983).
2. Morooka, S. and Y. Kato, J. Chem. Eng. Japan, 15 (1), 29 (1982).
3. Qin, Shao-zong and Liu, Guang-yuan, Fluidization, Conf. Papers China-Japan Symposium, 258 (1982).
4. Kato, Y. and S. Morooka, Kagaku Kogaku Ronbunshu, 6, 579 (1980).
5. Ellis, Paper B1-Symposium on two phase flow, EXETER (1965).
6. Linneweber, K.W. and E. Blass, Ger. Chem. Eng. 6(1), 28 (1983).
7. Stewart, P.S.B. and J.F. Davidson, Chem. Eng. Sci., 19, 319 (1964).
8. Østergaard, K., Chem. Eng. Sci. 20, 165 (1965).
9. Rigby, G.R. and C.E. Capes, Can. J. Chem. Eng., 48, 343 (1970).

Characterization of Bubbles in Three Phase Fluidized Beds Using U-Shaped Optical Fiber Probes

S.L.P. Lee and H.I. de Lasa
Department of Chemical Engineering
Faculty of Engineering Science
The University of Western Ontario
London, Ontario, Canada N6A 5B9

ABSTRACT

A novel U-shaped optical probe recently developed at the University of Western Ontario was used to investigate bubble behaviour in a co-current gas-liquid-solid fluidized bed (air, water and glass beads, 335 μm). Several U-shaped probes were employed to study bubble properties such as shape, included angle, bubble velocity, axial bubble length and the changes of bubble frequency with axial radial position.

SCOPE

There is a great need for direct measurement of bubbles in commercial units, operating at high temperature and pressure. This could be of particular importance in the case of hydrocracking reactors where very active catalysts are used and the conversion could be limited by the gas-liquid mass transfer of hydrogen. Similar concerns about mass transfer limitations exist in the Fischer-Tropsch process. In order to develop the required instrumentation for tracking bubbles in large three-phase reactors a multiprobe system constituted by several U-shaped fiber optic probes was developed (de Lasa et al. (1), Lee et al.(2)). In the present contribution several U-shaped sensors are further applied to characterize the bubble behaviour in a three-phase fluidized bed.

CONCLUSIONS AND SIGNIFICANCE

- The hydrodynamic characteristics of three phase fluidized beds were studied using a multiprobe fiber optic system constituted by several U-shaped probes.
- Relative bubble velocities with respect to the liquid phase were correlated against axial bubble length using a power law. Two different regions of bubble velocity were observed.
- The mean bubble size was found to decrease with increasing V_ℓ and to be also influenced by V_g. The bubble size distribution was affected by V_ℓ.

- The initial gas distribution, the redistribution of gas in the bed and the bubble breakage at the top of the bed were inferred by analyzing the average bubble frequency at different axial and radial positions.

Two typical approaches have been reported in the literature to predict the hydrodynamic parameters of three-phase fluidized beds: - the use of two-dimensional columns, where the bubbles are followed from the walls of the column by cinecamera films (Henrikson and Ostergaard (3), Kim et al. (4), Stewart and Davidson (5)),- the use of in-situ methods where electroconductivity probes are located inside three dimensional beds (Park et al. (6), Rigby et al. (7), Begovich and Watson (8), Linneweber and Blass (9),(10)). The two-dimensional column unduly constrains the bubbles and affects their behaviour. In-situ probes such as electroconductivity probes specifically designed to minimize flow disturbances, are preferable. These methods are, however, limited to cases where the approprite combination of electrical properties allow measurements to be made.

Due to the facts described, a multiprobe system constituted by several U-shaped fiber optic probes was developed (de Lasa et al. (1), Lee et al. (3)). The design and operation of these bubble sensor is based on the difference of refraction indices between the gas and the liquid-solid phase. In this respect this probe differs substantially from the two opposed fibers or the bundle of optical fibers proposed recently for two phase fluidized beds by Ishida et al. (11), Ishida and Hatano (12), Oki et al. (13). It is important to mention that the difference of refraction indices between gas and liquid phases was considered a few years ago by Miller and Mitchie (14), Danel and Delhaye (15), Abuaf et al. (16) for measuring gas hold-ups in two phase pipe flow. Recently Ishida and Tanaka (17) proposed the use of one of these probes in three-phase fluidized beds. However, special refinements of the probe were developed (de Lasa et al. (1), and Lee et al. (2)) in order to operate the U-shaped sensor in a three phase fluidized bed using a multiprobe configuration.

Experimental Apparatus

The experiments of the present study were carried out in a cylindrical bed of 80 cm diameter and 334 cm in height. The fluidized bed was instrumented with a system allowing the pressure at different bed levels to be recorded and windows allowing the U-shaped fiber optics to be positioned at different axial and radial positions.

The optical principle used in the design of these probes is the principle of total reflection. If this principle is applied to a fiber having a U-shape the fiber has to be bent as follows: the radius of curvature of the U has to be large enough so that the angle of incidence at the turning point is bigger than the angle of total reflection when the same fiber is exposed to air. If this condition is met, the source light is conserved in the fiber. However, the U has to be curved, as well, in such a way that the radius is small enough to secure an angle of incidence at the turning point smaller than the angle of total reflection for the same fiber dipped in water. Then, with this proposed U probe design the source light is totally or partially lost in water, depending on the specific probe dimensions, and conserved when exposed to air. Therefore, a significant difference can be seen at the detecting end of the fiber each time that a gas bubble contacts the U probe.

A single core silica optical fiber of 400μm diameter was bent in a U shape, according to the principles described above. The radius selected for the U was approximately 0.5 mm at the tip of the probe. The fiber was placed in a stainless steel tube so that only the top of the U was exposed to the fluidized bed.

The laser beam was focused on one of the ends of the U-shaped fibers and was guided through the U-shaped tip back to the other end. The perturbation of the returning light, generated when gas bubbles contacted the probes, was detected by a photodetector. The analog signal generated in each one of the photodetectors was digitized by the data acquisition system.

For the experiments addressed to define bubble characteristics such as bubble shape and bubble included angle, five U-shaped fiber optics of the kind described above were used. Two of them (Probes 1 and 2) were located with a 0.95 cm vertical separation. The other three (Probes 3,4 and 5) were positioned at the same level as Probe 2 which was located at the center, spaced 1.09 cm from the central probe with a 120° circumferential separation. For the runs directed to define bubble frequencies at different axial and radial positions only Probes 1 and 2 spaced 0.95 cm in the vertical direction were employed.

EXPERIMENTAL RESULTS

Bubble Velocity

The relative bubble velocity V_b was correlated against ℓ, the axial bubble length. The axial bubble length was defined as follows: - for bubbles measured by the five probes ℓ was estimated using a method similar to the one described by de Lasa et al (1), - for bubbles detected by the central probes and at least one lateral probe ℓ was defined with the longest axial bubble length.

Five different superficial liquid velocities, ranging from 0.389 to 1.945 cm/s were used while the superficial gas velocities considered were 1.219 and 1.896 cm/s. Figures 1 and 2 show how the various experimental conditions essentially overlap each other when a power law, $V_b = \alpha\ \ell^{\beta}$, is applied in the data analysis. For both superficial gas velocities it was observed that the data could be divided into two groups according to the axial bubble length. The velocities of the smaller bubbles were found to be less sensitive to the axial bubble length.

Details of the regression analysis performed for Vg = 1.214 cm/s and Vg = 1.896 cm/s are presented in Table 1.

For Vg = 1.214 cm/s, the following two correlations are proposed based on a sample of 228 bubbles:

$$V_b = (43.3 \pm 1.82)\ \ell^{0.342 \pm 0.05},\quad \ell \leq 1.56\ \text{cm} \qquad (1)$$
$$V_b = (36.44 \pm 7.4)\ \ell^{0.732 \pm 0.2},\quad \ell \geq 1.56\ \text{cm} \qquad (2)$$

- For Vg = 1.896 cm/s the following two correlations are considered based on a sample of 244 bubbles.

$$V_b = (40.59 \pm 3.37)\ \ell^{0.293 \pm 0.092},\quad \ell \leq 1.01\ \text{cm} \qquad (3)$$
$$V_b = (40.48 \pm 3.24)\ \ell^{0.598 \pm 0.096},\quad \ell \geq 1.01\ \text{cm} \qquad (4)$$

It is important to mention that Kim et al. (4) and Rigby et al. (7) found much higher β values in three-phase fluidized beds than the 2/3 predicted using the Davies and Taylor relationship for large isolated bubbles in an inviscid liquid. The reason of higher β values was attributed to the differences between a continuously bubbling three phase fluidized bed and a liquid fluidized bed with a single injected bubble. It was argued in that sense that bubbles in bubbling beds could rise faster than in single bubble systems. However, considering recent experimental results (de Lasa et al.(1), Lee et al.(2)) and the data of the present study, the experimental evidence seems to suggest that this effect is much smaller than originally thought. The velocity of bubbles with a characteristic length bigger than 1 cm - 1.5 cm could be correlated with a β parameter of 0.598 and 0.732 for Vg = 1.896 cm/s and Vg = 1.214 cm/s respectively. For bubbles smaller than 1 cm - 1.5 cm, β could be even smaller ranging between 0.293 for Vg = 1.896 cm/s and 0.342 for Vg = 1.214 cm/s. This behaviour of the smaller bubbles, that is bubble velocity less sensitive to the bubble length, reveals a certain similarity with observations carried out in gas-liquid flow (Clift et al. (18)).

Distribution and Mean Axial Bubble Length

The system constituted by several U-shaped probes was employed for measuring axial bubble length distribution and average axial bubble length inside the fluidized bed at 53.3 cm above the gas injector. It was observed that for Vg = 1.214 cm/s the axial bubble length distribution changed from normal type (V_ℓ = 0.389 cm/s) to the skewed type (V_ℓ = 0.778 - 1.94 cm/s)(Lee et al. (2)). Concerning the average axial bubble length it is shown in the present study (Figure 3) that an increase of liquid velocity has a tendency to significantly decrease the average axial bubble length. This is true for the two superficial gas velocities considered (Vg = 1.214 cm/s and Vg = 1.819 cm/s). Bubble size in a three phase fluidized bed is the result of the competition between bubble coalescence and splitting and the rate of bubble coalescence is related to liquid viscosity. Therefore, smaller bubbles could be expected at a higher liquid velocity as a result of a lower apparent viscosity in the fluidized bed. At the same time an increase of the superficial gas velocity produces, as could be predicted, a shift of the plot axial bubble length (ℓ) - liquid velocity (V_ℓ) towards the higher bubble sizes.

Bubble Shape

An important capability of the U-shaped multiprobe system is related to the definition of bubble shapes. The shapes of the bubbles are defined by analyzing the signals coming from the upper probes. The same type of technique was applied by us on both two dimensional three-phase fluidized beds (de Lasa et al. (1)) and cylindrical three-phase fluidized beds (Lee et al. (2)). For instance, for three dimensional bubbles, Lee et al. (2) observed included angles between 100° to 180° (slug type bubbles). Higher superficial liquid velocities and/or higher superficial gas velocities have a certain tendency to increase the bubble included angle. Moreover the coexistence of various shapes of bubble in the bed was also observed. Slug and disk-like forms were detected besides the expected spherical cap.

Bubble Frequency

A preliminary test was performed in this study to determine the minimum time required to ensure that the data taken were obtained under stationary conditions in terms of statistical analysis. This analysis was performed with two different sets of data taken with the same sampling time. The mean frequency, the root mean square and the standard deviation of each set of data were calculated and compared. It was shown that for the worst condition the mean frequency, the root mean square and the standard deviation vary only 0.08%, 0.1% and 3% respectively for a total sampling time of 90s. Then considering these results and for the frequency analysis reported in this study a total sampling time of 110 seconds was selected. It has to be mentioned that the analysis described above involves the determination of the baseline. A section of the signal without bubble peaks was considered and its mean value and standard deviation were calculated. The criterion for bubble counting was set to be ten times the standard deviation of the baseline. The recorded signal was divided into 55 sections with two seconds in each one of them. Bubble frequency in each section was then calculated. The frequencies obtained were classified as discrete frequency segments with a resolution of 0.5 s^{-1}. For example, N_f is the number of sections with average frequency f. Then the total number of frequency samples, N can be expressed as

$$N = \sum_{f=0}^{\infty} N_f = 55; \; f = 0, 0.5, 1, 1.5 \qquad (5)$$

The probability density function P(f) was then calculated

$$P(f) = \frac{N_f}{N} \; ; \; f = 0, 0.5, 1, 1.5 \qquad (6)$$

and the mean frequency was obtained from the following equation

$$f_{av} = \sum_{f=0}^{\infty} f \cdot P(f); \; f = 0, 0.5, 1, 1.5 \qquad (7)$$

The distribution of gas bubbles in the fluidized bed was investigated by measuring average bubble frequency at various locations in the bed. Only two probes, separated by 0.94 cm. in the axial direction, were used in these measurements to minimize the disturbance generated by the probes. These two probes were inserted into the fluidized bed from oppposite sides of the column and measurements were done at four different radial positions, r/R = 0, 0.25, 0.50 and 0.75. The experiments were carried out at several axial levels above the distributor: 7.5 cm, 22.5 cm, 37.5 cm and 52.5 cm. Some results of these experiments are shown in Figures 4, 5 and 6. Since gas was introduced through gas injectors, gas bubbles were concentrated around the injectors. A considerable higher frequency was observed in these zones. Very few bubbles were recorded in other areas at the grid level (z = 7.5 cm). However, the gas bubbles dispersed and redistributed quite rapidly across the bed and bubble frequencies were found rather uniformly distributed radially at higher axial positions (z = 37.5 cm). An interesting phenomenon is that even at the higher axial position, the bubble frequency decreases close to the wall due to a wall effect. It was also observed, analyzing f_{av} for different operating conditions, that the distribution of gas bubbles was influenced by V_g and V_ℓ and it was mainly controlled, for the cases considered in this study, by the superficial gas velocity. Another interesting phenomenon concerns the average bubble frequency at the very top section of the fluidized bed. The f_{av} at z = 52.5 cm increased noticeably. This can be explained by the fact that the solid porosity in the top section of the bed, close to the bed surface, is lower than in the rest of the fluidized bed. Consequently, the apparent viscosity at the top of the bed is reduced. This results in an increased tendency in bubble breakage, therefore higher bubble frequency can be expected. It should be noted that the f_{av} measured at z = 52.5 cm for the case shown in Figure 6 was taken in the two phase region just above the three-phase fluidized bed. The increased f_{av} is consistent with the other two cases reported in Figures 4 and 5 where the measurements were done inside the fluidized bed.

NOTATION

f	bubble frequency (1/s)
ℓ	axial bubble length (cm)
N	total number of frequency samples
N_f	number of sections with average frequency, f.
P(f)	probability density function
V_b	relative velocity of the bubble (cm/s).
Vg	superficial gas velocity (cm/s)
V_ℓ	superficial liquid velocity (cm/s)
z	axial position (cm)
α	coefficient of the bubble velocity correlation, Table 1
β	coefficient of the bubble velocity correlation, Table 1
av	average

LITERATURE CITED

1. de Lasa, H., S.L.P. Lee and M.A. Bergougnou, Can. J. Chem. Engng. 62, 165 (1984).
2. Lee, S.L,]., H.I. de Lasa and M.A. Bergougnou, AIChE Symp. Series, 241 (80), 110 (1984).
3. Henrikson, H.K. and K. Ostergaard, The Chem. Eng. J., 7, 141 (1974).

4. Kim, S.D., C.G.J. Baker and M.A. Bergougnou, Chem. Eng. Sci., 32, 1299 (1977).
5. Stewart, P.S.B. and J.F. Davidson, Chem. Eng. Sci., 19, 319 (1964).
6. Park, W.H., W.K. Kang, C.E. Capes and G.L. Osberg, Chem. Eng. Sci., 851 (1969).
7. Rigby, G.R., G.P. van Blockland, W.H. Park and C.E. Capes, Chem. Eng. Sci., 25, 1729 (1970).
8. Begovich, J.M. and J.S. Watson, AIChE Journal, 24, 351 (1978).
9. Λinneweber, K.W. and E. Blass, Proceedings Second World Congress of Chemical Engineering, Montreal, 3, 50-53 (1981).
10. Linneweber, K.W. and E. Blass, Ger. Chem. Eng., 6, 28 (1983).
11. Ishida, M., Nishiwaki and T. Shirai, in J.R. Grace and J. Matsen (Ed.) Fluidization, Pergamon Press (1980).
12. Ishida, M. and H. Hatano, Proceedings Fourth International Fluidization Conference, Japan, 1, 1-8-1 (1983).
13. Oki, K., M. Ishida and T. Shirai, in Fluidization, J.R Grace and J. Matsen (Ed.) Pergamon Press (1980).
14. Miller, M. and R.E. Mitchie, ASME, 82, (1969).
15. Danel, F. and J.M. Delhaye, Measures-Regulation-Automatisme, Aout-Sept. 99 (1971).
16. Abuaf, N., O.C. Jones and G.A. Zimmer, Rev. Sci. Instrum. 49 (8) 1090 (1978).
17. Ishida, M. and H. Tanaka, J. of Chem. Engng. Japan 15(2)389 (1982).
18. Clift, R., J.R. Grace and M.E. Weber, Bubbles, Drops and Particles, Academic Press, New York (1978).

Table 1

Regression Analysis of Bubble Rising Velocity Against Axial Bubble Length

$$V_b = \alpha\, \ell^{\beta}$$

V_ℓ(cm/s)	0.389 - 1.945	0.389 - 1.945
Vg(cm/s)	1.214	1.896
ℓ(cm_	$\leq$ 1.56 ($\geq$ 1.56)	$\leq$ 1.01 ($\geq$ 1.01)
Number of Data Points	173 (55)	97 (140)
α, 95% Confidence	43.3 ± 1.82 (36.44 ± 7.4)	40.59 ± 3.37 (40.48 ± 3.24)
β, 95% Confidence	0.342 ± 0.059) (0.732 ± 0.21)	0.293 ± 0.092 (0.598 ≠ 0.096)
Sample Correlation Coefficient	0.6615 (0.6853)	0.5478 (0.7231)
Standard Deviation cm/s	9.03 (16.96)	8.02 (15.67)
Overall Standard Deviation, cm/s for all data points	11.36	13.04

NOTE: Number in between parentheses correspond to the conditions $\ell \geq 1.56$ for Vg = 1.214 cm/s and $\ell \geq 1.01$ for Vg = 1.896 cm/s

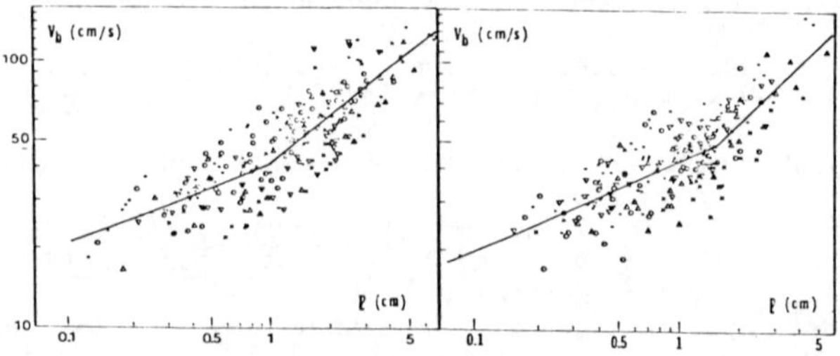

Fig. 1 V_g = 1.214 cm/s Fig. 2 V_g = 1.896 cm/s
(×) V_ℓ=0.389 cm/s, (△) V_ℓ=0.778 cm/s (▽) V_ℓ=1.167 cm/s
(○) V_ℓ=1.556 cm/s (●) V_ℓ=1.945 cm/s

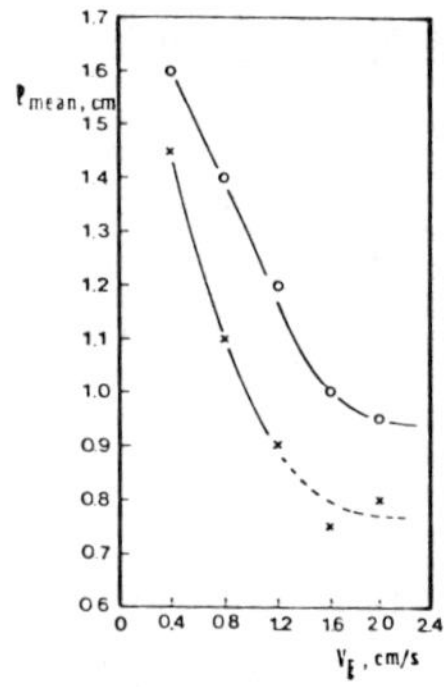

Fig. 3 Effect of V_ℓ and V_g on Mean Axial Bubble Length
(×)V_g=1.214cm/s,(○)V_g=1.896cm/s

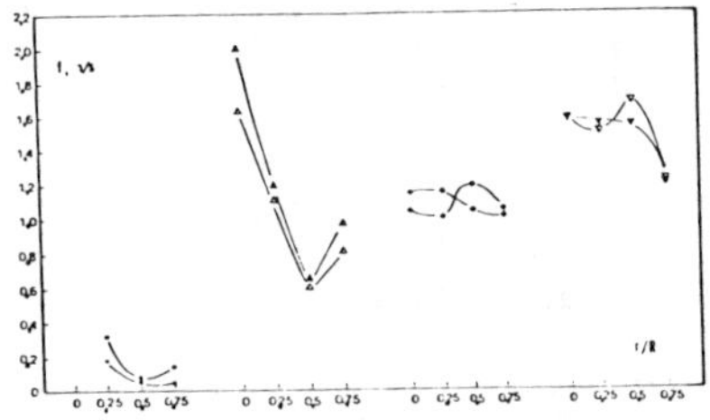

Fig. 4 Bubble Frequency in a TPFB
V_g=1.896 cm/s, V_ℓ=0.389 cm/s

● z=6.55cm, ○ z=7.5cm, ▲ z=21.55cm
△ z=22.5cm, ◆ z=36.55cm ◇ z=37.55cm
▼ z=51.5cm, ▽ z=52.5cm
Note: z=6.55cm, r/R=0 f=5.82
z=7.5cm, r/R=0 f=4.45

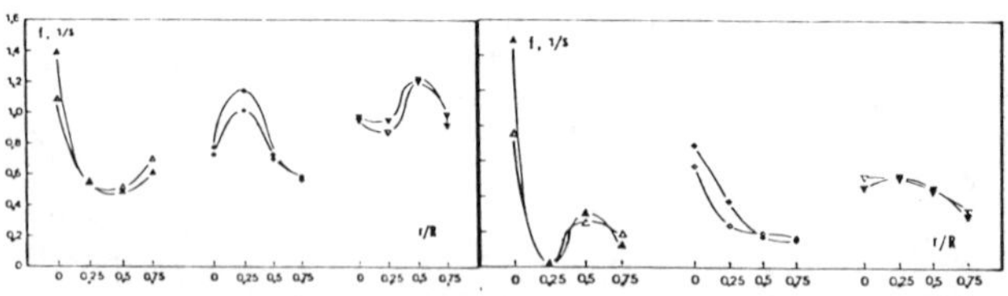

Fig. 5 Bubble Frequency in a TPFB
V_g=1.214cm/s; V_g=0.389cm/s

Fig. 6. Bubble Frequency in a TPFB
V_g=0.574cm/s; V_g=0.195cm/s

OVERALL GAS-LIQUID MASS TRANSFER COEFFICIENTS IN A DRAFT TUBE GAS-LIQUID-SOLID FLUIDIZED BED

Koichi Fujie and Liang-Shih Fan
Department of Chemical Engineering
The Ohio State University
Columbus, Ohio 43210 USA

ABSTRACT

Experiments were performed to study the gas-liquid mass transfer behavior in a draft tube gas-liquid-solid fluidized bed (DTFB) with zero liquid flow rate. A porous distributor plate was used as a fine bubble diffuser. The sodium sulfite oxidation method coupled with an off-gas concentration analysis was used to determine the overall gas-liquid mass transfer coefficient, in the DTFB. Six different particles were used including glass beads of 300 μm and 600 μm diameter, spherical activated carbon particles of 307 μm, 714 μm and 1500 μm diameter and bioflocs. The biofloc consists of a 307 μm activated carbon particle surrounded by a 50 μm immobilized biofilm. Distinct differences in gas-liquid mass transfer behavior between the DTFB with fine bubble diffuser and coarse bubble generator were observed.

SCOPE

The draft tube gas-liquid-solid fluidized bed (DTFB) involves the use of a draft tube coaxially situated inside the bed. The draft tube serves to provide bulk circulation of gas, liquid and solids between the draft tube and annulus and thereby achieve intimate contact among gas, liquid and solid phases in the bed. Applications of the DTFB have been available in the literature mainly in the area of biological wastewater treatment (e.g., 1,2,3,4). Fundamental work regarding the DTFB was begun, however, only recently (3,5,6,7,8,9).

With respect to the gas-liquid mass transfer, utilizing a perforated plate as gas injector, Fan et al. (3) reported that for spherical activated carbon particles of 714 μm, considerably higher K_La exists in the DTFB compared to the conventional cocurrent upward three phase fluidized bed (CTFB), draft tube gas-liquid bubble column (DTBC), inverse three phase fluidized bed (ITFB) and three phase spouted bed (TSB). Muroyama et al., (9) reported that for a DTFB with a single nozzle gas injector, K_La increases linearly on a log-log scale with superficial gas velocity, U_{g0}.

In this study, K_La in a DTFB operated under conditions of interest to biological processing was examined. A porous distributor plate was used as gas diffuser in the DTFB. Comparisons were made of K_La between the bio-active and bio-inactive systems. A comprehensive correlation was proposed for K_La in the DTFB.

CONCLUSIONS AND SIGNIFICANCE

In order to accurately evaluate K_La as a pure physical gas-liquid mass transfer coefficient, the inlet sodium sulfite concentration was limited to 1300 mg/ℓ. This was necessary due to the electrolyte effect of sodium sulfite which results in bubble size reduction. It was found that 2 to 4 vol.% loading of AC-700, GB-300 and GB-600 particles in the DTFB yields a higher K_La than that in the DTBC. Any loading of AC-300 particles in a DTFB, however, yields a lower K_La than in the DTBC. While K_La in the DTFB with fine bubble diffuser is higher than that in the CTFB, TSB, ITFB and DTFB with coarse bubble generator, it can be as low as about 1/2 of that in the conventional bubble column (CBC) with a fine bubble diffuser.

An empirical correlation expressed in terms of the mass transfer coefficient in a DTBC, particle Reynolds number, and solids loading was proposed for K_La in a DTFB. K_La obtained in the bio-active DTFB treating wastewater containing phenol was found to be slightly lower (within 23%) than that predicted by the proposed correlation, established based on the bio-inactive DTFB.

EXPERIMENTAL PROCEDURE AND DETERMINATION OF K_La

A schematic diagram of the experimental apparatus is shown in Fig. 1. The apparatus is made of plexiglas and is of 7.6 cm ID and 144 cm in height. A draft tube of 5.0 cm ID, 120 cm in height and 0.32 cm in wall thickness is coaxially situated inside the column. The effective liquid depth is 132 cm. A plastic porous plate made of polymethylmethacrylate (PMMA) with a nominal pore size of 30 μm is used as an air distributor and is located at the bottom of the conical section, which has an inclined angle of 45°. The cross sectional area of the distributor is 11.4 cm^2. Six types of particles including bioflocs were used. Their physical properties are given in Table 1. The bioflocs used were cultivated in the DTFB bioreactor using phenol as a substrate. The bioflocs were then placed in a pure water medium in a DTFB for the mass transfer experiments. Details of cultivation of bioflocs are given in Fan et al. (3). The biofloc consists of an AC-300 particle and an immobilized biofilm of 50 μm on the surface of the particle.

A 1300 mg/ℓ aqueous solution of sodium sulfite was introduced at the bottom of the column at a feed rate of 20 cm^3/sec. This low sodium sulfite concentration was experimentally verified in this study to be required to ensure negligible electrolytic effects on bubble size reduction and chemical absorption effects on mass transfer enhancement. No catalytic additives such as cobaltous ion were employed to enhance sodium sulfite oxidation. The superficial velocity of sodium sulfite solution in the column was a low 0.44 cm/sec, allowing its effect on the hydrodynamic behavior in the column to be considered negligible.

The dissolved oxygen (DO) concentration in the liquid was measured at the top of the draft tube by a polarographic oxygen probe. Measurement was also made of the DO concentration in the longitudinal direction of the draft tube and annulus. No significant longitudinal variation of the DO concentration was evident. Oxygen mole fractions in the sampling gas at the outlet of the column were continuously measured by a paramagnetic oxygen gas analyzer after removal of water mist in the gas by a silica gel column. The sampling gas velocity to the analyzer was maintained constant at 5 cm^3/sec. The temperature of water and air was controlled within 22 ±2°C.

K_La was determined by means of the off-gas (oxygen) analysis coupled with the liquid phase reaction involving sodium sulfite oxidation. Consider the state of complete mixing for the liquid phase and two extreme states of mixing for the gas phase, i.e., plug flow and complete mixing. For the plug flow pattern, the mass balance of oxygen is given by

$$G_0 \frac{d\eta}{d\ell} = K_La(C_0Z-C)S_a \qquad (1)$$

with the boundary condition:

$$\ell = 0 \; ; \; \eta = 0 \qquad (2)$$

C_0 in Eqn. (1) is given by

$$C_0 = C_0^* \{1 + \alpha(H - \ell)\} \qquad (3)$$

where $$\alpha = \rho_a/(10.33P^*) \qquad (4)$$

Z can be related to η by

$$Z = Z_0 \; (1 - \eta)/(1 - Z_0\eta) \qquad (5)$$

From Eqn. (5) we have

$$\eta = (Z_0 - Z)/Z_0(1-Z) \qquad (6)$$

Also, G_0 relates to the superficial gas velocity U_{g0}, by

$$G_0 = \frac{U_{g0}S_aZ_0}{v_m} \qquad (7)$$

To simplify the resulting first order non-linear differential equation for Eqn. (1), Eqn. (3) is approximated by

$$C_0 = C_0^* \; (1 + \alpha H/2) = \bar{C}_0 \qquad (8)$$

Here, ℓ in Eqn. (3) is replaced by H/2 which is the midpoint of the column. Equation (1) can then be integrated to give

$$K_La = \frac{U_{g0}[Z_0\eta_e(\bar{C}_0-C)-\bar{C}_0(1-Z_0)\ell n\{1-\eta_e[\bar{C}_0-C]/[\bar{C}_0-C/Z_0]\}]}{[\bar{C}_0-C]^2 v_m H} \quad (9)$$

For the complete mixing pattern, the mass balance of oxygen in the gas phase is given by

$$U_{g0}\eta_e = K_La^{\#}(\bar{C}_0 Z_e - C)\ Hv_m/Z_0 \quad (10)$$

With $\bar{C}_0$, given by Eqn. (8), $K_La^{\#}$ in Eqn. (10) can be expressed by

$$K_La^{\#} = \frac{U_{g0}\eta_e Z_0}{v_m H[C_0^*(1+\alpha H/2)Z_e - C]} \quad (11)$$

The mass transfer coefficient for the two gas mixing patterns, can be calculated from Eqn. (9) and (11) based on measured values of U_{g0}, C and Z_e. The calculated values of K_La are compared in Fig. 2 for three systems including a DTBC with a porous plate, DTBC with a perforated plate and DTFB with a porous plate and 6 vol.% of GB-600. It is seen that K_La calculated based on the completely mixed flow pattern is slightly larger than that based on the plug flow pattern. The difference between them, however, is not substantial. Based on visual observation in the operation of a DTFB, only a small fraction of gas bubbles are entrained into the annulus from the draft tube. Thus, in the present calculation of K_La, results based on the plug flow pattern for the gas phase are reported.

RESULTS AND DISCUSSION

Effect of Gas Velocity on $(K_La)_0$ for a DTBC. Figure 3 shows the variation of the mass transfer coefficient in a DTBC, $(K_La)_0$, with U_{g0}. Four flow regimes classified in Table 2 are noted in the figure. In Flow regimes I and II, $(K_La)_0$ increases sharply with an increase in U_{g0}. The rate of increase, however, decreases in Regimes III and IV. It is noted that the values of $(K_La)_0$ at low gas velocity are significantly higher than those reported by Fan et al. (3), Koide et al. (10) and Muroyama et al. (9) who used either a single nozzle or perforated plate as the gas distributor. At high gas velocities, the effect of the gas distributor on $(K_La)_0$ is reduced. Also shown in Fig. 3 are the mass transfer coefficients for a CBC based on the design given in Fig. 1, but without a draft tube, and CBC reported by Akita and Yoshida (11) using a single nozzle. It is seen that the mass transfer coefficient for the CBC with fine bubble diffuser is the highest for all U_{g0} among all gas-liquid contactors considered.

Effect of Gas Velocity and Solids Loading on K_La and ε_g for a DTFB. In the study of hydrodynamics of a DTFB, three modes of solids flow were identified by Fan et al. (6). They are packed bed mode, fluidized bed mode and circulated bed mode. K_La in the DTFB obtained in this study is restricted to the circulated bed mode operation.

Figures 4 and 5 show the effect of solids loading, L_s, on K_La and gas holdup, ε_g, in the DTFB for AC-700 particles. The mass transfer is expressed in terms of $K_La/(K_La)_0$. It is seen that at U_{g0} ranging from 99 to 270 m/hr, a maximum $K_La/(K_La)_0$ occurs at a solids loading of approximately 2 vol.%. It is noted that the maximum liquid circulation rate between the draft tube and the annulus was reported to occur at a similar solids loading, i.e., 1.5 vol.% (6). Conversely, at a gas velocity of 46 m/hr, $K_La/(K_La)_0$ decreases monotonically with an increase in solids loading. Apparently, changes in solids loading vary the bubble size, which in turn varies the mass transfer coefficient. $(\varepsilon_g)_0$ in Fig. 5 represents the gas holdup in the DTBC. It is seen that $\varepsilon_g/(\varepsilon_g)_0$ in the DTFB is practically invariant with solids loading.

Similar behavior is observed for GB-300 and GB-600 particles for which a maximum $K_La/(K_La)_0$ occurs at about 2 to 4 vol. % loading. For AC-300 particles and bioflocs, however, there is a monotonic decrease in $K_La/(K_La)_0$ with solids loading.

Correlation for K_La. Based on the experimental results obtained in this study, K_La for six types of particles can be satisfactorily correlated by

$$K_La/(K_La)_0 = 1 + F\ (Re_p, L_s) \tag{12}$$

where

$$F(Re_p, L_s) = (0.0029\ Re_p - 16.64\ L_s)(Re_p)^{-0.2521}(1-L_s)^{7.216} \tag{13}$$

A plot of experimental $K_La/(K_La)_0$ values vs. $F(Re_p, L_s)$ based on Eqn. (12) is given in Fig. 6. The correlation was found to fit all the data within an error of 20%. The ranges of operating variables covered in this study are: U_{g0}: 46 m/hr to 270 m/hr; Re_p: 5.88 to 163.12; L_s: 0 to 0.20. Values of $(K_La)_0$ used in Eqn. (12) can be satisfactorily correlated by

$$(K_La)_0 = A\ (U_{g0})^B \tag{14}$$

where A and B are empirical constants which vary with respect to the bubble flow regimes (see Table 2) as given by

A = 1.49, B = 1.12 for $U_{g0} < 65$ m/hr (Regimes I and II)

A = 81.5, B = 0.145 for $65 < U_{g0} < 129$ m/hr (Regime III)

A = 12.6, B = 0.523 for $U_{g0} > 129$ m/hr (Regime IV)

Table 3 shows a comparison of K_La obtained in a bio-active DTFB treating phenolic wastewater with values predicted from Eqn. (12). It is seen that the K_La obtained in a bio-active DTFB is slightly lower (within 23%) than that predicted by Eqn. (12). The discrepancy could be due to the effect of the substrate and microbial metabolites on the bubble properties. Despite this discrepancy, Eqn. (12) represents a reasonably good estimate for K_La in a bio-active DTFB.

It is noted that the values of K_La obtained in the present study are significantly higher compared to those reported by Muroyama et al. (9). Furthermore, in contrast to their study, K_La obtained here does not vary linearly on a log-log scale with U_{g0} over its entire range and the solids loading does have a distinct effect on K_La. Apparently, the gas distributor plays an important role in the determination of bubble behavior in the DTFB and hence the mass transfer behavior.

Acknowledgement

The work was supported bhy the U.S. Geological Survey under Grant 14-08-0001-G1150. The helpful assistance of Mr. T. S. Ramesh is gratefully acknowledged.

Notation

C	dissolved oxygen concentration in the liquid, kg/m^3
C_0	saturated value of C using pure oxygen, kg/m^3
$\overline{C}_0$	average value of C_0, given by Eqn. (8), kg/m^3
d_p	particle diameter, μm
G_0	mass flow rate of oxygen at the column inlet, kg/hr
H	liquid depth, m
K_La	volumetric mass transfer coefficient, hr^{-1}
$(K_La)_0$	K_La in a solids free column, hr^{-1}
L_s	volume fraction of particles in the column, -
ℓ	vertical distance from the air distributor, m
P^*	atmospheric pressure, atm
Re_p	particle Reynolds number, $d_p\rho_pU_{g0}/\mu$, -
S	phenol concentration in artificial wastewater, mg/ℓ
S_a	cross sectional area of the column, m^2
U_{g0}	superficial gas velocity at the column inlet, m/hr
v_m	specific volume of oxygen, m^3/kg
Z	mole fraction of oxygen in the gas phase, -
Z_0	mole fraction of oxygen in the gas phase at the column inlet, -

Greek

α	pressure variation factor, m^{-1}
δ	biofilm thickness on particle, μm

ε_g gas holdup, -
$(\varepsilon_g)_0$ gas holdup in a solids free column, -
η fraction of oxygen utilization, -
μ viscosity of liquid, kg/msec
ρ_a apparent bed density, kg/m^3
ρ_p particle density, kg/m^3

Superscripts
* value at atmospheric pressure

Subscripts
i influent
e effluent
value corresponding to complete mixing of gas phase

LITERATURE CITED

1. Sawada, S., Yosui to Haisui, 20, 1051 (1978).

2. Sawai, T., K. Sugata, and T. Miki, The Mitsubishi Heavy Industry Technical Report, 19, (5), 510 (1982).

3. Fan, L-S, K. Fujie, and T.R. Long, AIChE Symposium Series on Fluidization and Fluid Particle Systems: Recent Advances, No. 241, Vol. 80 (1984b).

4. W-T Tang and L-S Fan, paper presented at the ACS National Meeting, Chicago, Sep 8-14 (1985).

5. Fan, L-S, S.J. Hwang, and A. Matsuura, AIChE Symposium Series on Fluidization and Fluid Particle Systems, No. 234, Vol. 80 (1984a).

6. Fan, L-S, S.J. Hwang, and A. Matsuura, Chem. Eng. Sci., 39 (12) 1677 (1984c).

7. Hwang, S.J. and L-S Fan, paper presented at the 16th Annual Meeting of the Fine Particle Society, Miami Beach, April 22-26 (1985a).

8. Hwang, S.J. and L-S Fan, Chem. Eng. J. (in press, 1985b).

9. Muroyama, K., Y. Mitani and A. Yasunishi, Chem. Eng. Commun., 34, 87 (1985).

10. Koide, K., K. Kurematsu, S. Iwamoto, Y. Iwata and K. Horibe, J. Chem. Eng. Japan, 16, 413 (1983).

11. Akita, K. and F. Yoshida, Ind. Eng. Chem. Proc. Des. Dev., 12, 76 (1973).

12. Fan, L-S, K. Fujie, and T.R. Long, AIChE J., (in review, 1985).

Table 1. Physical properties of solid particles employed in the experiments.

Symbol	Material	Diameter (μm)	Wet Density (kg/m^3)	Terminal Velocity in Water (m/sec)
AC-300*	spherical	307	1.50×10^3	2.3×10^{-2}
AC-700*	activated	714	1.50×10^3	5.3×10^{-2}
AC-1500**	carbon particle	1500	1.45×10^3	10.48×10^{-2}
GB-300		300	2.5×10^3	4.67×10^{-2}
GB-600	glass beads	600	2.5×10^3	9.34×10^{-2}
Biofloc	activated carbon particle (AC-300) with immobilized living microbes	407	$1,22\times10^3$	1.73×10^{-2}

* Kureha Beads BAC®
** Takeda Pharmaceutical Co.

Table 2. Classification of bubble flow regimes in the DTFB

Regime	U_{g0} (m/hr)	Bubble Flow Regime in the Draft Tube	Bubble Flow Regime in the Annulus
I	<30	Dispersed bubble flow	No bubbles
II	30 to 70		Dispersed bubble, partial penetration
III	70 to 120	Transition from dispersed bubble flow to coalesced bubble flow	Dispersed bubble, full penetration
IV	>120	Coalesced bubble flow	

Table 3. Comparison of K_La obtained in the DTFB bioreactor treating phenolic wastewater with K_La predicted by Eqn. (12).

No.	U_{g0} (m/hr)	L_s (vol.%)	S_1 (mg/ℓ)	δ (μm)	C (mg/ℓ)	(K_La) bioreactor* (hr^{-1})	(K_La) Eqn. (12) (hr^{-1})
1	32.9	14.0	118	17.5	0.5	54.5	63.7
2	39.5	14.2	128	24.6	1.0	63.3	78.4
3	39.5	14.4	132	35.4	0.5	63.7	78.5

* (12)

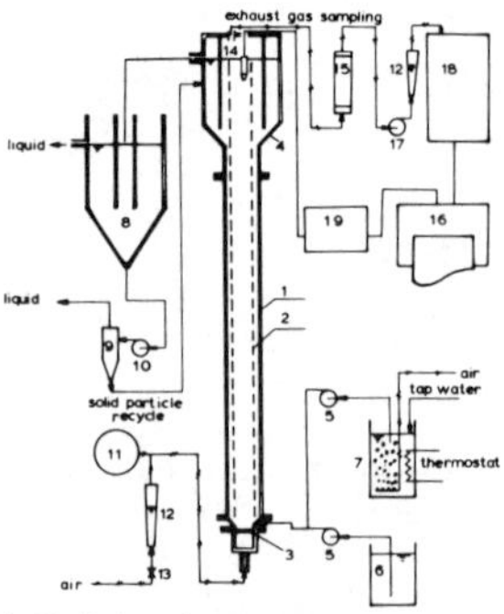

Fig. 1. Schematic diagram of experimental apparatus. 1. Fluidized bed; 2. Draft tube for DTFB (detachable); 3. Porous plate air diffuser; 4. Solid-liquid disengagement section; 5. Microfeeder; 6 Na_2SO_3 solution; 7. Head tank; 8. Settling tank; 9. Liquid cyclone; 10. Positive displacement pump; 11. Pressure gauge; 12. Air flow meter; 13. Air control valve; 14. DO probe; 15. Silica gel column; 16. Strip chart; 17. Air pump; 18. Oxygen gas analyzer; 19. DO meter (Thermometer).

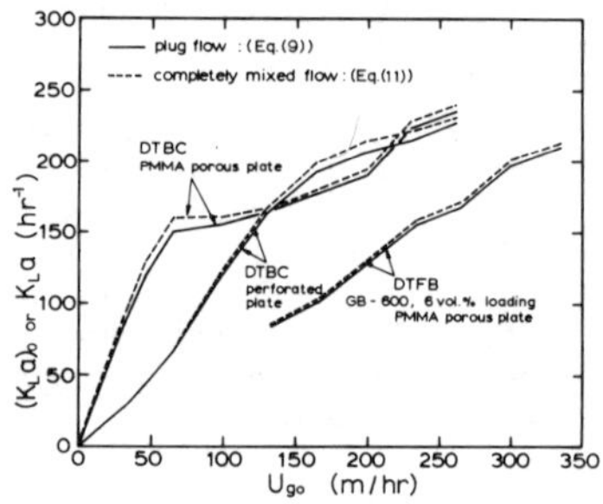

Fig. 2. Comparisons of $(K_La)_0$ or (K_La) obtained based on the plug flow and completely mixed flow patterns for the gas phase in the DTBC and DTFB.

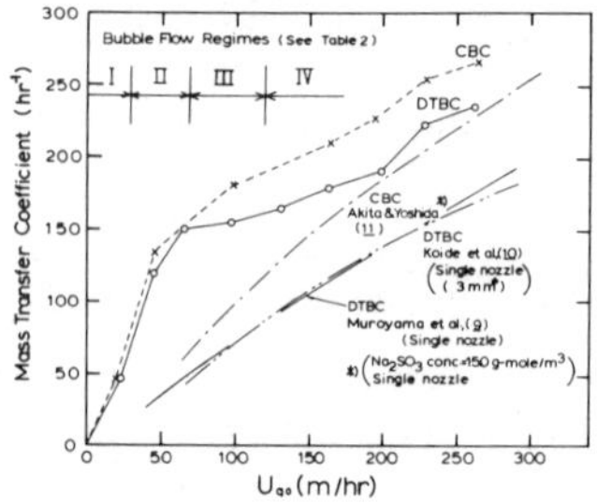

Fig. 3. Variation of mass transfer coefficient with U_{g0} for a CBC and a DTBC.

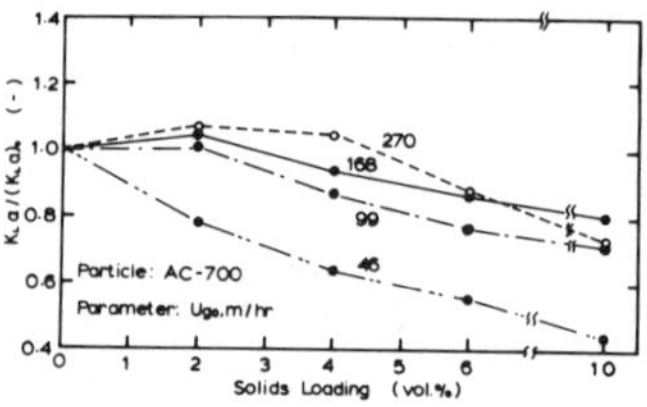

Fig. 4. Variation of $K_La/(K_La)_0$ with solids loading in a DTFB for AC-700 particles.

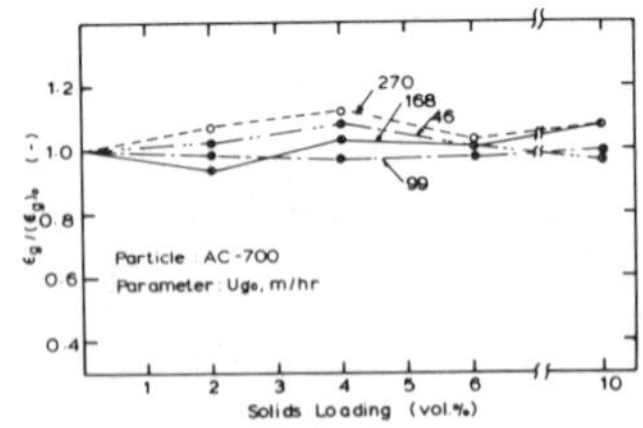

Fig. 5. Variation of $\varepsilon_g/(\varepsilon_g)_0$ with solids loading in a DTFB for AC-700 particles.

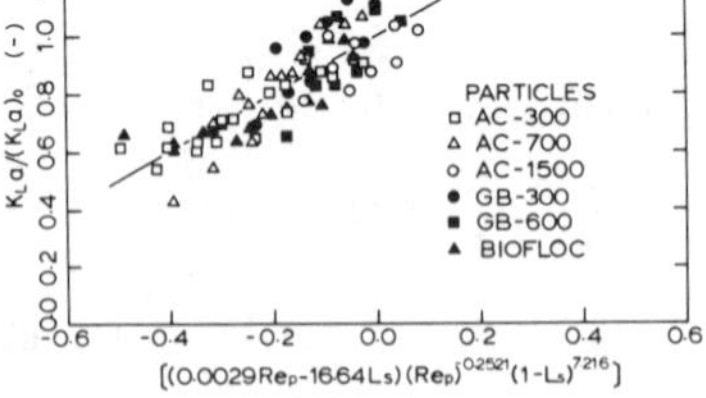

Fig. 6. Correlation for K_La for six types of particles considered in this study.

AXIAL MIXING AND INTERNAL RECIRCULATION FLOW IN THREE-PHASE FLUIDIZED BEDS

S. Morooka, M. Koyama and Y. Kato
Department of Applied Chemistry, Kyushu University
Fukuoka 812, Japan

ABSTRACT

Phase holdups and longitudinal dispersion coefficients of liquid in dense three-phase fluidized beds were measured and correlated by empirical equations. A wake model considering the radial distribution of gas holdup was effective for the explanation of the phase holdups and the internally recirculating liquid flow, but the longitudinal dispersion of liquid was further affected by the lateral exchange of liquid.

SCOPE

Mean phase holdups are important features of three-phase fluidization and are estimated by wake models (Østergaard, 15, Chern et al., 2, etc.). However, previous wake models assume that gas bubbles ascend uniformly at any radial position, and the solid fraction ratio in the wake region is sometimes speculated as zero or unity. Thus, the background of the wake model is still in need of hydrodynamic verification.

In gas-liquid bubble columns and gas-solid fluid beds, the radial distribution of gas holdup induces an internally recirculating flow of liquid or solid particles (Miyauchi et al., 11). A similar recirculation flow has been observed in three-phase fluidized beds (Morooka et al., 12). In this paper, radial distribution of gas holdup and liquid velocity in the three-phase fluidized beds are measured. A wake model with radially distributed parameters is newly developed to explain the experimental results. The longitudinal dispersion of liquid is also measured, and is discussed on the basis of the internal recirculation flow.

CONCLUSIONS AND SIGNIFICANCE

In dense three-phase fluidized beds, mean phase holdups and longitudinal dispersion coefficients of liquid were measured and correlated as functions of operating conditions and properties of solid particles. The value of local gas holdup was proportional to the square of the distance from the wall. The liquid flow pattern showed an up-flow in the central region and a down-flow in the peripheral region with respect to the interstitial mean liquid velocity.

The new wake model was successful to explain the radial distribution of gas holdup and the internal recirculation flow of liquid. The parameters of the model, k and x, were given by the empirical equations. However, the longitudinal dispersion coefficient of liquid was not in agreement with the proposed model, which did not include the exchange of liquid between the wake and the bulk liquid-solid region. More research is needed for the full description of flow properties in the three-phase fluidized bed.

EXPERIMENTAL APPARATUS AND PROCEDURE

The main columns were made of transparent acrylic resin pipes, which were 12 and 19 cm in diameter and ca. 250 cm in height. The fluidizing media were air and water at ambient conditions. Solid particles were glass spheres and porous alumina particles. The properties of the particles are listed in Table 1. The mean solid holdup was obtained from the bed height which was determined from the turning point of the axial distribution of static pressure. The mean gas holdup in the dense three-phase region was measured by the trap method (Kato et al., 6) with a horizontal shutter plate located at the upper surface of the dense region. The local gas holdup was measured with a fine electric conductivity probe.

The local liquid velocity was measured by using an electric conductivity probe illustrated in Figure 1 (Morooka et al., 12). The probe consisted of four needle electrodes and a tracer injection nozzle. When the tracer (an aqueous KCl solution) flowed near the sensing tip, the electric current between the tip and the counter electrode with a wide area varied. The signals of this variation were recorded for 30-50 s. The local liquid velocity was calculated from time differences between corresponding signals of adjacent electrodes.

The longitudinal dispersion coefficient of liquid in the dense three-phase region was measured by the impulse response method. The tracer was a saturated aqueous KCl solution and was injected instantaneously just above the distributor. The concentration of the tracer was measured by potentiometry (Kato et al., 7) near the upper surface of the dense region. The temperature was kept at 20-35°C during the experiment.

EXPERIMENTAL RESULTS

The mean gas holdup in the dense three-phase region was well correlated by the following equation (Kato et al., 6).

$$\bar{\varepsilon}_g = 0.3W^{1.3}/\{(1+1.1W^{1.5})(1+0.01Re_t^{0.62})\}$$

and

$$W = (gD_T^2\rho_l/\sigma)^{0.198}(gD_T^3\rho_l^2/\eta_l^2)^{0.035}(U_g/\sqrt{gD_T}) \tag{1}$$

Figure 2 shows the radial distribution of the local gas holdup. In the

Table 1. Properties of solid particles.

Particle	ρ_s/Mg·m^{-3}	d_p/mm	v_t/cm·s^{-1}
Glass	2.5	0.52	7.8
		1.2	17.6
		2.2	29.0
		5.2	50.0
Porous alumina	1.8	1.5	14.5
	1.7	3.2	24.4

dense three-phase region, ε_g is expressed as

$$\varepsilon_g = 2\bar{\varepsilon}_g(1-\phi^2) \tag{2}$$

The mean liquid holdup agrees with the equation of Kato et al. (8).

$$\left.\begin{aligned} &\bar{\varepsilon}_l = \{1-9.7(350+Re_t^{1.1})^{-0.5}K^{0.092}\}(U_l/v_t)^{1/n'} \\ \text{and}\quad &\{5.1(1+16.9K^{0.285})-n'\}/(n'-2.7) = 0.1(1+4.43K^{0.165})Re_t^{0.9} \end{aligned}\right\} \tag{3}$$

Figure 3 shows the radial distribution of time-averaged local liquid velocity. In the three-phase fluidized beds, an internal recirculation flow of liquid is induced. Figure 4 shows the liquid velocity along the

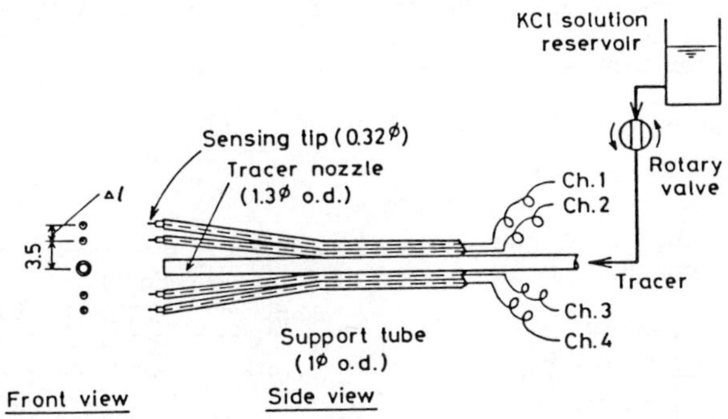

Figure 1. Probe for measurement of local liquid velocity. Δl = 2-3 mm.

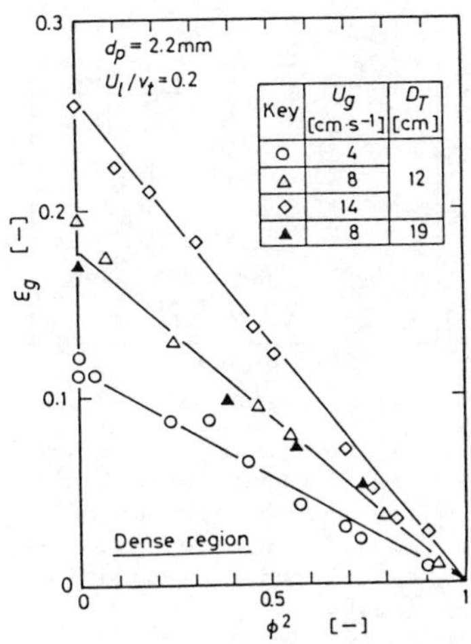

Figure 2. Radial distribution of gas holdup.

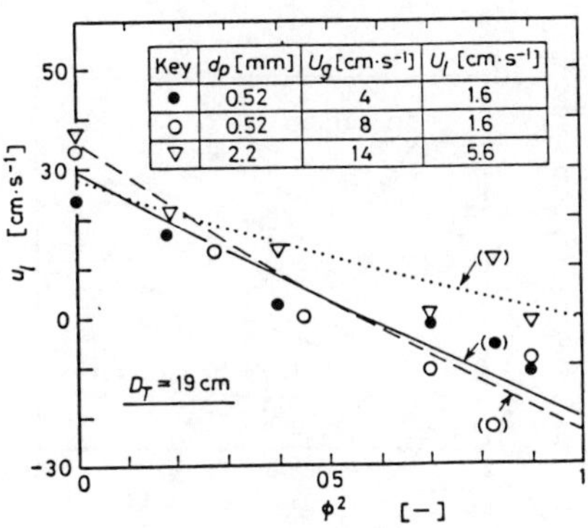

Figure 3. Radial distribution of liquid velocity.

central axis. The ordinate is the liquid velocity relative to the mean liquid velocity. The value of $(u_{l0}-U_l/\varepsilon_l)$ increases with increasing bed diameter and superficial gas velocity and decreasing particle diameter.

Figure 5 shows the longitudinal dispersion coefficient of liquid. The experimental data including those of Michelsen and Østergaard (10), Muroyama et al. (13) and Kim and Kim (9) are expressed by the following equation (Kato et al., 7).

$$U_g < 9\ \mathrm{cm\cdot s^{-1}};\quad Pe(TPFB) = Pe(BC)(1+v_t/U_l)^{0.40Bo^{0.225}} \tag{4}$$

$$U_g > 25\ \mathrm{cm\cdot s^{-1}};\quad Pe(TPFB) = Pe(BC) \tag{5}$$

$$U_g = 9\text{-}25\ \mathrm{cm\cdot s^{-1}};\quad Pe(TPFB) = Pe_1(U_g/9)^{2.3\log_{10}(Pe_2/Pe_1)} \tag{6}$$

Pe_1 and Pe_2 are the value of Pe(TPFB) of Equation (4) at $U_g = 9\ \mathrm{cm\cdot s^{-1}}$ and that of Equation (5) at $U_g = 25\ \mathrm{cm\cdot s^{-1}}$, respectively. Pe(BC) is the Peclet number for bubble columns and is given by

$$Pe(BC) = (\eta_l/\eta_w)^{0.07}\{13Fr/(1+6.5Fr^{0.8})\} \tag{7}$$

Table 2. Keys used in Figures 4, 5 and 8.

D_T [cm]	d_p [mm]	Key	U_l [cm·s⁻¹]
12	0.58	Φ	0.5
		○	1.8
		⊖	2.7
	1.2	△	3.6
	2.2	▽	3.4
		▽	5.5
		▽	7.6
19	0.52	●	1.6
	2.2	▼	5.8

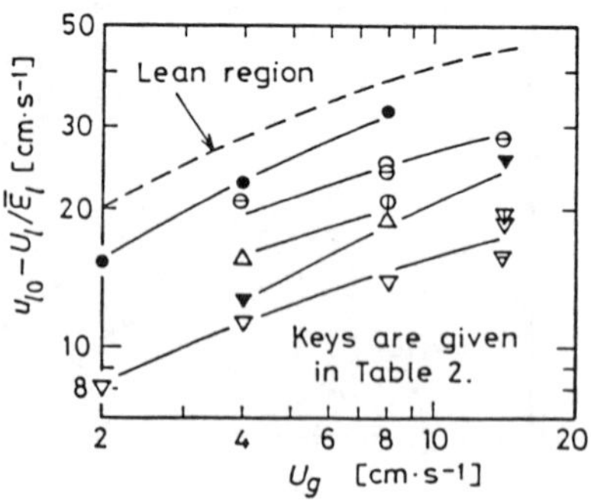

Figure 4. Liquid velocity at $\phi = 0$.

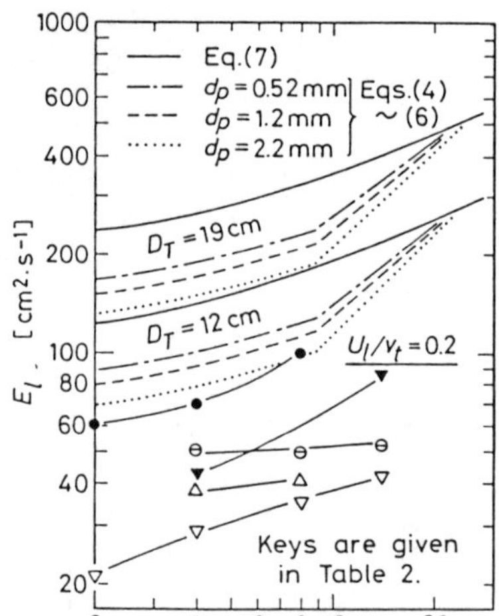

Figure 5. Longitudinal dispersion coefficient of liquid. Plotted data are calculated from Equation (A4).

DISCUSSION

Wake Model with Internal Recirculation Flow

In order to explain the results of Figure 3, the radial distribution of gas holdup has to be considered. The notation of various local phase holdups and velocities is given in Figure 6. The local phase holdups depend on radial position. From Equation (2), the local wake holdup can be expressed as

$$\varepsilon_w = 2k\bar{\varepsilon}_g(1-\phi^2) \tag{8}$$

The mass balances of liquid and solid particles are

$$\int_0^1 u_l\varepsilon_l\phi d\phi = \int_0^1 U_l\phi d\phi, \quad \int_0^1 u_s\varepsilon_s\phi d\phi = 0 \tag{9}$$

The wake velocity with respect to the wall, u_g, and the slip velocity between the wake and the liquid in the bulk liquid-solid region, (u_g-u_{lf}), are assumed to be radially unchanged. Then the local liquid holdup is given by

$$\varepsilon_l = 2\bar{\varepsilon}_g k(1-x)(1-\phi^2) + \{1-2\bar{\varepsilon}_g(1+k-kx)(1-\phi^2)\}(1-\varepsilon_{sf}) \tag{10}$$

where

$$1 - \varepsilon_{sf} = [\{U_l-U_g k(1-x)\}/(v_t\{1-(1+k)\bar{\varepsilon}_g\})]^{1/n} \tag{11}$$

The exponent n is given by Garside and Al-Dibouni (5). The cross-sectionally averaged liquid holdup is obtained as follows:

$$\bar{\varepsilon}_l = \int_0^1 \varepsilon_l\phi d\phi/\int_0^1 \phi d\phi = \bar{\varepsilon}_g k(1-x) + \{1-\bar{\varepsilon}_g(1+k-kx)\}(1-\varepsilon_{sf}) \tag{12}$$

The expression of $\bar{\varepsilon}_l$ is the same as the wake model where the gas holdup is radially unchanged. The local liquid velocity is

$$u_l = \frac{u_{lf}\{1-2(1+k)\bar{\varepsilon}_g(1-\phi^2)\}(1-\varepsilon_{sf})+U_g k(1-x\varepsilon_{sf})}{2\bar{\varepsilon}_g k(1-x)(1-\phi^2)+\{1-2\bar{\varepsilon}_g(1+k-kx)(1-\phi^2)\}(1-\varepsilon_{sf})} \tag{13}$$

The liquid velocity shows the maximum value at $\phi = 0$. u_{lf} is derived as follows:

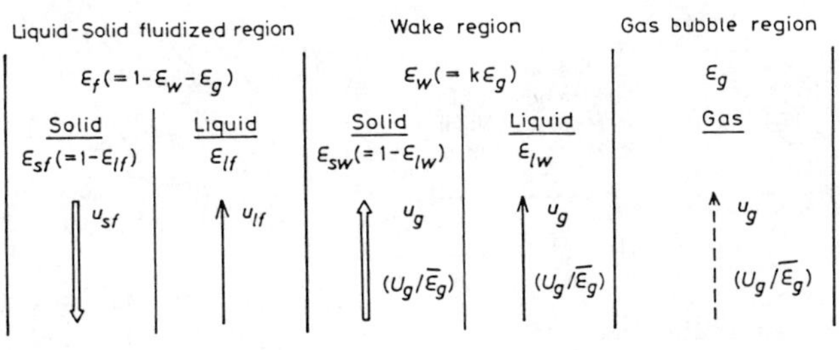

Figure 6. Notation of phase holdups and velocities.

$$u_{lf} = [U_l - U_g k(1-x\varepsilon_{sf})]/[\{1-(1+k)\overline{\varepsilon}_g\}(1-\varepsilon_{sf})] \quad (14)$$

When k and x are estimated from the experimental equations of Østergaard (15), Baker et al. (1) and Chern et al. (2), the calculated values of u_{l0} are only 1/2-1/20 of the experimentally measured values. The estimation with k and x based on the empirical equations of El-Temtamy and Epstein (4) gives relatively reasonable values of u_{l0} but very small values of $\overline{\varepsilon}_l$. Therefore, k and x have to be simultaneously evaluated by comparing Equations (12) and (13) with the data of $\overline{\varepsilon}_g$ and u_{l0}. The calculated values of (1-x) are independent of gas velocity and can be expressed by

$$(1-x) = 0.018U_l^{1.09}d_p^{-0.33} \quad (15)$$

where U_l and d_p are in the unit of $cm \cdot s^{-1}$ and mm, respectively. Equation (15) is applicable in the range of $U_g < 20\ cm \cdot s^{-1}$, $d_p \geqq 0.58$ mm and $U_l/v_t < 0.5$. The values of k are correlated by

$$k = 29U_g^{-0.985} \quad (16)$$

where the unit of U_g is $cm \cdot s^{-1}$. Equation (16) is applicable in the range of $U_g = 4\text{-}20\ cm \cdot s^{-1}$.

Figure 7 shows the experimental data of $\overline{\varepsilon}_l$ are in good agreement with Equations (12), (15) and (16). The data of Michelsen and Østergaard (10), Efremov and Vakhrushev (3) and Muroyama et al. (14) are also correlated within an error of ±10%. Figure 8 illustrates the data of u_{l0} agree with those calculated from Equations (14)-(16) within ±20%. The radial distribution of liquid velocity is also well explained by the proposed model as shown in Figure 3, where the lines are calculated from Equations (13)-(16).

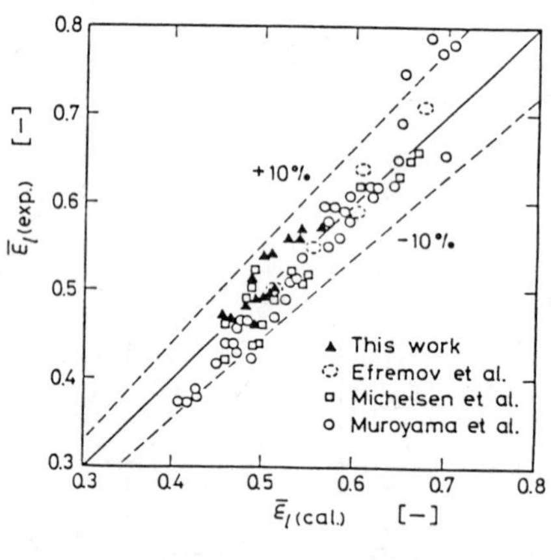

Figure 7. Comparison of experimental and calculated values of mean liquid holdup.

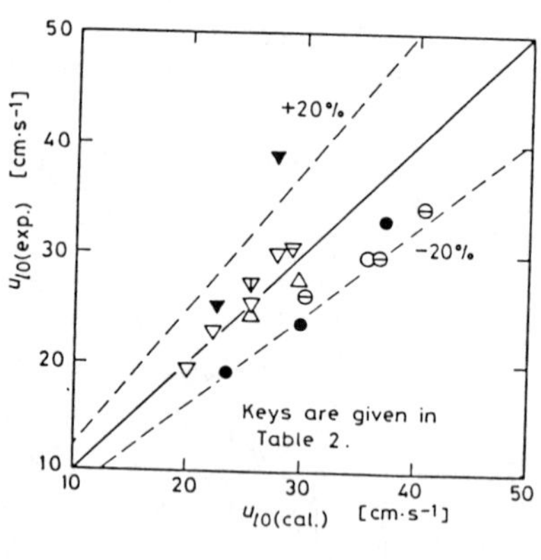

Figure 8. Comparison of experimental and calculated valued of liquid velocity at $\phi = 0$.

Longitudinal Dispersion Coefficient of Liquid

In the proposed wake model, the material exchange between the wake region and the bulk liquid-solid region is not taken into consideration. Liquid and solid particles in the wake are assumed to ascend without any resistance, and no axial distributions of chemical and physical properties exist in the bed. This means the axial dispersion of liquid and solid particles is infinitely large contrary to the experimental results.

In the bubble columns for gas-liquid systems, a momentum balance model for the internally recirculating liquid flow was presented by Miyauchi et al. (11). This model is based on the radial distribution of bouyancy force and the turbulent viscosity of gas-liquid mixture. The longitudinal dispersion coefficient of liquid is derived by applying the concept of Taylor dispersion to the radial velocity distribution. As shown in APPENDIX, the momentum balance model of Miyauchi et al. (11) is extended to the three-phase fluidized bed. However, Figure 5 indicates that the longitudinal dispersion coefficients of liquid calculated from Equation (A4) are still about 1/2-1/3 of the experimental values.

APPENDIX

In a sufficiently tall fluidized bed, the equation of motion for a steady state condition is given by

$$-(1/r)d(r\tau)/dr = d\overline{P}/dz + (1-\varepsilon_g)\rho_e g \tag{A1}$$

If the composition in the liquid-solid fluidized region is uniform radially and axially, the static pressure and the apparent density in the liquid-solid region are given by

$$d\overline{P}/dz = -(1-\overline{\varepsilon}_g)\rho_e g \quad \text{and} \quad \rho_e = (\rho_l\overline{\varepsilon}_l+\rho_s\overline{\varepsilon}_s)/(\overline{\varepsilon}_l+\overline{\varepsilon}_s) \tag{A2}$$

The shear stress is assumed to be radially uniform.

$$\tau = -(\nu_l+\nu_t)\rho_e du_m/dr \tag{A3}$$

Equations (A1) and (A3) can be solved similarly to the solution of Miyauchi et al. (11) for gas-liquid systems. Thus, the longitudinal dispersion coefficient of liquid in the three-phase fluidized bed is obtained:

$$E_l = 6.1\times10^{-3}\frac{q_r}{\overline{\varepsilon}_g}\left\{\frac{11}{15}+\frac{1}{(1-\overline{\varepsilon}_g)^2}\right\}\left\{\left(u_{m0}-\frac{U_l}{\overline{\varepsilon}_l}\right)\frac{1-\overline{\varepsilon}_g}{4-3\overline{\varepsilon}_g}\right\}^3 + q_z\nu_t \tag{A4}$$

If the slip velocity between liquid and solid particles in the liquid-solid fluidized region is not neglected, E_l becomes smaller than the calculated value from Equation (A4).

ACKNOWLEDGEMENT

This work was partially supported by Grant-in-Aid for Energy Research of the Ministry of Education, Science and Culture, Japan (No. 59045109). The authors are grateful to Mr. T. Kago, Mr. H. Kamano and Mr. H. Yabu for their assistance in the experiment.

NOTATION

Bo	$gd_p{}^2\rho_l/\sigma$	u_l	local liquid velocity
D_T	bed diameter	u_{l0}	u_l at $\phi = 0$
d_p	particle diameter	u_m	velocity of liquid-solid mixture
El	longitudinal dispersion coefficient of liquid	u_{m0}	u_m at $\phi = 0$
Fr	$U_g/\sqrt{gD_T}$	u_s	local solid velocity
g	accerelation of gravity	v_t	terminal velocity
K	$\rho_l U_g{}^4/g\sigma$	x	$\varepsilon_{sw}/\varepsilon_{sf}$
k	$\varepsilon_w/\varepsilon_g$	z	axial coordinate
n	exponent given by Garside and Al-Dibouni (5)	ε_g, ε_l, ε_s	local gas, liquid and solid holdup, respectively
n'	exponent in Equation (3)	η_l	liquid viscosity
$\overline{P}$	static pressure	η_w	water viscosity
Pe(BC)	$D_T U_g/E_l$ for bubble column	ν_l	kinematic liquid viscosity
Pe(TPFB)	$D_T U_g/E_l$ for three-phase fluidized bed	ν_t	turbulent kinematic viscosity
q_r, q_z	correction factor close to unity, given by Miyauchi et al. (11)	ρ_e	apparent density of liquid-solid mixture
Ret	$v_t d_p \rho_l/\eta_l$	ρ_l	liquid density
r	radial coordinate	ρ_s	solid density
U_g	superficial gas velocity	σ	surface tension
u_g	$U_g/\overline{\varepsilon_g}$	τ	shear stress
U_l	superficial liquid velocity	ϕ	dimensionless radial coordinate
		—	refers to mean value

LITERATURE CITED

1. Baker, C.G.J., S.D. Kim and A.M. Bergougnou, Powder Technol., 18, 201 (1977).
2. Chern, S.-H., L.-S. Fan and K. Muroyama, AIChE J., 30, 288 (1984).
3. Efremov, G.I. and I.A. Vakhrushev, Int. Chem. Eng., 10, 37 (1970).
4. El-Temtamy, S.A. and N. Epstein, Int. J. Multiphase Flow, 4, 19 (1978).
5. Garside, J. and M.R. Al-Dibouni, Ind. Eng. Chem. Process Des. Dev., 16, 206 (1977).
6. Kato, Y., S. Morooka, T. Kago, T. Saruwatari and S.-Z. Yang, J. Chem. Eng. Japan, 18, 308 (1985).
7. Kato, Y., S. Morooka, T. Kago, M. Koyama and S.-Z. Yang, J. Chem. Eng. Japan, 18, 313 (1985).
8. Kato, Y., K. Uchida, T. Kago and S. Morooka, Powder Technol., 28, 173 (1981).
9. Kim, S.D. and C.H. Kim, J. Chem. Eng. Japan, 16, 172 (1983).
10. Michelsen, M.L. and K. Østergaard, Chem. Eng. J., 1, 37 (1970).
11. Miyauchi, T., S. Furusaki, S. Morooka and Y. Ikeda, Adv. in Chem. Eng., 11, 275 (1981).
12. Morooka, S., K. Uchida and Y. Kato, J. Chem. Eng. Japan, 15, 29 (1982).
13. Muroyama, K., K. Hashimoto, T. Kawabata and M. Shiota, Kagaku Kogaku Ronbunshu, 4, 622 (1978).
14. Muroyama, K., M. Fukuma and A. Yasunishi, Can. J. Chem. Eng., 62, 62 (1984).
15. Østergaard, K., Chem. Eng. Sci., 20, 165 (1965).

RADIATIVE HEAT TRANSFER IN THE FREEBOARD REGION OF A FLUIDISED BED

J.J. Lindsay, W. Morton* and D.C. Newey

British Petroleum plc
Britannic House
Moor Lane
London EC2Y 9BU, England

ABSTRACT

Emittance and transmittance measurements were made above the surface of a large fluidised bed combustor. Heat transfer coefficients to tubes in this region were also investigated in this and a second combustor for bed and freeboard temperature ranges of 1050 to 1300K and 750 to 1150K respectively. The results were compared with isotropic scattering theory and good agreement was found.

SCOPE

The efficiency of many high temperature fluidised bed processes depends upon the amount of heat recovered in the freeboard region above the bed surface. Calculation of freeboard heat transfer coefficients, particularly the radiative component, is notoriously difficult - often necessitating the use of prototypes and/or computer simulation for design.

Several attempts have been made to adapt models of absorption in particle clouds to provide a theoretical description of radiation in a fluidised bed. Several early studies are described by Zabrodsky (1) and more recently there is the work of Kovenski (2), Makhorin et al (3), Baskakov (4) and related work by Borodulya and Kovenski (5). Most of the studies to date, however, have relied either upon the questionable assumption that the particles act as black body point emitters, or have reverted to situation specific computer models.

This work investigates the applicability of isotropic scattering theory to describing freeboard radiation - thus facilitating calculation of heat transfer coefficients over a wide range of conditions. Experimental measurements were made in two large diameter fluidised bed combustors: a 0.38 m diameter apparatus at Cambridge and a similar 0.3 m combustor at the BP Research Centre in Sunbury. Transmittance was measured by He/Ne laser and emittance was measured directly by pyrometer. Measurement of heat transfer coefficients to straight tubes were made on the 0.38 m apparatus and to a 'U'-tube on the 0.3 m apparatus. Bed and freeboard temperatures in the ranges of 1050 to 1300K and 750 to 1150K respectively were investigated at fluidising velocities between 0.5 and 2.5 ms^{-1}.

* University of Cambridge, Department of Chemical Engineering.

CONCLUSIONS AND SIGNIFICANCE

Freeboard transmittance and emittance and their variation with solids burden were predicted reliably by the theory. In the 0.38 m apparatus, at low solids burdens (S_b) under 0.1 kg m^{-3}, transmittance was found to be high, greater than 0.9. It dropped rapidly at higher values of S_b to under 0.5 at 1 kg m^{-3} as shown in Figure (5). Freeboard emittance was relatively low, less than 0.2 for S_b less than 0.1 kgm^{-3}, increasing to 0.6 at S_b values up to 1 kg m^{-3} as shown in Figure (6). In all cases, CO_2 and water vapour from the combustion of propane under near stoichiometric conditions were present.

Agreement between the experimental results and the theoretical predictions was of the order of ±20%. It is likely, however, that most uncertainty arose from experimental scatter because of unavoidable fluctuations in freeboard temperature and other random errors. Statistical analysis of the results indicated that the theory tends to underestimate in the case of transmittance and to overestimate in the case of emittance. This agrees with the neglection of secondary scatter terms when formulating the theoretical equations.

Reasonable agreement was also found between the measured and predicted values of the outside heat transfer coefficients to water and air carrying pipes. Heat transfer coefficients to the outside of the pipe were found to increase from 150 W $m^{-2}K^{-1}$ at virtually zero solids burdens up to 220 W $m^{-2}K^{-1}$ at solids burdens of 1 kg m^{-3}. The measured values from both fluidised bed combustors, together with the theoretical predictions for isotropic scatter by 0.2 mm particles in a stoichiometric off-gas mixture, are shown in Figure (7). The theoretical coefficient is broken down into its constituent gas convective and radiative components to illustrate their relative importance.

The advantages of isotropic scattering theory over previous models are greater flexibility and improved accuracy over a wide range of conditions. The theory can be further refined to consider the non-idealities of predominantly forward scatter (hard crystalline particles) and back scatter (dull amorphous particles) by considering secondary scattering effects.

ISOTROPIC SCATTERING THEORY

Freeboard transmittance

There are two components of transmittance in an absorbing/scattering system of grey particles. The first is the familiar direct component, T_d, described by analogy to Bouger-Lambert's law:

$$T_d = \exp[- (K_a + K_s) L_m/\cos \Theta] = \exp(-\kappa_1/\mu_0) \qquad [1]$$

The second conponent, T_s, describes the fraction of radiation eventually transmitted by scattering at the particle surfaces. This can be estimated from Chandrasekar's (6) equation:

$$T_s = \int_0^{\kappa_1} -\frac{\omega_o}{2} \Psi_2 \, [b(\kappa_1 - \kappa)] \, d[\exp(-\kappa/\mu_o)] \qquad [2]$$

Hence the total mean hemispherical transmittance is given by addition of Equations [1] and [2]:

$$\tau_s = \int_0^1 [T_d(\mu_o) + T_s(\mu o)] \, d(\mu_o^2) \qquad [3]$$

The predictions of Equation [3] are shown in Figure (1) for isotropic (ideal) scatter, the results for forward scatter, typical of hard semi-translucent crystalline particles and back scatter, typical of dull amorphous particles are also shown in Lindsay (7).

Freeboard emittance

By analogy to Equation [2] an equation can also be derived for the reflectance due to scatter by the particles:

$$R_s = \int_0^{\kappa_1} -\frac{\omega_o}{2} \Psi_2 \, (b\kappa) \, d[\exp(-\kappa/\mu_o)] \qquad [4]$$

Since the emittance, reflectance and transmittance of a grey system must sum to unity, the emittance is given by:

$$\varepsilon_s = \int_0^1 [1 - R_s(\mu_o) - T_s(\mu_o) - \exp(-\kappa_1/\mu_o)] \, d(\mu_o^2) \qquad [5]$$

The results of Equation [5] are shown for a range of scattering behaviour in Figure (2). In the presence of combustion products, CO_2 and water vapour, Equations [3] and [5] must be modified to account for the radiative properties of these gases as follows:

$$\tau_s' = \tau_s \, \tau_g \qquad [6]$$

$$\varepsilon_s' = 1 - (1-\varepsilon_g) \, \tau_s \qquad [7]$$

The correction factors τg and εg were derived by Hottel and Sarofim (8) and are widely reported in radiative texts.

Formulation of heat transfer coefficients

Designating the bed, the freeboard walls and the heat transfer surface 1,2 and 3 respectively, we can write equations for the radiosity W (i.e. the total emitted and reflected radiation) leaving each surface, typically for the bed surface, we have:

$$W_1 = \varepsilon_1 \sigma T^4{}_1 + \rho_1 (W_1 F_{11} \tau_{11} + \sum_{i=1}^{n} W_i F_{1i} \tau_{1i} + W_2 F_{12} \tau_{12} + W_3 F_{13} \tau_{13} + \varepsilon_{f1} \sigma T^4{}_g) \qquad [9]$$

We can write similar equations for all other surfaces in the system and finally an equation for heat transfer to the surface of interest:

$$Q_3 = A_3 \varepsilon_3 (W_3 - \sigma T^4{}_3)/\rho_3 = h_{rg} A_3 (T_g - T_3) \qquad [10]$$

If the temperatures of the surfaces and the gas are known Equations [9]-[10] can be solved simultaneously using linear matrix techniques to give Q_3 and hence the gas radiative component h_{rg}.

Figure 1.

Transmittance τ_s vs. optical depth for isotropic scatter albedo values ω_0

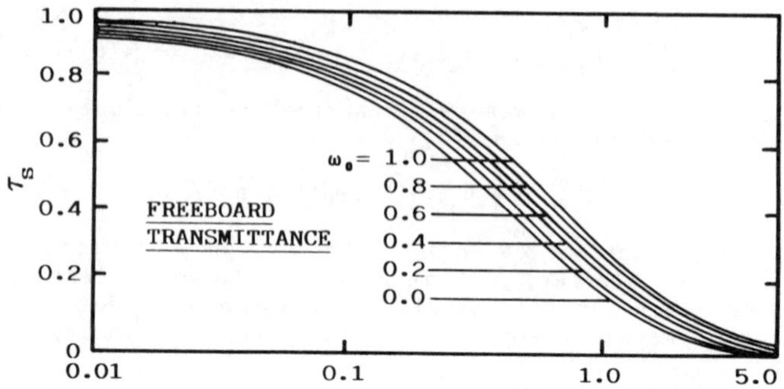

Figure 2.

Emittance ϵ_s vs. optical depth for isotropic scatter albedo values ω_0

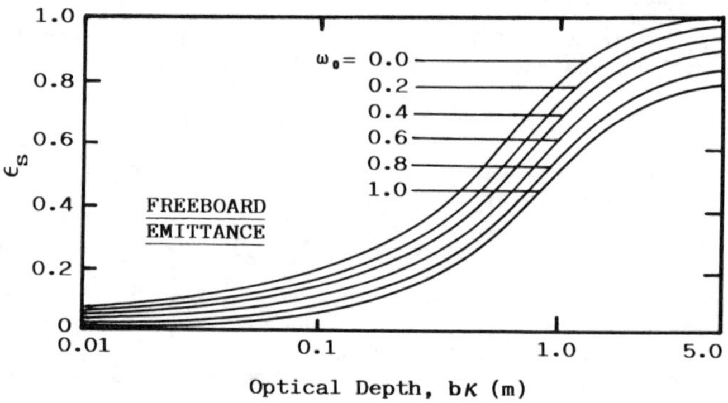

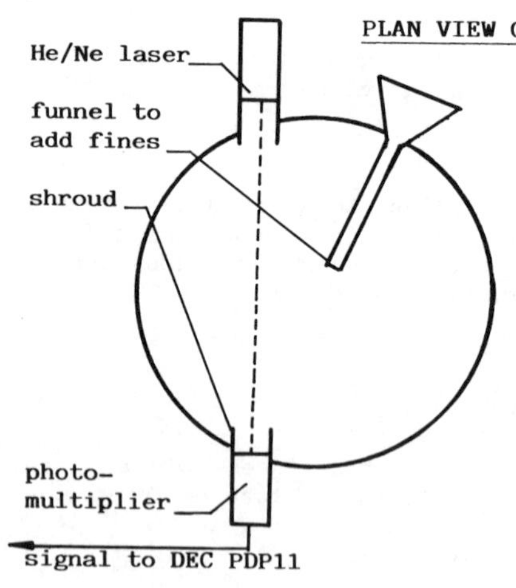

Fig. 3. Transmittance measurement.

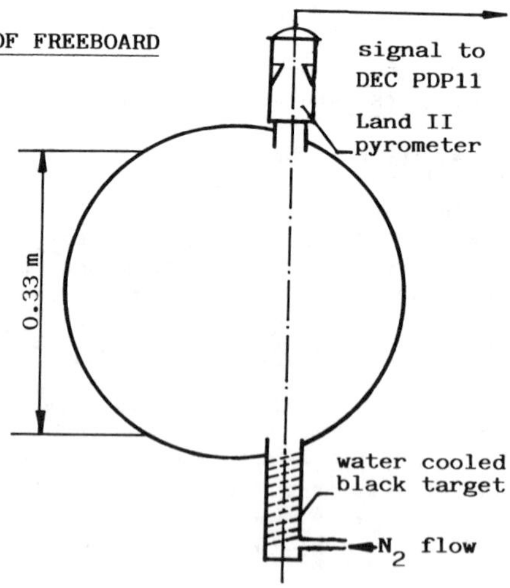

Fig. 4. Emittance measurement.

The freeboard convective component for low velocity, particle carrying streams was predicted reliably by the equation of Gorbis and Bakhtiozan (9).

Radiative and convective components in the freeboard, for the present work were assumed to be additive. This is because in the freeboard, unlike the bed, particle concentrations and gas velocities are relatively low (0.5 to 2.5 ms^{-1} at most in this work); therefore conditions approximating to steady state could be assumed. This approximation could still be used in larger industrial equipment by dividing the freeboard into quasi-steady state zones.

EXPERIMENTAL PROCEDURE

Transmittance and Emittance Measurements

These experiments were carried out in the 0.38 m apparatus, the experimental configurations are shown in Figures (3) and (4). Transmittance through the freeboard was inferred from the intensity of laser light falling upon a photomultiplier target from a source 0.33 m away. Automatic data logging via an analogue/digital (a/d) converter was carried out on a DEC PDP11 machine.

Emittance measurements were made by focussing a calibrated Land II pyrometer through the freeboard onto a cooled, black, nitrogen swept target. Emitted energy from the intervening gases could then be measured by the pyrometer directly. For both sets of experiments, the apparatus containing particles in the size range 1.2 to 1.4 mm was allowed to reach the desired operating conditions. A known quantity of fine (0.2 mm) material was then added via a funnel to the bed. Elutriation rates and the corresponding freeboard solids concentrations were then estimated from the models of Wen and Chen (10).

Elutriation data measured by isokinetic sampling probe were also available for a limited range of operating conditions from the 0.3 m apparatus at Sunbury. These were correlated and were used wherever possible because of their greater applicability to combustion conditions.

Heat Transfer Coefficients

Overall heat transfer coefficients in the freeboard were estimated from the temperature rise of water and air flowing through 0.0127 m (nominal ½ inch) steel and copper tubes. Precautions were taken to ensure adequate mixing at the inlet and outlet and the establishment of stable flow conditions in the tubes. Temperature measurement was by thermocouples connected via solid state amplifiers to a DEC PDP 11 computer.

Preliminary data indicated that thermal resistances inside the tubes could not be ignored, and so additional thermocouples were welded to the outside of the tubes on the Cambridge apparatus. Measurements on the Sunbury apparatus were also made using a stainless steel U-tube of internal and external diameters 0.0156 and 0.0213 m respectively. Here thermal resistance at the inside of the pipe were estimated.

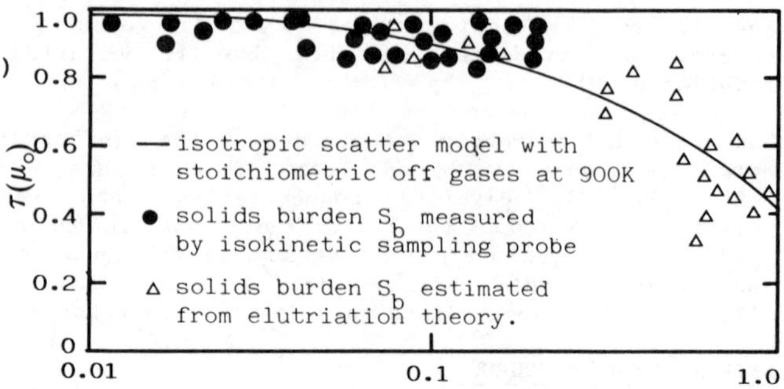

Figure 5. Transmittance $\tau(\mu_o)$ variation with solids burden S_b: comparison of experimental measurements and predictions of isotropic scatter theory.

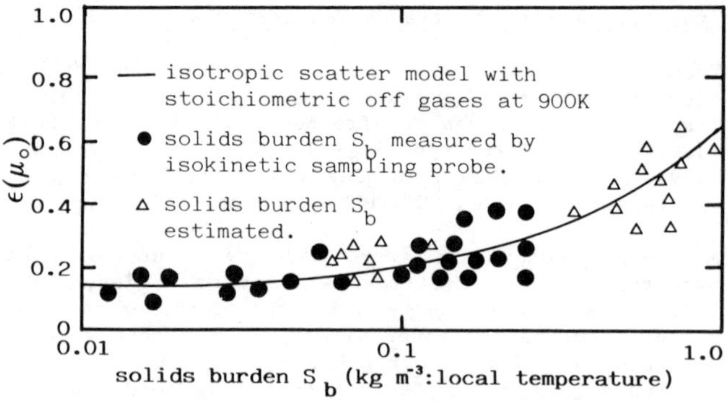

Figure 6. Emittance $\epsilon(\mu_o)$ variation with solids burden S_b: comparison of experimental measurements and predictions of isotropic scatter theory.

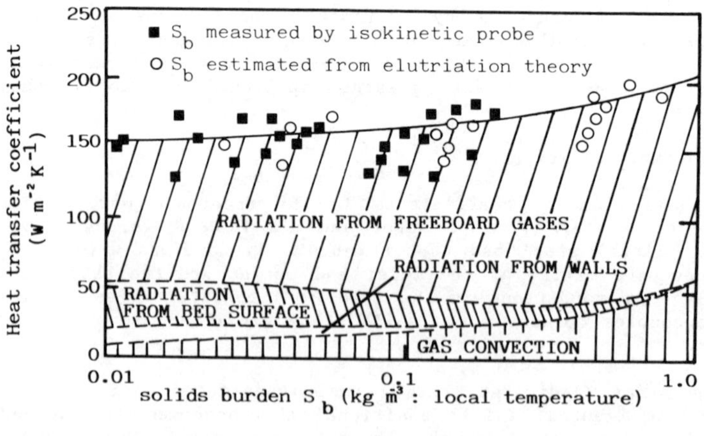

Figure 7. Comparison of experimental heat transfer coefficients to water cooled tubes and the predictions of isotropic scattering theory with stoichiometric off gases at 900K.

RESULTS AND DISCUSSION

The results of the transmittance and emittance experiments are shown in Figures (5) and (6). Without sophisticated sampling probes and particle size analysis in the freeboard, it was only possible to estimate solids burdens with ±20% reliability. Nevertheless, the marked decline in freeboard transmittance at solids burdens over 0.1 kg m^{-3} which occurred was predicted reliably by the theory as can be seen in Figure (5).

Figure (6) shows that the freeboard emittance is also described reasonably well by scattering theory for this range of conditions. Background CO_2 and water vapour emittance was found to lie in the range 0.14 to 0.18. It was found to increase progressively from this level at zero solids burden to just over 0.6 in the presence of 1.0 kg m^{-3} of solids. It is likely that at higher solids concentrations found under 'fast fluidisation' conditions, transmittance would tail off to zero, and the emittance would plateau at values similar to those found in the fluidised bed itself. For silica sands this value would lie between 0.8 and 0.85.

It should be noted that the theoretical predictions of the isotropic theory, modified to account for CO_2 and water vapour, do not sum precisely to unity. This is partly because of inaccuracies introduced in modifying the equations for combustion gases, and partly from the single scatter assumption implicit in the theory. In their analysis of scattering theory, Hottel and Sarofim (8) estimated mathematically that the single scatter approximation introduced errors of less than 10% for high emittance (ie. dark) particles and less than 40% for low emittance (ie. light) particles.

From figure (7), it can be seen that overall freeboard coefficients of the order 150-200 W $m^{-2}K^{-1}$ were measured. The theory gave a reasonable mean prediction tending toward a 5-10% overestimation. It can be seen from the breakdown of the theoretical coefficient in Figure (7) that the convective component changes in magnitude from about 10 W $m^{-2}K^{-1}$ under solids free conditions, to about 50 W $m^{-2}K^{-3}$ at 1 kg m^{-3}. The contribution of radiation from the bed surface and freeboard walls was for this geometry relatively small and was predicted to decline markedly at high solids concentration.

The dominant component predicted for this type of combustor was radiation from the freeboard gases. It must be noted that a major contributor to this component was the emittance from the combustion off-gases, almost 100 W $m^{-2}K^{-1}$. The particle contribution to the radiative component was, however, predicted to cause this to increase by over 50% as the solids concentration increased to 1 kg m^{-3}, more than compensating for the reduction in radiation from the bed surface and the freeboard walls.

ACKNOWLEDGEMENTS

Funding for this work in the form of a BP Research Studentship is acknowledged. The typing of Ms Susan Murray for this work is greatly appreciated.

NOTATION

$A_{1,2,3}$	Area of surface 1,2,3 etc (m^2)
A_t	Total surface area of particles per unit freeboard volume $6S_b/\rho_p d_p$ (m^2)
b	Constant with values of 1, $(1 - \omega_o)$, $(1 + \omega_o)$ for isotropic forward and back scatter respectively.
d_p	Particle diameter (m)
F_{ij}	View factor surface i to surface j
h_{rg}	Freeboard gas radiative component. ($W\ m^{-2}\ K^{-1}$)
$K_{a,s}$	Extinction coefficients for absorption and scatter respectively (m^{-1})
L_m	Mean beam length for gas radiation, related to optical depth via: $\kappa_1 = (\varepsilon_m + \rho_m)A_t L_m/4 = A_t L_m/4$ (m)
R_s	Reflectance due to scatter
Q_3	Heat flow to the heat transfer surface (Watts)
S_b	Solids Burden (at local temperature) ($kg\ m^{-3}$)
$T_{d,s}$	Transmittance direct and due to scatter respectively
$T_{g,3}$	Absolute temperature of gas and surface 3 respectively (K)
W	Radiosity (Watts)
$\varepsilon_{g,m,s}$	Emittance of combustion gases, particle material and the overall freeboard respectively
κ_1	Optical distance = $(K_a + K_s)/L_m$ (m)
μ_0	Cosine of incident radiation
ρ_m	Reflectivity of particle material
ρ_p	Density of particle material ($kg\ m^{-3}$)
σ	Stefan - Boltzman constant ($5.67 \times 10^{-8}\ W\ m^{-2}\ K^{-4}$)
$\tau_{g,s}$	Transmittance of combustion products and overall freeboard respectively
ω_0	Albedo for single scatter = $K_s/(K_a + K_s)$
ψ_2	The second exponential integral

LITERATURE CITED

(1) Zabrodsky S.S. "Hydrodynamics and Heat Transfer in Fluidized beds" MIT Press (1966).

(2) Kovenski V.I. "A Mathematical Model of Radiation in a Packed Bed" Inzjenerno-fizichesky zjurnal 38(6) pp 983-988 (1980)

(3) Makhorin K.E., Pikashov V.S. and Kuchin G.P. "Heat Transfer in a High Temperature Fluidized Bed" Naukova Dumka Kiev (1981)

(4) Baskokov A.P. "Heat and Mass Exchange Processes in a Fluidized bed" Metallurgia, Moscow (1978)

(5) Borodulya V.A. and Kovenski V.I. "Radiative Heat Transfer between a Fluidized Bed and a Surface" I.J.H.M.T. 26(2) pp 277-287 (1983)

(6) Chandrasekhar S. "Radiative Transfer" Oxford U.P. (1950)

(7) Lindsay J. "Radiative Heat Transfer in a Fluidised Bed" Ph.D. Thesis, University of Cambridge (1983)

(8) Hottel H.C. & Sarofim A.F. "Radiative Transfer" McGraw-Hill (1967)

(9) Gorbis Z.R. and Bakhtiozan R.A. "Experimental Investigation of Convective Heat Transfer in Dusty Streams" Materialy Soveshchaniya po Teplo - i Massobmenu (Doklad) Minsk (1961)

(10) Wen C.Y. and Chen L.H. "Fluidized Bed Freeboard Phenomena - Entrainment and Elutriation" A.I.Ch.E.J. August (1980)

APPARENT CHANGE IN HEAT TRANSFER MECHANISM AND CORRESPONDING BED BEHAVIOUR WITH CHANGE IN OPERATING TEMPERATURE

J.S.M. Botterill and G.D. Hawkes

Department of Chemical Engineering, University of Birmingham, PO Box 363, Birmingham. B15 2TT, U.K.

ABSTRACT

A sharp change in the dependence of maximum bed-to-surface heat transfer coefficient on operating temperature is associated with the change in dominant mechanism consequent upon a transition in bed behaviour. Evidence for this transition has been obtained from coincidental change in bubbling behaviour inferred from the cross correlation of pressure pulses.

SCOPE

In earlier work (1,2,3), we reported the existence of a sharp change in the dependence of the maximum bed-to-surface heat transfer coefficient on operating temperature. We ascribed this to a change in the characteristic mode of bed behaviour which engendered a change in the dominant mechanism of heat transfer. Thus, for a bed of sand or alumina of mean particle size about 1 mm, there is typically an abrupt change in the heat transfer behaviour when the operating temperature increases through the range of 500 to 600°C for a bed operating at atmospheric pressure. We attributed this to a transition from "D" to "B" type behaviour in terms of the classification suggested by Geldart (4). Over this temperature range, the decrease in gas density and increase in gas viscosity result in the reduction of the minimum fluidization Reynolds number from that characteristic of turbulent or transitional interstitial gas flow to laminar (3). At the higher temperature in the "B" mode of operation, the particle convective mechanism of heat transfer is dominant whereas the interphase gas convective component plays a significant role at the lower temperatures. The consequence of this is that, for example, a typical atmospheric fluid-bed combustor will behave in a different manner if tested under ambient conditions compared with if it is operated hot because of this discontinuity. In the hot condition, its heat transfer capability will be better and primarily so because of the particle convective mechanism of heat transfer rather than because of the increasing radiant heat transfer component.

The heat transfer experiments reported extend earlier work in order to verify that the change in heat transfer behaviour is not just a consequence of the calorimetric probe measurement technique used earlier. New measurements have been made of heat transfer between a bed of 982μm alumina and a single immersed horizontal tube of 35mm diameter and the bubbling behaviour has been inferred from the cross correlation of pressure pulses generated when bubbles rise past two vertically aligned probes inserted at the axis of the bed following the technique of Sitnai (5).

CONCLUSIONS AND SIGNIFICANCE

An abrupt change in the maximum heat transfer coefficient's dependence on operating temperature for beds of particles within the "D/B" group transition region has been further substantiated. Increasing contribution of radiant heat transfer as the reason for this change can be discounted and the causal mechanism is ascribed to the changing importance of the interphase gas convective and particle convective mechanisms of bed-to-surface heat transfer. A coincidental change in bubbling frequency has been observed associated with this change but its sharpness tends to reduce as the excess gas velocity is increased. Whilst, between different beds (1,2,3,6), the actual transition is observed to occur over a range of Archimedes and Reynolds numbers, the transition in behaviour is quite sharp and reproducible for a given bed. It is to be expected that particle shape, roughness and size distribution will be important variables and would need to be taken into account in the formulation of a criterion for the more precise demarcation of a "D/B" boundary.

EQUIPMENT AND EXPERIMENTAL TECHNIQUE

The experimental bed was contained within a 188mm diameter stainless steel cylinder. The range of operating temperatures (250°C to 800°C) was obtained by burning propane at a gas burner inserted into the air supply pipework just before it joined the plenum chamber beneath the distributor which was a stainless steel plate with 1.0mm holes drilled on a 6mm square pitch. The bed material used for the principal experiments (Table 1) was a narrow size distribution alumina of mean sieve diameter 982μm and density 3960 kg/m^3. The bed had a packed depth of 200mm and its minimum fluidizing velocity ranged from 0.81 m/s at 780°C to 0.91 m/s at 270°C, excess gas velocities were in the range 0.5 to 0.7 m/s at the operating temperature.

TABLE 1

Particle Properties

Material	Mean Particle Diameter (μm)	Density (kg/m^3)	U_{mf} 250°C (m/s)	U_{mf} 500°C (m/s)	U_{mf} 750°C (m/s)
Sand	380	2600	0.11	0.09	0.08
Sand	664	2600	0.26	0.23	0.23
Alumina	375	3640	0.15	0.11	0.10
Alumina	982	3960	0.91	0.87	0.82

The pressure fluctuations produced by rising bubbles were detected by two simple vertically aligned axial probes similar to those used by Zhang et al. (7). These consisted of lengths of 4mm i.d., 6mm o.d. stainless steel tube inclined downwards into the bed at 60° and their tips were 50mm apart with the lower one 50mm above the distributor. Each probe was connected to the high pressure port of a Furness Instruments 0 to 10mm w.g. micromanometer and, through a porous plastic snubber, to the low pressure port and operated against a small back pressure air bleed so that the system responded to the difference between the fluctuating pressure and the mean pressure at the tip of the probe. The signals from the two micromanometers were analysed by feeding them to a Hewlett Packard 3721A Correlator. By delaying the response from the lower probe and comparing it with the undelayed response from the upper probe, their degree of similarity could be compared using the cross correlation function. Provided that the bubbling behaviour was relatively steady, time averaged bubble properties including rise velocity, pierced height and point frequency could be estimated from the correlograms following Sitnai (5). Several correlograms were obtained for each set of experimental conditions based on data collected from bubbles over a total scan time of typically 8 minutes (see Figure 1). Some exhibited a high degree of damping because of excessive short term fluctuations in bubbling behaviour and these were discarded. This is a limitation of the technique which becomes more severe at higher gas velocities. The reliability of the technique was verified using visual observations of a 2-D bed.

The heat transfer measurements were made either using a small calorimetric probe or, to the immersed tube, by measuring the local radial heat flux through a thick walled water cooled tube. In the latter experiments, a 35mm o.d. stainless steel tube was used, positioned horizontally across the centre of the bed with its axis 150mm above the distributor. Details of the techniques used have been published elsewhere (1,2,3).

EXPERIMENTAL RESULTS AND DISCUSSION

The Heat Transfer Results

Calorimetric sphere (6) and immersed tube heat transfer results are presented for the 982μm alumina bed in Figure 2 together with other calorimetric sphere results for beds of smaller mean particle size (Table 1). Also indicated are the relevant predictions of the Zabrodsky (8) and Denloye and Botterill (9) correlations. It can be seen that there is a sharp change in dependence of the coefficient on bed operating temperature for the 982μm alumina at ∼550°C and that up to this temperature the results are well predicted by the Denloye and Botterill correlation (9) which was derived empirically for beds having "D" type characteristics. Such discontinuity is absent from the results for the beds of smaller particles which could be expected to display Group "B" type behaviour throughout the whole of the experimental range of operating temperatures. To estimate the possible contribution of radiant heat transfer in the higher temperature experiments, calorimetric sphere measurements were made using both fine gold and copper spheres. Whilst it is appreciated that there remains uncertainty over the precise surface emissivities (that of polished gold foil increases to 0.08 at 900°C; the fine gold sphere will have a higher emissivity than that but it will still be much lower than that of the oxidised copper surface) it is particularly notable that agreement between predictions of the Zabrodsky correlation and results for the gold sphere for the smaller particle beds continue to be close at the higher temperatures but there is increasing discrepancy as the operating temperature increases beyond 600°C with the measurements made with the copper sphere suggesting that the difference is a valid indication of the radiant heat transfer component. It can also be seen, Figure 2, that there is negligible difference between the results for the two calorimetric spheres in the 982μm alumina bed until in excess of 600°C and that the rate of increase with further increase in operating temperature is quite distinct from the sharp change in dependence of heat transfer coefficient noted at 550°C. Furthermore, the gold sphere results are closely parallelled by the results for heat transfer to the stainless steel tube whose surface can also be expected to preserve a low emissivity. From this evidence, we discount that the change in heat transfer dependence which we are discussing can have arisen as a consequence of the increasing contribution of radiant heat transfer. Rather, we think that this is a consequence of change in the mode of bed behaviour consistent with the bed reverting to "B" type behaviour from "D" as the operating temperature is increased and the fluidizing gas becomes more viscous and less dense. The calorimetric sphere technique averages the bed-to-surface heat transfer coefficient over the range of surface temperatures from its initial value before insertion into the bed to that of the bed when it has heated up. The bed-to-tube heat transfer measurements are carried out under "steady" state conditions (although there are minor local surface temperature fluctuations when fresh packets of particles first arrive at the surface from the bulk of the bed) and the stainless steel surface exposed to the bed operated at temperatures from 140°C for a bed at 350°C to 330°C for a bed operating at 775°C.

If the values for maximum bed-to-tube bank heat transfer coefficients are estimated from the data over a range of operating temperatures reported by Golan and Cherrington (10), the value for the lowest temperature (274°C) stands out from the line along which the others fall, Figure 2. On the assumption that it and the next highest at 400°C are representative of a bed of Group "D" characteristics (and the values for these two heat transfer coefficients are close to approximate predictions of the Denloye and Botterill (9) correlation), a sharp change in heat transfer mechanism could be envisaged as occurring at 425°C and a corresponding Archimedes number of ∿29000 for the wide distribution 1.3mm mean diameter dolomite bed, 0.6 x 1.5m area, used in their experiments. Thus, these pilot scale experiments are showing similar heat transfer behaviour.

Flow Conditions for Change in Bed Behaviour

We suggest that the sharp change in bed-to-surface heat transfer dependence is a consequence of an abrupt change in the dominant heat transfer mechanism consequent upon change in bed behaviour. Earlier we reported that the transition occurred at about the conditions for the minimum fluidization Reynolds Number, $Re_{mf} \sim 12.5$ and Archimedes Number, Ar, of about ∿26000 (2,3). We find that the transition condition is reproducible and sharp for a given bed but that it can occur over quite a wide range of Reynolds and Archimedes numbers between different beds (6). For the bed of 982μm alumina the corresponding value of Re_{mf} is 10.1 and Ar is

12000. The range of values, however, consistently lie within the transitional flow region as indicated by a plot of the relationship between Re_{mf} and Ar (3). A single precise value for a variety of beds is not to be expected because the interstitial flow conditions will be affected by particle shape, size distribution and roughness and no delineation of a boundary that fails to include these factors can be expected to be adequate. Recent work on local voidage variations around horizontal tubes (Decker and Glicksman, 11) and at immersed surfaces (Buyevich et al., 12) illustrate further the complexity of behaviour close to immersed surfaces in bubbling fluidized beds whilst Borodulya et al. (13) report that the transition in dependence of bubble-to-dense phase mass transfer occurs at a value of Ar as low as about 700 in a freely bubbling bed.

Bubbling Behaviour

Figure 3 presents data on bubble frequency, rise velocity and pierced height as derived from the cross correlation of pressure pulses generated by the rising bubbles in an unobstructed bed together with corresponding calorimetric sphere heat transfer measurements. Whilst it can be seen that there tends to be a steady increase in bubble rise velocity with increase in bed operating temperature for a given excess gas velocity, there is a discernible change in the rate of increase in the bubble frequency with increase in temperature coincident with the change in heat transfer coefficient dependence upon temperature. Within the scatter of the experimental estimates there is negligible change in the dependence of bubble rise velocity and pierced height on change in excess gas velocity. There is, however, a fall in the observed frequency of bubble passage with increasing excess gas velocity which is suggestive of increasing gas flow through the continuous phase, i.e. of some further increase in the continuous phase expansion if it is assumed that there is no change in the bubble throughflow. When the horizontal tube is inserted across the axis of the bed above the pressure detector probes, the change in bubbling behaviour detected is less marked and the rate of increase in bubble frequency tends rather to reduce with increase in bed operating temperature beyond that at which there was a change in the heat transfer coefficient dependence. The relationship between frequency and temperature over the whole temperature range (Figure 4) is now similar to that obtained beyond the transition point for the onset of "B-type" behaviour with the unobstructed bed (Figure 3). It is also noticeable that the bubble rise velocity and pierced height are higher when the tube is in place which is suggestive of the presence of the tube having affected the upstream bubbling behaviour. The work of Buyevich et al. (12), referred to above, illustrates the complex effect of an immersed slab on local bubbling behaviour. Insertion of a water cooled tube further affects the bed behaviour in that the gas adjacent to the tube will be at a sustained lower temperature than in the bulk of the bed whereas the unsteady state calorimetric sphere experiments involve measurements averaged over the full range of temperatures from ambient to bed. On this reasoning, it would seem that, whereas at the location of the probes behaviour is consistently "B" type with the tube in position because of its influence on the upstream bubble flow, adjacent to the tube, where the physical conditions are of necessity different, there is still a change in behaviour from "D" to "B" type at about 540°C. Few studies have been made of the effect of operating temperature on bubbling behaviour and no others close to this apparent transition point. Otake et al. (14) reported that the bubbling frequency increased with increase in operating temperature but that the rate of increase declined at higher temperatures. Frequency was little affected by gas flow rate from their experiments with "B" Group beds which is consistent with our findings. Sitthiphong et al. (15) worked with large particles and very much into the "D" regime at $Ar > 10^5$ and found an increase in bubble eruption diameter with increasing bed temperature under similar conditions of excess gas velocity and static bed height suggesting that bubble size increases with operating temperature for such "D" type beds. It is likely that wall effects were significant in their experiments but this is the opposite trend to that noted by Yoshida et al. (16), Tone et al. (17), and Stubington et al (18) who all noted a tendency for the bubble size to decrease in their Group "B" bed experiments.

ACKNOWLEDGEMENTS

The support of SERC for this work is gratefully acknowledged.

NOTATION

Ar	Archimedes number = $g d_p^3 \rho_g (\rho_p - \rho_g)/\mu_g^2$
d_p	particle diameter, m
f	mean bubble point frequency, H_z
g	gravitational acceleration = 9.81 m/s^2
h_B	mean bubble pierced height, m
h_{max}	maximum average heat transfer coefficient, W/m^2K
h_{pcmax}	maximum particle convective heat transfer coefficient, W/m^2K
k_g	gas thermal conductivity, W/m K
Re_{mf}	minimum fluidization Reynolds number = $d_p U_{mf} \rho_g/\mu_g$
u_y	mean bubble rise velocity, m/s
U	superficial gas velocity, m/s
U_{mf}	minimum fluidization velocity, m/s
ΔY_c	mean vertical bubble separation, m
ρ_g	gas density, kg/m^3
ρ_p	particle density, kg/m^3
μ_g	gas viscosity, Ns/m^2

Zabrodsky correlation (8),

$$h_{pcmax} = 35.8\ \rho_p^{0.2}\ k_g^{0.6}\ d_p^{-0.36}$$

Denloye and Botterill correlation (9),

$$h_{max} = \frac{0.843\ Ar^{0.15}\ k_g}{d_p} + \frac{0.86\ Ar^{0.39}\ k_g}{d_p^{0.5}} \quad ;\ 10^3 < Ar < 2 \times 10^6$$

LITERATURE CITED

1. Botterill, J.S.M. and Y. Teoman, "Fluid-bed behaviour at elevated temperatures", in Fluidization, Grace, J.R. and J.M. Matsen (Ed.), Plenum Press, New York, 1980.

2. Botterill, J.S.M., Y. Teoman and K.R. Yüregir, Powder Technol., 39, 117 (1984).

3. Botterill, J.S.M., Y. Teoman and K.R. Yüregir, "Factors affecting heat transfer between gas-fluidized beds and immersed surfaces", Paper 2.4, XVI ICHMT Symposium "Heat and Mass Transfer in Fixed and Fluidized Beds", Dubrovnik, Sept. 1984.

4. Geldart, D., Powder Technol., 7, 285 (1973).

5. Sitnai, O., Chem.Eng.Sci., 37, 1059 (1982).

6. Yüregir, K.R., "Temperature Effects on Fluidized Bed Behaviour", Ph.D. Thesis, Birmingham University, 1982.

7. Zhang, M.C., P.M. Walsh and J.M. Beér, "Determination of bubble size distributions from pressure fluctuations in a fluidized bed combustor", p.75, 7th International Conference on "Fluidized Bed Combustion", Philadelphia, October 1982.

8. Zabrodsky, S.S., Hydrodynamics and Heat Transfer in Fluidized Beds, p.234, M.I.T. Press, Mass (1966).

9. Denloye, A.O.O. and J.S.M. Botterill, Powder Technol., 19, 197, (1978).

10. Golan, L.P. and D.C. Cherrington, A.I.Ch.E.Symp.Series, No.208, 77, 374 (1981).

11. Decker, N.A. and L.R. Glicksman, "Local voidage around horizontal cylinders immersed in fluidized beds", Paper 2.1, XVI ICHMT Symposium "Heat and Mass Transfer in Fixed and Fluidized Beds", Dubrovnik, Sept. 1984.

12. Buyevich, Yu.A., V.N. Korolyov and N.I. Syromyatnikov, "Hydrodynamic conditions for the external heat exchange in granular beds", paper 4.2, ibid.

13. Borodulya, V.A., A. Buyevich and V.I. Dikalenko, J.Eng.Physics, BSSR Academy of Sciences, 41, 678 (1981).

14. Otake, T., S. Tone, M. Kawashima and T. Shibata, Jnl.Chem.Eng.Japan, 8, 388 (1975).

15. Sitthiphong, N., A.H. George and D. Bushnell, Chem.Eng.Sci., 36, 1259 (1981).

16. Yoshida, K., S. Fugii and D. Kunii, "Characteristics of fluidized beds at high temperature", in Fluidization Technology Vol.1. p.43, Keairns, D.L. (Ed.), Hemisphere Publishing Corporation, Washington (1976).

17. Tone, S., H. Seko, H. Maruyama and T. Otake, Jnl.Chem.Eng.Japan., 7, 44 (1974).

18. Stubington, J.F., D. Barrett and G. Lowry, Chem.Eng.Res. and Des., 62, 173 (1984).

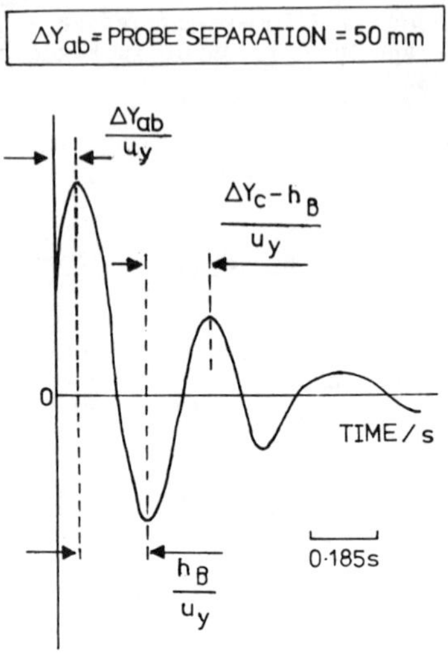

Figure 1 Correlogram.

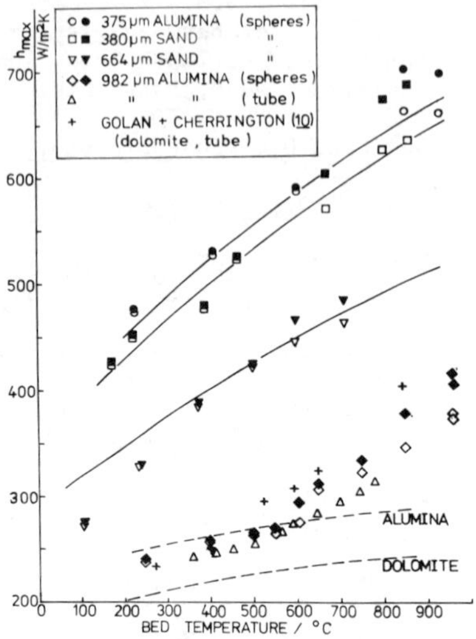

Figure 2 Calorimetric sphere and tube heat transfer data; ——— Zabrodsky (8), ----- Denloye and Botterill (9) correlations; open symbols gold surface, closed symbol copper surface.

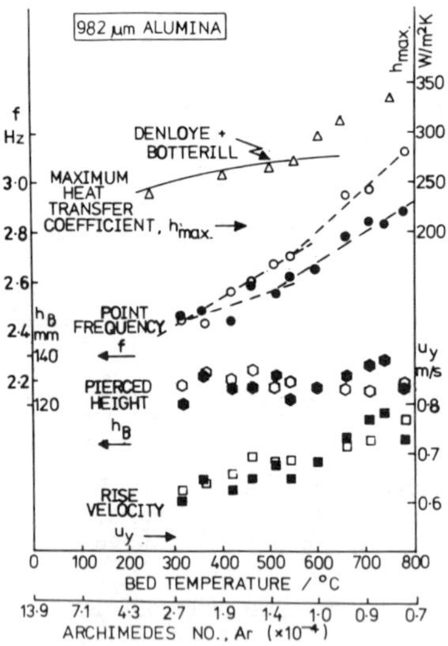

Figure 3 Temperature dependence of bubble properties and maximum heat transfer coefficient, calorimetric sphere experiments, open symbols $u-u_{mf}$ = 0.5 m/s, closed symbols $u-u_{mf}$ = 0.6 m/s.

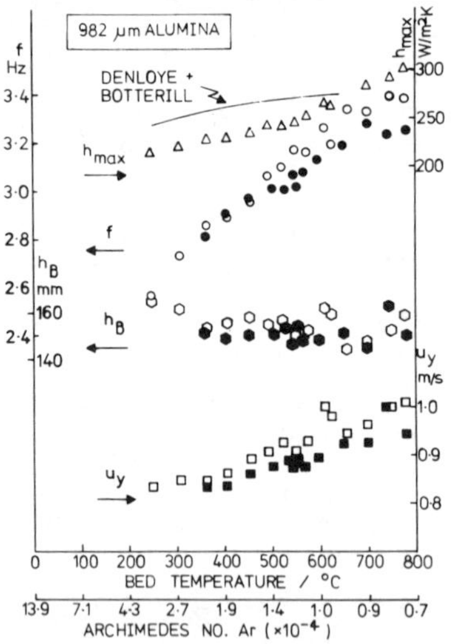

Figure 4 Temperature dependence of bubble properties and maximum heat transfer coefficient, horizontal tube experiments, open symbols $u-u_{mf}$ = 0.5 m/s, closed symbols $u-u_{mf}$ = 0.7 m/s.

SHALLOW-FLUID-BED TUBULAR HEAT EXCHANGER

Liu Dalu, Liu Juhua, Li Tonglan and Mooson Kwauk
Institute of Chemical Metallurgy, Academia Sinica
Beijing, People's Republic of China

ABSTRACT

Dynamics of a shallow fluid-bed is characterized by the presence of four consecutive zones: air cushion, dense bed, cloud and fog. Heat transfer to tubes immersed in a shallow fluid-bed maximizes itself against gas velocity, tube position, and settled bed height. Based on these findings, an activated solids shallow-bed heat exchanger is proposed for recovering heat from waste gases at minimal pressure drop.

SCOPE

Early experiments (1),(2) on heat transfer to tubes immersed in fluidized beds were carried out in beds seldom less than 500 mm in height. Enhancement of heat exchange from a gas to tubes by immersion in fluidized beds calls for minimal bed depth in order to reduce the pressure loss of the fluidizing gas. Shallow fluid-beds (3),(4),(5) are characterized, in addition, by small bubble diameters, absence of slugging, high solids expansion ratio, and low TDH. The shallow fluid-bed, together with fast fluidization and dilute-phase heat transfer, belongs to that class of gas-solid contacting designated as "bubble-less" (6), known to yield high transfer rates at relatively low energy expenditures. Design and scale-up for bubble-less Gas/Solids contacting devices will follow a course of its own, essentianlly divorced from the usual bubbling model, and possible aided by the relatively simple correlations of particulate fluidization (7).

The present investigation records the dynamic characteristics of a shallow fluid-bed and presents heat transfer rate to finned tubes immersed in a shallow fluid-bed as affected by gas velocity, position of heat exchange tubes in the bed, height of the settled solids bed, particle diameter, and the admixture of particles of a smaller size.

CONCLUSION AND SIGNIFICANCE

When a shallow bed of solids was fluidized, it was observed that the major solids form a dense phase buoyed on top of an air cushion next to the gas distributor, and capped by a cloud zone diminishing in solids concentration into a foggy constitution at the very top. The measured overall pressure drop of the solids bed was found to be less than that due to their weight.

Heat transfer rate to tubes immersed in a shallow fluid-bed was found to assume maximal values with respect to gas velocity, to the position of placement of the tubes and to the settled bed height. Heat transfer is favored by smaller particles, so much so, that the performance of a coarse- particle bed can be improved by admixture of fines. Also, high thermal conductivity of the solid particles, also results in better heat transfer.

These basic findings led to the proposed design of a cylindrically shaped actived solids shallow-bed heat exchanger (8), in which particles already suspended by a small activating gas stream are used to recover heat from a large stream of non-fluidizing waste gas stream at minimal pressure drop.

EXPERIMENTAL

Dynamic Characteristics of a Shallow Fluid-Bed

The dynamic behavior of shallow-bed fluidization was observed in a two-dimensional bed, measuring 40 x 150 mm, using ion-exchange resin

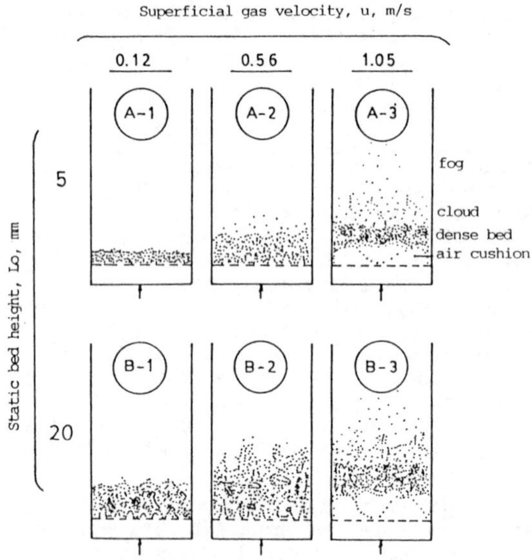

Figure 1 Shallow bed behavior, Resin, d_p=0.49 mm, u_{mf}=0.104 m/s

beads with an average diameter d_p=0.49 mm and fluidized with air. As shown in Figure 1, for a settled bed height of L_o=5 mm, the solids first acquired a vibrating surface at a gas velocity of u=0.12 m/s (Figure A-1) with a row of small jets registering with the apertures of the gas distributor. As the gas velocity was increased to u=0.56 m/s (Figure A-2), all solids were fully fluidized, with the row of jets now easily discernible With further increase in gas velocity to u=1.05 m/s (Figure

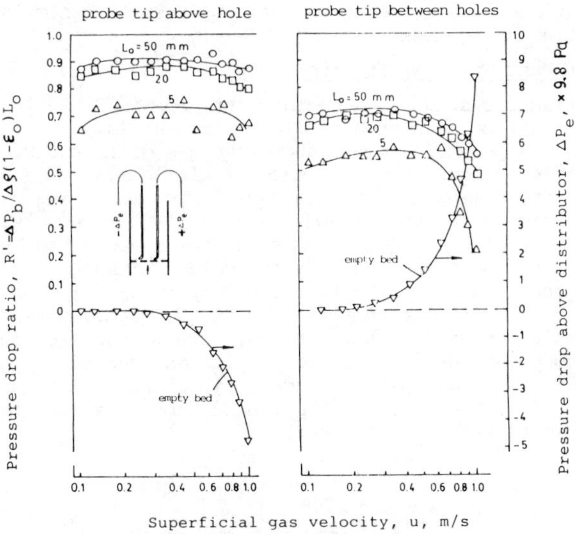

Figure 2 Pressure drop for shallow bed, Resin, d_p = 0.49 mm

A-3), the jets merged into an air cushion having very sparse particles population, supporting a dense bed of highly expanded solids, which was topped by a cloud layer with diminishing solids concentration until it lapsed into a foggy formation in the topmost region.

For a settled bed height of L_o=20 mm, full fluidization took place at the lower velocity level of u=0.12 s/m (Figure B-1), and the air cushioning phenomenon was equally evident at the high gas velocities (Figure B-3). For both the 5-mm and 20-mm beds, bubbles in their usual forms, were not identifiable. However, when a bed height of 50 mm was used, bubbles were observed through the gas velocity range adopted.

The four regions —— air cushion, dense bed, cloud and fog —— were found to be highly reproducible using the apparatus and the particles described.

The pressure drop ΔP of the shallow fluid-bed was measured with a probe inserted next to the top of the gas distributor. Two cases were differentiated: when the tip of the probe was above a hole of the air distributor (lefthand-side diagram of Figure 2), and when the tip was above the blank metal and between holes (righthand-side diagram). In the first instance, the measured pressure drop for the empty vessel was negative, while for the second, positive, as shown in the two diagrams of Figure 2. Then the vessel was filled with solids, and the pressure drop ΔP_b through the solids bed was measured at different gas velocities resulting in the family of three curves for the three settled bed heights of L_o=5, 20 and 50 mm. It should be noted that for both cases of probe placement, the measured pressure drop as compared to the solids weight, expressed as $R' = \Delta P_b / \Delta\rho(1-\varepsilon_o)L_o$, is always less than unity, indicating that for a shallow fluid-bed, its pressure drop ratio is always less than that due to the solids weight. The result is similar to that obtained by Liu and Squires (9).

Apparatus for Heat-Transfer Studies

For heat-transfer studies, a rectangular bed of 120 x 510 mm was used, in which a tube rack of a single horizontal layer of five tubes was immersed at adjustable levels above the gas distributor, as shown in Figure 3. Heat exchange tubes were made of aluminum, having a tube outside diameter of 14 mm with extruded radial fins measuring 2.5 mm in height and 0.6 mm in thickness, and placed at 5-mm intervals. Hot water, at a termperature of 95°C, from a constant temperature tank was supplied at known rates to the tubes, to heat the cold fluidizing air, and the rate of heat input was computed from the temperature rise of the water and its flow rate. Both the temperatures of the water, in and out, and of the bed, at three different locations, were measured by means of 0.2-mm bare copper- constantan thermocouples and recorded by a Thermal Logger VS-4300-A digital instrument on strip charts at an accuracy of 0.1°C. The total surface area of the finned tubes was used for calculating the heat transfer coefficient.

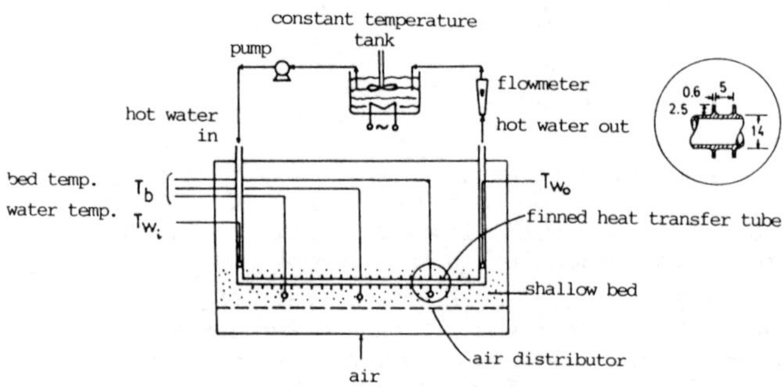

Figure 3 Experimental apparatus

The solid particles used were sand and copper beads having the following sizes with their corresponding minimum fluidization velocities:

Sand ρ_s=2.65 g/cm^3	$\bar{d}_p$, mm	0.085	0.125	0.175	0.225	0.283	0.335	0.425
	u_{mf}, m/s	0.008	0.012	0.030	0.035	0.058	0.092	0.12
Copper beads ρ_s=8.6 g/cm^3	$\bar{d}_p$, mm	0.08	0.14	0.175	0.25			
	u_{mf}, m/s	0.02	0.07	0.09	0.15			

Factors Affecting Heat-Transfer Rates

GAS VELOCITY. The curves in Figure 4 show that for any bed height, heat transfer coefficient passes through a maximum $(h_{max})_u$ as gas velocity increases. The maximum usually occurs at some common gas velocity, e.g., 0.175 m/s for sand and 0.37 m/s for copper beads.

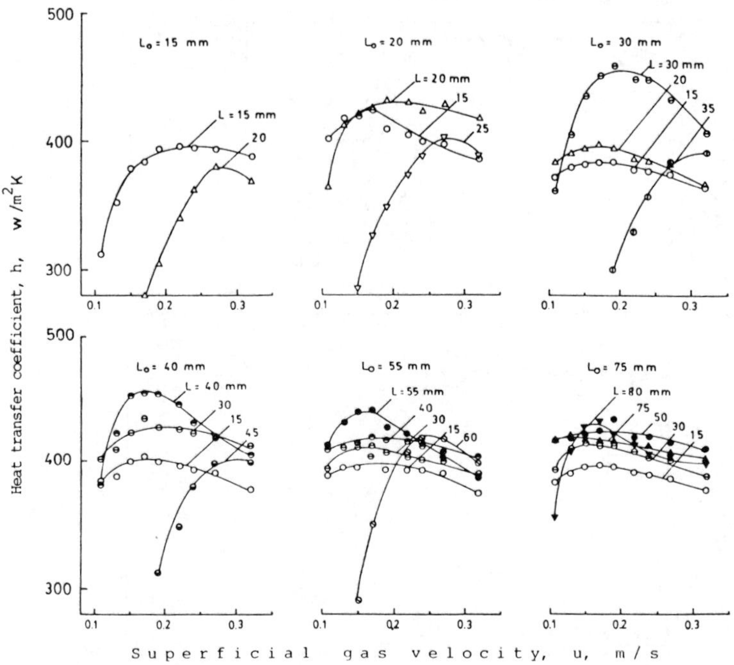

Figure 4 Variation of heat transfer coefficient with superficial gas velocity, sand d_p = 0.175 mm

TUBE PLACEMENT. Comparison of the members of any family of curves shown in Figure 4, indicates that heat transfer also maximizes itself on the position of tube placement above the gas distributor. Figure 5 plots the relative heat transfer coefficient to that at the velocity-maximum, $h/(h_{max})_u$, against the relative tube position to that of the settled bed height, L/L_o, and shows that for beds which are shallow, i.e., $L_o < 50$ mm, the tube-position maximum $(h_{max})_{u,L}$ occurs at a value of $L/L_o = 1$, indicating that for shallow fluid-beds, the optimum position of tube placement is at the top of the settled bed.

SETTLED BED HEIGHT. Figure 6 plots the velocity-and-tube-position maximum heat transfer coefficient $(h_{max})_{u,L}$ against settled bed height L_o for sand, and shows that the optimal bed height for this material is 30 mm, where a maximum value of 460 W/m^2K could be obtained.

MATERIALS. Figure 7 plots the thrice-maximum heat transfer coefficient, $(h_{max})_{u,L=L_o}$ against the particle size of the two materials, sand and copper beads, and shows that smaller particles are favored and that copper performs better than sand.

PARTICLE MIXTURES. While smaller particles give higher heat

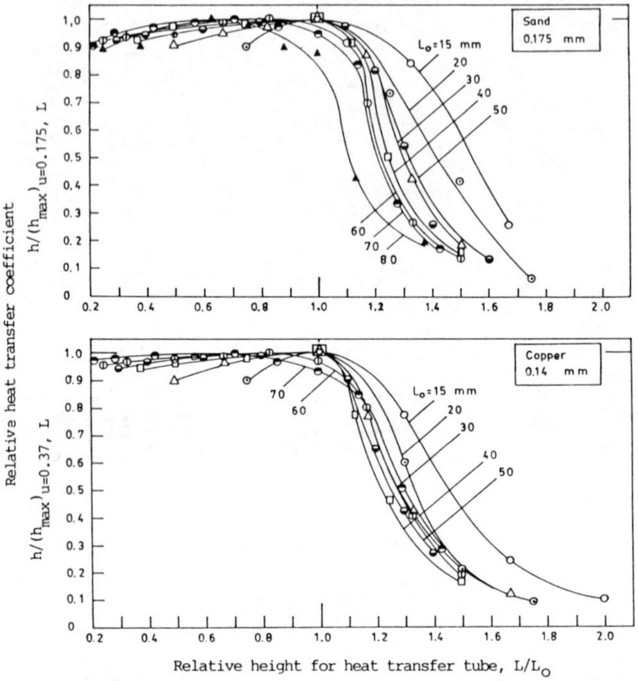

Figure 5 Variation of relative heat transfer coefficient with relative height of heat transfer tube

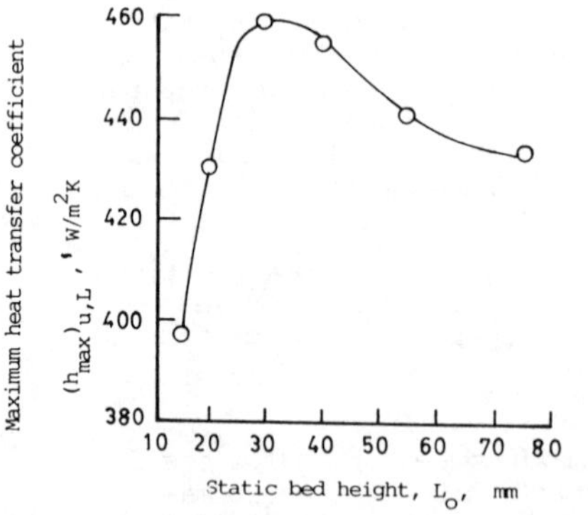

Figure 6 Variation of heat transfer coefficient with static bed height, sand d_p=0.175 mm

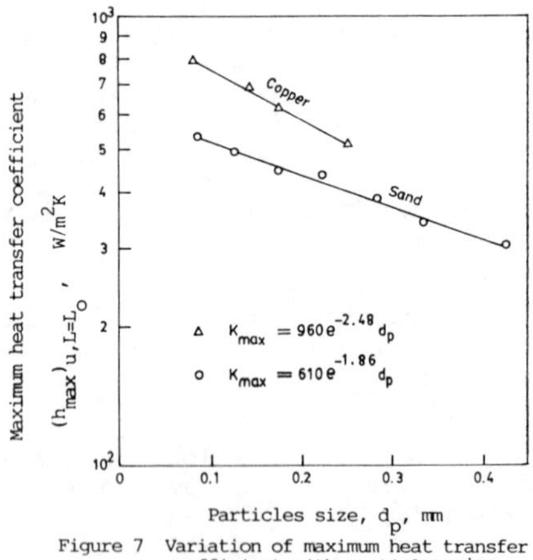

Figure 7 Variation of maximum heat transfer coefficient with particles size

transfer rates, Figure 8 shows the effect of intermixing a coarse particle, 0.283-mm sand, with a finer sand, 0.125. The salutary effect is evident from the ascending curves with increasing fines addition.

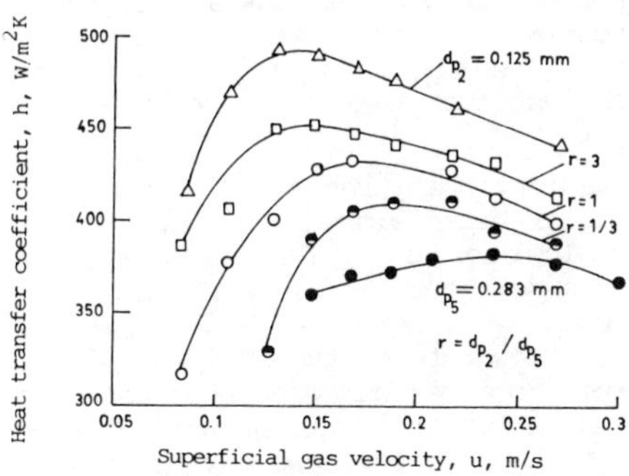

Figure 8 Variation of heat transfer coefficient with gas velocity, sand

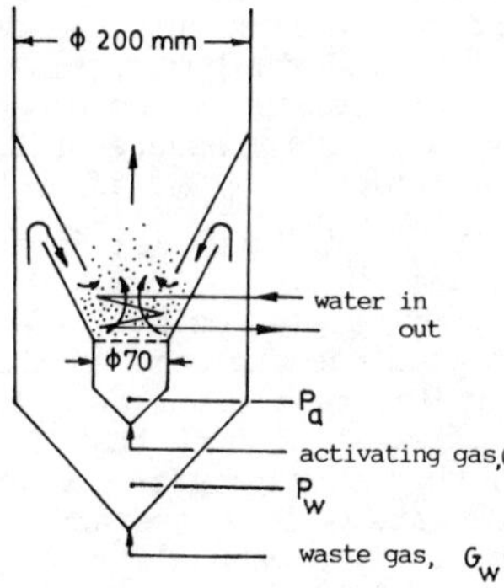

Figure 9 Activated solids shallow-bed heat exchanger

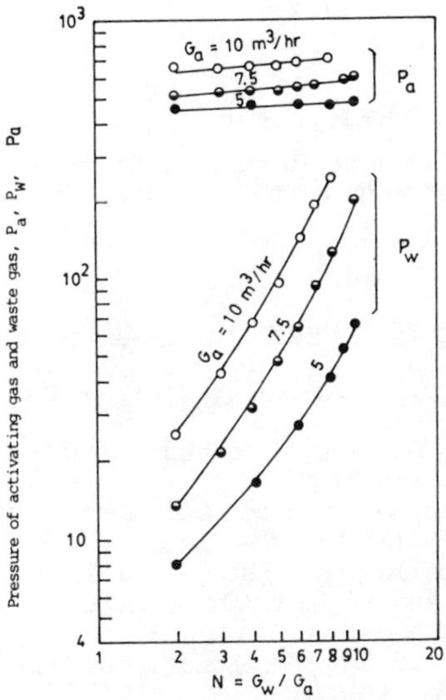

Figure 10 Variation of pressure of activating gas and waste gas with their flow ratio

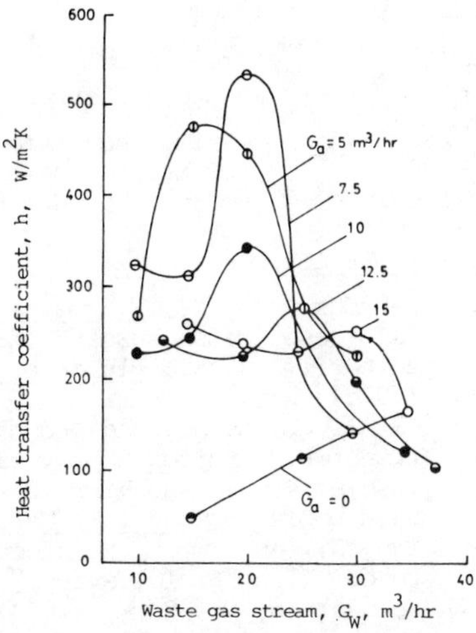

Figure 11 Variation of heat transfer coefficient with waste gas and activating gas, sand, d_p=0.175 mm

THE ACTIVATED SOLIDS SHALLOW-BED HEAT EXCHANGER

Shallow-fluid-bed tubular heat exchanger have been used in heat recovery from shipboard Diesel exhaust (4). For such an application, the exhaust gas has to overcome the full pressure drop of the fluid bed, and, besides, the gas distributor is subject to possible clogging due to the presence of carbonaceous materials in the hot gas. Figure 9 shows a cylindrically shaped shallow-fluid-bed tubular heat exchanger, designed to circumvent the above predicaments. Solid particles are fluidized with a small activating gas stream G_a, and the waste gas G_W, from which heat is to be recovered, is passed through the already suspended solid particles, thus experiencing a minimal pressure drop. Also, the waste gas does not pass through the gas distributor, in a manner similar to the heat tray described by Liu and Squires (9). Figure 10 shows the variation of pressure of both gas streams, G_a and G_W, with their flow ratio. Obviously the pressure of G_W was much less than that of G_a, even at a flow ratio $N=G_W/G_a$, as high as 10.

Due to difficulties in tube orientation, heat transfer studies were carried out in a 110 x 490 mm rectangular bed designed according to the same principle, using 12-mm diameter aluminum tubes. The hot water system used was the same as that in Figure 3. Figure 11 plots the heat transfer coefficient as affected by both gas streams, G_a and G_W. It is of interest to note that for any given flow rate of the activating gas G_a, the heat transfer coefficient rises to a maximum with the waste gas flow rate G_W. For the apparatus tested, a typical ratio of G_W/G_a for acceptably high heat transfer is about 4. Better design with the object of increasing this G_W/G_a ratio needs to be explored, such as recycling solids by jets or even by pneumatic transport.

NOTATION

G_a	activating gas stream, m^3/hr
G_W	waste gas stream (in Figure 11 air), m^3/hr
$(h_{max})_{u,L,L_o}$	maximum heat transfer coefficient in respect to gas velocity, to position of placement of tubes and to settled bed height, W/m^2K

LITERATURE CITED

1. Zabrodsky, S. S., Hydrodynamics and Heat Transfer in Fluidized Beds, MIT Press, Massachusetts (1966)
2. Botterill, J. F., Fluid-bed Heat Transfer, Academic Press, London (1975)
3. Al Ali, B. M. A., and Broughton, J., "Shallow fluidized bed heat transfer," Applied Energy, 3 (2), 101 (1977)
4. Cusdin, D. R., and Virr, M. J., "A marine fluidized bed waste heat boiler", Trans. I. Mar. E., 91, 1 (1979)
5. Yan, Z., and Kwauk, M., "Cocurrent shallow multi-stage fluid-bed reactor," in "Fluidization," D. Kunii and R. Toei (Ed.), Engineering Foundation, New York, (1984)
6. Kwauk, M., "Bubble-less G/S contacting," exchange-scholar lecture, Japan Society for Promotion of Science, (1983)
7. Kwauk, M., "Towards a unified hypothesis for fluidized systems," CHEMECA 80, 8th Australian Chemical Engineering Conference, Melbourne, Ausgust 24-27, (1980) Proceedings, p. 98-108
8. Liu, D. and Kwauk, M. "Recovering heat from waste gas by using multi-stage fluid beds" Chemical Metallurgy, (China), No.4 (1981)
9. Liu, Y. A., and Squires, A. M., "Fischer-Tropsch synthesis from a low H_2:CO gas in a dry fluidized-bed system," A technical progrees report, Virginia Polytechnic Institute and State University, (1983)

Shallow Fluid Beds of "Master Beads"

Jyh-Shing Yang, Y.A. Liu, and Arthur M. Squires

Department of Chemical Engineering
Virginia Polytechnic Institute and State University
Blacksburg, Virginia 24061, U.S.A.

ABSTRACT: Shallow fluid beds of spherical "Master Beads" (ca. 86% alumina) display nine "Types," of which seven may be useful. Coefficients of heat transfer from a heating tube are highest when the tube is situated near the surface of beds of "Types" that display a dense phase. A relatively high standard deviation of solid volume-fraction over time near the surface attests the greater solid "activity" in this region.

SCOPE: To study pressure and solid volume-fraction in shallow fluid beds, including those approaching zero solid inventory; to classify such beds in "Types"; and to study heat transfer in beds displaying a dense phase.

CONCLUSIONS and SIGNIFICANCE; Seven possibly useful "Types" of shallow bed are described. Heat transfer data for three sizes of "Master Beads" (ca. 86% alumina) conform, in general, with patterns established by earlier researches; but no consistent picture has emerged permitting the confident projection of shallow-bed heat-transfer behavior. The importance of shallow beds, especially for waste heat recovery, justifies further effort on their dynamic and heat-transfer behavior.

The late Douglas E. Elliott (*1*), with his students at Aston in Birmingham (*2*, *3*, *4*), pioneered study of "shallow" fluid beds (under 150 mm). This effort led quickly to commercial successes in both combustion of gases (*5*) and waste-heat recovery (*6*), where shallow beds offer the advantage of low gas pressure drop. Other shallow-bed studies have appeared (*7* - *11*), but no consistent picture for heat transfer in shallow beds has emerged. Data for sand of ca. 300 micrometers (μm) scatter by more than a factor of two (*2*, *9*, *10*). The efficiency of fins placed upon tubes in a shallow bed can be high (*6*), but two workers in Elliott's laboratory (*2*, *4*) obtained discrepant results.

We have undertaken studies to elucidate the dynamic and heat-transfer behavior of shallow beds. We give data here for "Master Beads," an inexpensive solid manufactured for use in grinding and as a packing in deep oil and gas wells, to prevent their closure through plastic deformation of the surrounding rock. Substantially spherical, particles of this solid are non-dusting in a fluid bed. Ca. 86% alumina, "Master Beads" contain 2-4% silica, 6-8% iron oxide, and 4-5% titania in solid solution; they have high body strength for impact and high wear resistance. Density is 3,680 kg/m^3. We have used three sizes: 12-20, 20-40, and 40-60 mesh (U.S. Standard); nominally, 1,260, 630, and 335 μm in diameter. Our experimental fluid bed is cylindrical and 101.6 mm in inside diameter.

Pressure in Shallow Beds

Fig. 1 gives representative data for pressure drop across shallow beds versus static bed height. We have data for three perforated-plate distributors (2.1 to 5.6% open area), two laminated plates (2.1 and 4.2%) purchased from

Dynapore (nos. 401430 and 407520), and three "micro-etched" screens (9.0 to 18.0%) from Buckbee-Mears Inc. (nos. 8-7, 6-6, and 14-2); also, for sand (nominally 359 and 630 μm), carbon (630 μm), and glass beads (353, 484, 705, and 842 μm). Normalized pressure drop, Δp, is linear in the reciprocal of static bed height, H: $\Delta p = a - b/H$, where a and b are functions of gas velocity, particle, and distributor. (The correlation coefficient is 0.984 for all of the data.) We have derived an equation of this form by equating the weight of fluidizing particles with the sum of supporting forces: (pressure drop + pressure recovery from high-velocity gas entering the bed) x (bed area) + (shear force between particles and bed wall). We will report elsewhere how we have correlated pressure recovery linearly with Reynolds and Froude numbers based upon gas velocity passing through the distributor.

Instantaneous Fluidized Density

We have modified Dutta and Wen's (<u>12</u>) optical probe to use infrared light, thereby permitting us to work in a lighted room and eliminating need for an amplifier. For solid volume-fractions smaller than about 0.03 (probe voltage $V \geq 6.305$), we use theory to deduce the volume fraction, ψ, from the voltage output, V, from the light probe: $\psi = (2/3)[1 - (0.1\ V)^{n}]$, where $n = D_p/L$, D_p is particle diameter, and L is the gap between infrared LED and photo transistor of the light probe (7.33 mm in our design). For higher ($V < 6.305$), two calibration procedures are in good accord:

(1) Fluidizing a known amount of 335 μm "Master Beads" at an air superficial velocity of 2.0 m/s, we obtained a vertical probe traverse of the bed. We varied the amount of solid from 10 to 250 g. Multiplying the above theoretical expression by a quadratic in V, we used a pattern-search technique to find coefficients in the quadratic giving minimum disagreement between known and calculated bed weights. The best quadratic was $(3.4 - 0.45\ V + 0.011\ V^2)$.

(2) We fabricated a chamber with transparent walls identical in size to the space between LED and photo transistor (4.13 x 7.33 x 7.62 mm). We placed a known quantity of "Master Beads" in the chamber and an identical amount into the space, enclosing the latter in black plastic. We placed both the chamber and the probe upon a vibrating table, permitting us to create vibrated beds (<u>13</u>) of the particles in both the chamber and the space. By adjusting a strobe light to flash in frequency with the vibration, we could observe the chamber to determine those frequencies and amplitudes that gave a uniform distribution of the particles. Frequencies between 20 and 28 Hertz and amplitudes providing a maximum upward acceleration greater than 3.5 times gravity were suitable. Fig. 2 gives calibrations obtained from vibrated beds of known quantities of solid. For 335 μm "Master Beads," disagreement between solid fractions determined by our two calibration procedures was always less than 4% of the value of the fraction.

Fig. 3 gives illustrative data from the probe, showing variability of solid fraction. Our probe obtains 600 data points per second, and we routinely acquired 3,000 points before determining average and standard deviation. Fan *et al*. (<u>11</u>) gave data showing frequencies of fluctuation in somewhat deeper beds (30 to 100 mm static bed height) comparable to the roughly 20 Hertz of Fig. 3.

"Types" of Shallow Fluid Bed

We identify the following "Types" of shallow fluid beds of Group B/D solids:

A. Similar to the familiar "deep" bubbling beds: clearly defined surface (although fluctuating owing to passage of bubbles) with spray of particles into the freeboard.

B. A shallower static bed than A, exhibiting a more diffuse surface. Elliott and his colleagues (3) spoke of a "cloud phase" interposed between a denser, bubbling zone below and a "spray zone" above.

C. A still shallower static bed, exhibiting a "dimpled" surface. The size of dimples increases with greater static bed height.

D. An even shallower static bed, but with swirling solids reaching much higher elevations than in C.

E. Portions are not fluidized; the risk of this is greater if the open area of the distributor is large.

F. A flatter surface than in C, and with a sharper drop in solid fraction at elevations above the bed.

G. No dense region; solids fly high, but without the swirling exhibited by D.

H. Sinuous "lines" of particles swirl back and forth upon the distributor; solids fly high, but with little swirl.

I. Solids are mostly "dead," but high-velocity gas from the distributor carry a few particles to high elevations.

Fig. 4 gives profiles of solid fraction for "Type A", "B", "C", "F", "G", and "H" beds together with sketches indicating their features.

Fig. 5 gives "maps" of the "Types." The location of the boundary between "Type I" and "Type E" and the seven "Types" that might be useful is a function of the distributor.

Heat-Transfer Data

We have used a "heat probe" developed in our laboratory (13) to determine heat transfer as a function of probe elevation. The probe is heated electrically and is designed to give substantially uniform temperatures over its entire surface even if local heat transfer were to vary markedly. The probe is cylindrical, 12.7 mm in diameter and 80 mm long.

Fig. 6 gives heat transfer coefficient determined at various probe elevations for a "Type B" bed of 335 μm "Master Beads." As others have observed, the coefficient is higher when the probe is placed in Elliott's "cloud phase," near the surface of the bed. Placing the probe near the distributor also leads to higher heat transfer, but this position might not be attractive in practice because of danger of erosion of a metal tube. Fig. 6 also gives profiles of pressure, solid fraction, and standard deviation of solid fraction. The latter data indicate zones of greater "activity" near the distributor and in the cloud phase, corresponding to regions of higher heat transfer.

Figs. 7 and 8 gives heat-transfer data for 335 μm "Master Beads" over a range of gas velocity and static bed height. For purposes of orientation, the figures show the curve of heat transfer coefficient drawn in Fig. 6. Although it would be confusing to draw curves through all of the data in Figs. 7 and 8, examination will reveal that each set of data shows higher heat transfer near the bed surface than in a "quieter" region below.

Figs. 9 and 10 compare heat transfer for three sizes of "Master Beads." The data display the usual trend toward higher heat transfer for smaller sizes.

Fig. 11 compares a typical set of our results for 335 μm "Master Beads" with heat transfer coefficients calculated from published correlations (*7*, *10*, *14*). Fig. 12 compares calculated coefficients with typical results for our three particles sizes. Clearly, more work will be needed before we can claim to understand shallow fluid beds.

Acknowledgment: The U.S. Department of Energy (Grant No. DE-FG-82P50791) and the Virginia Center for Coal and Energy Research (Project No. 2057090) supported our work. We are grateful to Norton Alcoa for supplies of "Master Beads" of its manufacture.

Literature Cited:

1. Elliott, D.E., "Possible Use of Fluidized-Bed Combustion and Heat Transfer," Paper 22, The Total Energy Conference, The Institute of Fuel, Brighton, England, November 1971.

2. Al-Ali, B.M.A., "Fin Spacing Interaction in Fluidised Bed Heat Exchangers," Ph.D. thesis (mentor: D.E. Elliott), Univ. of Aston in Birmingham, May 1976.

3. —— and J. Broughton, *Applied Energy*, 3 (1977), 101.

4. Atkinson, G.A., "Extended Surface Fluidised Bed Heat Transfer," Ph.D. thesis (mentor: D.E. Elliott), Univ. of Aston in Birmingham, October 1974.

5. Broughton, J., "Combustion of Gases in Fluidized Beds," in *Fluidized Beds: Combustion and Applications*, pp. 305-339, J.R. Howard (Ed.), Applied Science Publishers, New York (1983).

6. Virr, M.J., "Fluidized Bed Heat Recovery Systems," *ibid.*, pp. 341-366; M.J. Virr and H.W. Williams, *Chem. Eng. Prog.*, 81 (7), 50 (July 1985).

7. Abu Bakar, M.Y., "Heat Transfer Studies between a Horizontal Tube and Shallow Fluidized Bed," Ph.D. thesis (mentor: M.A. Bergougnou), Univ. of Western Ontario, June 1981.

8. ——, J.D. Tarasuk, M.A. Bergougnou and J.L. Sullivan, in *Fluidization*, pp. 355-362, D. Kunii and R. Toei (Eds.), Engineering Foundation, New York (1984).

9. Aerojet Energy Conversion Company, "Evaluation of Fluid Bed Heat Exchanger Optimization Parameters," Report No. 8100-FB-0001 to U.S. Dept. of Energy, Contract No. DE-AC22-78ET11343, March 1980.

10. Andeen, B.R., and L.R. Glicksman, ASME Paper 76-HT-67, presented at ASME-AIChE Heat Transfer Conference, St. Louis, August 1976.

11. Fan, L.T., T.C. Ho, N. Yutani, and W.P. Walawender, in *Fluidization* (*8*), pp. 15-24.

12. Dutta, S., and C.Y. Wen, *Can. J. Chem. Eng.*, 57 (1969), 115.

13. Sprung, R., B. Thomas, Y.A. Liu, and A.M. Squires, in *Proceedings*, Fifth Engineering Foundation Conference on Fluidization (this volume).

14. Vreedenburg, H.A., *Chem. Eng. Sci.*, 9, 52 (1958).

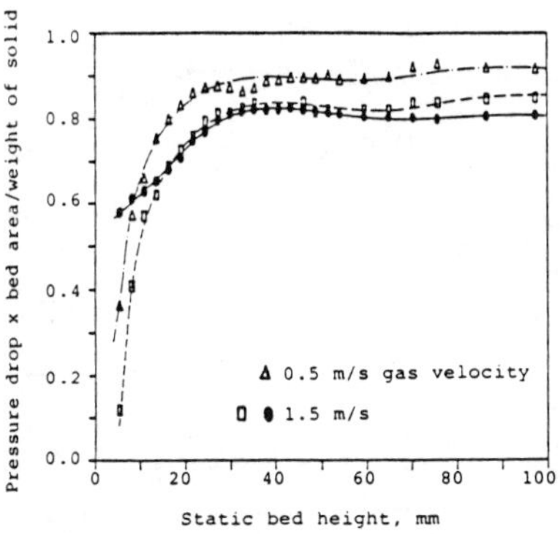

Fig. 1. Normalized pressure drop across shallow fluid beds of 335 μm "Master Beads"

Broken curves: perforated plate with 2.1% open area; 1.5 mm holes on 9.4 mm triangular pitch.

Solid curve: screen with 18.0% open area; 0.38 mm holes on 0.79 mm triangular pitch.

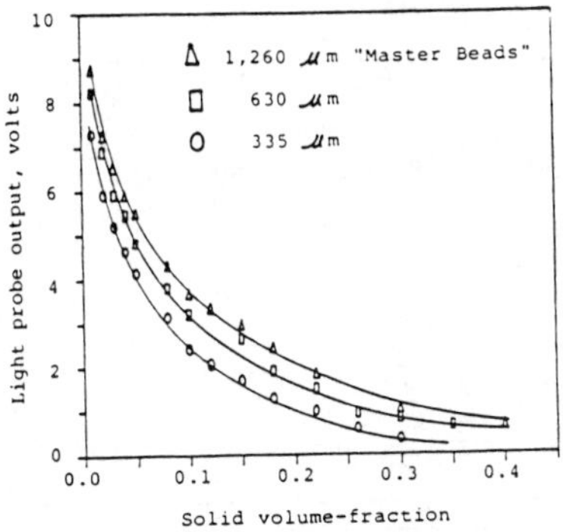

Fig. 2. Calibration of light probe for vibrated beds of three sizes of "Master Beads"

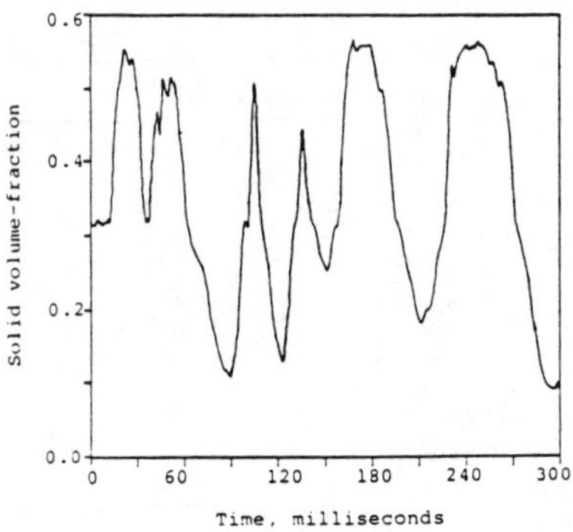

Fig. 3. Illustrative example showing variability of solid fraction

335 μm "Master Beads"
1.5 m/s superficial gas velocity
10.8 mm static bed height
Light probe at 5 mm elevation

Gas distributor as in Fig. 4(a).

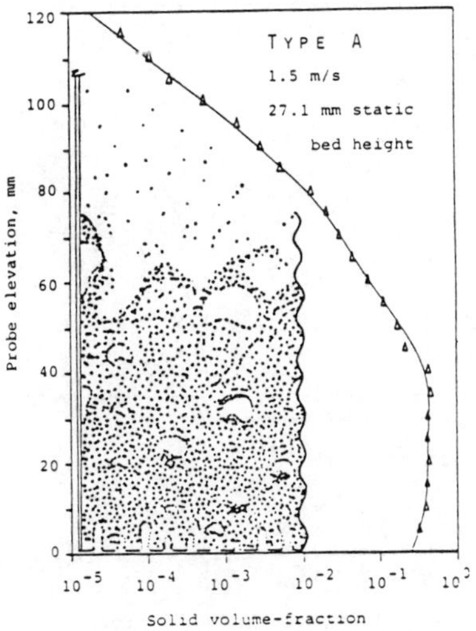

Fig. 4(a). "Type A" shallow fluid bed. Figs. 4(a) and 4(b) give data for 335 μm "Master Beads"

Gas distributor: 5.6% open area; 1 mm orifices on 4.01 mm triangular pitch.

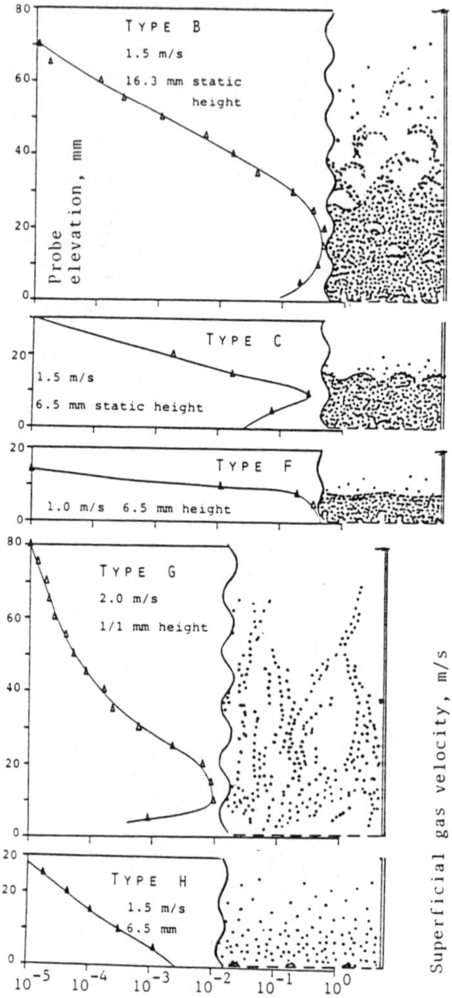

Fig. 4(b). "Type B", "C", "F", "G", and "H" shallow fluid beds.

Gas distributor as in Fig. 4(a).

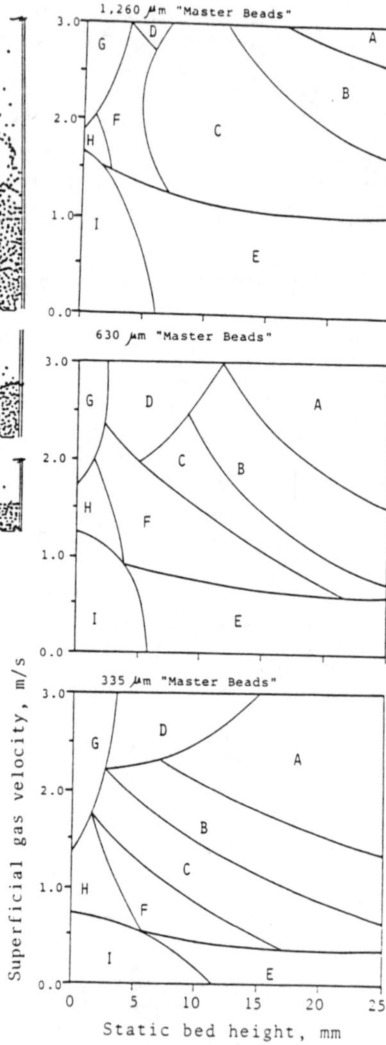

Fig. 5. "Types" of shallow fluid beds for three sizes of "Master Beads"

Gas distributor as in Fig. 4(a).

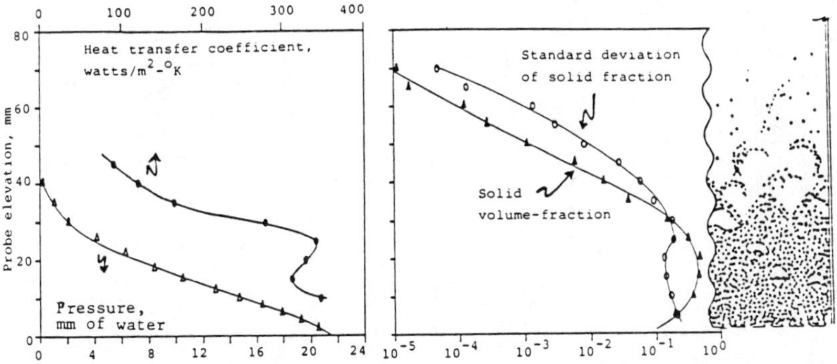

Fig. 6. Pressure, heat transfer coefficient, solid-fraction, and standard deviation of solid-fraction at various elevations in a "Type B" shallow fluid bed of 335 μm "Master Beads"

1.5 m/s superficial gas velocity; 16.3 mm static bed height; distributor as in Fig. 4(a).

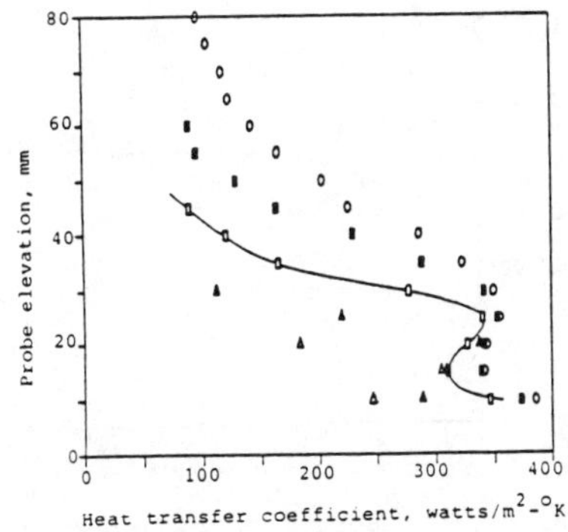

Fig. 7. Heat transfer for 335 m "Master Beads" at 16.3 mm static bed height

△ 0.5 m/s superficial gas velocity

▲ 1.0 m/s

□ 1.5 m/s

■ 2.0 m/s

○ 2.5 m/s

Distributor as in Fig. 4(a)

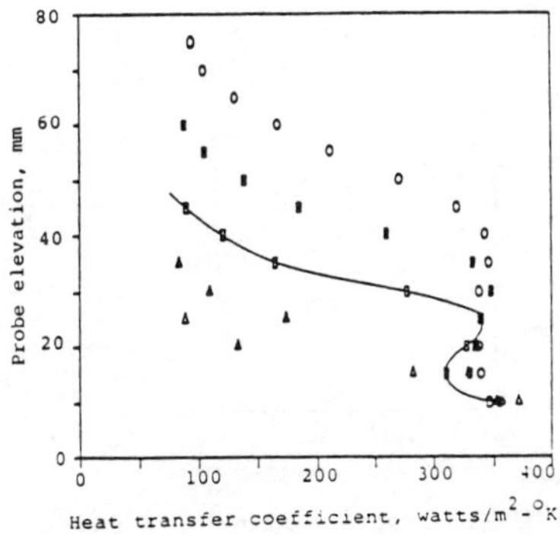

Fig. 8. Heat transfer for 335 m "Master Beads" at 1.5 m/s superficial gas velocity

△ 9.2 mm static bed height

▲ 12.5 mm

□ 16.3 mm

■ 21.7 mm

○ 27.1 mm

Distributor as in Fig. 4(a)

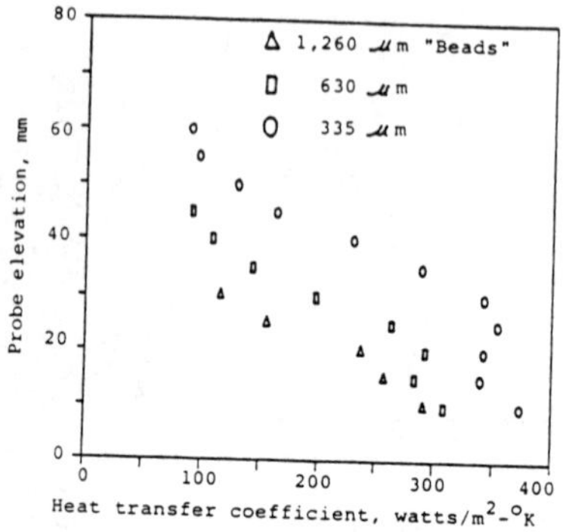

Fig. 9. Heat transfer for three sizes of "Master Beads" at 16.3 mm static bed height and 2.0 m/s gas velocity.

Distributor as in Fig. 4(a).

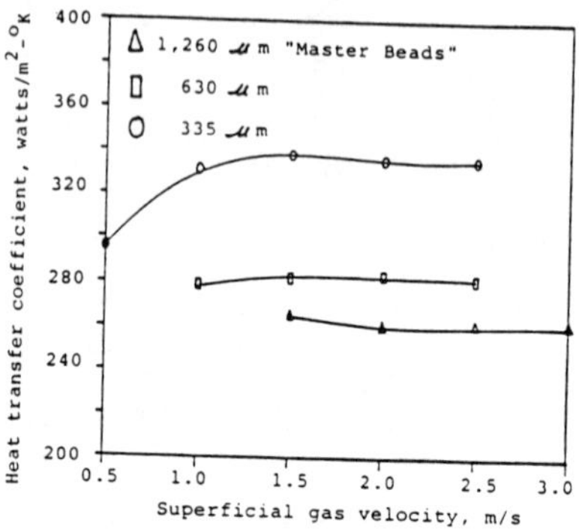

Fig. 10. Heat transfer for three sizes of "Master Beads" at 27.1 mm static bed height and 20 mm probe elevation.

Distributor as in Fig. 4(a).

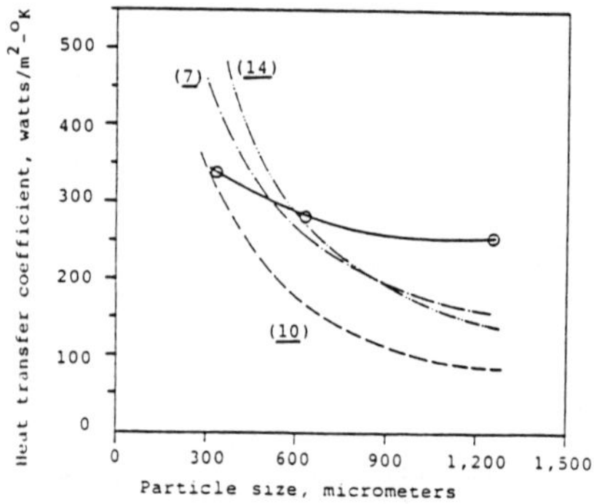

Fig. 11. Comparison of our data with published correlations.

335 μm "Master Beads"; 27.1 mm static bed height; probe at 30 mm elevation

-··—··—··— Vreedenburg (14)

—·—·—·—Abu Bakar (7)

– – – – – – Andeen and Glicksman (10)

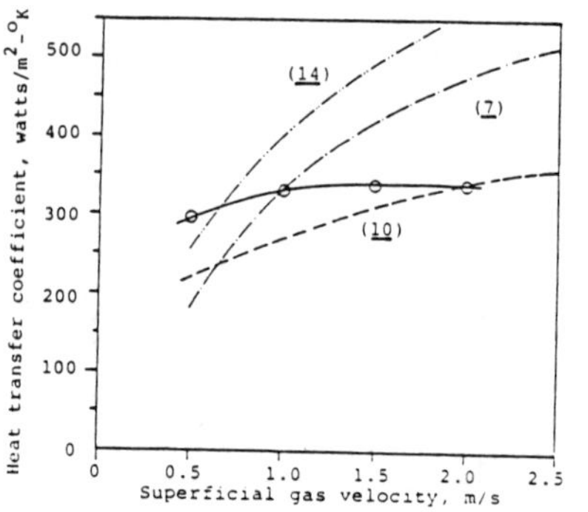

Fig. 12. Comparison of our data with published correlations.

1.5 m/s superficial gas velocity; 27.1 mm static bed height; probe at 30 mm

-··—··—··— Vreedenburg (14)

—·—·—·—Abu Bakar (7)

– – – – – – Andeen and Glicksman (10)

Shallow Vibrated Beds of "Master Beads"

Renato Sprung, Benku Thomas, Y.A. Liu, and Arthur M. Squires

Department of Chemical Engineering
Virginia Polytechnic Institute and State University
Blacksburg, Virginia 24061, U.S.A.

ABSTRACT: Arrangements for studying heat transfer, solid movement pattern, and instantaneous floor pressure in shallow "vibro-beds" are described; also, for phase-shift photography of such beds. At ca. 25 Hertz and a maximum upward acceleration of ca. 5 g's, transfer of heat from a single hot tube to 177 μm "Master Beads" can attain a value beyond 450 watts/m^2-K. This result can be understood qualitatively, in relation to results for coarser and finer "Master Beads," in light of (1) the fraction of the time the tube spends in contact with different solids and (2) patterns of movement in the solids.

SCOPE: To study heat transfer, solid movement, and instantaneous floor pressure in a wide, shallow vibrated bed fitted with a single heating tube.

CONCLUSIONS and SIGNIFICANCE: Transfer of heat between a tubular surface and a shallow vibrated bed can achieve levels beyond 400 watts/m^2-K. An application is visualized in which heat from horizontally flowing "supernatant" gas would move via a number of contact stages to a long, shallow vibro-bed and onward from the bed to cooling surface within the bed.

"Vibrated beds" are in wide use for drying granular solids. Pakowski *et al.* (*1*), who estimated that drying covers 90% of all applications, reviewed both vibrated beds and "vibrated fluid beds": the former comprises solids levitated solely by a vibrating solid floor; the latter, a bed into which gas flows from a vibrating porous floor. They cited 135 references, the preponderance from Poland and the U.S.S.R. Some authors speak of "vibro-fluidized" beds, and there is confusion over terminology. Vibrated beds little resemble gas-fluidized beds, and so we follow (*1*). But we will also speak of "vibro-beds," easier for the tongue; by vibro-bed we mean (*1*)'s vibrated bed.

The vertical displacement z of a plate vibrated vertically, sinusoidally, is $z_0 \sin \omega t$, where z_0 is the maximum displacement, ω is the angular frequency ($2\pi f$, where f is frequency), and t is time. Acceleration of the plate is $(-z_0\omega^2 \sin \omega t)$. A sufficient quantity of a granular solid forms a vibro-bed when a "vibration intensity parameter" $K = z_0\omega^2/g$ is greater than unity.

If one places just a very few coarse particles on a floor vibrating at 25 Hertz and K = 5, they bounce irregularly some 8 mm above the floor. The behavior can be understood simply by calculating trajectories of particles that, acting independently from one another, hit the floor at random times. If one slowly adds more particles to the plate, a point is reached at which some of the particles begin to move in concert. Bachman (*2*) reported that a layer six particles deep moves in a mass. A gap forms between the layer and the floor; above the layer is an "expanded zone" — particles looking a bit like the initial situation, before one had added sufficient solid to form a coherent bed, but expanding less than the earlier, freely bouncing particles. Gas moves vertically across the bed, reversing its direction of flow 25 times per second. "Supernatant" gas moves downward while the gap forms, and gas moves upward when the gap closes, maintaining an expanded condition in the surface zone.

Kroll (*3*) described a vibro-bed theoretically with the assumption that correlations for steady-state pressure drop in gas flowing through a packed bed would provide a good approximation for the transient forces acting in the dynamic situation. Gutman (*4, 5*) demonstrated that this assumption is valid for frequencies of interest and expanded Kroll's theory to include effects due to gas compressibility. For "narrow" vibro-beds of "coarse" solids, the Kroll-Gutman theory adequately predicts gap widths, gap durations (as a fraction of a vibrational cycle), and gas pressure within the gap versus time. In "wide" beds, solid circulation affects heat-transfer behavior. In beds sufficiently wide and deep, "bunkering" occurs — the piling up of solids at a preferred location. Buevich *et al.* (6) studied instabilities in beds to depths of 0.5 m and explained bunkering.

The heat-exchanger concept schematically indicated in Fig. 1 aroused our interest in wide, shallow vibro-beds. A horizontal flow of supernatant gas exchanges heat with the bed at a number of stages of contact created by vertical baffles descending a short distance into the bed. Preliminary study of the device has indicated that this distance can be "small" — i.e., creating but a low pressure drop at each contact stage — while providing a satisfactory approach between supernatant gas temperature and vibrated-solid temperature at each stage. We undertook the work reported here to elucidate the heat exchange between the solids and heat-transfer tubes placed within the vibro-bed.

We report data here for "Master Beads," a solid used in grinding and as a packing in deep oil and gas wells. Substantially spherical, the solid is non-dusting in a fluid bed. Ca. 86% alumina, "Master Beads" contain 2-4% silica, 6-8% iron oxide, and 4-5% titania in solid solution; they have high body strength for impact and high wear resistance. Density is 3,680 kg/m^3. We have used sizes ranging from 20 to 200 mesh (U.S. Standard); the figures note nominal sizes in micrometers (μm).

Experimental Arrangements

Figs. 2 and 3 sketch our two experimental rigs, providing cylindrical and "two-dimensional" vibro-bed spaces, respectively. A "heat probe" is mounted in the former for determining heat transfer coefficients; the latter has a dummy probe of the same diameter and at the same elevation. Side walls in Fig. 3 are transparent. We have taken pains to ensure that vibrations are strictly vertical; even a small deviation will cause bunkering. All data reported here are for a static bed height of 30 mm.

Fig. 4 shows our heat probe: within a copper tube is a heating coil wound on an aluminum support; the resistance is 10 ohm. At a heat release of 35 watts, no temperature difference around the tube's circumference should exceed 1.5°C. An electromagnetic vibrator, which drives the floor, requires vigorous air cooling. Heat from the probe is removed by heat transfer to the floor, cooled by the flow of air to the vibrator, and also by supernatant gas breathing up and down across the vibro-bed solids. Six thermocouples indicate bed temperature at locations that sample each major circulation pattern seen when the bed is viewed from above. The probe has four T-type thermocouples, in pairs at right angles to one another. In use, we place the probe so that one thermocouple faces upward; one, downward; and two, horizontally. Computer-controlled data acquisition monitors each thermocouple every 20 seconds. In general, the system requires roughly 20 to 40 minutes to reach a steady state (longer at lower K); we then gather data for 40 minutes, obtaining heat transfer coefficients at standard deviations typically about 1% of their value. Our reported values represent the heat-transfer performance based on the average of six bed-thermocouple readings. Mixing of solid appears to be more intense perpendicularly to the probe's axis than longitudinally. The two bed thermocouples closest to the probe (2 and 5 in Fig. 2) usually read a few degrees higher than the other four bed thermocouples. At the low end of the range of our measured coefficients, a value based only upon thermocouples 2 and 5 would, in general, fall about 15% higher than our reported value; at the high end, about 8% higher. We believe that thermocouples 1, 3, 4, and 6 reflect solid mixing longitudinal to the probe, and heat from the probe may not reach them as efficiently as it reaches 2 and 5. We will report elsewhere how deliberately induced solid circulation patterns affect heat transfer in vibrated beds.

Heat-Transfer Data

Fig. 5 gives heat transfer coefficients in vibro-beds of "Master Beads" at 25 Hertz. We have explored other solids and frequencies and will provide more data elsewhere. For the device of Fig. 1, frequencies between about 15 and 30 appear best; heat transfer at 50 Hertz is inferior. The best size of "Master Beads" appears to be about 177 μm.

Phase-Shift Photography

A velocity meter, attached to our vibrating system, gives the wave form of velocity, and electronic integration provides the wave form of displacement. A digital circuit locates the moment at which the displacement is zero and sends a signal, delayed by a desired phase angle (± ca. 8°), to a photographic flashgun, which illuminates the bed for 20 microseconds. In complete darkness, an open-lens camera photographs the two-dimensional bed of Fig. 3 at various phase angles.

Fig. 6 gives a series of pairs of phase-shift photographs, each pair at the same phase angle, for a vibro-bed of 707 μm "Master Beads" at 25 Hertz and vibration intensity parameter K = 4. Back-lit photos show gaps that form above and below the dummy heat probe. Front-lit photos give a better sense of the situation at the upper surface of the vibro-bed; the "dip" seen in back-lit photos above and a little to the left of the dummy probe is an artifact of the lighting.

For a vibro-bed of 177 μm "Master Beads" at 25 Hertz and K = 4, an arrangement of phase-shift photos in pairs loses information; Fig. 7 first presents back-lit photos and then front-lit ones at different phase angles.

Solid Motion and Instantaneous Pressures in Two-Dimensional Vibro-Bed

Fig. 8 illustrates solid movement patterns, and Fig. 9 gives instantaneous pressures at the floor of vibro-beds of three sizes of "Master Beads."

We judge solid motion from what we see through the two-dimensional bed's transparent wall — trying to remember that this may not always reflect accurately what is happening several particle layers away from the wall — and from the motion of a ca. 3 mm rubber "marker." From the motion of the latter, we believe that it quickly "discovers" a major solid recirculation path and takes up "residence" therein. We see it repeatedly to enter the bed at a particular location on the bed's surface; to remain out of view for an interval; to reappear repeatedly at a second particular location; and to slide quickly along the surface from the second to the first location, where it re-enters the bed. We will report elsewhere upon the times of its travels and what they tell us about the intensity of solid circulation. The time of disappearance declines with increasing K.

At K = 2.5, all three solids of Fig. 8 display mild bunkering, the bed being deeper above the probe. For the smallest solid, slight bunkering appears at the sides of the bed, increasing at K = 4. At a K beyond about 5, drastic bunkering occurs at one side wall or the other, with intense recirculation on a path up the bunkered wall and down the surface. The solid stream showers over the dummy probe and into a "well" that forms between the probe and the other wall. A lower "bunker" appears at the latter wall. At K = 4, the 177 μm "Master Beads" show slight bunkering at the sides. For the largest solid, the bed surface flattens as K increases from 2.5 to 4.

For all of the solids, a trapezoidal zone beneath the dummy probe tends to remain more or less quiet; this zone's width is close to the probe's diameter at the top and is about double this at the floor. Activity in this zone increases with particle diameter and with K. For 177 μm particles, an intense recirculation cell develops at K = 4. For 707 μm particles, this cell expands to include a strong downward current alongside the probe from the surface of the vibro-bed, and two intense recirculation cells develop just above and to the right and left of the dummy probe.

The broken curves of Fig. 9, giving instantaneous pressures at the floor midway between probe and side-wall, resemble predictions of the Kroll-Gutman theory; we will report elsewhere upon comparisons of experiment with theory. A negative pressure develops at the floor when the instantaneous acceleration falls below about -g. Shortly afterward, a gap forms at the floor; positive pressure develops as the gap is destroyed. As can be seen from Figs. 6 and 7, the "bump" in each of the solid curves of Fig. 9 occurs approximately when the gap beneath the dummy probe disappears.

Discussion

Features of the data in Fig. 5 can be understood qualitatively from Figs. 6 through 9. The leveling off of heat transfer coefficient for 707 μm solid at higher K-values is seen to reflect the growing fraction of time during which the probe is not in contact with the solid, because of large gaps that form above the probe and below. The higher heat transfer coefficients observed for 177 μm solid reflects, at least in part, the lesser fraction of time during which gaps above and below the heat probe interfere with flow of heat to the solid. The higher coefficients at 177 μm also reflect its more general pattern of solid circulation. The recirculation cell seen at the wall for 707μm solid at K = 4 does not help heat transfer very much, and circulation closer to the probe appears to be less vigorous. We comment that small circulation cells near the probe, above or below, probably contribute little, simply producing local regions of solid at a temperature closer to the probe than the average temperature of the bed.

The lower heat transfer to the 125μm and 88μm solids appears to relate to a lesser solid motion in these solids. Paths for gas flow into the solid appear to be "narrower" and to carry narrower currents of solid, the bed elsewhere being relatively stagnant. The trapezoidal region directly below the heat probe is especially quiet.

Action of gas leaving the vibro-bed provides a fascinating feature of our phase-shift photographs. This action is seen in Fig. 6 in front-lit photos for phase angles between about 260 and 0°, dying away between about 0 and 80°. Over this part of the vibration cycle, arrows in the photos indicate a clump of expanded solid about half way between side wall and dummy probe. This appears to be a "relic" of an expansion of the surface layer earlier in the cycle. We associate the clump with the circulation cell seen for 707μm solid in Fig. 8 at K = 4. For the finer solid of Fig. 7, bands of higher and lower density appear at the surface when the gas escapes (beyond about 200°, where Fig. 9 shows a reversal of pressure gradient across the solid). The bands are clearly seen in original photos but are harder to see in reproductions. We have seen marked banding of fluid cracking catalyst in a vibro-bed a fraction of a millimeter in depth. We will say more elsewhere about the history of the expanded surface zone through a vibration cycle.

Acknowledgments: The U.S. Department of Energy (Grant No. DE-FG07- 83ID12428) and the Virginia Center for Coal and Energy Research (Project No. 2057090) supported our work. We are grateful to Norton Alcoa for supplies of "Master Beads" of its manufacture. Scott Luco assisted in obtaining heat-transfer data, and Gilbert ("Chip") Mesec assisted in commissioning our vibration rig. Riley Chan devised and built phase-delay circuitry controlling time of flash for our photographs of vibro-beds and also assisted in design and fabrication of our heat probe. Renato Sprung received support from Maringa State University and the National Research Council (CNPq) of Brazil.

Literature Cited:

1. Pakowski, Z., A.S. Mujumdar, and C. Strumillo, "Theory and Application of Vibrated Beds and Vibrated Fluid Beds for Drying Processes," in *Advances in Drying*, vol. 3, pp. 245-306, A.S. Mujumdar (Ed.), Hemisphere Publishing Company, Washington, D.C. (1984).

2. Bachman, D., *Verfahrenstechnik Z.V.D.I. Beiheft*, no. 2, 43 (1940), cited in (4).

3. Kroll, W., *Forsch. Gebiete Ingenieurw.*, 20 (1), 2 (1954); *Chem. Ing. Techn.*, 27 (1), 33 (1955).

4. Gutman, R.G., "Vibrated Beds of Powders," Ph.D. thesis (mentor: John Davidson), Cambridge University, August 1974.

5. ——, *Trans. Inst. Chem. Engrs.*, 54, 174 and 251 (1976).

6. Buevich, Yu.A., A.F. Ryzhkov, and N.M. Kharisova, *Inzhenerno-Fizicheskii Zhurnal*, 37 (4), 626 (October 1979); *J. Eng. Physics*, 1980 (English translation, Plenum, New York), "Instability of a Vibrational Fluidized Bed," pp. 1162-1168.

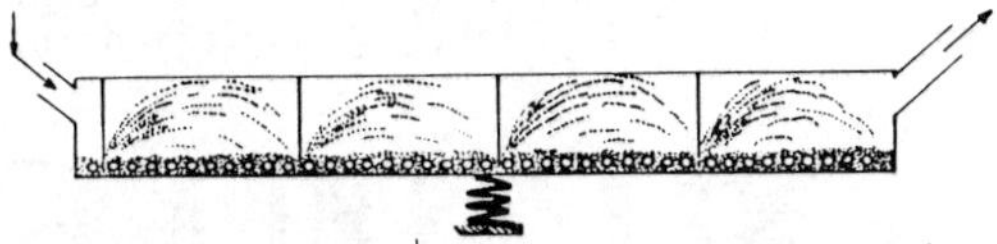

Fig. 1. Concept for heat exchanger with horizontal cooling tubes immersed in a shallow vibro-bed

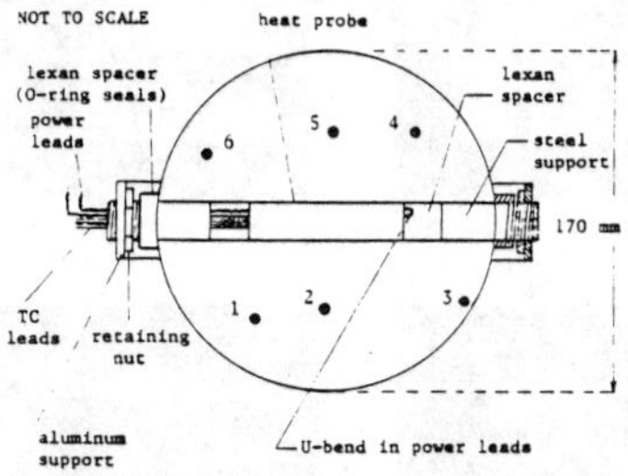

Fig. 2. Top view of cylindrical vibro-bed

Center line of heat probe is 20 mm above floor of vibro-bed.

Black circles show approximate positions of thermocouples in the bed: TCs 2 through 5 are level with center line of probe; TCs 1 and 6 are at upper and lower edges of the probe, respectively.

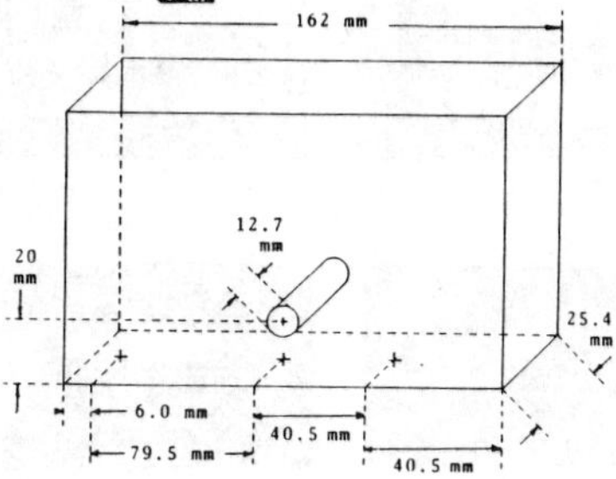

Fig. 3. Two-dimensional vibro-bed

X = pressure transducer on floor of bed.

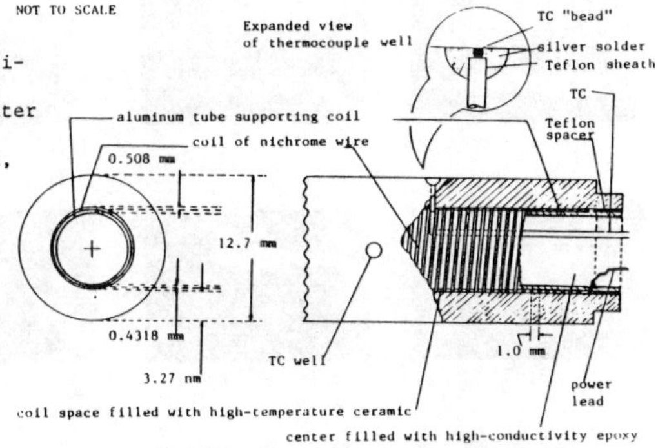

Fig. 4. Copper heat probe

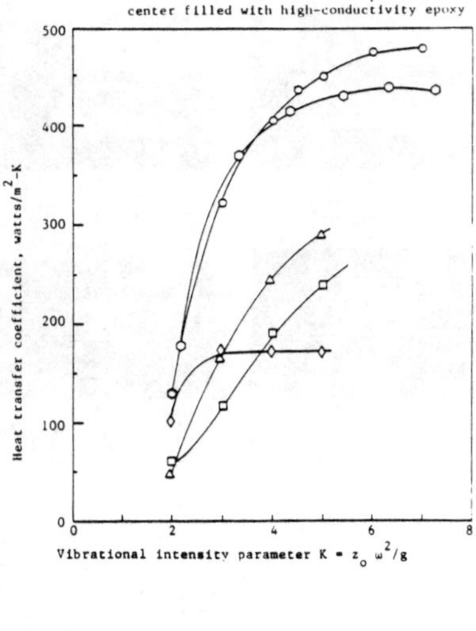

Fig. 5. Transfer of heat from probe to vibro-beds at 25 Hertz

"Master Beads" (3,680 kg/m^3)

◇ -20+30 mesh (707 μm nominal size)

⬡ -60+100 (193 μm) (50/50 mixture or -60 + 70 and -70 + 100)

○ -70+100 (177 μm)

△ -100+140 (125 μm)

□ -140+200 (88 μm)

Fig. 6. Phase-shift photographs of vibro-bed of -20 + 30 mesh "Master Beads" at 25 Hertz and vibration intensity parameter K = 4.

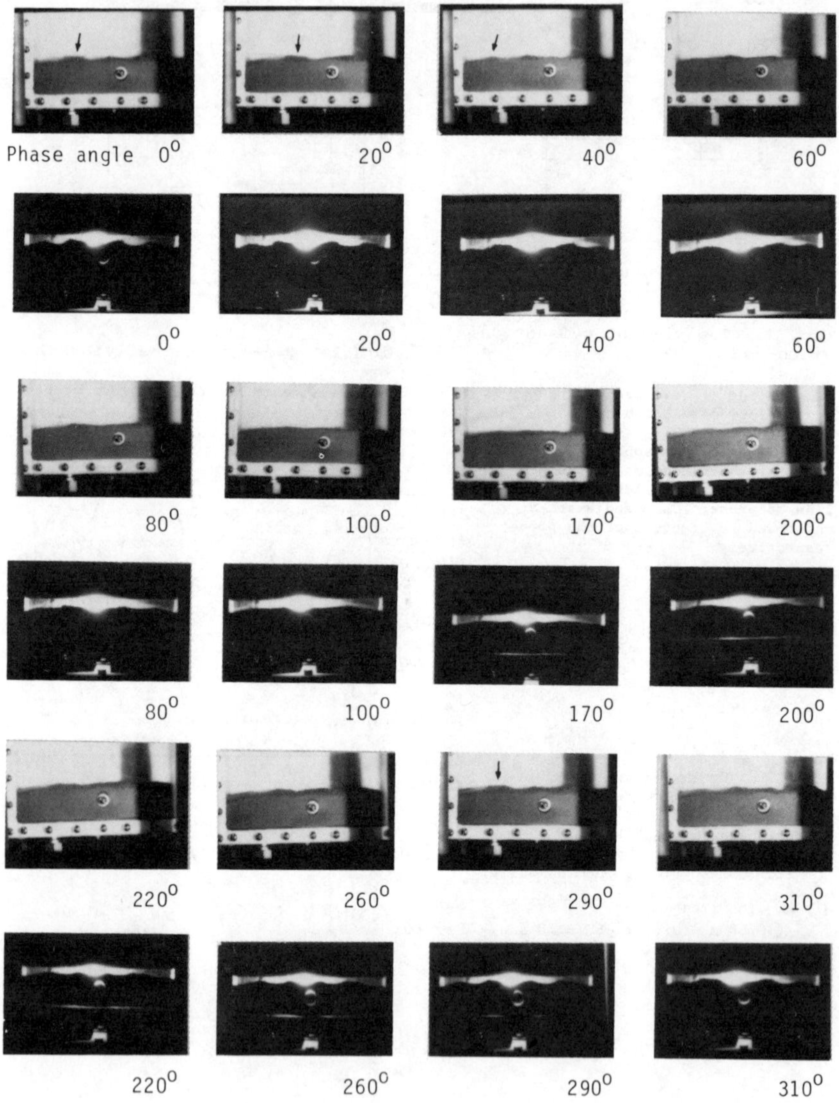

Fig. 7. Phase-shift photographs of vibro-bed of -70 + 100 mesh "Master Beads" at 25 Hertz and vibration intensity parameter K = 4.

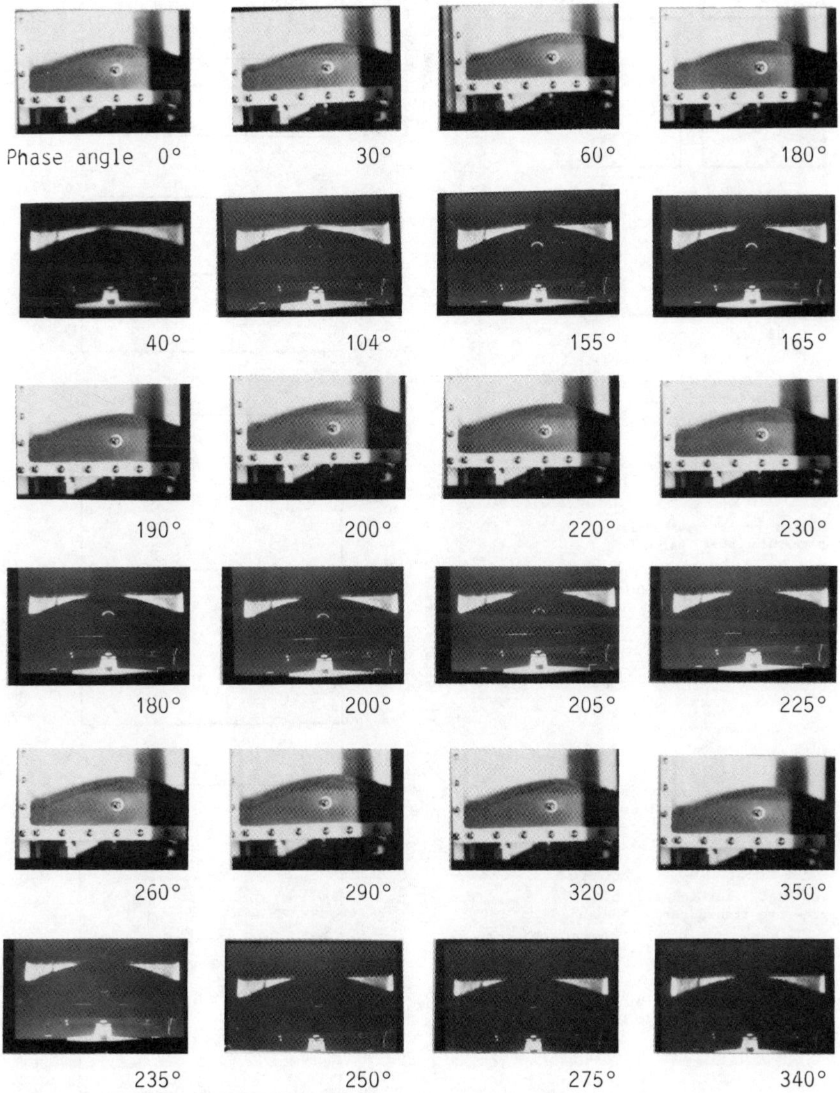

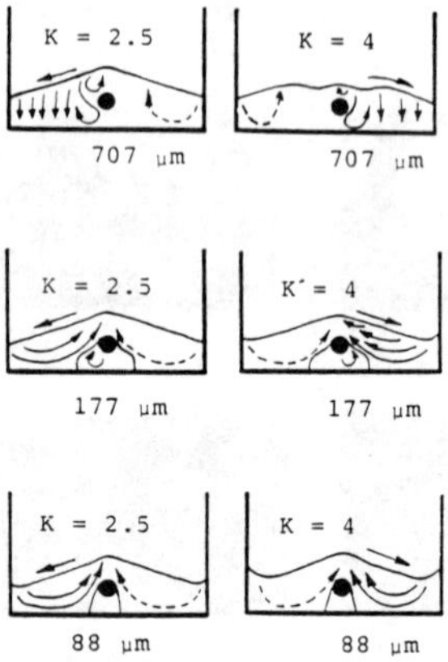

Fig. 8. Paths of solid circulation in vibro-beds of "Master Beads" at 25 Hertz

Solid curves show paths of particulates seen at wall of two-dimensional rig.

Dashed curves show probable paths of rubber "marker".

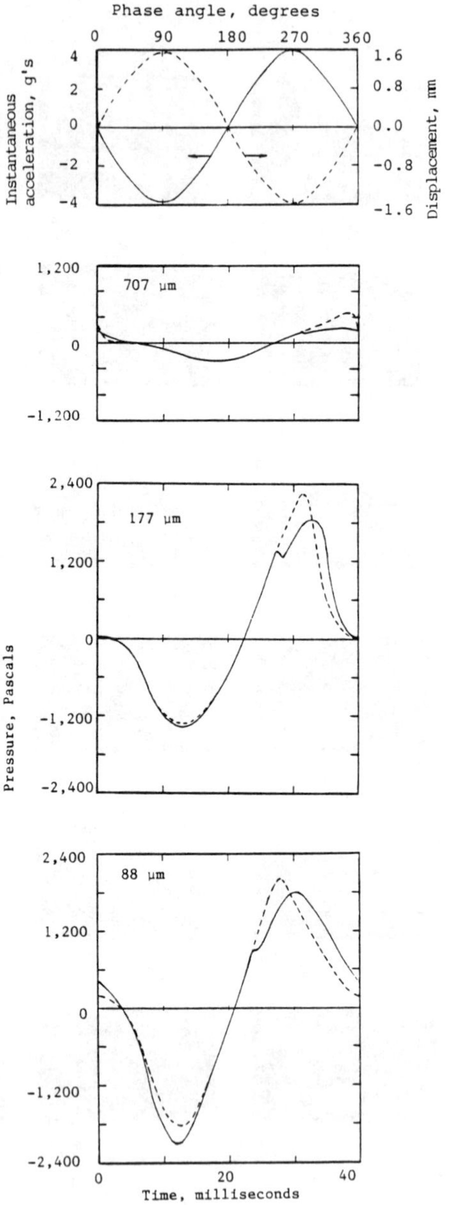

Fig. 9. Instantaneous pressures at floor of two-dimensional vibro-beds of "Master Beads" at 25 Hertz

See Fig. 3 for location of pressure transducers.

Solid curve = floor directly below dummy heat probe.

Broken curve = floor midway between probe and side wall.

HEAT AND MASS TRANSFER IN MAGNETICALLY STABILIZED FLUIDIZED BEDS

J. ARNALDOS, L. PUIGJANER AND J. CASAL

Chemical Engineering Department of the Universitat Politècnica de Catalunya Diagonal 647, 08028-Barcelona, Spain

ABSTRACT

Experimental results are presented on the heat transfer between a magnetically stabilized bed (MSB) and an immersed surface. The influence of operating variables is studied for both magnetizable particles and mixtures of magnetizable and non-magnetizable materials, and results are correlated using a previously developed mathematical model. Data on mass transfer (drying) are also presented and discussed.

SCOPE

The MSB was first studied by Filipov et al. in the early sixties (1, 2, 3). Some years later, more studies enlarged the knowledge of this subject: those of Katz and Sears (4) and Sonoliker et al. (5); an important contribution was the work of the bulgarian researchers (Ivanov, Grozev, Zrunchev), with more than eighty articles published from 1964 to 1985, studying the hydrodynamical behaviour and the application to ammonia synthesis (6, 7).

Both theoretical and experimental investigation on the MSB have been carried out in recent years (8 to 16). Some interesting applications also emerge : filtration (18, 19), separation of solids and chromatographic separation (20). A quite different application is the magnetic valve, studied by Levenspiel et al. (21), which is related to a device used by Piccard (22) to control the flow of particulate solids from a container.

However, very little work has been devoted to the investigation, modelling and correlation of heat transfer (Rosensweig (8); Neff and Rubinsky (14); Kellog et al. (15); Rubinsky and

Starnes (16)) and mass transfer in these beds.

The objective of this study is to enlarge the knowledge and understanding on these transfer phenomena, providing useful information for the application of MSB to processes in which heat and mass transfer take place. The study has been extended to the partially stabilized bed, which in some cases can offer more flexibility for practical applications (Jovanovic et al. (17)).

CONCLUSIONS AND SIGNIFICANCE

The influence of magnetic field intensity, fluid velocity and mass fraction of magnetizable material on the heat transfer has been studied. The experimental results show that the heat transfer coefficient increases slightly with fluid velocity in the MSB, and considerably in the partially stabilized bed (bubbling bed). As the magnetic field intensity increases, the heat transfer coefficient decreases. When the mass fraction of magnetizable material increases, the stabilization increases and the coefficient diminishes.

The plot of these experimental results shows the existence of three regimes: I) fixed bed, in which the influence of gas velocity is negligible; II) stabilized bed, where an increase in velocity causes a certain increase of heat transfer coefficient; III) partially stabilized bed, in which a sudden increase of h is observed when the transition velocity is exceeded, with a strong influence of gas velocity on the heat transfer coefficient.

The results obtained indicate a certain degree of movement inside the MSB, as well as an important influence of the degree of bubbling on the heat transfer coefficient, for a given fluid velocity. Data have been correlated using a mathematical model previously proposed by Arnaldos et al. (12). Equations (4,5) and (6,7) allow the calculation of j_h as a function of the operating conditions and magnetical properties of the fluidized material.

Finally, initial results on mass transfer (air draying) are also presented. The experimental data show that operation efficiency increases with H, as the by-pass due to bubbling is reduced, while for a fixed value of magnetic field intensity, the increase in fluid velocity diminishes the efficiency.

EXPERIMENTAL ARRANGEMENT

Heat transfer was studied in a stainless-steel column, 70 mm I.D. Good gas distribution was obtained with a fixed bed of glass particles and a perforated plate (1 mm holes). The gas was always air at atmospheric pressure, filtered and dried.

An external coil coaxial with the column provided the magnetic field (10). The heating system consisted in a stainless-steel tube, 140 mm long and 6 mm O.D., electrically heated; the following heat transfer ratings were used: 35, 95 and 188 W.

Bed temperature was measured with four thermocouples situated at 30, 60, 90 and 120 mm from distributor, respectively. The temperature of the heating surface was measured with two thermocouples. The characteristics of the particles used are summarized in Table 1, where a and b are magnetic hardness constants (at 25 ºC).

HEAT TRANSFER RESULTS AND DISCUSSION

The heat transfer coefficient (wall-to-bed) was calculated having taken into account the temperature gradient in the bed:

$$h = \frac{Q}{\Sigma\ S_i\ \Delta T_{lm}}$$

Table 1

material	density (kg/m^3)	shape factor	a	b (At/m)$^{-1}$
steel	7500	0.9	40	0.0377
nickel	5900	0.8	46.2	0.0243
silica	2720	0.6	-	-
alumina	2080	0.7	-	-

In the absence of magnetic field, or in the case of MSB operating above transition conditions, temperature gradient was practically inexistent and this expression adopted its simplest form.

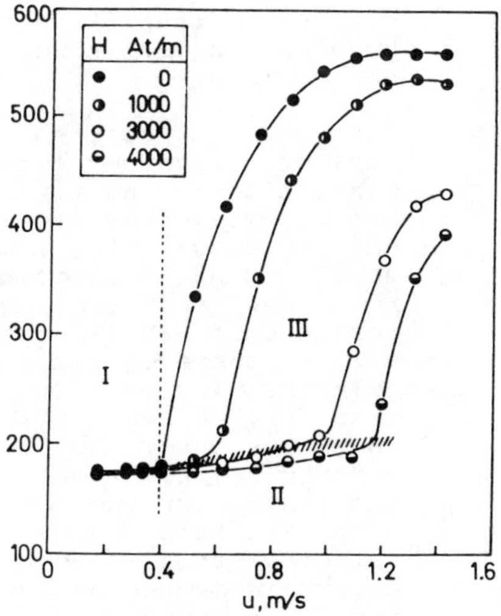

Figure 1. Steel particles (420 < d < 500 μm).

A set of typical results can be seen in Fig.1, showing the change of heat transfer coefficient as a function of gas velocity, at different values of magnetic field intensity. The existence of three clearly distinct regimes can be observed: I) fixed bed ($u < u_{mf}$), in which the influence of gas velocity is negligible; II) stabilized bed ($u_{mf} < u < u_b$), in which an increase in velocity causes a certain increase in h; and III) partially stabilized bed or bubbling bed ($u > u_b$), where a sudden increase of h is observed when the transition velocity is exceeded, with a strong influence of gas velocity on the heat transfer coefficient.

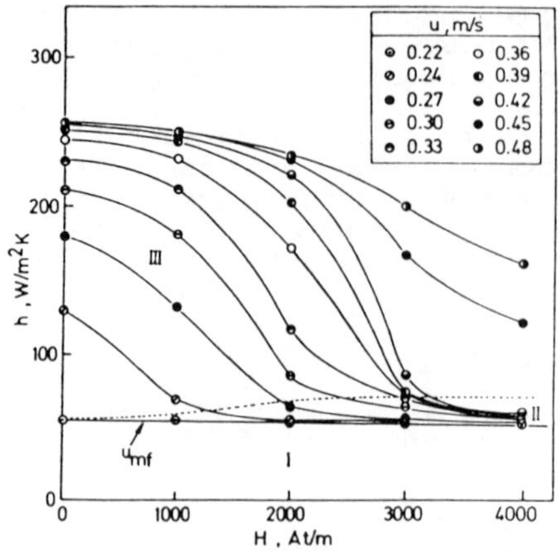

Figure 2. Variation of h as a function of H and u (nickel, 250 < d < 400 μm).

A similar behaviour is observed in Figure 2, where the change of h as a function of magnetic field intensity has been plotted at different gas velocities; the three regimes are again obvious. It can be seen that h decreases as the magnetic field intensity increases. These results agree with those obtained by Neff and Rubinsky (14).

The same experiments were repeated with mixtures of magnetizable and non-magnetizable materials (nickel-silica and steel-silica). A similar behaviour can also be observed (Figure 3) but showing shifts in the transition zones due to the non-magnetizable material. The influence of the mass fraction of this material is as follows: the increase in the percentage of magnetic material clearly increases the stabilization of the bed, moving to higher gas velocities the transition to bubbling regime (sudden change in h).

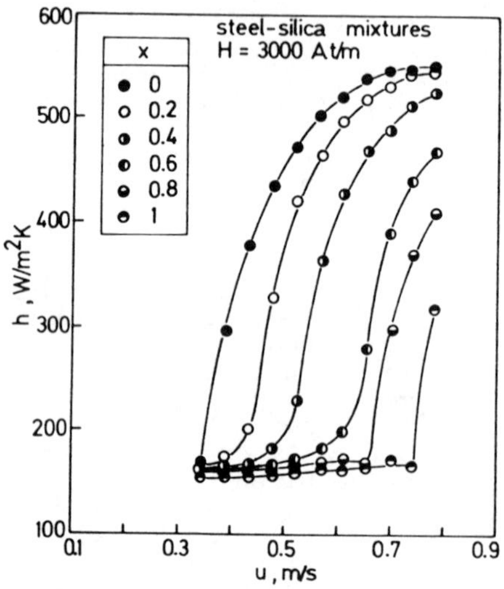

Figure 3. Mixtures of steel (350 < d < 420 μm) and silica (630 < d < 890 μm).

This was, in fact, the behaviour already expected. Concerning to the fixed bed regime, experimental data agree with those usually found in literature. A new regime, characteristic of MSB (zone II), exists however. From the point of view of heat transfer, it seems to be an extension of the fixed bed regime, although it shows a higher increase of h as a function of gas velocity: this indicates a certain degree of movement in the bed. Finally, the presence of bubbles, which adds agitation to the system, causes an important increase in h. It is also noticeable the effect of the magnetic field which, at diminishing bubling, decreases the value of h.

DATA CORRELATION

Correlation of the experimental data has been achieved through the relationship between the j_h factor, the Reynolds number and the bed voidage function. Gupta and Thodos (23) have suggested the following expression:

$$j_h \, f(\varepsilon) = f(\mathrm{Re}) \qquad (1)$$

Following the hydrodynamical model proposed in a previous work (12), this expression should be:

$$j_{h_{mf}} \left(\frac{\varepsilon_{mf}^3}{1-\varepsilon_{mf}} \cos^2 \Psi_o \right) = f(\mathrm{Re}_{mf}) \qquad (2\text{-a})$$

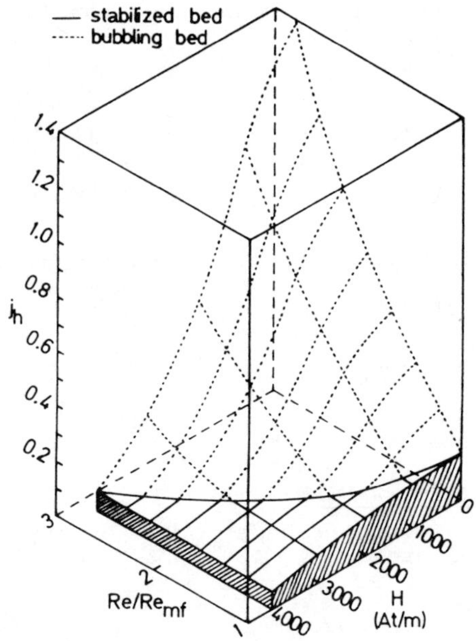

Figure 4. j_h - Re/Re$_{mf}$ - H relationship for steel particles (420 < d < 500 μm; lines following experimental data).

$$j_{h_s} \left(\frac{\varepsilon_b^3}{(1-\varepsilon_b)} \cos^2 \Psi \right) = f(\mathrm{Re}) \qquad (2\text{-b})$$

where

$$\frac{\cos^2 \Psi \; \varepsilon_b^3}{(1-\varepsilon_b)} = \frac{\cos^2 \Psi_o \; \varepsilon_{mf}^3}{(1-\varepsilon_{mf})} \; e^{(a+bH)\mu_o H} \qquad (3)$$

Introducing this relationship in Equation (2), the following expression is obtained:

$$j_{h_s} = j_{h_{mf}} \; e^{-\mu_o H (a+bH)} \left(\frac{\mathrm{Re}}{\mathrm{Re}_{mf}} \right)^{0.2} \qquad (4)$$

where a and b are specific constants for a given magnetic material, and the value 0.2 has been obtained from the correlation of experimental data. This expression is valid for the MSB; at $u < u_b$ the behaviour is different (Figure 4). Having taken into account Equation (1), as well as the hydrodynamical model (12),

$$Re_b = Re_{mf}\ e^{\mu_o H(a+bH)}$$

the following expression is obtained:

$$j_{h_F} = j_{h_b}\ e^{-2\mu_o H(a+bH)} (Re/Re_{mf})^2 \quad (5)$$

in which the value 2 has been obtained from experimental data. Similar equations have been obtained for mixtures of magnetizable and non-magnetizable materials:

$$j_{h_s} = j_{h_{mf}}\ e^{-\mu_o H(a+bH)X} (Re/Re_{mf})^{0.2} \quad (6)$$

$$j_{h_F} = j_{h_b}\ e^{-2\mu_o H(a+bH)X} (Re/Re_{mf})^2 \quad (7)$$

MASS TRANSFER

The study was performed by drying air in a MSB with mixtures of steel (350-420 μm) and alumina (630-890 μm). The bed was situated in a 51 I. D. column, sorrounded by the coil as described before. The air was previously humidified to saturation and then mixed with dry air. The temperature at the entrance of the bed was 30ºC and its relative humidity 90%. Air humidities were determinated using gas chromatografy, provided with thermal conductivity detector and a column of Porapak Q. The carrier consisted in a mixture of $He\text{-}H_2$.

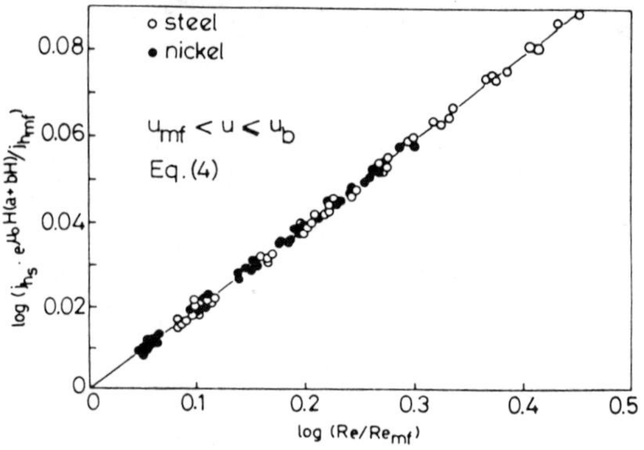

Figure 5. Correlation of Equation (4).

The operation was effectuated at different values of magnetic field intensity: 0, 1500, 3000 and 4500 A.turn/m, and the influence of gas velocity was studied.

For a given value of the magnetic field intensity, the efficiency of the operation decreases (the humidity in the exit air increases) due to the fall in contact time (Figure 6); this effect masks partially the magnetic field influence. This influence can be better observed in Figure 7, in which the humidity of exit air has been plotted versus the time of contact for a given gas velocity, at different values of

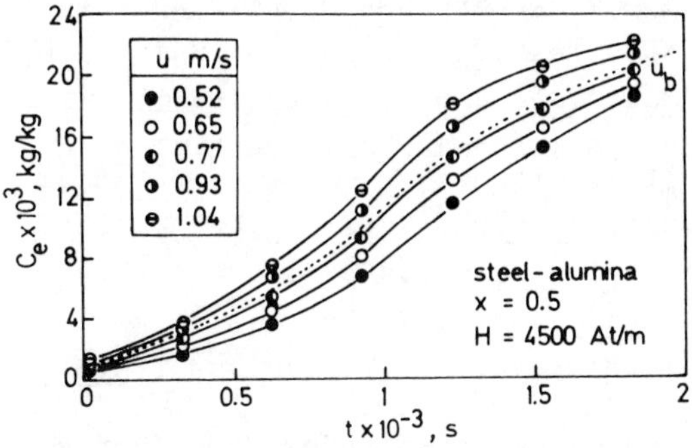

Figure 6. Draying of air in a MSB.

magnetic field intensity. The efficiency of mass transfer decreases as H increases, i.e., as the stabilization increases. At H = 0, a certain degree of bubbling exists, with the consequent bypass of gas; at the value of H corresponding to transition conditions, bubbling is practically inexistent, and the improvement in the contact solid-gas increases the rate of mass transfer. A later increase of H causes still a decrease in C_e, due to the disapearance of the sporadic bubbles that may exist at transition conditions.

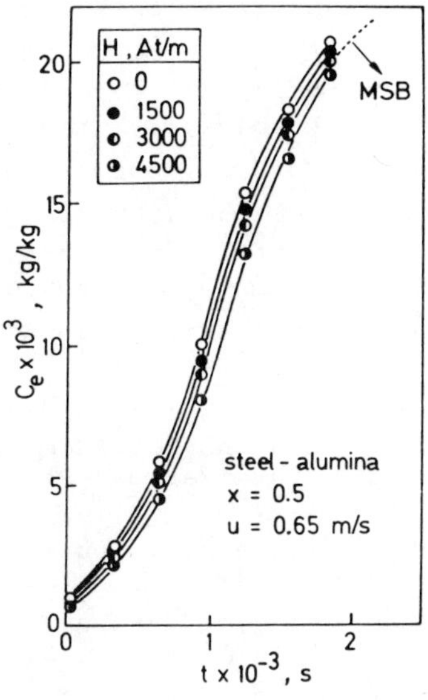

Figure 7. Mass transfer as a function of H.

NOTATION

a magnetic hardness constant
b magnetic hardness constant
C_e exit humidity
d particle diameter
H magnetic field intensity
h heat transfer coefficient
j_h (St $Pr^{2/3}$)
Pr Prandtl number
Q heat transfer rate
Re Reynolds number (particle)
S heat transfer surface
St Staton number
T temperature
t time of contact
u fluid velocity
X mass fraction of magn.material

Greek symbols

ε bed voidage
Ψ velocity angle
Ψ Ψ value for random bed
μ_o permeability free space

Subscripts

F bubbling bed
mf minimum fluidization
b transition conditions
s stabilized bed.

LITERATURE CITED

1. Filipov, M.V., Prik.Magnitogidrodin.Tr.Inst.Fiz.Akad. Nank.Latv.SSR (USSR), 12, 215 (1960).

2. Filipov,M.V.,Izv.Akad.Nank.Latv.SSR(USSR)12,47 (1961).

3. Kirko,J.,Filipov,M.Zh.Tekh.Fiz.(USSR),30,1081 (1960).

4. Katz,H.,Sears,J.,Can.J.Chem.Eng.,47,50 (1969).

5. Sonoliker,R.L.,Ingle,S.,Giradkar,J.,Meue,P.,Indian J. Technol.,10, 377 (1972).

6. Ivanov,D.,Grozev,G.T.,Acad.Bulg.Sci.,23,787 (1970).

7. Zrunchev,I.A.,Acad.Bulg.Sci.,16,285 (1983).

8. Rosensweig,R.E.,Science,204,57 (1979).

9. Rosensweig,R.E.,Siegell,J.H.,Lee,W.K.,Mikus,T.A.I.Ch. E.,Symp.Ser.,77,8 (1981).

10. Casal,J.,Ripol,J.,Puigjaner,L."Journées Européennes sur la Fluidisation.Actes",1,7,1,Toulouse (1981).

11. Lucas, A.,Casal,J.,Puigjaner,L.,in "Fluidization",D. Kunii-R.Toei (Eds.),New York (1984).

12. Arnaldos,J.,Casal,J.,Lucas,A.,Puigjaner,L.Powder Technol.,44,57 (1985).

13. Resnick,W.,Zimmels,Y."Proc. 8th CHISA",p.1213 ,Praha (1984).

14. Neff,J.,Rubinsky,B.Int.J.Heat Mass Transfer,26,1885 (1983).

15. Kellog,K.,Rubinsky,B.,Greif,R.,Int.J.Heat Mass Transfer,26,151 (1983).

16. Rubinsky,B.,Starness,G.,J.Heat Transfer,4, 809 (1983).

17. Jovanovic,G.,Jovanic,P.,Vukovic,D.,Colakian,M.,"Proc. 8th CHISA",767,Praha (1984).

18. Albert,R.V.,Tien,C.,A.I.Ch.E.J.,31,288 (1985).

19. Geuzens,P.L."Some aspects of magnetically stabilized fluidization".PhD thesis,Techn.Hogeschool,Eindhoven (1985).

20. Siegell,J.,Coulaloglou,C.,A.I.Ch.E.Symp.Ser.,80, 129 (1984).

21. Yang,W.,Jaraiz,J.,Guo-Tai,Z.,Chan,R,Levenspiel,O.,in "Fluidization",D.Kunii-R.Toei (Eds.),New York (1984).

22. Piccard,A."Au fond des mers bathyscaphe".Arthaud.Paris (1954).

23. Gupta,A.S.,Thodos,G.,A.I.Ch.E.J.,9,751 (1963).

THE INFLUENCE OF SOLIDS SPECIFIC INTERFACIAL AREA ON GAS-SOLID MASS TRANSFER IN GAS FLUIDISED BEDS

J.J. de Kok, N.L. Stark and W.P.M. van Swaaij
Twente University of Technology, Department of Chemical Technology,
P.O. Box 217, 7500 AE Enschede, The Netherlands.

ABSTRACT

In a 30 cm ID air fluidised bed the overall linear gas-solid mass transfer coefficient has been measured using a bed consisting of inert glass beads and a varying mass fraction of subliming naphthalene beads as a model system.

At constant bed height, inert particle size and superficial gas velocity this coefficient was found to be constant across a wide range of naphthalene specific interfacial area for a given naphthalene particle size. However, beyond a critical value of the interfacial area all coefficients decreased.

SCOPE

In most gas fluidisation processes particle Reynolds number is well below 100. In this range a reliable estimate of mass transfer from dense phase gas to (single) particles is difficult to make. Most experimental evidence (1,2) indicates that for decreasing Re the mass transfer coefficient in the fluidised bed may become much smaller than the corresponding coefficient for a single sphere. Both for fixed and fluidised beds correlations and models have been developed to account for the experimental results (1-8). A critical evaluation of model assumptions was given by Joos (5) and Glicksman and Joos (6).

The problem described above became of more interest since in more recent applications of fluidised beds (such as combustion and gasification of coal and other carbonaceous materials) only a (varying) fraction (active particles) of the bed is involved in a particular process while the remainder serves as diluent (inert particles). Dissimilarity of particle size and density between inert and active particles, which is often encountered in such cases, poses an additional complication. Several workers (9-12) investigated mass transfer to/from one active particle in a matrix of inert particles experimentally, while others (13,10) also presented a model for this particular case. In three papers experiments have been described where mass transfer was measured using a larger fixed (14) or variable (15,16) fraction of active particles. The influence of dilution of active material has been accounted for in models proposed by Beek (2), Joos (5), Glicksman and Joos (6) and Resnick (7).

The objective of this study was the experimental determination of the overall linear dense phase gas to particles mass transfer coefficient for a fluidised bed consisting of inert particles and a varying fraction of active particles using the sublimation of naphthalene as a model system. Conditions were selected such that size and density of active and inert particles were comparable to those during fluidised combustion of coal.

CONCLUSIONS AND SIGNIFICANCE

When linear plotted against the logarithm of naphthalene specific interfacial area (a_n) the overall mass transfer coefficient (k_{ov}) was found to be constant and of the same order of magnitude as reported for a single freely moving naphthalene particle in a fluidised bed of inert particles. Within this range the coefficient increased with decreasing naphthalene particle size. However, beyond a critical value of a_n all coefficients became smaller, the decrease being stronger the smaller the naphthalene particle size. The decrease of the mass transfer coefficient as observed in this work supports the model predictions of Schlünder (3), Joos (5) and Resnick and Golt (7) and agrees with a similar decrease shown with experiments in fixed beds (17,7,18).

Results of adsorption experiments indicated that in general non-porous glass beads showed superior inertness compared to some types of polymer beads.

As for the typical case of fluidised bed coal combustion, the relation between coal hold-up and excess air as measured and modeled by amongst others Donsi et al. (23) probably reflects the relation between the character of the controlling mass transfer resistance and a_n emerging from this work. Similar reasoning applies to the relation between coal charge and overall combustion rate/burn-out time for batch fluidised bed coal combustion.

EXPERIMENTAL APPARATUS AND PROCEDURE

Mass transfer experiments. The 30 cm ID fluidised bed column was made from stainless steel and equipped with a porous plate distributor. Experiments were carried out at a bed temperature of approximately 63 °C. Temperature could be measured within 0.1 °C. In order to reduce temperature gradients the fluidising gas (air) was pre-heated and the exterior of the column insulated. Non-porous glass beads ($0.4 < d_{p,i}$ (mm) < 0.6) were used as inert particles. Superficial gas velocity was kept constant at 0.33 m/s (1.8 $u_{mf,i}$) during all experiments. Expanded bed height was 0.30 m.

Naphthalene vapor concentration at the column outlet was determined discontinuously in gas samples withdrawn from a small continuous gas sample flow. All gas sampling lines and valves were heated to prevent naphthalene condensation/adsorption. A stirrer, located above the bed surface, was used to reduce concentration fluctuations and mix the gas flow from bubble and dense phase. Gas samples were analysed quantitatively using a flame ionisation detector, which had been calibrated withsolutions of naphthalene in carbon tetrachloride. Within the range of interest detector response was essentially linear.

Practically spherical naphthalene particles were prepared according to a method outlined by Kato et al. (19). Naphthalene mass fraction and particle diameter were independently varied between 0.001 and 1% and 0.95 and 8.76 mm respectively.

The sublimation of naphthalene as it was applied in this work differed slightly from the way this technique was utilized by previous authors. After the stirrer, gas sampling and analysis system, air flowrate and bed temperature had been adjusted to the selected operating conditions the air flow was suddenly stopped. A weighed amount of naphthalene with known particle size was quickly dropped onto the bed and the air flow restarted as soon as possible. The first gas sample was withdrawn 30 s after addition of the naphthalene. Subsequent samples could be analysed at a maximum rate of one sample per two minutes. Gas analysis was stopped as soon as detector response fell below 1% of the maximum response during the experiment.

Adsorption experiments. These were carried out using a double-channel gas chromatograph equipped with a thermal conductivity detector. One of the columns was packed with the particles of which the adsorption characteristics had to be measured. Helium was used as carrier gas. In a typical transient experiment the test column was first subjected to a step change from the carrier gas to a non-adsorbing gas mixture (helium/argon) until steady-state was reached. Subsequently, a step change was initiated from helium to helium saturated with naphthalene vapor. In a preliminary experiment with the non adsorbing gas the concentration of argon was selected that would match at its steady-state value with the steady-state response of the adsorbing gas. Identical operating conditions of temperature and volumetric flow rate were maintained. Column temperature was varied between ambient and 65 °C. The total amount of naphthalene adsorbed was calculated by integrating the area between the response curves of the non-adsorbing and adsorbing gas.

RESULTS AND DISCUSSION

Selection Of Inert Particles

Adsorption experiments. It has been shown (15,20,12) that higher mass transfer coefficients are obtained if the "inert" particles adsorb the component being transferred to/from the active particles. Although the results of Van Heerden (15) and Hsiung and Thodos (16) show a significant effect of dilution on the dense phase gas to particles mass transfer coefficient it is not entirely clear (15) to what extent the results were affected by adsorption.

We selected non-porous glass beads from two sources, polystyrene beads and styrene/40% divinylbenzene co polymer beads for our inertness test. In Figure 1 the adsorption capacity is shown as a function of temperature. Each symbol represents the average of two determinations. The adsorption capacity of the co-polymer beads (not shown) was only determined at 65 °C and amounted to 45 μg/g. As it was anticipated to be higher at lower temperature no determinations in this range were made. It is clear that the glass beads show essentially negligible adsorption independent on temperature whereas the adsorption capacity of polystyrene beads strongly depends on temperature. By virtue of their structure polymers are known to be capable of dissolving molecules. This potential may be reduced by an increasing degree of corsslinking e.g. by copolymerisation of styrene with divinylbenzene. This, and the sensitivity of final product properties on the method of preparation probably explains the disagreement between our results and those of Hsiung and Thodos (16). This point may be further illustrated by our finding that a rapid gas chromatographic analysis of naphthalene, which we carried out on a collumn packing made from the same copolymer from a different source, required a relatively high column temperature.

Both the relative inertness and higher (than naphthalene) particle density made us decide to use glass beads for the mass transfer experiments. A higher than ambient temperature was chosen, because it has been found (20), that the contribution of attrition and erosion of naphthalene to overall mass transfer rate decreases with increasing temperature.

Solid mixing experiments. It has been found that the mass transfer coefficient is affected by the freedom of movement and segregation of active particles (11,14,12). Because the remaining aspect yet to be considered was the possibly segregating tendency of the chosen model system we carried out some solid mixing experiments. A single naphthalene particle sized 7, 5, 3 or 2 mm and ink blackened, was dropped onto the bed surface and during 5 min. the total time was measured that the particle was present at the surface. At a superficial gas velocity slightly lower than used in the mass transfer experiments this time was 7 s for the 7 mm particle and decreased rapidly with decreasing particle size. Also, for each particle size this time decreased rapidly with increasing superficial gas velocity. From these observations and results in literature (11,14) dealing with the influence of restricted movement and segregation on the mass transfer coefficient we concluded that under the prevailing experimental conditions our fluidised bed was well mixed.

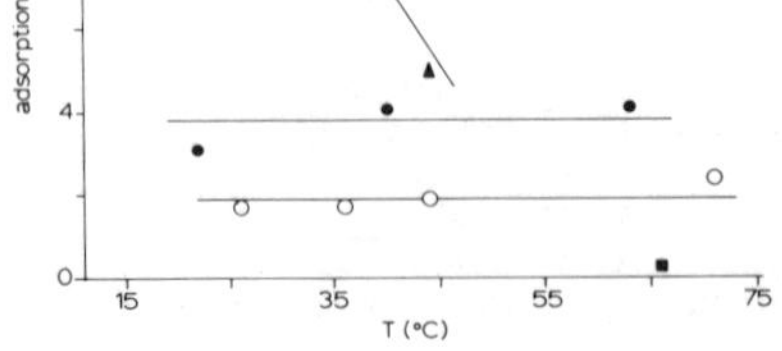

Figure 1. Influence of temperature on adsorption capacity. Particle size: ● 0.35-0.40 mm; O,▲ 0.42-0.50 mm.

Mass Transfer Experiments

A typical plot of the measured dimensionless naphthalene vapor concentration at the column outlet against time after addition of a

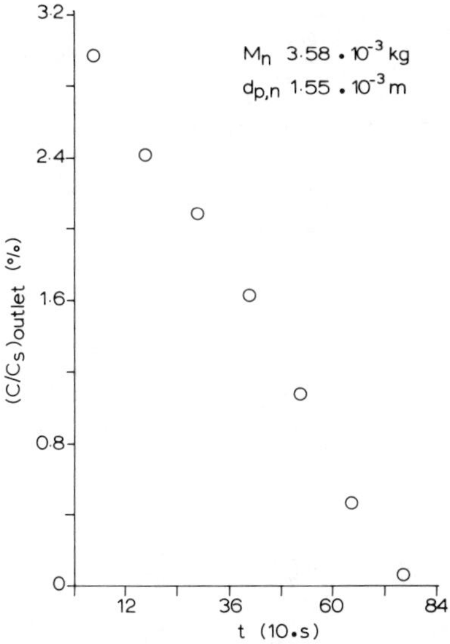

Figure 2. Dimensionless naphthalene vapor concentration at column outlet during a mass transfer experiment.

batch of naphthalene particles is shown in Figure 2. The saturated napththalene vapor concentration (c_s) was calculated using the following vapor pressure relationship reported in literature (21) and the ideal gas law:

$$^{10}\log p_s \text{ (mm Hg)} = \frac{-3729.75}{T} + 11.45$$

$$(291.15 < T\ (°K) < 353.15)$$

Naphtalene particle surface temperature was assumed to be equal to bed temperature. As the specific naphthalene interfacial area (a_n) was most accurately known at the initial stage of the batch experiment, the initial value of the dimensionless naphthalene vapor concentration at the column outlet and the initial naphthalene mass fraction were used for calculation of the mass transfer coefficient. In Figure 3 this initial concentration has been plotted against the initial naphtahalene mass fraction. Each symbol represents one batch experiment. Integration of the concentration vs. time curve allowed calculation of the mass balance. Only data points within ± 15% are presented. A worst case calculation shows that the contribution of naphthalene mass loss due to entrainment is negligible. Also, naphthalene could not be detected in the very small amount of solids separated by the cyclone. It can be seen in Figure 3 that the outlet concentration increases proportionally with the mass fraction for a given $d_{p,n}$. Beyond a mass fraction of ~ 0.5% the increase tends to level off for al $d_{p,n}$. All particle sizes show the same rate of increase. At fixed initial mass fraction small naphthalene particles produce a higher outlet concentration than larger ones, but this difference becomes smaller beyond a mass fraction of ~0,5%. Collection of data at much higher mass fractions than indicated was not considered useful for two reasons: (i) the gradual change to entirely different hydrodynamic behaviour as a result of the dissimilarity of density and size between inert and active particles (ii) the range of interest from the point of view of fluidised coal combustion is not significant wider. Only few experiments reported by Hsiung and Thodos (16) were carried out at fixed bed height and gas velocity. These data have been included in Figure 3 (solid lines) and show the trend expected from our data. An increase of fractional saturation with increasing naphthalene content of the bed was also observed by Van Heerden (15), but his data were affected by adsorption.

The overall linear mass transfer coefficient (k_{ov}) was calculated from the experimental data assuming plug flow in the gas phase and ideal mixing of the solid phase. A steady-state mass balance yields the following familiar equation for the mass transfer coefficient:

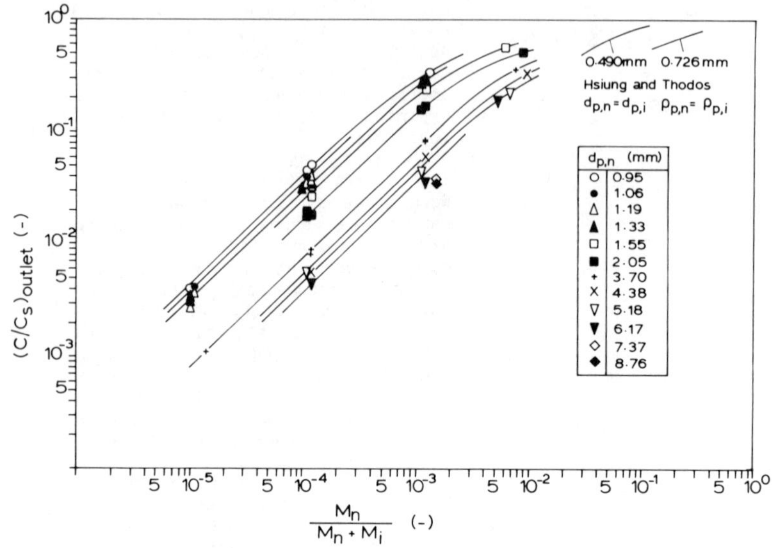

Figure 3. Influence of naphthalene mass fraction and particle size on dimensionless naphthalene vapor concentration at column outlet.

$$k_{ov} = -\frac{u}{a_n H} \ln\left(1 - \frac{c}{c_s}\right) \tag{1}$$

where a_n is the initial specific naphthalene interfacial area defined as:

$$a_n = \frac{6(1-\varepsilon)}{d_{p,n}} \left[\frac{M_n/\rho_{p,n}}{(M_n/\rho_{p,n}) + (M_i/\rho_{p,i})}\right] \tag{2}$$

Effects due to adsorption, volume change, attrition and entrainment were neglected. The mass transfer coefficient calculated according to Equation (1) has been plotted against a_n in Figure 4. For a given $d_{p,n}$ the coefficient is essentially constant up to a value for $a_n \simeq 2\ m^{-1}$. Within this range k_{ov} increases with decreasing $d_{p,n}$. If $a_n \gtrsim 2$ the coefficients decrease for all $d_{p,n}$. The absolute error in k_{ov} was calculated to be approximately 15%.

At infinite dilution the initial naphthalene mass fraction is one active particle. It is anticipated therefore that the limiting value of k_{ov} with decreasing a_n will be of the same order of magnitude as the (local overall) coefficient for a single particle. In order to test this hypothesis the correlation proposed by Prins et al. (12) was used to calculate the single particle mass transfer coefficient and a_n defined by Equation (2). The results are plotted in the left hand part of Figure 4, where the upper and lower series of data points refer to an inert particle size of 0.6 mm and 0.4 mm respectively. The agreement is best for intermediate $d_{p,n}$, although the correlation proposed by Pal'chenok and

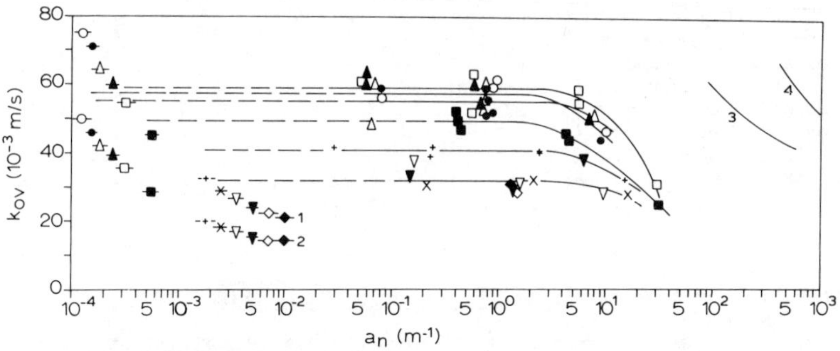

Figure 4. Dependence of overall mass transfer coefficient on naphthalene specific interfacial area and particle size compared with data from literature. Symbols as used in Figure 3. Curve 3 ($d_{p,i} = d_{p,n} = 0.490$ mm) and 4 ($d_{p,i} = d_{p,n} = 0.726$ mm): data from (16), see Figure 3. -Symbol- : calculated from (12): 1 (upper series) $d_{p,i} = 0.6$ mm, 2 (lower series) $d_{p,i} = 0.4$ mm.

Tamarin (11) gives better agreement for the larger particle size. The combined results nevertheless indicate that all active particles behave as individual particles provided $a_n \leq 2\ m^{-1}$. The experimental data from Hsiung and Thodos (16) as presented in Figure 3 served to draw curve 3 and 4 in Figure 4. Although both curves show a decrease for high a_n the absolute values for k_{ov} are too high relative to a_n. This might be due to the similarity of active and inert particles and/or lower gas residence times.

In order to ascertain whether the observed k_{ov} solely represents the dense phase gas to solid mass transfer coefficient the corresponding mass transfer resistance, $1/(k_p.a_n)$, has been compared with the calculated bubble to dense phase resistance, $1/(\overline{k_b.a_b})$. k_p was calculated from the correlation given by Prins et al. (12). The results showed (22), that under the prevailing experimental conditions the bubble to dense phase mass transfer resistance (range: 0.125 - 0.071 s) was negligible compared to dense phase gas to solid mass transfer resistance (range: 0,8 - 1390 s). It means that when a_n increases beyond the highest value applied in this work this situation could gradually reverse.

An elaborate description of the work described in this paper is given elsewhere (22).

ACKNOWLEDGEMENTS

This investigation was commissioned by the Project Office for Energy Research within the framework of the Dutch National Coal Research Programme.
The reading of the manuscript by dr. J.J. van Deemter is appreciated.

NOTATION

a specific interfacial area (m^2/m^3 bed)

c/c_s dimensionless naphthalene vapor concentration at column outlet
d_p particle diameter (m)
H expanded bed height (m)
M (initial) weight of particles (kg)
u superficial gas velocity (m/s)
ε overall bed porosity (-)
ρ_p particle density (kg/m^3)
b bubble, i inert, n naphthalene, ov overall

LITERATURE CITED

1. Kunii, D. and Levenspiel, O., Fluidization Engineering, Wiley, New York, Ch. 7 (1969).
2. Beek, W.J., "Mass transfer in fluidized beds", in Fluidization, Davidson, J.F. and Harrison, D. (Eds.), Academic Press, London (1971).
3. Schlünder, E.U., Chem. Eng. Sci., 32, 845 (1977).
4. Wakao, N. and Funazkri, T., Chem. Eng. Sci., 33, 1375 (1978).
5. Joos, F.M., "Mass transfer in fluidized beds", M.Sc thesis, Mass. Instit. of Technol., Cambridge (1978).
6. Glicksman, L.R. and Joos, F.M., Trans. ASME, J. Heat Transfer, 102, 736 (1980).
7. Resnick, W. and Golt, M., Int. J. Heat Mass Transfer, 24, 387 (1981).
8. Fedkiw, P.S. and Newman, J., Int. J. Heat Mass Transfer, 25, 935 (1982).
9. Tamarin, A.I., Galershteyn, D.M., Shuklina, V.M. and Zabrodskiy, S.S., Heat Transfer - Sov. Res., 14 (3), 88 (1982).
10. La Nauze, R.D., Jung, K. and Kastl, J., Chem. Eng. Sci., 39, 1623 (1984).
11. Pal'chenok, G.I. and Tamarin, A.I., Inz. Fiz. Zh., 47 (2), 235 (1984).
12. Prins, W., Casteleijn, T.P., Draijer, W. and van Swaaij, W.P.M., Chem. Eng. Sci., 40, 481 (1985).
13. Tamarin, A.I., J. Engng. Phys., 41 (6), 1346 (1981).
14. Cobbinah, S., Laguérie, C. and Gibert, H., "Simultaneous heat and., mass transfer between a fluidized bed of fine particles and immersed large particles", presented at the 16th Int. Symp. Heat Mass Trans., ICHMT, Dubrovnik (1984).
15. Van Heerden, C., J. Appl. Chem., 2, Suppl. Iss. 1, S7 (1952).
16. Hsiung, T.H. and Thodos, G., Chem. Eng. Sci., 32, 581 (1977).
17. Petrovic, L.J. and Thodos, G., Ind. Eng. Chem., Fund., 7 (2), 274 (1968).
18. Brunello, G., Terron, L.R., Batistela, S.A. and Shima, R.Y., Proc. 3rd Pac. Chem. Eng. Congr., 1, 377 (1983).
19. Kato, K., Kubota, H. and Wen, C.Y., Chem. Eng. Progr. Symp. Ser.,66 (105), 87 (1970).
20. Baskakov, A.P. and Suprun, V.M., Sov. Chem. Ind., (9), 61 (1970).
21. Int. Critical Tables, Vol. 3, p. 208, McGraw-Hill, New York (1928).
22. De Kok, J.J., "Reactor engineering aspects of fluidised bed coal combustors", Dr. Thesis, Twente Univ. of Technology, Enschede, The Netherlands (1986).
23. Donsi, G., Massimilla, L., Miccio, M., Russo, G. and Stecconi, P., Comb. Sci. Technol., 21, 25 (1979).

DEVELOPMENT OF A DENSE PHASE FLUIDISED BED FISCHER-TROPSCH REACTOR

R.W. Silverman[1], A.H. Thompson[1], A. Steynberg[1], Y. Yukawa[2] and T. Shingles[1]

[1]Sasol Technology (Pty) Ltd.,
Sasol 1,
Sasolburg,
South Africa

[2]Badger Engineers, Inc.,
Cambridge,
Mass.,
USA

ABSTRACT

Cold model work was carried out to determine the effect of catalyst properties and reactor scale-up on fluidization in a dense phase Fischer-Tropsch reactor. The results indicated that the use of a fine powder would avoid problems previously encountered with this reactor system. A demonstration reactor is now in operation.

SCOPE

Circulating Fluidised Bed (CFB) Synthol reactors are employed at Sasol One, Two and Three on a commercial scale to convert CO and H_2 to liquid fuels over an iron catalyst via the Fischer-Tropsch reaction (1). An alternative reactor system to the CFB is the dense phase fluidised bed, the in-house terminology for this type of reactor is a fixed fluidised bed (FFB). The use of an FFB rather than a CFB promises reduced capital, maintenance and operating costs.

FFB reactors were used 30 years ago at Brownsville, Texas with unsatisfactory results. Those reactors were designed by Hydrocarbon Research Inc., and their history has been fully outlined by Squires (2). There were many problems with the Brownsville reactors particularly with regard to poor conversions. Despite these problems it was decided that the FFB reactor system was worth re-investigating. Cold model work on three tube sizes was initially carried out to optimise catalyst particle size, determine scale-up criteria and in general, to determine if the quality of fluidization was acceptable. The results of the cold model studies would, to a large extent, be responsible for deciding whether or not to proceed with the project and build a semi-works synthesis reactor.

CONCLUSIONS AND SIGNIFICANCE

Cold model work showed that the catalyst bed voidage was much higher than that at Hydrocarbon Research due to Sasol's use of smaller catalyst particles. The higher total voidages, which are postulated as being due to higher dense phase voidages, should result in better gas to catalyst contact. Initial synthesis runs on the demonstration reactor have shown that conversions and product selectivities are in line with those obtained under equivalent gas flow conditions on the commercial Circulating Fluidized Bed reactors. If the results from further extensive synthesis runs confirm the initial optimism, then further development would be in the construction of a commercial size reactor at Sasol One. Success at this stage would then greatly reduce capital, operating and maintenance costs for future synthetic fuel plants employing Synthol reactors.

DESIGN COMPARISONS OF CFB AND FFB REACTORS

The FFB reactor is potentially a less expensive reactor than the CFB and requires considerably less mass of metal. This is best illustrated by comparing Figure 1, which shows a Sasol Two and Three Synthol CFB reactor, with Figure 2 which represents, on the same scale, an equivalent gas capacity FFB reactor. Since the FFB reactor is self-supporting, that is standing on a skirt, the structure is also much less complex and therefore less expensive than that required by a CFB reactor. The nett result is that the installed capital cost of an FFB type reactor is less than three quarters than that of an equal capacity CFB reactor. Additionally, although the voidage is lower and the pressure drop across the grid plate higher for an FFB reactor, the nett result is a lower pressure drop across the FFB reactor by up to 50 %. This will result in a saving in both the capital and operating costs of the gas compressors.

COLD MODEL WORK

Experimental

Cold model work was carried out in transparent tubes of 5 to 6 m in length and internal diameters of 0.05, 0.14 and 0.64 m diameter. All three models were used to study the effect of catalyst particle size and density on the quality of fluidization and bed voidage at gas superficial velocities up to 0.6 m/sec. Full experimental details and results on the effect of particle size and density on dense phase voidage have been given by Dry (3). The larger model was, in addition, used to determine the effect of reactor internals and distributor grid design.

The all important but somewhat nebulous expression "quality of fluidization" is difficult to quantify. Low pressure fluctuations (4) and a bed pressure drop close to the catalyst weight per unit area were taken as indications of good fluidization quality but the comment by Squires (2) that the "eyeball is the first teacher", still holds.

Results

Extensive tests with magnetite powder lead to the development of a particle size distribution that exhibited the best fluidization characteristics in that the powder could be readily fluidized whilst large bubbles or slugs were minimised. On Geldart's classification (5) the powder type is transitional between types A and C exhibiting properties of each type. Dry et al (3) and Wong (6) have termed this class of powder as type AC, the significance of this will be dealt with later. Catalyst of similar size but lower density, taken from a running commercial reactor, exhibited better fluidization characteristics but at ambient conditions the waxy products on the catalyst would not manifest their potential detrimental effects.

The averaged results with magnetite powder plotted in Figure 3, taken at bed aspect ratio above which tube length no longer influenced the results, show that voidage decreases, bed density increases, going from the slugging 0.05 m tube to the bubbling or turbulently fluidized 0.64 m tube fitted with vertical tubes. This trend was observed in cold model studies carried out by Hydrocarbon Research (2) and indicates the absence of slugs or very large bubbles in the larger tubes. Figures 4, 5 and 6 show that voidage is higher with the highest density catalyst. Since density has been shown to yield no significant correlation with voidage in the dense phase (3), the high total voidage with the denser catalyst is probably due to the presence of more large bubbles or slugs. Direct observations indicate that this is in fact the case but it is difficult to eliminate subjectivity. The voidage given in Figures 3 to 6 was used to predict catalyst requirements and to design vessels, such as storage hoppers, for the demonstration reactor.

The overall conclusion from the cold model work was that the catalyst exhibited sufficiently good fluidization characteristics to warrant the construction of a demonstration reactor.

Cold model results comparison between Hydrocarbon Research Inc. and Sasol

Hydrocarbon Research carried out cold model experiments using tubes of different diameter. Their results at a gas superficial velocity of 0.30 m/s, given by Squires (2) and Volk (7), have been plotted as voidage against tube diameter in Figure 7 along with Sasol results from Figure 3. The Hydrocarbon Research results, obtained with a typical catalyst of 110 microns average size, gave voidages similar to Sasol's in the small diameter slugging tubes but much lower voidages with the larger diameter tubes operating in the bubbling or turbulent regime. Since small particles are less prone to slugging than larger particles it is postulated that the extra voidage of the Sasol system is due to a higher dense phase voidage rather than to the presence of more large bubbles or slugs. It is further postulated that the poor conversions achieved by Hydrocarbon Research on their larger non-slugging reactors was due, in part at least, to poor catalyst to gas contact brought about by the relatively low dense phase voidage of their Geldart class B/D catalyst. The effect of fines on dense phase voidage has been fully covered by Dry et al (3). Hydrocarbon carried out extensive work with reactor internals to improve the fluidization quality of their class B/D catalyst. With class A materials, internals are probably not as critical for good fluidization. Volk (7) did work on small particles but the size range was too narrow, which lead to the advantages of small particles being missed. Under fixed process conditions the use of large particles does minimise problems such as excessive elutration but success with the type of reactor system in question must be approached from the angle of first achieving the necessary syn gas conversion then solving the mechanical problems.

In cold model work there is always a problem in deciding what minimum tube diameter will give realistic scale-up data. It is suggested that data be plotted as per Figure 7 and the point at which the voidage asymptotically approaches the abscissa be taken as the minimum diameter.

SEMI-WORKS DEMONSTRATION REACTOR

System and operation description

The demonstration reactor, see Figure 8, is operated in parallel with a commercial CFB reactor at the Sasol One Plant and it can thus only be run whilst the commercial reactor is in operation. It can however be started-up and shut-down whilst the commercial reactor is in operation. The fresh feed and the recycle gas are taken from the commercial reactor streams, in this way the need to install a recycle compressor was avoided. The recycle stream is not therefore a true recycle for the demonstration reactor. The total effluent gas, apart from a small sampling stream, flows into the commercial CFB reactor outlet line. A relatively new feature is that after leaving an internal cyclone, the effluent stream passes through a filtration system for final catalyst and effluent separation. If the system works, it could lead to significant steam cost savings for present and future Plants and it would remove the need for the current catalyst and heavy oil separation plant.

Operational problems

Cyclones

The system was designed with two internal cyclones in series with diplegs discharging into the bed. This is common practice on FFB reactors but unlike the cyclones in the Synthol CFB reactor hoppers which discharge into a lean phase region. Special features of type AC solids are that they flow well when fluidized, the type A influence, but, if allowed to, they will permanently defluidize on any horizontal surface and block any narrow pipe, the type C influence. The presence of sticky products on the solids exacerbates the type C influence. Severe problems with dipleg blockage lead to the design of a proprietary single stage system. This cyclone has performed satisfactorily since its installation.

Distributor grid

The main anticipated problem prior to start-up of the reactor was leakage, or weepage, of the catalyst through the distributor nozzles and orifices. This is a well-known distributor problem (8). This could lead to blockage of the nozzles or chemical reaction, leading to hot spots, on stagnant catalyst in the plenum chamber beneath the distributor. At the present moment in time the grid is the main area receiving development attention as far as mechanical design is concerned.

Initial performance results

Bed voidage

From beginning of run to relatively steady state conditions, the catalyst density and hence the bed density drops to about 70 % of initial level. Over this density range the bed voidage remained fairly constant at a superficial gas velocity of 0.4 m/s. The voidage was a little higher than those given in Figures 3 and 6 for the 0.64 m diameter cold model despite the fact that the diameter of the semi-works reactor is larger. From the results given in Figures 3 to 6 the voidage in the semi-works reactor may have been expected to be lower than those reported for the 0.64 m cold model due to the diameter increase. The fact that voidages are higher in the synthesis unit points to an increase in dense phase voidage, rather than an increase in bubbles, at the higher pressure. This has been reported previously by other workers (9, 10 and 11) and explains the improvement in quality of fluidization, referred to in the literature, as pressure increases.

Synthesis performance

Conversions comparable to commercial CFB performance have been achieved (1). Product selectivities of CH_4 to condensed oil are also in line with expectations.

Future action

At the present moment in time process optimisation studies are proceeding hand in hand with the design of a Sasol One size commercial FFB reactor.

LITERATURE CITED

(1) Dry, M.E., Applied Industrial Catalysis, Academic Press Inc., 2, 5, 167 (1983)

(2) Squires, A.M., Proceedings, Joint Meeting of Chemical Engineering, Chemical Industry and Engineering Society of China, American Institute of Chemical Engineers, Beijing, 1982 September 19 to 22

(3) Dry, R.J., Judd, M.R., and Shingles T., Powder Technol., 34, 213 (1983)

(4) Jones, D.H., Shingles, T., Kelly, F.T., and Dry, M.E., U.S. Patent No. 4,225,531 (1980)

(5) Geldart, D., Powder Technol., 7,285 (1973)

(6) Chi-Ying Wong, A., "Fluidization of Cohesive Powders", Ph.D Thesis, Univ. of Bradford, England (1983)

(7) Volk, W., Johnson, C.A., and Stotler, H.H., Chem. Eng. Progr., 58(3), 44 (1962)

(8) Briens, C., Bergougnou, M.A., and Baker, C.G.J., "Fluidization", Plenum Press, p413 (1980)

(9) Subzwari, M.P., Clift, R., and Pyle, D.L., "Fluidization", Cambridge University Press, p51 (1978)

(10) Varadi, T., and Grace, J.R., "Fluidization", Cambridge University Press, p55 (1978)

(11) Guedes de Carvahlo, J.R.F., Chem.Eng. Sci., 36,413 (1980)

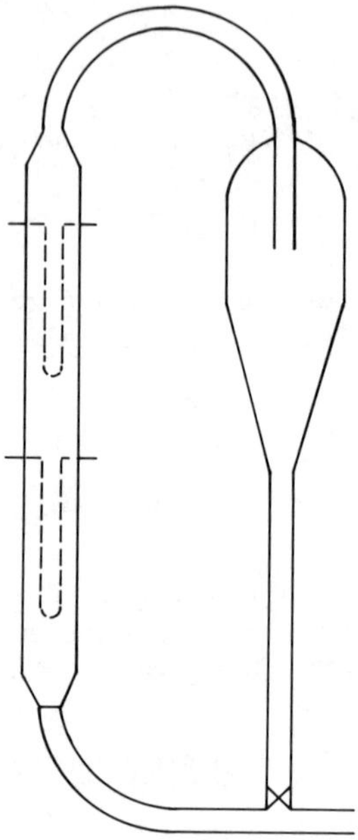

FIG.1 SASOL 2 AND 3 SYNTHOL CFB REACTOR

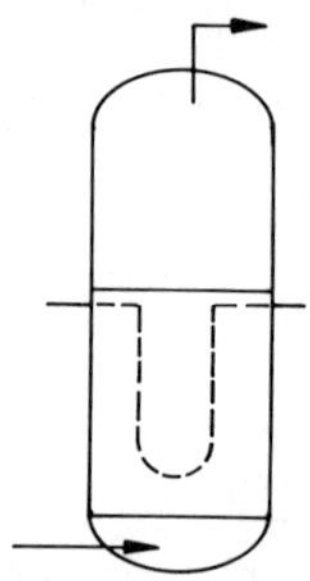

FIG.2 EQUIVALENT CAPACITY FFB REACTOR

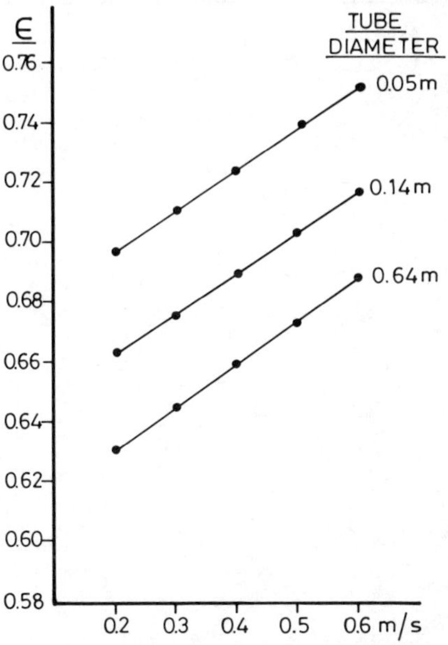

FIG. 3 FIXED ρ_p OF 5 190 Kg/m^3

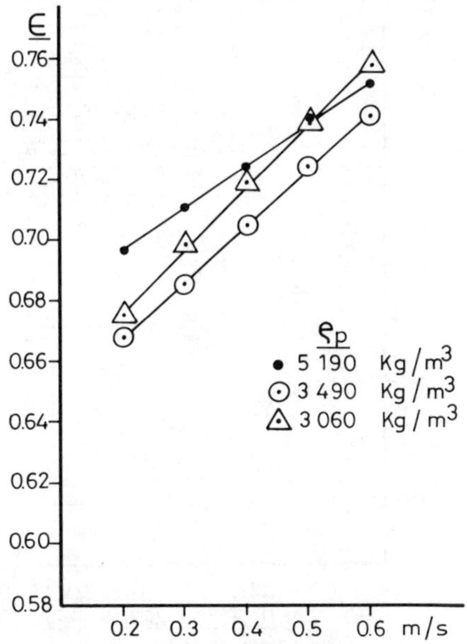

FIG. 4 0.05 m I.D. TUBE

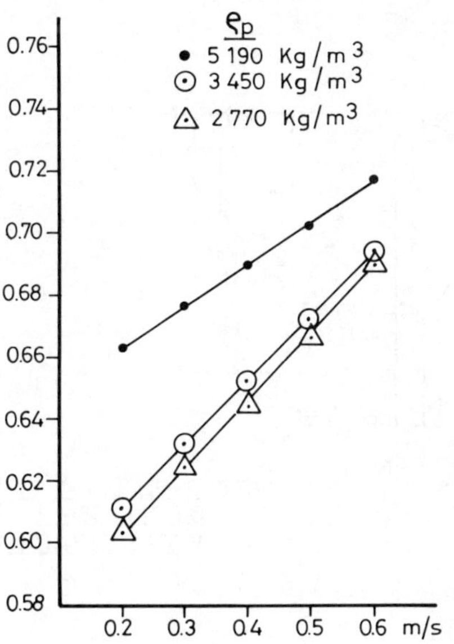

FIG.5 0.14 m I.D. TUBE

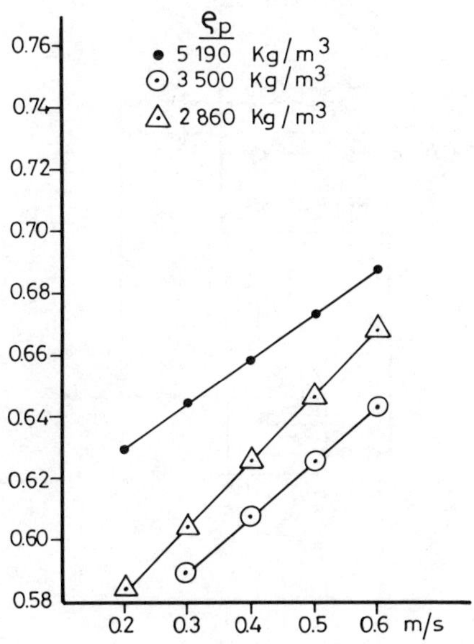

FIG. 6 0.64 m I.D. TUBE

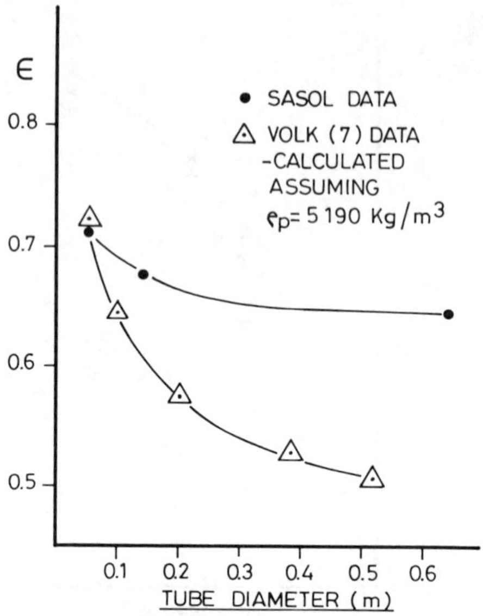

FIG. 7 TUBE DIAMETER AND BED VOIDAGE (ϵ) AT A GAS VELOCITY OF 0.30 m/s

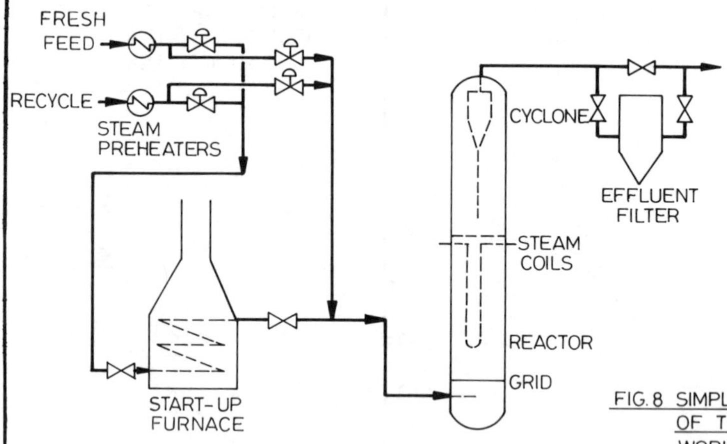

FIG. 8 SIMPLIFIED PFD OF THE SEMI-WORKS REACTOR

FLUID-BED REACTION WITH HORIZONTAL TUBES AS INTERNALS

Q.M. MAO and O.E. POTTER

Department of Chemical Engineering
Monash University
Clayton, Victoria, Australia, 3168

ABSTRACT

The long-term objective of the study is to establish design criteria for internals in fluidized beds. In this paper data are presented for closely packed horizontal coils in a 0.61 m diameter reactor covering conversion and pressure profiles for a sand of mean size 117 μm and for spent cracking catalyst of mean size 65 μm.

SCOPE

What happens when internals are fitted in a fluidized bed? Is it possible to increase conversion by adding internals? Answering these and related questions is the long-term objective of this study. The equipment comprises a 0.61 m diameter reactor fitted with cyclones and bag-filter, a horizontal coil, or a vertical tube bundle - on which no data are reported here, and an ozone generator for catalytic reaction. In this paper concentration profiles and pressure profiles are presented covering fine sand, and to some extent FCC. The horizontal tube bundle is of rectangular arrangement, 5.6 cm horizontal pitch and 3.75 cm vertical. The velocities employed are in the commercial range and extend from 0.1 to 1.2 m s^{-1}.

Internals may be fitted purely for heat-transfer purposes and a knowledge of the effect of installing internals whether vertical or horizontal is necessary if adequate design procedures are to be developed. Examples of catalytic reactions which could possibly benefit from properly designed internals are the Mobil gasoline synthesis from methanol, and acrylonitrile production. Indeed, increased conversions in fluidized-bed reactors are very much to be desired. It is known that such internals have been used and the effects have been ascribed to a bubble-diameter dependent in some way on the vertical and horizontal separations. Literature results are scanty.

CONCLUSIONS AND SIGNIFICANCE

The first significant results relate to the appearance of static voids in the bed in both sand and FCC. These static voids are inferred from pressure-gradients as shown in Figures 1 and 2. If the inference is correct, such static voids are similar to the dilute zone found under horizontal baffles. From pressure profiles the height of a static void assumed to extend over the whole cross-section can be approximately evaluated and Figure 3 shows how the height of the static void varies with velocity in sand and FCC.

The second significant result relates to the conversion profiles. It seems difficult to escape the conclusion that models will need two-dimensions rather than one as the conversion profiles would seem to be closely dependent on bubble paths. In Figure 4 radial concentration profiles at different heights are plotted for the sand bed without internals on the right hand side and with the horizontal tube bundle on the left hand side, for a velocity of 0.8 m s^{-1}. The internals-free bed is compatible with bubbles rapidly coalescing to form a large central bubble. The horizontal bundle shows an annular region in which bubbles rise - possibly channelling.

FLUIDIZED-BED MODELS

The modelling of fluid-bed reactors incorporating the effect of internals, wether vertical or horizontal, has not been adequately developed although there is good evidence that internals serve to reduce bubble-size and there are some preliminary methods of predicting the effect of internals on effective vessel diameter. Therefore, modelling with internals must clearly take as starting point modelling without internals.

Existing Models

Two models which are respectively diffusive and convective are:

Dense phase diffusion model (DPDM). This diffusion-type model has been developed by May and van Deemter.

Counter-current backmixing model (CCBM). This is a convection-type model which has been developed by Kunii and Levenspiel, van Deemter, Fryer and Potter, etc.

These models have been reviewed (1) and are essentially one-dimensional. Miyauchi et al (2) has developed a model based on gas-liquid bubbling flow. This is two-dimensional.

Bubble Size

The knowledge of bubble size and distribution is of primary importance in carrying out model predictions. So far as fluidized beds without internals and particles > 100 μm are concerned, Mori and Wen (3) have developed an approximate correlation, as also have Werther (4,5), Rowe (6) and Darton (7). These correlations have not been adequately tested at commercial velocities.

For the case of fluidized beds with internals, few correlations are available (8, 9) and the applicability is uncertain. Ikeda (10) has defined a generalized reactor equivalent diameter, considering that the commercial reactors have not only vertical internals but also horizontal baffles. The proposed equivalent diameter is

$$\frac{1}{D_{eq}} = \frac{1}{D_{eV}} + \frac{1}{D_{eH}} \quad \text{... (1)}$$

where D_{eV} and D_{eH} are equivalent diameters based on the vertical surfaces and the horizontal surfaces respectively. The calculated equivalent diameter of a large reactor then will give the same conversion as a small reactor of the same actual diameter, or so it is argued.

Doraiswamy (11) adopted a similar concept of equivalent diameter for the case of a vertical tube bundle. He suggested that the bubble diameter used can be taken as equal to this equivalent diameter.

An attempt has been made to compare the average concentration profiles predicted by CCBM for such bubble sizes as might occur with internals or without internals (12).

EXPERIMENTAL EQUIPMENT

The experimental fluidized bed reactor system used is the same as described previously (1). Figure 5 is a schematic diagram of the equipment. The ozone decomposition reaction is conducted in a 61 cm dia. reactor, in which a vertical or a horizontal tube bundle can be inserted. The horizontal tube bundle used was constructed with coils made from 2.54 cm dia. aluminium tube. The bottom of the bundle is supported by a 4 cm wide stainless steel ring on the wall about 30 cm above the distributor. The distributor is of bubble-cap type with removable nozzles of different diameter fitted in the riser to adjust grid pressure drop to the gas flow rate. The catalyst used is 117 µm sand coated with ferric oxide. Further studies with FCC are under way. Six sampling probes were installed in the bed along a radius from central axis, with a pitch of 5.6 cm. They can be moved axially along the bed height. Gas samples can be withdrawn from these probes and analyzed using six UV absorption photometers. These probes are also used to measure pressures.

RESULTS AND DISCUSSION

Concentration Profiles

The average axial concentration profiles measured for the bed with and without horizontal tubes at U = 80 cm s^{-1} are shown in Figure 6. The data might suggest that the tubes did not make a significant difference. However, the radial profiles tell a different story. Figure 4 shows the radial concentration profiles at different heights above distributor, plotted from the same sets of data as Figure 6. The dense bed surface was about 1.38 meters above the distributor for the tube-free bed and about 1.45 meters for the bed with tubes. The shape of the curves in the case of the tube-free bed is quite similar to the flow pattern of the bubble phase measured by Werther (13), suggesting a similar flow pattern. That is, bubbles appear mainly in an annular area near the wall at distributor level, then gradually move toward the centre of the vessel with increased height as a consequence of the rapid increase in bubble size due to coalescence. However, in the bed with horizontal tubes the shape of the concentration profiles shows a rather different flow pattern. The presence of the tube bundle not only limited bubble size but also inhibited lateral movement of bubbles. The bubble streams in the annular area seemingly do not coalesce to form a central bubble stream.

Comparison of concentration profiles in Figures 4 indicates that the presence of the tube bundle has reduced bubble size, as found by various authors (14, 8, 9). But to estimate the bubble size for modelling work is still not simple because of the confusion and disagreement in reported data and correlations. For example, Sitnai et al. (15) reported that in their fluidized bed containing a horizontal tube bank, measured bubble diameter is about 40 cm instead of 17 cm estimated using Bar-Cohen and Hughes' formulae (8).

Gas mixing is another factor which influences chemical reaction. In the case of an internals-free bed, solids descending along the wall carried gas flow downwards, while with horizontal tubes, the downflow stream also occurred in the centre of the column. The much higher conversion (about 82%) near the wall at 80 cm above the distributor in the latter case is most likely caused by the fast downflow solid stream in the gap between the outermost tubes and the wall, as shown by Sitnai (16), in a solid mixing experiment. Potter et al. (17), reported that gas backmixing was suppressed by horizontal tubes but enhanced by vertical tubes in the wall region at higher velocity. The maximum fluidizing velocity used in their work is about 12.6 times incipient fluidizing velocity while in the present case this value is about 47. The different tube bundle geometry in the present work could be responsible for different bed behaviour. More thorough investigations are needed.

Another fact which should not be overlooked is that the overall conversion was not improved by the insertion of the tube bundle and the bed expansion is less than expected. This may suggest that some channelling has occurred.

Pressure Profiles

In order to detect the channelling caused by these densely packed tubes, two pressure transducers have been installed and pressures in both axial and radial directions in the bed, have been measured. Figure 1 shows the average pressure gradient versus bed height at U = 10, 40, 80 and 120 cm s^{-1}. The big decrease of pressure gradient at about 20 to 30 cm above the distributor indicates the probable existence of a big gas pocket caused by the bottom of the tube bundle. The tube bundle starts at 30 cm above the distributor and the free cross-section at the narrowest section is 0.60. The bundle rests on a cross with free space 0.78. In the case of U = 120 cm s^{-1}, the second minimum at 85 cm may be caused by a second supporting cross, which sits at about 100 cm above distributor.

Figure 2 shows the pressure gradient for FCC at U = 10, 20 and 30 cm s^{-1}, displaying a similar situation as in the case of sand.

In order to verify the gas pocket, the so called static void height was calculated and the result was shown in Figure 3. The increasing of the slope at the gas velocity higher than 0.8 m s^{-1} in the case of sand is no doubt due to the appearance of the second gas pocket caused by the second supporting cross. It is also interesting to notice that in the case of FCC, the height of static void increased more rapidly than sand.

Analogy with Flooding in Gas-Liquid Packed Columns

In principle flooding in a gas-liquid packed column commences when the liquid cannot escape readily from the bottom of the packing or the distributor, as a result of the upwards drag of gas on liquid.

If the horizontal tubes of our reactor were immersed in glycerine, or for that matter water, and gas velocity is increased a velocity will be reached at which the liquid cannot escape readily from the packing so that a gas space is created underneath the tubes. Flooding correlations can be extrapolated to zero liquid flow rate. Using the correlation of Lobo et al (18) it will be noted that when the liquid flow rate is small a limiting value is reached such that:

$$\frac{U^2 a}{g\varepsilon^3} \frac{\rho_G}{\rho_L} \left(\frac{\mu_L}{\mu_W}\right)^{0.2} \simeq 0.30 \qquad \ldots (2)$$

as a first approximation take $\varepsilon = 0.60$ and $a = 0.38\ cm^{-1}$ then for a 10-poise liquid flooding occurs at $U = 1.93$ m s^{-1} and for a 1-poise liquid at $U = 2.43$ m s^{-1}. These approximate values for flooding velocity are 2-3 times the values employed in the fluidized bed. On the other hand the model of a packed bed needs to be re-considered for the horizontal tube geometry. Some measurements are in hand. The sand employed could be expected to have an apparent viscosity of the order of 10-poise.

ACKNOWLEDGEMENTS

The authors wish to thank the Australian Research Grants Committee for their continuous financial support to this project.

NOTATION

a	surface of packing/unit volume of tower
g	gravitational acceleration
U	superficial velocity of fluidizing gas
ε	fractional voids
μ_L	liquid viscosity
μ_W	water viscosity
ρ_G	gas or vapour density
ρ_L	liquid density

LITERATURE CITED

1. Mao, Q.M. and Potter, O.E., A.I.Ch.E. Symp. Series, 241 (80), 65 (1984).

2. Miyauchi, T., Furusaki, S., Morooka, S. and Ikeda, Y., Advances in Chemical Engineering, 11, ed. Drew T.B. et al., Academic press, New York, pp 276-448 (1981).

3. Mori, S. and Wen, C.Y., A.I.Ch.E. J., 21, pp 109-115 (1975).

4. Werther, J., Chem. Ing. Tech., 49, No. 11, pp 901-903 (1977).

5. Werther, J., Fourth International Conference on Fluidization, Kashikojima, Japan, Engineering Foundation & Society of Chemical Engineers, Japan, Paper 1-12 (1983).

6. Rowe, P.N., Chem. Eng. Sci.,31, pp 285-288. (1976).

7. Darton, R.C., La Nauze, R.D., Davidson, J.F. and Harrison, D., Trans. Inst. Chem. Engrs., 55, No. 4, pp 274-280 (1977).

8. Bar-Cohen, A. and Hughes, R.W., Israel J. Technol, 18 137(1980)

9. Jodra, L.G. and Aragon, J.M., Int. Chem. Eng., 23(1), p 18(1983)

10. Ikeda, Y., quoted in: Miyauchi, T., Furusaki, S., Morooka, S. and Ikeda, Y., Advances in Chemical Engineering, 11, ed. Drew, T.B. et al., Academic Press, New York, 307 (1981).

11. Doraiswamy, L.K. and Sharma, M.M., HETEROGENEOUS REACTIONS: Analysis, Examples, and Reactor Design. Vol. 1, John Wiley & Sons, pp 352 (1984).

12. Mao, Q.M. and Potter, O.E., Proceedings of 3rd Australasian Conference on Heat and Mass Transfer, (Melbourne) 355 (1985).

13. Werther, J., A.I.Ch.E. Symp. Series, 141 (70), 53 (1975).

14. Nguyen, H.V., Potter, O.E. and Whitehead A.B., Chem. Eng. Sci., 34, pp 1163 (1979).

15. Sitnai, O., Dent, D.C. and Whitehead, A.B., Chem. Eng. Sci., 37 (9), pp 1429 (1982).

16. Sitnai, O., Ind. Eng. Chem. Process Des. Dev., 20, pp 533 (1981).

17. Potter, O.E., Whitehead, A.B. and Nguyen, H.V., Chem. Eng. in Australia, 6, pp 42-44 (1981).

18. Lobo, W.E., Friend, L., Hashmall, F. and Zenz, F., Trans. Amer. Inst. Chem. Engs., 41, pp 693-710 (1945).

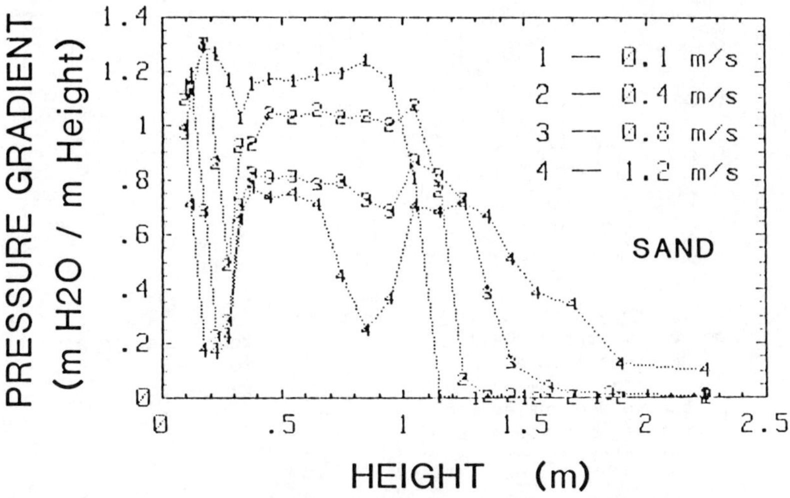

Figure 1. Pressure gradient measured in the bed with sand at different gas velocities.

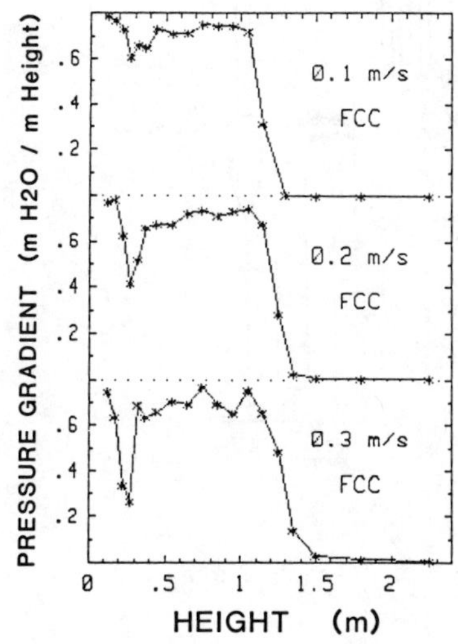

Figure 2. Pressure gradient measured for FCC.

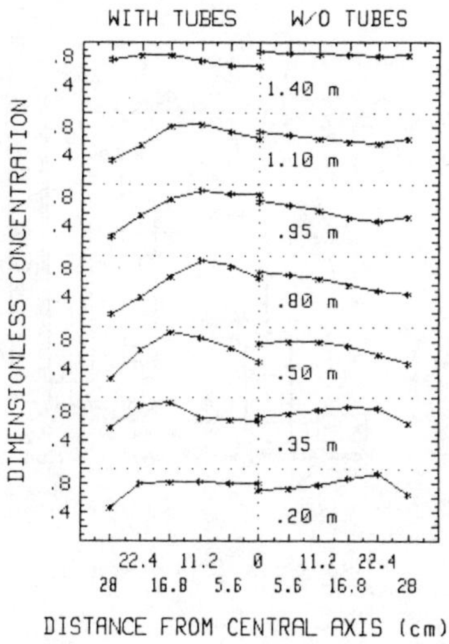

Figure 4. Radial concentration profiles for sand at 0.8 m/s gas velocity.

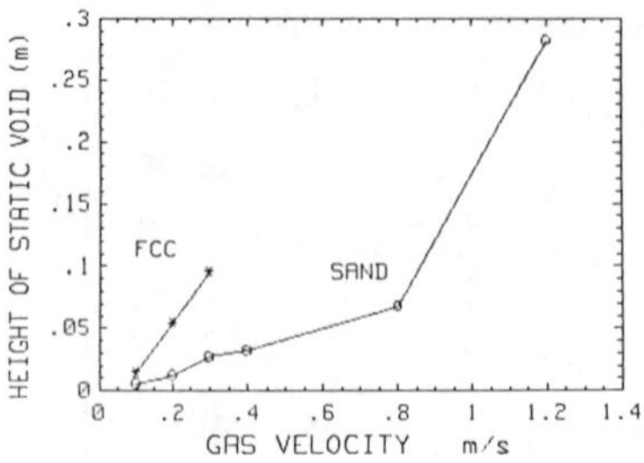

Figure 3. The height of static void in the bed with horizontal tube bundles, assuming that the void occupies the entire cross-section area of the bed.

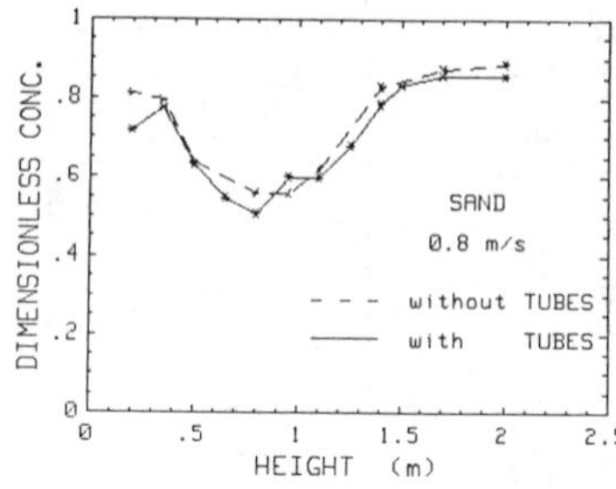

Figure 6. Average axial concentration profiles for sand at 0.8 m/s gas velocity.

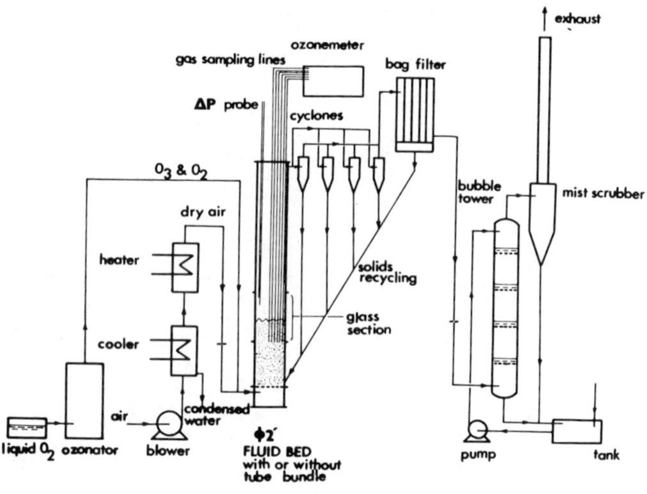

Figure 5. Experimental fluidized bed reactor system.

MODELING AND SCALE UP OF MOBIL'S FLUID-BED MTG PROCESS

A. Avidan and M. Edwards

Mobil Research and Development Corporation
Paulsboro Research Laboratory
Paulsboro, New Jersey 08066 U.S.A.

ABSTRACT

Mobil's fluid-bed Methanol-To-Gasoline (MTG) process was scaled up from bench-scale, through a pilot plant, to a 0.6 m diameter demonstration plant. Nearly 100% conversion, and excellent yields, were maintained. Bubble suppression was achieved by operation in the turbulent regime with a minimum of 15% < 40 μm fines. Reactor efficiency was improved further by horizontal baffles. The homogeneous axial dispersion model was used. Gas tracer response yielded Peclet numbers which were coupled with apparent kinetics.

SCOPE

Designers of fluid-bed reactors have always faced the problem of maintaining the yield and conversion of a bench-scale pilot plant as reactor diameter is increased. Scale up of fluid-bed reactors has been more risky than fixed-bed reactors.(1) A major part of scale up is using a model to predict the performance of a commercial-scale reactor based on the performance of a pilot plant. The model includes a description of fluid-bed hydrodynamics and gas tracer studies have long been used to characterize them.

Most fluid-bed reactor models are based on the two-phase approach which typically requires 2 to 6 parameters. This communication describes the use of the simpler axial dispersion model to analyze scale up effects.

CONCLUSIONS AND SIGNIFICANCE

The axial dispersion model can be used to predict conversion in a turbulent fluid-bed reactor under certain conditions. The model assumes overall homogeneity, which requires bubble suppression. This is achieved in a turbulent bed of group A powder by operation at superficial velocities above 0.3 m/s, with a minimum of 15% less than 40 μm fines, and a sufficient bed height. Staging the bed with horizontal baffles improves conversion further. Implementation of these design parameters permitted scale up of the fluid-bed

MTG process from bench-scale, through a pilot plant, to a 0.6 m diameter, 20 tonnes/day, demonstration plant without loss of conversion efficiency, as predicted by the model.

THE MTG PROCESS

In the early 1970's Mobil scientists discovered a novel process for converting methanol to gasoline using a selective zeolite catalyst. Bench-scale development was conducted in both fixed and fluid-bed reactors. The fluid-bed reactor offers several advantages. The heat of reaction can be easily removed and the fluid bed can be maintained at any desired catalyst activity level. Gasoline yield in the fluid bed is higher, and the gasoline has a higher octane rating.

The fluid bed was successfully scaled up to a 0.1 m diameter, 7.6 m tall, 0.6 tonnes/day ("4 B/D") pilot plant.(2) In 1980, the New Zealand government selected fixed-bed MTG technology because it was ready for commercialization. At the same time, a five-year international project to demonstrate scale up of the fluid-bed MTG process in a 0.6 m diameter, 20 tonnes/day plant in Germany was initiated. The design rate corresponds to 160 B/D. The demonstration plant has been referred to as "100 B/D". The plant ran for 360 days with methanol. Complete methanol conversion (99.9% +), excellent gasoline yields (over 92%, including alkylate and excess olefin recycle) and quality were demonstrated. Catalyst make-up rate was low, and depended on the desired catalyst activity.(3)

FLUID-BED SCALE UP

The potential for loss in reactor efficiency upon scale up is illustrated schematically in Figure 1.(4) The major culprit for this loss is gas bypassing in the form of large bubbles. Yet surprisingly, bubble properties, rather than their suppression, have been studied extensively. Bubble suppression can be achieved by using these three design parameters:

- Increasing gas velocity - When gas velocity is increased for a group A powder, the bed undergoes transition from bubbling to turbulent fluidization. The transition velocity decreases with increasing bed diameter. The bed assumes a more homogeneous appearance: bubbles are short lived, and gas backmixing is reduced. The change in slope in the expansion curve, Figure 2, upon transition to turbulence has been interpreted to show more homogeneity by use of the Richardson-Zaki equation.(5)

- Increasing fines level - For a typical group A powder, the less than 40 μm fines fraction suppresses bubble size.(6)

- Bed height and staging - Some of the lowest reactor efficiencies were observed in shallow beds. Increasing the bed height in commercial reactors reduces gas bypassing.(7) Staging the bed reduces bypassing and backmixing.

FLUID-BED REACTOR MODELING

Shinnar (8) classifies chemical reactor models into learning and predictive models. We consider models which use bubble properties as learning models. They are useful in improving our fundamental understanding of bubbling fluid-beds. However, they are not used in the scale up of turbulent bed reactors. Bubbles coalesce and break up in a complex fashion in the turbulent fluidization regime.

Predictive design models typically use the two-phase approach. Model parameters include solids fractions and velocities in the two phases, dispersion in each phase, and interchange coefficients. This model has been used to interpret adsorptive and non-adsorptive tracer data in our demonstration plant CFM.(9)

The homogeneous appearance of the turbulent fluid-bed led us to the axial dispersion model. Such models fell into disrepute because they cannot predict conversions lower than an ideally mixed reactor. Reactant bypassing, e.g. in a bubbling bed, has led to such low conversion levels.(1) We avoided this situation by employing the three design parameters cited previously.

EXPERIMENTAL

We scaled up the fluid-bed MTG process from a 0.04 m diameter, bench-scale reactor, through a 0.1 m pilot plant, to a 0.6 m diameter demonstration plant. Full-scale CFM's were used to study hydrodynamics by using a gas tracer technique. Sulfur hexaflouride (SF_6) tracer was injected below the reactor distributor. The sample probe was 1.6 mm diameter to minimize contribution to lag time. It was located just above the catalyst bed. (Experience from earlier work in the pilot plant CFM had shown that a sample probe located in the catalyst bed would yield a noisy signal). A commercial leak detector, with a Ni-63 source, was used to measure the response signal.

Pulse injections in the demonstration plant CFM were conducted at isuperficial gas velocities of 0.3 - 1 m/s. Moisture level in the system was varied by steam addition. We evaluated the fluid bed without baffles, with vertical baffles, and with horizontal baffles. The solids used in these experiments were typical FCC catalyst. Additional details are given elsewhere.(9)

AXIAL DISPERSION MODEL

The fluid-bed axial dispersion model yields by mass balance (11):

$$c(k) = \frac{4\,a\,\exp\left(^{Pe}/_{2}\right)}{(1+a)^2\exp(aPe/2) - (1-a)^2\exp(-aPe/2)} \qquad (1)$$

$$\text{where} \qquad a = \left(1 + \frac{4k\tau}{Pe}\right)^{1/2} \qquad (2)$$

The kinetics used were obtained in laboratory reactors, using Equation (1) and the Pe number from gas tracer studies in these reactors. The apparent activation energy measured in the bench unit is 50 kJ/mol with a first order fit. (11) The frequency factor was related to catalyst activity by the "reaction index", RI, which is the product propane/propene ratio. As the catalyst is deactivated (reversibly by coke deposition, and irreversibly by steaming), more olefins are produced. RI was found to be a convenient index of catalyst activity. (2) RI varies with temperature, space velocity, and pressure. It was corrected to 0.27 MPa, 680 K, and 1 SV.

The Peclet number is calculated from tracer data. The equations for one-dimensional dispersion are:

$$\frac{\partial C}{\partial \theta} = \frac{1}{Pe}\frac{\partial^2 C}{\partial Z^2} - \frac{\partial C}{\partial Z} \qquad (3)$$

$$\text{where:} \quad C = c(t)/\int_o^\infty c(t)\,dt \qquad (4)$$

The system is divided into two regions: the bed and the freeboard, and b is the ratio of the bed to freeboard axial dispersion coefficients. Because of the high gas velocity in the reactor riser we assumed that $D_{Z<O}=0$.

For a pulse injection, the solution of Equation (3) is:

$$\text{variance:} \quad \sigma_\theta^{\;2} = 2(e^{-Pe}(1-b) + Pe-1+b+1.5b^2)/Pe^2 \qquad (5)$$

$$\text{mean:} \quad \theta = 1+b/Pe \qquad (6)$$

Axial Peclet numbers were calculated from Equations (5) and (6) by matching the response curve to a theoretical curve. A comparison of the model to an experimental curve is shown in Figure 3.

The Peclet number obtained in the demonstration plant CFM is plotted versus catalyst moisture content in Figure 4. Horizontal baffles have a higher Peclet number than vertical baffles and the unbaffled bed. This implies lower dispersion and a higher methanol conversion efficiency for horizontal baffles. The Peclet number increased from 7 to 10 - 12. This is caused by the staging effect and the lower catalyst density in the horizontally baffled bed (Figure 2). The results of the MTG conversion model are illustrated in Figure 5 at low activity. Bed efficiency (compared to plug flow) decreases as conversion increases.

FLUID-BED MTG SCALE UP

The MTG conversion model predicted similar methanol conversion from the bench-scale reactor to the 100 B/D demonstration plant. The units did exhibit similar conversion performance (Figure 6). Methanol breakthrough (conversions of less than 99.5%) occurs in all units in the same activity range. The bubbling-bed bench-scale unit exhibits higher methanol breakthrough than the other units. Small amounts of gas bypassing, not detected in tracer studies, result in the higher methanol breakthrough despite the unit's higher Pe number. The model assumption of a homogeneous flow is not completely satisfied in a bubbling bed.

The pilot plant operated in the slugging/turbulent regime with less opportunity for methanol bypassing (Pe=10). The demonstration plant reactor data includes breakthrough curves from three reactor configurations: "Phase I" operation with baffles and external cooling (Pe=12); "Phase II" operation with baffles and internal cooling coils (Pe=12); and "Phase II" operation without baffles, but retaining the cooling coils which provided some baffling (Pe=8 to 10). Thus, the pilot plant breakthrough curve is in agreement with the Phase II unbaffled bed performance.

Demonstration plant performance for the three reactor configurations was:

	Aqueous Phase Methanol, ppm		
	Actual	Predicted	Pe
Phase I (baffled)	80	90	12
Phase II (baffled)	115	90	12
Phase II (unbaffled)	195	380	7
		200	8-10

Gasoline yield was also scaled up successfully from the bench-scale to the demonstration plant. High gasoline quality was maintained throughout (95 - 96 Research Octane number). Additional details of the demonstration plant operation were given by Gierlich et al.(3)

ACKNOWLEDGEMENTS

The 100 B/D Project was a joint effort by Union Rheinische Braunkohlen Kraftstoff AG, Uhde GmbH, and Mobil Research and Development Corporation, supported in part by the U.S.A. Department of Energy and the Federal Republic of Germany Bundesministerium fur Forschung und Technologie. We thank C. K. Lee and M. N. Lo for developing the tracer methods.

NOTATION

c(k)	:	Unconverted reactant fraction
c(t)	:	Tracer concentration
C	:	Nondimensional tracer concentration
D	:	Gas axial dispersion coefficient, m^2/s
k	:	Apparent reaction rate constant, h^{-1}
L	:	Bed length, m
n	:	Richardson-Zaki exponent (Figure 2)
Pe	:	Peclet number, $U_g L/D$
RI	:	Reaction Index, propane/propene yield weight ratio at 680 K, 0.27 MPa, 1 SV
SV	:	Space velocity ($1/\tau$), kg/h feed/kg catalyst
t	:	Time, s
U_g	:	Superficial gas velocity, m/s
U_t^*	:	Particle cluster terminal velocity, m/s
Z	:	Fractional axial length
ϵ	:	Average catalyst bed voidage
θ	:	Tracer mean residence time/theoretical residence time
σ_θ^2	:	Tracer variance
τ	:	Methanol space time, h

LITERATURE CITED

1. Squires, A. M., Proc. Joint Meeting Chem. Ind. Eng. Soc. China and AIChE, Beijing, 322 (1982).
2. Kam, A. Y. and Lee, W., Fluid Bed Process Studies on Conversion of Methanol to High Octane Gasoline, Final Report, FE-2490-15 (1978).
3. Gierlich, H.-H., K.-H. Keim, N. Thiagarajan, E. Nitschke, N. Daviduk, and A. Y. Kam, "Successful Scaleup of the Fluid-Bed Methanol-to-Gasoline (MTG) Process to 100 BPD Demonstration Plant". 2nd EPRI Conf. (Coal Gasification and Synthetic Fuels for Power Generation), April 15-19, San Francisco (1985).
4. Squires, A. M., Kwauk, M. and Avidan, A. A., "Fluid Beds: At Last, Challenging Two Entrenched Practices," to be published, Science (1985).
5. Avidan, A. A. and Yerushalmi, J., Powder Technology, 32, 223 1982).
6. de Vries, R. J., van Swaaij, W. P. M., Mantovani, C., and Heijkoop, A., Proc. 5th European Symp. on Chem.

Reaction Eng., B9-59, Elsevier Pub., Amsterdam (1972).
7. van Deemter, J. J., "Mixing Patterns in Large-Scale Fluidized Beds," Fluidization, Plenum Press, New York (1980).
8. Shinnar, R., "Chemical Reactor Modeling-The Desirable and the Achievable" in Chemical Reaction Engineering Reviews - Houston, ACS Symp. Ser. 72,1-36, American Chemical Society, Washington, D.C. (1978).
9. Krambeck, F. J., Avidan, A. A., Lee, C. K., and Lo, M., "Predicting Fluid-Bed Reactor Efficiency Using Adsorbing Gas Tracers," Presented at the AIChE Annual Meeting, Chicago (November, 1985).
10. Wehner, J. F. and Wilhelm, R. H., Chem. Eng. Sci., 6, 89 (1956)
11. Liederman, D., Jacob, S. M., Voltz, S. E., and Wise, J. J., Ind. Eng. Chem. Proc. Des. Dev., 17, 340 (1978).

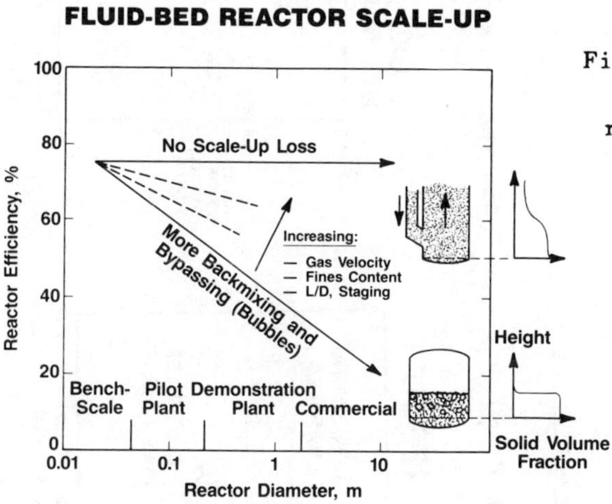

Figure 1: Fluid-bed reactor scale up

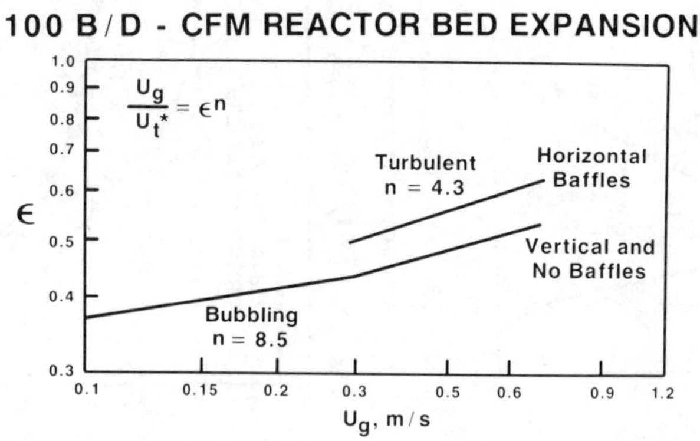

Figure 2: Demonstration Plant CFM reactor bed expansion

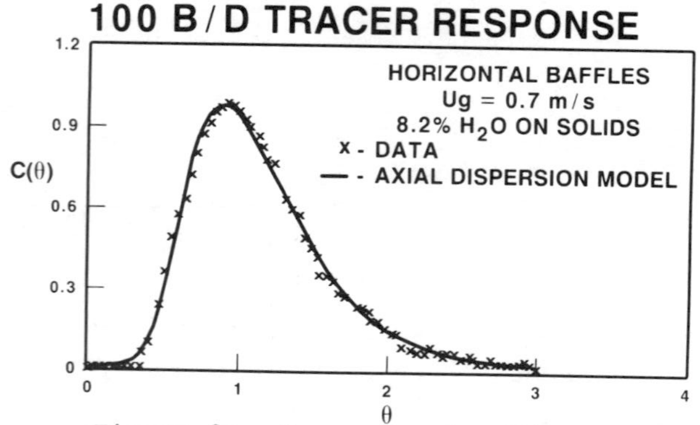

Figure 3: Demonstration Plant typical tracer response

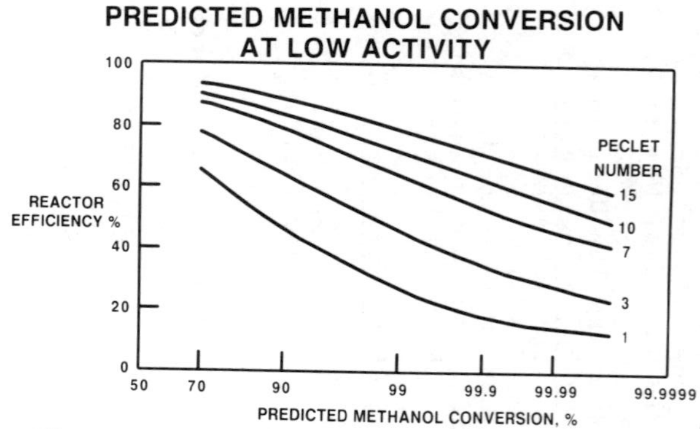

Figure 5: Predicted methanol conversion at low activity

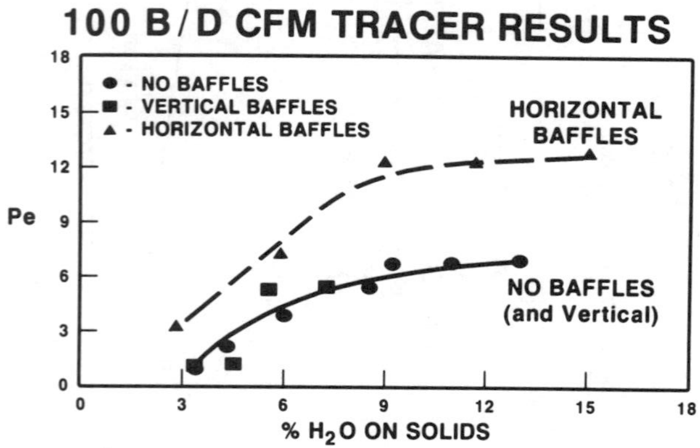

Figure 4: Demonstration Plant CFM tracer results

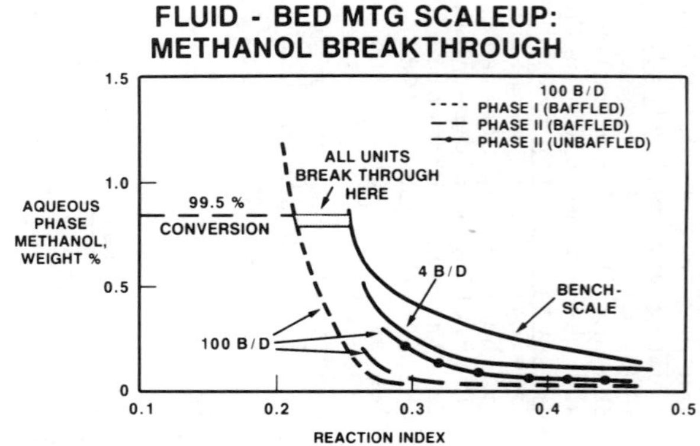

Figure 6: MTG scale up - methanol breakthrough

A Computer Simulation of Propane Cracking in a Spout-Fluid Bed Reactor with a Draft Tube

by

Rudolf K. Stocker, Ashutosh Rastogi, Leo A. Behie, and William Y. Svrcek
Department of Chemical and Petroleum Engineering,
The University of Calgary, Calgary, Alberta, Canada

and

Maurice A. Bergougnou, Faculty of Engineering Science,
The University of Western Ontario, London, Ontario, Canada

ABSTRACT

Predictions from a two-dimensional computer model of propane cracking in a spout-fluid bed reactor with a draft tube are compared to experimental results from an 11 cm diameter bench-scale bed. Although the reactant conversion predicted by the model is low, the product distribution is reasonably good. Reactant contact times in the reactor are very short, ranging from 5 to 20 ms.

SCOPE

Spouted beds are a relatively recent development in fluidization technology, consisting of a bed of medium to coarse particles vertically penetrated by a fluid jet. A spout-fluid bed combines the windbox and gas distributor of a fluidized bed with the nozzle and central spout of a spouted bed. The action of the fluidizing gas is to reduce particle-particle interaction in the entrainment region at the base of the bed such that the rate of particle entrainment into the spout gas stream is enhanced, and to promote radial heat heat transfer in the fluidized region of the annulus.

A computer model for heat transfer in spout-fluid beds previously developed by us (Stocker *et al.* (8,9)) was adapted to simulate the propane cracking reactions in such a bed. Product distributions from the simulation are compared, in this paper, to the results of a propane cracking test in a bench-scale experimental spout-fluid bed.

The significance of this study is that, for the first time, the spout region of a spouted or spout-fluid bed is considered to be significant in terms of reactant conversion. Previous models have always assumed the spout to be a region of reactant by-passing because of the high voidage relative to the annulus. For a fast homogeneous gas-phase reaction, however, the short residence times and high heat-up rates in the spout region can be a tremendous advantage compared with the longer and more distributed residence times seen in the annulus. Enhancement of the bed with the inclusion of a draft tube to confine the spout gases beyond the entrainment region promotes this effect. Moreover, the spout-fluid bed permits rapid heat transfer from particles to gas in the spout region while allowing the particles a much longer regeneration period in the slower moving annulus.

CONCLUSIONS AND SIGNIFICANCE

This study has clearly demonstrated the potential for utilization of spout-fluid beds as reactors where fast (millisecond time scale) endothermic or exothermic reactions are concerned. The combination of efficient heat transfer to the spout region and spout gas confinement with a draft tube enables manipulation of the reaction selectivities and selection of the principal product species in situations where competing reactions take place. There are remaining problems in the modelling of chemical reactions in such beds; a definitive solids entrainment model has not yet been proposed and kinetic rate equations based on data for time scales similar to those seen in the spout have not yet been generated for most reactions. It has been demonstrated, however, that a relatively simple model is adequate to describe the hydrodynamic and thermal processes taking place in a bed of this configuration.

Spout-fluid bed technology is one of the newest derivatives of the spouted bed unit process, which itself dates only from the mid 1950's. Nonetheless, it shows significant promise in terms of potential applications in many of the fields of chemical and process engineering. Efforts to model chemical reactions in spouted and spout-fluid beds, however, have largely been confined to simplified cases: isothermal beds, one-dimensional models, and invariant spout voidage. Moreover, virtually all of the models published to date have neglected reaction in the spout region entirely or have considered it to be of negligible extent.

As described by Mathur and Epstein (1), Uemaki was the first to attempt the modelling of chemical reactions in spouted beds in connection with his work on pyrolysis of heavy oils in the late 1960's. Little information concerning that early model is available. Subsequent efforts include those of Mathur and Lim (2), Piccinini *et al.* (3), Littman *et al.* (4), Kumpinsky and Amundsen (5), and Viswanathan (6), all of which assume that the bed is isothermal and that the spout region voidage is constant and reaction in the spout region can be neglected. The isothermal bed assumption may hold reasonably well for mild bed conditions and reactions which are neither highly exothermic or endothermic. Where these are not the prevailing conditions, variations in gas density and reactant concentration, gas and solid velocities, and heat and mass transfer coefficients and the existence of large radial thermal gradients may cause significant deviation from the predictions of these models.

Smith *et al.* (7) recently modelled the ozone decomposition reaction in an adiabatic spouted bed without presuming the bed to be isothermal. The model was largely based on the previous one-dimensional models with the addition of the heat transfer equations for transfer between spout and annulus. Radial gradients within the annulus, however, were neglected, even in predicting the effect of bed diameter on reactant conversion.

The model described below was initially developed in the course of a study on heat transfer in spout-fluid beds (Stocker *et al.* (8), Stocker *et al.* (9)), a crucial precursor to futher study on their application as chemical reactors. A bench-scale spout-fluid bed used for that study was modified to permit the cracking of propane, as described below. Figure 1 illustrates the regions of a spout-fluid bed schematically.

COMPUTER MODEL

The axisymmetric two region model originally proposed by Lefroy and Davidson (10) was modified in the following respects in order to better describe the actual bench-scale bed:

(1) The inclusion of a draft tube in the bed configuration has eliminated the crossflow of solids and gases between spout and annulus above the entrainment region. Heat transfer in that region between spout and annulus is limited to conduction across and radiation from the draft tube wall.

(2) Heat flux through the outer wall of the externally heated bed is based on a constant external wall temperature.

(3) The annular region is described with a two-dimensional (axial-radial) representation of a slowly-moving packed bed. Bed conductivity is thus based on a packed bed model accounting for all forms of heat transfer, including radiation. Radial fluxes into and out of the annulus are based on overall transfer resistances at the packed bed boundaries. The two phases are treated together thermally as a pseudo-homogeneous phase.

(4) A one-dimensional model is used to describe the spout region, neglecting potential radial thermal and velocity gradients because of the extreme turbulence in that region. The two flowing phases are separately treated hydrodynamically and thermally, rigorously accounting for acceleration of particles in the spout. Solids circulation rates are input to the program based on those inferred from the heat transfer work because of the limited reliability and regions of applicability of current correlations for solids entrainment into fast-flowing fluid streams.

Given these amendments, the equations for the major regions of the bed can be written as follows. The spout solids mass balance is given by:

$$\frac{\partial F_p}{\partial z} = \frac{\partial}{\partial z}[(1 - \varepsilon)\,\rho_p\,u_p\,A_s] \tag{1}$$

where the solids entrainment rate $\partial F_p/\partial z$ is finite within the entrainment region and zero elsewhere.

Rearranging and carrying out the differentiation of the right hand side of Equation (1), one obtains:

$$\frac{\partial \varepsilon}{\partial z} = \frac{(1 - \varepsilon)}{u_p}\frac{\partial u_p}{\partial z} - \frac{1}{\rho_p u_p A_s}\frac{\partial F_p}{\partial z} \tag{2}$$

a relation for the axial variation of spout voidage with height in the spout.

A force balance on a single solid particle in the spout region with the assumption of minimal particle-particle interaction yields:

$$\frac{\partial u_p}{\partial z} = \frac{1}{u_p}\left[\frac{3}{4}\, d_p\, C_D\, \frac{\rho_g}{\rho_p}(u_g - u_p)|u_g - u_p| - g\,\frac{(\rho_p - \rho_g)}{\rho_p}\right] \quad (3)$$

The drag coefficient C_D is obtained at the spout conditions through a number of empirical correlations fitted to the Reynold's number Re. The correlations, too lengthy to include here, were drawn from Clift, Grace and Weber (11); they are described in greater detail in Stocker (12).

A similar mass balance on the spout gases yields the relation:

$$\frac{\partial F_g}{\partial z} = \frac{\partial}{\partial z}[\varepsilon\, \rho_g\, u_g\, A_s] = 0 \quad (4)$$

which reduces to:

$$\frac{\partial u_g}{\partial z} = -\frac{u_g}{\rho_g}\left(\frac{\partial \rho_g}{\partial z}\right) - \frac{u_g}{\varepsilon}\left(\frac{\partial \varepsilon}{\partial z}\right) \quad (5)$$

The rate of energy transfer to the spout gases is obtained from an energy balance on an individual solid particle in the spout:

$$\frac{\partial T_p}{\partial z} = \frac{1}{u_p}\,\frac{6}{\rho_p C_{pp} d_p}\left[-h_p(T_p - T_g) + \sigma F_{oi}(T_a^4 - T_p^4)\right] \quad (6)$$

The first term on the right hand side of Equation (6) constitutes the rate of heat transfer by convection to the spout gases, where h_p is the convective heat transfer coefficient. The second term is the rate of energy gain by the particles through radiation from the draft tube wall. The overall interchange factor F_{oi} is set equal to the emissivity of the spout particles because the assumption is made that the draft tube wall constitutes a black body completely surrounding the gray body spout particles.

Thus, an energy balance on the spout gases yields:

$$\frac{\partial T_g}{\partial z} = -\frac{(1-\varepsilon)}{\varepsilon}\,\frac{\rho_p}{\rho_g}\,\frac{u_p}{u_g}\,\frac{C_{pp}}{C_{pg}}\left(\frac{\partial T_p}{\partial z}\right) + \frac{2R_w}{r_s \rho_g C_{pg} u_g \varepsilon}(T_g - T_a) \quad (7)$$

The first term on the right hand side of Equation (7) accounts for the rate of heat transfer from the spout particles to the gas. The second term denotes the rate of heat transfer by convection from the spout wall, where R_w is the sum of three resistances, including a term for the rate enhancing effect of the particles in the spout stream. The spout gases are assumed to be transparent to infrared radiation.

Finally, an energy balance on an element of the pseudo-fluid in the annulus yields the following partial differential equation:

$$\frac{\partial T_a}{\partial z} = \frac{k_e}{G\, C_p}\left(\frac{1}{r}\frac{\partial T_a}{\partial r} + \frac{\partial^2 T_a}{\partial r^2}\right) \quad (8)$$

Thus far, the model described is identical to the heat transfer model previously detailed in Stocker et al. (8,9). Simulation of propane cracking entailed the inclusion of the reaction mechanism proposed by Sundaram and Froment (13,14). The reaction mechanism consists of eleven series and parallel elementary reaction steps, represented with first order rate equations, the parameters for which were derived from a large number of short duration (0.2 - 2.0 s) cracking experiments. The equations describe the rates of consumption and formation of a number of molecular species, from H_2 to C_6+ in competing reactions. A solid phase coke formation reaction which is surface area dependent is also included.

Details of the method of solution of the above equations are described in Stocker (12).

EXPERIMENTAL

The bench-scale spout-fluid bed (11.0 cm diameter) constructed for the heat transfer study was used in this phase of the work as well. Figure 2 shows the configuration of the bed and ancillaries. The spout-fluid bed itself was of semi-cylindrical configuration, with a thick (12.7 mm) sheet of Johns-Manville Marinite 'I' between the flanged carbon steel shell halves to act as a thermal mirror. This configuration enabled the mounting of an array of shielded K-type (chromel-alumel) thermocouples longitudinally in the spout and annulus. The thermocouples in the annulus were staggered between sides to better yield the average annular mid-point temperature as well as give an additional indication as to the state of the bed. Thermocouple leads and tubing for manometer connections occupied the inactive shell side. A flat, perforated grid (70 0.8 mm holes equally spaced on a diagonal grid pattern) above the windbox formed the base of the active shell half, with the spout nozzle protruding slightly into the bed above it for enhanced spout stability. A coarse screen, set 5 cm below the bed entrance, supported the bed inside the 1.91 cm diameter nozzle. The draft tube (2.54 cm diameter) was mounted directly on the insulating sheet such that its distance from the inlet nozzle could be varied between 5.0 and 12.0 cm.

Temperature readings from the thermocouple arrays were monitored with a TAURUS ONE data logger and transmitted to an IBM-PC/XT microcomputer for data conversion, display, and storage on a periodic basis (typically every two minutes). This made it possible to monitor the state of the bed and determine when steady-state was achieved. A bank of manometers yielded differential pressure profiles along the spout and annulus, enabling determination of the state of flow inside the bed, which was completely enclosed in an electric furnace (rated at 2450 W with only one side heating).

For the propane cracking experiments, nitrogen was supplied from a cylinder and split into spouting and fluidizing gas streams, with individual stream rates controllable and measured with rotameters. Propane gas was supplied from a cylinder at rates predetermined not to exceed 1.0 % of the total inlet gas flowrate by volume and measured with a rotameter. Clean silica sand (mean particle diameter: 1.034 mm) was used as bed solids; the bed depth in the quiescent state was 30 cm. Exhaust gases were sampled from a tap just above the bed outlet with a vacuum bottle and analyzed on a Shimadzu GC-8APT gas chromatograph using a TCD detector with a current of 100 mA, detector and injector temperatures of 260°C, a ramp temperature program from 50°C to 250°C at 20°C/min, and helium as the carrier gas flowing at 30 ml/min in a 1.83 m 80-100 mesh Poropak 'Q' column of 3.2 mm diameter.

ANALYSIS OF RESULTS

Figure 3 shows an experimental steady-state axial spout gas temperature profile from the earlier heat transfer work (Stocker *et al.* (9)) and the model profile generated for the same conditions. The large heat-up rates in the lower spout region are indicated by the rapid initial increase in gas temperature and the flat profile over the remainder of the spout length. The spout gas temperature profile actually measured during the propane cracking test is shown in Figure 4.

Figures 5 and 6 present the axial and exit selectivity profiles generated by the model for propane and the major product species expected according to the reaction mechanism proposed by Sundaram and Froment (13). It is apparent that, while there is agreement for some of the species (C_3 to C_6+), there are significant differences for the lighter gases (H_2, CH_4, and C_2's). The reason for such differences, given the species involved, may be due to differences in the time scales between the conditions in the spout region of the spout-fluid bed and the conditions of the experiments originally conducted for the kinetic studies.

ACKNOWLEDGEMENTS

This work was supported by the Natural Sciences and Engineering Research Council of Canada.

NOTATION

A	cross sectional area of flow, m^2
C_D	particle drag coefficient
C_{pg}	gas specific heat, J/kg K
C_{pp}	solids specific heat, J/kg K
d_p	mean solids diameter (smallest cross-section), m
F_g	inlet gas mass flowrate, kg/s
F_{oi}	overall radiative interchange factor
F_p	solids mass circulation rate, kg/s
g	gravitational acceleration, m/s^2
G	total annular mass flowrate, kg/s
h_p	particle heat transfer coefficient, W/m^2 K
k_e	effective annular thermal conductivity, W/m K
r_s	spout radius, m
R_w	overall heat transfer coefficient, W/m^2 K
Re_p	particle Reynold's number
T_a	annular temperature, K
T_g	spout gas temperature, K
T_p	spout solids temperature, K
u_g	spout gas velocity, m/s
u_p	spout solids velocity, m/s
z	axial distance from base of bed, m
α_p	particle absorptivity
ε	voidage
ε_p	particle emissivity
ρ_g	gas density, kg/m^3
ρ_p	solids density, kg/m^3
σ	Stefan-Boltzmann constant, $W/m^2 K^4$

LITERATURE CITED

1. Mathur, K.B., and Epstein, N., Spouted Beds, p. 185, Academic Press Inc., New York, N.Y. (1974)

2. Mathur, K.B., and Lim, C.J., "Vapour Phase Chemical Reaction in Spouted Beds: a Theoretical Model", Chem. Eng. Sci., 29, 789 (1974)

3. Piccinini, N., Grace, J.R., and Mathur, K.B., "Vapour Phase Reaction in Spouted Beds: Verification of Theory", Chem. Eng. Sci., 34, 1257 (1979)

4. Littman, H., Narayanan, P.V., Tomlins, A.H., and Friedman, M.L., "A Complete Theoretical Model for a First Order Isothermal Catalytic Reaction in a Spouted Bed", AIChE Symp. Ser., 77 (205), 174 (1981)

5. Kumpinsky, E., and Amundsen, N.R., "Reactor Model Assessment for the Combustion of Char in Spouted Beds of Sand", I.E.C. - Proc. Des. Dev., 23 (4), 611 (1984)

6. Viswanathan, K., "A Semi-Compartmental Model for Spouted Bed Reactors", C.J.Ch.E., 62 (7), 623 (1984)

7. Smith, K.J., Arkun, Y., and Littman, H., "Studies on Modelling and Control of Spouted Bed Reactors - I: Reactor Modelling", Chem. Eng. Sci., 37 (4), 567 (1982)

8. Stocker, R.K., Behie, L.A., Svrcek, W.Y., and Bergougnou, M.A., "A Two-Dimensional Computer Simulation of Heat Transfer in Spout-Fluid Beds", presented at the 34th Canadian Chemical Engineering Conference, Quebec, P.Q., Canada (1984)

9. Stocker, R.K., Behie, L.A., Svrcek, W.Y., and Bergougnou, M.A., "Heat Transfer in a Spout-Fluid Bed Equipped with a Draft Tube", to be presented at the 35th Canadian Chemical Engineering Conference, Calgary, Alta., Canada (1985)

10. Lefroy, G.A., and Davidson, J.F., "The Mechanics of Spouted Beds", Trans. Inst. Chem. Eng., 47, T120 (1969)

11. Clift, R., Grace, J.R., and Weber, M.E., Bubbles, Drops, and Particles, p. 112, Academic Press Inc., New York, N.Y. (1978)

12. Stocker, R.K., "Ultrapyrolysis of Alberta Heavy Oil in a Spout-Fluid Bed Reactor", PhD thesis (in preparation), Univ. of Calgary, Calgary, Alta., Canada (1985)

13. Sundaram, K.M., and Froment, G.F., "Modeling of Thermal Cracking Kinetics - I: Thermal Cracking of Ethane, Propane, and their Mixtures", Chem. Eng. Sci., 32, 601 (1977)

14. Sundaram, K.M., and Froment, G.F., "Kinetics of Coke Deposition in the Thermal Cracking of Propane", Chem. Eng. Sci., 34, 635 (1979)

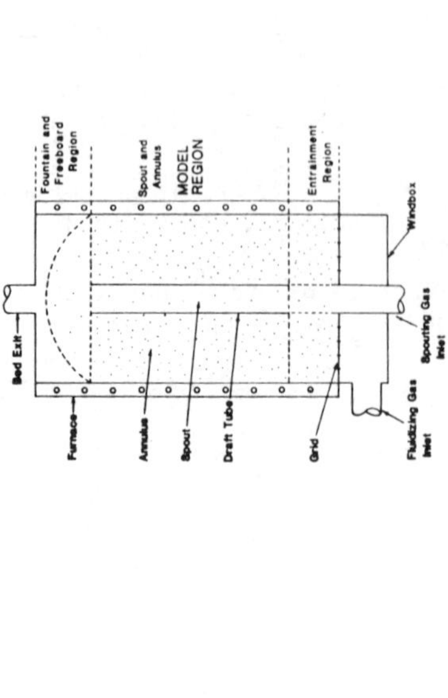

Figure 1: Spout–Fluid Bed Schematic

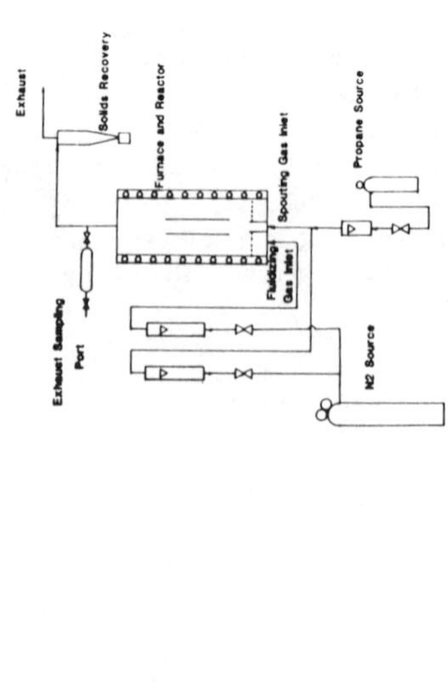

Figure 2: Experimental Equipment

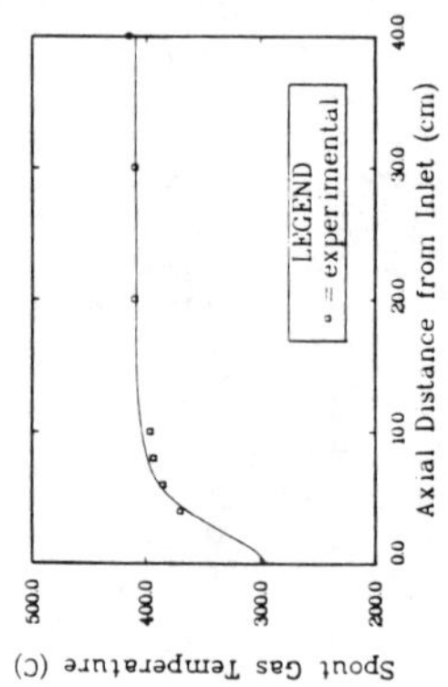

Figure 3: Axial Spout Gas Temperature Profile – Heat Transfer Study

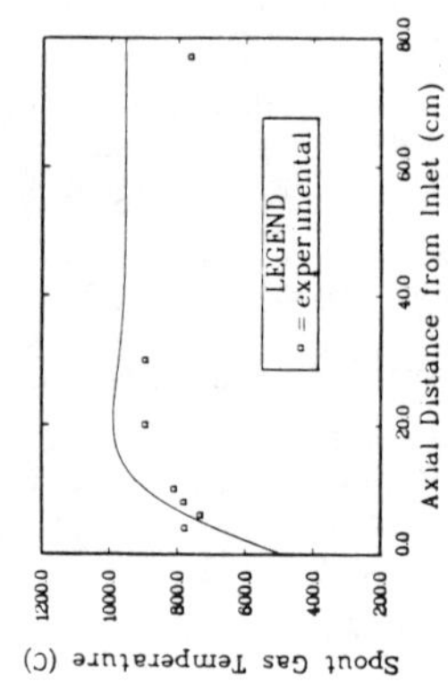

Figure 4: Axial Spout Gas Temperature Profile – Propane Cracking Run

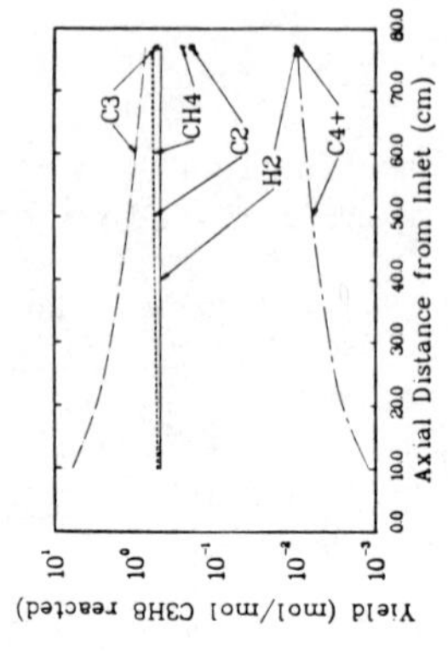

Figure 5: Axial Product Yield Profiles

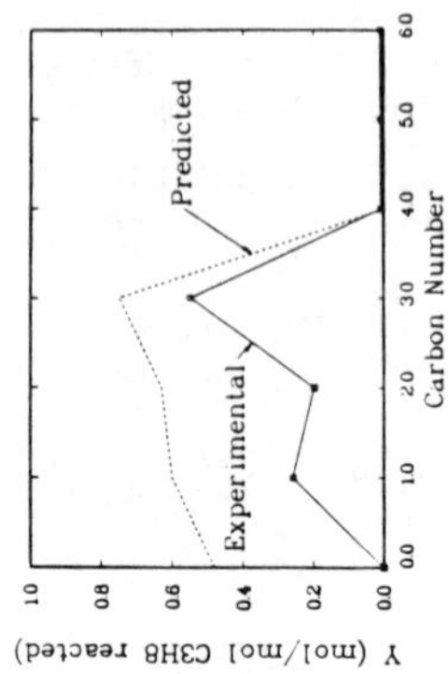

Figure 6: Exit Product Yield as a Function of Carbon Number

THE ULTRA-RAPID FLUIDIZED (URF) REACTOR: APPLICATION TO DETERMINE THE KINETICS OF THE FAST PYROLYSIS (ULTRAPYROLYSIS) OF CELLULOSE

R.G. Graham, B.A. Freel, R.P. Overend*, L.K. Mok and M.A. Bergougnou

Faculty of Engineering Science
The University of Western Ontario
London, Ontario, Canada N6A 5B9

*National Research Council
Ottawa, Ontario, Canada K1A 0R6

ABSTRACT

A major new reactor system, the ultra-rapid fluidized (URF) reactor is being developed at the University of Western Ontario using cold models and has been applied to study fast pyrolysis kinetics in a mini-pilot plant. The results of cellulose pyrolysis experiments conducted at 800 C fit a first order kinetic model.

SCOPE

A major new reactor system, the ultra-rapid fluidized (URF) reactor is being developed to exploit the field of high-temperature, extremely fast reactions. It involves the rapid contacting and separation of a hot finely divided solid heat carrier (thermofor) with, for instance, a carbonaceous feedstock. Cold modeling work is in progress to define optimal contacting and separation modes. The scheme is being applied, at the mini-pilot plant level, to the fast pyrolysis of cellulose in order to exploit the high heating rates, short residence times, high temperature and rapid quenching which are required to produce valuable non-equilibrium chemical intermediates (i.e. ethylene, acetylene, BTX, etc.) from carbonaceous feedstocks. The primary objective of the first phase of the work was to develop a novel reactor system which could accommodate the rapid and severe process requirements and which could also serve as a conceptual design for a continuous commercial process. The equipment evolved considerably from the first concept to the current design with the aid of extensive cold model testing.

In the mini-pilot plant, hot solids and/or inert gas are used to carry and transfer heat to particulate carbonaceous feedstocks (i.e. wood, cellulose, lignite), in a very turbulent vortical contactor, termed a thermovortactor. The turbulence creates an ideal environment for fast thorough mixing and extremely rapid heat transfer. These features, combined with the quasi plug-flow nature of the reactor and the fast quenching of the products, are integrated into a system which exhibits excellent residence time control (on a scale of tens of ms) and is, therefore, uniquely suited to investigate the isothermal kinetics of fast pyrolysis reactions.

Kinetics experiments using Avicel pH102 microcrystalline cellulose and an inert nitrogen heat carrier were conducted at temperatures between 750 and 900 C. In this paper, the mass yields of the important gaseous products are reported as a function of residence time at 800 C. The first-order kinetic parameters are given for the rate of production of total gas, hydrogen, carbon dioxide, ethylene, methane and carbon monoxide.

CONCLUSIONS AND SIGNIFICANCE

Using the unique features of the ULTRAPYROLYSIS system as outlined in the previous section, the mass yields of total gas (i.e. the sum of the gaseous species which are non-condensible at room temperature), carbon dioxide, carbon monoxide, hydrogen, methane, ethylene and acetylene were measured as a function of residence time for the fast pyrolysis of cellulose at 800 C. These data are represented by the curves in Figures 2 to 7. It is evident from the curves that they all exhibit the same general shape which reflects an initial rapid rate followed by a tapering off to a maximum yield. For example, the total gas production rate (Figure 2) is quite fast up to about 150 ms and then moderates until a maximum mass yield of 78.8 percent by mass is evident at 250 ms. Corresponding values for the individual gas components are illustrated in Figures 3 to 7 and are summarized in Table 1.

A first order kinetic model for the appearance of the total gas and individual gas species provides an excellent fit for the experimental data. This first order expression is given as Equation (1). The two kinetic parameters, V^* and K, are estimated to be 78.8 percent (by mass) and 17.5 s^{-1}, respectively, for total gas production. The parameters for the individual gases are reported in Table 1. The results seem to indicate that primary pyrolysis consists simply of depolymerization of the cellulose followed by rapid thermal fragmentation of the characteristic depolymerized units. The water-gas shift reaction ($H_2O + CO \leftrightarrow H_2 + CO_2$) and thermal cracking of the gases do not appear to significantly occur over the range of residence times investigated.

ULTRAPYROLYSIS (UP) PROGRAM REVIEW

The fast pyrolysis of biomass in an ultra-rapid fluidized (URF) reactor has exhibited promising yields of high quality chemical intermediates (i.e. ethylene, acetylene, light organic liquids). Fundamental research has indicated that olefin yields of 10 to 15% by mass of the biomass feedstock can be realized and that little or no char need be formed under conditions of short vapour residence times (<500 ms), high temperatures (>700 C), rapid heating rates (>1000 C/s) and rapid quenching of the reaction products (1,2,3,4,5). A complete characterization of fast pyrolysis and an outline of its advantages with respect to product quality and selectivity over the more conventional conversion processes, have been presented previously (1,2).

At the University of Western Ontario, the first phase of a fast pyrolysis program (Ultrapyrolysis) was conceptualized with the intention of engineering a continuous process which would successfully demonstrate the chemistry of fast pyrolysis. In effect, a successful demonstration would require that valuable non-equilibrium thermochemical conversion products (i.e. olefins, BTX and reactive intermediates) and synthesis gas (CO/H_2) be generated at steady state in equipment that could be used as a conceptual basis for eventual commercial scale-up. The heart of this primary objective was the development of a new ultra-rapid fluidized (URF) reactor design in order to exploit and optimize fast pyrolysis product yields. Preliminary results, using wood, cellulose and lignin feedstocks are reported, along with details of the design, in a report to the Canadian Government (1). Upon completion of the first phase of the project, the system was modified to enhance operation for kinetic modeling. At the time of writing, nearly 300 kinetic experiments had been conducted using Avicel cellulose at temperatures between 700 and 900 C and at residence times of 40 to 900 ms. This paper briefly outlines the current Ultrapyrolysis process flow scheme and the experimental operating procedure. Quantitative results are summarized for cellulose pyrolysis trials at 800 C and kinetic parameters are reported for a first order model.

ULTRAPYROLYSIS FLOW SCHEME

The major components of the current Ultrapyrolysis (UP) process are illustrated in Figure 1. Rapid mixing and heat transfer are carried out in two conical vessels known as vortical contactors or Vortactors. The first has been termed a Thermovortactor and allows heat to be transferred from a hot "thermofor" stream (i.e. gaseous nitrogen, suspended particulate solids, or a combination of the two) to the biomass. The second is a Cryovortactor and allows fast quenching of the products by the direct transfer of heat to "cryofor" (i.e. cryogenic nitrogen). The Thermovortactor has two opposing tangential inlets for the thermofor. One tangential stream effectively destroys the momentum of the other causing severe turbulence. Biomass feedstocks are then injected from the top of the Thermovortactor through an air cooled tube into the turbulent region where mixing occurs within 30 milliseconds. The hot gaseous product is rapidly cooled (i.e. in less than 30 ms) by the injection of a single tangential stream of cryogenic nitrogen.

Mechanical table feeders are used to supply biomass and thermofor solids to the reactor system. Solids pass from sealed hoppers (which have a sufficient inventory of solids for several experiments) through a double funnel system and are thereby metered onto a rotating table. Two fixed armatures sit near the surface of the round table and plough the solids off the outer circumference. From the table, the solids then fall into a conical chamber where they are picked up and carried into the transport line by nitrogen carrier gas. The overall range of the feed rate of the solids is controlled by setting the gap between the lower funnel and the table. Fine control is exercised by the rotation speed of the table. Currently the biomass feeder is set in the range of 0 to 20 grams per minute and the thermofor feeder is set to deliver 0 to 200 grams per minute. The biomass feeder is mounted on a 60 kg balance to facilitate the determination of the mass balance. When thermofor solids are required to supply the process heat (i.e. either in place of gaseous nitrogen thermofor or as a supplement), the thermofor feeder sends cold (room temperature) particulate solids to a double preheater coil assembly where they are heated to the reaction temperature. These solids are then sent on to the Thermovortactor. Thermofor nitrogen can be sent directly to the preheater coil.

The fast pyrolysis of biomass is initiated in the Thermovortactor and continues in a plug-flow entrained bed downflow reactor. The reactor is simply a one meter length of Inconel pipe in an electrical oven. The mixture of hot gases and biomass passes from the Thermovortactor through the reactor to the Cryovortactor. With the insertion of cylindrical inserts to reduce the reactor volume and by manipulating thermofor/biomass flowrates, the residence time (i.e. the time from the heating of the biomass to the exit from the reactor) can be set between 30 and 900 ms. Reactor temperatures can be set in the range of 700 to 1000 C. The UP flow scheme is described in detail in a previous publication (6).

EXPERIMENTAL

In preparation for an experiment, the system is purged with nitrogen. The furnaces are then turned on and allowed to reach their set point temperatures. When the desired temperatures are reached, cryogenic nitrogen is injected into the Cryovortactor. Thermofor (i.e. transported silica sand and/or inert nitrogen gas) is then introduced into the system and the cryogenic nitrogen flowrate is adjusted to ensure that the Cryovortactor temperature is less than 350 C. Gaseous nitrogen was used as the thermofor for all of the experiments reported in this paper. During the startup period, the stream exiting from the Cryovortactor is sent to a bypass filter assembly and vented. The biomass feeder is activated and its feedrate is verified by directing the biomass flow to a collection device mounted on a balance.

Immediately prior to the actual steady state experimental run, the thermofor/cryofor stream (which is exiting from the Cryovortactor) is switched from "bypass" to the mass balance filter. The biomass is directed to the Thermovortactor and the actual fast pyrolysis experiment is thereby initiated. Products are initially quenched in the Cryovortactor while additional cooling (approaching room temperature) is carried out in a water-cooled coiled tube heat exchanger. Solids and condensibles are deposited in the filter, and the entire non-condensible product gas is collected in the gas collection bags. For extremely short residence time trials (when the yield of "condensibles" increases), an electrostatic precipitator unit can easily be integrated in the downstream gas line. In addition, an efficient cyclonic condenser (which is submerged in an ice bath) has been developed to increase the yield of pure (i.e. non-solvent extracted) liquid sample. The steady state experiment continues for 3 to 10 minutes. Static samples are taken from the gas collection bags and analyzed after the run is completed. Gas analysis is accomplished using a Carle (Model 111-H/197A) gas chromatograph. Condensible products are extracted with a solvent and recovered in a rotary evaporator.

FLUIDIZATION AND TRANSPORT PARAMETERS

Typical velocities in the reactor system during steady-state operation at 800°C are given as follows:

Thermofor Injector (using sand thermofor):	50 m/s
Thermofor Injector (using gaseous thermofor):	30 m/s
Biomass Injector:	1 m/s
Downflow Transport Reactor:	4 m/s
Downflow Transport Reactor: (region of the inserts)	15 m/s

Solids loadings (i.e. the mass ratio of the solid to its carrier gas) were 1.5 for the biomass (nitrogen carrier) and 5 for the sand thermofor (nitrogen carrier). The thermofor to biomass mass ratio was in the range of 10 to 15 for both types of thermofor, and the average particle size was 100μm for the biomass and the particulate sand heat carrier.

RESULTS AND DISCUSSION

The results of the fast pyrolysis experiments conducted at 800 C using Avicel pH102 cellulose (a 100 μm crystalline powder supplied by FMC) as the biomass feedstock and nitrogen gas as the heat carrier, are summarized in Figures 2 to 7. The solid lines on the curves represent the first order kinetic model (discussed later) which has been derived from the indicated experimental points. Table 1 is a summary of the kinetic parameters for the model.

The plot of the mass yield of the total non-condensible gas product vs reactor residence time (ms) is given in Figure 2. The general shape of this curve indicates an initial rapid production rate followed by a tapering off to a constant maximum yield, and is representative of all of the individual component yield curves (i.e. Figures 3 to 7). For example, the total gas production rate (i.e. the slope of the curve) is very rapid up to about 150 ms and then tapers off giving a maximum yield of 78.8% (by mass) after approximately 250 ms (Figure 2). Carbon monoxide (Figure 3) and carbon dioxide (Figure 4) yields also reach maximum values of 56.5 and 5.6 percent by mass, respectively, after 250 ms. The yields of hydrogen (Figure 5) and methane (Figure 6) both seem to exhibit their constant values of 1.7 and 5.7 percent, respectively, at a much longer residence time of about 600 ms, indicating perhaps that their appearance may result from a similar reaction mechanism or a common parent compound. Finally, the yield of the major unsaturated component, ethylene (Figure 7) reaches a constant value of 6.0 percent at 250 ms (which is similar to that reported for the total gas, CO and CO_2).

The optimum yield of total unsaturates (i.e. ethylene, acetylene and propylene) is about 10 percent by mass at 800 C while the corresponding value for total hydrocarbons is 16 percent.

A consideration of the equilibrium composition of the water-gas shift reaction ($H_2O + CO \leftrightarrow H_2 + CO_2$) at 900 C, and the fact that the CO, CO_2 and H_2 yields remain constant, leads to the conclusion that the shift reaction does not play an important role in the early pyrolysis reaction mechanism (i.e. at least up to 900 ms at 800 C). In addition, the yield curves for "crackable" hydrocarbons (i.e. such as ethylene, acetylene, ethane) indicate that no net thermal decomposition of these hydrocarbons occurs over the residence times studied. Therefore, primary pyrolysis appears to consist simply of the "unzipping" of the biomass polymer followed by rapid thermal fragmentation of the characteristic depolymerized units. There seems to be a lag time before additional gas phase reactions, such as the inevitable water-gas shift or thermal cracking of the hydrocarbons, occur to any extent.

The mass balance procedure and results are not detailed in this paper. However, after the liquid/solid recovery procedure (i.e. solvent extraction, and subsequent filtration and rotary evaporation) was carried out for about thirty (30) selected experiments, between 95 and 98% of the input mass was accounted for in the product mass. The systematic deficiency in the mass balance closure was assumed to be due to slight losses of the light volatile liquids with the solvent in the rotary evaporation procedure. The non-gaseous products were essentially all liquids with only trace amounts of char observed.

A simple and convenient approach to kinetic modeling assumes that cellulose decomposes directly to each individual gas component by a single independent pathway and that the kinetics of the decomposition can be represented by a unimolecular first order reaction. This approach has been widely used in representing the kinetics of the fast/flash pyrolysis of biomass and coal (4,7). The following rate equation is thereby derived:

$$V = V^* \left[1 - \exp(-Kt) \right] \tag{1}$$

Where V represents the mass yield of a component at time 't' and temperature 'T', V* is the maximum attainable yield of the component at temperature 'T' and long residence times, and K is the Arrhenius rate constant $[K = K_o \exp(-E/RT)]$. Best fit values for the kinetic parameters (V* and K) were estimated for all major components at 800 C using a least squares non-linear regression procedures. These parameters are summarized in Table 1 and the corresponding regression curves are shown as the solid lines in Figures 2 to 7.

The fact that the non-gaseous products are liquids rather than solid char or unpyrolyzed cellulose (even at extremely short residence times) would seem to suggest that the solid phase decomposition to primary liquids is practically "instantaneous" in the ultrapyrolysis reactor system. This is likely a result of a combination of the inherent high heat transfer rates and the small cellulose particle size. Therefore the kinetic model is best described as representing the rate of appearance of the gas and its components from the primary pyrolytic liquids. Upon completion of the analysis of the experimental data over the entire temperature, the isothermal kinetics will be published for temperatures of 750, 800 and 900 C. The activation energies (Ea) and pre-exponential constants (Ko) will then be estimated for the rate of appearance of each gaseous component during the fast pyrolysis of cellulose.

ACKNOWLEDGEMENT

Substantial grants-in-aid of research from the ENFOR (EMR) biomass program, NRC, NSERC and the Imperial Oil Company made this work possible and are here gratefully acknowledged.

REFERENCES

1. Bergougnou, M.A., H.I. de Lasa, L.K. Mok, R.G. Graham, B.A. Freel and J.D. Hazlett. "ULTRAPYROLYSIS of Cellulose and Wood Components", Enfor C-147 Final Report, Environment Canada,Ottawa, Canada (1983).
2. Graham, R.G., M.A. Bergougnou and R.P. Overend, "Fast Pyrolysis of Biomass", Journal of Analytical and Applied Pyrolysis, 6:95-135 (1984).
3. Diebold, J.P., "Gasoline from Solid Wastes by Non-Catalytic Thermal Process" In Thermal Conversion of Solid Wastes and Biomass, Jerry L. Jones and Shirley B. Radding,editors, American Chemical Society Washington, DC (1980).
4. Hajaligol, MR., J.B. Howard, J.P. Longwell and W.A. Peters, "Product Compositions and Kinetics for Rapid Pyrolysis of Cellulose", IE&C Proc. Des. Dev. 21:457-465 (1982).
5. Diebold, J.P. Editor, Specialists' Workshop on Fast Pyrolysis of Biomass, Solar Energy Research Institute, Golden, CO. (1980).
6. Mok, L.K., R.G. Graham, B.A. Freel, M.A. Bergougnou and R.P. Overend. "Fast Pyrolysis (ULTRAPYROLYSIS) of Cellulose and Wood Components", Journal of Analytical and Applied Pyrolysis 8:391-400 (1985).
7. Antal, M.J. Jr., H.L. Friedman and F.E. Rogers, "Kinetics of Cellulose Pyrolysis in Nitrogen and Steam", Combustion Science and Technology, 2:141-152. (1980).

TABLE 1 FAST PYROLYSIS OF CELLULOSE AT 800 C: SUMMARY OF KINETIC PARAMETERS V* AND K

Component	Parameter (Regression Value) V*	K
Total Gas	78.8	17.5
Carbon Monoxide	56.5	18.6
Carbon Dioxide	5.6	16.3
Hydrogen	1.7	10.5
Ethylene	6.0	18.3
Methane	5.7	9.0

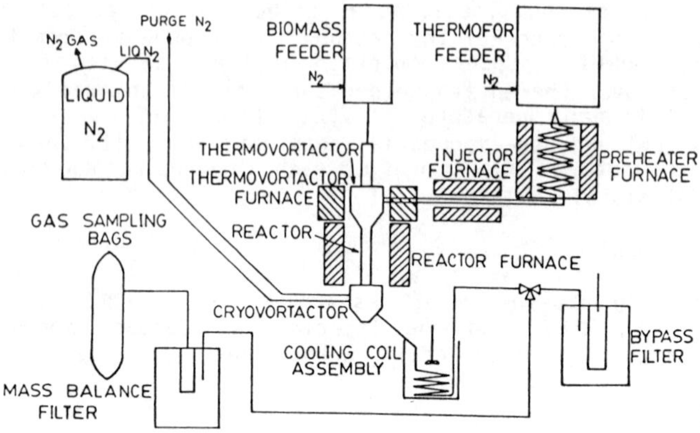

FIGURE 1 : ULTRAPYROLYSIS FLOW SCHEME

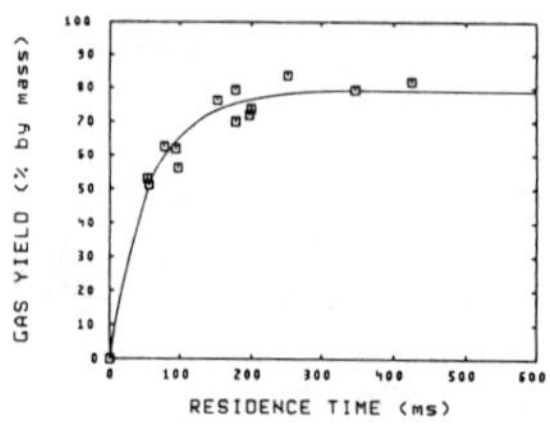

FIGURE 2 GAS YIELD VS RESIDENCE TIME

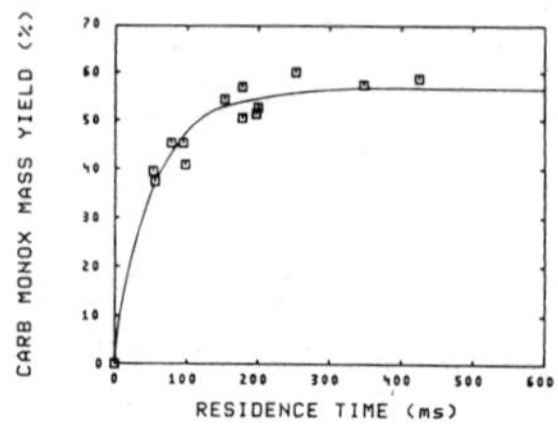

FIGURE 3 CARBON MONOXIDE YIELD VS RESIDENCE TIME

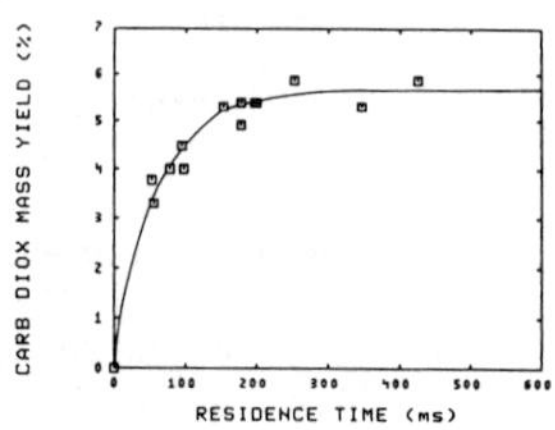

FIGURE 4 CARBON DIOXIDE YIELD VS RESIDENCE TIME

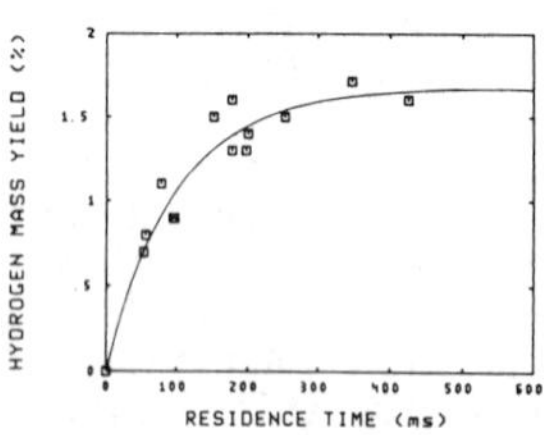

FIGURE 5 HYDROGEN YIELD VS RESIDENCE TIME

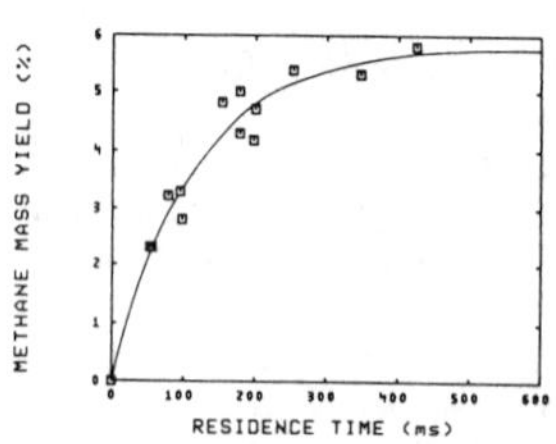

FIGURE 6 METHANE YIELD VS RESIDENCE TIME

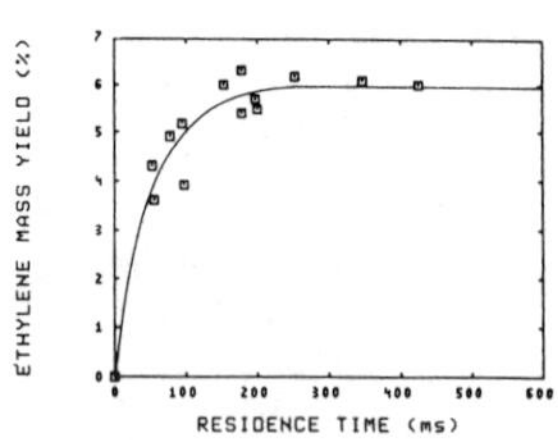

FIGURE 7 ETHYLENE YIELD VS RESIDENCE TIME

"Development status of the High-Temperature Winkler (HTW)-Process"

by J. Deinert, J. Lambertz, Chr. Nassenstein and L. Schrader
Rheinische Braunkohlenwerke AG, Stüttgenweg 2, D-5000 Cologne 41
West Germany

Abstract

The High-Temperature Winkler (HTW)-Process for generation of synthesis gas or low-BTU gas is suited to gasify low-ash and high-ash renish brown coal with spec. synthesis gas output more than tripled in comparison to the elder atmospheric Winkler gas generators. The HTW-Process is also able to gasify subbituminous coal, coke, wood and peat. Fluiddynamic measurements in a cold model with 2,75 m in diameter show, that a large commerical-scale HTW-gasifier, which is now under construction and will go on stream in 1985, can be fluidized sufficiently to reach good solid mixing.

Scope

The Rheinische Braunkohlenwerke AG (Rheinbraun) develops the High-Temperature-Winkler (HTW)-Process for the generation of synthesis gas as well as low-BTU gas. The heart of a 25 t/day pilot plant, which has run since 1978 up to 1985, consists of a fluidized-bed gasifier with an operating pressure up to 10 bar and an operating temperature up to 1 100 °C. Based on the high efficiency of the pilot plant, a commercial-scale demonstration plant is now under construction. For scale-up considerations a large cold model on a 1 : 1 scale of the demonstration plant gasifier was built to study the fluiddynamics of the fluidized bed. By use of a Constant-Temperature-Anemometer (CTA)-System and a Capacitance-Measurement-System characteristic bubble parameters and heat transfer coefficients within the fluidized bed were measured, which were dependent on size and shape of the nozzles for introduction of the fluidizing gas.

Conclusion and Significance

Under optimized operating conditions the HTW pilot plant is capable of reaching a high synthesis gas output and a high synthesis gas yield for low-ash and high-ash renish brown coals. For an oxygen input of 0,39 m^3STP/kg coal, maximum values of 5 500 m^3STP/h . m^2 for the spec. synthesis gas output and of 1 400 m^3STP/t $coal_{maf}$* for the spec. synthesis gas yield were measured.

The C-conversion is about 90 % to 95 % renish brown coals. The synthesis gas contains no tars or CS_2. Specific catalytic converters were tested to reduce the concentration of COS and HCN in the product gas. NH_3 and C_6H_6, which can be concentrated in gas-cleaning units, were recirculated to the gasifier. By this means the total amount of these compounds were significantly reduced.

The fluiddynamic measurements in the cold gasifier model of the demonstration plant showed that good fluidization is possible although the fluidized-bed was strongly scaled up to 2,75 m in diameter and although the fluidization only by few nozzles is difficult in principle.

*) maf = moisture and ash free

Description of the HTW pilot plant

A flow sheet of the HTW pilot plant including the maximum efficiency data is shown in fig. 1. As feedstock, predried brown coal with an 18 % moisture is used. The coal is dried down to 12 % moisture content and pneumatically conveyed to a lock-hopper system and then fed into the reactor by a screw-conveyor. In the brick-lined fluidized bed gasifier the coal first reacts at about 700 °C to 800 °C with steam/oxygen-mixtures or air, injected at different heights in the fluidized-bed. In this stage, in addition to partial coal and gas combustion for heat generation, the devolatilization reactions take place and they produce mainly CH_4, CO, H_2 and CO_2. The semi-coke produced reacts endothermally with steam and with CO_2 under a shrinking particle mechanism. Elutriated coke particles are reconverted in the freeboard region where steam/ oxygen-mixtures are also injected. Entrained bed material is recycled to the gasifier by means of the first cyclone.
In order to test parts of the gas-cleaning units, required for synthesis gas production, a partial raw gas stream is branched off to a venturi scrubber where the residual dust is washed out while in a downstream converter various sulphur resistent catalysts are tested for the Shift-reaction and for the hydrolysis of COS and HCN.

General operating experience with the HTW pilot plant

By the end of 1984 the pilot plant had been in operation for about 34,500 hours including 24,600 hours of operation under gasification conditions. About 18,300 t of dried brown coal had been processed in various experimental programs. The average availibility of the unit was between 65 % to 75 % during the last years. The longest continuous operating period was seven weeks.
Compared to the older atmospheric Winkler gasifier it is possible to triple the spec. synthesis gas output to about 7,700 m^3 STP per m^2 of gasifier cross-section an hour by increasing the pressure up to 10 bar and by using an optimal spec. O_2-supply and an optimal oxygen distribution along the reactor height.
The HTW gasifier shows a favourable load behaviour. No trouble occurred when at 10 bar operating pressure the coal throughput was varied from about 470 to 1,860 kg/h, where the upper limitation was only caused by the minimum cycle time of the lock-hopper system. When a constant oxygen supply of 0,5 m^3 STP per kg coal is maintained the synthesis gas yield considerably increases with a rise in coal throughput up to a maximum value of 1,600 m^3 STP per t coal (maf). On the other hand, by raising the oxygen supply, a drop in the spec. synthesis gas yield, caused by a reduced, throughput can be compensated.

Efficiency data of the gasification of low-ash and high-ash brown coals

Fig. 2 shows the spec. synthesis gas output and the spec. synthesis gas yield as a function of the coal throughput on a maf basis for low-ash and high-ash renish brown coals. A relatively low spec. oxygen supply of 0,39 m^3 STP per kg coal (maf) was used here. The low-ash coal contains about 5 % ash and 12 % moisture; the lower heating value is about 21 - 23 MJ/kg. The high-ash coal respectively contains about 15 - 20 % ash and the same moisture; the lower heating value is about 19 - 21 MJ/kg. The average feed size of the coals was about 2 - 3 mm. The oxide analysis of the two coals show that except for SiO_2 all other components have relatively nearly the same concentration.

The spec. synthesis gas output increases significantly with a rise in the coal throughput for both coals. The synthesis gas yield for the high-ash coal also increases considerably up to nearly 1,300 m^3 STP per t coal (maf). The synthesis gas yield for the low-ash coal runs through a slight maximum of about 1,380 m^3 STP/t_{maf} and tends to fall insignificantly with increasing coal throughput. At the high load region the efficiency data for both coals reach an identical level. At the low load region the spec. synthesis gas output also becomes similar for both coals, however a large difference in the spec. synthesis gas yield did occur.

There are three aspects to consider for the explanation of these effects:

a) The different mean residence time of both coals in the fluidized bed
b) The different mean density and heat capacity of the bed material
c) The heat generation in relation to the heat losses.

At low coal throughput the heat losses of the system in relation to the heat generation by partial combustion increases considerably for low-ash and high-ash coals. The high-ash coal particles contain inert material (SiO_2) with a high heat capacity. Therefore the mean surface temperature of the reacting high ash coal particles is lower than that of the low-ash coal particles, correspondingly a lower reaction rate results for the heterogeneous water vapour gasification. Another reason for the difference in the spec. synthesis gas yield is affected by the different mean density of the bed material in relation to the low synthesis gas output. The high-ash coals won't be fluidized as well as the low-ash particles for constant fluidization conditions, especially in the upper parts of the bed.

At high coal throughput these effects, described here, are not important. The heat generation increases, the fluidization conditions are excellent, and correspondingly the synthesis gas yield of low-ash and high-ash brown coal become similar.

The reason for the similarity of the synthesis gas output for both coals in the low load region, although there is a greater difference in the synthesis gas yield, is that the amount of CO and H_2, generated by devolatilization, becomes important in relation to the amount of CO and H_2 generated by the gasification reactions. The devolatilization of both coals are not significantly affected by the three points mentioned above.

The influence of the spec. oxygen supply on the efficiency of the HTW-gasifier at a constant coal throughput of 1,000 kg/h maf low- and high-ash brown coal is shown in fig. 3. There is a considerable increase of the spec. synthesis gas output and of the spec. synthesis gas yield for both coals. For low-ash coals and for the parameter range considered a yield of about 1,450 m^3 STP/t coal (maf), respectively an output of about 5,100 m^3 STP/m^2 . h were reached. The synthesis gas output for the gasification of high-ash coals were about 10 % lower at constant oxygen supply because of high heat capacity of the inert materials in the coal. The synthesis gas yield of high-ash coals were about 5 % lower than those of the low-ash coals.

The data for the C-conversion in relation to coal throughput are relatively scattered. The C-conversion for low-ash coals reaches values up to 95 % but the data for high-ash coals are always lower under the same operating conditions, because to maintain a constant bed height it is necessary to remove more ash per unit of time and therefore the solid residence time drops. By process or design changes, for instance total dust and/or partial ash recirculation or modifications of the removal system, the C-conversion can be improved.

Minor trace compounds and gas-cleaning for synthesis gas production

Because of the complexity of the gasification process and the feedstock, gas-cleaning is necessary for utilizing the raw gas as a synthesis gas. H_2S is removed during the gas scrubbing and will run through a stretford unit to produce pure sulphur. NH_3 and C_6H_6 are enriched in special scrubbers. Both compounds can be recirculated to the gasifier resulting in a 99,8 % conversion for C_6H_6 an a 67 % conversion for ammonia. Both substances can also be further purified to reach the sales standards. Various catalysts have been tested for the hydrolysis of COS and HCN. In a single converter high conversions up to 90 % have already been reached. By three catalytical converters, planned for the HTW demonstration plant, conversions greater 99 % are attainable. CS_2 and tars are not measurable in the raw gas stream.

Description of the cold model of the HTW demonstration plant

In order to test the fluiddynamics in the 2,75 m in diameter gasifier of the HTW demonstration plant, a cold atmospheric fluidized-bed model on a 1 : 1 scale was developed. Fig. 4 shows a sketch of this model. Bed material of the HTW pilot plant was used; the fluidization medium was air. There are two ash removal systems in the lower part which are fluidized by gas injectors at the level N 9. The gas for the basic fluidization of the bed is fed into the reactor by the nozzles at level N 10. The fluidization characteristics are investigated at the levels M and L of the reactor.

Measurement techniques and results

For the direct measurement of bubble parameters in relation to different size and shape of the nozzles a capacitance measurement system was developed. This system has an extraordinary high signal/noise-ratio. For a precise analysis of the signals an AD-converter and a magnetic tape were connected to the system to transport data to the host computer. Another condition for safe operation of the HTW demonstration plant gasifier is a high and uniform heat transfer within the bed. The characteristic heat transfer coefficient has been determined by a modern constant temperature anemometer (CTA) system, which is on-line-coupled with a small computer system for direkt estimation of the coefficient.

Fig. 5 shows the behaviour of the heat transfer coefficient in the vinicity of the bubbling point at level L. With an increasing gas stream only to the nozzle level N 9 the bed is fluidized earlier in the middle parts of the bed. But at a fluidization velocity greater than 2 u_{mf} to 3 u_{mf} a uniform heat transfer coefficient over the whole cross-section is present. The bubble frequency in a relation to the heat transfer coefficient over the whole cross-section at the two levels L and N is presented in fig. 6/7. The fluidization gas streams to the N 9 and N 10-nozzles, used here, are calculated via the similarity theory to represent the same conditions as in the high-temperature and pressure fluidized bed of the HTW demonstration gasifier. At level N the bubbling frequency shows a minimum at the center of the bed and a maximum at the outer regions. The heat transfer profile is nearly uniform with a slight minimum at the same place where most of the bubbles rise. The jets of the nozzles at N 10 are directed vertical, so that at level M there must be a maximum of the bubble frequency.

The heat transfer coefficient, measured here, is composed of one part related to the gas esp. the bubbles and of another part related to the solid's motion. The heat capacity of the solid is much higher than that of the bubbles. This means that although there is a minimum in bubble frequency at the center of the bed, the solid's motion is very high resulting in a high heat transfer coefficient of about 420 $W/m^2 \cdot k$. Because of the high fluidization number of about 10 - 15, used for standard operating conditions, the fluidized bed is relatively highly expanded. A great part of the gas flows through the emulsion phase and only a minor part forms distinct bubbles. Under these circumstances the bubbles are not so important for solid's mixing as at low fluidization numbers. Another aspect is that, as already mentioned in the literature, bubbles in beds of coarse solid particles have no cloud or wake. This explanation is underlined by the fact that at level L (fig. 7) the profile of the heat transfer coefficient remains uniform although there is a remarkably strong drift of the bubbles to the center of the bed.

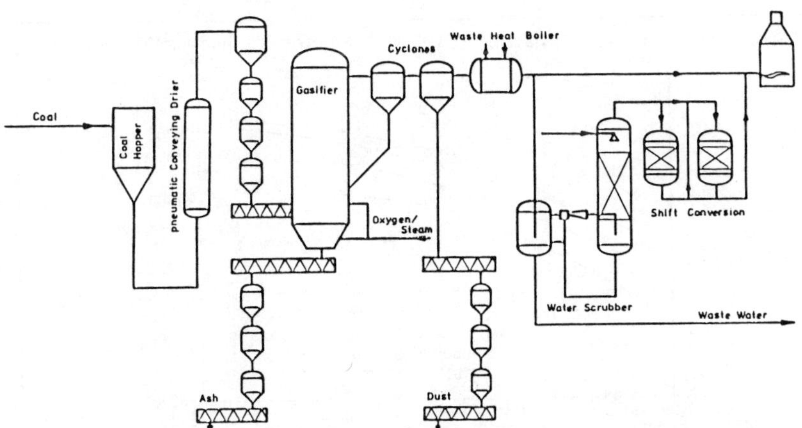

throughput (dry lignite) : up to 1800 kg/h; gasification agents : oxygen/steam or air
raw gas output : up to 4000 m^3/h; pressure : up to 10 bar
temperature : up to 1100 °C ;

Fig. 1 Flow sheet HTW pilot plant RHEINBRAUN

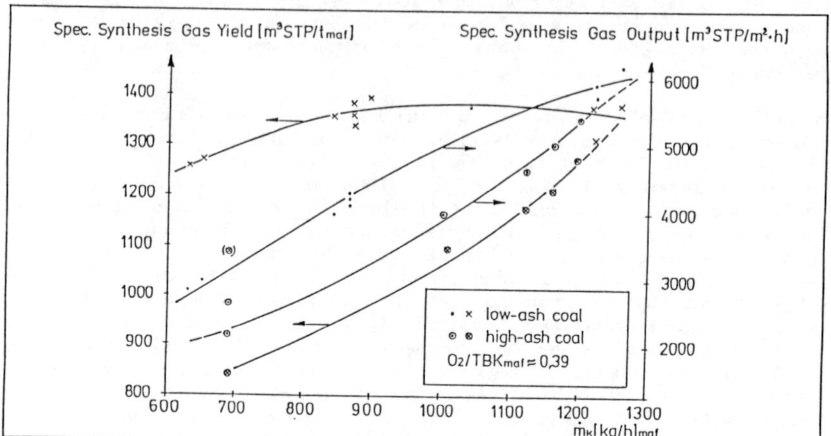

Fig. 2 | Low-ash and high-ash brown coal gasification efficiency data | RHEINBRAUN

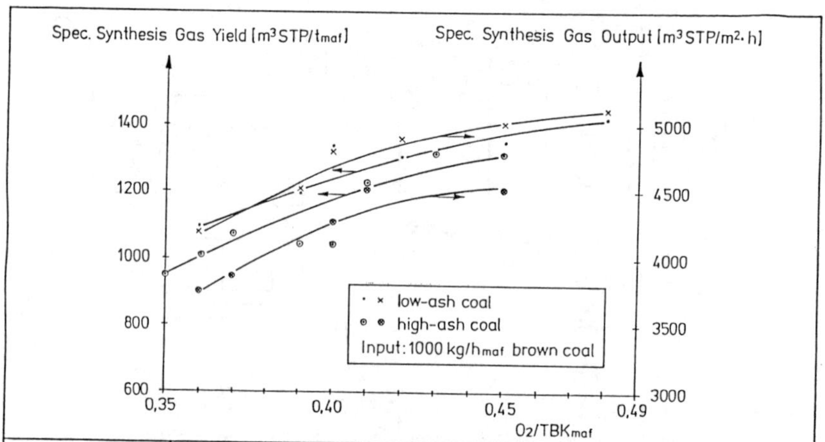

Fig. 3 | Low-ash and high-ash brown coal gasification efficiency data | RHEINBRAUN

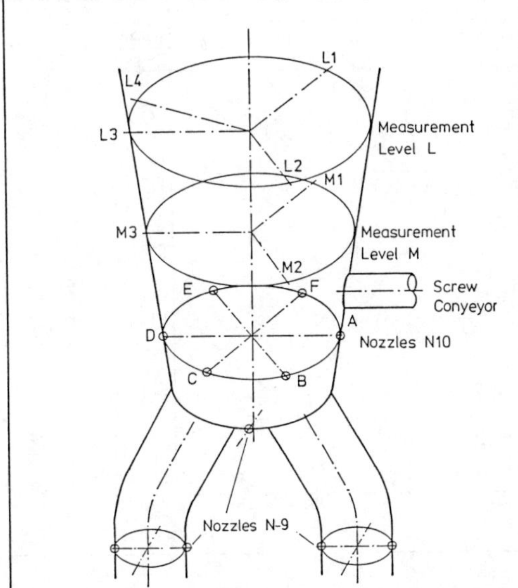

Fig. 4 | Geometrics of the cold model | RHEINBRAUN

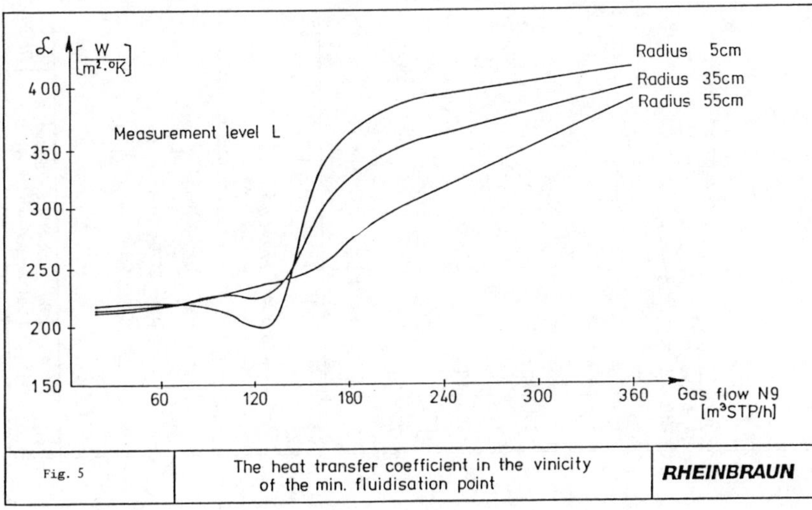

Fig. 5 | The heat transfer coefficient in the vinicity of the min. fluidisation point | RHEINBRAUN

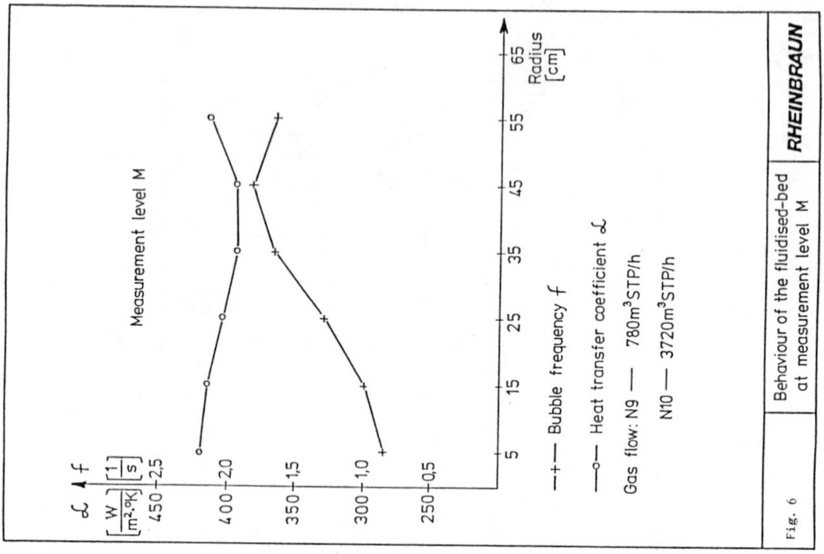

Fig. 6 Behaviour of the fluidised-bed at measurement level M

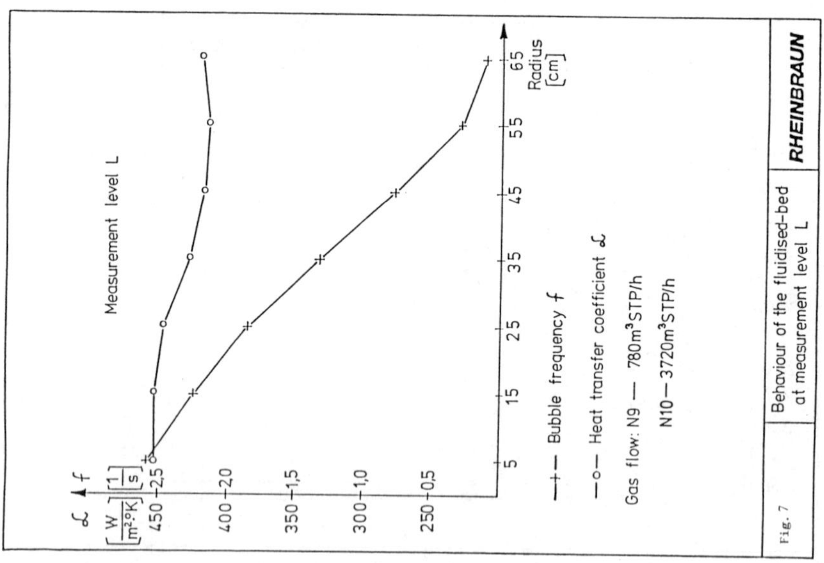

Fig. 7 Behaviour of the fluidised-bed at measurement level L

THE COMFLUX-PROCESS FOR SUBSTITUTE NATURAL GAS (SNG)

Hans-Joachim Bock, Otto Molerus, Werner Lommerzheim*, Gregor Friedrichs*

Lehrstuhl für Mechanische Verfahrenstechnik, University Erlangen-Nürnberg, Martensstr. 9, D 8520 Erlangen, FRG

*Thyssengas GmbH, Postfach 11 05 62, D 4100 Duisburg, FRG

ABSTRACT

The Comflux-process is a single-stage process for the combined methanation and partial shift conversion of gases rich in CO and H_2 in a catalytic pressurized fluidized bed reactor.

SCOPE

Methanation is an important step during the conversion of coal and oil products into SNG. The key problem in the technical realisation lies in the removal of the reaction heat and the temperature control of the reactor. Two or multi-stage processes in a fixed bed are the state of the art. In fixed beds the temperature is controlled by additional feed gas or recycle gas and/or intermediate gas coolers. Because of the high heat transfer coefficient between a fluidized bed and immersed cooling surfaces, the process presented here was developed on the basis of the fluidized bed technique.
Each step of process development was tested in full-scale models at ambient conditions.
With these tests the optimization of the internal components of the reactor, like the cooling surfaces, gas distributors and product gas clean-up systems, was achieved. The results obtained served as the basis for the detailed design of the high pressure test plants.

CONCLUSIONS AND SIGNIFICANCE

In the first stage of process development a semi technical test plant with an internal diameter of 0.4 m was designed for a pressure in the range of 3 to 8 MPa, a fluidized bed temperature of 300° C to 500° C and a capacity of 100 to 200 m^3/h SNG. The semi technical test plant ran for a total of 14,000 hours and proved the feasibility of the COMFLUX-concept. In order to minimize the development risks, in a second stage of process development a pilot plant with an internal diameter of 1 m and a capacity of 2,000 m³ SNG/h was designed. During the operation period of two years the step-by-step improvement of the plant technique led to a high degree of availability. This was particularly demonstrated by the concluding test run of approximately 2,200 hours. Meanwhile the preliminary planning for the commercial application of the process has been started. The COMFLUX-process is the result of almost ten years of development work, started in 1974 by Thyssengas in cooperation with Didier Engineering, Essen, with substantial financial support by the Ministry for Research and Technology of the FRG.

INTRODUCTION

The standard equations for the fast and catalytic controlled reactions of methanation and shift conversion are:

$$CO + 3H_2 \rightarrow CH_4 + H_2O \qquad \Delta H_{R\ 298} = -206.7 \text{ kJ/mol} \quad (1)$$

$$CO + H_2O \rightarrow CO_2 + H_2 \qquad \Delta H_{R\ 298} = -41.2 \text{ kJ/mol} \quad (2)$$

The key problems in their technical realization lie in the removal and profitable utilization of the reaction heat and in the temperature control of the reactor.

From the results of Werther (1) it was known that the hydrodynamics of fluidization depends strongly on the bed diameter for those smaller than 0.2 m which are normally used for laboratory scale experiments. It follows from these investigations that a minimum bed diameter is needed for fluidized beds if the results of the experiments are to be used for scaling-up. In order to assume the transformability of the results a fluidized bed reactor was designed with an internal diameter of 0.4 m.

Preliminary tests were initiated in 1975 with studies of flow mechanics and measurements of the heat transfer coefficients on a full-scale model at ambient conditions at the University of Erlangen-Nürnberg. The model reactor was used for the optimization of the internal components, such as cooling surfaces, gas distributors and solids clean-up. The results obtained served as the basis for the detailed design of a semi technical test plant. The design data of the semi technical test plant are given in Table 1.

Reactor pressure	:	3 to 8 MPa
Reactor height	:	8 m
Steam drum pressure	:	8 MPa
Bed temperature	:	300 to 500° C
Steam temperature	:	295° C
Fluidized bed height	:	1.5 to 3 m
Catalyst charge	:	100 to 300 kg
SNG production	:	100 to 200 m³/h

Table 1: Design data of the semi technical test plant

EXPERIMENTAL RESULTS AND DISCUSSION

The semi technical test plant was run from the end of 1977 until early 1981 with some longer interruptions for a total of 14,000 hours from which approximately 7,000 hours were in synthesis operation. More details of the process design and operating conditions of the semi technical test plant have already been published - see for example Lommerzheim (2) - but the following results are to be emphasized:

The catalytic fluidized bed is stabilized by the vertical tube bundle. Neither any erosion nor vibration problems were observed. The heat transfer to the immersed tube bundle exceeded the extrapolated data from the model tests. In particular, the SNG-production could be increased up to 350 m³/h. The expected SNG-quality could also be reached, even when the H_2/CO - ratio was less than stoichiometric. The recycle ratio could normally be kept below 0.5 vol./vol.. The particle size distribution of the catalysts used varied in the range from 10 to 400 microns.

For a commercial application of the COMFLUX-process the SNG-production is expected to exceed 20,000 m³/h. This means in comparison with the results from the semi technical test plant a scaling-up by a factor of 50 to 100.

In order to minimize the scale-up risks the process development was continued with a scale-up factor of 7 to 10. In a second stage of process development a pilot plant was designed with an internal diameter of 1 m of the fluidized bed.
The diameter of 1 m additionally fulfilled the requirement that all key components should be tested in a large design which could be scaled up. The scale-up concept of a commercial COMFLUX-reactor is based on the partition of the reaction space into so-called modules. The partition wall consists of thin plates. The cross section of each module almost equals that of the pilot reactor. Figure 1 shows the COMFLUX-reactor development.

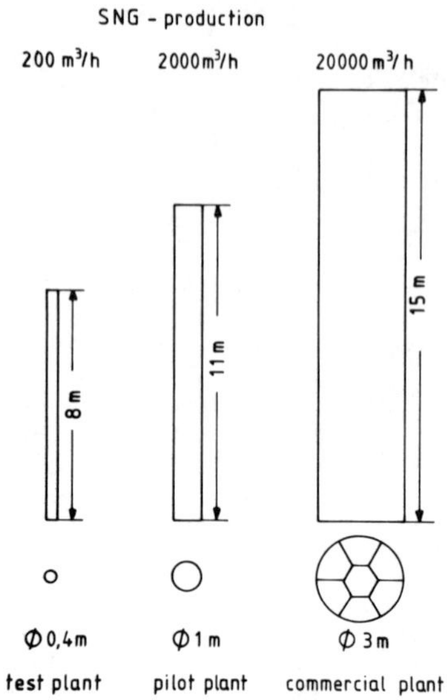

Figure 1: COMFLUX-reactor development

Before final construction of the pilot plant was begun the modified reactor design was tested in a full-scale model at ambient conditions. Measurements of local bubble behaviour and local heat transfer coefficients at various positions in the bed have shown that, in fluidized beds of small particles, it is impossible to obtain uniform conditions throughout the bed cross-section (see Bock (3), (4)).

For example, figures 2 and 3 present the results of local heat transfer measurements in the full scale model reactor of 1 m diameter, equipped with a vertical tube bundle. The heat transfer coefficients were measured with electric heating elements, 0,15 m long, which were mounted in the tube.At a height of 0.3 m above the distributor a reduction in the heat transfer coefficient occured in the wall region. This effect is more dominant for the fluidized bed with a nozzle plate distributor. Figure 3 also shows, that at low gas velocities, there was a drastic reduction in heat transfer at tubes which were not located over a nozzle. This effect disappeared at increasing gas velocities but the measured values in the wall region were very low at all the investigated gas velocities. The results indicate, that a change in bed geometry leads to variation in the fluid dynamic properties and in the heat transfer. With these results the final design data of the fluidized bed were found.

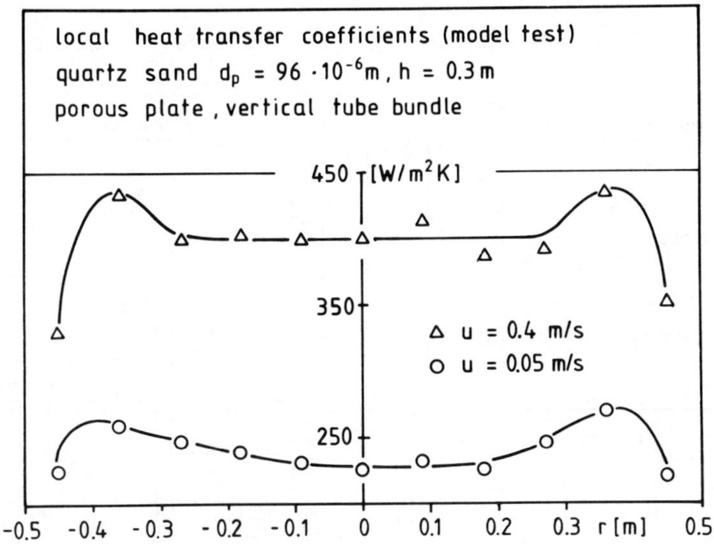

Figure 2: Local heat transfer coefficients, porous plate

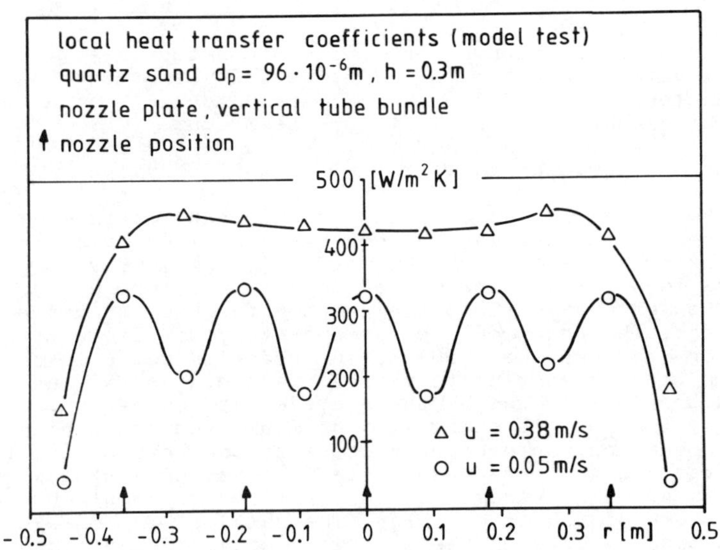

Figure 3: Local heat transfer coefficients, nozzle plate

PROCESS DESCRIPTION

The flow diagram of the pilot plant is shown in Figure 4.

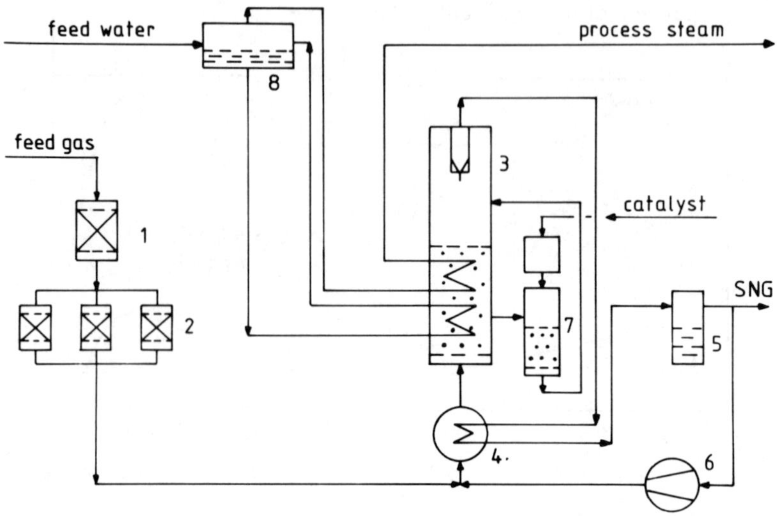

1. Zinc oxide desulphurizer
2. Activated charcoal
3. Reactor
4. Heat exchanger
5. Condensate vessel
6. Recycle compressor
7. Activator
8. Steam drum

Figure 4: Flow diagram of the COMFLUX-pilot plant

Because nickel is the main active component of the catalyst used here, the feed gas is thoroughly purifed by means of zinc oxide and activated charcoal filters to prevent deactivation. After being heated to a temperature of 200° C the feed gas passes into the reactor through the gas distributor. The reaction gas is completely dedusted at the reactor head and then cooled. The reaction water is collected in the gas condensate vessel. Part of the SNG may be added to the feed gas by the recycle gas compressor, particularly in order to stabilize the fluidized bed during low load operation. Before being fed into the reactor the fresh catalyst has to be reduced in the activating system where it is treated with hydrogen at temperatures of about 400° C and a pressure of 0.2 to 0.5 MPa. The catalyst is conveyed pneumatically between the storage vessel, the activator and the reactor. Even during the operation the catalyst can be withdrawn from and returned to the pressurized reactor. The heat removal during methanation is done by evaporative cooling. The boiled water at 7 to 12 MPa is fed to the vertical field tube bundle and the steam/water mixture returns into the steam drum by natural convection. A part of the tube bundle is used to superheat the steam up to 475° C; at specific conditions it is possible to superheat the steam up to 475° C.

RESULTS

The general and main operational data of the pilot plant are listed in Table 2.

Engineering start up	Oct. 1979
Construction period	Nov. 1980 to Dec. 1981
Test period	Jan. 1982 to Oct. 1983
Reactor pressure	1.3 to 6.0 MPa
Steam drum pressure	7.0 to 12.0 MPa
Bed temperature	450 to 550° C
Feed gas	3,000 to 10,800 m^3/h
H_2/CO ratio	2.0 to 3.0 Vol./Vol.
Recycle gas ratio	0 to 0.3 Vol./Vol.
SNG production	1,200 to 3,000 m^3/h
Steam production (netto)	1.0 to 5.2 t/h
Steam temperature	370 to 475° C
Bed height	2.0 to 4.0 m
Catalyst charge	0.8 to 1.6 t

Table 2: COMFLUX-pilot plant data

During the operation period of almost two years the step-by-step improvement of the plant technique led to a high degree of availability. This was particularly demonstrated by the concluding test run of approximately 2,200 hours. In three months of continous operation 14 millions m^3 feed gas were converted into 4.3 million m^3 SNG.

The pilot plant runs have shown that the SNG capacity could be increased up to 3,000 m^3 SNG/h in the whole pressure range from 2 to 6 MPa without problems in the heat exchanger system. The change in the height/diameter ratio from approximately 6/1 for the test plant to 3/1 for the pilot plant did not have any influence on the SNG-capacity and on the degree of conversion in the pressurized fluidized bed. Part-load runs of the reactor down to 30 % of the maximum output could be realized. The partition of the heat exchanger into an evaporative cooler and a superheater led to a higher steam quality and a more flexible temperature control of the reactor. With the catalyst feeding and removal system it is possible to adjust to any required bed level and/or to change part of the catalyst charge for reactivation.

Meanwhile thermic and flow mechanic studies have been carried out in a full scale model of a reactor compartment. The results agreed with those achieved in the pilot plant. Together with the optimized pilot plant results they serve as a sound basis for the design of commercial COMFLUX-plants.

The successful development of the COMFLUX-process has justified the usefulness of cold model tests. In comparison with the investment costs for the high pressure plant the expenses for the model tests are negligible.

LITERATURE CITED

1. Werther, J.,: "Zur Maßstabsvergrößerung von Gas/Feststoff-Wirbelschichten", Aufbereitungs-Technik, 12, 670 (1974).

2. Lommerzheim, W.,: "Methanisierung von CO/H_2-Gemischen im Wirbelschicht-Reaktor", Das Gas- und Wasserfach, Gas/Erdgas, 118 (10), 417 (1977).

3. Bock, H.J.: "Dimensioning of vertical heat transfer surfaces in gas/solid fluidized beds, German Chemical Engineering, 4 (6), 356 (1981).

4. Bock, H.J., Molerus, O.: "Heat transfer in gas/solid fluidized beds", German Chemical Engineering, 6 (1), 57 (1983).

GASIFICATION OF A SUB-BITUMINOUS COAL IN A TWO-STAGE, JET-SPOUTED BED REACTOR

Osamu Uemaki and Toshiro Tsuji

Department of Applied Chemistry
Hokkaido University
Sapporo, 060 Japan

ABSTRACT

The application of the spouted bed technique to a coal gasifier has been studied in laboratory scale equipment. The experiments were carried out using a jet-spouted-bed reactor and a conventional spouted-bed reactor. Both these reactors had two particulate beds, one above the other. The fractional conversions of carbon were 0.60 to 0.88 over a temperature range of 1000K to 1300K at atmospheric pressure. Gases produced had calorific values up to 11 Mj/m^3.

SCOPE

The spouted bed technique has already been accepted widely as a solid-fluid contacting method. Recently, the use of spouted beds for the thermal cracking of petroleum and the gasification of coal has also received attention (1,2,3). The advantages of the spouted bed for coal gasification are the absence of a gas distributor plate and the forced solids mixing in the bed by the spouting gas. The jet-spouted bed is particularly characterised by high velocity gas jets and shallow bed depth (4). Even in the absence of inert solids, a jet-spouted-bed coal gasifier can be operated at higher temperature without incurring sintering or clinkering troubles compared with a conventional spouted-bed gasifier and a fluidized-bed gasifier.

However, the spouted bed has a disadvantage, namely gas bypassing due to the characteristics of fluid flow. This diadvantage is overcome by having an additional bed of particles in the reactor, to prevent tar trouble as well as ensuring a high steam decomposition. These features of the two-stage spouted-bed reactor result not only in sufficient rates of the gasification reactions but also high thermal efficiency. In this work two types of two-stage reactors were used ; one containing a jet-spouted bed and the other, a conventional spouted bed. The gasification tests were first conducted by use of the conventional spouted-bed reactor with air-steam mixtures (5). After that the jet-spouted bed reactor using oxygen-steam mixtures was developed to improve carbon conversion and gas quality.

The objectives of the present study are to prove successful application of the spouted bed to a coal gasification reactor and to compare the results obtained in the jet-spouted-bed gasifier with those of the conventional spouted-bed gasifier. The effects of the operating conditions on gas composition, gas calorific value, steam decomposition and carbon conversion were examined. A mathematical model is used to predict the steam decomposition rate in the reactor.

CONCLUSIONS AND SIGNIFICANCE

The gasification of a sub-bituminous coal has been carried out successfully using a two-stage jet-spouted-bed reactor and a two-stage conventional spouted-bed reactor. The effects of oxygen feed rate, steam feed rate, coal feed rate and other operational characteristics on carbon conversion, steam decomposition and product gas quality were explored.

The gases produced had calorific values up to 11 Mj/m^3 for the jet-spouted-bed reactor in which oxygen-steam mixtures were used as the gasification medium. A gasification temperature of as much as 1300K can be attained without agglomeration problems in the jet-spouted-bed reactor. Thereby, a high carbon conversion up to 0.88 as well as good quality gas with a cold gas thermal efficiency of 0.80 were obtained.

A comparison of results in the jet-spouted-bed reactor with those of the conventional spouted-bed reactor, which contained sand particles and used air not oxygen, shows improvements in carbon conversion and gas quality for the jet-spouted-bed reactor. The carbon conversion and cold gas thermal efficiency are considerably affected by the coal-to-oxygen feed ratio.

The effectiveness of an additional particulate bed in promoting high steam utilization in the two-stage reactor was confirmed by estimation of the steam conversion by use of the simplified two-region spouted bed model.

EXPERIMENTAL APPARATUS AND PROCEDURES

A schematic diagram of the gasification unit is shown in Figure 1. The jet-spouted-bed reactor is 0.21 m inside diameter × 1.24 m high with a 40-degree conical base and a 0.015 m diameter inlet pipe. The column is lined with refractory to an inside diameter of 0.114 m to tolerate high temperature. Temperatures and pressures at various points in the reactor are measured to show the bed depth as well as to detect ash sintering in the beds.

As shown in Figure 1, the gasification reactor is

divided into two stages by a gas distributor with a conical cap. The lower stage contains the jet-spouted bed and the upper stage contains the fluidized bed in which the particles move up and down at random. The opening area ratio of the gas distributor installed between these beds is 16.3 %. This value is five to ten times that of the ordinary gas distributors used in practical fluidized beds, in order to avoid blockage of the distributor with agglomerated ash. The solids loading in each stage can be controlled by the spouting gas velocity. The bed heights of the spouted bed and of the fluidized bed are about 0.25 m and 0.22 m,respectively.

Steam for the gasification reaction is superheated up to about 870K and then mixed with preheated oxygen at the inlet pipe section of the gasification reactor. Crushed coal is fed continuously into the lower stage of the reactor by means of a rotary feeder. The partial combustion and gasification of coal take place over a temperature range of 1000 to 1300K in the present work. Most of the ash is discharged from the reactor through a withdrawal pipe and the rest is elutriated. The product gas passes through a direct water scrubber in which condensable hydrocarbons and steam are condenced.

A small amount of the product gas is sampled and then analyzed by gas chromatography. The composition of the bed is measured to determine the relative amounts of carbonaceous material and ash. Proximate analysis is carried out on the solids caught in the cyclone to determine the extent of reaction of the carryover.

The conventional spouted-bed reactor is the size of the jet-spouted-bed reactor except for a 60-degree conical base. In the conventional spouted-bed reactor, inert silica sand is used as the spouting solids and an air-steam gas mixture as the gasification medium to prevent the ash from sintering.

The coal used in this work is Taiheiyo coal. The properties of Taiheiyo coal and of the silica sand such as average particle size and density are given in Table 1. The experimental conditions are given in Table 2.

EXPERIMENTAL RESULTS AND DISCUSSION

Gasification in the jet-spouted-bed reactor

Some typical results are shown in Figures 2 to 4. Figure 2 shows the effect of oxygen feed rate on the bed temperatures, carbon conversion and cold gas thermal efficiency (C.G.T.E) at the steam-to-coal ratio of 1.12. As can be seen from Figure 2 the temperatures at various points in the reactor show a steady increase with oxygen feed rate. The highest temperature(T_2) is near the bottom of the jet-spouted bed, where the exothermic partial combustion reaction takes place predominantly. The temperature(T_5) at the top of the jet-spouted bed is 50 to 100K lower than that near the bottom. This indicates that the endothermic steam gasification reaction takes place near the top of the jet-spouted bed. The temperature (T_6) of the fluidized bed in the upper stage is about 200K lower than that at the top of the jet-spouted bed in the lower stage. This is also due to the endothermic steam gasification reaction. As Figure 2 also demonstrates, the carbon conversion and cold gas thermal efficiency increase with increasing oxygen feed rate. As may be expected, higher values of the carbon conversion are attained under the condition of higher gasification temperature. It was found that a carbon conversion of more than 0.85 was obtained when the reaction temperature(T_2) exceeded of 1250K in the jet-spouted reactor.

Figure 3 shows the effect of oxygen feed rate on the product gas composition, the gas calorific value and the gas yield. The trends of the gas composition profiles suggest that increasing oxygen rate results in increasing carbon monoxide and in decreasing carbon dioxide. The contents of hydrogen and methane are nearly constant over a temperature range of 1100K to 1300K. The gases contain typically 40 to 45% H_2, 20 to 30% CO, 20 to 27% CO_2 and about 4% CH_4. The gas calorific values are 9.5 to 11 Mj/m^3.

Figure 4 shows the effect of steam feed rate on the carbon conversion and on the gas composition at the oxygen rate of 3.5 (kg/h). The carbon conversion decreases gradually with an increase in steam feed rate which is accompanied by a decrease in the bed temperature at a given oxygen feed rate.

In the jet-spouted-bed reactor fines were carried over with the product gas and were collected in the cyclones. The weight loss of the fines on ignition indicated that the carbon content of the carryover was about 30%.

Gasification in the conventional spouted-bed reactor

The experimental results are shown in Figures 5 to 7. Figure 5 shows the effect of air feed rate on carbon conversion, steam conversion and bed temperatures at the steam-to-coal feed ratio of 0.55. In the conventional spouted-bed reactor, the temperature(T_4) near the top of the spouted bed in the lower stage is the highest. As Figure 5 demonstrates the carbon and steam conversions increase with increasing air rate. Figure 6 also shows the effect of air feed rate on the gas composition, the gas calorific value and the gas yield. The gas contains typically 15 to 20% H_2, 8 to 14% CO, 12 to 15% CO_2 and about 3% CH_4. The gas calorific values are 4 to 5 Mj/m^3 in the present runs since the gases contain about 50% nitrogen.

Figure 7 shows the effect of coal feed rate on the carbon conversion, the steam conversion and the bed temperatures at the air-to-coal feed ratio of 2.0. Although the temperatures in the reactor show a steady increase with the coal feed rate, the steam conversion does not have a definite tendency. This would be due to segregation of the coal particles in the reactor. The carbon conversion decreases with coal feed rate up to 5 kg/h and then begins to keep a constant value at higher coal feed rate in the present runs. This indicates that at a given air and coal rate the coal feed rate would be an important operating parameter for this reactor.

There was no need to remove the ash from the particulate beds by discharge through the withdrawal pipe since all the ash was ground by the inert sand particles and was carried over to the cyclones. The carbon content of the ash collected in the cyclones was about 50% due to the high carryover rate of unreacted carbon. Agglomeration of ash sometimes occurred in the bottom part of the reactor due to the slow movement of the particles in the reactor.

Estimation of steam conversion

The simplified two-region spouted bed model shown in Figure 8, which represents the conventional spouted bed reactor, was used to estimate the steam conversion in the lower stage of the conventional spouted-bed reactor (6). Although more sophisticated models have been proposed, their use did not seem justified in the present work since many of the required parameters were not available.

According to this model (6) the fractional conversion of steam in the spouted bed is represented by Equation (1).

$$\eta = 1 - [\,\overline{F}_{sr}\exp(-\frac{6\alpha R}{\overline{F}_{sr}}) + \overline{F}_{ar}\exp(-\frac{6(1-\alpha)R}{\overline{F}_{ar}})\,] \qquad (1)$$

The values of the reaction rate constant in Equation(1) were obtained by another experiment in which the steam gasification reaction of Taiheiyo coal was conducted using a fixed bed reactor. The values of the steam conversion calculated by Equation (1) are represented graphically in Figure 9, in comparison with the steam conversions obtained in the conventional spouted-bed reactor in the present runs. Figure 9 indicates that the steam conversion in the lower stage of the spouted bed is about one third of that in the upper stage of the fluidized bed. Consequently, it is suggested that the additional particulate bed in the upper stage brings about a remarkable increase in the staem conversion over and above that due to the spouted bed.

ACKNOWLEDGEMENT

This work was supported by the Science Research Foundation of the Ministry of Education Japan under Grants No. 58045003 and 59045005.

NOTATION

$\overline{F}_{ar}$ = dimensionless average gas flow rate in the annulus
$\overline{F}_{sr}$ = dimensionless average gas flow rate in the spout
G_a = air feed rate in the gasification reactor kg/h
G_c = coal feed rate kg/h
G_o = oxygen feed rate kg/h
G_s = steam feed rate in the gasification reactor kg/h
R = dimensionless reaction rate group
α = fraction of particles existing in the spout
η = fractional overall conversion of reactant

LITERATURE CITED

1. Uemaki, O. and M.Kugo, Jr.Japan Petr. Inst., 20, 410 (1977)
2. Foong,S.K., Lim,C.J. and A.P.Watkinson, Can.J.Chem.Eng., 58, 84 (1980)
3. Foong,S.K., Ceng,G. and A.P.Watkinson, Can.J.Chem.Eng., 59, 625 (1981)
4. Markowski,A. and W.Kaminski, Can.J.Chem.Eng., 61, 377 (1983)
5. Takegami,Y. et al.(Eds), "Research on Coal Liquefaction and Gasification" SPEY 12, p.145 (1984)
6. Uemaki,O. and M.Kugo, Bull. Faculty of Eng., Hokkaido Univ., No.88, 59 (1978)

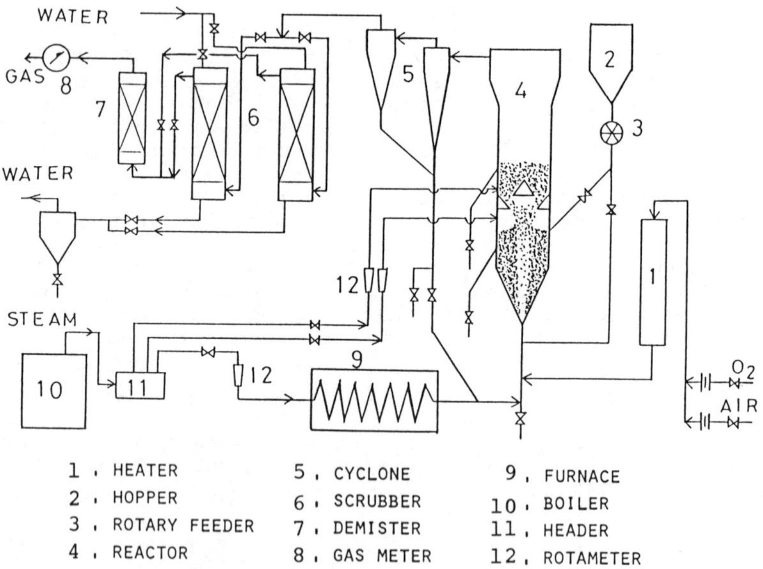

Figure 1 Schematic diagram of spouted bed coal gasifier

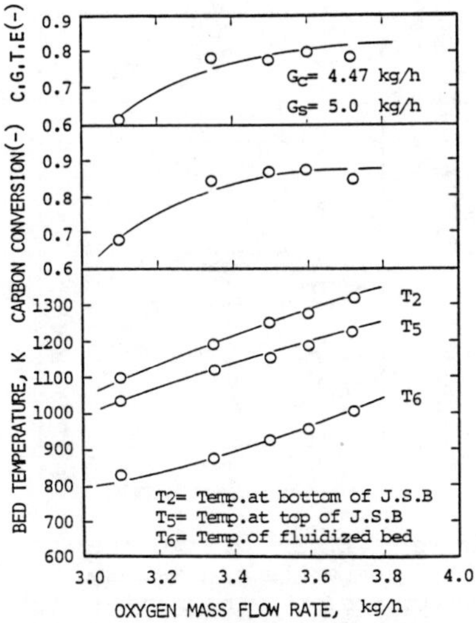

Figure 2 Effect of oxygen feed rate on bed temperature, carbon conversion and C.G.T.E. with the jet-spouted-bed reactor

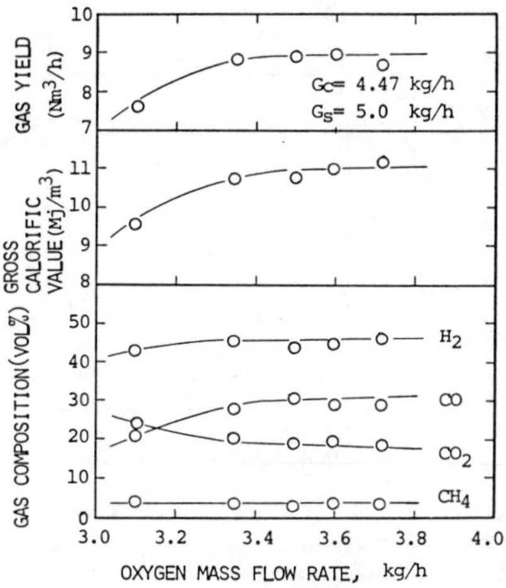

Figure 3 Effect of oxygen feed rate on gas composition, calorific value and yield for dry gas with the jet-spouted-bed reactor

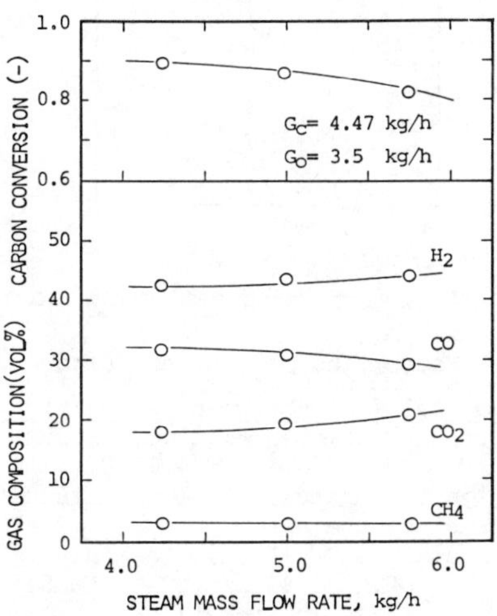

Figure 4 Effect of steam feed rate on gas composition and carbon conversion with the jet-spouted-bed reactor

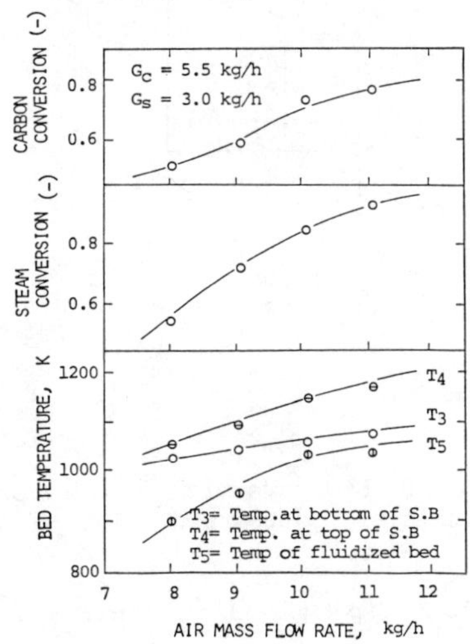

Figure 5 Effect of air feed rate on bed temperature, steam and carbon conversion with the conventional spouted-bed reactor

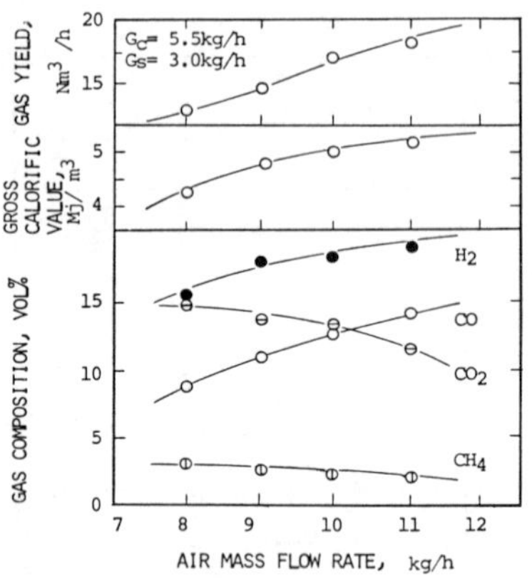

Figure 6 Relationship between air rate and gas quality for dry gas with conventional spouted-bed reactor

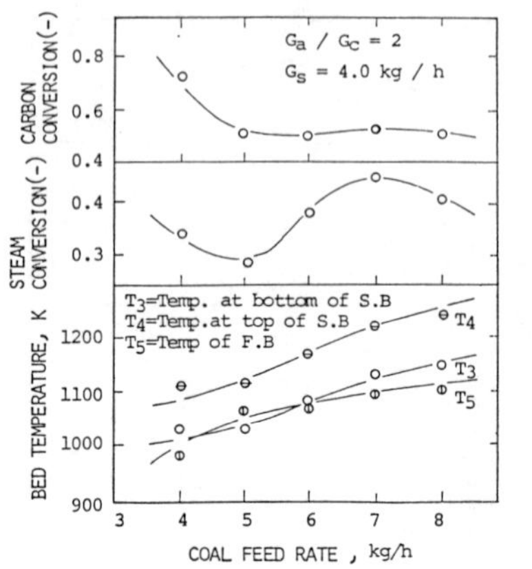

Figure 7 Effect of coal feed rate on bed temperature, steam and carbon conversions with conventional spouted-bed reactor

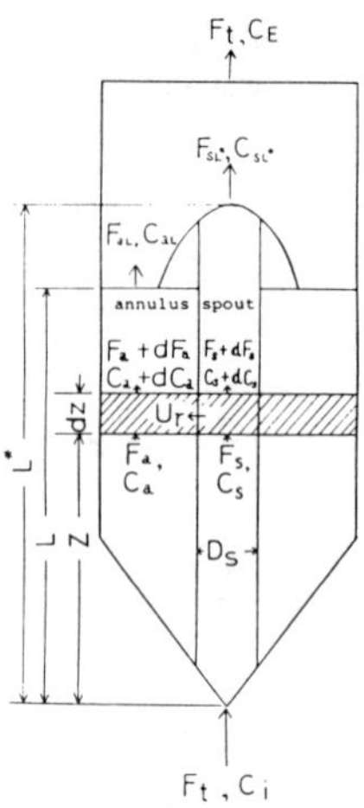

Figure 8 Hydrodynamic features of spouted bed

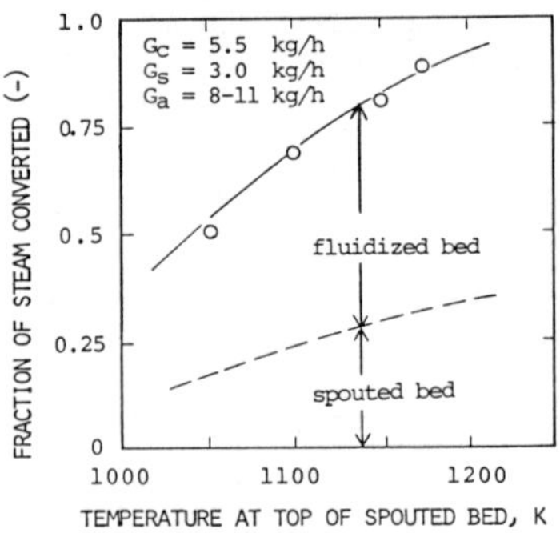

Figure 9 Comparison of steam conversion

Table 1 Properties of coal and sand

Taiheiyo coal (Sub-bituminous coal)			
Mean particle diameter			1.27mm, 1.95mm
Density			1290 kg/m^3
Ash fusion temperature			1600K
CaO content of the ash			2.4%
Ultimate analysis		Proximate analysis	
% C	65.6	% moisture	5.4
H	5.4	% volatiles	41.1
O	15.6	%fixed carbon	40.4
Ash	13.1		
Silica sand			
Mean particle diameter			1.52mm
Density			2630 kg/m^3

Table 2 Experimental conditions

Temperature	1000- 1300 K
Feed rate of coal	4- 8 kg/h
Flow rate of oxygen	3- 4 kg/h
Flow rate of air	8-16 kg/h
Flow rate of steam	2- 8 kg/h

GASIFICATION OF COAL IN A FLUIDISED BED WITH A DRAFT TUBE

M.R. JUDD* and V. RUDOLPH**

* Dept. Chemical Engineering, University of South Africa, Durban, 4001, South Africa.

** Engineering Management Services Limited, Box 585, Bedfordview, 2008 South Africa.

ABSTRACT

Synthesis quality gas, containing over 80% $CO + H_2$, balance CO_2, has been produced in a mini-pilot scale fluidised coal gasifier containing a draft tube, using air and steam as the gasifying agents. The H_2/CO ratio can be varied from about 1 to 3 by changing the operating conditions.

SCOPE

The overall aims of the project are to develop a coal gasifier suitable for producing synthesis quality gas

- from a wide range of coals and high ash discard materials;
- capable of accepting a wide size distribution of feed particles and in particular a heavy fines load;
- free of methane and tarry by-products;
- capable of operation under pressure;
- using air rather than oxygen as the combustion gas.

The work is of particular significance in South Africa where there are accumulating stockpiles of high ash and fine coal discards.

CONCLUSIONS AND SIGNIFICANCE

A fluidised bed containing a draft tube can be effectively divided into two separate reaction compartments; one within the draft tube and the other in the annulus surrounding it, without interfering with the bulk solid circulation between these zones.

Applying this to coal gasification, a synthesis quality gas has been produced by steam gasification in the draft tube and using air for the combustion reaction in the annulus; the syn- and wastegas streams produced by these reactions may be taken off separately from the reactor.

The H_2/CO ratio of the syngas can be varied from about 1 to 3, by changing the reactor operating conditions.

Despite the separation of the reactions into discrete zones, solid circulation, which provides the energy transfer from the combustion to the gasification reactions, can be maintained.

The key features of the process have been demonstrated on a mini-pilot scale gasifier processing about 40 kg/h of coal.

FLUIDISED GASIFIER WITH DRAFT TUBE

The many inherent advantages of draft tube fluidised beds, especially in processing fine and poor quality coals to yield hydrocarbon-free gasification products (1,2), are further extended by the ability to operate the bed as two separate and independent reaction zones defined by the regions inside and outside the draft tube, while still allowing free circulation of solids around the draft tube annulus loop (3).

In our apparatus, shown in Figure 1, the draft tube operates as a slugging bed with the annulus at, or near, incipient fluidising conditions, causing a strong bulk solid flow up the draft tube and down the annulus. Steam is supplied to the draft tube through drilled, bolt-type tuyeres on the bottom distributor plate. Air for combustion is supplied some way up the annulus through a double-pipe sparger arrangement.

The partition of fluidising gas between the draft tube and annulus legs is determined by the equalisation of the pressure drop over these legs. In order to achieve zone separation (such that there is no crossflow of gas between the draft tube and the annulus) this pressure balance must be achieved by appropriate bed geometrical configuration and operating conditions.

Separation of the gases leaving the bed can be achieved by a suitable manifold over the draft tube (4), as shown in Figure 1, again by careful pressure balancing. This arrangement has not yet been demonstrated under hot gasification conditions, but has been proven in cold simulations.

It is particularly advantageous to operate the draft tube in square-nosed slug flow (5). In this fluidisation regime the bubble rise velocity is low and the dense phase gas flow rate is high, resulting in improved gas-solid contacting.

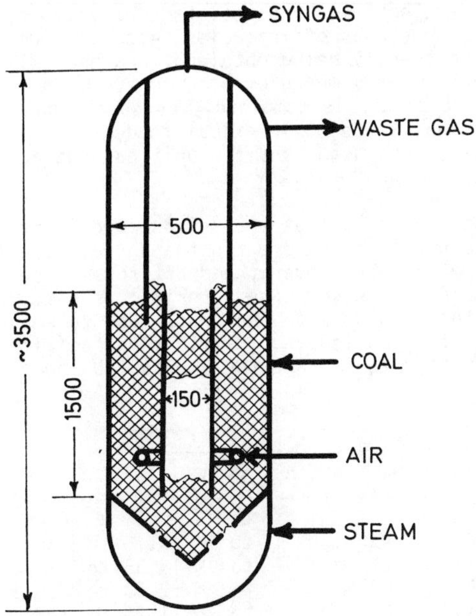

FIGURE 1 PILOT PLANT GASIFIER

GASIFICATION RESULTS

Sasol II and III coal was used in these experiments, but of discard quality due to the size range and high fines load. The properties are given in Table 1. Silica sand, with mean particle size of 640 μm and density 2480 kg/m^3, makes up about 95% of the bed material.

Typical gas analyses (sampled directly from the draft tube and annulus, because the draft tube hood arrangement was not in hot operation) are given in Table 2. The draft tube gas is of comparable quality to syngas produced by other oxygen-blown processes. A small proportion of H_2 and CO was detected in the annulus gas, indicating a slight degree of crossflow of steam, supplied for the draft tube, to the annulus.

Some crossflow is unavoidable because of pressure fluctuations resulting from the slug flow in the draft tube, although this may be expected to become relatively less important when the process is operated under pressure. Process requirements will dictate whether it is preferable to tolerate crossflow in the form of dilution of the draft tube syngas with combustion flue gas (e.g. in heating or reductant applications), or as an efficiency loss of product gas to the waste stream (e.g. in syngas production).

Material balances have not yet been satisfactorily closed, and gas production rates, cold syngas efficiencies, etc., are not known exactly. These are key data presently being obtained, and they will be published in due course. Preliminary measurements, however, indicate that better than 97% C is converted, air is completely consumed, but steam bypass may be rather high (perhaps 30 to 40%). Coal feedrates of between about 40 and 100 kg/hr have been used, but the unit has not been taken near maximum throughput.

An interesting observation is that the solids circulation rate varies in such a way as to counteract pressure fluctuations in the system. If, for example, the pressure drop over the draft tube is suddenly reduced (on eruption of a slug of dense phase material), the tendency is for gas to crossflow from the annulus into the draft tube to equalise the pressure drops over the two legs. This in turn facilitates particle flow into the draft tube to restore the pressure balance.

Table 1 Coal Analysis

Proximate Analysis		Ultimate Analysis		Size Range	Mass Fraction
H_2O (tot)	5 to 15%	H_2O	5 to 15%	> 6,30 mm	-
Ash	17 to 24	Ash	17 to 24	4,75 to 6,30	0,054
VM (DAF)	25 to 37	C	77 to 80	3,35 to 4,75	0,101
FC (DAF)	63 to 75	H	3,7 to 5,1	2,63 to 3,35	0,121
CV (DAF)	29,4 to 31,4	O	1,2 to 1,7	1,70 to 2,63	0,117
		N	1,3 to 2,5	1,18 to 1,70	0,148
		S	0,4 to 2,6	0,85 to 1,18	0,115
				0,425 to 0,85	0,189
				< 0,425	0,155

Table 2 Product Gas Analysis (dry basis)

	Draft tube		Annulus
Temp	840°C	905°C	850°C
H_2	59%	43%	2,5%
CO	23%	35%	4,2%
CO_2	18%	15%	17,2%

GAS COMPOSITION

Figure 2 shows that the gas composition is a strong function of temperature. The H_2/CO ratio can apparently be varied over a range of about 1 to 3, by changing the operating temperature of the reactor.

The overall chemistry of the process seems well described by the simple reaction scheme

$$C + O_2 = CO_2 \quad (1)$$

$$C + xH_2O = (2 - x)\,CO + (x - 1)\,CO_2 + xH_2 \quad (2)$$

with x in the range 1 to 1,5.

Since the gasification reaction (2) is confined to the draft tube, some shift of CO to CO_2 and H_2 must be occurring for the measured condition of $x > 1$. The equilibrium of this shift is displaced towards the H_2 side with lower temperature, and this could partly explain the variation of the H_2/CO ratio with the operating temperature of the reactor.

The ability to tailor the H_2/CO ratio of the product gas to the specific downstream requirements, without an additional shift step, is a significant bonus for this process.

At this time we have no accurate analysis for methane or higher hydrocarbons. Simple Draeger tube checks suggested very low levels of hydrocarbon, and, when the gas was bubbled through water, there were only traces of liquid organics. More work needs to be done to quantify these crude observations. Moisture content is high and not known with certainty yet. Preliminary measurements suggest that it may be about 0,8 gm/ℓ dry gas.

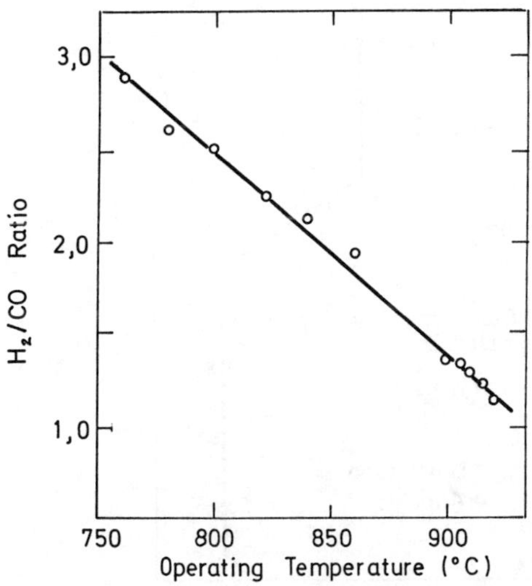

FIG. 2 EFFECT OF TEMPERATURE ON THE H_2/CO RATIO OF PRODUCT GAS

BED TEMPERATURES

The temperature profile within the bed was found to be very uniform, as shown in Figure 3. Even though the endothermic gasification reactions are confined within the draft tube and the combustion reactions in the annulus, all the bed temperatures are within 10°C, demonstrating the good energy transfer achieved by the strong bulk solids circulation. Free board temperatures are typically 100 to 150°C lower than the inbed temperatures, as has also been observed for fluidised bed combustion (6). Under distressed operating conditions, the freeboard temperatures sometimes rose above the bed temperatures, presumably due to oxygen bypass and burning of gas above the bed.

The coals studied have high ash fusion points (1200 to 1300°C), and we have not been concerned about sintering except under upset conditions.

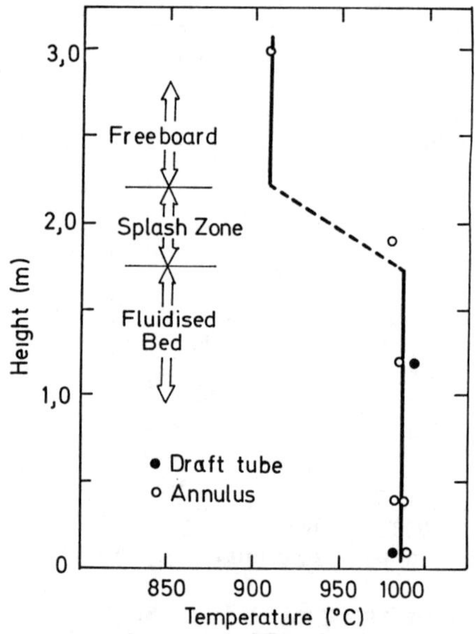

FIG. 3 TYPICAL BED TEMPERATURE PROFILE DURING GASIFICATION

SOLIDS CIRCULATION

The circulation rate of solids within the bed was monitored by tracking the progress of a radioactive particle. This particle is made up of a Co.60 core of about 4 m Ci, coated with alumina refractory. The refractory reduces the particle density to about 2000 kg/m^3 which is comparable with the bed material, and serves to protect the metal core against the ~ 1000°C operating temperature and alternating oxidising and reducing conditions encountered as the particle passes through the combustion and gasification zones in the bed. The particle is about 6 mm in diameter - the same size as the larger coal and ash particles in the bed.

For the bed geometry used in these experiments, the bed is turned over entirely about every minute. The precise rate of bulk solids circulation around the draft tube - annulus circuit is determined by the balance of the driving force - due to the density difference between the two zones and the frictional resistance to flow (7). Typically solids spend about 80% of their time (i.e. about 40 to 50 secs.) in the annulus and 20% of their time in the draft tube; whilst in the draft tube there may be quite a wide spread in residence time depending on the very complicated way in which solids are transported up the tube in the slugging mode. The radioactive particle would generally move up and down the draft tube in the rising plugs of solids several times before finally being tossed over into the annulus. A piece of coal probably circulates around the bed several times before being combusted/gasified to extinction.

From independent measurements of plug rise velocity and frequency, as well as overall voidages in the draft tube (5,7), it appears that there is effectively only one plug of solids at any one time in the draft tube. Our circulation data is well modelled by assuming that it is the frictional resistance associated with the rise of this plug that dominates the resistance to solids circulation, and we find that a simple dry friction model (with a fitted constant friction factor of 0,4) is adequate to describe the system.

ACKNOWLEDGEMENT

This work is funded by the Foundation for Research and Development of the CSIR (South Africa). The gasifier development program is supported by Engineering Management Services Limited.

LITERATURE CITED

1. Judd, M.R., Henri Masson and Meihack, W.F. "Solid circulation and gasification experiments in a fluidised bed with a draught tube" in Fluidization, Kunii, Daizo and Ryozo Toei (Eds), Engineering Foundation, New York (1983).
2. Judd, M.R. and Victor Rudolph, "Fluidised bed combustion and gasification of coal, R/D program at Natal University", 9th RSA/ Israeli Binational Symposium on Coal Technology, Beersheva, May 1984.
3. Rudolph, Victor, "Zonal separation and solids circulation in a draft tube fluidised bed applied to coal gasification", Ph.D.Thesis, University of Natal, South Africa (1984).
4. Judd, M.R., South African Prov. Pat. No. 84/8067 to SAIDCOR (1984).
5. Rudolph, Victor and Judd, M.R., "Circulation and slugging in a fluid bed gasifier fitted with a draft tube", submitted to 1st International Conf. on Circulating Fluidised Beds, Halifax, Canada (1985).
6. Yavuzkurt, Savas, Chaim Gutfinger and Joshua Dayan, "Fluidised combustion of oil shale", in Fluidization, Grace, J.R. and Matsen, J.M. (Eds), (1), Plenum Press, New York (1980).
7. La Nauze, R.D., Powder Technol., 15, 285 (1976).

Fluidized Bed Gasifier Model Taking Account of Dynamic Change of Surface Concentrations, Experimental and Theoretical Study

Takehiko Furusawa, Tadafumi Adschiri,
Toshinori Kojima and Hideo Kawaguchi

Department of Chemical Engineering
The University of Tokyo

ABSTRACT

The time of the intensive char circulation between the oxidizing zone and the reducing zone within the gasifier was of similar order of magnitude to the time of the dynamic change in surface oxide complex. The CO_2 gasification of char particle within the experimental fluidized bed gasification was analyzed by the new model taking account of the dynamics of the surface oxide at surfaces of char particles.

SCOPE

Fluidized bed gasification of coal char is generally analyzed assuming that the gasification rate is determined by the local concentrations of gaseous reactants and products around the solids, i.e. that the concentrations of the reactants and the products on the char surface exist in an instantaneous equilibrium with the concentrations in gas phase(1). However , it was suggested that the intensive char circulation between the oxidizing atmosphere and the reducing atmosphere within the gasifier might affect the reactivity of the coal char surface(2,3,4,5), thus the direct application of the thermogravimetrically obtained gasification kinetics to the analysis of fluidized bed gasification may not be always justified.

CONCLUSION AND SIGNIFICANCE

The time of solids circulation within the bed was of similar order of magnitude to the time of surface oxide dynamics. Furthermore the experimentally observed response against the step change of carbon dioxide verified the importance of the new model. The performance of the experimental gasifier of char by carbon dioxide was numerically analyzed. It was suggested that the intensive solid circulation between the oxidizing zone and the reducing zone may result in the reduction of the occupied fraction and consequently the decreased conversion of carbon dioxide.

IMPORTANCE OF SURFACE DYNAMICS

Theory

Several mechanisms have been proposed so far for the char-CO_2 reaction. The following mechanism proposed by Ergun(6) is widely accepted.

$$C + CO_2 \underset{i_2}{\overset{i_1}{\rightleftharpoons}} C(O) + CO \qquad C(O) \xrightarrow{i_3} CO \tag{1}$$

$$d\theta/dt = i_1(1-\theta)P_{CO2} - i_2\theta P_{CO} - i_3\theta = 0 \tag{2}$$

$$\text{rate} = i_3\theta = i_1 P_{CO2}/(1+i_2P_{CO}/i_3+i_1P_{CO2}/i_3) \tag{3}$$

Thus the time constant of the dynamic change in the fraction is

$$\tau_\theta = (i_1P_{CO2} + i_2P_{CO} + i_3)^{-1} \tag{4}$$

Experimental

The char used for this investigation was prepared by carbonizing Taiheiyo coal at 623-673 K, which is a typical Japanese non-caking subbituminous coal. The physical properties and ultimate analysis of the employed char are shown in Table 1. All of the experiments were conducted by use of an isothermal thermogravimetric apparatus at atmospheric pressure. The details of the apparatus and the procedure were found elsewhere(1).

Table 1 Scope of the Experiment

Temperature : 1173 K					
Diameter of the char : 0.50-0.59 mm					
Particle density of the bed material : 0.95 g/cm^3					
u/u_{mf} at the inlet : 4.9					
Ultimate analysis of the initial char used(Dry basis %)					
C	H	N	S	O	Ash
66.2	2.3	1.3	0	7.9	22.3
Heating rate of the bed				473-773 K	30 K/min
				773-1073 K	13 K/min
				1073-1173 K	5 K/min
Concentration of the step input CO_2 = 50.4% N_2 = 49.6%					

Results

The thermogravimetrically obtained data were analyzed in term of the volume reaction model as follows

$$dX/dt = (k_1[CO_2]/(1+k_2[CO_2]+k_3[CO]))(1-X) \tag{5}$$

The rate constants were measured in the various concentration atmosphere of reactant gas. The retardation effect of carbon monoxide was tested by conducting the experiments in the mixture of carbon dioxide and carbon monoxide. The rate constants were obtained as,

$$\begin{aligned} k_1 &= 1.52\times10^{9}\exp(-348000/RT)\ [cm^3/mol/s] \\ k_2 &= 8.11\times10^{-5}\exp(\ 52700/RT)\ [cm^3/mol/s] \\ k_3 &= 1.64\times10^{-9}\exp(\ 84300/RT)\ [cm^3/mol/s] \end{aligned} \quad \text{at } X = 0.62 \tag{6}$$

Also the different rate constants were obtained at the different conversion level of the conversion-time curve.

Estimation of the Time of Surface Dynamics

The rates of fundamental reaction steps were determined experimentally by use of the following equation

$$C_0 S(i_3 \theta) = (dX/dt)(F_0/M_0) \quad (7)$$

The rates of elemental reaction steps determined for X = 0.62 are given as follows

$$i_1 = 1.85\text{x}10^7/T \exp(-34800/RT) \quad [atm^{-1}s^{-1}]$$

$$i_2 = 3.74\text{x}10^2/T \exp(-3170/RT) \quad [atm^{-1}s^{-1}]$$

$$i_3 = 1.83\text{x}10^{13} \exp(-87500/RT) \quad [s^{-1}] \quad (8)$$

The typical time constant of surface dynamics for a following experimental condition: T = 1273 K, P_{CO2} = 0.26 [atm], P_{CO} = 0.52 [atm] was τ_θ = 16.4.

ESTIMATION OF CIRCULATION RATE OF SOLIDS

A bubble in a fluidized bed carries particles upward in a wake and cloud. Rowe proposed a calculation procedure of solids circulation rate caused by the bubble motion(7). Let the upward transport in a cloud volume be f_c and also let the wake volume be approximated by f_w, then the upward particle velocity and the downward particle velocity can be given by

$$U_{up} = ((u_0-u_{mf})/\delta)(f_w+f_c) \quad (9)$$

$$U_{down} = (f_w+f_c)(u_0-u_{mf})/(\delta -1) \quad (10)$$

Thus the order of magnitude of solids circulation time in a bubbling bed can be estimated as(8)

$$\tau_c = L_0/(u_0-u_{mf})(f_w+f_c) \quad (11)$$

where

$$f_w = 0.25 \quad (12)$$

and

$$f_c = 3u_{mf}/\varepsilon_{mf}/(0.711(gD_b)^{1/2}-u_{mf}/\varepsilon_{mf}) \quad (13)$$

The order of the time for solids circulation was estimated for following cases

1) Experimental fluidized bed employed for the present investigation(I.D. = 79 mm, L = 196 mm); τ_c = 2.1 s for d_p = 0.55 mm, τ_c = 20 s for d_p = 0.26 mm
2) Fluidized bed coal gasification pilot plant(I.D. = 3000 mm, L = 6000 mm); $\tau_c \cong$ 15 s for d_p = 1 mm.

PROPOSED MODEL TAKING ACCOUNT OF DYNAMIC CHANGE OF SURFACE CONCENTRATIONS

As mentioned above, the time for the circulation of the char particles within the gasifier was of similar order of magnitude to time constant of the dynamic change in surface oxide complex. This suggested that the performance of the gasifier cannot be expressed by the conventional models which assume the instantaneous equilibrium between the concentrations in the gas surrounding the char and at the char surface.

A new model was developed modifying the conventional Bubbling Bed Model(9). It is assumed in this investigation that all the gases travel through the bed only as bubbles(1). The bubble volume fraction was defined as u_0/u_b. The effect of particles dispersed within the bubble was also ignored. The critical assumptions employed were (1) that the all particles within the wake and cloud regions were transported with bubbles from the bottom of the bed to the surface of the bed at the rate u_b, thus the particles in the emulsion were transported downwards at the rate $u_b(f_w+f_c)/f_c$ and (2) that the exchange of particles between the cloud-wake phases and the emulsion phase was ignored. The detailed calculation procedure which follows the transient changes in the gas phase concentrations through the gasifier after starting the supply of the reactant gas will be discussed elsewhere.

RESULTS AND DISCUSSION

Importance of the Proposed Model-Experimental Verification

The importance of the proposed model was verified by use of the step responses of gases in the batch fluidized bed gasification. The details of the apparatus used were reported elsewhere(1). The scope of the present experiment is shown in Table 1. The fluidized bed(I.D. = 79 mm, Height = 19.6 mm) packed by the char was heated in a nitrogen stream with the rate indicated in Table 1 to 1173 K. When the desired temperature was attained, the gasification was conducted until the conversion of solids attained 0.62. The feed composition was switched to carbon dioxide diluted by nitrogen. The dynamic responses of gases were followed intermittently. A typical result is shown in Figure 1. The calculated ratio of the time constant of the surface dynamics to the time for solids circulation τ_c/τ_θ was 0.14. The step response of the bed calculated by the conventional model which reached the steady state within 5 s could not explain the experimental result. Furthermore, the overshoot of carbon monoxide concentration could not be expressed by the conventional model but by the proposed model. This strongly suggests the importance of the present model.

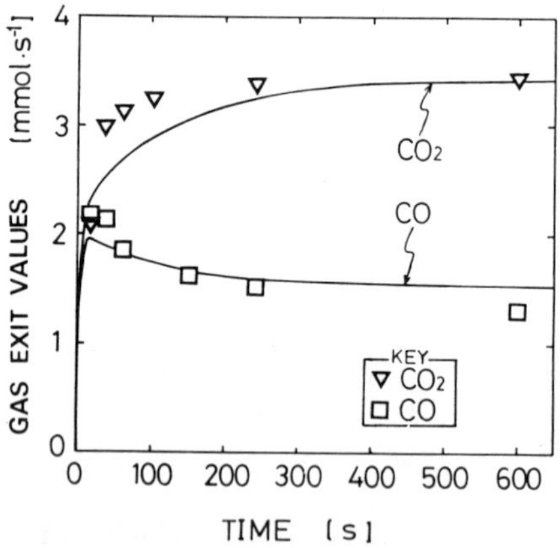

Figure 1 Dynamic response of carbon monoxide and carbon dioxide against step change of inlet composition in batch fluidized bed of char.

Data Used for Numerical Calculation

The significant difference between the results calculated by the proposed model and these calculated by the conventional model was expected over the high range of conversions of gas and solids. Thus the physical properties and the operating conditions listed in Table 2 were employed for the numerical analysis. The rates employed for the present analysis are given by following equation. The details of the derivation will be given elsewhere.

$$i_1 = 3.51\times10^8 \exp(-71900/RT)/T \quad [atm^{-1}s^{-1}]$$

$$i_2 = 0.26 \quad \exp(13600/RT)/T \quad [atm^{-1}s^{-1}]$$

$$i_3 = 6.58\times10^5 \exp(-45400/RT) \quad [s^{-1}] \qquad (18)$$

Table 2 Scope of Numerical Experiment

Mean diameter of char*	0.219mm(X=0.86)
Apparent density of char*	$0.64 g/cm^3$ (X=0.86)
Exposed active carbon site	0.0085mol/g
diffusion coefficient of CO	$3\times10^{-3}T^{-1.43}cm^2/s$
diffusion coefficient of CO_2	$2.3\times10^{-3}T^{-1.08}cm^3/s$
Void fraction of bed at minimum fluidization	0.5
Bubble diameter	1-3 cm
Bed height at fluidization	19.6 cm (10-40 cm)
Partial pressure of inlet CO	0 atm
Partial pressure of inlet CO_2	0.704 atm
Temperature	1273 K
Gas flow rate	$27\ cm^3/s$ at 290 K

* determined experimentally for Taiheiyo coal char (1)

Effects of Bed Height on Calculated Conversion of Carbon Dioxide

The typical conversion of carbon dioxide calculated by both the proposed model and the conventional model are plotted against the bed height in Figure 2. The difference between the two calculated conversions was not significant in a shallow bed. However, the difference is enhanced with the increase in the bed height, since the conversion calculated by the present model did not increase so much as that calculated from the conventional model. The decreased sensitivity of the bubble diameter with increased conversion is found in the present model. The results suggested that the validity of the present model can be examined by the bed of increased height.

Longitudinal Profiles of the Reactant, the Product and the Occupied Fraction of the Sites

The significant difference between the two results was observed with the increased bed height. The profiles of partial pressure of gases calculated for the proposed model and the conventional model are respectively shown in Figure 3 and Figure 4. The bed height and the effective bubble diameter were respectively 39.4 cm and 3 cm. The significant difference between the two calculated results was that the longitudinal profiles of carbon dioxide from the proposed model did not decrease monotonously over the whole range of conversion. The carbon dioxide partial pressure in each phase shown in Figure 3 first decreases to a minimum and then slightly increase, while the partial pressure of carbon dioxide in each phases shown in Figure 4 decreases monotonously. The order of the carbon dioxide partial pressure in each phase was reversed over the upper part of the bed. This is more clearly shown in the enlarged portion of Figure 3. The longitudinal profiles of the occupied fraction of sites calculated from the proposed model and the conventional model are shown respectively in Figure 5 and Figure 6. In case of the present model, the occupied fractions in emulsion phase has a minimum and that in cloud-wake phase has a maximum. This indicates that the consumption of the surface oxide complex due to either desorption or reduction of surface oxide by carbon monoxide are predominant in upper part of emulsion phase. The reversed profile of carbon dioxide in the upper part of the bed indicates that the surface oxide formed in the lower part of the bed was transported upwards by upflowing motion of the cloud-wake solids and was released in the upper part of the bed. The profiles of the occupied fraction for the conventional model are shown in Figure 6, where the particle pressure profiles of carbon monoxide and carbon dioxide changed monotonously. The averaged value of the occupied fraction for the present model, which is far below the level of the occupied fraction in the conventional model was attributable to the upward transport of solids. For the proposed model, the promoted reduction of surface oxide by carbon monoxide in the upper part of the bed resulted in the decreased conversion of carbon dioxide. For the conventional model, the rather large value of the occupied fraction in the lower part of the bed increased the rate of carbon monoxide production. The high sensitivity of bubble diameter, observed for the conventional model, can be also explained by increased rate of gasification, i.e. the increase in the relative importance of mass transfer resistance.

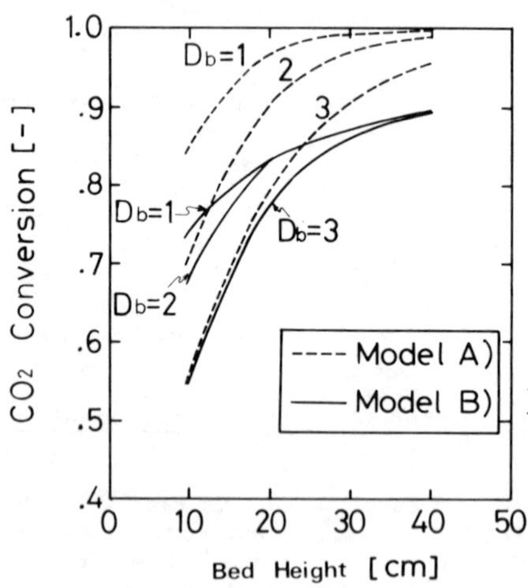

Figure 2 Variation of gas phase conversion with bed height. ModelA:The conventional model, Model B:The proposed model.

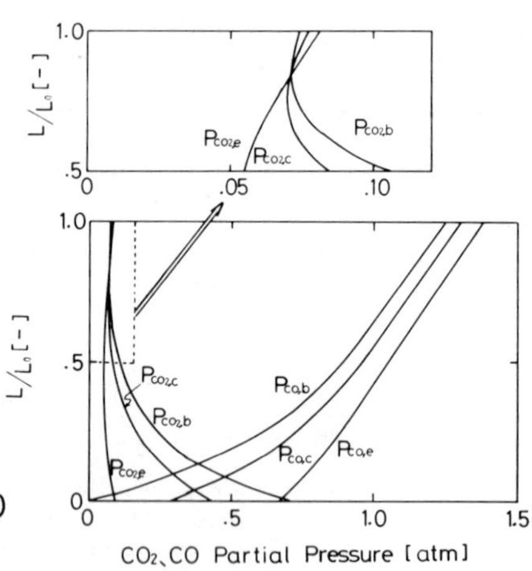

Figure 3 Longitudinal profiles of carbon monoxide and carbon dioxide of bubble, cloud and emulsion phase for the proposed model.

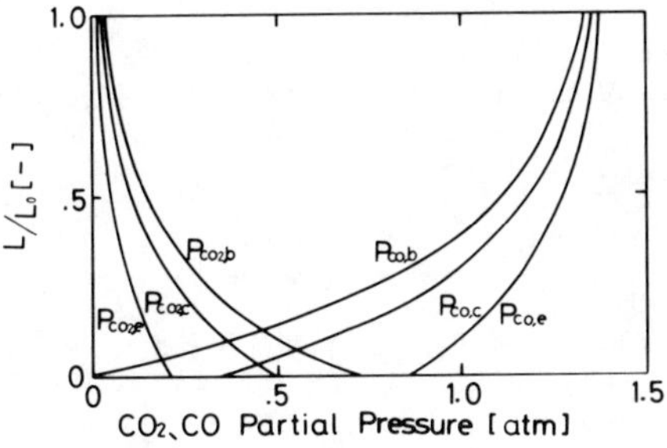

Figure 4 Longitudinal profiles of carbon monoxide and carbon dioxide of bubble, cloud and emuldion phase for the conventional model.

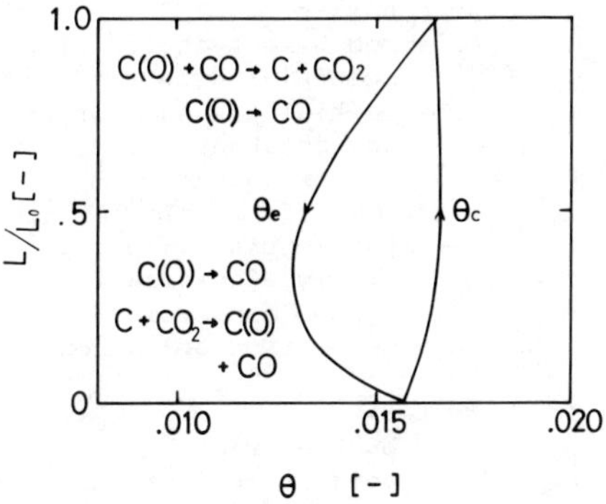

Figure 5 Longitudinal profiles of the occupied fraction of the sites for the proposed model.

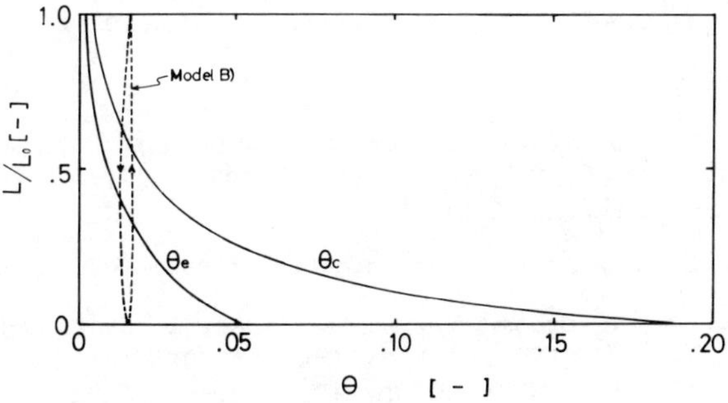

Figure 6 Longitudinal profiles of the occupied fraction of the sites for the conventional model.

ACKNOWLEDGEMENT

The author(T.F.) wishes to express his thanks to Grants in Aid for Energy Research (59045201 and 59040033) of the Ministry of Education Science and Culture, Japan.

NOTATION

C_0S = exposed active site [mol/g-initial char]
D_b = bubble diameter [cm]
f_c, f_w, f_e = ratio of cloud, wake or emulsion volume to bubble volume [-]
d_p = diameter of particle [cm]
F_0 = fraction of carbon within char [-]
g = acceleration of gravity = 980 [cm s^{-2}]
i_j j=1-3 = rate of elemental reaction steps [dimension in text]

k_j j=1-3 = rate constants for C-CO_2 reaction [dimension in text]
L = height [cm]
L_0 = bed height at fluidization [cm]
M_0 = molecular weight of carbon [g mol^{-1}]
P_i = partial pressure of reactant and product [atm]
R = gas constant [cal mol^{-1} K^{-1}]
T = temperature [K]
u_0 = superficial gas velocity [cm s^{-1}]
u_b = bubble rise velocity [cm s^{-1}]
u_{mf} = minimum fluidization velocity [cm s^{-1}]
X = carbon conversion [-]
X_{CO2} = conversion of carbon dioxide [-]

Greek Letters
δ = bubble fraction [-]
ε_{mf} = void fraction at minimum fluidization [-]
θ = occupied fraction of site [-]
τ_θ = time of surface reaction [s]
τ_c = time of circulation [s]

Subscripts
b = bubble phase
c = cloud-wake phase
e = emulsion phase
i (i=1,2) = CO_2, CO

LITERATURE CITED

1. Matsui,I., D.Kuniiand T.Furusawa, Study of Fluidized Bed Steam Gasification of Char by Thermograbimetrically Obtained Kinetics, J. of Chem. Eng. Japan, 18, 105 (1985)

2. Squires, A.M., Private Communication (1982)

3. Squires, A. M., Steam-Oxyugen Gasification of Fine-Sizes of Coal in a Fluidized Bed at Elevated Pressure Part I, Reaction of carbon with Hydrogen, Trans. Inst. Chem. Engrs., 39, 3 (1961) Part II. Reaction of Carbon with Steam, ibid., 39, 10 (1961)

4. Squires, A.M., Role of Solid Mixing in Fluidized Bed Reaction Kinetics, AIChE Symp. Ser. No.128, 69, 8 (1973)

5. Dent, F.J., Steam-Oxygen Gasification of Fine Sizes of Coal ina Fluidized Bed at Elevated Pressure-Discussion, Trans. Inst. Chem. Engrs., 39, 22 (1961)

6. Ergun, S., Kinetics of the Reduction of Carbon Dioxide with Carbon, J. Phys. Chem., 62, 480 (1956)

7. Rowe, P.N., Estimation of Solids Circulation Rate in a Bubbling Fluidized Bed, Chem. Eng. Sci., 28, 980 (1973)

8. Muchi,I.,S. Moriand M. Horio,Ryudoso no Hanno Kogaku, Baihukan, p114 (1983)

9. Kunii, D. and O. Levenspiel, Fluidization Engineering, John Wiley & Son, p.110-114 (1977)

MODELING OF A FLUIDISED BED WOOD GASIFIER

F.G. van den Aarsen*, A.A.C.M. Beenackers**, W.P.M. van Swaaij; Twente University of Technology - Laboratory of Chemical Reaction Engineering, P.O. Box 217, 7500 AE Enschede, The Netherlands

ABSTRACT

We developed a "learning model" for a fluidised bed wood gasifier, which predicts the product gas composition, the char hold-up in the reactor and the unit capacity as function of the reactor bed temperature and the fluidised bed height. In this model we incorporate the specific processes which occur in wood gasification with the hydrodynamics of a bubbling fluidised bed. The results of our model calculations are compared with our experimental data, that were obtained in a 30 cm diameter pilot plant reactor.

SCOPE

The thermochemical conversion of solid fuels into a combustable gas can be accomplished in several reactor types. Especially small particles such as sawdust and high ash containing materials (e.g. refuse derived fuels and rice hulls) cause flow and sintering problems in reactors of the fixed bed type. For such difficult feedstocks a fluidised bed gasifier may turn out to be a prospective alternative because of its rather omnivorous character with respect to feedstock specifications. In a fluidised bed wood gasifier, the following processes occur simultaneously:

- partial oxidation of the wood
- fast pyrolysis of the wood
- gasification of pyrolysis char by means of reaction with CO_2 and H_2O
- conversion of pyrolysis tar
- the possibility of chemical enhanced mass transfer of the reacting gases from the bubble into the dense phase.

The complexity of this integrated process asks for a mathematical model of the gasifier in order to correlate the effects of process variables on the product gas composition and the dimensions of the gasifier. Therefore a "learning model" of a fluidised bed wood gasifier was developed which incorporates the wood gasification process and the hydrodynamic features of a bubbling fluidised bed.

---

* Present address: Biomass Energy Consultants and Engineers - BECE b.v., P.O. Box 498, 7600 AL Almelo, The Netherlands

** Present address: Groningen State University - Industrial Chemistry Department, Nijenborgh 16, 9747 AG Groningen, The Netherlands

CONCLUSIONS AND SIGNIFICANCE

We developed a "learning model" on fluidised bed wood gasification. To test the results of the model, a 30 cm diameter, continuously fed fluidised bed pilot gasifier was designed, constructed and successfully operated with beech wood particles as a feedstock.
Analysis of the computed results and comparison with the performance of a pilot fluidised bed gasifier revealed that:

- the mass transfer rate between the bubble and the dense phase turned out to be of less importance in a fluidised bed gasifier. The gasification rate per m^3 dense phase is relatively slow as compared to the possible mass transfer rate. This equally holds at elevated temperatures as then the char hold-up becomes very small.
- the steady state char hold-up is fairly well predicted by the model. The product gas composition is equally well predicted by the model but mainly depends on the pyrolysis gas composition and the water-gas shift equilibrium.

MODEL DEVELOPMENT

In our model (figure 1) we distinguish two separate regions:

a. The oxidation zone, which is the lower part of the bed immediately above the distributor plate, where part of the incoming wood is combusted under consumption of all the oxygen that is supplied with the air.
b. The gasification zone, which is that part of the bed above the oxidation zone, where the uncombusted wood is pyrolysed, the char is gasified and tar is converted.

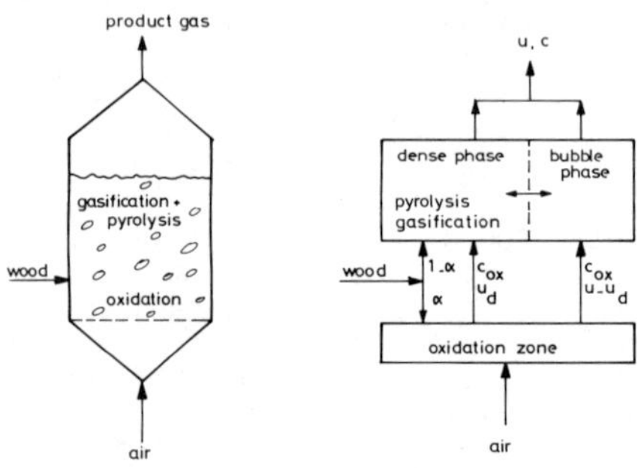

Figure 1 : SCHEME OF A FLUIDISED BED WOOD GASIFIER.

Oxidation Zone

The air-wood ratio determines which part of the incoming wood is combusted in the oxidation zone. The thermodynamic model of Schläpfer [1, 2] is applied to calculate the optimum air-wood ratio for gasifying wood of a certain moisture content at a fixed reactor temperature. The gas composition at the exit of the oxidation zone is determined by the complete combustion of the pyrolysis gases (including tar) resulting from the wood fraction that is assumed to be combusted. Pyrolysis char on the other hand will not combust completely, but be converted into carbon monoxide and carbon dioxide as a result of the gas-solid combustion reaction mechanism. The CO-CO_2 ratio resulting from the char combustion is dependent on the combustion temperature [3]:

$$\frac{CO}{CO_2} = 10^{3.4} \exp(-12400/RT) \qquad (1)$$

After calculating the CO-quantity on basis of the total amount of char converted in the oxidation zone, the CO_2 and H_2O quantities follow from component balances for H, O and C respectively. Then the final gas omposition at the exit of the oxidation zone is obtained by assuming the equilibration of the homogeneous water gas shift reaction.

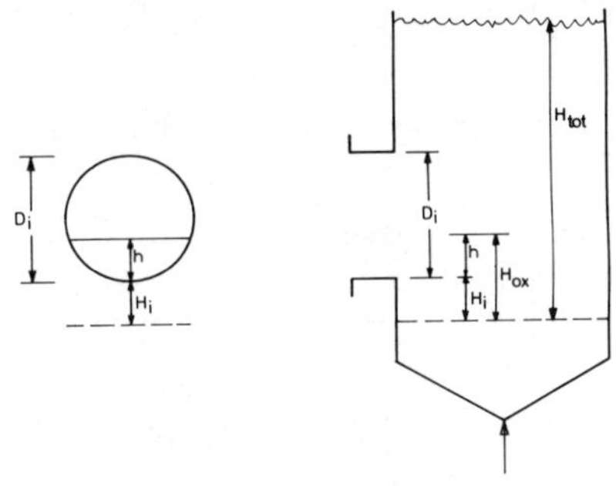

Figure 2 : HEIGHT OF THE OXIDATION ZONE

The height of the oxidation zone is thought to depend on both the biomass feed port location and on the fraction of wood required for consuming all the oxygen by combustion, as follows: assume the cross sectional area of the biomass feed port to be divided into two surfaces, A_1 and A_2, in a ratio equivalent to the ratio of the biomass that is converted by combustion to the biomass that is left for gasification. Then the imaginary embarcation line between the two areas is supposed to indicate the end of the oxidation zone (figure 2).

Gasification Zone

The section of the fluidised bed gasifier above the oxidation zone is called the gasification zone. The performance of this part of the reactor is determined by the rate at which a number of physical and chemical processes occur:

- pyrolysis of the biomass remaining after partial combustion;
- gasification of the char by CO_2 and H_2O;
- (thermal) conversion of the tar;
- mass transfer of the reactive gases from the bubble phase into the dense phase.

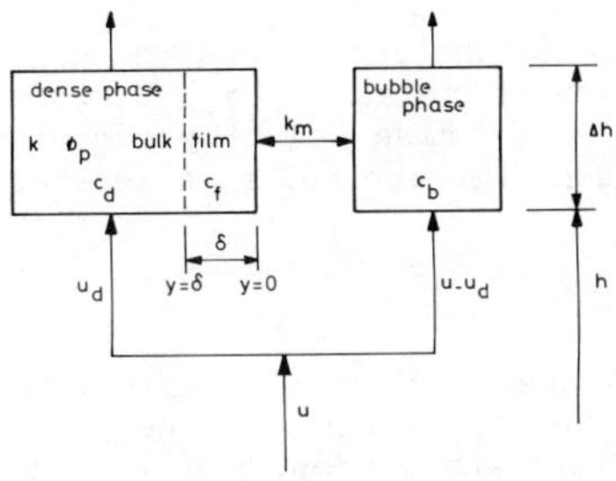

Figure 3 : REACTOR MODEL OF GASIFICATION ZONE

These processes are incorporated in our model for a fluidised bed wood gasifier; the basic reactor model is an extended two phase model, which takes into account the possibility of chemical enhanced mass transfer, analogous to gas-liquid systems (Danckwerts [4], Westerterp, van Swaaij and Beenackers [5]) and a throughflow of the dense phase above minimum fluidisation quantities. The basic scheme of the model is given in figure 3.

Two phases are distinguished, a bubble phase and a dense phase, which are separated by a film of thickness δ, through which mass transfer occurs, possibly enhanced by chemical reaction.

The following assumptions are made:

1. Gas in both the bubble and the dense phase is in plug flow.
2. No solids in the bubble phase.
3. Interfacial area and bubble hold-up are a function of bed height, but not of the radial position in the bed.
4. The bubble phase flow is given by:

$$u_b = 0.48\, u_{mf} \left(\frac{u-u_{mf}}{u_{mf}}\right)^{1.15} \quad (2)$$

which correlation is based on experiments in 3-dimensional systems [2].

The differential equations for the material balances, for stationary conditions and n-th order chemical reaction are [2]:

Bubble phase:

$$(1-\gamma)\frac{dc_b}{d\zeta} + N_{k_m}\{(c_b - c_d + \frac{k\delta^2}{4D}(c_b^n + c_d^n)\} + (1-C_v)\, N_p\, (c_b - c_d) = 0 \quad (3)$$

Dense phase:

$$\gamma \frac{dc_d}{d\zeta} + N_r c_d + N_{k_m}\{c_d - c_b + \frac{k\delta^2}{4D}(c_b^n + c_d^n)\} + N_p\,(c_d - c_p) = 0 \quad (4)$$

The chemical enhancement of mass transfer is represented in equation (3) and (4) by the term:

$$\frac{k\delta^2}{4D}(c_d^n + c_b^n)$$

which is an approximate solution for reaction order n [2].

MODEL PARAMETERS

Bubble to Dense Phase Mass Transfer Kinetics

For calculating the mass transfer rate between the bubble and the dense phase, we make use of Davidson's mass transfer relation [6], Darton's bubble growth relation [7] and Werther's bubble rise velocity correlation [8], of which the applicability has been demonstrated in a separate study [9].

Pyrolysis

The products of the wood pyrolysis process are in yield and composition dependent on the pyrolysis temperature, the particle size and the particle heating rate. The pyrolysis product yield and composition were determined in a bench scale fluid bed reactor [2, 10].

With respect to pyrolysis we assume the following in our model calculations:

- pyrolysis occurs in the dense phase only
- a gas quantity equivalent to the gas volume as produced by pyrolysis, is flowing into the bubble phase; this gas has the same composition as the local dense phase gas
- the reactor height across which pyrolysis occurs, is dependent on the particle pyrolysis time and the solids mixing time. We derived an empirical expression for the pyrolysis height on basis of literature and experimental data in case biomass is fed near the bottom of a fluidised bed gasifier [2]:

$$H_{pyr} = 300\ d_p \tag{5}$$

where,
d_p : characteristic dimension of a wood slab $< 3{*}10^{-3}$ m
H_{pyr} : pyrolysis height measured from the top of the oxidation zone.

Char Gasification Kinetics

Knowledge on the reactivity of char as it is present in a fluidised bed wood gasifier is essential for proper modeling of this reactor. We determined the gasification rate of char, originating from our pilot plant in a bench scale reactor for various carbon dioxide concentrations and temperatures [2], yielding:

$$R''_{CO_2} = k_{CO_2}\ [CO_2]^{0.83}\ \text{mol c/(m}^2\ \text{char.s)} \tag{6}$$

and

$$k_{CO_2} = 7.2\ \exp\ (-)\ \frac{166156}{RT}\ \frac{\text{mol}}{\text{m}^2\text{s}}\ (\frac{\text{mol}}{\text{m}^3})^{-0.83} \tag{7}$$

The surface reaction rate is rewritten as a volumetric reaction rate for convenient use in our model:

$$R_v = R''\ A_o\ \rho_c\ 10^3\ \frac{\varepsilon_c}{(1-\varepsilon_b)} \tag{8}$$

Rensfelt [11] measured a char reactivity for H_2O which was twice the reactivity for CO_2 and found the same dependency of the reaction rate on the water concentration as for the carbon dioxide concentration. We applied these findings in equation (6) for calculating the char reactivity with respect to water:

$$R''_{H_2O} = 2\ k_{CO_2}\ [H_2O]^{0.83} \tag{9}$$

Homogeneous water-gas shift equilibrium

We assume in our model calculations the establishment of the homogeneous water-gas shift equilibrium in both the dense phase and the bubble phase, after each differential calculation step Δh.

Tar Conversion

Tar cracking which takes place in both the dense and bubble phase is included in the model calculations. Details on the kinetics and experimental results have been presented in [2].

EXPERIMENTAL

The performance of a fluidised bed wood gasifier was studied in a 0.3 m diameter pilot scale reactor system (see [2] for details of this plant). The operating variables are given in table 1.

Table 1: Operating variables of the 0.3 m diameter fluidised bed wood gasifier

Fuel	: beed wood particles ($CH_{1.48}O_{0.7}$)
Particle size	: 1 - 3 mm
Moisture content	: 5% wt
Fuel feed rate	: 10 - 50 kg/hr
Air feed rate	: 10 - 60 Nm^3/hr
Reactor temperature	: 700 - 950 °C
Reactor pressure	: 1.1 - 1.3 bara
Bed material	: dense alumina
Mean particle diameter	: 325.10^{-6} m
Particle density	: 3500 kg/m^3
Minimum fluidisation velocity	: 0.21 m/s
Bed porosity at min. fluid. vel.	: 0.43 m^3/m^3
Bed diameter	: 0.3 m
Bed height	: 0.2-1.0 m

Experimental And Calculated Results

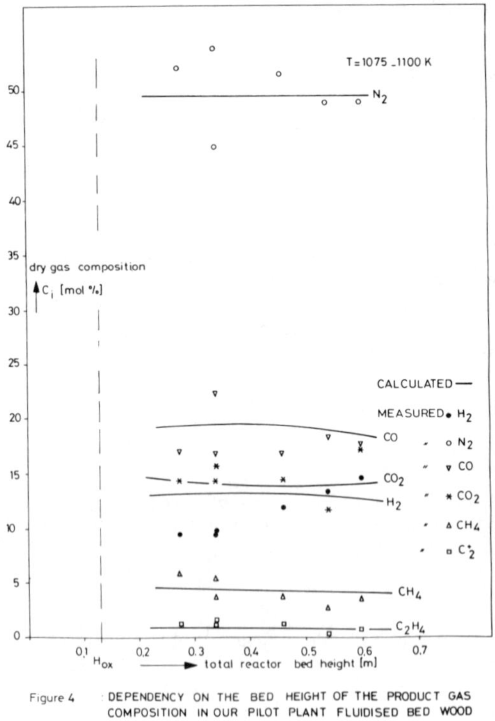

Figure 4 DEPENDENCY ON THE BED HEIGHT OF THE PRODUCT GAS COMPOSITION IN OUR PILOT PLANT FLUIDISED BED WOOD GASIFIER : MEASURED (POINTS) AND CALCULATED (DRAWN LINES)

Operating temperature, bed height and biomass feed rate were the variables in our pilot scale experiments, while gas production rate, gas composition, gas heating value and char hold-up were measured. We used these data for comparison with our model calculations (figure 4 and 5).

The gas composition and the product gas heating value that follows from our computations is in good agreement with the experimental data and is relatively insensitive to the operating temperature and bed height. This is the result of the relatively small amount of char left for gasification once the pyrolysis process is completed. The relatively small amount of gas from the gasification reaction of char with H_2O and CO_2 is thus hardly effecting the product gas composition. However, this gasification reaction determines the maximum char quantity that can be converted per unit time at a given char hold-up in the fluidised bed gasifier. This char

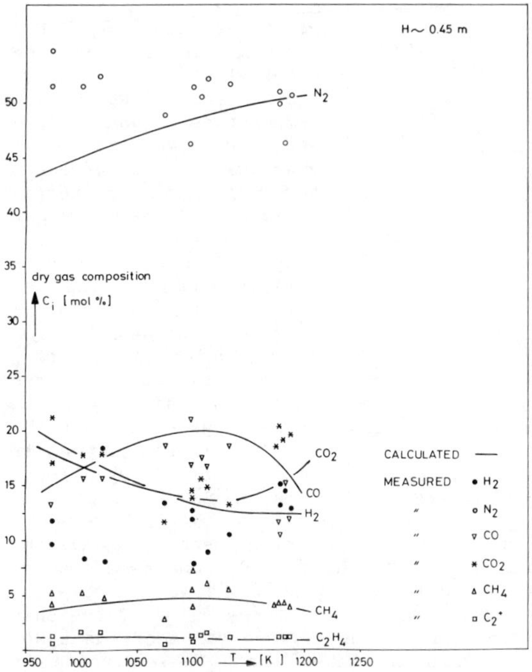

Figure 5 : DEPENDENCY OF FLUIDISED BED PRODUCT GAS COMPOSITION ON THE REACTOR BED TEMPERATURE: MEASURED (POINTS) AND CALCULATED (DRAWN LINES)

hold-up is dependent on the char reactivity, operating temperature and bed height. The value of the char hold-up, that results from the model calculations, satisfies the condition: char production rate by pyrolysis = char consumption rate by gasification in the fluidised bed reactor.

The experimental data on the char hold-up and the calculated values are summarised in table 2 and figure 6.
Apparently, the discrepancy between measured and calculated values of the char hold-up is relatively large for a number of experiments. The reason is a low accuracy in measuring the char hold-up because of the relatively small amount of char that has to be separated from a relatively large amount of bed material. However, the order of magnitude is well predicted, which is important in relation to predicting the

Table 2: Experimental data and calculated values for the char hold-up

	Ø wood [kg/hr]	Ø air [m^3/hr]	T K	H [m]	$\varepsilon_{measured}$ [m^3 char/ m^3 bed]	$\varepsilon_{calculated}$ [m^3 char/ m^3 bed]
1	23.84	34.7	1078	0.25	0.211	0.123
2	23.7	33.0	1113	0.42	0.045	0.018
3	21.75	33.0	1098	0.46	0.033	0.024
4	22.73	28.5	1108	0.47	0.025	0.017
5	21.83	29.3	1098	0.54	0.033	0.016
6	28.73	36.5	1073	0.54	0.024	0.036
7	33.07	39.0	1133	0.54	0.011	0.014
8	28.9	30.0	1093	0.22	0.125	0.114
9	23.3	31.0	1100	0.34	0.051	0.032
10	29.1	33.6	1103	0.44	0.063	0.023
11	27.9	34.5	1148	0.22	0.019	0.042
12	25.2	33.5	1158	0.30	0.026	0.016
13	23.4	34.0	1153	0.45	0.009	0.008

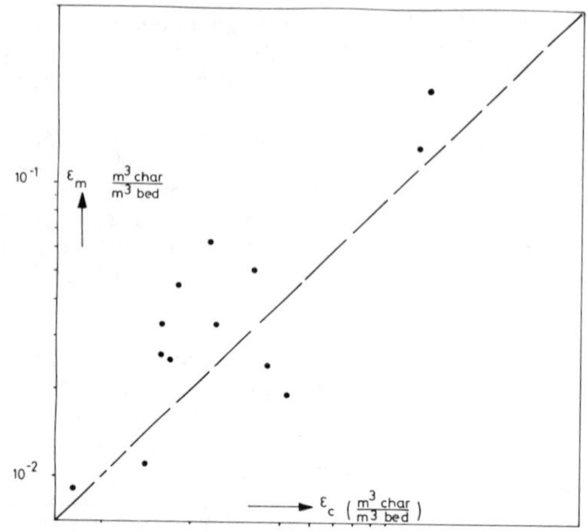

Figure 6 : COMPARISON OF MEASURED AND CALCULATED VALUES OF THE CHAR HOLD-UP IN A FLUIDISED BED WOOD GASIFIER

maximum capacity per unit reactor-volume. The char hold-up determines the reactivity of a certain reactor volume, at a given temperature (equation (8)) and thus the amount of char that can be converted per unit time. The temperature and bed height dependency is given in figures 7 and 8. The calculated char hold-up decreases with increasing reactor temperature (figure 7), which is according to expectations, as the increased reactivity due to temperature and a higher CO_2 concentration, is counterbalanced by a lowering in char hold-up at a constant overall rate of char gasification (kg/s). Again we notice a discrepency between the measured and calculated values of the char hold-up for the same reasons as mentioned previously, but the order of magnitude is well predicted. Figure 7 also shows that the logarithm of the inverse value of the char hold-up, is nearly a linear function of the inverse value of the absolute gasification temperature. This indicates that the influence of fluidised bed hydrodynamics on the char conversion rate is small.

The calculated char hold-up decreases with increasing height of the gasification zone (figure 8), which is a result of the larger residence time of the gas, thus requiring a lower bed reactivity to obtain the same char conversion. In figure 8 we also see that the char hold-up is nearly proportionally dependent on the inverse value of the height of

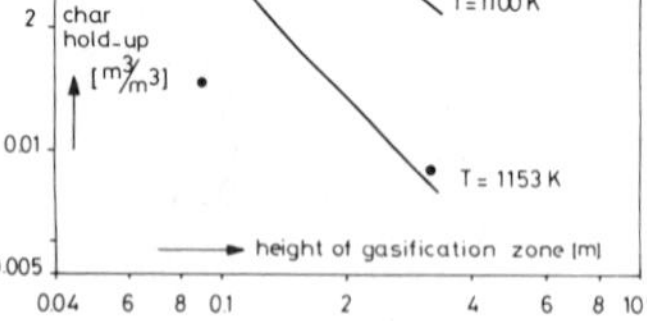

Figure 8 :
CHAR HOLD-UP AS FUNCTION OF THE HEIGHT OF THE GASIFICATION ZONE
o EXPERIMENTAL DATA AT T = 1100 K - - - CALCULATED
• EXPERIMENTAL DATA AT T = 1153 K —— CALCULATED
WOOD FEED RATE = 23.4 kg/hr

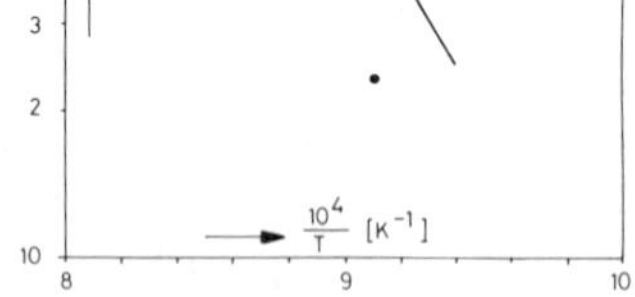

Figure 7 :
CHAR HOLD-UP IN A FLUIDISED BED WOOD GASIFIER AS FUNCTION OF THE REACTION TEMPERATURE
• EXPERIMENTAL DATA AT A GASIFICATION ZONE HEIGHT OF 0.41m, WOOD FEED RATE = 28.7 kg/hr
— DATA RESULTING FROM MODEL CALCULATIONS AT THE SAME CONDITIONS

the gasification zone. This again illustrates the minor influence of fluidised bed hydrodynamics on the char conversion rate. For the same reason as mentioned in the previous paragraphs we find a random deviation between the calculated and measured char hold-up.

LIST OF SYMBOLS

Ar	Archimedes number $\frac{d_p^3\ \rho_g(\rho_s-\rho_g)g}{\eta_g^2}$	[-]
A_o	catchment area of distributor plate per orifice	[m^2]
a	specific surface area of bubbles	[m^2/m^3]
a_o	initial specific surface area of solids	[m^2/g]
α	fraction of biomass that is combusted	[kg/kg]
c_b	gas concentration in the bubble phase	[mol/m^3]
c_d	gas concentration in the dense phase	[mol/m^3]
c_f	gas concentration in the film	[mol/m^3]
C, C_v	constant	[-]
c_{ox}	gas concentration at the exit of the oxidation zone	[mol/m^3]
γ	dense phase through flow ratio = u_d/u	[-]
d_p	particle dimension	[m]
D_e	effective bubble diameter	[m]
δ	film thickness (= D/α)	[m]
D	gas diffusion coefficient	[m^2/s]
D	reactor diameter	[m]
E_a	activation energy	[kJ/mol]
ε_c	char hold-up	[m^3/m^3]
ε_b	gas hold-up in the bubble phase	[m^3/m^3]
ε_{mf}	gas hold-up at minimum fluidization	[m^3/m^3]
ε_o	initial particle porosity	[m^3/m^3]
ϕ_{air}	air feed rate	[m^3/hr]
ϕ_p	volumetric production rate of pyrolysis gases	[$m^3/m^3.s$]
g	gravitational acceleration (= 9.81)	[m/s^2]
h	height	[m]
h^*	height up to which bubbles grow	[m]
H	height	[m]
H_{pyr}	height of pyrolysis zone	[m]
Ha	Hatta number (= $\sqrt{\frac{k\ D}{\alpha}}$ for first order reaction)	[-]
K(T)	equilibrium constant for homogeneous water gas reaction	[-]
k	reaction rate constant	[$(\frac{mol}{m^2 s})(\frac{mol}{m^3})^{-0.83}$]
k_m	bubble to dense phase mass transfer coefficient	[m/s]
k_o	frequency factor	[s^{-1}]
N_{k_m}	number of mass transfer units (= $\frac{k_m a H}{u}$)	[-]
N_p	number of pyrolysis units (= $\frac{\phi_p H}{u}$)	[-]
N_r	number of reaction units (= $\frac{k_v(1-\varepsilon_b-a\delta)c^{(n-1)}H}{u}$)	[-]

n	reaction order	[-]
η_g	gas viscosity	[Ns/m^2]
R	gas constant (= 8.134)	[J/mol.K]
R"	surface reaction rate	[mol/m^2.s]
R_v	volumetric reaction rate	[mol/m^{3}dense phase .2]
ρ_g	gas density	[kg/m^3]
ρ_c	char density	[kg/m^3 char]
S_o	initial specific surface area of solids	[m^2/m^3]
T	absolute temperature	[K]
u	superficial gas velocity	[m/s]
u_d	superficial gas velocity in the dense phase	[m/s]
u_b	superficial gas velocity in the bubble phase	[m/s]
u_A	bubble rise velocity	[m/s]
u_{mf}	minimum fluidization velocity	[m/s]
ζ	dimensionless height (= h/H)	[-]

LITERATURE

1. Schläpfer, P. and Tobler, J.; "Theoretische und Praktische Untersuchungen über den Betrieb von Motorfahrzeugen mit Holzgas", Bern (1933) - Bericht Nr. 3 (Separatabdruck aus "Der Motorlastwagen").
2. Van den aarsen, F.G.; Ph.D.Thesis, Twente University, Enschede, The Netherlands (1985).
3. Arthur, J.R.; Trans. Faraday Soc.; 47 (1951) 164.
4. Danckwerts, P.V.; "Gas-Liquid Reactions"; McGraw-Hill, London (1970).
5. Westerterp, K.R., van Swaaij, W.P.M. and Beenackers, A.A.C.M.; "Chemical Reactor Design and Operation"; John Wiley & Sons Ltd. (1984).
6. Davidson, J.F., Harrison, D., Darton, R.C. and La Nauze, R.D.; In "Chemical Rector Theory: A Review" (L. Lapidus and N.R. Amundsen, eds.) (1977) 583; Prentice Hall, Eaglewood Cliffs, New Jersey.
7. Darton, R.C., La Nauze, R.D., Davidson, J.F. and Harrisson, D.; Trans. Inst. Chem. Eng. 55 (1977) 274.
8. Werther, J.; Chem. Ing. Techn. 49 (1977) 11, 777.
9. Van den Aarsen, F.G., Beenackers, A.A.C.M. and van Swaaij, W.P.M.; Proc. of the 16th ICHMT Symp. "Heat and Mass transfer in fixed and fluidized beds"; W.P.M. van Swaaij, N.H. Afgan (eds.); Hemisphere Corp., Washington (1985), 397.
10. Van den Aarsen, F.G., Beenackers, A.A.C.M and van Swaaij, W.P.M.; Proc. of the Int. Conf. "Fundamentals of Thermochemical Biomass Conversion"; R.P. Overend (ed.), Elsevier Appl. Science Publ. (1984) p.691.
11. Rensfelt, E., et. al.; Proc. I.G.T. Conference "Energy from Biomass and Waste", Washington D.C. (1978).

PERFORMANCE OF FLUIDIZED BED GASIFIERS AND COMBUSTORS WITH BROWN COAL OF DIFFERENT ORIGINS. CALCULATED AND MEASURED GAS COMPOSITION PROFILES IN FLUIDIZED BED AND FREEBOARD.

M. Mittmann, J. Deinert, K. Schütte, R. Wingelsdorf and K. Schügerl
Institut für Technische Chemie, Universität Hannover, Callinstr. 3, D-3000 Hannover

ABSTRACT

In a laboratory fluidized bed, brown coal was gasified and/or burned, and the concentrations of H_2, N_2, O_2, Ar, CO, CO_2, NO, H_2S, hydrocarbons and tar were measured in the outlet gas and calculated by means of a model. In a bench-scale fluidized bed, brown coal was burned, and the spatial distributions of gas components (O_2, CO, CO_2, NO, SO_2) were measured under different operational conditions and compared with model calculations.

SCOPE

Gasification and/or combustion of three different types of brown coal from the main brown coal districts of the Federal Republic of Germany (Hesse, Helmstedt and Rhine) in a laboratory-scale fluidized bed is investigated in an oxygen-free atmosphere and/or in air. In the bench-scale equipment, only the brown coal from the Hesse district was used to evaluate the longitudinal concentration profiles in the bed and in the freeboard at three different radial positions. The measured flue gas compositions are compared with the model calculations.
The aim of the present investigations is to compare the model predictions with the measurements, and, in the long range, to develop models for the adaptive control of fluidized bed reactors.

CONCLUSIONS AND SIGNIFICANCE

Under the operational conditions used in the investigations presented, the concentrations of gaseous components are not constant in the bed and in the freeboard. The assumption of ideal gas mixing cannot be sustained.
The radial and longitudinal concentration profiles of the gas components allow to conclude that by feeding the coal from the top, the coal degassing occurs at the fluidized bed surface and in a thin layer below that.
The investigations indicate that SO_2 is formed mainly in that degassing layer and NO in the entire bed and the free-

board.
Since the volatile components are mixed fairly well in the bed/freeboard system, one can assume that SO_2 is formed mainly from nonvolatile and NO from the volatile components. The flue gas contains several nonvolatile, multinucleate aromatic compounds, which were analyzed by infrared and nuclear magnetic resonance spectroscopy. It was not possible to identify these compounds using standard compounds of known identity.

In the previous paper (Howind et al. 1984), some results were published on the flue gas composition of brown coal combustion in laboratory- and bench-scale fluidized beds. In the present paper, more components of the flue gas were measured and, also, the spatial distributions of the gas components in the fluidized bed and the freeboard are examined. The results are compared with model predictions. The long-term aim is to develop dynamical models for the adaptive control of fluidized bed combustors in cooperation with the Institute of Automatic Control of the University of Hanover and the Deutsche Babcock AG, Krefeld, Germany.

1. Experiments

1.1 Brown coals

The brown coal used in these investigations is characterized in Table 1.

Table 1 Composition of brown coal

composition		weight %	
		coal	
	I (Hesse)	II (Helmstedt)	III (Rhine)
H_2O (raw)	47.3	-	-
ash (dry)	8.9	10.0	3.1
volatiles (dry)	46.7	46.6	-
carbon (waf)+	65.0	71.5	71.7
hydrogen (waf)+	5.1	5.6	5.1
sulphur (waf)+	4.2	2.0	0.3
nitrogen (waf)+	0.9	2.9	-
oxygen (waf)+	24.9	18.0	-
SiO_2 ash composition	34.2	2.9	8.0
Fe_2O_3 ash composition	22.2	27.2	7.5
Al_2O_3 ash composition	15.0	3.5	4.5
CaO ash composition	11.2	28.6	50.5
MgO	1.4	2.2	14.4
Na_2O	0.4	0.3	1.6
K_2O	0.3	0.2	0.5
SO_3	13.6	34.0	12.2
MnO	0.1	-	-

+ waf = water- and ash-free

The laboratory reactor (2.3 cm in diameter and 1 m high) and bench-scale reactor (30 cm in diameter and 1 m high) have already been described by Howind et al. (1984) and Schütte et al. (1985).
In the laboratory reactor, sand particles of 0.2 to 0.3 mm diameter formed a fluidized bed of 3 cm height. In the bench-scale reactor, sand particles of 0.3 to 0.4 mm in diameter were used. The bed height was 30 cm. In both steady-state-operated fluidized beds the coal was fed from the top to the bed surface. The coal feed in the bench-scale reactor was usually controlled by keeping the oxygen concentration in the flue gas at a constant level. The particles were separated from the flue gas in a hot cyclone and the condensable components (water, tar) in a water-cooled cyclone. The flue gas was sampled for the chemical analysis after the second cyclone.
The analyser system was described by Schütte et al. (1985).

Results

Investigations in the laboratory reactor. In the laboratory reactor, the three brown coal types of 0.25 to 0.5 mm in diameter were treated either in a nitrogen atmosphere at temperatures from 425 to 900 °C at a coal feed rate of 330 mg/min and at a nitrogen flow rate of 110 L/h, or in air at temperatures from 425-950 °C at coal feed rates from 100 to 1200 mg/min and at an air flow rate of 110 N L/h. The composition of the volatile components in the nitrogen gas was analysed by a gas chromatograph (Mittmann, 1984).
In Figures 1a and 1b, the concentrations of CH_4 and C_2H_4 in nitrogen gas are plotted as a function of the temperature in the bed. One can recognize that there is only a small difference between the three different coal types with regard to the gas composition. In the presence of oxygen in the air, the difference between the three different coal types with regard to the gas composition is larger (Figs. 2a and 2b).

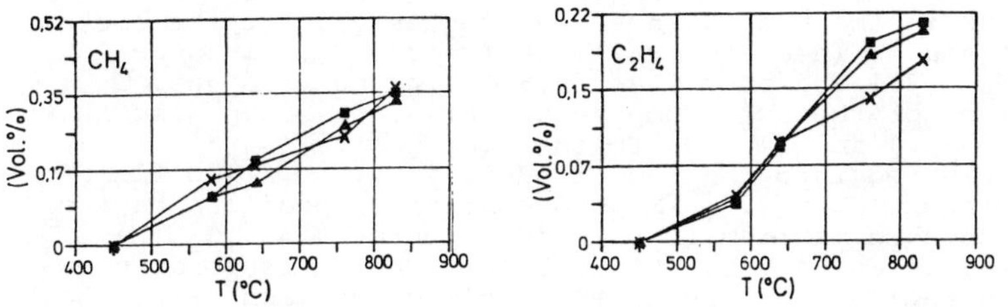

Figs. 1a and 1b: Concentrations as functions of the temperature in the laboratory reactor at nitrogen flow rate 110 N L/h and coal feed rate 330 mg/min with different coal
■ Hesse ▲ Helmstedt x Rhine
a) CH_4, b) C_2H_4

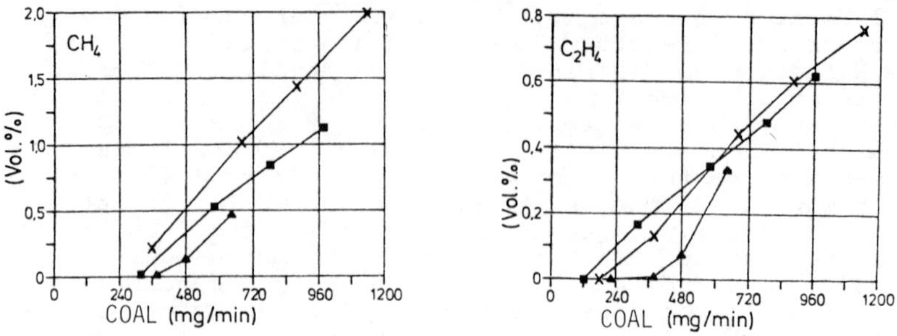

Figs. 2a and 2b: Concentrations as functions of the coal feed in the laboratory reactor at air flow rate 110 N L/h and T = 800 °C with different coal (for symbols, see Figs. 1a and 1b)
a) CH_4, b) C_2H_4

The measured CH_4-concentrations were compared with those calculated by using the model of Rajan and Wen (1980) in the absence of oxygen. The agreement is good. The comparisons of measured and calculated CO_2-, CO- and CH_4-concentrations in the presence of oxygen show a satisfactory agreement (Schütte et al. 1985).

3. Investigations in the bench-scale reactor.

3.1 In the bench-scale reactor only coal of type I was used. The combustion was carried out at 670 and 835 °C at oxygen concentrations in the flue gas of 6.0, 9.6 and 14 % and at three fluidization index values: 2.1, 2.5 and 3.1.
The longitudinal concentration profiles were measured in three different radial positions in the fluidized bed and freeboard.

3.2 Some of the CO_2, SO_2 and NO profiles have already been discussed by Schütte et al. (1985). Some other SO_2- and NO-profiles are taken into consideration here.
In Figures 3, 4 and 5, longitudinal SO_2- and NO-profiles are presented which were measured in three different radial positions in the bed: 3 (near the center), 9 (at the coal feed) and 13 (near the wall) at T = 835 °C, fluidization index U_s = 2.1 and/or 3.1 and 9.6 O_2 in the flue gas. The profiles of the same components (Figs. 5a,b are also shown at U_s = 2.1 and 14 % O_2 in the flue gas).
Coal was fed into the bed at position 9. At a low fluidization index (U_s = 2.1), SO_2-concentration profiles have a considerable maximum in this position at 9.6 % O_2 (Figs.3a,b) as well as at 14 % O_2 (Figs. 5a,b). The corresponding NO-concentration profiles exhibit no such maximum (Figs. 3a,b and 5a,b). At a higher fluidization index (U_s = 3.1), this maximum of So_2-profiles disappears (Figs. 4a,b). The NO-profile in this position has a minimum in the bed (Figs.4a and 4b).
The variation of SO_2- and NO-concentrations in the freeboard above (30 cm) is significant, especially at high

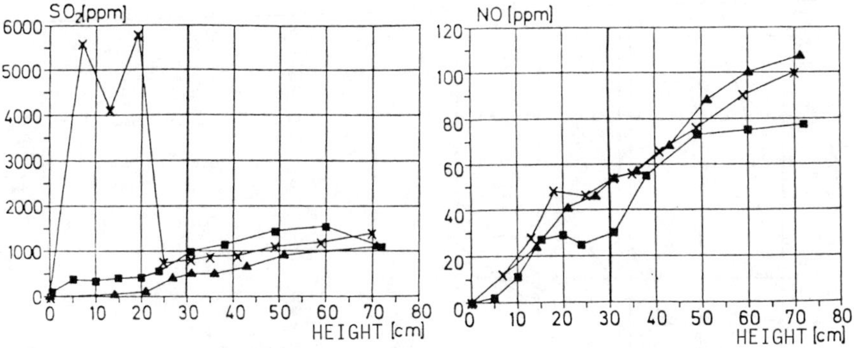

Figs. 3a and 3b: Longitudinal concentration profiles in the bench-scale reactor at T = 835 °C, U_s = 2.1, 9.6 % O_2
■ position 13 ▲ position 3 x position 9
a) SO_2, b) NO

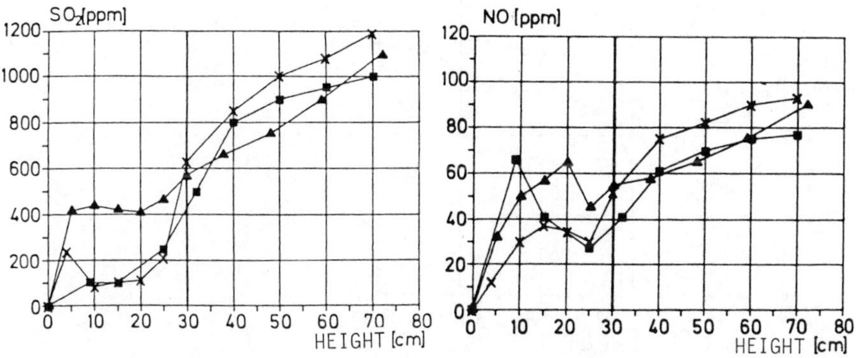

Figs. 4a and 4b: Longitudinal concentration profiles in the bench-scale reactor at T = 835 °C, U_s = 3.1, 9.6 % O_2
(for symbols, see Figs. 3a and 3b)

fluidization index values (Figs. 3 to 5).
It is possible to fit the calculated SO_2- and NO-profiles to the measured ones, but with a different set of parameters. This indicates that they have different formation sources. This assumption is supported by the considerable differences in their profiles.
One can assume that the gas mixing in the bed is already high at U_s = 2.1. With an increasing fluidization index, the gas mixing does not change significantly. However, the solid mixing is low at U_s = 2.1 and is much higher at U_s = 3.1. At low fluidization index values, at which the solid mixing intensity in the bed is low, significant concentration maxima of SO_2 occur at the coal feed position. At a higher fluidization index, this maximum disappears. This indicates

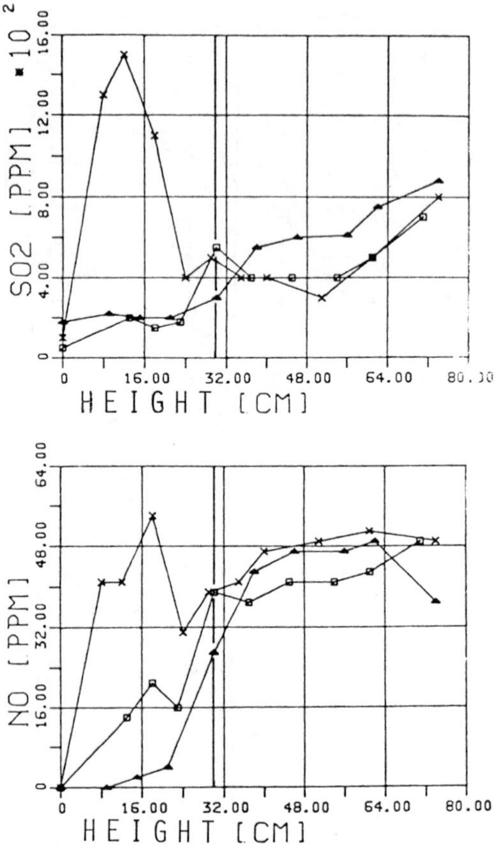

Figs. 5a and 5b:

Longitudinal concentration profiles in the bench-scale reactor at $T = 835\ ^{o}C$, $U_s = 2.1$, 14 % O_2 (for symbols, see Fig.3a,b) a) SO_2, b) NO

that SO_2 is formed mainly in the solid phase. Since the NO-profiles do not exhibit such behavior, one can assume that it forms mainly in the gas phase.
With increasing fluidization index, the concentration variation of SO_2 and NO in the freeboard becomes more significant. This indicates that a considerable fraction of the reaction is shifted into the freeboard.
No comparable investigations with brown coal have been published yet. According to our investigations, coal and anthracite behave fairly different (Schütte 1986). A comparison with NO-Profiles evaluated with peat (Hampartsoumian, Gibbs, 1984) indicates a different behavior. In brown coal systems at 850 ^{o}C, the NO-concentration increases in the bed and freeboard with the height. Only close to the coal feeding position a maximum occurs. At low temperatures (670 ^{o}C), NO-profiles were found which are somewhat similar to the profiles which were measured with peat. More investigations are

necessary to explain these differences.

Condensable organic components in the flue gas. In the hot cyclone, the solid particles were recovered; in the water-cooled second cyclone, water and volatile organic components and aerosol were retained. In order to analyze the organic components, the emulsion (with low pH) was extracted by methylene chloride to recover the acidic components. Then the pH of the remaining aqueous solution was treated with an NaOH-solution, and the basic components were extracted again by methylene chloride.
The acidic and alkaline extracts were analyzed by IR- and NMR-spectroscopy after the separation of their components by thick-layer chromatography.
This analysis indicates that these organic components are mainly multinucleate aromatic compounds. Their concentration in the flue gas lies in the range of several ppm.
It was not possible to identify these components using standard compounds of known identity. Several multinucleate aromatic compounds are known to be carcinogenic.

Acknowledgement

The authors gratefully acknowledge the financial support of the Ministry of Research and Technology of the Federal Republic of Germany, Bonn, and the support by Prof. M. Thoma, Institut für Regelungstechnik der Universität Hannover, and Mr. Poersch, Deutsche Babcock AG, Krefeld.

Literature cited

1. H.R. Howind, J. Deinert, M. Mittmann, H. Kröger, K. Schügerl (1984)
Fluidized bed combustion of lignite, in "Fluidization", D. Kunii, R. Toei, Eds., Engng. Foundation Conferences, May 29-June 3, 1983 in Kashikojima, Japan, 411-418

2. K. Schütte, J. Deinert, M. Mittmann, K.D. Weluda, W. Wittler, K. Schügerl (1985)
Concentration profiles in fluidized bed combustors. "La Fluidisation", Toulouse, 18-19. Septembre 1985

3. M. Mittmann (1984)
Untersuchung zur Abgaszusammensetzung bei Verfeuerung von Braunkohle in Wirbelschichten.
Dissertation, University of Hanover

4. R.R. Rajan, C.Y. Wen (1980)
A comprehensive model for fluidized bed coal combustors.
A.J.Ch.E. Journal 26, 642-655

5. E. Hampartsoumian, B.M. Gibbs (1984)
NO_x-formation and reduction in fluidized bed combustors.
J. of the Inst. of Energy 57, 402-410

6. K. Schütte (1986)
ongoing dissertation

THE COMBUSTION OF HYDROCARBON GAS, PRE-MIXED WITH AIR, IN A FLUIDISED BED.

D.R. van der Vaart and J.F. Davidson,
University of Cambridge, Department of Chemical Engineering,
Pembroke Street, CAMBRIDGE CB2 3RA, England.

ABSTRACT

Pre-mixed hydrocarbon gas and air fluidised an inert bed of particles at 600 to 900°C. Observations are reported as to the nature and position of the combustion, i.e. whether it was in the bed or the freeboard. The results are relevant to the combustion of volatiles released from coal burning in a fluidised bed.

SCOPE

The objective was to elucidate the mechanism of combustion, about which there is controversy. Questions to be answered were: (i) does ignition occur in the bed and by what means; (ii) are the reactions explosive, or steady and limited by heat transfer or termination of free radicals on the surface of the bed material; (iii) how are the reactions divided between the bubble phase (favoured by Sadilov and Baskakov (1)) and particulate phase (favoured by Broughton (2) and Pillai (3))? Special attention was given to the transition, as the bed temperature is raised, from (a) noisy over-bed burning of erupting fuel/air bubbles to (b) quiet in-bed burning at higher temperatures.

Three separate experiments were performed as follows. (i) A pressure transducer measured freeboard explosion intensity as a function of bed temperature, particle diameter, excess fluidising velocity and bed height. (ii) Observations of the bed temperature when (a) heating up with fuel on, and (b) cooling down with fuel off, gave an energy balance from which the fraction of fuel burned on the bed could be deduced. (iii) Chemical analysis of rapidly quenched samples from a probe gave the species present in and above the bed.

CONCLUSIONS AND SIGNIFICANCE

From the pressure transducer results, a critical bed temperature T_{cr} was defined, above which pressure pulses ceased. It was found that: (i) increase in bed height reduces T_{cr}; (ii) increase in $U-U_{mf}$ increases T_{cr}; (iii) increase in U_{mf}, at given $U-U_{mf}$, increases T_{cr}.

The chemical analysis showed that T_{cr} was not merely the bed temperature above which bubble ignition receded into the bed, but, rather, the temperature for in-bed conversion to be complete enough to preclude over-bed ignition.

The heat and mass balances showed counteracting effects so that only

60 to 85% of the heating value of the gas was retained in the bed throughout the temperature range 600 to 900°C. At low bed temperatures there was little combustion in the bed, unreacted fuel gas entered the freeboard and ignited giving high temperatures and consequent heat loss, though some of the freeboard heat was returned to the bed via the splash zone. At high bed temperatures there was more in-bed combustion, so that the fuel leaving the bed was too lean to ignite in the freeboard, giving a loss of unburnt fuel.

A tentative reactor model for propane combustion predicted the reduction of peak over-bed temperature, as bed temperature increased, in accordance with measurements.

INTRODUCTION

Fluidised combustion of coal or oil involves gas-phase oxidative reactions of the volatiles released when the fuel enters the bed. An important question is: do these combustibles burn wholly in the bed or partly above it? The optimal conditions for in-bed combustion of hydrocarbon gases occur when the fuel and air are pre-mixed below the reactor and are influenced by: (i) bed temperature; (ii) residence time in the reactor; (iii) heat transfer within the bed; (iv) surface effects on the particles.

When inert particles are fluidised by air pre-mixed with combustible gas, burning is mainly above the bed if its temperature is below about 650°C. As the bed temperature is increased, burning occurs nearer the surface in an explosive manner as each bubble leaves the bed. The loud explosions become less intense as the bed temperature increases; when it reaches about 850°C burning is wholly in the bed.

Questions about the combustion process are: (i) in which phase (bubble or emulsion) does combustion occur; (ii) why is the required bed temperature for in-bed reaction so high; (iii) what is the combustion efficiency; (iv) what is the chemistry; (v) how does the state of fluidisation affect performance? There is some published work on these questions, but few experiments. Sadilov and Baskakov (1) considered the freeboard explosions due to bubble ignition at bed temperatures below 800°C. Broughton (2) and Pillai (3) treat the combustion as an emulsion phase phenomenon, using catalytic reactor theory, allowing for bubble by-passing.

EXPERIMENTAL

A 76 mm i.d. transparent quartz reactor 800 mm high was used. The distributor was of Inconel 800 sheet, with 48 x 0.5 mm holes in an 8 mm triangular pitch. The bed temperature was measured by a 3 mm dia bare Chromel-Alumel thermocouple 3 cm above the distributor plate. Propane and air were separately metered through rotameters, at 10 psig, followed by a mixing chamber. The quartz tube was spring-loaded from above into a groove around the perforated part of the distributor plate. The bed, but not the plenum chamber, was lagged with 15 cm thick Kaowool. The bed could be operated (a) with an electrical heating coil wrapped round the lower part of the bed, (b) without the coil, permitting visual observations. A stoichiometric propane/air mixture was usually used except where the effect of fuel composition was studied.

Three sets of experiments were performed, as follows: (i) measurement of freeboard pressure pulsations; (ii) total heat balance; (iii) total material balance of chemical species.

To measure the transition from over-bed to in-bed burning, a high speed pressure transducer was placed in the freeboard. The bed was then heated gradually and the pulses due to freeboard ignition were recorded as a function of bed temperature.

The following technique was used to measure the thermal efficiency of the fluidised combustor, i.e. find out how much of the heating value of the fuel was retained in the bed. The relation between bed temperature and time, the heating curve, with propane and air reacting is governed by

$$dT/dt = c(R - Q), \text{ constant } c \text{ depending on bed properties} \quad (1)$$

where T is bed temperature and R represents the heat of reaction retained in the bed; Q represents the heat loss through the walls and by heating up the fluidising gas. With fluidising air alone

$$dT/dt = -cQ \quad (2)$$

assuming that the cooling term Q in Equations (1) and (2) is similar after allowing for the heating of the incoming propane in the first case. Figure 1 shows the form of the curves. At a given bed temperature, the difference in slopes between the heating and cooling curves gives R which is compared with ΔH, the heat of combustion of the fuel. The thermal efficiency $\eta = R/\Delta H$.

For the chemical analysis, dried gas samples from the reactor, via an air-cooled probe, were analysed with a calibrated Phillips Pye Unicam PU 4500 gas chromatograph; to measure light hydrocarbons, CO_2 CO and O_2, a two-column technique (column isolation, Cowper and de Rose (4)) was used.

RESULTS AND DISCUSSION

Pressure Pulsations

To analyse the results, explosions were defined as pulses whose measured pressure amplitude was greater than ½ mm H_2O. The frequency of the explosions, Figure 2, was calculated from the number of pressure peaks over a 6 sec. period. An average bed temperature was determined in the middle of the 6 sec. period. To compare the results in Figure 2(a) and (b) a critical temperature T_{cr} is defined as the bed temperature at which over-bed ignition ceases. Thus in Figure 2(a) $T_{cr} = 825^{\circ}C$; in Figure 2(b) $T_{cr} = 840^{\circ}C$. From Figure 2, and from similar experiments, the effect of operating variables on T_{cr} was as follows, (i) T_{cr} increases with $U-U_{mf}$ as reported by Sadilov and

Baskakov (1); (ii) T_{cr} increases with an increase in U_{mf}, for given $U-U_{mf}$; (iii) T_{cr} diminishes with an increase in bed height.

Heat Released in the Bed

Figure 3 shows energy balance results for different bed masses of sand with $U-U_{mf}$ = 26 cm/s. It is evident that $\eta = R/\Delta H$ is constant for the shallow bed, but increases with bed temperature for the two deeper beds. Values of T_{cr} are shown: no significant difference in η, above and below T_{cr}, is apparent. The energy balance, Figure 3, shows a decrease in η as bed height is increased; the pressure pulsation experiments showed that an increase in bed height facilitates in-bed ignition. This apparent contradiction will be discussed below.

The effects of the operating parameters on combustion efficiency η may be summarised as follows. (i) Shallow beds afford the highest thermal efficiency (about 85%) regardless of bed temperature. (ii) An increase in bed temperature usually increases η. (iii) $U-U_{mf}$ has little effect on η. (iv) T_{cr} and η are not related. (v) Large particles give higher η. (vi) Experiments showed that low and high heating value fuels can be burned with comparable efficiencies.

Chemical Analysis from Sample Probe

Figure 4 shows the effect of bed temperature on fuel conversion. At 750°C, the bed is in the unstable over-bed ignition regime, a significant part of the fuel passing through the bed unreacted and other combustible species, especially CO, enter the freeboard and ignite there. The effect of ignition is to give a marked drop in oxygen concentration just above the bed surface. At 850°C, Figure 4, more of the fuel and oxygen are consumed in the bed: combustion is stable, without freeboard explosions, the temperature being above T_{cr}. Should raising the bed temperature above T_{cr} cause in-bed bubble ignition, the sudden chemical conversion in the freeboard at 750°C should be observed lower down in the bed. Instead, a gradual conversion takes place, the formation of intermediates and CO being more rapid at 850°C than at 750°C. It is evident that T_{cr} does not denote a transition from over-bed to in-bed ignition; T_{cr} is, rather, the bed temperature at which in-bed conversion via steady reactions becomes sufficient to reduce the fuel concentration at the bed surface so that ignition is not possible. The consequence is a dichotomy: (a) at temperatures below T_{cr}, much of the fuel is explosively converted in the freeboard, giving high freeboard temperatures; (b) at temperatures above T_{cr}, most of the fuel is converted in the bed at a steady rate, pre-empting over-bed ignition. No hot spot, due to ignition, exists anywhere in the reactor, so complete conversion is not possible: it follows that when the bed is hotter than T_{cr}, combustion efficiency may be almost as low as when the bed is cooler than T_{cr}. This is discussed later.

The results of the chemical analysis may now be summarised. (i) over-bed ignition was characterised by a reaction zone within 3 cm above the bed surface. (ii) In-bed conversion increased at a gradual rate with increase of bed temperature. (iii) Higher fluidising velocity caused (a) an increase in the amount of CO which escaped after burning in the freeboard and (b) a taller, i.e. more diffuse, over-bed reaction zone. (iv) For fixed $U-U_{mf}$ and bed temperature, larger particles gave less in-bed conversion. (v) Complete combustion (i.e. bed and freeboard) was observed for fuel-lean mixtures; as the mixture was made fuel-rich, more CO and some CH_4 were observed and did not burn in the freeboard.

Heat and Mass Balances Reconciled

The heat and mass balances need to be reconciled to answer the question: Why is the heat released in the bed (R) less than the calorific value of the fuel (Figure 3)? There are two causes of the low efficiency, namely (a) unburnt species, especially CO but also CH_4, see Figure 4, escape into the freeboard; (b) the bubble gas entering the freeboard is hotter than the bed; the analysis using Equations (1) and (2) presupposed equality of bed and freeboard temperatures.

The magnitude of the effects is best illustrated by an example. For a 250 gm bed of 0.25 to 0.425 mm sand at 750°C, R = 410 kcal/mol. The bed height H_{mf} being 5 cm, it is assumed that combustible species detected more than 8 cm above the distributor represent a loss of energy as unburnt species, item (a) above. With $U-U_{mf}$ = 26 cm/s, 0.25% CO was measured at 8 cm height, representing a heat loss of 4.2 kcal/mol propane fed. Hence the heat loss via the sensible heat in the flue gas is 480 - 410 - 4.2 = 68.5 kcal/mol, which when converted to a freeboard temperature rise gives 346°C, i.e. the freeboard temperature should be 750 + 346 = 1096°C, a plausible value. In view of the well known difficulty of measuring hot gas temperatures, this value could not be verified experimentally.

Summary of Effects

Table 1.

Change in System Parameter	T_{cr}	Thermal Loss	Chemical Loss	Thermal Combustion Efficiency
Bed Temperature ↑	-	↓	↑	↑
Bed Height ↑	↓	↑	↑	↓
$U-U_{mf}$ ↑	↑	↓	↑	unchanged
U_{mf} ↑	↑	↑	↓	↑

The table shows the effects of system parameters on the three measurements namely T_{cr}, thermal loss and chemical loss. A full discussion is given by van der Vaart (9). The most important finding is that as the temperature is raised, the conversion within the bed increases but there is less chance of unburnt species igniting in the freeboard: consequently the chemical loss increases with bed temperature but the thermal loss decreases.

THEORY

For combustion of hydrocarbon gas in a fluidised bed of inert particles above 600°C, the assumptions are as follows. The thermal theory of ignition is used: the work of Potter and Berlad (5) and of Friedman and Johnston (6) indicates that the thermal theory is more appropriate than the mass transfer of free radicals for describing high temperature processes like flame quenching. In addition, the physical differences between bubble and emulsion phases mean that the nature of combustion, ignition versus steady reaction, could be different in two phases. For ignition, the residence time of the gas must exceed the characteristic ignition delay time of the fuel/air mixture. At temperatures above 700°C, Brokaw and Jackson (7) reported delay times for propane of about 1 sec., the order of magnitude of the bubble residence time. Another mechanism limiting in-bed ignition is heat transfer which, if rapid, can cause transition from explosive to steady oxidation (Rice et al, (8). The emulsion phase should be quiet because of especially good heat transfer. The bed material is here assumed to be chemically inert, though with some materials, e.g. platinum, there are catalytic effects.

Using these assumptions, tentative theory, treating (a) each bubble as an unsteady-state CSTR and (b) the emulsion phase as a steady-state CSTR has been developed (van der Vaart (9)): this theory predicts the over-bed temperature rise, due to ignition in the freeboard. Results are shown in Figure 5 which gives the bubble temperature profiles in the bed and freeboard, assuming that each bubble retains its identity in the freeboard. The predictions are in qualitative agreement with what is observed, i.e. the maximum temperature in the freeboard. The results also explain the 'superadiabatic temperatures' reported by Sadilov and Baskakov (1). Part of the mixed fuel and air, initially at room temperature, is heated to bed temperature prior to ignition in the freeboard: thus its post-ignition temperature may exceed the conventional adiabatic flame temperature because of the pre-heating in the bed.

LITERATURE CITED

1. Sadilov, P.V. and A.P. Baskakov, Int. Chem. Eng., 13, 448 (1973).

2. Broughton, J., Applied Energy, 1, 61 (1975).

3. Pillai, K.K., J. of Inst. of Fuel, Dec., 200 (1976).

4. Cowper, C.J. and A.J. de Rose, The Analysis of Gases by Chromatography, Pergamon, Oxford (1983).

5. Potter, Fr., A.E. and A.L. Berlad, J. of Physical Chem., 60, 97 (1956).

6. Friedman, R. and W.C. Johnston, J. of Applied Physics, 2, 791 (1950).

7. Brokaw, R.S. and J.L. Jackson, 5th Symp. (Int.) of Combustion, The Combustion Institute, 563 (1955).

8. Rice, O.K., A.O. Allen and H.C. Campbell, J. Amer. Chem. Soc., 57, 2212 (1953).

9. van der Vaart, D.R., "The Combustion of Gases in a Fluidised Bed", Ph.D. thesis, Cambridge University, Cambridge (1985).

NOMENCLATURE

ΔH	heat of combustion
Q	heat loss
R	heat of reaction retained in the bed
t	time
T	bed temperature
T_{cr}	temperature at which freeboard explosions cease
U	superficial fluidising velocity
U_{mf}	minimum fluidising velocity
η	thermal efficiency

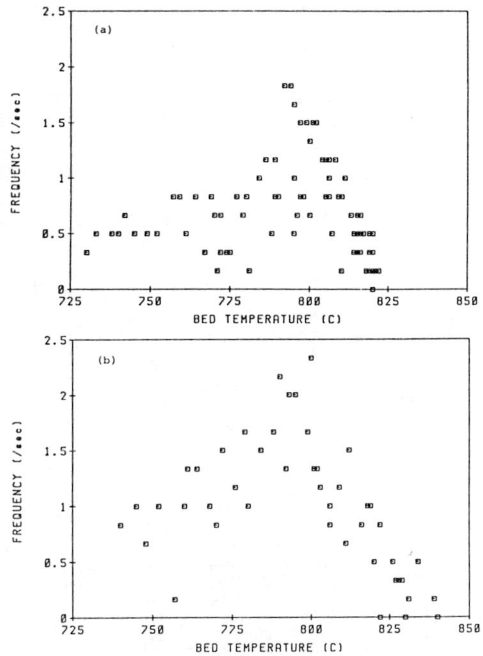

FIGURE 2. FREQUENCY OF BUBBLE EXPLOSIONS MEASURED BY PRESSURE TRANSDUCER IN THE FREEBOARD. 500 gm OF 0.25 - 0.425 mm SAND (a) $U-U_{mf}$ = 13 cm/s, (b) $U-U_{mf}$ = 26 cm/s.

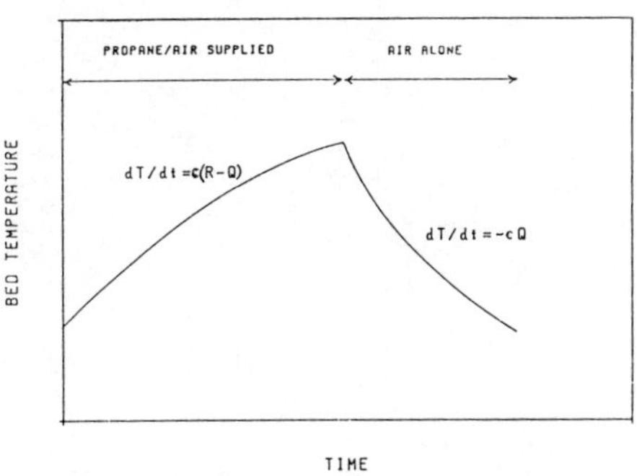

FIGURE 1. HEAT TRANSFER EXPERIMENT.

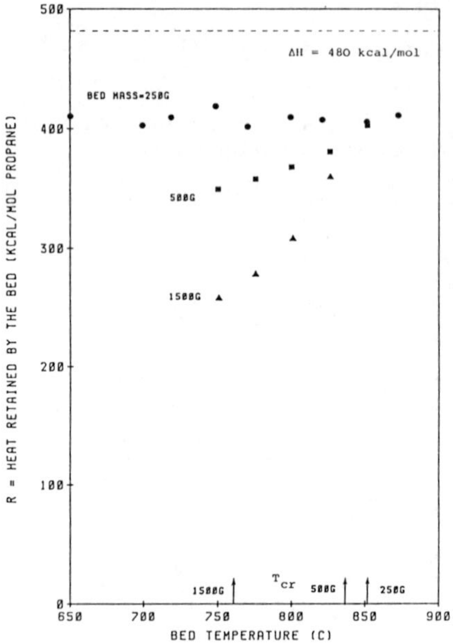

FIGURE 3. EFFECT OF BED TEMPERATURE FOR VARIOUS BED HEIGHTS FOR 0.3 - 0.6 mm SAND AND $U-U_{mf}$ = 26 cm/s.

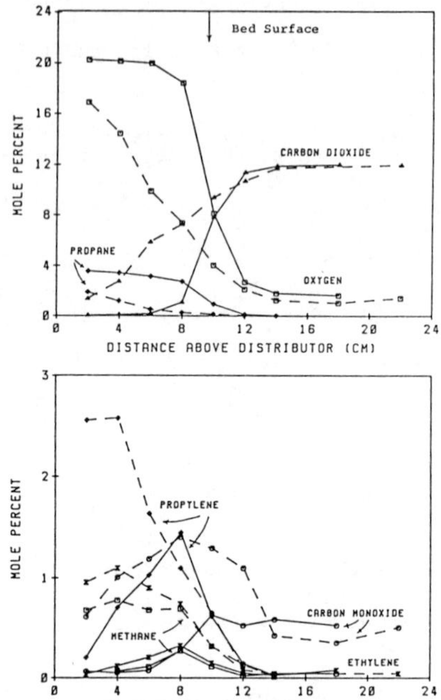

FIGURE 4. CHEMICAL SPECIES FROM IN-BED PROBE MEASURED BY CHROMATOGRAPH,

500 gm OF 0.25 - 0.425 mm SAND.

$U-U_{mf}$ = 26 cm/s BED TEMPERATURES, —— 750°C, --- 850°C

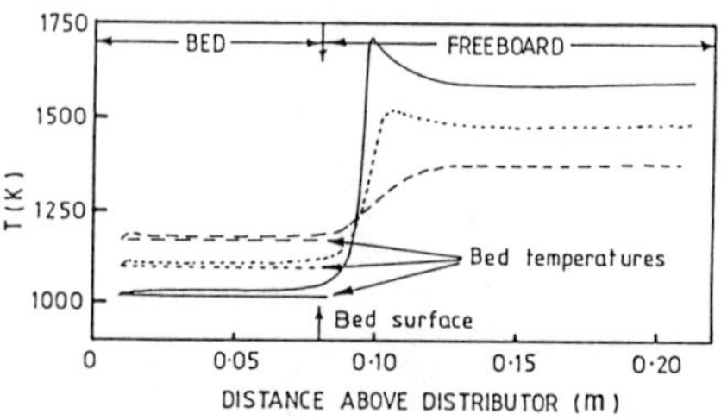

Figure 5. Theoretical predictions of bubble temperature profile.

Bed temperatures (K) : ———— 1025

: ············ 1100

: - - - - - - - 1175

U_{mf} = 5 cm/s., $U - U_{mf}$ = 30 cm/s.

AN INVESTIGATION OF CHANGES IN FLUIDISATION AND HEAT TRANSFER CHARACTERISTICS DURING COMBUSTION OF HIGH ASH COALS IN SHALLOW SAND FLUIDISED BEDS

BY M.A. VICKERS and R.K. CHAKRABORTY
NATIONAL COAL BOARD
COAL RESEARCH ESTABLISHMENT
STOKE ORCHARD, NR. CHELTENHAM
GLOS. GL52 4RZ
ENGLAND

ABSTRACT

Investigations, into the effect of burning high ash coals in shallow sand fluidised beds, have shown that accumulation of oversize ash particles causes segregation and defluidisation. Bed regrading systems can remove such particles but not the smaller ash particles which result in changed in-bed heat transfer characteristics.

SCOPE

Investigations have been carried out on a purpose built test rig to obtain detailed heat transfer characteristics for two representative bed samples obtained from industrial boilers which had given reduced heat outputs arising from the accumulation of ash particles in the bed (originally graded sand). The coals burnt in these boilers, were of ash contents 10% and 15% respectively.

A second series of tests was carried out with an air classifier bed regrading unit fitted to the test rig. This unit was operated at different conditions to assess its effectiveness in reducing the ash accumulation rates in the bed. The coal burnt during the tests had an ash content of 18%.

The test data is relevant to industrial boiler operation. In particular, knowledge is required concerning (i) the operating characteristics required to prevent bed segregation and defluidisation, (ii) the associated changes in bed heat transfer characteristics. No previous data was available.

CONCLUSIONS AND SIGNIFICANCE

The two beds sampled from industrial boilers had reduced heat transfer coefficients compared to those formed from the original sand. For the first bed the maximum heat transfer coefficient was 273 $Wm^{-2}K$ compared to the initial 339 $Wm^{-2}K$ for the virgin 0.7mm sand. For the second bed (originally 0.8mm sand) the corresponding heat transfer coefficients were 268 and >300 $Wm^{-2}K$. Also the bed expansion for the ash beds was less than that for the sand.

Analysis of both beds showed the presence of a high proportion (>30%) of ash particles in the size range 1mm to 4mm. The increased mean particle size of the beds accounted for some of the reduction in heat transfer coefficient. Changes in particle properties for ash beds were also expected to contribute towards the reduced heat transfer coefficients.

The accumulation of ash in the bed also resulted in a reduced bulk density and hence a reduced pressure gradient across the bed.

Combustion of the high ash content coal (18%) in the bed (originally 0.8mm sand) showed that 23% of the input ash accumulated in the bed without a bed cleaning device in operation. A large proportion (50%) of the accumulated ash had a greater size than 4mm and resulted in segregation and defluidisation of the bed which was detected by large variations in the vertical bed temperature gradient.

By incorporating the air classifier bed regrading unit, most of the particles larger than 4mm could be removed from the bed. Particles in the size range 1.2mm to 4mm were not readily removed from the bed and it was predicted that an array of units (one 51mm air classifier per $0.5m^2$ of bed area, (See Fig. 1)) would not be capable of separating ash in this size range from the original sand.

The changes of both heat transfer and bed pressure drop characteristics arising from accumulation of smaller ash particles must be considered when designing tube arrangements and bed level control systems for industrial boilers.

INTRODUCTION

The National Coal Board is currently involved in the commercial application of fluidised bed technology to the design of coal fired boilers (1). There are 75 units (giving a total of 600 MW heat output capacity) either operating or being installed at commercial sites in the United Kingdom.

Initially most of the boilers were designed to operate by burning washed graded (size up to 25mm) coals of low ash content (<10%) in shallow beds of graded silica sand. Most of the ash formed during combustion of washed coals is in the form of fine powder which is elutriated from the bed. A small quantity of oversize ash particles accumulate in the bed but these are readily removed by periodic operation of bed regrading systems (2). Addition of sand to the bed is required to make up for losses by elutriation. Therefore when burning washed coals the bed remains substantially in the form of sand particles.

During the past two years there has been an increasing requirement to burn coals of higher ash content (>10%). In such cases, even with bed regrading systems in operation, an excess of ash particles has accumulated in the bed such that material has to be drained from it to maintain the required level. Addition of sand is not required to the bed and eventually it consists almost entirely of ash particles. The changes in the composition of the bed have resulted in changes in heat output and the reasons for this have been investigated in a test rig at the Coal Research Establishment. Tests have also been carried out to assess the change in bed characteristics when operating an air classifier bed regrading system at different conditions. The test programme and results are described in this paper.

THE TEST RIG

The detailed design features of the rig have been described elsewhere(3). A brief description is given below.
The combustion chamber is refractory-lined and of internal dimensions 1m long by 0.5m wide and 2m in height to give a bed area of $0.5m^2$. For the current tests either six or nine water-cooled tubes were positioned in the expanded and splash bed regions. The tubes, 1m long with an outside diameter of 51mm, are arranged in horizontal rows comprising three tubes per row. (Figure 1.). Data on heat transfer can be determined separately for each row of tubes. Air from a forced draught fan is supplied to the bed through a standpipe air-distributor. Facilities are provided for firing the bed by coal and pre-mixed gas. An internal air classifier bed regrading unit (2) was installed for the second series of tests.

TESTS TO ASSESS THE EFFECT OF ASH ACCUMULATION ON BED CHARACTERISTICS

Experimental

In order to illustrate the phenomena and investigate the reasons for changed bed heat transfer characteristics with ash accumulation in the bed, the test results for two different coals (designated 1 and 2) are presented. The coals were burnt on two different 5 MW industrial boilers.

The original bed material in the first boiler was 16/30 grade sand (mean size 0.7mm). Changes in heat output were noted during a period of five weeks over which 280 tonnes of Coal 1 were burnt in the bed. An air classifier bed regrading system was in operation during part of this period but it was not possible to obtain separation between the sand and smaller ash particles. The materials appeared to have similar aerodynamic properties but different heat transfer properties.

The original bed material in the second boiler was 14/25 grade sand (mean size 0.8mm). Changes in heat output were noted during a period of 100 hours over which 60 tonnes of Coal 2 were burnt in the bed. Bed regrading equipment was not in operation, however material was periodically removed through drain ports to maintain the bed level reasonably constant.

Sufficient quantities of bed material were extracted from the boilers to enable an assessment of the heat transfer and fluidisation characteristics to be made in the test rig. For comparison tests were also carried out on the original graded sand beds. Two rows of tubes (A and B) were used when testing the bed from the first boiler; a third row (C) was added for testing of the bed from the second boiler. The fuel burnt was pre-mixed natural gas and the static bed heights were between 160 and 168mm. The bed temperature was nominally 880°C and the range of fluidising velocities between 0.6 and 2.6 $m.s^{-1}$. Each steady-state test was of about one hour duration.

Results and Discussion

The ash bed formed from Coal 1 had a mean size of 0.95mm, with 48.7% by weight >1mm and 4.2% >4mm. The original 16/30 grade sand had a mean size of 0.7mm with 4.1% >1mm. The bulk density of the ash and sand beds were 1247 and 1475 $kg.m^{-3}$ respectively.

The ash bed formed from Coal 2 had a mean size of 1.07mm with 56.4% by weight >1mm and 20.9% >3.35mm. The original 14/25 grade sand had a mean size of 0.8mm with 7% >1mm. Bulk densities of the ash and sand beds were 977 and 1478 $kg.m^{-3}$ respectively.

The ash beds formed from Coals 1 and 2 had reduced heat transfer coefficients compared to those formed from the original sand. For the first bed the maximum heat transfer coefficient was 273 $Wm^{-2}K$ (to tube row B at a fluidising velocity of 2.2 $m.s^{-1}$) compared to the initial 339 $Wm^{-2}K$ for the original 16/30 grade sand at the same operating conditions (Figure 2). For the second bed the maximum heat transfer coefficient was 268 $Wm^{-2}K$ (to tube row C at fluidising velocities above 2 $m.s^{-1}$) compared to the initial value of >300 $Wm^{-2}K$ for 14/25 grade sand (Figure 3). The reduced heat transfer coefficients for the ash beds are partly accounted for by the increase in mean particle size. Changes in particle properties such as density, shape, heat capacity and thermal conductivity are also expected to contribute towards reduced heat transfer coefficients.

With reducing fluidising velocity the heat transfer coefficients (particularly for tube rows A and B) gradually decreased as the tubes became uncovered from the expanded bed (Figures 2 and 3). The heat transfer data has been used in a mathematical model (4) to derive correlations of bed expansion with fluidising velocity. The values of bed expansion obtained for the ash beds were less than those for the original sand.

During the tests on the ash bed formed from Coal 2 a vertical temperature gradient of 50°C occurred in the bed (higher at the base). At the end of the test some small clinkers were obtained from the base of the bed. It was evident that some bed segregation had occurred because of the presence of a high proportion of particles (21%) >3.35mm which were not properly fluidised.

TESTS ON A HIGH ASH COAL WITH BED REGRADING UNIT IN OPERATION

Experimental

Tests were carried out with the coal of 18% ash content (designated Coal 3) to assess the performance of the fluidised bed with the air classifier bed regrading system operating under different conditions. Four series of tests were conducted. The bed material at the start of each test series was 14/25 grade sand (mean size 0.8mm) with a static height of approximately 160mm. Series 1, 2 and 3 were carried out at a high fire condition (fluidising velocity 2.5 $m.s^{-1}$ and bed temperature of 900 - 925°C) to assess the respective effects of non-operation, intermittent operation and continuous operation at a low classifying velocity. Series 4 was carried out at a low fire condition (fluidising velocity 1 m/s at a bed temperature of 850°C) with the classifier operating continuously at different velocities. Operating parameters for the classifier are shown in Figures 4 to 6.

Results and Discussions

Comparison of the performance of the classifier at different operating conditions, can be conveniently made in terms of a parameter given by the average accumulation rate of oversize ash in the bed expressed as a percentage of the ash feed rate in the coal. Oversize ash was defined as being greater than 1.2mm, the top size of the sand. The accumulation of oversize ash in the bed was 23.4% of the ash feed without the classifier operating (series 1). This accumulation reduced to 9.4% of the ash feed (series 2) by removing oversize ash through the classifier operating intermittently at 8.2 $m.s^{-1}$; there was a small residue of sand loss with the ash. A further reduction in ash accumulation to 5.7% of the ash feed was achieved (series 3) by operating the classifier continuously at a lower velocity of 5.5 $m.s.^{-1}$, however the sand loss rate with the ash was greater. The changes in static bed height and mean particle size in the test series 1 to 4 are shown in Table 1.

A relevant practical issue is the variation in oversize ash concentration in the bed as a function of time for different size fractions. These are shown in Figures 4, 5 and 6 for series 2, 3 and 4. Values of oversize ash concentration were evaluated at intervals of 30 minutes from a mass balance on the bed. The rate of ash entering the bed and the elutriation rates of sand from the bed were assumed to be constant during each series. From the measured ash and sand removal rates the quantity of ash in the bed could be calculated. Figures 4, 5 and 6 show that:

(i) Most of the +4mm particles were removed from the bed by the classifier. The +4mm material remaining in the bed attained a nominally steady concentration level depending upon the mode of operation of the classifier. The concentration levels were approximately 12.5% for series 2 (intermittent classifier operation at 8.2 $m.s^{-1}$) and between 2.5% and 3.5% for series 3 and 4 (continuous classifier operation at velocities between 5.5 and 7.8 $m.s^{-1}$).

(ii) Material between 1.2 and 4mm size progressively accumulated within the bed. It was more difficult to separate this material from the sand particles. It is expected that in most cases the bulk of the bed material will eventually be in the size range 1.2 to 4mm. An exception must be made for operation at low coal feed rate and a low classifying velocity (series 4, Figure 6). In this case the material between 1.2 and 4mm reached a steady concentration in the bed. In order to obtain the same equilibrium concentration in the bed for this size fraction at higher coal feed rates it would be necessary to supply additional classifiers to serve the 0.5 m^2 bed area. However, cost considerations will limit the number of classifiers that can be supplied per unit bed area for industrial boilers.

At the end of series 1 segregation of bed material and clinker formation occurred. Immediately prior to this the heat transfer coefficients (bed to tube) had reduced by only 5% from the values obtained with the sand bed. However a large temperature gradient (>200°C) was noted vertically through the bed (temperature lower at base of bed). In series 2 to 4 segregation of bed material did not occur and the temperatures indicated in a vertical array of thermocouples were similar. In these latter tests there were few oversize particles greater than 4mm in size but in series 1 nearly 33% of the bed was above 4mm. This indicates that the removal of +4mm ash particles by the classifier is fairly critical in preventing segregation/defluidisation occurring.

ACKNOWLEDGEMENTS

The authors wish to thank the National Coal Board for permission to publish this paper. Any views expressed are those of the authors and not necessarily those of the NCB.

LITERATURE CITED

1. Highley J. "The Development of Fluidised Bed Combustion for Industry". Fluidised Bed Combustion and Applied Technology Symposium, Beizing (Peking), 22-25 August 1983.

2. Cooke, M.J., Rogers, E.A., Dando, R.L., and Gauld, D.W., "Ash Management in the Fluidised Bed Combustion of Coal". 3rd International Conference on Fluidised Bed Combustion, Institute of Energy, London, 16/17 October 1984.

3. Chakraborty, R.K. and Vickers, M.A., "Heat Transfer Characteristics of Cooling Tubes in the Expanded and Splash Regions of a Fluidised Combustor", Ibid.

4. Moodie, J. and Vickers, M.A., "Some Heat and Mass Transfer Considerations in the Design of Fluidised Bed Boilers", Eighth International Symposium on Chemical Reaction Engineering, Edinburgh, September 1984.

TABLE 1

Data on static bed height and bed particle mean size for four series of tests burning Coal 3.

Test series number	1	2	3	4
Static bed height				
(i) at the start of the test	160 mm	160 mm	160 mm	160 mm
(ii) at the end of the test	258 mm	161 mm	133 mm	183 mm
Bed particle mean size				
(i) at the start of the test	0.8 mm	0.8 mm	0.8 mm	0.8 mm
(ii) at the end of the test	1.54 mm	1.05 mm	0.83mm	0.88mm

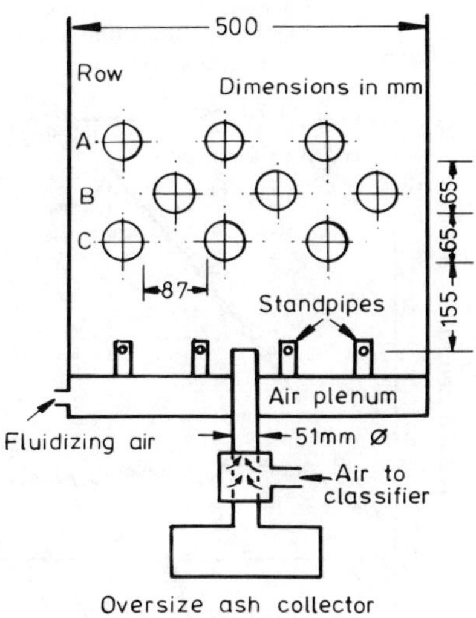

Figure 1: Tube arrangement and air classifier.

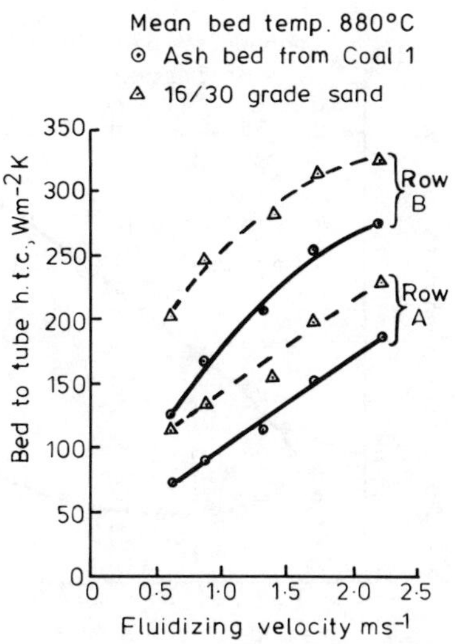

Figure 2: Heat transfer coefficients for ash (Coal 1) and sand.

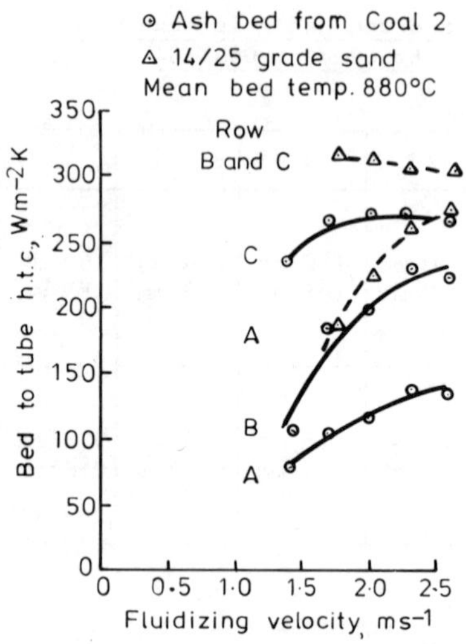

Figure 3: Heat transfer coefficients for ash (Coal 2) and sand

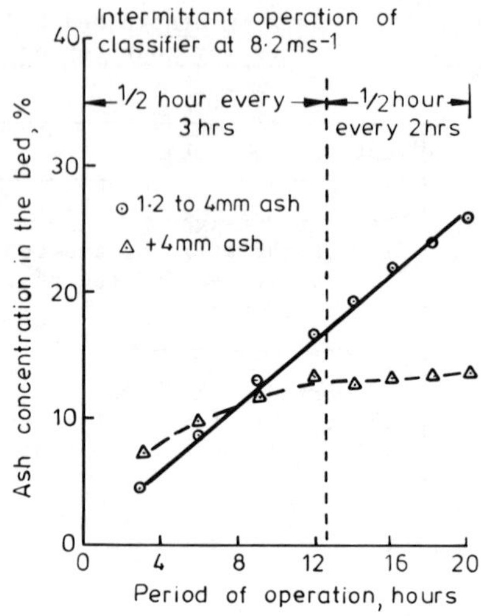

Figure 4: Oversize ash accumulation in test series 2

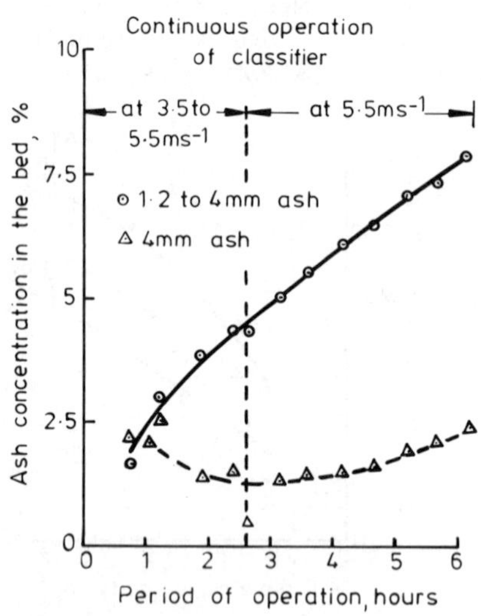

Figure 5: Oversize ash accumulation in test series 3

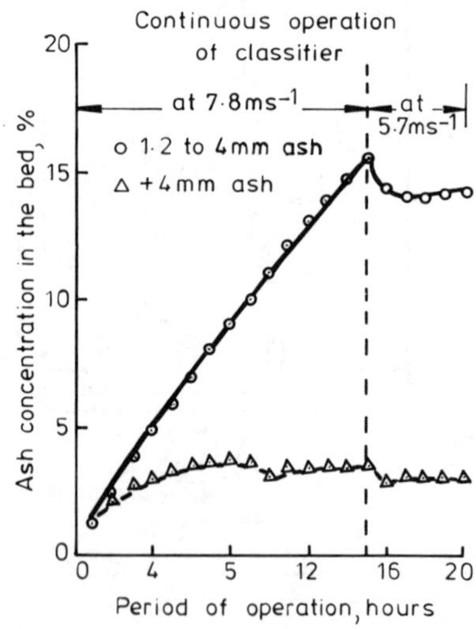

Figure 6: Oversize ash accumulation in test series 4

COMBINED ENTRAINED-BED/MOVING-BED COMBUSTOR FOR LOW-GRADE SOLID FUELS

Masanobu Hasatani, Ichiro Naruse,
Hitoki Matsuda, Hiroaki Sugita, and Norio Arai

Department of Chemical Engineering
Nagoya University
Furocho, Chikusa-ku, Nagoya, 464, JAPAN

ABSTRACT

A new type of two-stage solid fuel combustor with a combined entrained-bed and a moving-bed was proposed. The flow behaviors of both the solid particles and the air through the combustor were examined by using two- and three-dimensional cold-models.

The chars produced by the pyrolysis of raw solid fuels were found to be completely burned in the moving-bed with the formation of NO_x being depressed.

SCOPE

In the combustion of such low-grade solid fuels as brown coal, lignite, and surplus active sludge, which usually contain considerable amounts of S and N compounds and ash component, special attention should be paid for the development of the effective combustor with a high efficiency as well as with an air pollution control.

A FBC (Fluidized-Bed Combustor) so far has been considered as one of the potent techniques for effective coal combustion. In the application of fluidized-bed coal combustion, two different combustion modes are possible; one with complete and the other with partial combustion of coal. In the second combustion mode, called two-stage combustion, a stoichiometric deficiency of air is added to the fluidized-bed and the resulting char, H_2, CO, and hydrocarbons are burned by providing additional oxygen into the second-stage of combustion. In the first-stage of combustion, in general, the nitric and the sulfur compounds are released from the coal in the form of such volatile-N as NO, NH_3, HCN, and of H_2S and SO_2, respectively. The mixture of H_2S and SO_2 may be removed more effectively with CaO pellets in a reducing atmosphere of the first stage (1 and 2). It was also reported that the NO is removed with CaO (3) and chars (4 and 5). The two-stage fluidized-bed combustion is considered to be potential for increasing the efficiency of CaO utilization and also in terms of NO_x reduction.

In view of the above, in the present study, as one of promising candidates of efficient combustor for low-grade solid fuels, a new type of two-stage coal combustor with a combined entrained-bed and a moving-bed is proposed (Figure 1). As a first step in the development of such a combustor, the fundamental behaviors of both the solid particles and the air flowing through the combustor were observed under various solid/air ratios by using two- and three-dimensional cold-models. The chars produced by the pyrolysis of raw surplus active sludge, which is considered as a typical example of low-grade solid fuels, were burned at the second-stage of the moving-bed combustion zone in a laboratory-scale combustor. In the combustion experiments in the moving-bed, the conditions for achieving complete combustion and the ultimate level of NO_x emission were examined.

CONCLUSION AND SIGNIFICANCE

The results obtained in this study were summarized as:

(1) It was observed from the two-dimensional cold-model experiment that there exists a minimum gas velocity to form the apparent steady state of a fast- or an entrained-bed for a continuous feed of solid particles into the inner part (the partial combustion and gasification zone) of the double-piped combustor. The velocities of solid particles, in both the inner and the annular parts, were found to be increasing in proportion to the mean gas velocity almost independently of the particle feed rate. The particle hold-up and the particle residence time were increased with the particles feed rate, and were inversely proportional to the gas mean velocity. With the increases of the bed height and particle feed rate, the pressure drop in the annular part of the moving-bed increases.

(2) From the active sludge combustion experiments, it was observed that in the moving-bed the combustion of the chars produced by the first stage were completely accomplished. By adding NH_3 as a source of volatile N to the char combustion zone, the NO_x emission was strongly reduced when the air-fuel ratio was close to one. With increasing air-fuel ratio, however, the NO_x emission was increased. Therefore, it was considered that there is an optimum in the air-fuel ratio and the NH_3 concentration for low NO_x emission.

These cold-model and the combustion experiment result may present the useful guideline for the development of an effective coal combustor with a combined entrained-bed and a moving-bed to a practical use.

EXPERIMENTAL

CONCEPTUAL DESIGN OF THE COMBINED ENTRAINED-BED/MOVING-BED COMBUSTOR

A conceptual drawing of the solid fuel combustor with a combined entrained-bed and a moving-bed is schematically shown in Figure 1. The combustor is composed of a double-piped combustion chamber so as to accomplish the two-stage combustion effectively. The inner pipe is the coal gasification and partial combustion zone, and the annular part is the char combustion zone. The coal mixed with CaO are fed into the combustor from the lower part of the inner tube accompanied by the primary air. The secondary air is injected radially near the top of the inner pipe where the chars are spouted out. The char combustion is accomplished when falling down the annular part of the moving-bed, and, at the same time, NO_x and SO_x released during the coal combustion are removed by char as well as CaO particles. The ash particulates separated from the gas are captured in the ash box equipped with baffles. In this combustor, the combustion heat is transferred to steam pipes at the top of the apparatus. In addition, the sensible heat of the burned coal particles in the moving-bed is transferred to the inner pipe so that the coal gasification and partial combustion might be enhanced there.

EXPERIMENTAL APPARATUS AND PROCEDURES

COLD MODEL The two-dimensional cold-model apparatus was made of glass plates except the side wall made of acrylic resin plates, the inner part, 950 mm in height and 40 mm in width. A 400 mesh stainless-steel wire net was employed as a distributor, and was set below the solids feed inlet port. Five pressure taps were equipped at each position 1 through 5 as shown in Figure 1. Small glass beads of 37 to 74 μm, 177 to 250 μm, and 420 to 590 μm were used and fed into the inner part with a screw feeder at a constant rate. Primary air was introduced into the inner part through the distributor so that a fast fluidized-bed or an entrained-bed might be formed in the inner part. In the two-dimensional model, however, the secondary air was not introduced. In order to form a moving-bed in the annular part, relatively larger glass beads were packed in the lower section. The velocities of the particles both in the inner and in the annular part were determined by a high-speed photograph. A part of the experiments was carried out by employing the three-dimensional apparatus.

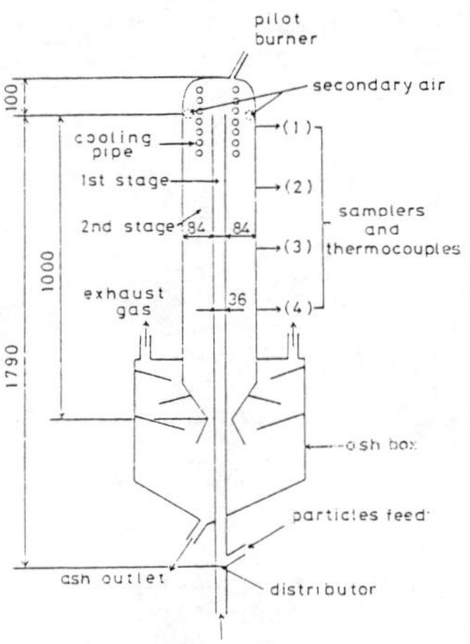

Figure 1 Experimental Apparatus of the Combustor

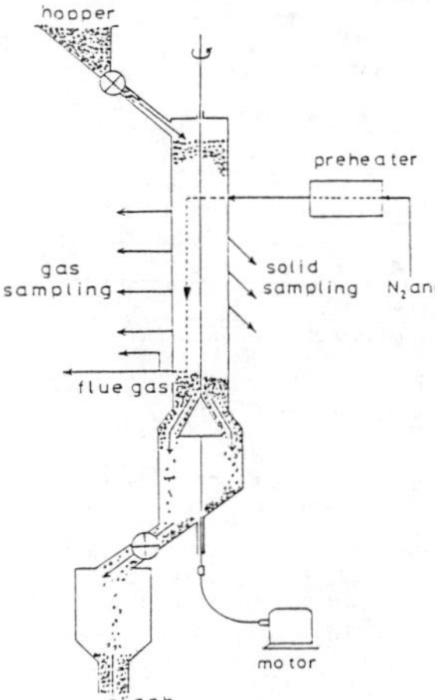

Figure 2 Schematic Diagram of the Moving-Bed Combustor

HOT-MODEL In order to obtain some of the information concerning the actual combustion in the annular part of a moving-bed in the hot-model, the chars produced by the pyrolysis of raw surplus active sludge were burned in the moving-bed combustor. Figure 2 shows a schematic diagram of the moving-bed combustor employed in the experiment. The reaction tube was a stainless-steel tube, 610 mm in height and 83 mm in diameter. A rotating shaft attached to a cone in the center of the tube was installed so that the flow of solid particles in the moving-bed might be as near plug flow as possible. The reaction tube was heated from the surrounding by electric heaters to control moving-bed temperature. A mixture of N_2 and O_2 was used as oxidizing gas. In order to examine the NO possible reduction, NH_3 was mixed with the oxidizing gas at the moving-bed inlet. The flow rate of the oxidizing gas was kept constant at 1.0 $Nm^3 \cdot h^{-1}$ and the air-fuel ratio m was varied from 1.06 to 2.5 with the O_2 concentration. The chars flow rate and the inlet gas temperature were 0.35 to 0.38 $kg \cdot s^{-1}$, and 673.2 K, respectively.

RESULTS AND DISCUSSION

COLD-MODEL TEST RESULTS

At first, the dynamic behavior of the gas and solid particles mixed flow in the inner part of the two-dimensional cold-model was examined. The typical examples of the experimental results are shown in Figure 3. The data obtained with 550 mm height of the inner part and also with the three-dimensional cold-model are included. The lines in Figure 3

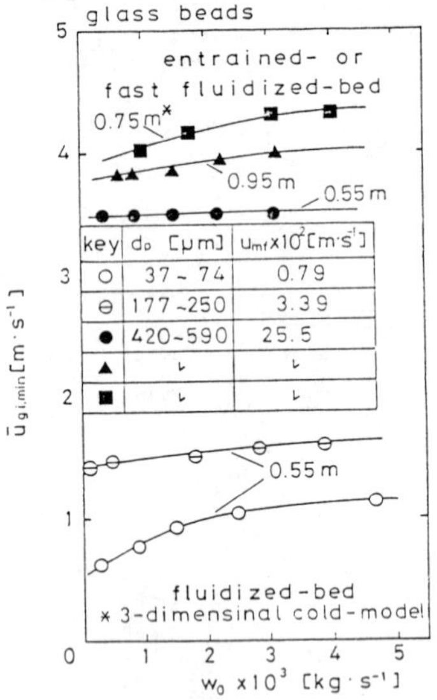

Figure 3 Relation between the Minimum Gas Velocity and the Particle Feed Rate

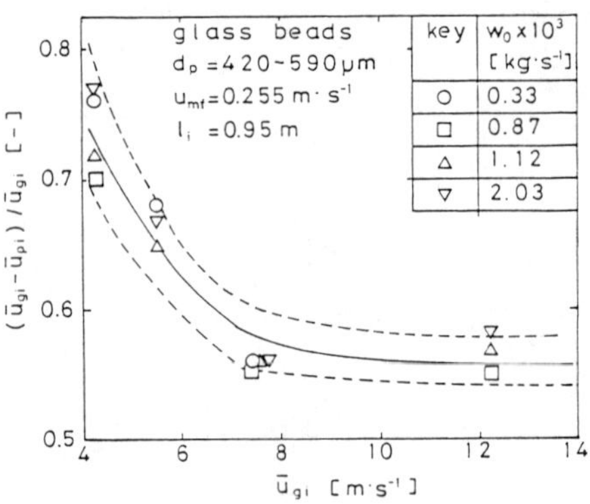

Figure 4 Slip Ratio in the Inner Pipe

indicate phase boundary across which an entrained-bed or a fast fluidized-bed (above the line) and a fluidized-bed (below the line) are formed in the inner part. These phase boundaries were determined as the points where the pressure drop ΔP_{1-2} began to increase suddenly when the gas flow rate was throttled. As it can be seen from Figure 3, there is a minimum gas velocity $\bar{u}_{gi,min}$ to form a stable entrained-bed and the $\bar{u}_{gi,min}$ tends to increase gradually with the increase in the feed rate of glass particles. The bigger the particle size and the higher the inner part, the greater is the minimum gas velocity $\bar{u}_{gi,min}$ which is required for a stable entrained-bed.

After the particles velocities u_{pi} in the inner pipe had been measured by high-speed photography, the slip ratio $(\bar{u}_{gi}-\bar{u}_{pi})/\bar{u}_{gi}$ was plotted against the mean gas velocity $\bar{u}_{gi}$ (Figure 4). As it can be seen from Figure 4, as the mean gas velocity $\bar{u}_{gi}$ increases, the slip ratio decreases, and this trend is stronger when the total height of the inner pipe l_i becomes greater.

The hold-up of the particles $(1-\varepsilon_i)$ and $(1-\varepsilon_a)$ in the inner part and in the annular part are shown in Figures 5 and 6, respectively. The void fraction ε_i and ε_a were determined by the following correlations.

$$1-\varepsilon_i = \frac{w_0}{\rho_p \bar{u}_{pi} W_i t} \quad (1)$$

$$1-\varepsilon_a = \frac{w_0}{2\rho_p \bar{u}_{pa} W_a t} \quad (2)$$

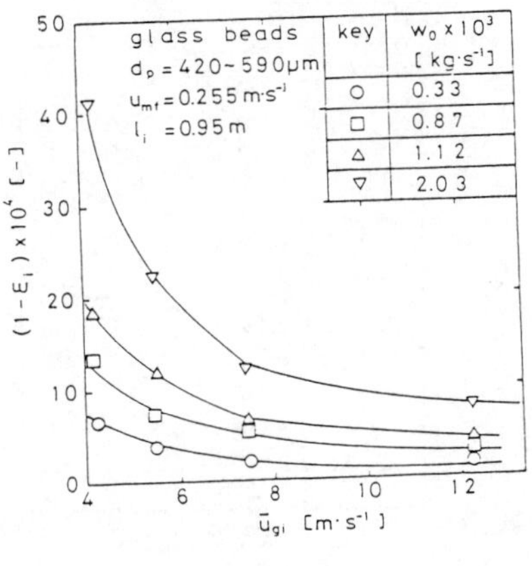

Figure 5 Particle Hold-up in the Inner Pipe

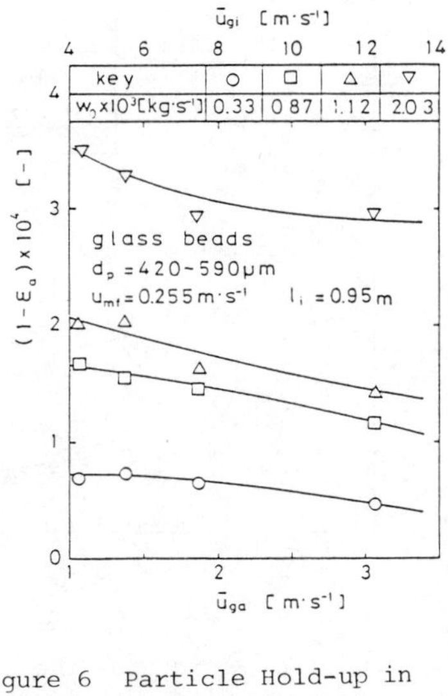

Figure 6 Particle Hold-up in the Annular Part

Where ρ_p denotes the particle density, W_i the width of the inner part, and t the depth. W_a is the width of the annular part. As it can be seen from the above figures, the hold-up is inversely proportional to the mean gas velocity $\bar{u}_{gi}$ as well as to particles feed rate w_0.

The mean residence time in the inner part θ_i and the total residence time in the whole apparatus θ_t are plotted against $\bar{u}_{gi}$ and $\bar{u}_{ga}$ in Figure 7. One can see from Figure 7 that both θ_i and θ_t decrease with increase in $\bar{u}_{gi}$ and $\bar{u}_{ga}$. Despite the different particle feed rate w_0, the results are located within the region surrounded by the broken lines and the chained lines. These experimental data may provide a valuable information in designing the hot-model apparatus and an actual solid fuel combustor.
of the packed bed l_b and the particle feed rate w_0. As it is shown in Fi gure 8, the pressure difference between the position 4 a nd 5 (see Fig As it is shown in Figure 8, the pressure d ifference between the position 4 and 5 (see Figure 1) increases with inc rease in both the height of the packed bed l_b and the particle feed rat e w_0.

HOT-MODEL TEST RESULTS

For the steady state operation, the test results of the axial distribution of the bed temperature t_B, gas composition, and the char conversion in the parallel-flow moving-bed are shown in Figure 9. As it

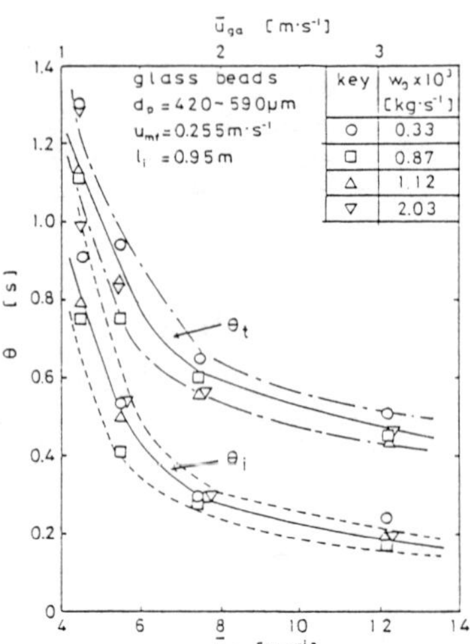

Figure 7 Residence Time of Solid Particles

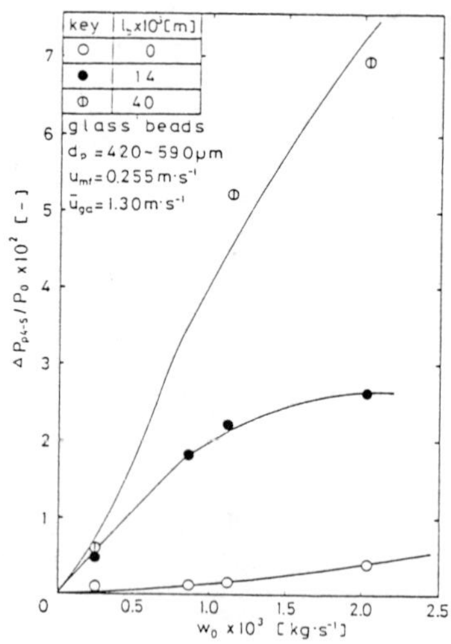

Figure 8 Pressure Drop at the Baffles

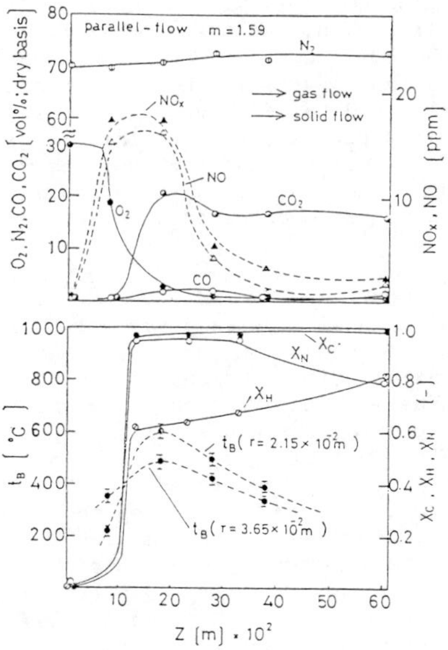

Figure 9 Combustion Experimental Data Obtained with a Parallel-Flow Moving-Bed

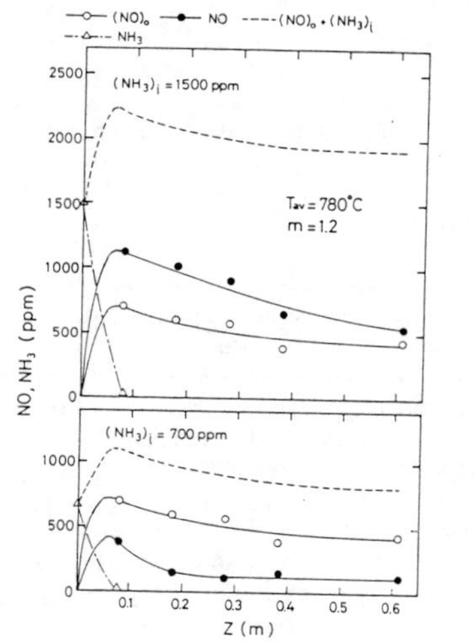

Figure 10 Axial Profiles of NO and NH_3 (m=1.2)

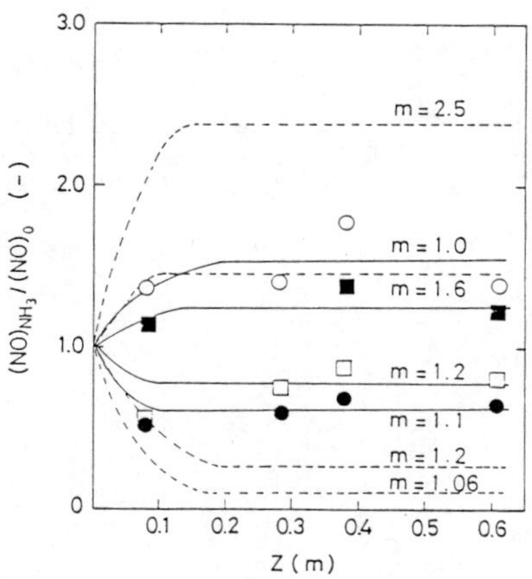

Figure 11 NO Formation and Reduction with NH_3 ($(NH_3)_i$=700 ppm)

can be seen from Figure 9, the oxygen is rapidly consumed in 0.2 m diatance from the bed inlet, forming CO, CO_2, NO, and NO_x. The maximum temperature is developed in this region. Because of the high-rate of combustion in this region, such components as carbon, hydrogen, and nitrogen contained in the chars, the conversions being represented as X_C, X_H, and X_N, respectively, are found to be suddenly increased along through this point in the bed. It is noteworthy that in this experiment, any unburned carbons in the chars as well as ash particulates were not detected at the bed outlet, and that as seen in Figure 9, the concentrations of NO and NO_x were reduced enough to an acceptable extent.

In order to examine the above-mentioned NO reduction effect in the moving-bed more clearly, NH_3 which is considered as a typical volatile-N source, was fed at the inlet of the bed accompanied by NO and chars. Figure 10 shows the obtained axial distributions of the NO and NH_3 concentrations. $(NO)_0$, which is the NO concentration in the absence of NH_3, decreased gradually after passing through a peak around the bed height of z=0.1 m. This decrease in $(NO)_0$ was considered due to the reaction with CaO in the ash (5). In the presence of NH_3, the NO formation in the bed showed different trends depending on the inlet NH_3 concentration. When the air-fuel ratio, m=1.2, the addition of 700 ppm of NH_3 strongly depressed the NO formation, and the outlet NO concentration was reduced to one-third of $(NO)_0$. Under the other conditions shown in Figure 11, however, the NO formation was rather increased because of the conversion of the excess of NH_3 into NO. Consequently, it seemed that there is an optimum in the air-fuel ratio and the NH_3 concentration for low emission of NO.

LITERATURE CITED

1. Hasatani, M., M. Yuzawa, and I. Yajima: Kagaku Kogaku Ronbunshu, 9, 166, (1983)
2. Hasatani, M., M. Yuzawa: Kagaku Kogaku Ronbunshu, 10, 300, (1984)
3. James, N.J. et al.: Environ. Sci. and Tech., 11, 1191 (1977)
4. Furusawa, T., et al.: "ACS-Division of Fuel Chem.", 27, 1, 262, (1982)
5. Hasatani, M., N. Arai, and S. Sugiyama: Kagaku Kogaku Ronbunshu, 7, 651,(1981)

THE CALCINATION OF LIMESTONE PARTICLES IN FLUIDISED BEDS.

J.S. Dennis and A.N. Hayhurst,
Department of Chemical Engineering, University of Cambridge,
Pembroke Street, CAMBRIDGE CB2 3RA, England.

SCOPE

There is little information on the rates of calcination of limestone particles in the size range used for desulphurisation, especially at elevated pressures. The actual rate-controlling step has been subject to considerable controversy. This study elucidates the mechanism of calcination by monitoring CO_2 in the off-gas, when batches of limestone are added to an electrically-heated bed of sand fluidised by air containing CO_2. The rate of calcination affects the final pore size distribution and the capacity of the sorbent for SO_2.

ABSTRACT

The rate of calcination of limestone (diam. 0.4 to 2.0 mm) in fluidised beds is controlled by chemical reaction at a sharp interface between $CaCO_3$ and CaO for $1 < P < 18$ bar and $1098 < T < 1198$ K. The reaction temperature is just below that of the bed. At 1123 K, the rate constant is $1.0 \pm 0.2 \times 10^{-2}$ mol/m^2 s, with an activation energy close to ΔH.

CONCLUSION

Calcination times increased with pressure even with no CO_2 present. This indicates the rate of calcination is $\bar{k}\ (p_e - p_i - Py_I)$, where p_e and p_i are, respectively, the partial pressures of CO_2 at equilibrium and during the experiment, $\bar{k}$ is the rate constant for carbonation, P is total pressure and y_I is a mole fraction of CO_2. It is difficult to justify the y_I, but it describes well the results for $p_i \neq p_e$. At high pressures, whenever $p_e \approx p_i$, the rate expression becomes non-linear in $(p_e - p_i)$.

INTRODUCTION

The fluidised bed combustion of coal has an intrinsic means of reducing sulphurous emissions by adding limestone. A combustor would normally operate between 1073 and 1273 K, at which temperatures CaO can always react with SO_2 and O_2 to yield $CaSO_4$, in $CaO + SO_2 + \frac{1}{2}O_2 = CaSO_4$ (1). Limestones have negligible natural porosity, so a necessity for effective SO_2 capture is the development of internal pores by calcination. Baker (2) investigated the calcination equilibrium: $CaCO_3 = CaO + CO_2$ and found that the equilibrium pressure of CO_2, p_e (in atmos.), obeys:

$$\log_{10} p_e = \frac{8308}{T} + 7.079 \qquad (1)$$

Using Equation (1), for, typically, 15 mol% CO_2 in the off-gas (1), $CaCO_3$ calcines at 1 bar if T > 1051 K, but at 18 bar only above 1250

K. Previous studies (1) (3) have shown that the overall conversion of Ca in uncalcined stone to $CaSO_4$ decreases, if the calcination time is comparable with the sulphation time. Little is known on rates of calcination of particles of the size usually used for desulphurisation (i.e. diam. 0.4 to 2.5 mm), especially at pressure.

PREVIOUS WORK

Some workers (4 to 7) conclude that thermal decomposition occurs at a definite boundary between the CaO and $CaCO_3$, which moves at a constant rate. The actual rate-controlling step is the subject of controversy. For example, Azbe (4) suggested that control is by heat transfer through the layer of calcined CaO. However, subsequent workers (5) (6) observed that, since the thickness of the calcined layer increases linearly with time, the rate is governed by chemical reaction. Hills (7) disputed this and concluded that the decomposition of large spheres (10 mm diam.) was controlled by heat and mass transfer in the calcined layer. Beruto and Searcy (8) have shown that the decomposition of single crystals *in vacuo* is controlled by chemical kinetics. Borgwardt (9) has reached a similar conclusion for small (10 to 90 μm diam.) particles.

EXPERIMENTAL

The calcination of pre-sized particles was studied by adding them in batches to a bed of sand (sieved to +425, −500 μm), held at a fixed temperature between 1098 and 1198 K and fluidised by air containing up to 20 vol% CO_2. Experiments at 1 atmos. were carried out in a ceramic bed (78 mm diam.), constructed from recrystallised alumina. The required temperature was maintained by external electric heaters. A perforated plate distributor was used and bed height at minimum fluidisation, H_{mf}, was 80 mm. The flowrate of fluidising gas was fixed at 0.02 mol/s. The crossflow factor, X, was measured to be ∿ 3.0. The apparatus has been described elsewhere (1) (3). For work at pressures up to 18 bar, a 100 mm bed (10) was used. Here, silica sand (+500, −600 μm) was fluidised with H_{mf} = 150 mm. The ratio U/U_{mf} was always greater than 5.0, with X > 2.5. Only Penrith limestone was studied; it contained 98 wt% $CaCO_3$. The mole fraction, y_o, of CO_2 in the off-gas was measured continuously, by an infra-red analyser, following the addition of a batch of limestone. A typical curve of y_o, as a function of time, is depicted in Figure 1. Here 3.5 g of limestone (sized to +710, −850 μm) was calcined in air at 1 atmos. It was found that the gas sampling system for both beds behaved as a C.S.T.R. (with mixing time constant, τ), in series with a plug-flow element. At 1 atmos. τ = 3.5 s, whilst for the 100 mm bed, τ was 10 to 20 s, depending on pressure. Thus, the true $y_o = x_o + \tau \dot{x}_o$, where x_o is the observed mole fraction at time t. Hills (7) has shown that r_c, the radius of the sharply defined interface between CaO and $CaCO_3$ in a calcining particle is given by:

$$r_c/r_o = [\,(m_t - m_c)\,/\,(m_o - m_c)]^{1/3} = r^* \qquad (2)$$

Here, r_o is the radius of the particle (assumed constant) (7) and m_o, m_t and m_c are the masses of the particles before, during and after complete calcination. The quantity $(m_t - m_c)$ is obtained from: (i) the area under an off-gas curve (e.g. Figure 1), (ii) the mass of particles in the batch and (iii) the molar flowrate of fluidising gas. It was assumed that particles were spherical with diameter $d_p = 2r_o$ equal to the mean of their successive sieve sizes. Equation (2) thus enables r* to be obtained at various times during calcination.

RESULTS

Figures 2 and 3 show r* plotted against t at 1 atmos. and 1148 K. Both plots are linear. The former illustrates the effect of varying the mole fraction of CO_2 in the fluidising gas, y_i. Figure 3 shows the effect of varying d_p^2 at two concentrations of CO_2. It is clear that an increase in d_p or y_i increases t_c, i.e. the time required for complete calcination, which occurs when r* = 0. Figure 4 shows the effects of pressure for $y_i = 0$, 1148 K and various d_p for pressures up to 18 bar. Similar results are shown in Figure 5 for 1098 K. The rate of calcination evidently falls at high pressure, even when there is no CO_2 in the fluidising gases. Finally, Figure 6 shows that plots of t_c against r_o are linear and pass close to the origin.

THEORY

A model must include some or all of the following resistances: (i) chemical reaction at the interface between $CaCO_3$ and CaO, (ii) diffusion of CO_2 through the CaO away from reaction zone, (iii) conduction of heat to the interface, (iv) diffusion of CO_2 away from the particles into the particulate phase, (v) heat transfer from the particulate phase to the particle, (vi) interphase mass transfer from the particulate phase to the bubbles.

Chemical Reaction Control

For reversible calcination, the net rate is: $Q_c = \bar{k}\,(p_e - p_r)$ mol m^{-2} s^{-1} which assumes reaction is confined to a surface. Here, $\bar{k}$ is the rate constant for the backward, carbonation, step and p_r is the actual partial pressure of CO_2 at the interface. Assuming shrinking core kinetics controlled by reaction, r* varies with t according to:

$$\frac{r_o}{\bar{k}RT}(1 - r*) + \frac{WV\,(1 - r*^3)}{AM_{CaCO_3}\,[U - (U-U_{mf})e^{-X}]} = (p_e - p_i)\,\frac{V}{RT}\,t \qquad (3)$$

Here, V is the molar volume of $CaCO_3$, M_{CaCO_3} is the molar mass of $CaCO_3$, W is the mass of the batch, A is the bed area, X is the crossflow factor for a superficial velocity, U, and p_i is the partial pressure of CO_2 in the fluidising gas. The temperature of the bed is T. Figures 2 to 5 show that r* varies linearly with t, as found by others (5 to 7). This means that the second term in Equation (3), which describes interphase mass transfer, is negligible. To substantiate this, t_c was calculated from Equation (3) for $\bar{k} \to \infty$, when the interphase resistance is controlling. For $y_i = 0$, T = 1148 K and 1 atmos., Figure 3 gives $t_c \approx 60$ s for $d_p = 0.40$ mm. The batch mass, W, was at most, 0.01 kg, so that Equation (3), with $\bar{k} \to \infty$, and conditions in the 78 mm bed, gives $t_c = 2.6$ s. This shows that the term in Equation (3) representing inter-phase mass transfer is unimportant. Equation (3) thus gives $t_c \approx r_o/\bar{k}V\,(p_e - p_i)$. Figure 6 confirms that t varies linearly with r_o. There are, however, two problems. The first can be seen by considering Figure 5. Now, at 1098 K, p_e = 0.33 bar. Thus t_c for limestone calcined in the presence of 15% CO_2 should be ~ 1.8 times t_c for CO_2 not added. In practice, the difference is greater, as seen in Figure 5. This could occur if the interfacial temperature is lower than that in the bed. Thus, there might well be a contribution from heat transfer to the interface. The second problem occurs with Figure 4, where the total pressure, P, is increased for y_i = 0. Since p_e is effectively independent of P, plots of r* against t for different P should lie on the same straight line. This is clearly not so from Figure 4.

Intraparticle Temperature Gradients

To estimate the temperature difference, $\Delta T = (T_r - T)$, between the reaction interface and the bed, an estimate of ΔT can be obtained (11) by relating the flux of CO_2 from the reaction interface to the heat transferred, thus:

$$D \frac{dC}{dr} \cdot \Delta H = k \frac{dT}{dr} \tag{4}$$

Here C is the concentration of CO_2 in the CaO, D is the diffusivity, k the thermal conductivity of CaO and ΔH is the heat of calcination. Integration of this expression yields:

$$\Delta T = T_r - T_s = \frac{\Delta H\ D}{k} (C_s - C_r) \tag{5}$$

Here, T_r and C_r are the temperature and concentration of CO_2 at the reaction front. Similarly, T_s and C_s refer to the external surface. The maximum value of ΔT is obtained by taking T_S equal to the bed temperature, T, and C_r to be its maximum value, C_e, i.e. when there is equilibrium at the interface, and also C_s = 0. For T = 1148 K, ΔH = 169.4 kJ/mol (7) and k = 0.46 W/m K (12). D was calculated using the approximation of Wakao and Smith (13), i.e. D is related to the Knudsen diffusivity, D_K, the molecular diffusivity, D_m, and the porosity of the calcine, ε_o, by $1/D = \varepsilon_o^2 (1/D_K + 1/D_m)$. Here, D_K is estimated from $D_K = 97\ (2\varepsilon_o/S_o)\ \sqrt{(T/M)}$, where S_o is the B.E.T. area of the calcined stone and M is the molar mass of CO_2. It is estimated that $D_M = 1.66 \times 10^{-4}$ m^2/s. S_o was measured for limestone calcined at 1148 K in the absence of CO_2 as 2.93×10^{-7} m^{-1}. Using ε_o = 0.54, one obtains $D_K = 1.80 \times 10^{-5}$ m^2/s and $D = 4.8 \times 10^{-6}$ m^2/s. The value of C_e is related to T_r by Equation (1). Equations (1) and (5) were solved simultaneously for T_r and C_e. ΔT turns out to be about 11 K; D and ΔH do not change over such a small ΔT. Hills (7) estimates that k might be as high as 1.0 W/m K, so it is likely that ΔT < 11 K. Thus ΔT is small. The only parameter which is sensitive to such small temperature changes is p_e in Equation (1).

Combination of Resistances

A model was derived to take account of all the resistances to calcination listed earlier. However, from the foregoing, both heat transfer and interphase mass transfer are unimportant. Also, detailed calculations (14) show that external mass transfer of CO_2 is rapid. In this case the general equation for t_c reduces to:

$$t_c = (RTr_o^2/D + 6r_o/\bar{k}) \ /\ (6V\ (p_e - p_i) \tag{6}$$

where p_e is now evaluated at the calculated value of T_r. A plot of t_c against r_o should, therefore, initially be linear, but acquire a component in r_o^2, at high r_o. Figure 6 presents such plots; they are linear, pass close to the origin and do not curve upwards; indicating that $RTr_o/D \ll 6/\bar{k}$. This is confirmed for 1 atmos. and 1148 K by estimating the diffusional contribution to t_c, i.e. $RTr_o^2/6V\,D(p_e - p_i)$, which is 3.8 s for d_p = 0.78 mm, p_i = 0 and ΔT = 11 K. However, from Figure 2, $t_c \approx 100$ s. Diffusion is predominantly Knudsen up to 3 bar. Thus, there should be no change in D or t_c (for y_i = 0) up to at least this pressure - even if the diffusive resistance were significant. This is not actually so, because t_c increases with pressure up to 3 bar, as shown by a comparison of Figures 3 and 4. This pressure effect is discussed below but it can be noted that Equation (5) gives ΔT independent of pressure. Also, the dominant term in Equation (6) is usually the kinetic one, which makes it identical with t_c from Equation (3), except that p_e is now based on the temperature of the interface, rather than the bed. The rate of decomposition of $CaCO_3$ has been studied (15) as a function of $(p_e - p)$, where p is the partial pressure of CO_2 near a decomposing particle. The rate, Q_c, was not proportional to $(p_e - p)$ (15). Instead, for values of p somewhat less than p_e, Q_c varied as $(p' - p)$, where p' ($< p_e$) was an effective equilibrium partial pressure. For decomposition *in vacuo*, however, it was found (15) that $Q_c \propto p_e$. These observations suggest that Q_c is, in fact, proportional to $(p_e - p - Py_I)$. Here, y_I is a constant mole fraction at a fixed temperature. The effective equilibrium partial pressure of CO_2, p', is thus $p' = p_e - Py_I$, so that *in vacuo* $Q_c \propto p_e$, as expected. The reaction interface thus behaves as if it is subject to a constant mole fraction of CO_2, y_I, which depends on temperature only. In this case, Equation (6) becomes, for kinetic control:

$$t_c = r_o/\{\bar{k}V(p_e - p_i - Py_I)\} \tag{7}$$

Hence, if p_i = 0, t_c will vary with total pressure, P, because y_I is a constant. Also, a plot of $(p_e - p_i)/P$ against r_o/t_cP should be linear with intercept y_I and slope $1/V\bar{k}$ for a given temperature. The calculation of p_e was again based on the interfacial temperature estimated using Equation (5). The plots shown in Figure 7 are for calcination at T = 1098 and 1148 K, and various P and p_i. The linearity of the plots is striking and their slopes are equal. This suggests that we now have an adequate description of a calcining particle. Finally, the intercept y_I increases with temperature, but is typically 0.05 to 0.10.

DISCUSSION

The plots in Figure 7 have equal slopes. This was true at 1198 K and implies that $\bar{k}$, the rate constant, is not affected by temperature. Thus the activation energy and energy barrier for carbonation are close to zero. This corresponds to the activation energy for calcination being equal to ΔH for the process, which is about (1) (7) 169 kJ/mol, as deduced by others (5), (6), (15). From Figure 7, $\bar{k}$ = 2.1×10^{-1} mol/bar m^2 s. Borgwardt (9) suggests the rate constant for calcination is $1.35 \pm 0.15 \times 10^{-2}$ mol/m^{2}s at 1123 K. The value of $\bar{k}$ derived here gives the rate constant for calcination as $\bar{k}\,p_e = 1.0 \pm 0.2 \times 10^{-2}$ mol/m^2 s. The agreement is good.

It is now possible to calculate r* at various t using the simplified Equation (3), along with (5), and the new rate expression. The results are compared with experimental ones in Figures 2 to 5. There is good agreement at 1 atmos, but errors arise when the difference between p_e and $(p_i + Py_I)$ is small. Part of this lies in the difficulty of estimating p_e; an error of 5 K alters the value considerably. Also, when $(p_i + Py_I) \rightarrow p_e$, the rate does not vary linearly (15) with $(p_e - p_i - Py_I)$. Thus Equation (7) predicts calcination times well when $(p_i + Py_I)$ is not close to p_e. It is difficult to justify insertion of the constant mole fraction, y_I. However, very complex processes (15) must occur at the interface, including the breakdown of the rhombohedral $CaCO_3$ lattice, followed by the desorption of CO_2 and the nucleation and growth of cubic CaO.

LITERATURE CITED

1. Dennis, J.S., Ph.D. Dissertation, University of Cambridge, (1985).
2. Baker, E.H., J.Chem.Soc., 464 (1962).
3. Fieldes, R.B., Ph.D. Dissertation, University of Cambridge,(1979).
4. Azbe, V.J., Rock Prod., 56(2), 100 (1983).
5. Satterfield, C.N. and Feakes, F., A.I.Ch.E.J., 5, 113 (1959).
6. Ingraham, T.R. and Marier, P. Can.J.Chem.Eng., 41, 170 (1963).
7. Hills, A.W.D., Chem.Eng.Sci., 23, 297 (1968).
8. Beruto, D. and Searcy, A.W., J.Chem.Soc.Far.Trans.I., 70, 2145 (1974).
9. Borgwardt, R.H., A.I.Ch.E.J., 31, 103 (1985).
10. Turnbull, E. et al., Chem.Eng.Res.Des., 62, 223 (1984).
11. Prater, C.D., Chem.Eng.Sci., 8, 284 (1958).
12. Turkdogan, E.J., "Physical Chemistry of High Temperature Technology" Academic, (1980).
13. Wakao, N. and Smith, J.M., Chem.Eng.Sci., 17, 285 (1962).
14. Dennis, J.S. and Hayhurst, A.N., submitted to Chem.Eng.Res.Des., (1986).
15. Zawadzki, J. and Bretsnajder, S.,Z.Elektrochem. 41, 221 (1935).

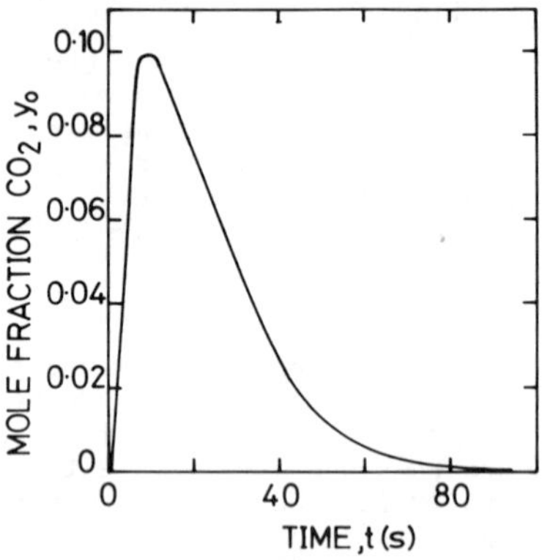

FIGURE 1: Variation of CO_2 mole fraction in off-gas following the addition of 3.5 g limestone to the 78 mm bed fluidised by air at 1148 K.

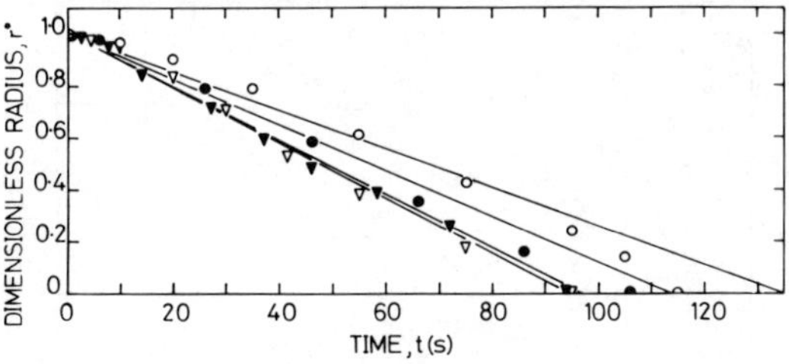

FIGURE 2: Radius of shrinking core, r*, versus time for 1148 K, 1 bar and d_p = 0.78 mm: ▽ 0% CO_2; ▼ 4% CO_2; ● 8% CO_2; O 15% CO_2.

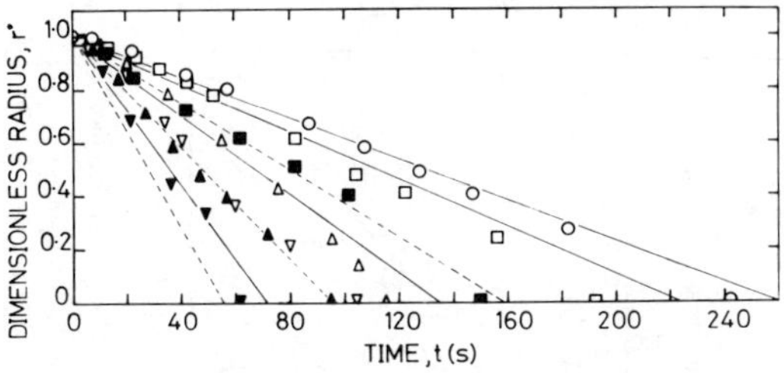

FIGURE 3: Radius of shrinking core, r*, versus time for 1148 K and 1 bar: O, □, △, ▽ = d_p = 1.55, 1.29, 0.78 and 0.4 mm, 15% CO_2; ■, ▲, ▼ d_p = 1.29, 0.78 and 0.4 mm, 0% CO_2. Model fits:—— 15% CO_2; ---- 0% CO_2.

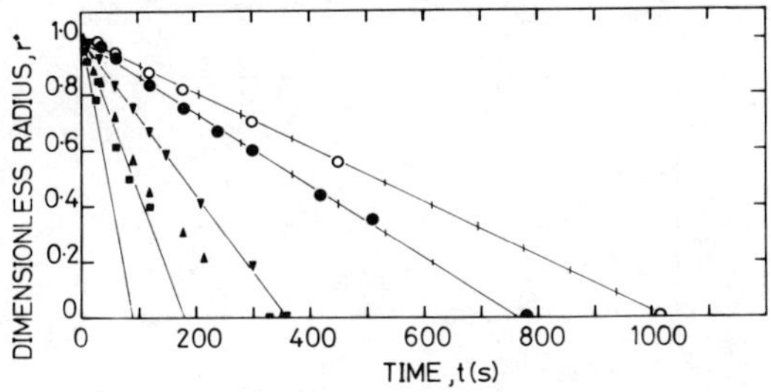

FIGURE 4: Radius of shrinking core, r*, versus time for 1148 K and 0% CO_2: O d_p = 1.55 mm, 12 bar; ● d_p = 0.78 mm, 12 bar; ▼ d_p = 1.55 mm, 3 bar; ▲ d_p = 0.78 mm, 3 bar; ■ d_p = 0.4 mm, 3 bar. Model fit;——: trend line, —+— .

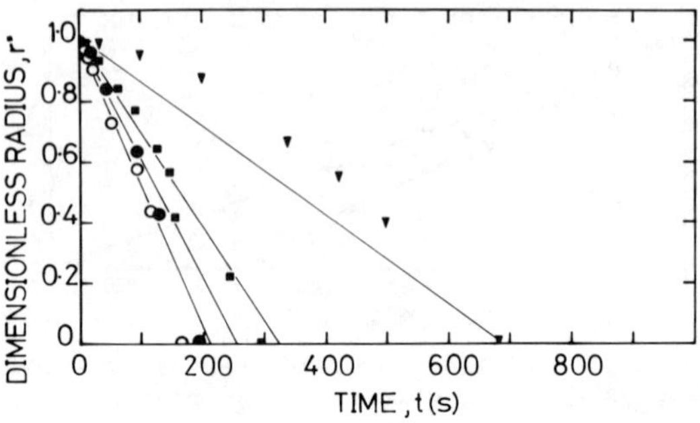

FIGURE 5: Radius of shrinking core, r*, versus time for 1098 K, 1 bar and d_p = 0.78 mm: ∇ 15% CO_2; ■ 8% CO_2; ● 4% CO_2; 0 0% CO_2. Model fit——.

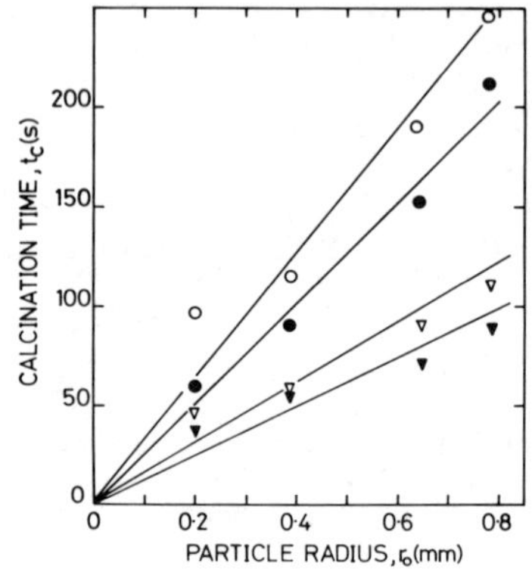

FIGURE 6: Calcination time versus particle radius: ●, 0 T = 1148 K, y_i = 15 or 0 mol%; ▼, ∇ T= 1198 K, y_i = 15 or 0 mol%. P = 1 bar.

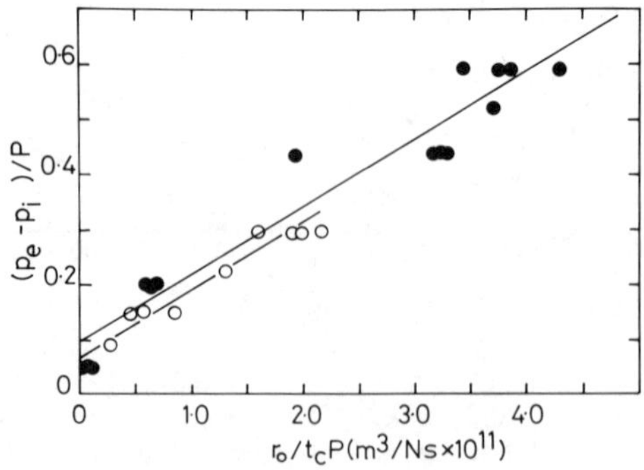

FIGURE 7: Plot to show validity of Equation (7): ● T = 1148 K; 0 1098K

KINETICS OF THE REACTION BETWEEN SULPHUR DIOXIDE AND LIMESTONE PARTICLES IN FLUIDIZED BEDS

Kim Dam-Johansen and Knud Østergaard
Department of Chemical Engineering
Technical University of Denmark
Lyngby, Denmark

ABSTRACT

18 different limestones were examined in a laboratory fluidized bed and found to have widely differing capacities for reaction with sulphur dioxide. Temperature, particle size, and sulphur dioxide concentration influence the degree of conversion of limestone. It is shown by energy dispersive x-ray analysis that the rate of the intraparticle reaction is influenced by at least two parallel steps.

SCOPE

The objectives of the present work have been:

- the evaluation of locally available limestones as desulphurization agents in fluidized beds
- the study of the effect of operating variables such as temperature, particle size, and gas composition on the rate and extent of the desulphurization process
- the development of theoretical models for the reaction between sulphur dioxide and limestone particles.

Experiments have been carried out in three reactor systems:

- a laboratory recirculation (differential) reactor
- an electrically heated laboratory fluidized bed
- a coal fired, pilot plant sized fluidized bed boiler.

Limestone particles at different stages of conversion have been examined by energy dispersive x-ray analysis.

The work is part of a continuing research programme in this Department. Further details about this programme, the experiments and results reported in the present paper are available in (1).

CONCLUSIONS AND SIGNIFICANCE

18 different limestones were evaluated at standardized reaction conditions in a laboratory fluidized bed. In Fig. 2 are shown degrees of conversion of the calcium oxide content as a function of time. It is apparent that the limestones differ widely with respect to their capacity for reaction with sulphur dioxide. Four of the limestones have been evaluated in a coal fired pilot plant fluidized bed. The ranking of the limestones is qualitatively identical to that obtained in the laboratory fluidized bed.

For a reactive limestone (No. 7) the influence of reaction variables such as temperature, limestone particle size, gas composition a.o. has been studied in a series of experiments in a laboratory fluidized bed. Influences of limestone particle size, temperature, and sulphur dioxide concentration have been demonstrated (Figs. 3 and 4).

Cross sections of limestone particles at varying stages of conversion have been examined by energy dispersive x-ray analysis, and local sulphur and calcium concentrations have been measured. The local degree of calcium oxide conversion has been calculated on the basis of these measurements. Results of such calculations are shown in Fig. 7. It is apparent that the rate of the overall process must be influenced by at least two parallel steps.

EXPERIMENTAL AND RESULTS

Evaluation of limestones in laboratory fluid bed reactor

The laboratory fluid bed reactor was electrically heated and of 5.6 cm internal diameter. The fluidizing gas, at atmospheric pressure, was nitrogen containing carbon dioxide, oxygen, and sulphur dioxide. Experiments were carried out by instantaneous addition of an amount of limestone and subsequent monitoring of temperature and gas composition. The types of limestone are described in Table 1. A characteristic particle size distribution is shown in Fig. 1.

Experiments were carried out at the following conditions:

T	U	M^{o}_{CaO}	$X^{o}_{SO_2}$	$X^{o}_{CO_2}$	$X^{o}_{O_2}$
850±2	.562±.003	.066±.002	1502±7	10.C±.2	4.02±.05

The amount of sulphur dioxide removed from the fluidizing gas was determined from the measured gas compositions and flow rates, and the result was used for calculation of the conversion of the calcium oxide content. Corrections were made for the formation of sulphur trioxide (cf. Fieldes et al. (2)) and for the elutriation of calcium oxide. (Elutriated calcium sulphate was neglected, the maximum error thus introduced on the calculated calcium oxide conversion being 0.03.)

Fig. 2 shows the degree of conversion of the calcium oxide content as a function of time for the types of limestone studied in this series. It is apparent that the limestones differ widely with respect to their capacity for reaction with sulphur dioxide. The most reactive types (in agreement with observations of Burdett (3)) are chalks of a porous, fine grained texture and of relatively young geological age (see Fig. 5), whereas the least reactive are geologically old, crystalline calcites of very low porosity. Types of intermediate reactivity are limestones of intermediate geological age and of a porous, coarse grained texture. It is noted that in this intermediate group a content of iron oxide appears to correlate with a relatively low reactivity.

Pilot plant experiments

A series of 42 experiments were carried out in a coal fired fluidized bed pilot plant of 0.50 m internal diameter, fired with bituminous coal of narrow size distribution (median size 5.5 mm) containing 0.95 % (weight) sulphur. The relative ranking of the 4 limestones examined in this series was qualitatively identical to that found in the laboratory experiments.

Influence of process variable

One of the most reactive types of limestone, No. 7, was studied in a series of experiments in the laboratory fluid bed reactor, under variation, one at a time, of the following reaction variables:

Variables	Values in Reference Experiment	Values in Other Experiments
T	850	800 825 875 900
d_p	.779	.815 .328 .550 .925 2.000
U	.56	.4 .8
M^o_{CaO}	.07	.034 .21
$X^o_{SO_2}$	1500	500 1000 2000
$X^o_{CO_2}$	10	0 5 15
$X^o_{O_2}$	4	1.7 8
Amount of inerts, cm^3	200	100 300
d_{pi} (inerts)	.800-1.500	0-.200 .200-.400 .400-.800

Analysis of experiments in which bed height and inert particle size were varied (and of experiments by Hartman et al. (4)) led to the conclusion, supported also by theoretical consideration, that bubble phenomena do not cause a significant deviation from plug flow in this reactor system. Furthermore, the greater part of the literature on the reaction between sulphur dioxide and limestone lends support to an assumption that the reaction is of first order with respect to sulphur dioxide concentration, e.g. Borgwardt (5), Marsh and Ulrichson (6). Thus, based on these two assumptions regarding gas flow and reaction order, experiments may be interpreted in terms of an initial first order reaction rate constant given by the following expression:

$$k = -\,(UA/M^o_{CaO})\ \ln\,(x_{SO_2}/x^o_{SO_2})$$

(It may be noted that a survey of the literature on the reaction between sulphur dioxide and calcium oxide has led to the conclusion that the reaction should be described by a less simple rate expression with a reaction order - taking into account the likely reaction mechanisms - varying between 0 and 2, dependent on temperature and sulphur dioxide concentration. Details of this will be published elsewhere.)

Influence of inlet carbon dioxide and oxygen concentrations

Variation of inlet carbon dioxide partial pressure had no significant effect on the observed initial reaction rate or conversion. Nor was an effect expected, since the equilibrium partial pressure over calcium carbonate at 850 °C is approximately 0.56 atm. compared to 0.15 atm., the maximum in these experiments.

Oxygen concentration was observed to be of little importance. A very slight increase in the reaction rate constant was observed when the oxygen concentration was increased from 4 to 8 %, cf. that the data of Fieldes et al. (2) indicate a 7 % increase of the rate constant when oxygen concentration is increased from 1.6 % to 21 %.

Influence of sulphur dioxide concentration (Fig. 3)

The plot of calcium oxide conversion after 15 min. can be fitted to a straight line through the origin, in agreement with the assumption of first order reaction kinetics. After longer periods of reaction, conversions at the higher values of sulphur dioxide concentrations are lower than would be expected from first order kinetics. It is assumed that this is caused by internal mass transport limitation.

Influence of temperature

At reaction times lower than 90 min., the solids conversion exhibits a maximum a 825-850 °C. This is in general agreement with observations by other authors, e.g. O'Neill et al. (7). However, at reaction times of 120 min. and above, solids conversion was observed to increase steadily with increasing temperature. The initial reaction rate constant exhibits a local maximum at 825-850 °C, and it increases with increasing temperature in the range 875-900 °C. Thus, both positive and negative activation energies are observed. In general, the influence of temperature is slight.

Influence of particle size (Fig. 4)

The effect of increasing particle size is a lower reaction rate indicating gas transport limitations.

Intraparticle concentration profiles

Limestone particles reacted at 850 °C at standardized reaction conditions in a differential laboratory reactor were withdrawn after varying times of reaction. Particle cross sections were examined in a scanning electron microscope by energy dispersive x-ray analysis, and sulphur and calcium concentrations were determined along a line bisecting the particle cross section. Calcium oxide conversion profiles were calculated on this basis, an example being shown in Fig. 7. The intraparticle reaction appears to be controlled by diffusion through the porous structure and by a parallel process, e.g. diffusion in the solid phase. Further analysis of these observations will be reported elsewhere.

ACKNOWLEDGEMENT

Pilot plant experiments were carried out at the Risø National Laboratory, and scanning electron microscopy and energy dispersive x-ray analysis were done at the Department of Metallurgy, Technical University of Denmark. The valuable assistance received from these institutions and their staffs is acknowledged with gratitude.

NOTATION

A	Empty bed cross sectional area (25 cm^2)
d_p	Particle size of limestone (mm)
d_{pi}	Particle size of inerts (sand) (mm)
F	Particle size distribution function
k	Initial first order reaction rate constant (m^3/s mole)
M^o_{CaO}	Amount of added calcium oxide (mole)
T	Reaction temperature (°C)
U	Superficial fluidization velocity (m/s)
X_{CaO}	Degree of conversion of limestone calcium oxide content
$x^o_{CO_2}$	Inlet molefraction of carbon dioxide (%)
$x^o_{O_2}$	Inlet molefraction of oxygen (%)
$x^o_{SO_2}$	Inlet molefraction of sulphur dioxide (ppm)
x_{SO_2}	Outlet molefraction of sulphur dioxide (ppm)

LITERATURE CITED

1. Dam-Johansen, K., "Sulphur Dioxide Retention on Dry Chalk" (in Danish), Ph.D. thesis, Technical University of Denmark, Lyngby, Denmark (1986).

2. Fieldes, R.B., N.A. Burdett, and J.F. Davidson, Trans.Instn. Chem.Engrs., 57, 276 (1979)

3. Burdett, N.A., in "Fluidized Combustion: Systems and Applications", p. V-1-1, Institute of Energy, London (1980)

4. Hartman, M., K. Svoboda, and O. Trnka, EFCE Publication Series No. 37 (ISCRE 8), Instn.Chem.Engrs., 509 (1984)

5. Borgwardt, R.H., Environm.Sci. and Techn. 4 (1), 59 (1970)

6. Marsh, D.W. and D.L. Ulrichson, Chem.Eng.Sci., 40, 423 (1985)

7. O'Neill, E.P., D.L. Keairns, and W.F. Kittle, Thermochim. Acta, 14, 209 (1976)

Table 1 Chemical composition of the evaluated limestones (wt %)

No.	Name	Origin	CO_2	CaO	MgO	SiO_2	Fe_2O_3	Trace[1] Elements
1	Faxe Industri	Denmark	43.6	55.1	0.56	0.5	0.11	0.16
2	Faxe Bryozo	Denmark	43.6	55.1	0.43	0.5	0.08	0.24
3	Faxe Bryozo[2]	Denmark	43.4	54.8	0.31	0.6	0.58	0.30
4	Faxe Koral	Denmark	43.7	55.2	0.46	0.2	0.07	0.15
5	Faxe Koral[3]	Denmark	43.7	55.0	0.48	0.2	0.32	0.20
6	Katholm	Denmark	39.9	49.3	0.50	9.3	0.29	0.95
7	Stevns chalk	Denmark	43.6	55.2	0.29	0.4	0.04	0.22
8	Stevns chalk[4]	Denmark	37.3	51.5	0.34	0.6	0.04	0.24
9	Stevns chalk[5]	Denmark	42.7	54.1	0.34	2.2	0.13	0.27
10	Aggersund chalk	Denmark	41.7	52.1	0.38	4.3	0.26	0.96
11	Aggersund chalk[6]	Denmark	41.4	51.6	0.43	4.9	0.28	1.06
12	Aggersund chalk[7]	Denmark	41.6	51.7	0.39	4.7	0.29	1.18
13	Gotland	Sweden	42.3	52.0	1.15	2.5	0.50	1.48
14	Ignaberga	Sweden	40.1	49.8	0.55	8.3	0.22	0.87
15	Förby	Finland	41.3	52.5	1.26	3.4	0.36	1.07
16	Carmeuse	Belgium	40.3	50.8	1.01	5.3	0.26	1.68
17	Malaga calcite	Spain	43.7	54.3	1.20	0.4	0.11	0.25
18	Malaga dolomite	Spain	47.4	30.9	21.4	0.1	0.05	0.13

[1] Al_2O_3, MnO, TiO_2, P_2O_5, K_2O, Na_2O.
[2] Similar to no. 2, but containing Fe_2O_3.
[3] Similar to no. 4, but containing Fe_2O_3.
[4] Similar to no. 7, but containing $CaSO_4$.
[5] Waste from production of "wash whiting".
[6] As no. 10, agricultural quality.
[7] As no. 10, oven dryed.

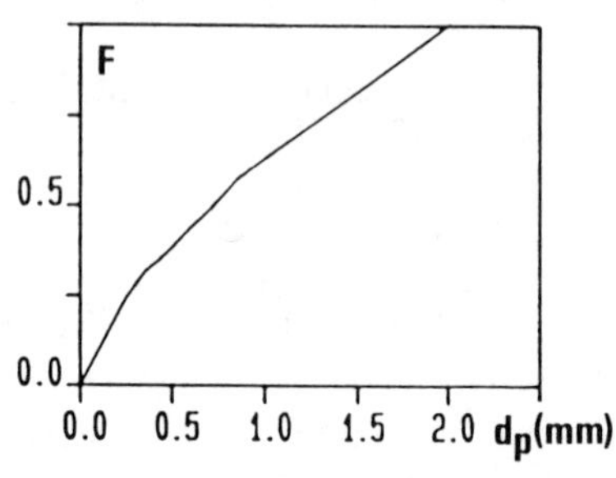

Fig. 1. Characteristic particle size distribution for the limestones listed in Table 1.

Fig. 2. Conversion of limestone calcium oxide content as function of time for limestones listed in Table 1.

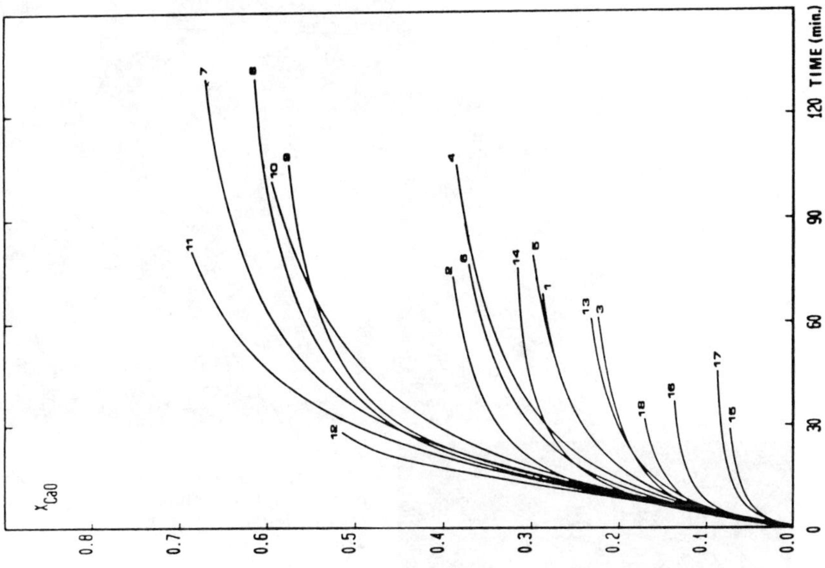

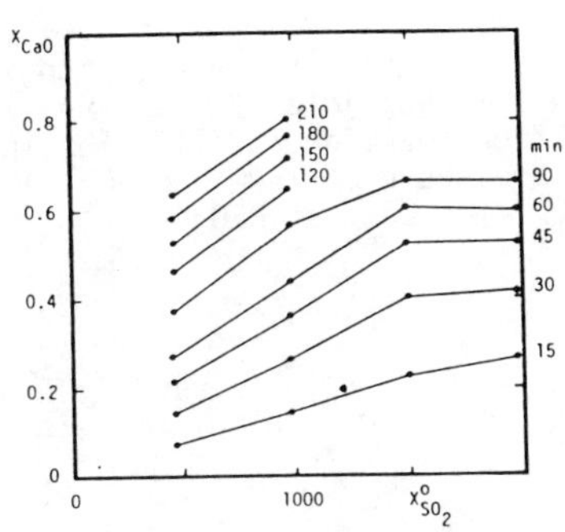

Fig. 3. Conversion of calcium oxide in chalk no. 7 as function of inlet sulphur dioxide concentration and time.

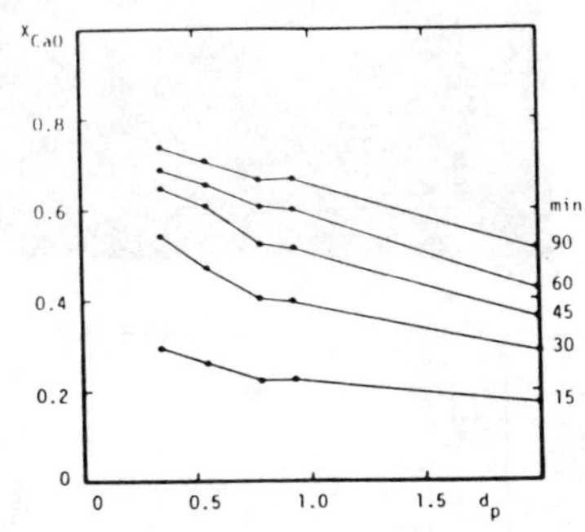

Fig. 4. Conversion of calcium oxide in chalk no. 7 as function of particle diameter and time.

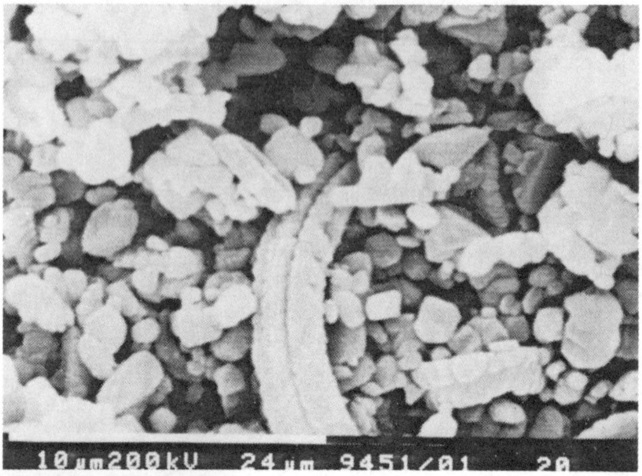

Fig. 5. Scanning electron microphotograph of chalk no. 7, characterized by a highly porous, fine grained texture.

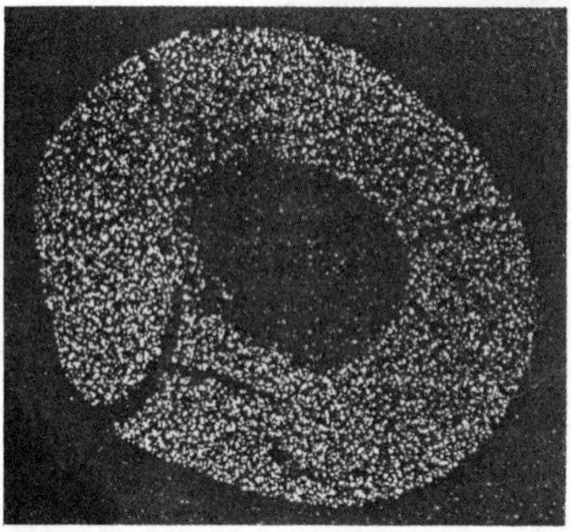

Fig. 6. Sulphur distribution in a 2 mm particle of chalk no. 7 after reaction for 112 min. in atmosphere of 1500 ppm SO_2, 4% O_2, 10% CO_2, at 850 °C.

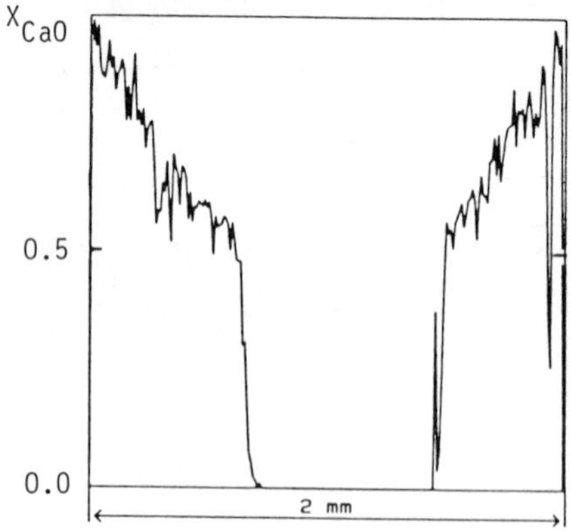

Fig. 7. Calcium oxide conversion in particle shown in Fig. 6. Conversion rate limited by two parallel processes (pore diffusion and solids diffusion).

THE INFLUENCE OF GAS TO PARTICLES MASS TRANSFER ON THE RETENTION OF SULFUR OXIDES DURING FLUIDIZED BED COMBUSTION OF COAL

I.W. Noordergraaf,* J.J. van Deemter,+ C.M. van den Bleek,
P.J. van den Berg,
Laboratory of Chemical Technology, Delft University of Technology,
Julianalaan 136, 2628 BL Delft, The Netherlands

ABSTRACT

Comparison of results on mass transfer in fluidized beds of particles in the range of 0.5 - 3 mm shows that, contrary to results with small particle fluidization, mass transfer coefficients become larger for larger bed diameter.
For fluidized bed combustion of coal and the capture of sulfur oxides by sorbent particles the controlling resistance to mass transfer must be located near individual reactive coal and sorbent particles.
From available data it may be derived that under certain conditions the neglect of mass transfer may result in predicted emission values which are low by a factor 2.

SCOPE

Fluidized bed combustion offers possibilities to capture sulfur oxides by the addition of sorbent particles containing calcium oxide.
The transport of sulfur dioxide from the area of generation (a coal particle) to sorbent particles may be a factor limiting the effectiveness of sulfur retention. The chance for sulfur dioxide to escape from the bed increases the more entrainment by the main vertical gas flow dominates compared to mass transfer to the sorbent.
Experimental data on mass transfer phenomena in large particle fluidized beds are limited. Nevertheless an analysis of available information may throw some light on the question of how critical mass transfer of sulfur dioxide may be in fluidized bed combustion.

CONCLUSIONS AND SIGNIFICANCE

In fluidized beds with particles around 2 mm diameter volumetric mass transfer coefficients, describing the transport of components from the dilute phase to dense phase particles, are smaller in a 0.1 m diameter bed than in 0.4 m diameter beds for equal superficial gas velocity. This difference can be ascribed to differences in flow regimes, namely square nose slugs in the small diameter bed as against turbulent fluidization or slow bubbles for large diameter.

* Koninklijke/Shell-Laboratorium, The Netherlands

+ Twente University of Technology, The Netherlands

Furthermore mass transfer coefficients appear to be nearly proportional to the amount of active particles in the bed. This can be explained as a resistance to mass transfer in a small region around the single active particles.
Particularly in the case of rapid SO_2 generation in fluidized bed combustion the effect of mass transfer limitation may be a doubling of sulfur emission compared to a situation where no mass transfer limitation would exist. This factor can, therefore, not be dismissed in design calculations.

CAPTURE OF SULFUR DIOXIDE

In the path of SO_2 through a fluidized bed combustor (FBC) three steps can be distinguished:
1. SO_2 generation as a result of combustion of coal.
2. Transport of SO_2 from a coal particle to a sorbent particle.
3. SO_2 capture by the sorbent particle.

In a previous paper (1) it was shown that each of these three rate steps may have an effect on the overall sulfur capture efficiency. Here the mass transfer step will be discussed in somewhat more detail.

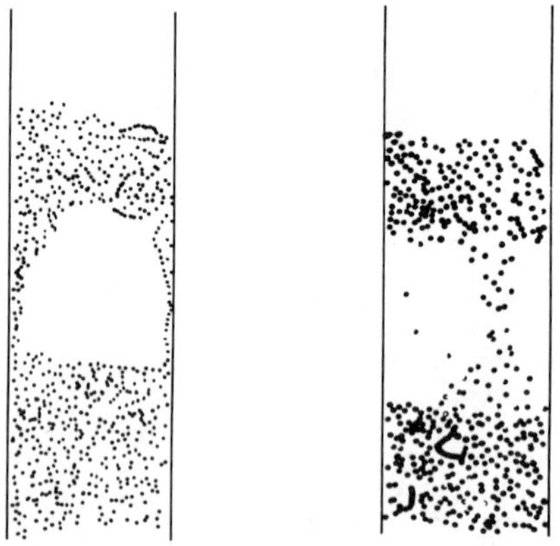

ROUND - NOSE
- small particle systems
- particles descend alongside
- slug length can be observed
- slug rise velocity higher than gas velocity

SQUARE - NOSE
- large particle systems
- particles rain through void
- slug length hardly defineable
- slug rise velocity lower than gas velocity

Figure 1. Slugging types

Since mass transfer has to do with convection and eddy currents one should realize that in an FBC with its relatively large particles (0.5 - 3 mm) the fluid dynamics differ appreciably from that of a fluidized bed with small particles (0.05 - 0.2 mm). At velocities below the turbulent regime "slow" bubbles will be present in the bed. This means that the bubble rise velocity is smaller than the dense phase interstitial velocity. Under the latter conditions particles may "rain" through bubbles. G. Boelens (local meeting, Delft, 1983) recommends to regard such a bubble as a void of locally higher porosity.
In practical FBC systems horizontal or vertical tube bundles are present and affect the flow regime. Horizontal bundles may cause that bubble behaviour resembles a two-dimensional fluidized bed (2). It may well be that an FBC with

closely packed vertical internals can be regarded, at least as a first approximation, as a system of parallel small diameter beds with little or no exchange between them. If this were so a large unit could then experimentally be simulated by a laboratory unit of a diameter representative of the geometry of the internals. The small unit would operate in a slugging fashion; in particular with the large particles the slugs would be of the "square nose" type (Figure 1).
In the comparison of beds of various diameters, with or without internals, phenomena as summarized above play a role in mass transfer effects and sulfur capture.

MASS TRANSFER MODEL

The mass transfer coefficients used in this paper are defined by the model presented in Figure 2.
K_1, K_2, k_1 and k_2 are volumetric quantities expressed as m^3 component per second transported or per second converted per m^3 of fluidized bed volume. K_1 refers to the transport of O_2 from the dilute phase (bubbles) through a layer of inert particles to the burning coal particles. The SO_2 generated is assumed to move in the first instance to the main gas stream; it may transfer from there to sorbent particles (K_2).
When the SO_2 released transfers directly to sorbent particles rather than first entering the main gas stream the model can still be applied, be it that a lower K_2 value will then be found because SO_2 has a lower chance to escape with the main gas flow.
In the model it is left open for the moment where major resistances in the path of O_2 and SO_2 are located.
A considerable simplification can be introduced when particle mixing is rapid compared to the time of release of SO_2. In that case the SO_2 generation is uniform over the bed. Otherwise a more complicated pattern of SO_2 formation must be assumed or calculated. The principle of the approach remains, however, the same. In this paper the complete mixing assumption will be used. The mathematical equations corresponding to this model are linear and first order; they lead to solutions in terms of exponentials (1). Important parameters are the apparent rate constants

for SO_2 generation $$\beta_1 = k_1K_1/(k_1+K_1) \qquad (1)$$

for SO_2 capture $$\beta_2 = k_2K_2/(k_2+K_2) \qquad (2)$$

Experimental values for k_1 and k_2 can be obtained from separate experiments under more ideal conditions such as in a fixed bed reactor, although one should be careful to translate fixed bed data to a fluidized bed situation that gives rise to particle attrition. K_1 and K_2 can be found by determining β_1 and β_2 in fluidized beds.

MASS TRANSFER DATA COMPARED

In two previous papers (1, 3) data on mass transfer under FBC conditions have been published. In the first data could be derived from steady state coal combustion in a 0.39 m diameter fluidized bed

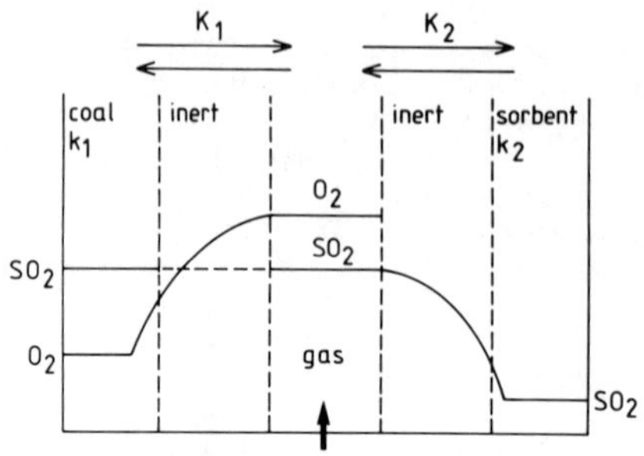

Figure 2. Mass transfer model. K_1, K_2 and k_2 are volumetric quantities (m^3/s of component transported or converted per m^3 of fluid bed volume)

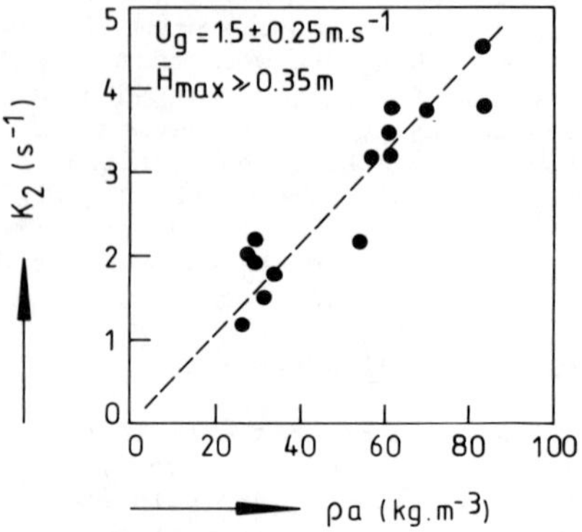

Figure 3. The mass transfer coefficient as a function of the mass concentration of sorbent particles

to which dolomite was added batchwise. The volumetric mass transfer coefficients for both O_2 and SO_2 transport were found to be of the order 10 /s. From Table 1 it appears that for coal combustion K_1 increases for increasing coal content in the bed. The variations in k_1 are due only to different bed carbon content, not to temperature which was kept constant in the experiments. Discussions on the influence of temperature on k_1 and k_2 can be found in (11). The experiments with batches of dolomite gave values of K_2 varying from 7 - 20 /s, where the higher values were found for larger batches.

Table 1

% w carbon	0.8	2.0	2.3
k_1 (/s)	5.4	13	16
K_1 (/s)	4.6	7.1	9.0

The second paper describes experiments in a 0.1 m diameter fluidized bed with model sorbent particles where SO_2 is injected below the distributor. Values of K_2 varied from 1 - 5 /s more or less proportional to the mass of sorbent material in the bed (Figure 3). There was no influence of bed height and only a slight decrease of K_2 with an increase of the superficial velocity from 1 to 2 m/s. The fluidized bed was in the (square nose) slugging regime.

Earlier preliminary data on mass transfer during coal combustion in a small diameter bed were published by Avedesian and Davidson (4), and data from a 0.37 m diameter combustor by Kolenko and Collin (5).

In none of the units referred to above tube bundles or other internals were present.

In Figure 4 the available results are compared in terms of the height of a transfer unit, U_g/K_1 and U_g/K_2 where U_g denotes the superficial gas velocity. The shaded area covers the range of data for small particle fluidized beds (6), all other data are for fluidized beds of large particles (d_p in the order of 1 mm).

For bed diameters larger than 0.1 m the differences between mass transfer for small and large particles become considerable. Contrary to small particle systems the contacting between gas and large particles tends to improve when going to larger diameter vessels. In addition calculation procedures based on small particle fluidization results do not provide any leads. For instance, according to (7) and (8) predicted values for the height of a transfer unit are 0.1 m for the experiments in the 0.1 m diameter unit, and 0.014 m for the results in the 0.39 m vessel, i.e. factors 2 - 10 too small.
The height of a transfer unit may be associated with the scale of the mass transfer mechanism, for instance the dimension of an eddy current, the size of the circulation around a bubble. In small particle fluidized beds it is conceivable that the largest eddies increase in size to more than 1 m with increase in diameter. But 0.05 - 0.2 m would seem to be rather small for the size of eddies in a large particle system unless one has to do with a "quiet" turbulent flow regime. On the other hand these values are much larger than the average distance between active particles in FBC. With 1 - 2 % active particles this distance is equal to about 5 d_p, i.e. roughly 0.01 m. This indicates that other approaches to understand the mechanism of mass transfer for these systems should be explored.

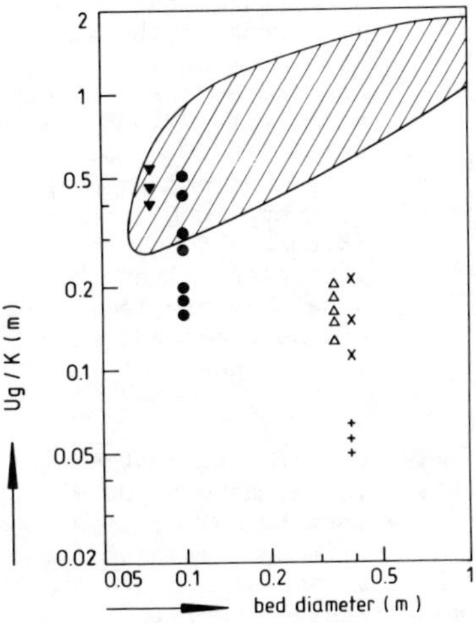

Figure 4.
Comparison of mass transfer results with literature data

Data source:

● Model sorbent

X Coal combustion } 0.39 m bed

\+ Dolomite

Δ Kolenko & Collin (5)

▼ Avedesian & Davidson (4)

Shaded area: data summarized by Van Deemter (6) for small particles

MASS TRANSFER NEAR ACTIVE PARTICLES

The proportionality between K_2 and ρ_a (Figure 3) is a strong indication that the resistance to mass transfer could be located near single active particles and not so much near a bubble surface or in the bulk of the dense phase.
To relate the volumetric coefficients K_1 and K_2 to a film mass transfer coefficient k_g use will be made of the relations

$$K = 6\ \rho_a k_g/(\phi\ d_p \rho_p) \tag{3}$$

$$k_g d_p/\mathbb{D} = Sh(Re,\ Sc) \tag{4}$$

where $Re = ud_p/\nu$ and $Sc = \nu/\mathbb{D}$ (see also notation). A possible correlation for equation (4) is (9)

$$Sh = 2 + 0.69\ Re^{0.5} Sc^{0.3} \tag{5}$$

Eliminating k_g from equations (3) and (4) gives

$$K/\rho_a = 6\ \mathbb{D}Sh/(\phi\ d_p^2 \rho_p) \tag{6}$$

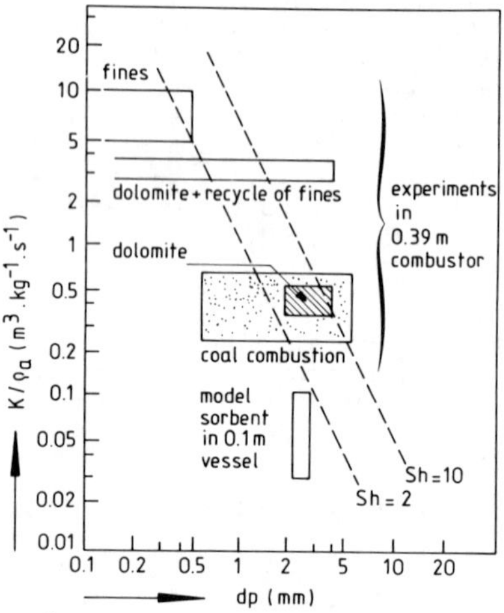

Figure 5. Mass transfer results compared to Sherwood numbers for individual particles

In Figure 5 experimental values K_1/ρ_a and K_2/ρ_a have been plotted as a function of d_p. As a comparison Equation (6) has been drawn for two values of the Sherwood number Sh (2 and 10) assuming a reasonable set of values for $\mathbb{D}$ and ρ_p. It appears that Equation (6) predicts the right order of magnitude of coefficients K. The experiments with "fine" dolomite particles ($d_p < 0.45$ mm) show that equation (5) is less applicable for mass transfer of SO_2 to smaller particles, although in this case they are less than 1% of the total bed mass (mainly sand, $d_p = 0.85$ mm).

Mass transfer coefficients in the 0.1 m vessel are somewhat smaller than predicted by Equations (5) and (6). This may have to do with the slugging flow in this unit compared to the more turbulent flow regime observed in the large vessel. A large unit with heat exchange tubes may on the other hand also operate in a slugging fashion and should then also be associated with a lower mass transfer coefficient.

MASS TRANSFER AND SULFUR EMISSION

Assuming completely mixed particles the percentage emission, E, of SO_2 in an FBC becomes according to the model of Figure 2 (1)

$$E = \frac{\beta_1}{\beta_2 - \beta_1} \quad \frac{\exp(-\beta_1 H/U_g) - \exp(-\beta_2 H/U_g)}{1 - \exp(-\beta_1 H/U_g)} \tag{7}$$

As an example an FBC of 0.6 m bed height will be considered for two cases: $\beta_1 = 4.6$ /s and $\beta_1 = 1.1$ /s. Both values were derived from

experiments in fluidized beds, (1), (10). Values of k_2 were obtained from time-averaged reaction rates of dolomite in a fluidized bed, including the influence of particle attrition (1). Both k_2 and K_2 are dependent on the Ca/S ratio via ρ_a, and k_2 also via the mean residence time of the dolomite particles. In (11) this procedure is described in more detail. K_2 values used were calculated from the conservative value $K_2/\rho_a = 0.05$ in Figure 5.
The SO_2 retention is calculated both for K_2 values as indicated above and for $K_2 = \infty$. The results are presented in Figures 6 and 7.

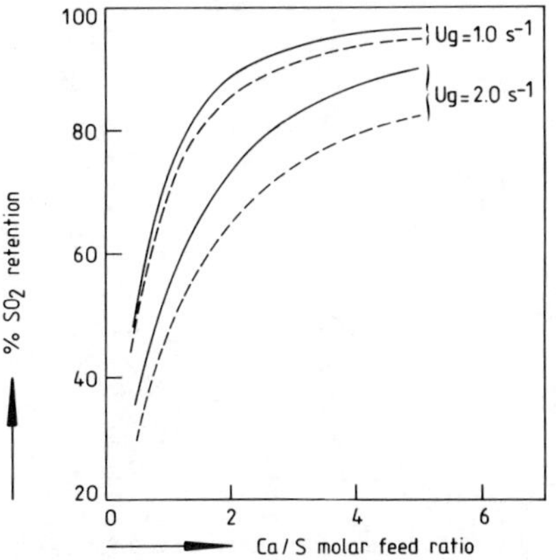

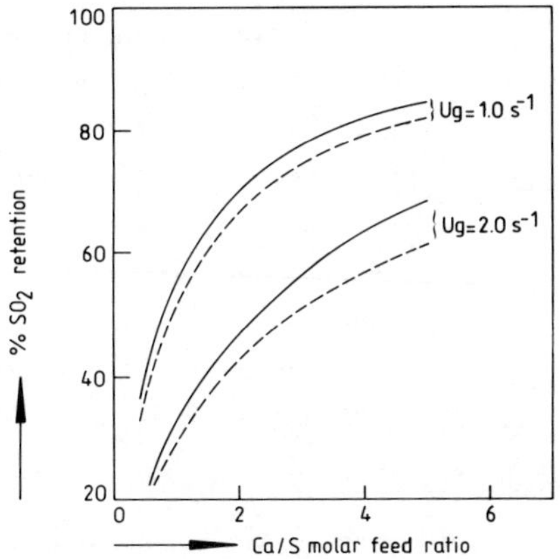

Effect of mass transfer of SO_2 on sulfur retention projected by model.
Solid lines: mass tranfer limitations ignored.
Broken lines: mass transfer limitations incorporated.

Figure 6.
Rapid coal combustion $\beta_1 = 4.6$ /s

Figure 7.
Low coal combustion $\beta_1 = 1.1$ /s

The mass transfer effect on SO_2 emission becomes larger at higher gas velocities because at the same coal combustion rate more coal is burnt and more sorbent must be added. Then the sorbent residence time is smaller, but its average reactivity larger, so that a mass transfer limitation becomes more pronounced.
At a lower combustion rate ($\beta_1 = 1.1$ /s) a larger proportion of SO_2 is generated in the upper regions of the bed where the chance to escape is increased. The mass transfer effect on emissions is then of less importance. The oxygen utilization is quite low, namely 48 and 28 % for $U_g = 1$ and 2 m/s respectively. This can, however, be improved by various means: addition of more coal, higher bed height, lower gas velocity, etc. At the same time the SO_2 retention would improve.

The examples show that "learning" models of the type proposed in this study, combined with preliminary data on mass transfer may provide indications on the role of important design factors. Where needed more accurate calculations can be performed after introduction of more sophisticated models in combination with additional experimental information on factors as mass transfer and mixing.

NOTATION

d_p	particle diameter	m
$\mathbb{D}$	diffusion coefficient	m^2/s
k_g	linear mass transfer coefficient	m/s
$k_{1,2}$	volumetric 1st order rate constant (see Figure 2)	/s
K, K_1, K_2	volumetric mass transfer coefficient	/s
U_g	superficial gas velocity	m/s
u	characteristic gas velocity around an active particle	m/s
$\beta_{1,2}$	combined volumetric rate parameter (Eq. 1 & 2)	/s
ν	kinematic viscosity	m^2/s
ρ_a	mass concentration of sorbent (coal) in reactor	kg/m^3
ρ_p	apparent particle density	kg/m^3
ϕ	shape factor	-

LITERATURE CITED

1. Noordergraaf, I.W., J.J. van Deemter, J. van Kooten and P.J. van den Berg, 8th Int. Symp. Chem. React. Eng., Edinburgh, UK (1984).
2. Lord, W.K., G. McAndrews, M. Sakagami, J.A. Valenzuela, L.R. Glicksman, 7th Int. Conf. Fluidized Bed Combustion, Philadelphia, USA (1982).
3. Noordergraaf, I.W., C.M. van den Bleek, P.J. van den Berg, 8th Int. Conf. Fluidized Bed Combustion, Houston, USA (1985).
4. Avedesian, M.M., J.F. Davidson, Trans. I. Chem. Engs., 51, p. 121 (1973).
5. Kolenko, T., R. Collin, 2nd Int. Conf. Fluidization, Cambridge, UK (1978).
6. van Deemter, J.J., 3rd Int. Conf. Fluidization, Henniker, USA (1980).
7. Kunii, D., O. Levenspiel, "Fluidization Engineering", Wiley, New York, USA (1969).
8. Davidson, J.F., D. Harrison, "Fluidized Particles", Cambridge Univ. Press, UK (1963).
9. Olofsson, J., "Mathematical modeling of fluidized bed combustors", IEA report nr. ICTIS/TR14, Paris, France (1980).
10. Walsh, P.M., C. Li, A. Dutta, J.M. Béer, 8th Int. Symp. Chem. React. Eng., Edinburgh, UK (1984).
11. Noordergraaf, I.W., "Sulfur capture during fluidized bed combustion of coal. External mass transfer phenomen and the fluid-dynamical behaviour of large particles", PhD thesis, Delft University of Technology, Delft, The Netherlands (1985).

FREEBOARD SULFUR CAPTURE IN A FLUIDIZED BED COMBUSTOR

Sadriye Küçükbayrak*, Dick Boersma**, Pieter J. van den Berg***

*İstanbul Technical University, Faculty of Chemical-Metallurgical Engineering, İstanbul, Turkey
**Delft University of Technology, Laboratory for Thermal Power Engineering, Delft, The Netherlands
***Delft University of Technology, Laboratory for Chemical Technology, Delft, The Netherlands

ABSTRACT

The purpose of this study is to examine the sulfur capture potency of the freeboard and to show how it is influenced by the important operation variables in an atmospheric fluidized bed combustor. The results show that the freeboard region of a fluidized bed combustor can be very effective in overall sulfur capture.

SCOPE

In the freeboard zone of a fluidized bed combustor solid particles entrained from the bed surface cause further reactions. The understanding of the reactions occurring in this region is very important for modeling of a fluidized bed combustor. One of the favourable conversions of solid particles in the freeboard is absorption of sulfur dioxide by sorbent particles. Several operating variables are affecting the sorbent particle concentration and thereby the amount of sulfur dioxide captured in the freeboard region of a fluidized bed combustor.

In the present work the effects of sorbent particle size, calcium to sulfur molar ratio in the feed, initial sulfur dioxide concentration and superficial gas velocity on the sulfur capture potency of the freeboard were examined.

CONCLUSIONS AND SIGNIFICANCE

On the basis of the experimental data, it is proposed that the freeboard region of a fluidized bed combustor can be very effective in reducing sulfur dioxide. In some cases a large amount of sulfur dioxide is captured in the freeboard region of the combustor.

The size distribution of feed dolomite has a large effect on the amount of sulfur dioxide captured in the freeboard. Decreasing the particle size of the sorbent material increases the freeboard particle hold-up and therefore the freeboard sulfur reduction increases.

The amount of sulfur dioxide captured in the freeboard is a strong function of the calcium to sulfur molar-ratio in the feed to the combustor. An increase in calcium to sulfur molar ratio in the feed increases the effectiveness of the freeboard as a result of an increase of sorbent material concentration in the bed and in the freeboard.

For higher initial sulfur dioxide concentrations the freeboard of the combustor is more effective for sulfur reduction.

Increasing the gas velocity can also improve the desulfurisation efficiency of the freeboard versus the bed, but up to a limited value, because residence times of gas and solid particles must be long enough.

The fines produced from slow attrition are highly sulfated, but the fines originating from the feed or produced by the fast attrition process have short residence times in the bed and are therefore active to sulfur dioxide capture in the freeboard region.

INTRODUCTION

The importance of the freeboard region in the overall performance of the fluidized bed reactor has been reported(1,2,3,4),but there are only a little quantitative data available to understand the complex physical and chemical processes occurring in this region. It is important to study the extent of the reactions in the bed and in the freeboard separately.

The most important solid reactants in the freeboard of a fluidized bed combustor are char, ash and sorbent particles. The gases leaving the top of the bed are contacting sorbent particles entrained into the freeboard, therefore the concentration of sorbent particles in this region can affect the sulfur absorption in a fluidized bed combustor significantly. There are two sorbent particle fluxes in the freeboard region of a fluidized bed combustor. Fines with terminal velocities less than the superficial gas velocity have a constant flux in the freeboard and they will be elutriated. Other entrained particles which have greater terminal velocities than the gas velocity will reach a maximum height within the freeboard and then return ultimately back into the bed. The mass concentration of this second sorbent flux decreases along the freeboard with height(6).

The total amount of elutriated sorbent material is the sum of fine components due to feed,fragmentation, fast attrition and slow attrition (5).The thermal shock, the calcination of raw sorbent material and the mechanical stress results in production of fines before sulfation (fast attrition). Collisions cause further fines production after sulfation (slow attrition).

The objective of this study is to examine the sulfur capture potency of the freeboard and to see how it is influenced by the important operation variables in a fluidized bed combustor. The effects of the following variables were examined: sorbent particle size distribution, calcium to sulfur molar ratio in the feed, initial sulfur dioxide concentration and superficial gas velocity.

Extreme conditions were set to see how fast and how much the freeboard region can be effective in the overall sulfur reduction. Sulfur dioxide was supplied from a cylinder into the freeboard. Usually the sulfur dioxide concentration in the freeboard region will be less than in our case and sorbent particles entrained from the bed surface will be less active to capture sulfur dioxide in the freeboard.

THE EXPERIMENTAL COMBUSTOR

The experiments were carried out with an atmospheric fluidized bed combustor which had been built in the Laboratory for Thermal Power Engineering of the Delft University of Technology. The main dimensions are: the fluidized bed and freeboard diameter are both 0.39 m, freeboard height 2.75 m, expanded bed height 0.6 m.

Fresh dolomite particles are continuously fed into the bed, 0.1 m above the distributor, by means of a screw feeder and a chute. Sulfur dioxide supplied from a pressurized cylinder is fed centrally into the freeboard via a nozzle. Two consecutive standard cyclones, with diameters of 0.3 m and 0.2 m, respectively, remove elutriated solids from the exit gases. The gas samples pass an analysis system after being cooled to $150^{\circ}C$ and cleaned. The sulfur dioxide analyser is a type 400 Photometric Analyser (Do Pont Instruments). More details about the experimental combustor and also a sketch of it can be found in a recent study(6).

THE EXPERIMENTS

In all experiments, the bed temperature was kept at 1103K. The bulk of the bed material was silica sand with a surface mean diameter of d_{32} = 0.66 mm. The sorbent material used was dolomite originating from Marche les Dames (Belgium). The chemical composition of fresh dolomite as analysed is shown in table 1. The size distributions of the dolomite as determined by sieving are shown in figure 1. Feed1 is finer than 1.0 mm (d_{32} = 0.125 mm); feed2 is the dolomite as received and is finer than 4.0 mm (d_{32} = 1.2 mm); feed3 is classified between 2.0 mm to 4.0 mm (d_{32} = 2.8 mm). The combustor was fired with natural gas. The steady state oxygen content in the flue gas was about 8%. After the combustor reaches steady state conditions (combustion as well as feed sulfur dioxide) the experiment was started by switching on the dolomite feed.

In the experiments which have been carried out sulfur dioxide was supplied centrally into the freeboard at about 0.1 m height above the bed surface and the concentration was measured at 2.4 m height above the bed surface.

In the experiments which were carried out to determine the effects of the particle size distribution of sorbent material and of the initial sulfur dioxide concentration on the sulfur dioxide captured in the freeboard, three different size distributions of dolomite were used (figure 1). In these experiments the superficial gas velocity was constant (1.0 m/s), the calcium to sulfur molar ratio changed (1, 3 or 5) and the initial sulfur dioxide concentration in the exit gas was chosen at 1000 ppm, 2000 ppm or 3000 ppm.

To see the influence of the calcium to sulfur molar ratio on sulfur dioxide captured in the freeboard of the combustor, the superficial gas velocity was kept constant at 1.0 m/s. At calcium to sulfur molar ratios of 1, 3, 5 or 7 two different size distributions of sorbent material (feed1 and feed2) were used and the initial concentration of supplied sulfur dioxide was about 1000 ppm or 2000 ppm.

The effect of the superficial gas velocity was examined at velocities of 0.5, 1.0 and 1.5 m/s using dolomite from feed2 at a calcium to sulfur molar ratio of 3 and at an initial sulfur dioxide concentration of 2000 ppm.

RESULTS AND DISCUSSION

Effects of the Sorbent Particle Size Distribution

The entrainment rate of sorbent particles from the surface of the bed and the available surface area per mass unit for reaction are related strongly to the particle size distribution. This is caused by the decreasing terminal velocity of the particles, which was caused in turn by the decreasing particle size, thereby increasing the driving force. As to be seen from figures 2, 3 and 4, decreasing the particle size increases the sulfur retention in the freeboard. In experiments which are carried out using dolomite from feed1, the fines from the feed have a large effect on the sulfur dioxide captured in the freeboard. Therefore, just after starting feeding sorbent material into the bed, the sulfur dioxide concentration level in the exit gas is decreasing sharply (figures 2 and 4). With dolomite particles below 1.0 mm (feed1) and after 3000 s about 50% of the dolomite fed is elutriated. But, with particles below 4.0 mm and with particles between 2.0 mm and 4.0 mm, this value is 29% and 14% respectively(6). For small particles the rate of accumulation of sorbent material in the bed is less, therefore the slow attrition is not effective as it is for the bigger particles on the sulfur retention in the freeboard and it is easier to reach the steady state condition. The chemical analyses of cyclone materials showed that the elutriated sorbent particles from experiments which were carried out using bigger particles contain more sulfate. The cyclone material from an experiment, with dolomite particles below 1.0 mm at a calcium to sulfur molar ratio of 5 contains about 44% less sulfate than the one from an experiment using dolomite particles between 2.0 mm and 4.0 mm and 28% less than the one with dolomite particles below 4.0 mm. For a calcium to sulfur molar ratio of 3 in the feed these differences are 38% and 20% respectively. In these experiments the initial sulfur dioxide concentration was about 2000 ppm and the superficial gas velocity 1 m/s. The differences are reasonable, because feeding bigger sorbent particles into the bed causes an increase in the slow attrition rate and the dust produced from the surfaces of the partly sulfated larger particles contains a large proportion of sulfate.

Effects of the Calcium to Sulfur Molar Ratio in the Feed

As can be seen from figures 2 and 5 sulfur dioxide captured in the freeboard is increasing with an increase in calcium to sulfur molar ratio in the feed as a result of an increase of sorbent material concentration in the bed and in the freeboard (freeboard holdup). The data from the experiments which have been carried out with different calcium to sulfur molar ratios using different particle sizes of sorbent material showed that a decrease in particle sizes of sorbent increases the effectiveness. For higher feeding rates of sorbent the fines from feed and fast attrition are more effective on sulfur dioxide reduction. The bed material from experiments with higher calcium to sulfur molar ratios in the feed contains less sulfate. Therefore these sorbent particles are less resistant to size degradation and as a result the slow attrition is starting later. The chemical analyses showed that the sulfate content of elutriated sorbent material is decreasing with an increase in calcium to sulfur molar ratio in the feed. The cyclone material from the experiment which was carried out with a calcium to sulfur molar ratio of 3 (using dolomite from feed2) contains about 20% less sulfate than the one from the molar-ratio of 1.

Effects of Initial Sulfur Dioxide Concentration

It is obvious from figures 4 and 6 that for higher initial sulfur dioxide concentrations the freeboard of the combustor is more effective for sulfur reduction. For smaller particles the influence of this variable is larger, because higher sulfur dioxide concentrations result in a more rapid pore-blocking for coarse particles. The pore plugging time is inversely proportional to the surrounding concentration of sulfur dioxide(7). The results from chemical sulfate analyses indicate that the sorbent particles which are returning after being entrained from the surface of the bed and the sorbent particles in the cyclones contain more sulfate as the initial sulfur dioxide concentrations increase. For bigger particles the accumulation rate of sorbent in the bed is high and the slow attrition starts late when the sulfation degree is high. Attrition rates are decreasing with sulfation(8). To reach steady state conditions of sulfur capture in the freeboard is difficult if the sorbent material contains coarse particles and the ambient sulfur dioxide concentration is low. For an experiment with initial sulfur dioxide concentration of 1000 ppm the sorbent material from the cyclones contains about 15% less than the one which is carried out at 2000 ppm and about 39% less than the one at 3000 ppm. In these experiments dolomite particles below 4.0 mm were used at a calcium to sulfur molar ratio of 3.

Effects of Superficial Gas Velocity

At low velocities the freeboard sorbent particle holdup is small and therefore the sulfur reduction in the freeboard is less (figure 7), but the gas and solids residence times are longer. At higher superficial gas velocities the entrainment from the bed surface is large, but the residence time is short. Our experimental data showed that the bed material from the experiment which has been done at a superficial gas velocity of 0.5 m/s contains about 41% less sulfate than the one from the experiment at 1.5 m/s. But the cyclone material from the experiment at 1.5 m/s contains about 37% less sulfate than the one from 0.5 m/s. For a superficial gas velocity of 1.0 m/s the residence time is longer than the one at 1.5 m/s and the entrainment from the bed surface is larger than the one at 0.5 m/s. As a result of the combination of these effects the experiment which was carried out with a superficial gas velocity of 1.0 m/s showed the best sulfur retention in the freeboard under the examined velocities. In the experiment with a superficial gas velocity of 0.5 m/s after 3000 s about 4% of input dolomite was elutriated. For a superficial velocity of 1.0 m/s the elutriation increases to about 29%, and for 1.5 m/s to about 47% of input dolomite(6). Because of the longer residence time the elutriated material from the experiment at 0.5 m/s should contain more sulfate than the one at 1.0 m/s, but the elutriated sorbent particles from both experiments contain almost the same amount of sulfate. The reason for this could be the smaller sulfate content of the bed material and the less dust production from the surfaces of the partly sulfated bigger particles at low velocity.

LITERATUR CITED

1. Wen,C.Y., Chen,L.H., AIChE J., Vol. 28, No.1, 117 (1982).

2. Yates,J.G., Rowe,P.N., Trans. Instn. Chem. Engrs., Vol.55, 137 (1977).

3. Rajan,R.R., "Simulation of fluidized bed coal combustors," Dissertation, West Virginia Uni., Morgantown, W-VA, USA (1978).

4. Martens,F.J.A., "Freeboard phenomena in a fluidized bed coal combustor," Dissertation, Lab. for Thermal Power Eng., Delft Univ. of Tech., Delft, NL (1984).

5. Küçükbayrak,S., Martens,F.J.A., van den Berg,P.J., presented at the 8th Int. Conf. on FBC in Houston, Texas (18-21 March 1985).

6. Küçükbayrak,S., Boersma,D., van den Berg,P.J., presented at 12. Deutscher Flammentag, Düsseldorf, BRD (10-11 Sept. 1985).

7. Chrostowski,J.W., Georgakis,C., ACS Symposium Series, Vol.65 (1978).

8. Franceschi,J., Kolar,A., Miller,G., Zakkay,Y., Ho,C., Skelley,W., Hakim,S., Proceedings of the 6th Int. Conf. on FBC, Nat. Tech. Inf. Ser., Springfield, VA, USA, 1028 (9-11 April 1980).

9. Horio,M., Taki,A., Hsieh,Y.S., Muchi,I., Fluidization, Grace and Matsen (eds.), Plenum Press, New York, (1980).

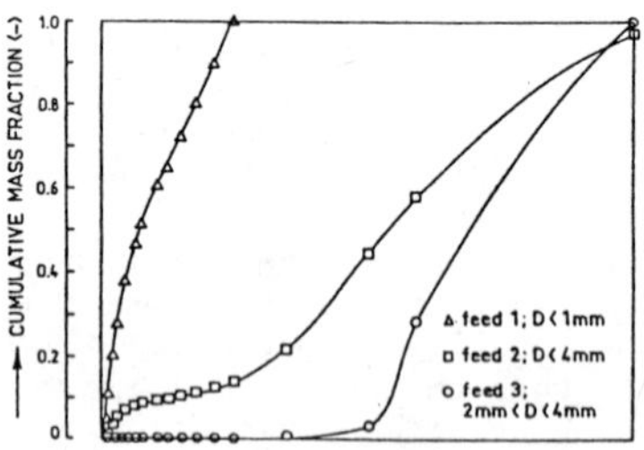

Figure 1. Size distribution of feed dolomite.

Table 1. Chemical composition of fresh dolomite (weight %)

$MgCO_3$	45.54%
$CaCO_3$	54.06%
SiO_2	0.14%
Al_2O_3	0.05%
Fe_2O_3	0.21%

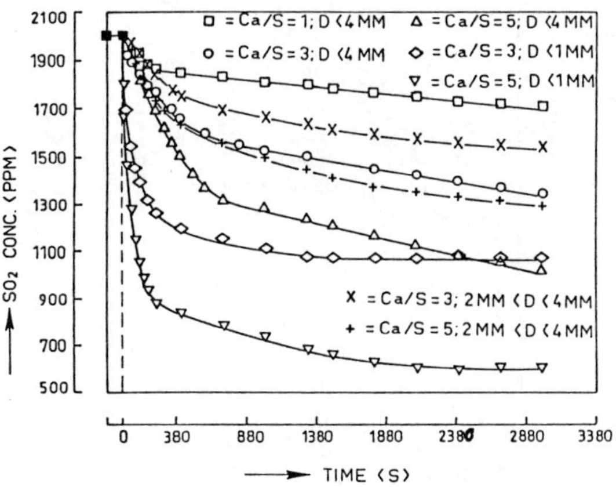

Figure 2. Effect of the sorbent feed rate and particle size distribution on the sulfur dioxide concentration in the exit gas (initial sulfur dioxide concentration 2000 ppm).

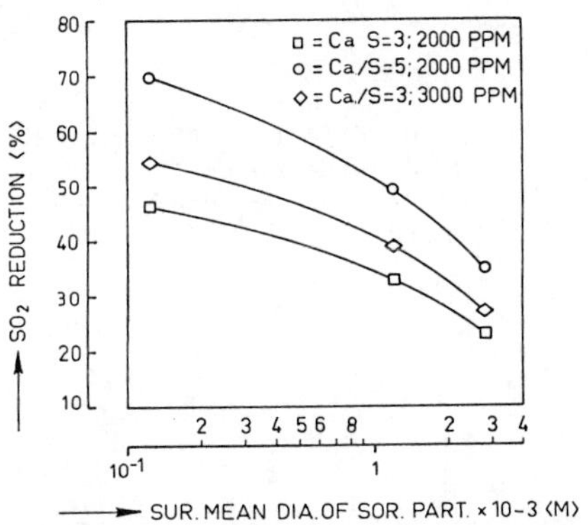

Figure 3. Effect of the surface mean diameter of sorbent particles in the feed on sulfur dioxide captured in the freeboard region.

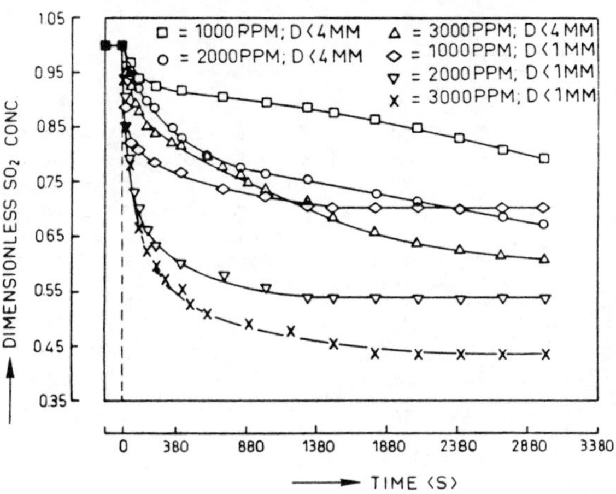

Figure 4. Effect of the initial sulfur dioxide concentration and particle size distribution of sorbent material on the sulfur dioxide concentration in the exit gas (Ca/S = 3).

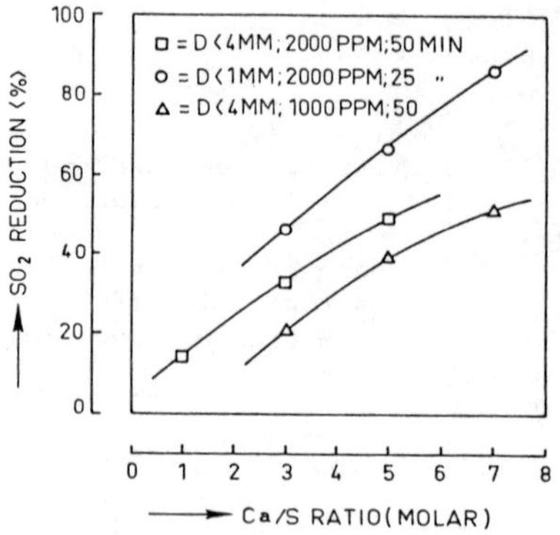

Figure 5. Effect of the feed rate of sorbent material on sulfur dioxide captured in the freeboard region.

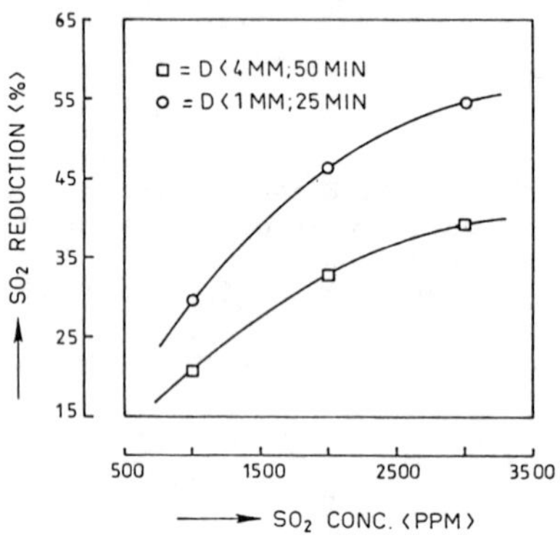

Figure 6. Effect of the initial sulfur dioxide concentration on sulfur dioxide captured in the freeboard region (Ca/S = 3).

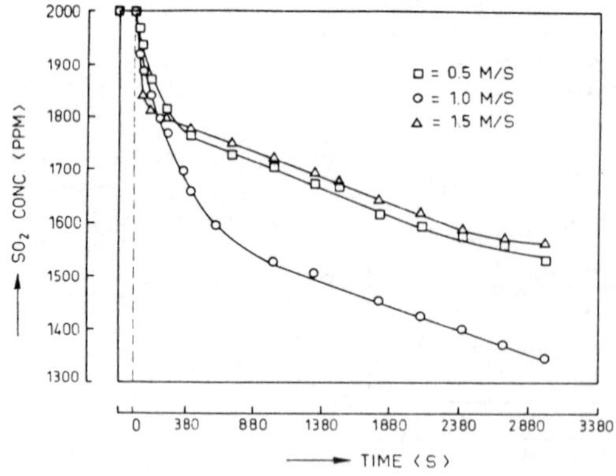

Figure 7. Effect of the superficial gas velocity on the exit gas sulfur dioxide concentration (initial sulfur dioxide concentration 2000 ppm; Ca/S = 3).

STEAM ACTIVATION OF COAL IN A FLUIDIZED BED

K.E. Makhorin

Institute of Gas, Ukrainian Academy of Sciences
Kiev, USSR

V.A. Borodulya

Heat and Mass Transfer Institute, BSSR Academy of Sciences
Minsk, USSR

ABSTRACT: The present study deals with the technology of steam activation of anthracite in fluidized bed furnaces. This process is studied in the pilot and laboratory reactors. The technology is tested in the demonstration reactor. In the USSR, the active anthracite production is developed. A new type of carbon sorbent is used for additional decontamination of sewage from organic admixtures.

SCOPE: The production of active coals is a promising branch of chemical industry. To diminish their cost, anthracite, brown coal and other non-traditional carbon-containing materials are used as raw materials. The steam processing of raw materials within 850-950°C (e.g. Wilson (1), Mathorin (2)) is a perspective tool to produce active coal from anthracite.

CONCLUSIONS AND SIGNIFICANCE: Physico-chemical characteristics of active anthracite depending on the content of H_2O mixed with nitrogen are specified. Means of coal heating in a fluidized bed up to an activation temperature are considered. Preference is given to steam activation of anthracite mixed with natural gas combustion products. Test indices of pilot and laboratory reactors are given. Operational data of industrial installations involving batch and continuous coal charge and discharge are presented.

Laboratory studies. Laboratory studies were begun of the utilization of the technology of anthracite activation. Anthracite was activated in the tube, through the wall of which heat was supplied. The Donetsk anthracite (moisture was 5%, ash was 6.8-8%, sulphur was 1.2-1.7%, volatile matter was 5.8-7.0%, particle size was 0.25-0.1 mm) was activated in the vertical 67 mm dia tube in the electric furnace. The feed of 200 g of anthracite corresponded to the 75 mm high deposited bed. The steam was blown through the bed in the amount ensuring the fluidized state. The steam flow rate was decreased from 1.8 to 1.2 kg/hr to preserve a constant fluidization particle velocity. When the coal was fed into the reactor the temperature was decreased up to 450-550°C, and 5-6 minutes were needed to heat the coal up to 900-950°C. In each 30 min, the coal samplings were made. Figure 1 shows adsorption isotherms for anthracite-activated benzene vapours in the vacuum device. A typical Langmuir adsorption isotherm is observed at a high anthracite activity at small P/P_s. The curves have no hysteresis loop, which points to a microporous coal structure. For 2 hr activation the combustion loss attained 65-75% and the micropore volume was 0.58-0.66 cm^3/g (Fig. 2). With the activation time being 2.5 hr, the micropore volume was decreased up to 0.43-0.63 cm^3/g. Similar adsorption isotherms for methyl alcohol and n-hexane vapours are obtained.

Table 1 displays the physico-chemical characteristics of active coal from 1-2 mm size grain anthracite (moisture is 2-3%; ash, 3-5%; volatile matter, 5%). For a temperature of 850°C, time of 2.5 hr and vapour velocity of 0.55-0.75 m/s, anthracite combustion loss was 55-60%; active coal contained 0.29-0.32 cm^3/g of micropores. When the steam was diluted with nitrogen amounting to more than 50%, the carbon gasification rate diminished. When the particle size was decreased less than 1 mm, this rate did not depend on the concentration of the steam mixed with nitrogen, starting from H_2O> 65%.

Table 1

Physico-Chemical Characteristics of Active Anthracite Versus Nitrogen Concentration in the Steam

Concentration, %		Combustio loss, %	Density, g/cm^3		Micropore volume cm^3/g	Surface m^2/g	Mean pore radius Å
H_2O	N_2		Bulk	apparent			
14	86	17	0.80	-	0.09	145	-
37	63	32	0.67	-	0.12	270	-
45	55	34	0.66	1.42	0.14	360	8.0
50	50	42	0.60	1.25	0.31	540	11.5
55	45	48	0.53	1.15	0.26	670	7.6
66	34	54	0.51	1.28	0.25	600	10.0
100	-	55	0.50	1.24	0.29	745	7.8
100	-	60	0.40	1.20	0.32	700	9.2

With anthracite combustion loss being 55-60%, the abrasion resistance is 90% according to the standard methods.

Pilot reactor. Figure 3 shows possible schemes of heat supply into the bed. When the steam is used as an activator, it must be heated up to 1000-1100°C. This superheating is performed in ceramic regenerators or the steam is generated by hydrogen combustion. In metal recuperators, the steam may be superheated up to 500-600°C. The additional amount of heat is supplied by fuel combustion products (Fig. 3b,c,d). Preference is given to vapour-gas activation of anthracite (Fig. 3c). This technology is verified in pilot reactors when an optimum regime of activation was achieved.

The utilization of pilot reactors has shown that heat losses are large because of their small sizes. In the chamber beneath a gas distributor in the 170 mm dia reactor, the temperature was kept within 1000-1200°C, at a height of 100 mm above the gas distributor it was decreased up to 800-950°C and at the reactor exit it was 600°C. The temperature gradient between the chamber beneath the gas distributor and the activation chamber was increased with decreasing particle size and with growing coal feed. At a temperature of 1150-1120°C beneath the gas distributor, anthracite ash was softened, thus promoting slagging. Therefore, in the pilot reactors the reaction gas temperature did not exceed 1100°C. The fuel was burnt in the external burner at a stoichiometric air flow rate. The fuel and air were preliminarily mixed and burnt in the no-flame burner in order to completely burn hydrocarbons and to withdraw oxygen from combustion products. The steam was added to the flue gases at the tunnel exit. The available free oxygen accelerated the activation and deteriorated the quality of active anthracite. Experiments were made in the gas amtosphere: H_2O=40-50%, CO_2=5-10%, N_2= =55-40%.

Anthracite starts activation at 750°C. The best samples of the active 0.75-2 mm size coal were obtained in the 170 mm dia reactor within 850-900°C, and the activation time was 2-2.5 hr. The coal combustion loss was 55-60%. The ultimately adsorption volume of pores was 0.33-0.35 cm^3/g and the absorptivity of phenol and methylene blue from the aqueous solution amounted to 10-14% of the coal mass. At 0.3-0.6 mm particle activation, the time was increased up to 5 hr (due to an increase in the coal heating time), and the reaction gas flow rate was decreased from 40-45 to 20-25 m^3/hr. For 60÷65% combustion loss, the phenol absorptivity of active anthracite was increased up to 16% of the coal mass.

In the 400 mm dia pilot reactor, the anthracite-containing moisture equal to 2.5%, ash equal to 7.5% and volatile matter being 5% was activated. The gas was filtrated through the 0.25-2.0 mm grain bed at a velocity of 0.6-0.75 m/s while in the 170 mm dia reactor, the velocity was equal to 1.8-2.5 m/s. The fluidized bed temperature varied from 900 to 950°C, the activation time was 4-4.5 hr and the combustion loss was 60-70%. In active anthracite, the amount of 0.25-1.0 mm particles was 75-85%. The reactor output was 60-75 kg of active coal a day. One tone of active coal having the following physico-chemical characteristics is manufactured:

Table 2

Quality of Active Anthracite under Different Vapour-Gas Activation Conditions

Flow rate gas m^3/hr	air m^3/h	vapour kg/h	Temperature, ^oC	Time hr	Combustion loss, %	Bulk density, g/cm^3	Pore volume cm^3/g	Surface m^2/g	Pore radius, Å	Temperature sampling ^oC
20	185	150	800	5.7	50	0.70	0.214	-	-	2.2
22	200	100	860	5.0	64	0.62	0.24	564	11	3.1
22	200	90	860	6.3	62	0.65	0.28	582	12	4.3

Table 3

Indices of Vapour-Gas Anthracite Activation in Batch Single-Chamber Fluidized Bed Reactors

Reactor	Activation chamber diameter, mm	Feed, kg	Particle size, mm	Temperature, ^oC	Time, hr heating	process	Coal mass loss, %
Laboratory	28	0.02	1-2	850	0.05-0.1	2.5	55-60
	67	0.1-0.2	0.25-1.0	900-950	0.1	2-2.5	65-75
Pilot	170	5	0.75-2.0	850-900	0.5	2.0-2.5	55-60
	400	40	0.25-2.0	900-950	1.0	4-4.5	60-70
	500	40-50	0.25-1.5	860-890	1.25	3.3-4.5	60-65
Demonstration	800	160-180	0.25-2.0	830-880	0.5-0.7	4.5-60	50-65
Industrial	2270	930	0.4-3.0	900	1.5-2.0	4-5	50-65
	2440	1500	0.25-2.0	850-870	2-3	5-6	60
	3570	5000	<3		4	7	65

Bulk density, g/cm^3	0.54-0.56
Apparent density, g/cm^3	1.17-1.25
True density, g/cm^3	1.94-1.99
Ultimately-adsorption pore volume, cm^3/g	0.31-0.34
Surface, m^2/g	760-930
Phenol adsorptivity, mg/g	150-170
Methylene blue, mg/h	140-160
Temperature sampling, ^oC	3.2-5.0

The pilot reactors did not reproduce the better samples of active coal under vapour gas activation conditions, as compared to those obtained in the steam atmosphere.

Demonstration unit. The developments of the Institute of Gas of the Ukrainian Academy of Sciences were adopted to design and test a demonstration unit. Lump anthracite served as the raw material and was ground on the jaw breaker, then on the rollers, achieving less than 6 mm grains. It was screened on the slieve having 0.25 and 2 mm holes. More than 2 mm grains were again ground. The dust coal (0-0.25 mm) formed due to grinding in the number of 10% was not activated. The fraction (0.25-2.0 mm) was transported by the air into the furnace. Here, the coal was deposited in the bunker and was periodically injected by the air into the fluidized bed reactor, shaped as a vertical cylindrical metal vessel lined with chamotte bricks, thus forming the inner 800 mm diameter. The shaft along the height was partitioned by the arch refractory gas distributor, into the holes of which the chamotte sieve caps were inserted. The useful gas distributor cross-section was 1.25%. In the chamber beneath the gas distributor 20-22 m^3/hr of a natural gas were burnt. 90-150 kg/hr of the steam were added to combustion products. The reaction gas within 1050-1100^oC was directed into the activation chamber. Here, anthracite was activated within 830-880^oC and at gas velocities of 0.7-0.95 m/s. The height of the fluidized bed was 600-800 mm and of the deposited bed, 350-400 mm. Activation gases were used for thermal regeneration of active coal and, when purified from dust, within 500-550^oC, were thrown into atmosphere. The one-time coal feed was 160-180 kg. The coal was kept in the reactor for 4-5 hr. 65-80 kg of active anthracite was removed from the furnace. When cooling, hydrogen and carbon oxide were evolved from active anthracite. The steam was accumulated in the freeboard area. The bunker was periodically blown-through by the steam to protect the gas explosion. During activation the coal mass and apparent density are gradually decreased. At a constant vapour-gas flow rate, the fluidized bed height in the reactor is decreased, and the fluidization velocity is increased, thus resulting in the elevated entrainment of coal dust from the fluidized bed. For one operation, 20-30 kg of coal dust can deposit in the cyclone. Despite the fact that anthracite has a very high thermal stability, the particles were ground in the fluidized bed. The quality index of active anthracite is given in Table 2.

The demonstration unit operated for 4 years. The production of active anthracite from 1 m^2 of a gas distributor was 0.5-1.6 tone a day. The production of one tone of active coal required 3 tones of anthracite, 1600 m^3 of a natural gas, 10 tones of the steam and 960 kW-hr of electrical energy.

Industrial reactor. Table 3 displays the indices of anthracite activation in industrial reactors, as compared to the laboratory, pilot and demonstration ones. In order to increase thermal efficiency of the industrial reactor, the activation gases are used for steam production and thermal regeneration of active anthracite. The technological scheme of one of the industrial installations is shown in Fig. 4. Donetsk anthracite of a 25+13 mm class is ground up to the grains of less than 1.6 mm. The 0.2-1.6 mm fraction coal is transported into the furnace compartment where it is deposited in the bunker. The 6.3 mm high reactor is lined inside with the chamotte bricks. The shaft is partitioned by a hole hearth into the chamber beneath the gas distributor 2.04 m dia and into the 2.27 m long activation chamber. The four--section arch gas distributor of heat-resistant concrete serves as a hearth. The chamotte caps having a 2-3% useful cross-section are inserted into the gas distributor. Anthracite in the amount of 200-237 kg is fed to the fluidized bed surface. Active coal (70-80 kg/hr) is fed through the opening at the gas distributor level into the tubular heat exchanger where the coal is cooled up to 50-80°C in the moving packed bed. 55-60 m^3/hr of a natural gas is burnt in the burner of the chamber beneath the gas distributor. Combustion products are mixed with 0.45-0.5 tone/hr of the steam and injected into the reactor within 1000-1050°C. In the fluidized bed, the temperature is maintained within 870-900°C. Three burners are mounted under the reactor arch, and 25 m^3/hr of a natural gas are burnt in them. A temperature beneath the arch is 1000-1100°C. The dust entrainment does not exceed 10%. About 50% of the anthracite dust are trapped in the cyclone. The remainder dust is deposited in the coal regeneration furnace chamber beneath the gas distributor. The particle size is mainly less than 200 microns. The production of one tone of active coal in the operating installation requires 3.3 tones of anthracite, 1100 m^3 of a natural gas and 6 tones of the steam.

LITERATURE CITED

1. Wilson, J., Fuel, 60 (9) 823 (1981).
2. Makhorin, K.E., Khim. Tekhnolog., 6, 7 (1982).

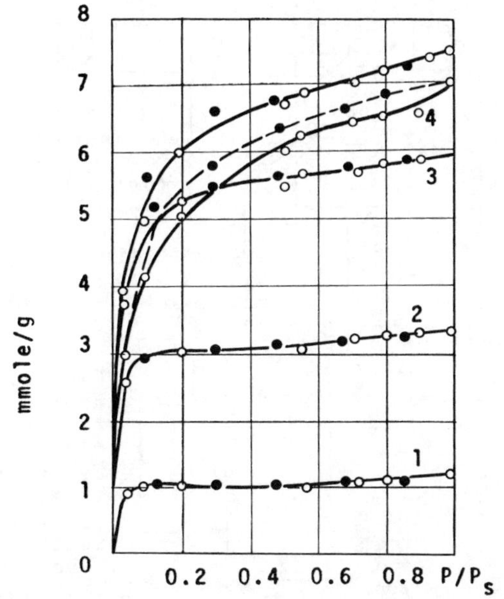

Fig. 1. Isotherms for active anthracite adsorption-desorption of benzene vapours: o-o, adsorption; ●-●, desorption; activation temperature is 950°C; time activation is in hours; 1, 0.5; 2, 1.0; 3, 1.5; 4, 2.0; 5, 2.5.

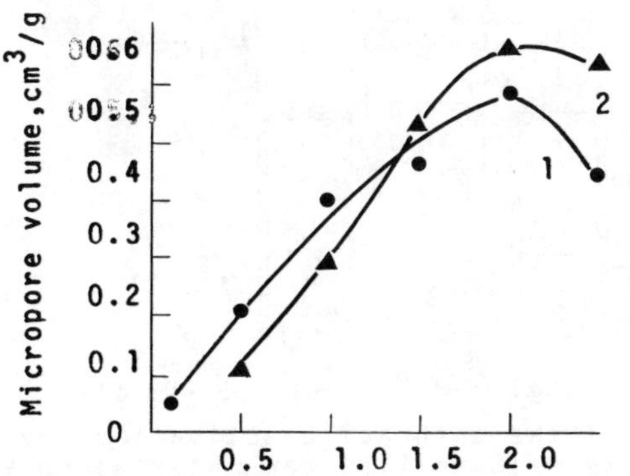

Fig. 2. Change in the micropore volume in activated anthracite vs time process, °C; 1, 900; 2, 950.

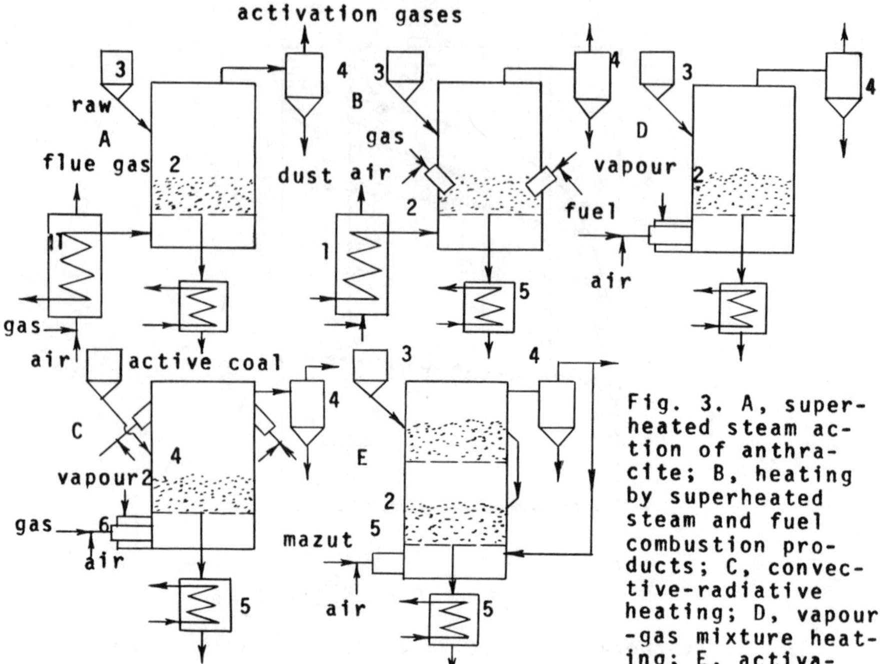

Fig. 3. A, superheated steam action of anthracite; B, heating by superheated steam and fuel combustion products; C, convective-radiative heating; D, vapour-gas mixture heating; E, activation gas heating; 1, steam superheater; 2, reactor; 3, bunker; 4, cyclone; 5, refrigerator; 6, bunker.

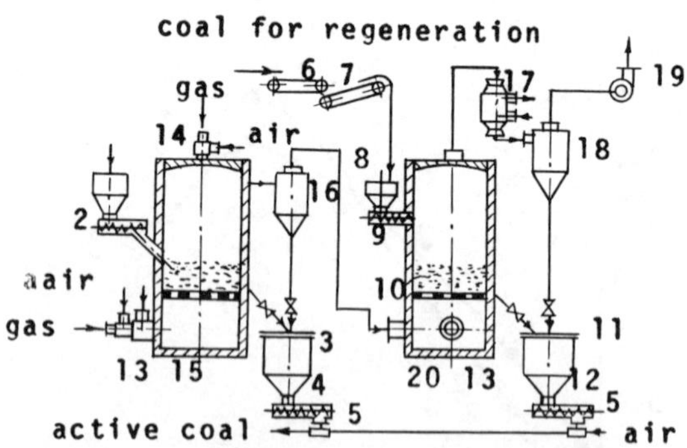

Fig. 4. Schematic of the anthracite activation device developed at the Institute of Gas: 1,8, raw material bunker;2,4,9, 12,feeder; 3,11, refrigerator; 5, pneumotube; 6, vacuum belt filter;7, conveyer belt; 10, reactor for active coal regeneration; 13,14, bunker; 15, reactor for anthracite activation; 16,18, cyclone; 17, air heater; 19, exhaust fan.

THE ABSORPTION OF LIGHT ENERGY IN A FLUIDIZED PHOTOREACTOR

P L Yue

School of Chemical Engineering, University of Bath, Bath, BA2 7AY, England.

ABSTRACT: The transmittance and reflectance of light energy in a flat gas-solid photoreactor containing fluidized catalyst particles were studied by an optical technique. Mean bed transmittance increases with bed expansion and the square root of particle diameter. Light reflectance was found to decrease with gas flow rate. These measurements provide the basis for the modelling of light absorption in fluidized photoreactors.

SCOPE: Research efforts in the conversion of solar energy to chemical energy are based on photocatalytic or photo-assisted reactions. Pseudo-homogenous and heterogeneous photocatalysis have been found to be effective in promoting the desired transformations (1) to (8). Photocatalysts in solid or dispersion form were employed in gas-liquid, gas-solid and gas-liquid-solid reactors. Before these photoreactions can be turned into feasible industrial processes, the modelling and design of heterogeneous photoreactors must be studied.

The complexity of modelling multiphase photoprocesses has long been pointed out (9). Some useful analyses have been conducted for homogeneous systems eg (10), (11) and have been reviewed recently (12), (13). Very few attempts have been made to model and design heterogeneous photoreactors. One type of flow reactor which has been used for performing gas-solid photocatalytic reactions is a fluidized bed (3), (8). The fluidized bed has the advantage of good and frequent contact between reactants, photons and catalysts. In one of these studies (8), a fluidized bed photoreactor with two transparent flat walls, ie a two-dimensional bed configuration, was used. Fluidization provided continuous exposure of the catalyst particles to light and improved light penetration into the bed. The high surface area to volume ratio of the flat bed enhanced the irradiation of solids and the absorption of the photons. In order to analyze such types of fluidized photoreactors, it is necessary to model the absorption of light energy by the fluidized photocatalyst. Previously work has been done on the measurement of light transmission through an irradiated fluidized bed containing gamma-alumina particles (14). The present investigation yields results of the first attempt to study the absorption of light energy by zeolite particles in a flat fluidized bed photoreactor.

The fluidized bed with an internal diameter of 2.1 mm was irradiated by a line or point source of light directed by an optical fibre bundle. The fraction of light which passed through the reactor and that which reflected from the walls were measured at different bed heights over a range of reactor conditions. Mean particle diameter varied from 0.114 to 0.165 mm. The range of gas flow rate used covered from minimum fluidization to eight times the minimum. Both transmitted and reflected light have been correlated with reactor and fluidization variables.

CONCLUSIONS AND SIGNIFICANCE: Light transmission through a flat fluidized photoreactor increases with bed height and gas flow rate. The fraction of light energy lost through the entire reactor varies linearly with bed expansion and the square root of particle diameter. The fraction of reflected light decreases with increasing flow rate, but is not a strong function of bed height at low flow rates. The flat fluidized bed is an efficient photoreactor because of its high absorbance.

INTRODUCTION

Recent research on heterogeneous photocatalysis has raised the hope of directly converting solar energy to chemical energy. However, fundamental studies of photoreactor design are required if these photocatalytic processes are to be exploited. In the modelling and design of photoreactors, the transport phenomena and reaction kinetics must be analyzed simultaneously with the absorption of light radiation energy.

The absorption of light energy in a gas-liquid photoreactor can be determined by performing a test reaction such as the photoreduction of potassium ferrioxalate, the kinetics of which are well established. There is, as yet, no suitable test reaction which can be easily performed in gas-solid photoreactors. The optical method described in the present study does not rely on a test reaction of known kinetics.

In a homogeneous medium light absorption follows the Beer-Lambert law which is expressed by the simple exponential extinction functional form. In the case of a non-stationary heterogeneous medium such as fluidized solid photocatalyst particles, the scattering of photons in every direction pervades the entire reactor. The absorption of radiant energy by an absorbing-scattering medium is therefore considerably more complex (15). The present study is an attempt to measure experimentally the two key variables, namely transmittance and reflectance, which are needed for the determination of light absorption in a fluidized photoreactor by the optical method. Although no reaction was conducted in the fluidized bed the catalyst and fluidization conditions were typical of those used for photocatalysis.

The photoreactor used in this study is of the parallel-plate geometry. Such a configuration is, of course, that of a two-dimensional fluidized bed with transparent walls. The parallel-plate (or flat bed) geometry, in contrast with the cylindrical one increases the fraction of solids simultaneously exposed to irradiation. Fluidization provides the continuous exposure of catalyst particles to light and improves light penetration into the bed, thus enhancing photocatalysis (3), (8).

EXPERIMENTAL

The flat fluidized bed used has the dimensions of 250 x 100 x 2.1 mm.

The reactor was masked at the sides to eliminate lateral loss of light. The catalyst was ground and screened to give different size fractions. The particles were supported on a porous distributor which gave a sufficient pressure drop to ensure uniform gas distribution and were fluidized by high purity nitrogen. The total static bed height was about 96 mm. Bed characteristics at incipient fluidization, ie U_{mf} and H_{mf} were carefully determined from the measurements of pressure drop and bed height variation with gas flow rate. Gas flow rates were measured by two calibrated rotameters.

The light source was a 3 mm diameter bifurcated fibre-optic probe which was used in two different ways depending on whether transmittance or reflectance was being measured. For the former measurements the probe was inserted into a metal housing where the light was converted into a thin horizontal beam by means of an optical fibre bundle converter. The thin light beam was collimated by a lens to produce a 0.1 mm wide by 1.5 mm thick line source at the focus of the lens. The reactor was placed at this focus; thus only a narrow strip of the reactor was irradiated. The light transmitted through the reactor was collimated and collected by another set of lens and optical fibre converter operating in the opposite direction, and was collected and measured by a photodiode. The arrangement is schematically represented in Figure 1a.

For the reflectance experiments, the optical fibre converters and collimators were not used. Instead, the fibre-optic light probe was placed directly against the front wall of the reactor giving effectively a point light source. The reflected light was directed by means of a bifurcation in the probe to a second photodiode, as shown in Figure 1b. The signals of the photodiodes were fed to a photometer which gave digital displays as well as analogue voltage outputs. The photometer gave measures of either illumination (lumens m^{-2}) or irradiance (w m^{-2}) and was callibrated by taking readings of light through a series of apertures of known area.

The voltage outputs were amplified and recorded onto tapes by means of a Racal 'Store 4' instrumentation tape recorder with four channels. The tapes were replayed at a slower speed and the signals digitized by the analog-to-digital converter of a MINC-23 minicomputer at a sample frequency of 100 Hz. The signals were then analyzed to produce time mean values under different conditions. Since a point source was used for reflectance measurements, space averages were obtained for any given bed height by traversing the light probe across the width of the bed. Thus it was possible to measure light transmittance and reflectance at different bed heights for a chosen particle size and gas flow rate.

RESULTS AND DISCUSSION

In all the experiments performed, the particles were observed to fluidize uniformly throughout the bed, with bubble size distribution characteristic of a freely bubbling bed. The particles belong to the Group A classification.

Local light transmittance E_T is a function of bubble voidage e_b and the percentage of particles within a given bubble γ_b. Bubble voidage, with respect to light transmission through a two dimensional fluidized bed can be treated as having two portions. One portion (e_{b1}) permits net transmission of light but not the other (e_{b2}). The proportion of e_{b1} to e_{b2} depends onbubble sizes relative to the gap between the transparent walls of the bed. As bubble size increases due to increasing flow rate or coalescence, e_{b1} increases because more bubbles bridge the internal walls of the bed. Particles present in the bubbles cause light scattering and reduce light transmittance. γ_b is a function of particles raining through the bubbles, particle wall bridging and the content of fines.

Figure 2 shows the variation of light transmittance E_T with gas flow rate and bed height. E_T is expressed as a fraction of that transmitted through the empty reactor. At the lowest rate of 1.3 U_{mf}, light transmittance was less than 1% with little variation over the entire height of the bed. As the flow rate was increased to 2.3 U_{mf}, bubbling became more vigorous giving higher transmittance at the upper part of the bed. Further increase in flow rate caused rise in transmittance between 50 to 70 mm above the distributor. Beyond this bed height, the transmittance levelled off and for most cases began to drop by a small amount as the top of the bed was approached. The slight reduction was most probably due to an increase in γ_b. Bubbles appeared to be cloudier as they moved up the bed because of a small degree of particle erosion and size segregation. No significant trend of reduction of light transmittance was observed with fluidized gamma-alumina at the top of the bed because there was less particle erosion. Otherwise the present results are consistent with the previous ones (14) obtained for gamma-alumina.

The results of light reflectance as functions of flow rate and bed height are illustrated in Figure 3. The values of reflectance are given in terms of the fraction of the reflectance obtained when the particles were stationary. Reflectance was found to decline with increasing flow rates for all three particle sizes. It can be argued that the major part of light reflectance is due to that reflected or refracted by the particles in the vicinity of the reactor front wall. Incident light on a void is either transmitted through the reactor or scattered from the bed-void interface (see Figure 4). Hence increasing gas flow rate increases bubble voidage and reduces reflectance. From Figure 3 it can be seen that at low flow rates (ie less than $3U_{mf}$) reflectance is essentially independent of bed height as there are not many voids which span across the thickness of the bed. However, at higher flow rates there is an initial increase in reflectance below the first 50 mm of the bed followed by a decline. This indicates that although bubble voidage is increased with higher flow rates, the voids only span the reactor walls beyond a certain bed height (about 50 mm). Below this bed height bubble voidage enhances the scattering of light, thus increasing light reflectance.

The average light transmittance $\bar{E}_T$ of the entire reactor were obtained by integrating the area under the E_T versus h curves and converting E_T into percentage incident light. Of course, light scattering has an effect on local transmittance measurements at any given bed height. However, the use of a 20 mm diameter collimating lens on the detection side has largely included the scattering effect in the vicinity of the bed height under consideration.

The mean bed transmittance was correlated with bed expansion and particle diameter resulting in the following equation:

$$\bar{E}_T = a + b\,\varepsilon_b\, d_p^{\frac{1}{2}} \qquad (1)$$

The values of a and b were -0.131 and 7.689 respectively, with a regression coefficient of 0.96. Bed expansion was evaluated by Geldart's correlation (16). Equation (1) therefore confirms that transmittance is directly proportional to bed expansion and to the square root of particle diameter (14).

The mean bed reflectance was obtained by similar integration and correlated in the same way to give equation (2).

$$\bar{E}_R = \left(\frac{3.83}{d_p} - 5.0\right) - \left(\frac{5.92}{d_p} + 0.87\right)\varepsilon_b \qquad (2)$$

The above equation shows that reflectance decreases with increasing bed expansion and particle diameter.

Since only one reactor was used in the present study, it was not possible to obtain an expression to correlate light absorption with distance of penetration. However, the study reported here provides the methodology and a quantitative description of the two variables which can be used for the modelling of light absorption in flat fluidized beds.

Finally, it should be noted that the flat fluidized bed has a significantly high absorbance, as the maximum transmittance is in the order of 10%, while reflectance is no more than 4% of incident light. At flow rates of 2 to 3 times U_{mf}, bed transmittance is only about 2%. It would appear that a fluidized photoreactor is most efficient when the reactor conditions are such that there is significant bubble voidage, but the size of which is below the reactor thickness. Further studies are required to produce the optimum design of such photoreactors.

ACKNOWLEDGEMENT

The author wishes to acknowledge the award of a Science and Engineering Research Council grant and the assistance of Mr P Speller and Professor L. Rizzuti.

NOTATION

a	coefficient in equation (1)	
b	coefficient in equation (2)	
d_p	mean particle diameter	mm
E_T	transmittance at a given bed height	
$\bar{E}_T$	mean bed transmittance	

$\underline{E}_R$	reflectance at a given bed height	
$\bar{E}_R$	mean bed reflectance	
e_b	bubble voidage	
e_{b1}	that fraction of bubble voidage which permits net light transmission	
e_{b2}	that fraction of bubble voidage which does not permit net light transmission	
h	bed height	mm
H	total bed height	mm
H_{mf}	bed height at incipient fluidization	mm
U	fluidization velocity	mm s^{-1}
U_{mf}	minimum fluidization velocity	mm s^{-1}
ε_b	bed expansion	
γ_b	percentage of particles present within a given bubble	

LITERATURE CITED

1. Sato, S. and J.M. White, J. Catal., 69, 128 (1981).
2. Kwai, T. and T. Sakata, Chem. Phys. Lett., 72, 87, (1980).
3. Augugliaro, V., Lauricella, A., Rizzuti, L., Schiavello, M. and A. Sclafani, Int. J. Hydrogen Energy, 7 (11), 845 (1982).
4. Schrauzer, G.N. and T.D. Guth, J. Am. Chem. Soc., 99, 7189 (1977).
5. Borgarello, E., Kiwi, J., Pelizzeti, E., Visca, M. and M. Grätzel, J. Am. Chem. Soc., 103, 6324 (1981).
6. Pichat, P., Hermann, J.-M., Courbon, H., Disdier, J. and M.-N. Mozzanega, Can. J. Chem. Eng., 60, 27 (1982).
7. Khan, F., Yue, P.L., Rizzuti, L., Augugliaro, V. and A. Burcato, Ind. Eng. Chem. Prod. R. & D., 22, 238 (1983).
8. Yue, P.L., Khan, F. and L. Rizzuti, Chem. Eng. Sci., 38 (11), 1893 (1983).
9. Cassano, A.E., Silverston, P.L. and J.M. Smith, Ind. Eng. Chem., 59, 18 (1967).
10. Jacob, S.M. and J.S. Dranoff, A.I.Chem.E.Symp. Ser., No. 89, 64, 54 (1968).
11. Skarbo, R. and J.A. Williams, Chem. Eng. Sci., 28, 83 (1973).
12. Yue, P.L., NATO ASI Ser., C., 146, 527 (1984).
13. Augugliaro, V., NATO ASI Ser. C., 146, 605 (1984).
14. Rizzuti, L. and P.L. Yue, Chem. Eng. Sci., 38 (8), 1241 (1983).
15. Santarelli, F., NATO ASI Ser. C., 146, 549 (1984).
16. Geldart, D., "The Expansion of Fluidized Beds", in Fluidization Technology, Keairns, D.L. (Ed.), McGraw-Hill, New York (1975).

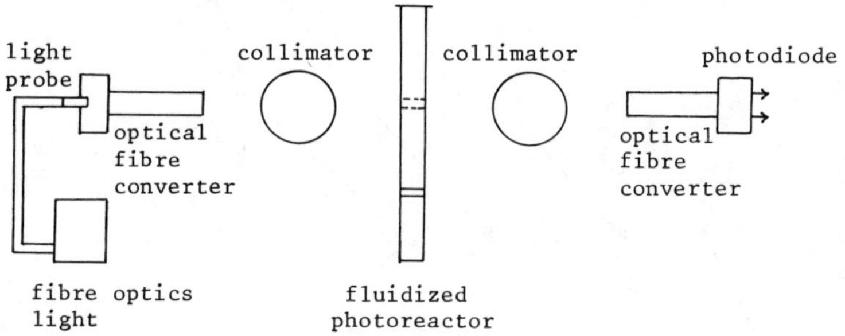

Figure 1a. Measuring Transmittance

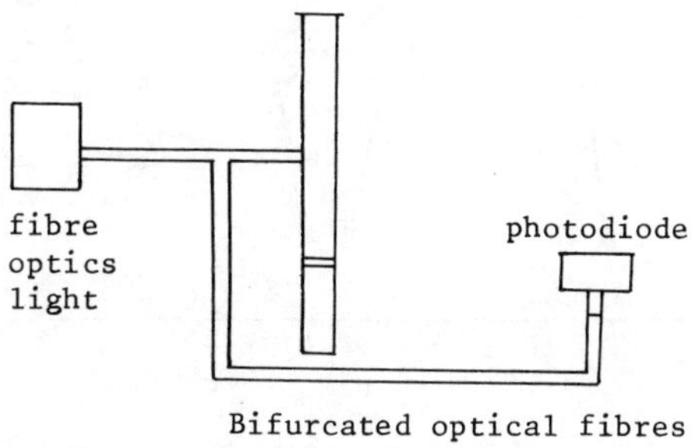

Figure 1b. Measuring Reflectance

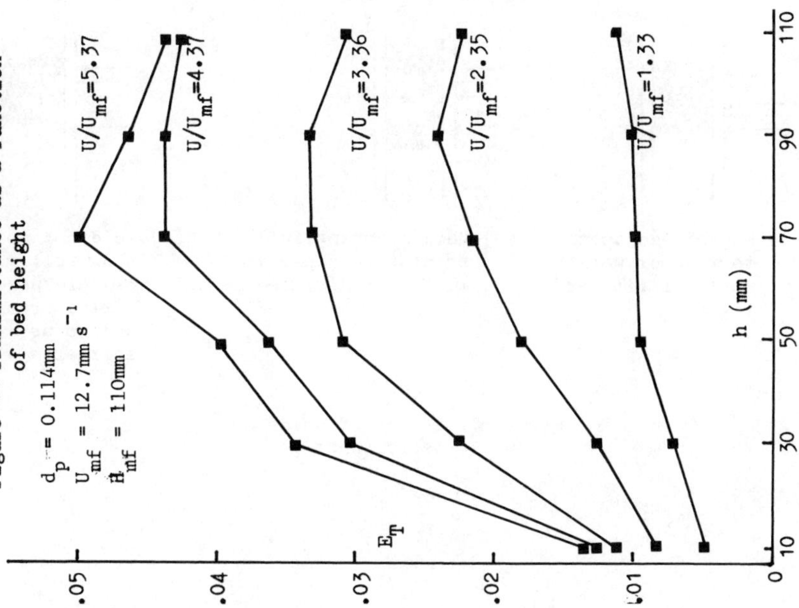

Figure 2. Transmittance as a function of bed height

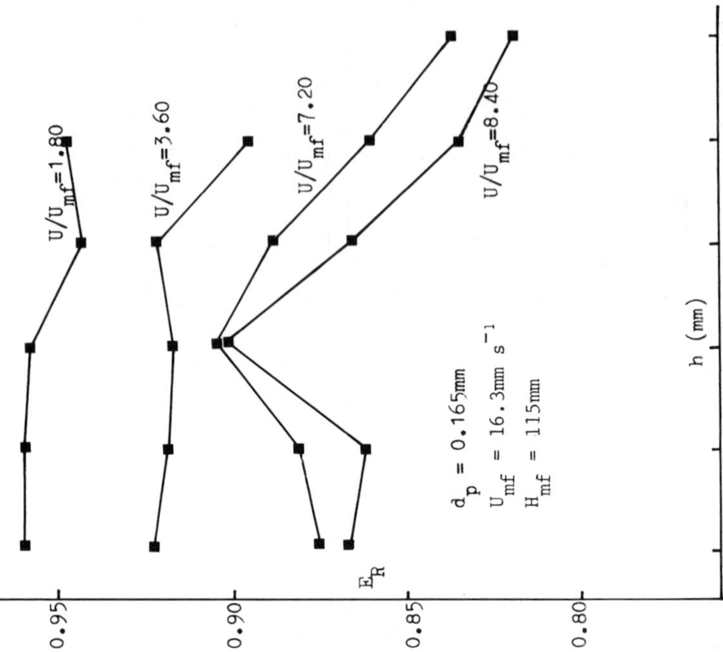

Figure 3. Reflectance as a function of bed height

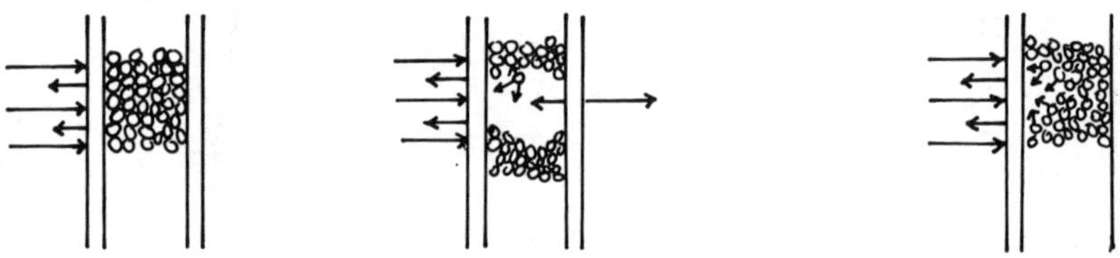

(a) No void adjacent to reactor wall: mostly reflected light

(b) Void spanning full reactor thickness: mostly transmitted light

(c) Void not spanning full reactor thickness: partial reflectance plus scattering

Figure 4. Effect of voids on reflectance

SIMULATION AND OPTIMIZATION OF A CONTINUOUS FLUIDIZED BED BIOREACTOR PRODUCING THE ANTIBIOTIC PENICILLIN-G

Nicolas Kalogerakis and Leo A. Behie
Department of Chemical and Petroleum Engineering
and
G. Maurice Gaucher
Division of Biochemistry
The University of Calgary, Calgary, Alberta T2N 1N4

ABSTRACT

A 1.2 L three phase fluidized bed bioreactor has been operated continuously producing the antibiotic penicillin-G. Maximum specific rates of reaction (q_p^{max}) and yields (Y) were measured at 13.5 mg Pen-G(K+)/g protein/h and 0.068 mol Pen-G/mol lactose respectively which are comparable with present industry operations. Further, the performance of the system was simulated under computer control and the expected productivity gains are presented.

SCOPE

The development of three phase fluidized bed bioreactors is presently having an enormous impact on the pharmaceutical industry. This is particularly true for antibiotics which are the most important products produced by biological processes. For example, the current world production of Penicillin-G exceeds 20 000 tonne/year and bioreactor vessels up to 100 000 L are not uncommon. Despite the large scale of production, batch and fed-batch submerged free cell fermentations remain the predominant modes of operation in industry.

The main problem of free cell batch or fed-batch fermentations is the severe limitations in oxygen transfer (from gas bubbles to microbes dispersed in the liquid medium) since, as the amount of biomass in the reactor increases, the liquid phase becomes very viscous and difficult to agitate. On the other hand, in a continuous operation when the growth rate of cells is less than their removal rate, the cells are washed out and thus the antibiotic production is gradually ceased. Unless these problems are overcome, no economic benefit can be expected by going to larger scale continuous operation.

Immobilized cell technology provides a simple and promising alternative to free cell fermentations, the principle difference being the capability to run the fermentation in a continuous fashion. Recently, our biotechnology group at the University of Calgary has successfully developed a three phase fluidized bed bioreactor where the limitations of oxygen transfer are completely alleviated (Berk et al., 1,2,3; Jones et al., 4,5). The growing cells containing the necessary biocatalyst (enzymes) are immobilized inside inert support matrices. This makes up the solid phase. With concurrent air, the fluidized bed can be described as having bubble supported solids in which the liquid (medium) flow is far below its minimum fluidization velocity.

Besides new strain development, proper selection of the feeding medium and state-of-the-art computer control of the fermentation process are the only means to achieve high production yields which could make the continuous operation an economic success. This paper presents data on mass transfer effects in the fluidized bed fermentor, specific rates of reaction for immobilized cell penicillin-G production, and a mathematical model used to evaluate the performance of the new process under computer control.

CONCLUSIONS AND SIGNIFICANCE

An immobilized cell fluidized bed bioreactor has been run continuously for over 10 days using two different support matrices. With our control system in place, the productivity of the bioreactor can be increased substantially to make continuous fermentations economically attractive.

The advent of immobilized whole cell technology has generated considerable interest in the field of biotechnology and several applications have been extrapolated from the laboratory to the pilot plant or industrial scale (Chibata and Tosa, 6). A number of advantages are inherent in the use of immobilized cells. These include the increased lifetime of the cells in the immobilized state, increased productivity, and from a chemical point of view, a means of overcoming mass transfer problems usually associated with growth at high cell densities. This change in catalyst conformation from either single or mycelial cells to that of larger spherical or irregularly shaped particles has also lead to the development of specialized bioreactors, for example bubble columns, fluidized beds or airlift reactors to accomodate these biocatalysts.

Fungi are an important source of a variety of industrially important products e.g. antibiotics, steroids and enzymes. However, an unfortunate feature of the growth of these mycelial organisms is the high viscosity of the culture broth. This results in a decrease in the mass transfer capabilities of conventional stirred tank reactors for oxygen and to a lesser degree for nutrients such as carbon and nitrogen sources. To overcome these problems the agitation and aeration rates can be increased but this often results in severe shear stresses being imposed on the cells. Consequently cell viability decreases and thus productivity declines. An alternative to mycelial growth is that of pelleted growth. The latter can be induced by manipulation of the growth medium or environmental factors such as pH (Kim et al., 7; Konig et al., 8; Wittler et al., 9). Alternatively, the rheology of the broth can be similarly changed by using immobilized cells, which can be regarded as mycelial pellets since growth is often restricted to the external surface of the beads. Thus, immobilized fungal cells can serve as a convenient method of directed pellet formation which does not suffer from the often unpredictable media or environmental manipulations necessary for induced pelletization in normal cultures.

Our studies have been directed to exploring the potential uses of immobilized cells for the production of antibiotics in three phase fluidized beds and to an understanding of the engineering aspects of bioreactor design for the use of such biocatalysts. As model systems we have examined the production of the mycotoxin, patulin (Berk et al., 2,3) and the industrially important antibiotic, penicillin-G (Berk et al., 1). Several aspects for the cells of Penicillium Chrysogenum immobilized in carrageenan have been examined and a medium for the semi-continuous or continuous production of penicillin-G over a period of 10-15 days has been developed. As an extension of these studies, attention has now been turned to the use of Celite beads as an immobilization matrix (Jones et al., 5). Celite, which is available commercially in a variety of particle sizes, is also used as a support in heterogenous catalysis systems. Its use as a support matrix for Penicillium Chrysogenum was described by Gbowenyo and Wang (10, 11). These authors showed that the cells would successfully produce penicillin in a tower loop reactor. Of interest is their finding that normal cultures had a much reduced capacity for antibiotic production under similar conditions of growth.

This paper presents data on two similar types of cell support matrices used in a 1.2 L three phase fluidized bed bioreactor running continuously. Futhermore, a model of the system is presented and is subsequently used to simulate the performance of the bioreactor under computer control.

EXPERIMENTAL MATERIALS AND METHODS

Immobilization in Carrageenan

Two cell support matrices were used in this study. The first, κ-carrageenan, is an inert polysacharide derived from seaweed and sold commercially by the FMC Corporation. It had many attributes as a support. Not only were the carrageenan beads very robust but they allowed the fungi to grow inside the beads without affecting the morphology of the cells. Furthermore, the spherical beads, having a diameter of either 3.0 mm or 1.8 mm, had a very narrow size distribution (standard deviation 0.15 mm). The carrageenan beads were casted in a device developed specifically for this study (Berk et al., 2; Jones et al., 4). About 20 000 beads were used in the reactor. The culture of Penicillium Chrysogenum E15 (ATCC 26818) was obtained as freeze-dried ampoules from Dr. S.W. Queener of the Eli Lilly Co. Ltd.).

Immobilization in Celite Beads

Celite is basically diatomaceous earth which comes in different mesh sizes and different pore size distributions. In this study, Celite R-630 (Johns Manville Co. Ltd), was used with an initial mean diameter of 410 µm, pore diameter of 8.0 µm, specific pore volume 1.70 cm^3/g and bulk density of 0.27 g/cm^3. All preparation details can be found in Jones et al. (5).

Three Phase Fluidized Bed Reactor

The reactor had a diameter of 7.5 cm and a working volume of 1.2 L. At the reactor top, the diameter increased to provide a solids disengaging section. Saturated air was introduced uniformly at the bottom through a high pressure drop porous distributor at a superficial velocity of 1.4 cm/s. The reactor was enclosed in a constant temperature jacket that maintained the temperature at 26 C.

The liquid was fed through a tube just above the distributor. In the production medium all necessary nutrients were present, including: KH_2PO_4 (6.8 g/L), K_2SO_4 (3.95 g/L), phenylacetic acid (0.3 g/L) and finally, NH_4Cl and lactose as nitrogen and carbon source respectively. All microbes were confined in the beads at all times with the medium essentially free of cells. Further, the medium pH was kept in the range 6.2 to 6.8 in order to minimize penicillin-G hydrolysis to 6-APA (Pirt, 12). Finally, the mean residence time of the medium in the reactor was in the range 24 to 50 h. More details on the design and operation of the reactor as well as all measurement techniques can be found in Berk et al (2,3).

EXPERIMENTAL RESULTS AND DISCUSSION

Carrageenan Beads

In the continuous runs presented here, three different levels of lactose concentration in the feed (10, 5 and 2 g/L) were used. The penicillin concentration in the reactor effluent versus time, at a liquid residence time of 24 h, is shown in Figure 1 for a bead size of 3.0 mm. Also shown is the effect of reducing the nitrogen source in the feed stream from 22 mM to 11 mM (NH_4Cl). The results show that when the lactose concentration in the feed is 2 g/L, the penicillin concentration in the effluent stays constant at 150 mg/L for the first 100 h and then starts declining slowly until it reaches 75 mg/L by 300 h. On the other hand, a rapid increase in the production of penicillin is observed during the first 40 to 60 h when the higher feed concentrations of lactose (10 and 5 g/L) are used. Decreasing the concentration of the ammonium ion from 22 to 11 mM results in a decrease in the penicillin concentration.

The effluent concentrations of ammonia and lactose were also measured and are shown in Figures 2 and 3. When the lactose content in the feed was low (5 or 2 g/L), its residual concentration in the effluent was zero at all times indicating that it was the limiting substrate. This was never the case with the high lactose concentration feeds. On the other hand, there was a nitrogen limitation in the reactor when the ammonium ion was reduced to 11 mM.

An unfortunate characteristic of the *P. Chrysogenum* for maintaining its capability to synthesize penicillin is that it requires the formation of new cells. The effect of lactose concentration in the feed on the increase in the total protein content in the reactor is shown in Table 1. As expected, higher lactose concentrations in the feed resulted in a higher biomass increase in the reactor. The highest biomass concentration was obtained when neither lactose nor ammonium were in limiting concentrations. It is interesting to note that the highest penicillin concentration was also obtained for the same conditions in the feed. As the biomass was reduced due to nutrient limitations, the penicillin production was also seen to decrease.

In figures 4 and 5 the specific reaction rates (q_p) and yield coefficients (Y) are shown versus time for two different bead sizes (3.0 and 1.8 mm). The curves were generated from a material balance on the reactor taking into account the accumulation terms as well. Note that, although the reactor was running continuously, it never achieved steady state conditions.

The dominant mass transfer resistance in the reactor was found to be diffusion within the solid support matrix. Mass diffusivities in carrageenan have been measured and confirm this conclusion (Berk et al., 2,3). However, most of the cell growth occurs on the surface of the beads (final layer thickness was less than 60 μm). Thus, by going from 3.0 mm beads to 1.8 mm, the reaction rates and yields remained the same as shown in Figures 4 and 5. Hence, the bead mass transfer resistance was not significant. In other words, the reaction rates were very slow and the reactor was under kinetic control. Although cell spores were initially uniformly dispersed throughout the bead, only the growing cells on the surface produced penicillin.

Celite Beads

In a recent study, Jones et al. (5) found that Celite beads were five times more productive on a reactor volume basis. This result is not surprising since Celite beads were able to support much higher cell densities (more than 30 g/L), and hence, they perform much better in a three phase fluidized bed. Nevertheless, the maximum specific reaction rates (q_p^{max}) observed for both carrageenan and celite were practically the same at about 13.5 mg Pen-G(K+)/g protein/h, a value comparable to that found in submerged free cell fed-batch fermentations.

INCREASED PRODUCTIVITY UNDER COMPUTER CONTROL

Recently, we designed a simple noninteractive control system which allows the user to control the specific growth rate (μ) and the penicillin concentration in the effluent stream, to any desired values by manipulating the feed flowrate and the substrate concentration in the feed (Kalogerakis et al., 13).

The only limitation of the system is that, as the cells in the reactor grow continuously at the specified growth rate, there will come a certain point in time where the bed will defluidize. The way around this problem is to remove beads and add new ones in the reactor continuously or in an intermittent fashion so that the average cell concentration in the reactor remains constant. This will allow the fermentations to run for extended periods of time until other factors come into play (eg. contamination). The controller has two feedback loops and thus posesses self correcting capabilities to modelling errors (which can be viewed as slowly varying model parameters).

The cell concentration in the reactor can easily be kept constant by introducing an independent P+I controller which will manipulate the bead removal rate. Such a modification to the reactor will allow the system to reach steady state conditions where the average cell concentration, specific growth rate and penicillin concentration remain constant to the values selected by the process engineer.

The underlying idea here is to select the desired specific cell growth rate (0.010 to 0.015 h^{-1}) where the penicillin production rate is maximum and at the same time keep the average cell concentration at a constant but high value (35 to 50 g/L) so that the penicillin productivity ($q_p x$) is maintained very high at all times during the fermentation. Furthermore, the penicillin concentration in the effluent will be kept constant at the specified concentration level by the control system so that the potency of the product stream is high. The actual value will be the result of experimental investigations so that any possible inhibition effects or side reactions are avoided. In summary, the system will be maintained under tightly controlled environmental conditions which will correspond to the best operating conditions for penicillin production. It is assumed here that pH and temperature will also be kept constant at optimally selected values by conventional PID controllers. Dissolved oxygen is not a problem for our reactor since we work with high air flows and the broth remains clear of cells at all times.

The dynamic behavior of the fermentation process is modelled by considering unsteady state component mass balances, namely

$$dx/dt = (\mu - F)x \quad (1)$$

$$ds/dt = -\sigma x + D(s_f - s) \quad (2)$$

$$dp/dt = q_p x - Dp \quad (3)$$

$$dz/dt = -bq_p x + D(z_f - z) \quad (4)$$

where x,s,p and z are the cell, limiting substrate (glucose), penicillin and precursor (PAA) concentrations (g/L) in the reactor. D is the dilution factor (volumetric feedrate over liquid volume, h^{-1}), s_f and z_f are the substrate and precursor concentrations (g/L) in the feed stream and F is the bead removal rate expressed as a volumetric cell removal flow over liquid volume (h^{-1}).

The specific growth rate, μ, is assumed to follow the usual Contois type kinetics, i.e.,

$$\mu = \mu_{max}\, s/(K_s x + s) \quad (5)$$

The specific uptake rate, σ, is related to the growth rate by (Pirt, 12)

$$\sigma = \mu/Y_G + m + q_p/Y_p \quad (6)$$

where Y_G is the cell growth yield, m is the maintenance energy and Y_p is the penicillin from substrate yield (carbon and ATP source).

The specific penicillin production rate, q_p, is related to the average cell age, λ, by (Fishman and Biryokov, 14; Holmberg and Ranta, 15)

$$q_p = q_p^{max} \, a \lambda \exp(1 - a \lambda) \tag{7}$$

where $$d\lambda/dt = 1 - \mu \lambda \tag{8}$$

A typical transient response of the system under computer control and the behaviour of the manipulated variables is shown in Figures 6 and 7 respectively. As seen, after some time we reach and maintain steady state conditions where penicillin is produced at a high rate. The amount of penicillin collected in the exit stream for the above run, as well as, for other penicillin concentrations in the effluent, is shown in Figure 8. As seen, there is a trade off between penicillin concentration in the exit stream and amount produced.

ACKNOWLEDGEMENTS

This work was supported by Natural Science and Engineering Research Council of Canada Strategic Grant No. G-0639.

LITERATURE CITED

1. Berk, D., L.A. Behie, A. Jones, B.H. Lesser and G.M. Gaucher, proc. of 34th Can Chem Eng Conf, Quebec City, Canada, 279(1984).
2. Berk, D., L.A. Behie, A. Jones, B.H. Lesser and G.M. Gaucher, Can J Chem Eng, 62, 112(1984).
3. Berk, D., L.A. Behie, A. Jones, B.H. Lesser and G.M. Gaucher, Can J Chem Eng, 62, 120(1984).
4. Jones, A., R. Siqueira, D. Berk, T. Razniewska, B.H. Lesser, L.A. Behie and G.M. Gaucher, Can J Microbiol, 30, 475(1984).
5. Jones, A., D.N. Wood, T. Razniewska, G.M. Gaucher and L.A. Behie, proc. of 35th Can Chem Eng Conf, Calgary, Canada, Oct., (1985).
6. Chibata, I. and S.J. Pirt, Ann Rev Biophys Bioeng, 10, 197(1981).
7. Kim, J.H., J.M. Lebeault and M. Reuss, Europ J Applied Micro Biotech, 18, 11(1982).
8. Konig, B., K. Schugerl and C. Seewald, Biotech Bioeng, 24, 259(1982).
9. Wittler, R., R. Matthes and K. Schugerl, Europ J Applied Microb Biotech, 18, 17(1983).
10. Gbewonyo K. and D.I.C. Wang, Biotechn Bioeng, 25, 967(1983).
11. Gbewonyo K. and D.I.C. Wang, Biotechn Bioeng, 25, 2873(1983).
12. Pirt, S.J., Kem Ind, 34, 13(1985).
13. Kalogerakis,N., T. Linardos, L.A. Behie, W.Y. Svrcek and G.M. Gaucher, proc. of 35th Can Chem Eng Conf, Calgary, Canada, Oct. (1985).
14. Fishman V.M. and V.V. Biryokov, Biotechn Bioeng, 16, 647(1974).
15. Holmberg, A. and J. Ranta, Automatica, 18, 181(1982).

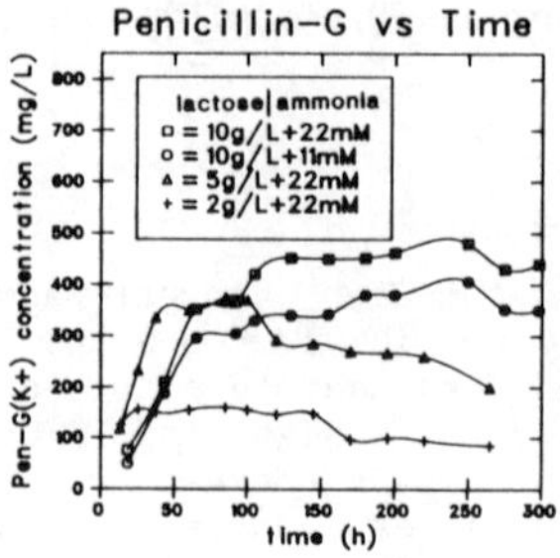

Figure 1 Concentration of Penicillin-G(K^+) in the Fluidized Bed Effluent for Various Liquid Mediums (3.0 mm Carrageenan beads)

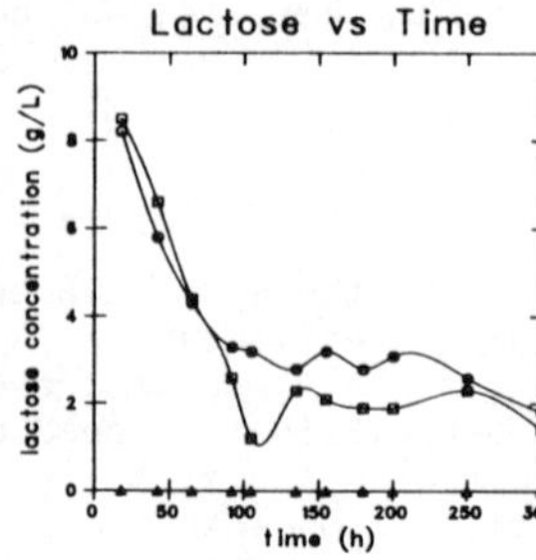

Figure 2 Concentration of Lactose in the Reactor Effluent During the Continuous Production of Penicillin-G (same legend as Figure 1)

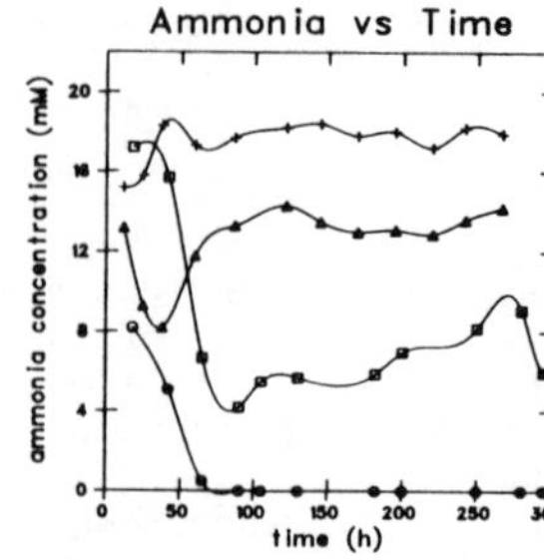

Figure 3 Concentration of Ammonia in the Reactor Effluent During the Continuous Production of Penicillin-G (Same legend as Figure 1)

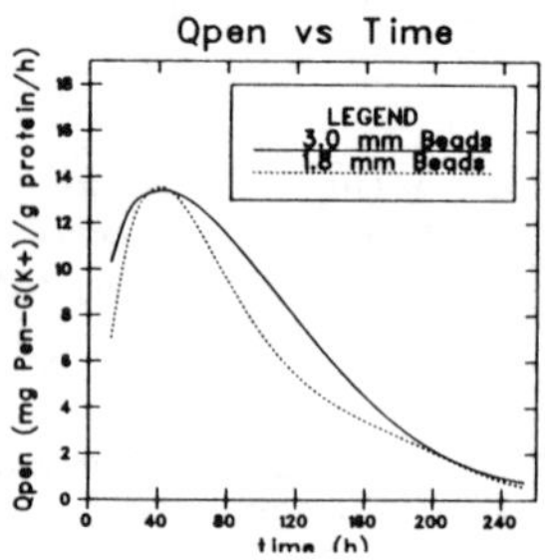

Figure 4 Specific Rates of Reaction for Penicillin-G Production Using Two Carrageenan Bead Sizes

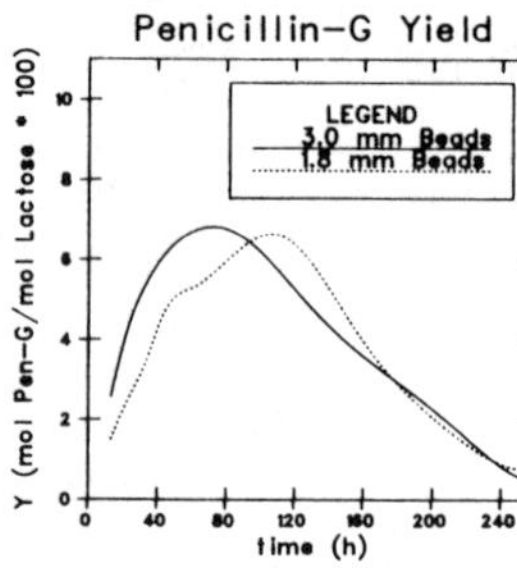

Figure 5 Variation of Penicillin-G Yield Coefficients (Y) For Two Carrageenan Bead Sizes

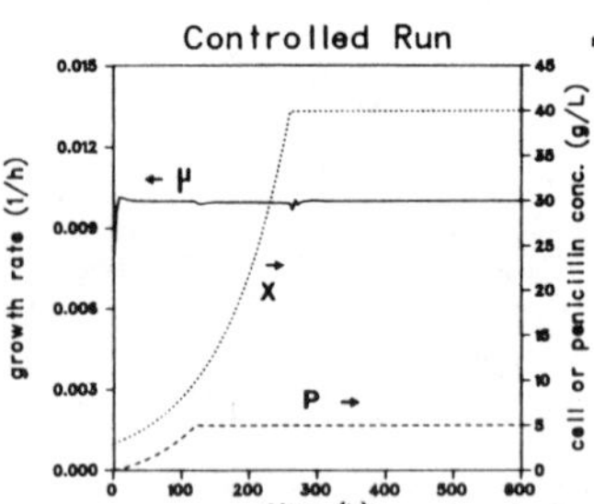

Figure 6 Cell Growth Rate, Penicillin-G Concentration and Cell Concentration Transients for a Typical Continuous Fermentation Run Under Computer Control

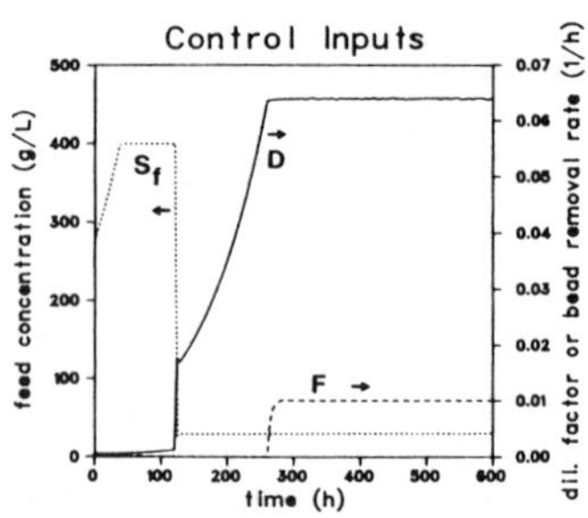

Figure 7 Behaviour of the Manipulated Variable for the Typical Continuous Fermentation Run Shown in Figure 6

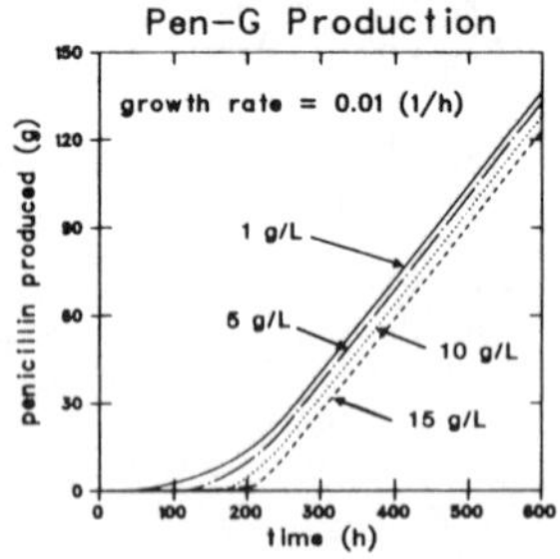

Figure 8 Total Penicillin Produced at Different Penicillin Concentrations in the Effluent Stream Under Computer Control

TABLE 1

INCREASE OF THE TOTAL AMOUNT OF PROTEIN IN THE FLUIDIZED BED REACTOR

SUBSTRATE CONCENTRATION		TOTAL PROTEIN CONTENT (9)		PROTEIN PERCENT CHANGE
Lactose g/L	Ammonia mM	Initial	220 h	
10.0	22.0	1.10	4.98	+353
10.0	11.0	1.04	2.88	+177
5.0	22.0	0.82	1.53	+ 87
2.0	22.0	0.88	0.82	-6.8

THE USE OF GAS /SOLID FLUIDIZATION FOR BIOCATALYSED REACTIONS: EXPERIMENTS AND MODELLING

W. J. Bauer
Fraunhofer- Institute of Food Process Engineering and Packing at the Technical University of Munich,
Schragenhofstrasse 35, 8000 München 50, F. R. G.

ABSTRACT

Gas/solid fluidization is shown to deliver favourable fermentation conditions for biocatalysed reactions at low water content. For two cases i.e. production of ethanol and glutathione with means of S. cerevisiae experiments at different scales are performed and compared to results from mathematical modelling taking into account complex biokinetics and fluid bed hydrodynamics.

SCOPE

Fermentations are performed in submerged culture techniques as well as in solid state, resp. solid substrate, techniques, the latter being applied mainly in East-Asia i.e. for the Koji process /1/ or for the production of food colourants /2/. Conventionally plate fermenters, fixed beds and rotary drums are used for solid state fermentation. These are difficult to control in temperature, moisture content and oxygen supply. The aim of this work was to apply the principle of gas/solid fluidization to biocatalysed reactions as fluidized bed reactors deliver good mass and heat transfer properties.
The processes choosen were i) ethanol production, ii) synthesis of glutathione, a secondary metabolite, with means of small yeast particles of 400 to 800 µm diameter fluidized with air resp. inert gas. Experiments and modelling mainly were performed with respect to scale-up and control of the process.
Moebus /3/ first showed that gas/solid fluidization may be used for growth of yeast in solid state mode. Akao /4/ applied this system to produce enzymes with Aspergillus sojae fluidized on wheat bran.
Our studies on the fluidization behaviour /5/, control of substrate feed resp. water content in the fermenter /6/, kinetics of bioreactions in solid state mode and comparison to submerged culture for ethanol /7/ and glutathione production /8/ permitted the development of mathematical models for interpretation of experimental data /9/ as a function of the fluid bed properties and kinetics as well as for the scale-up of the process /10/. The aim of this work is to show how the product itself - volatile or intracellular accumulating - influences the design and the control of the fluid bed fermenter.

CONCLUSION AND SIGNIFICANCE

The mathematical modelling of the unsteady state biocatalysed reaction combining mass transfer data, the data for phase-equilibrium between liquid phase of microorganisms and the fluidizing gas with the reaction kinetics and the mixing behaviour of the fluid bed allows to calculate the productivity of the fluid bed fermenter for the production of ethanol at steady state. On the other hand for the design of processes with intracellular accumulation products like glutathione a competitive feedback inhibition kinetics has to be taken into account and combined with the control of substrate feed to achieve maximum yield with respect to the precursors used, as well as to achieve maximum product concentrations in the process.

PRINCIPLE OF GAS/SOLID FLUIDIZED BED FERMENTATION

In the fluid bed fermenter small microorganism pellets resp. microorganisms on a carrier material are fluidized in a gas stream with air or inert gas depending on the process condition needed. The scheme of a laboratory scale fermenter of 0.2 m diameter is given in figure 1.
The substrate is either the fluidizing material itself for the solid substrate fermentation or is fed in aqueous solution to the fluidized microorganisms by two phase spraying nozzles. The composition of the fluidizing gas with respect to oxygen, CO_2, EtOH and water is continuously controlled by means of infrared-analysers.
A system of optical fibre probes allows to control the fluidization behaviour, as for the microorganisms relatively high water content (φ = 0.55 to 0.65) has to be maintained in the fermenter thus bearing the risk of defluidization, when free water occurs /6/.
Figure 2a) and b) show how the moisture content φ in the fermenter influences flowability and fluidization behaviour indicating that flow conditioners have to be applied to achieve good fluidization at fermentation conditons (φ = 0.55 to 0.65). The yeast pellets used here with diameters of d_p = 400 to 800 µm can be classified according to Geldart B powders /5/. For the production of volatile products, i.e. ethanol, the fermenter is run continuously, the ethanol being stripped from the microorganisms by the fluidizing gas. For the production of nonvolatile products, i.e. glutathione, the fermenter has to be run batchwise and the microorganisms removed from the fermenter when maximum product accumulation is reached.

CONTINUOUS PROCESS : ETHANOL

Reaction Kinetics

The production of ethanol in the fluid bed fermenter with the yeast S. cerevisiae is performed anaerobic and aerobic (i.e. when regeneration of yeast is needed) /9/. For mass balance the following reactions have to be considered.

i) Alcohol production from glucose:

$$C_6 H_{12} O_6 \longrightarrow 2\, C_2 H_5 OH + 2\, CO_2$$

ii) Growth on glucose:

$$\alpha_1 C_6 H_{12} O_6 + \alpha_2 NH_3 + --- \longrightarrow \beta_1 C_6 H_7 O_3 NH_2 + \beta_2 H_2O + ---$$

iii) Growth on ethanol:

$$\alpha_3 C_2 H_5 OH + \alpha_4 NH_3 + --- \longrightarrow \beta_3 C_6 H_7 O_3 NH_2 + \beta_4 H_2O + ---$$

iv) Oxidation of ethanol:

$$C_2 H_5 OH + 3 O_2 \longrightarrow 2 CO_2 + 3 H_2O$$

The oxidation of glucose is considered by combination of i) and iv). The specific reaction rate is given by /7/ :

$$\gamma (t) = \gamma_{max,o} \exp \left(- \frac{E_\gamma}{R \cdot T} \right) \cdot \frac{C_G(t)}{K_M + C_G(t)} \left(1 - \frac{C_E}{C_E^*} \right)$$

showing the strong dependance upon ethanol concentration in the yeast pellets C_E resp. on the maximum tolerable concentration C_E^* and the Michaelis-Menten-type dependance upon glucose concentration.

Fermenter Modelling

As the substrate is fed to the fermenter by spray nozzles and as on the other hand the yeast pellets are mixed continuously in the fluid bed, each yeast particle after being supplied with substrate behaves discontinuously with respect to product formation. The efficiency of the fermenter thus depends as well on the mixing characteristics of the fluid bed and therefore to a large degree upon the scale of the fermenter as can be seen from figure 3. The actual reaction rate in the single pellet $\gamma(t)$ for different reaction times τ_R yields different effective reaction rates γ_{eff}.

The reaction time itself is a function of the probability for the single yeast pellet to be supplied with substrate and thus is a function of solids mixing and spraying intensity, the calculation of which is given in /9/. The overall reaction rate γ_{eff} then is calculated for fermenter steady state conditions and compared to the maximum possible reaction rate γ_{max} at optimum reaction conditions. Figure 4 shows calculated results with $K \cdot \tau_R$ the modified kinetic parameter. Increasing the bed height decreases fermenter efficiency, while an increase in solids mixing and feed intensity markedly improves reactor efficiency.

Results and Discussion

Figure 5 gives results from experiments performed in a laboratory (d_t= 0.2 m) and a technical scale (d_t=0.55 m) fermenter. Productivities up to 40 kg EtOH/$m^3 \cdot h$ were reached at aerobic conditions. These results are compared to calculations according to the proposed model (figure 6, data in table 1). They show that i) the model describes the influence of bed height and bed diameter well and ii) that there is still the possibility of increasing fermenter efficiency, as even the results with high fermenter productivities were obtained at an efficiency of about 30 %.

BATCH PROCESS : GLUTATHIONE

Reaction Kinetics

For batch processes in the fluid bed fermenter, the synthesis of glutathione (GSH) by the yeast S. cerevisiae with means of the precursors cysteine, glutamate and glycine was choosen. Glutathione (L-γ-glutamyl-L-cysteinyl-glycine) plays an important role in biology and medicine /11/. It is produced intracellularly by a two-step enzyme-catalysed reaction (figure 7). In the first step GC-synthetase [EC 6.3.2.2.] combines L-glutamata and L-cysteine to γ-glutamyl-cysteine. The second step with glycine as presursor and the enzyme glutathione-synthetase [EC 6.3.2.3.] yields GSH. Both reactions depend on ATP as energy source, so in the process ATP has to be continuously regenerated e.g. by the glycolytic pathway or by respiration of the intact yeast cells affording C-substrate and oxygen.
The third reaction step yields GSSG which - when the fermentation is performed in submerged culture - is excreted in the fermentation broth and lost. In solid state fermentation this can be avoided.
Our experiments in submerged culture and solid state mode /12/ showed the marked influence of various substrates and of the precursors itself upon reaction rates and GSH-maximum /8/. The GSH-accumulation at given substrate conditions reaches a maximum at temperatures of T=27 to 31 °C as with increasing temperature the third reaction step producing GSSG is prefered (figure 9). The GSH-concentration in the cells has a strong inhibitory effect of the feedback inhibition type.

$$\text{cys} + \text{glu} \xrightarrow{[\text{EC } 6.3.2.2.]} \gamma\text{-glu-cys} + \text{gly} \xrightarrow{[\text{EC } 6.3.2.3.]} \text{GSH} \; - - -$$

(dashed feedback line from GSH back to the first reaction arrow)

With cysteine being the limiting precursor concentration, the reaction rate is

$$\gamma_{GSH} = \gamma_{GSH,max} \cdot \frac{C_{cys}}{K_M \cdot \left(1 + \frac{C^*_{GSH}}{K_i}\right) + C_{cys}}$$

Fermenter Modelling

The aims of fermenter design for the process are i) to achieve high product concentrations in order to reduce the costs for downstream processing and ii) to reach high yields with respect to the expensive precursors applied. This means that, on the one hand, the process has to be performed discontinuously and, on the other hand, that the substrate feed rate has to be controlled according to the reaction kinetics to avoid loss of precursor. The integration of reaction rate to yield the actual intracellular GSH-concentration and the parallel calculation of the actual precursor-concentration, leads iteratively to a program to control the feed rate in order to achieve maximum product concentrations and optimum yield at the same time.

Results and Discussion

Figure 8 shows the schematic of the fluid bed laboratory fermenter and the probe preparation that was used for the glutathione experiments. The experiments were run under aerobic conditions with a strict control of moisture content in the fermenter. At constant time intervals probes were taken from the fermenter and prepared for enzymatic determination of GSH and GSSG.
Figures 10a) and b) show how the control of the feed rate influences the yield Y calculated as glutathione, produced at time t with respect to cysteine fed to the fermenter and the relative GSH-concentration with respect to the initial concentration $C_{GSH}(t)/\ C_{GSH}(t_o)$.

In run no. 9 with controlled feed, maximum yield and maximum GSH-accumulation were brought to coincidence, while in run no.7, without control of feed according to the proposed model, the maxima are too far apart for reaching optimum fermentation conditions. Further results with a large scale fluidized bed (d_t = 1.0 m, m_s = 42 kg are published elsewhere /12/.

ACKNOWLEDGMENT

This work was supported by Glatt Apparatebau, Binzen, and Deutsche Hefewerke Wandsbek, Hamburg.
I wish to express my sincere thanks to H. G. Kirk for helpful advice and B. Heuer and L. Röttenbacher for the careful experimental work.

NOTATION

C_{cys}	cysteine concentration	mol/l
C_{GSH}	glutathione conc. with resp. to dry matter	mol/g dm
C^*_{GSH}	glutathione conc. with resp. to intracell. water content	mol/l
C_E	ethanol concentration	mol/l
C^*_E	inhibitory ethanol concentration	mol/l
C_G	glucose concentration	mol/l
ΔC_p	glucose conc.-augmentation for single yeast particle	mol/l
d_p	particle diameter	µm
d_t	fluid bed diameter	m
E_γ	activation energy for ethanol production	J/mol
ff_c	flowability factor	-
K_M	Michaelis-Menten-Constant	mol/l
K_i	inhibition-constant for GSH production	mol/l
m_s	mass of wet solid in fermenter	kg
$\dot{n}_{cys}$	specific molar flowrate of cysteine with respect to dry matter	mol/kg dm/h
P_{stat}	fermenter productivity at steady state	kg EtOH/h/l

R	gas constant	J/mol/K
T	temperature	K, °C
t	time	min
u_{mf}	minimum fluidization velocity	m/s
Y	yield	mol
α, β	stoichiometric constants	-
φ	water content	kg water/kg solid
γ	reaction rate	h^{-1}
γ_{eff}	overall reaction rate	h^{-1}
γ_{max}	maximum possible reaction rate	h^{-1}
τ_R	reaction time	s
τ_{rel}	relative reaction time	-

LITERTURE CITED

1. Hesseltine, C.W., Biotechn. Bioeng. 14, 517 (1972)
2. Beuchat, L.R. "Indigeneous Fermented Food", in Biotechnology Vol. 5, Food and Feed Production with Microorganisms, Rehm H.J. ed., Verlag Chemie, Weinheim (1983).
3. Moebus, O., Teuber, M. and H. Reuter, "Pneumatischer Wirbelschicht - Bioreaktor zur Herstellung von Einzellereiweiß". Dechema-Monographien Vol. 95, Verlag Chemie (1984).
4. Akao, T. and Y. Okamoto, "Cultivation of Microorganisms in an Air-Solid Fluidized Bed" Proc. IVth Int. Conf. Fluidization, Kashidojima, 29.5 - 3.6. 1983.
5. Egerer, B., Zimmermann, K. and W. Bauer, "Flow and Fluidization Behaviour of Yeast in Gas/Solid Fermentation and Drying". I. Chem. E. Symp. Ser. No. 91, p. 157, (1983).
6. Bauer, W. and L. Röttenbacher, Zeitschrift für Lebensmittelverfahrenstechnik (ZFL) 1, 18 (1984).
7. Röttenbacher, L., Behlau L. and W. Bauer, J. of Biotechn. 2, 137 (1985).
8. Bauer, W. and H. G. Kirk, Biotech-Forum (BTF) 1, 3,(1985).
9. Röttenbacher, L., Schössler , M. and W. Bauer, "Mathematical Modelling of Alcoholic Fermentation in a Gas/Solid Bioreactor : Combined Effects of Solid Mixing and Non-Steady-State Kinetics" Proc. 1st IFAC Symp. on Modelling and Control of Biotechnological Processes, 11.-13. Dec 1985, Noordwijkerhout.
10. Bauer, W., Zeitschrift f. Lebensmittelverfahrenstechnik (ZFL) 3, 154 (1983).
11. Meister, A., Tate, S., Annual Review of Biochemistry, Vol. 45, 559, (1976)
12. Bauer, W., "Comparing Submerged Culture and Solid State Fermentation for the Production of Glutathione", Proc. 35th CSChE Conf. 6.-9. Oct. 1985, Calgary.

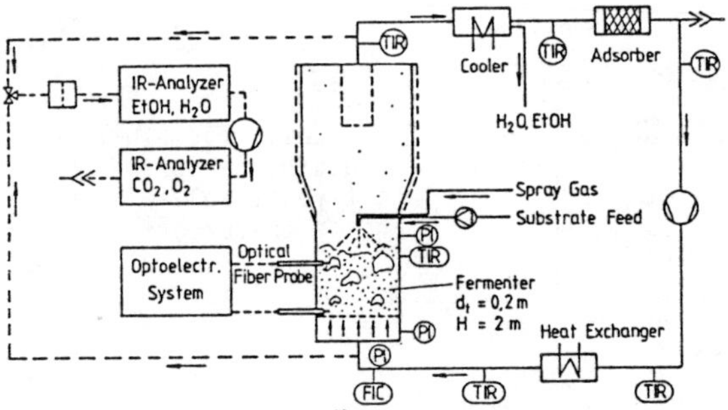

fig.1. Scheme of laboratory fluid bed fermenter for ethanol production.

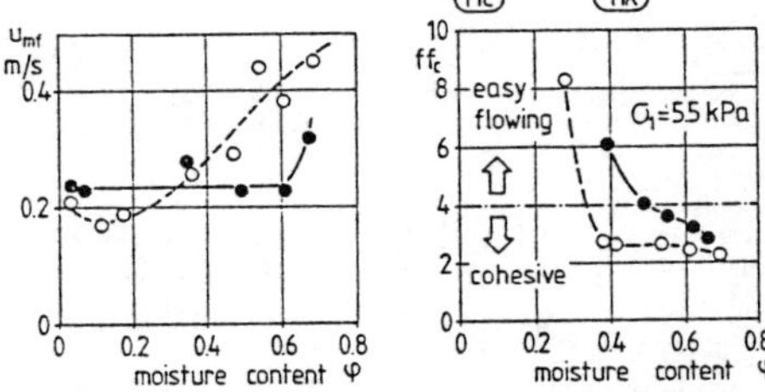

○ yeast without conditioner ● yeast with conditioner

fig.2. Influence moisture content on minimum fluidization u_{mf} (2a) and flowability ffc (2b).

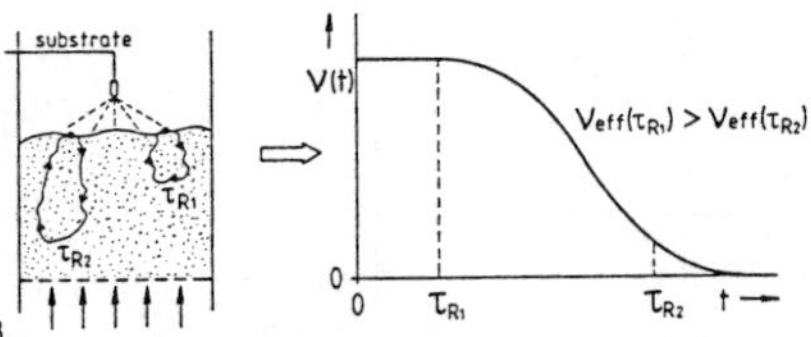

fig.3. Reaction rate γ as function of reaction time τ_R. Influence of fluid bed characteristics on effective reaction rate γ_{eff}.

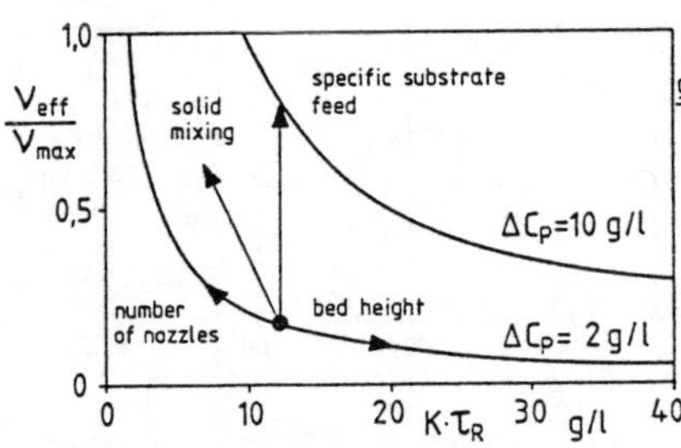

fig.4. Calculated fermenter efficiency $\gamma_{eff}/\gamma_{max}$ as function of reaction time τ_R and feed intensity Δc_p.

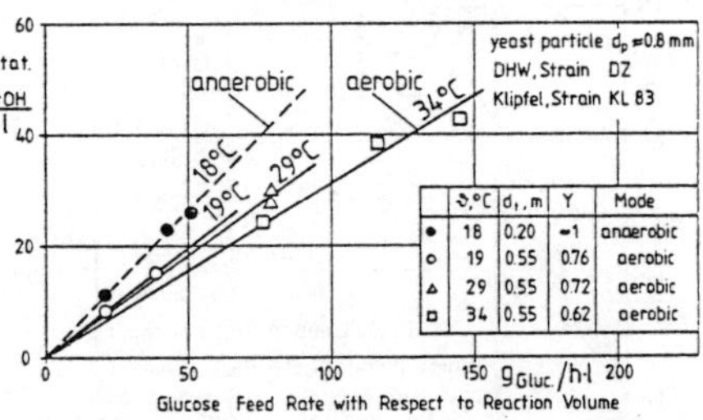

fig.5. Experimentally determined fermenter productivity at steady state P_{stat}.

	d_t m	T °C	M kg	$\dot{m}_{Gl}$ g/s	V_{eff} h^{-1}	τ_{rel}
■	0,2	20	1,2	0,030	0,144	1,0
●	0,2	22	1,5	0,047	0,127	1,2
◆	0,2	22	1,5	0,04	0,107	1,2
▲	0,2	22	2,25	0,049	0,096	1,9
▼	0,2	20,5	2,25	0,021	0,045	1,9
×	0,2	18	2,4	0,028	0,069	1,9
△	0,55	34	20	0,85	0,163	4,1
○	0,55	34	20	1,29	0,257	4,1
□	0,55	29	20	0,74	0,173	4,1
▽	0,55	29	20	0,67	0,109	4,1
◇	0,55	34	20	1,34	0,239	4,1
⊕	0,55	34	20	1,17	0,200	4,1

tab.1 Experimental results.

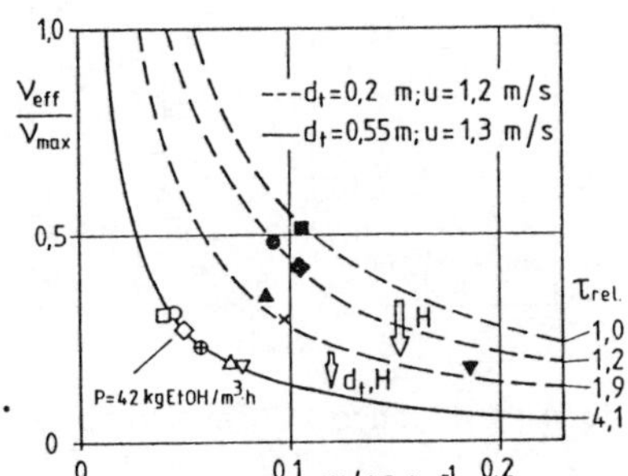

fig.6 Comparision of experimental data with calculated fermentation efficiency.

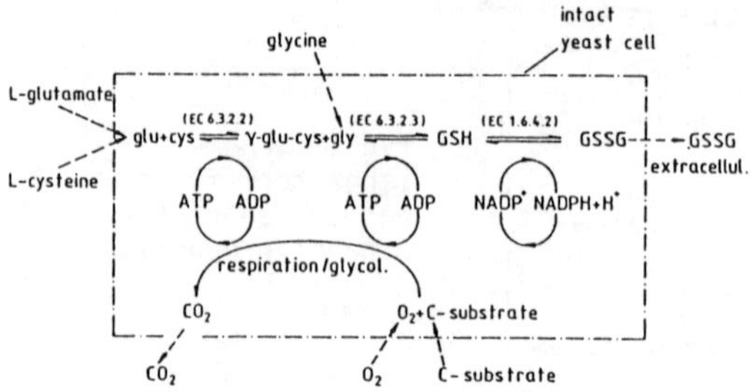

fig.7. Enzymatic synthesis of glutathione in an intact yeast cell with means of precursors.

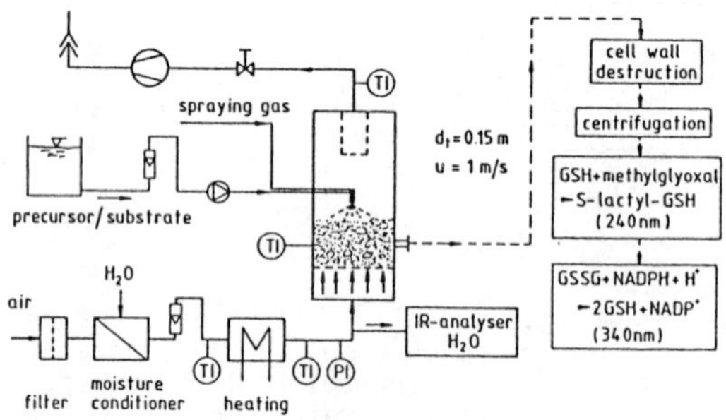

fig.8. Laboratory scale fluid bed fermenter for glutathione production.

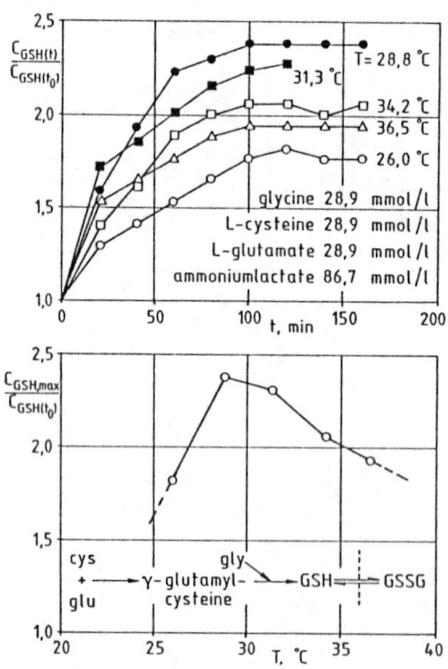

fig.9. Influence of fermentation temperature on GSH-accumulation

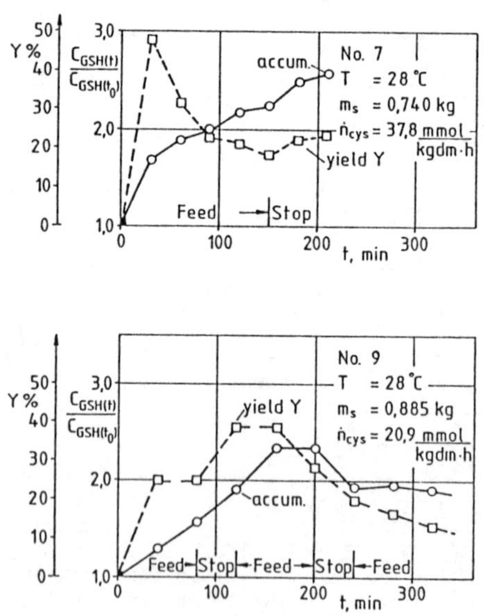

fig.10. Experimental results fluid bed fermenter; accumulation factor $C_{GSH}(t)/C_{GSH}(t_0)$ and yield Y as funktion of time.

OXYGEN TRANSFER AND ORGANIC CARBON ABATEMENT IN A BIOLOGICAL HIGH COMPACTING MULTIPHASIC REACTOR

S. Elmaleh, S. Papaconstantinou, G.M. Rios and A. Grasmick

Laboratoire de Génie des Procédés
Université des Sciences et Techniques du Languedoc
34060 Montpellier Cedex, France.

ABSTRACT.

The High Compacting Multiphasic Reactor (HCMR) is a three-phase spouted bed where a regular solid circulation is ensured by a special gas distributor. Oxygen transfer is relatively intense and little energy is required. As an aerobic attached-cell reactor, the HCMR can treat a polluted influent with higher loading than classical fluidized reactors.

SCOPE.

The HCMR is a new device that enables fluidized type contacting of a gas, a liquid and dispersed solids in some cases - specially with solid and liquid having nearly the same density, sticky surface particles and/or high solid hold-up - where classical fluidization is not possible due to agglomeration/floatation tendencies of particles. Basically, the HCMR is a spouted bed reactor where regular solid circulation results from a specially designed gas distributor which ensures a tight and smooth contact between all phases. In the field of wastewater treatment,aerobic fluidized reactors allow significant reduction of the units but the fulfilment of the oxygen requirement raises some technical and economical problems. The objectives of this work are the quantification of the oxygen transfer in a HCMR and the evaluation of the contactor as a basic wastewater purifier when a biofilm is attached to the spouted particles.

CONCLUSIONS AND SIGNIFICANCES.

The HCMR overall oxygen transfer coefficient values are situated between those corresponding to small particle and large particle fluidized beds. The energy required for good operating conditions is fairly low and the HCMR compares then with the most efficient air injection systems with mechanical stirring operated in slurry reactors. As particle-particle attrition is important, the biomass hold-up is lower than in the classical aerobic fluidized bed ; however, the volumetric pollution abatement rate is higher. The HCMR is therefore promising in the field of wastewater treatment but further research is needed to optimize particle biomass hold-up.

BACKGROUND

As compared to fixed beds - used at present in most industrial processes involving gas-liquid-solid contact - fluidized reactors enable to over come difficulties arising from a strong exothermic reaction or from the requirements of continuous solid extraction. Lately their development has allowed to succesfully investigate new applications where a good phase circulation is needed.

Ordinary fluidized beds have also proved to be unsuitable in some special cases, e.g. in food processing and biotechnology where the limitations of the technique are due to :

1°) flat shaped solid particles having nearly the same density as the liquid phase ;
2°) particles with sticky surfaces which tend to agglomerate into a cake ;
3°) liquid-phase volume strongly reduced in an effort to minimize extraction of soluble substances.

The HCMR is designed to ensure good contact and reaction pattern of such particles with a gas phase and a liquid phase (1, 2). The HCMR is characterized by the original configuration of its bottom plate built up from two components (Figure 1) :

1°) a horizontal grid uniformly distributing the gas and ensuring then a local upward particle transport ;
2°) an inclined deflector which brings back the solid phase towards the bubbling zone.

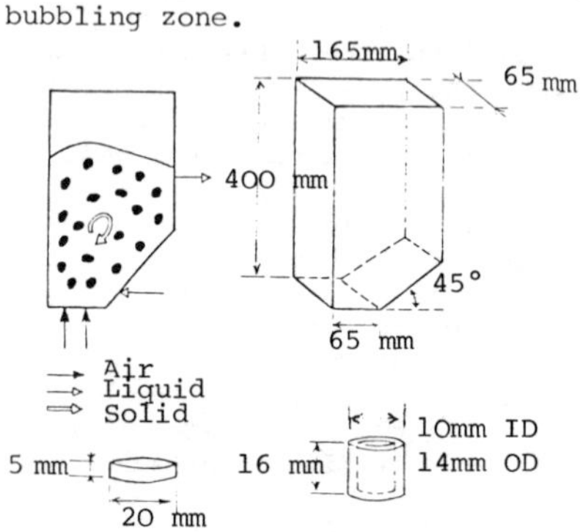

Figure 1 - Apparatus

A whirling movement of the solid phase and of the dragged liquid is therefore induced around a horizontal axis, setting up a real vortex. Besides, a low amplitude and high frequency movement is promoted and superimposed to the overall motion.

A previous work mainly concerned with reactor hydrodynamics (3) has shown that :

1°) by analogy with traditional fluidization, a minimum gas velocity inducing the whirling movement may be defined ; this velocity is related to particle shape and solid hold-up ;
2°) gas hold-up is lower than in fluidized bed ;
3°) the liquid phase is perfectly macromixed ;
4°) the solid phase is not uniformly distributed ; solids heavier than the liquid tend to segregate at the bottom and lighter particles tend to float.

This work intends to evaluate the HCMR applicability to the treatment of biodegradable wastewater after cell immobilization on support particles. In the field of wastewater treatment, aerobic fluidized reactors have promoted significant process intensification (4). In such reactors, the required oxygen is currently supplied by one of the methods listed in Table I. Let's note that each one raises some technical and economical problems which limit industrial development.

Table 1

Oxygen transfer into fluidized reactor

Method	Technical and economical problems
Air injection through a three-phase fluidized reactor (5)	Only fluidized bed of large particles - 4 to 6 x 10^{-3} in diameter as an example - possess good aeration characteristics (8). Fluidization of such particles requires high energy input.
External oxygen transfer from a device located on a recycle line (6)	Stoechioemetry implies high values of recycle rate and therefore high energy input.
Injection of an oxygen liberating chemical, e.g. hydrogen peroxyde (7)	Such chemicals are expensive.

In the following, the HCMR oxygen transfer will be evaluated and then its performance in biochemical oxydation of organic pollution.

MATERIAL AND METHODS

Reactor

One single unit has been tested with particles of different shapes (Figure 1) made of polyamide (disks, 1140 kg.m^{-3} in density) and nylon (cylinders, 1100 kg.m^{-3} in density).

Oxygen transfer runs

Oxygen transfer was evaluated by the classical method using clean water. Dissolved oxygen was brought down to zero by adding sodium sulfite Na_2SO_3 and Co^{++} as catalyst. During the reoxygenation step, dissolved oxygen concentration was recorded.

Organic carbon abatement runs

Wastewater. The unit was operated with a molasse and ammonium hydrogenophosphate solution in tap water in order to get a TOC/N ratio of 10.

Microorganisms. Cell immobilization was obtained after 10 days of batch operation with settled urban wastewater at low air flowrate. The reactor was then operated with synthetic wastewater till a steady state

Table 2

Oxygen transfer results

Liquid phase volume $10^{-3}m^3$	Solid phase volume $10^{-3}m^3$	u_G $10^{-2}m.s^{-1}$	k_La s^{-1}	OC $10^{-4}kgm^{-3}s^{-1}$	AE $kg.kWh^{-1}$
		5.0	0.022	2.23	3.53
30	1.5	7.5	0.028	2.91	3.06
		10.0	0.032	3.10	2.63
		6.0	0.022	2.31	2.92
30	3	7.5	0.026	2.68	2.63
		10.0	0.029	3.10	2.45
		7.5	0.023	2.36	2.56
30	4.5	10	0.027	2.88	2.27
		11.5	0.029	2.96	2.20

Table 3

Experimental variation of k_La : ↑ increase -↓ decrease

Parameter	Observation	Comments
u_G ↑	k_La ↑	When u_G increases, gas hold-up increases and bubble size decreases (3)
solid hold-up ↑	k_La ↓	In spite of higher gas hold-up (3), an increase of solid hold-up favours bubble coalescence
particle shape	$(k_La)_{cylinders}$ higher than $(k_La)_{disks}$	Turbulence in the center of the reactor is better when cylinders are spouted (3) ; bubble breakage is improved due to internal particle voids.
reactor size	k_La is higher in the larger unit	Wall effects are less in the larger unit

was reached.

Pollution concentration. Pollution concentration was determined with a TOC-meter. All the determinations were made in conformity with the Standard Methods (9).

RESULTS AND DISCUSSION.

Oxygen transfer

Assuming that the oxygen partial pressure in the bubbles is uniform, the oxygen transfer rate in the batch reactor is quantified by :

$$\frac{dC}{dt} = k_L\, a\, (C^* - C) \qquad (1)$$

The overall mass transfer coefficient k_La is easily calculated from Equation (1) or from its integrated form :

$$\text{Log}\,(C^* - C) = \text{Log}\,C^* - k_La\,t \qquad (2)$$

In wastewater technology, the performance of an aeration device is currently characterized by the following parameters :

1°) Oxygenation capacity (OC), i.e. oxygen mass flowrate transferred per unit volume while zero oxygen concentration is maintained ;
2°) aeration efficiency (AE), i.e. oxygen mass transferred per unit of required energy.

It is easily derived from the definitions that :

$$OC = k_L\, a\, C^* \qquad (3)$$

and

$$AE = OC/P \qquad (4)$$

Mixing being provided by the gas only, the dissipated power in the HCMR is estimated by (neglecting the pressure drop in the grid) :

$$P = Q_G\, \rho_L\, g\, H \qquad (5)$$

Some typical results are given Table 2 ; Table 3 summarizes the conclusions on k_La when different parameters are varied.

The HCMR performance is compared to that of classical three-phase fluidized beds in Table 4. It follows that the HCMR k_La values are situated between those corresponding to small particle and large particle fluidized beds. The AE values are comparable with the ones that characterize the most efficient air injection systems with mechanical stirring operated in slurry reactors (1.5 to 2.5 $kg.kWh^{-1}$); they are even better than the values obtained in classical surface aerators used in activated sludge units.

Table 4

Mass transfer performance of fluidized bed and HCMR

Reactor	Liquid velocity u_L $10^{-2} m.s^{-1}$	Gas velocity u_G $10^{-2} m.s^{-1}$	Overall oxygen transfer coefficient k_La s^{-1}	References
6 mm glass bead fluidized bed	16.2	4	0.06	(8)
0.6-1.0 mm crushed brick granules fluidized bed	0.83	0.88	0.007	(5)
HCMR	-	5	0.022	

Organic carbon abatement

Biomass hold-up. The high level of turbulence in the reactor, which induces significant particle-particle attrition, tend to limit the biofilm thickness. In fact, biomass is retained only on the internal surface of the cylindrical particles. Biomass hold-up depends then solely on the turbulence conditions promoted by the air flowrate. The values of biomass hold-up are between 2 and 5 $kg.m^{-3}$ which are lower than the values generally observed in aerobic fluidized bed (30-50 $kg.m^{-3}$). Biomass hold-up can probably be improved by using particles with high internal surface.

Organic carbon abatement efficiency. Pollution abatement efficiency is plotted against liquid phase space-time Figure 2 : a plateau is reached after 2 hours space-time. This result shows that an influent loaded up to 18 kg $BOD_5.m^{-3}.day^{-1}$ can be treated ; the HCMR appears then to perform better than a classical three-phase fluidized reactor where the maximum loading rate is 10 kg $BOD_5.m^{-3}.day^{-1}$(5). The biofilm attached to the spouted particles of the HCMR is significantly more active.

Organic carbon abatement kinetics

The organic carbon abatement rate per unit volume of the perfectly mixed liquid phase is :

$$r = \frac{Si - Se}{\tau} \qquad (6)$$

The plot of the experimental results in the (S,r) plane leads to only one curve of hyperbolic shape (Figure 3). The intraparticle biofilm

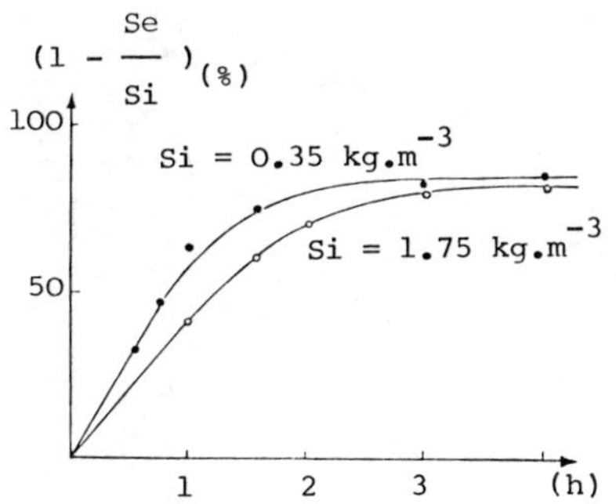

Figure 2 - Pollution abatement efficiency versus space time

being fairly thin, there is no rate limitation by diffusion through a solid phase (10).

A Monod type relationship can model the rate equation :

$$r = r_o \frac{S}{S + S_M} \qquad (7)$$

Its validity is checked by plotting 1/r against 1/S which gives a straight line and numerical values of r_o and S_M (Figure 4).

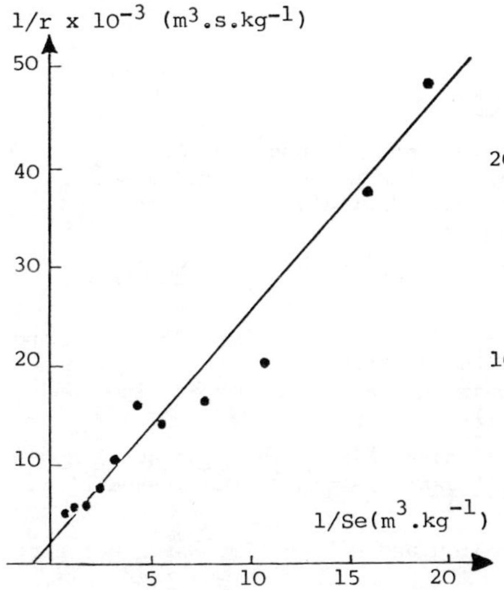

Figure 4 - 1/r versus 1/Se

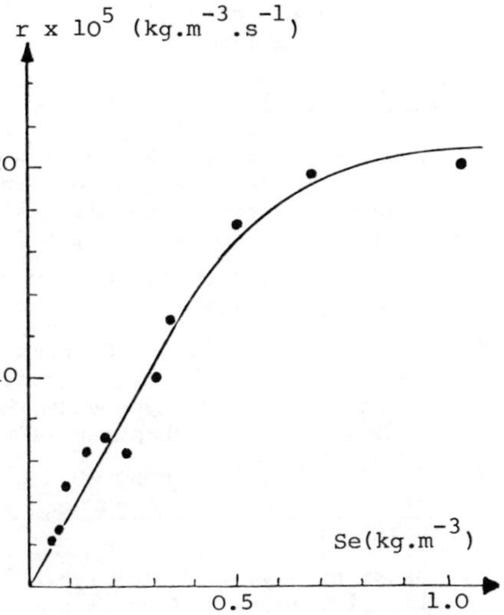

Figure 3 - r versus Se

ACKNOWLEDGEMENT

Financial assistance from CNRS through grant n° 001229 is fully acknowledged.

NOTATION

a	gas-liquid interfacial area	L^{-1}
AE	aeration efficiency	$L^{-2}T^{2}$
BOD_5	5-day biochemical oxygen demand	
C	oxygen concentration	ML^{-3}
C^*	saturation oxygen concentration	ML^{-3}
g	gravity	$L^{2}T^{-1}$

H	bed height	L
k_L	oxygen transfer coefficient	LT^{-1}
OC	oxygenation capacity	$ML^{-3}T^{-1}$
P	power	ML^2T^{-3}
Q_G	gas volumetric flowrate	L^3T^{-1}
r	organic carbon abatement rate	$ML^{-3}T^{-1}$
r_o	maximum organic carbon abatement rate	$ML^{-3}T^{-1}$
S	organic carbon concentration	ML^{-3}
S_e	outlet concentration	ML^{-3}
S_i	inlet concentration	ML^{-3}
S_M	saturation constant	ML^{-3}
t	time	T
TOC	total organic carbon	
u_G	superficial gas velocity	LT^{-1}
u_L	superficial liquid velocity	LT^{-1}

Greek letter

ρ_L	liquid density	ML^{-3}
τ	space time	T

LITERATURE CITED

1. Rios, G.M., J.L. Baxerres and H. Gibert, "Procédé de mise en contact d'une phase gazeuse, d'au moins une phase liquide et d'au moins une phase solide à l'état divisé", U.S. Patent n° 4.334.998 to ANVAR (June 15, 1982).

2. Jacquet, M., J.C. Vincent, G.M. Rios and H. Gibert, Thé, Cacao, Café, 1, 45, (1981).

3. Papaconstantinou, S., S. Elmaleh and G.M. Rios, "Hydrodynamics and mass transfer in a High Compacting Multiphasic Reactor" presented at the 16th ICHMT Symposium, Heat and Mass Transfer in Fixed and Fluidized Beds, Dubrovnik (1984).

4. Elmaleh, S., "Les Réacteurs à Biomasse Fixée" in Point sur l'Epuration et le Traitement des Effluents, Martin G. (Ed), Lavoisier, Paris (1982).

5. Grasmick, A., B. Chatib, S. Elmaleh and R. Ben Aīm, Water Research, 15, 719, (1981).

6. Sutton, P.M. , W.K. Shieh, P. Kas and P.R. Dunning, "Dorr-Oliver's Oxitrŏn System fluidised-bed water and wastewater treatment process" in Biological Fluidised Bed Treatment of Water and Wastewater, Cooper P.F. and Atkinson B. (Ed), Ellis Horwood Limited, Chichester (1982).

7. Yahi, H., A. Grasmick, S. Elmaleh and G.M. Faup, Environ. Technol. Letters, 3, 281 (1982).

8. Ostergaard K. and P. Fosbol, Chem. Eng. J., 3, 105 (1972).

9. Standard Methods for the Examination of Water and Wastewater, American Public Health Association, 13rd ed. (1971).

10. Rodrigues, A.E., A. Grasmick and S. Elmaleh, Chem. Eng. J., 21 (3), 1339 (1983).

CHEESE-WHEY PROTEINS RECOVERY BY ION EXCHANGE IN FLUIDIZED BEDS

B. BISCANS and J.P. COUDERC
Laboratoire de Génie Chimique
U.A. CNRS 192 - ENSIGC
Chemin de la Loge
31078 TOULOUSE Cédex (France)

ABSTRACT

Cheese-whey proteins can be recovered by ion exchange in fluidized beds, avoiding the usual fixed beds clogging. Two kinds of multistage fluidized columns have been developed, one without solids circulation, alternating fixation and elution stages and a second with solids circulation and a continuous fixation.

SCOPE

The recovery of biological products and particularly of proteins from biological solutions involves important but difficult operations. Among the extraction methods, ion exchange offers promising potentialities linked to its selectivity and also to the fact that it does not denaturate the products. However, its industrial development has still been hindered by the many technical problems encountered with the fixed beds equipment generally used.

The aim of this study is to analyse the performances of a new process for ion exchange, in fluidized beds, in the particular case of the recovery of cheese-whey proteins.

The usual treatment of cheese-whey, with a fixed-bed of resins and alternating fixation and elution stages (1), (2), is limited by the phenomenon of bed clogging, caused by the fine solid particles suspended in the solution, casein, fats or precipitated salts. This constitutes a serious drawback because the efficiency of the equipment decreases progressively. Moreover, the different treatments necessary for the declogging of the bed and the washing of the particles reduce the time of life of the resins. Finally, any pretreatment of cheese-whey to clear it, would complicate the plant and decrease its profitability.

In this work, the performances and limits of a three stages fluidized system have been examined. The study can be divided in two parts :

- a theoretical modelling of the phenomenon, based on a mass transfer mechanism analysis.

- an experimental study with a small scale pilot plant demonstrating the feasability of the process and enabling a comparison with the model predictions.

CONCLUSIONS AND SIGNIFICANCE

Two types of fluidization equipment and procedures have been studied in this work :

- the first one, without circulation of the resin from stage to stage, involves a cyclic operation, alternating fixation and elution steps,

- the second one is a continuous system with resin flow through down-comers from stage to stage.

For the cyclic procedure, in a three stage pilot plant, the mass transfer results have been similar to those obtained in fixed-beds but without any clogging problem. So, it has been shown that, at a small scale, the recovery of cheese-whey proteins by ion-exchange in fluidized beds, works conveniently. Moreover, the model results have been shown to predict satisfactory experimental data even if the precision could be improved by a better knowledge of equilibrium data.

For the continuous process, with resin circulation through down-comers, it has been shown that the recovery of cheese-whey proteins is possible, but the efficiency remained low and additional analysis are still necessary to optimize the resin movements from stage to stage and in every fluidized bed.

It is important to recall that the results and conclusions obtained could be extended to the treatment of other biological solutions than cheese-whey.

Cheese-whey is a mixture of lactose, mineral salts and proteins. Table 1 gives the proportions of these different elements. It must be observed that the amount of proteins is very small, which makes their recovery difficult. Among the various kinds of proteins in cheese-whey, the extraction of the two major ones, αlactalbumin and βlactoglobulin, by ion exchange with a Spherosil QMA resin, has been particularly analyzed.

FLUIDIZATION PLANTS

Figure 1 is a schematic view of the pilot plant, without resin circulation. The column is made of one or several stages (generally three) of 5cm in diameter and 50cm in height. The liquid distributor at the bottom of each stage is made of a perforated plate. The cheese-whey flow rate has been measured by a rotameter.

WATER	94	
NON PROTEIN NITROGEN	0,03	
PROTEIN SUBSTANCES	0,55 to 0,58	β-lactoglobulin 53,6 α-lactalbumin 21,4 serum-albumin 5,3 immunoglobulins 9 proteoses-peptones 10,7
FATS	0,05	
LACTOSE	4,5 to 4,8	
MINERAL SALTS	0,45 to 0,5	

TABLE 1 - Massic percentages of cheese-whey components

For the case with a continuous resin circulation, the previous equipment has been modified (figure 2). The resin is introduced at the top of the column by hydraulic transport, then circulates from stage to stage through downcomers and is removed from the bottom through a lock chamber.

The variations in concentration of the liquid have been followed during the experiments, by taking samples of the solution at the outlet of each fluidized stage. The analysis have been made by zone electrophoresis.

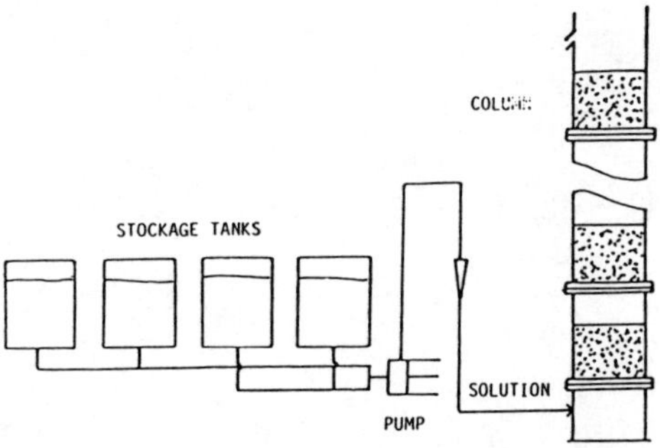

Figure 1 : Cyclic fluidization system without resin circulation

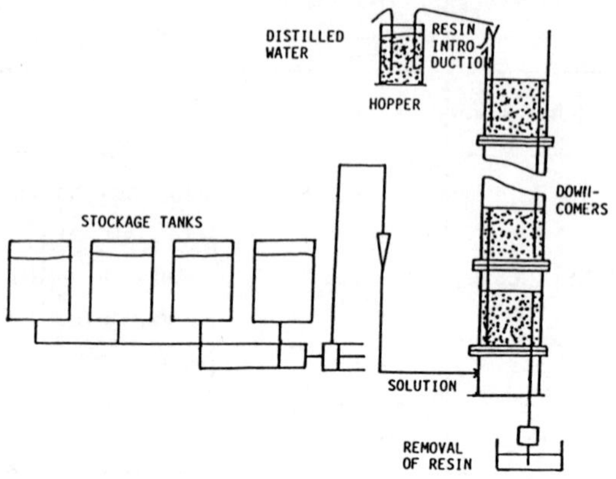

Figure 2 : Continuous fluidization system with resin circulation

MODELLING

The equations usually used to describe the ion exchange operations in a fixed bed have been modified to take into account the fluidized beds properties, adding to the classical assumptions (plug flow of the liquid, no internal diffusion inside the particles), new hypotesis concerning the solids movements, for each kind of fluidization system (3)

- For the cyclic operation it has been supposed that the solids are perfectly mixed in each fluidized bed

- For the continuous procedure, it has been supposed that, as soon as the particles enter the bed, they are perfectly mixed with the particlesat different concentration in the bed, and that they immediatly reach the mean concentration value.

Equations

Cyclic system

$$C(t,h) = C^*(t) + (Co - C^*(t)) \exp\left[- \frac{Sk_L h}{v_L} \right]$$

$$\frac{dq}{dt} = \frac{v_L}{(1-\varepsilon)h} \left[1 - \exp\left(- \frac{Sk_L h}{v_L}\right)\right] \left[Co - C^*(t) \right]$$

Continuous system

$$\frac{dC}{dx} = - \frac{Sk_L}{v_L} (C - C^*)$$

$$Q_S (q - q_o) = v_L A (C_o - C(h))$$

For the two cases, the distribution isotherm of the different ionic species at equilibrium must be added to the preceeding equations, i.e.

$C^*(t) = f(q)$

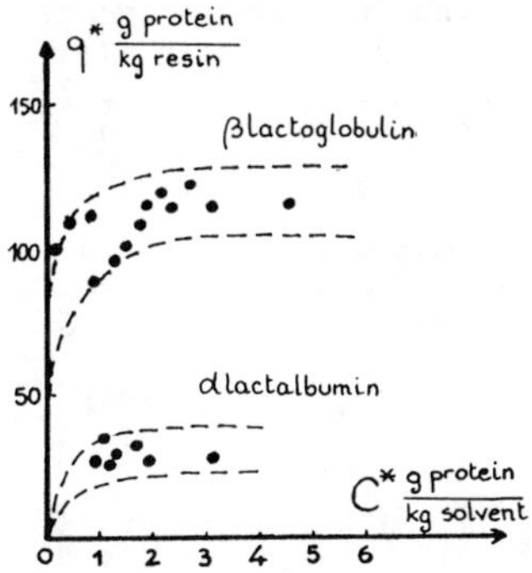

Figure 3 : Equilibrium isotherms (25°C)

The corresponding data for this equilibrium have been experimentaly obtained by agitating during 24 hours, at constant temperature, various mixtures of resin and cheese-whey, then analysing the liquid phase. Figure 3 shows the results obtained for αlactalbumin and βlactoglobulin ; owing to difficulties, in particular in the electrophoretic analysis, the precision is not very good.

EXPERIMENTAL RESULTS - COMPARISON WITH MODEL PREDICTIONS

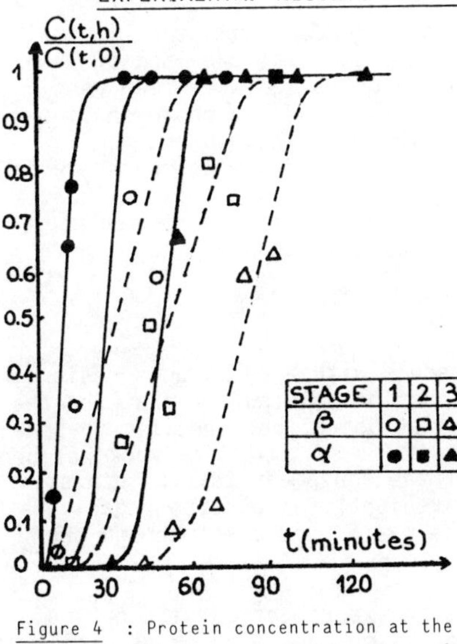

Figure 4 : Protein concentration at the outlet of a three-stage fluidization column

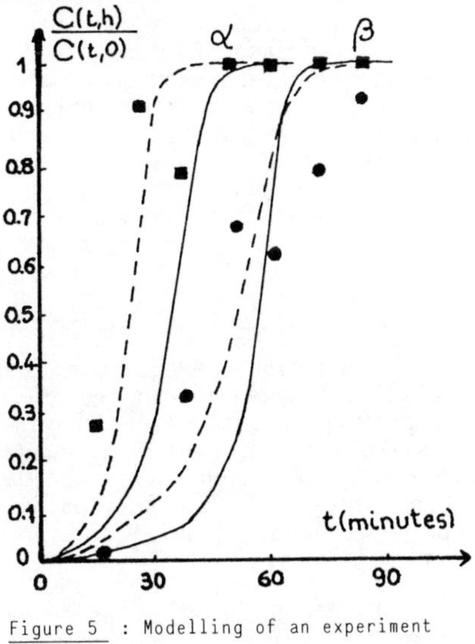

Figure 5 : Modelling of an experiment

System Without Resin Circulation

Figure 4 is an example of the results obtained with a three stage fluidized column. It presents the protein concentration variations at the outlet of each fluidized stage versus time. The experimental contions were 1.3 10^{-3} ms^{-1} for the liquid flow-rate and 1.32 10^{-4} m^3 for the volume of solid in each stage. The initial solution Ph was 6.6 at the inlet of the column. The operation has been carried out at 25°C. At the beginning of the experiment, the protein concentration at the outlet of all the stages is zero. It increases, then, rapidly at the exit of the lowest stage in the column and more slowly for the higher stages. It is

It is interesting to observe that the phenomena are different for the two proteins analysed During approximately 30 minutes, both proteins are completely recovered from the cheese-whey, but after this time the beds become saturated in αlactalbumin, the concentration of which increases in the liquid. Between 30 minutes and 45 minutes after the beginning of the experiment, βlactoglubulin is still completely adsorbed on the resin, while αlactalbumin remains in the liquid phase. This result opens the possibility of recovering quite pure αlactalbumin by treating the corresponding liquid in a second column.

The exchange capacity of the resin is the quantity of protein adsorbed per gram of resin and which can be recovered after alution. For all the experiments performed, the capacity has been about 20 mg per gram of resin for αlactalbumin and 110 mg per gram of resin for βlactoglobulin. These values appear to be similar to those obtained in a fixed bed.

Finally it is important to note that no clogging problem has been observed, neither in the fluidized bed itself nor for the distributors at the bottom of the stages and that the efficiency of the fluidized process has not decreased during the twenty cycles that have been performed.

Figure 5 compares the predictions of the model with experimental data in the case of a one stage experiment. Owing to the lack of precision of the equilibrium data, two series of values have been used to perform these calculations (the dotted line and the unbroken line), showing the important influence of the equilibrium data on the model results. Taking into account all the incertainties of the concentration measurements and of the model calculations, it can be concluded that the agreement is satisfactory and that in the future, improvements would result from increases in the analysis precision.

System with resin circulation

Concerning the equipment and procedure with a continuous resin circulation through downcomers, first it must be pointed out that for that kind of apparatus, several difficulties of operation, specific to Spherosil QMA resin, have been encountered. A lot of empirical work has been necessary to design particular parts of the equipment (distributors, downcomers...etc) and to find out a convenient procedure to organize a continuous and steady circulation of the resin particles through the system (4).

Figure 6 presents, as an example, an experiment carried out in a three stage fluidized column. A transient step can be observed, followed by a stationnary period. It can be noted that, in reduced concentration αlactalbumin is fixed in less important quantities than βlactoglobulin. It can also be observed that, with the solid and liquid flow rates used, the first stage, at the bottom of the column is useless, which means that the particles of Spherosil QMA are quite completely saturated when they leave the second stage. Nevertheless, these experimental results show clearly that the continuous recovery of cheese-whey proteins in a multistage fluidized bed with a counter-current flow of resin is possible.

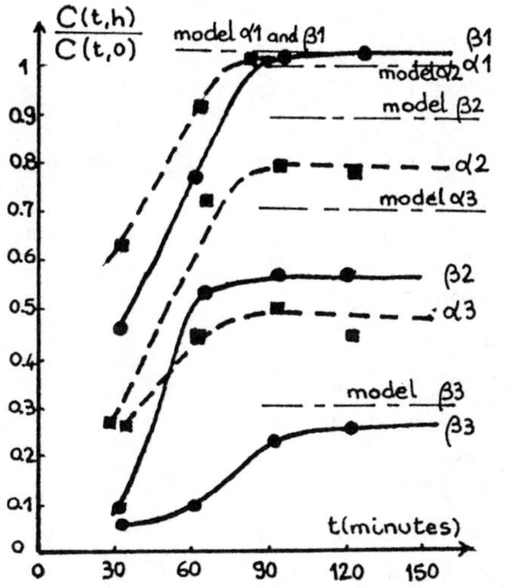

Figure 6 : Continuous process trial

Figure 6 shows clearly that the agreement between experimental data and the model results is far from being satisfactory. Without entering, here, a detailed discussion, it can be said that additional experiments have demonstrated that these discrepancies result from a very bad modelling of the real solid motion in the fluidized beds. Two kinds of improvements are still necessary :

- for the experimental procedure, the volume of dead zones in the beds must be reduced, so, increasing the mass transfer efficiency,

- from a modelling point of view, it is necessary to improve the real solid motion descriptions.

CONCLUSIONS

As a conclusion, it can be said that the cyclic equipment and procedure are now being developed at a larger scale, as well for the cheese whey proteins recovery as for the treatment of other biological products of pharmaceutical interest. For the continuous equipment and procedure, additional laboratory works are going on, first to improve our knowledge of the solids motions and the influence of convenient baffles, then to increase the efficiency of the ion exchange process.

NOTATIONS

$C(t,x)$	-	proteins concentration in the liquid phase ($g\ Kg^{-1}$)
$C(t,h)$	-	proteins concentration in the liquid phase at the outlet of the fluidized bed ($g\ Kg^{-1}$)
$C(t,o)$	-	proteins concentration in the liquid phase at the inlet of the column ($g\ Kg^{-1}$)
$C^*(t,x)$	-	proteins concentrations in the liquid phase, at equilibrium ($g\ Kg^{-1}$)
f	-	function describing the isotherm curve
h	-	height of the fluidized bed (m)
k_L	-	mass transfer coefficient in the liquid phase ($m\ s^{-1}$)
$q(t,x)$	-	proteins concentration in the solid phase ($g\ Kg^{-1}$)
$q^*(t,x)$	-	proteins concentration in the solid phase at equilibrium ($g\ Kg^{-1}$)
Q_s	-	volumetric flow rate of the solid ($m^3\ s^{-1}$)
S	-	volumetric specific area of the resin ($m^2 m^{-3}$)
t	-	time (s)
v_L	-	superficial flow rate of the liquid ($m\ s^{-1}$)
x	-	axial coordinate (m)
ε	-	bed void fraction

LITERATURE CITED

1 - MIRABEL, B., "Nouveau procédé de valorisation du lactosérum", Informations Chimie, 175, p. 105 (1978),

2 - MIRABEL, B., "Séparation des protéines par chromatographie préparative", L'actualité Chimique, p. 39 (juin-juillet 1980)

3 - BISCANS, B., RIBA J.P., COUDERC J.P., "Mise en oeuvre de résines échangeuses d'ions en couche fixe et en couche fluidisée multi-étagée - Comparaison des deux procédés", The Chem. Eng. Journal, 30(2), p. 81 (1985).

4 - BISCANS, B., SANCHEZ V., COUDERC J.P., "Procédé et dispositif pour la séparation chromatographique de macromolécules biologiques", Brevet d'invention FR 85.00463, Déposant : Brevet d'invention FR 85.00463, Déposant : ANVAR 04-01-85.

FLUIDIZED BED DRYING OF FINE PARTICLES WITH INTERNAL DIFFUSION LIMITATION

A.H.M.Verkooijen
Eindhoven University of Technology
Department of Chemical Engineering
Eindhoven, The Netherlands

ABSTRACT

A model for continuous fluidized bed drying of solids is proposed in which diffusional moisture transport in the dense phase particles is combined with exchange between gas in the bubbles and in the dense phase. In one variant a resistance in the cloud around the bubbles is introduced, in the second not. Drying experiments with solids of various particle size indicated that the resistance of the cloud can be neglected if the particles are larger than 120 µm, for smaller particles the cloud must be taken into account.

SCOPE

Fluidized beds are very well suited for drying solids in which moisture transport rate is determined by diffusion. Where other drying equipment provides only small contact times, in the order of seconds, the fluidized bed poses no restriction on residence time.

When fine particles are dried in a fluidized bed a significant proportion of the drying gas flows as bubbles through the bed and exchange of moisture between gas in the bubble and in the dense phase must be taken into account. For fluidized beds used as chemical reactors this exchange received ample attention and many different models have been proposed which take this exchange into account. For fluidized bed drying until recently most models assumed thermal and thermodynamical equilibrium between outgoing gas and the solids.

Palancz [1] proposed a three phase model with mass transfer between the bubble phase, the dense phase gas and the solid particles. Transport limitation of heat and mass inside the particles was assumed to be negligible. Results show that heat and mass transfer resistance between particles and gas must be taken into account. Palancz indicates that in case mass transfer inside the particle is rate limiting this model needs improvement.

In a study on batch fluidized bed drying Hoebink [2] proposed a model for drying particles with internal diffusion limitation. The bed is divided into three zones: -bubble phase -cloud phase -dense phase. In the dense phase mass transfer is assumed to be so fast, due to the large surface area, that there is equilibrium between the interstitial gas and the solid surface. However, drying in the particles is a slow process. In the cloud phase particles come into contact with relatively dry gas that flows from the bubble into the cloud and start a fast drying process of short duration. Thus the moisture concentration in the cloud will be higher than in the bubble and the moisture transfer from the dense phase will decrease. For this model no experimental verification has been given.

In this contribution a model is presented for continuous fluidized bed drying of solids with internal diffusion limitation. The drying process is thought to take place in three steps. First the moisture is transported by diffusion to the surface of the particle then transferred into the dense phase gas. Finally the moisture is transferred to the bubble phase. With respect to the latter two cases are considered: either this mass transfer is modelled using boundary layer theory, as was done by Chiba and Kobayashi [3] or Hoebink's model is used in which the presence of and the drying process in the cloud is taken into account. In order to be able to verify the model a number of refinements are necessary. The growth of bubbles as they rise in the bed influences many parameters in the model and is therefore included in the model. Also a dependence of the sorption isotherm on the moisture concentration is used.
An overall moisture balance links the slow diffusion inside the particles and the mass transfer to the bubbles.
A number of continuous drying experiments were performed to validate the model. Silicagel was used as a solid which exhibits severe diffusion limitation. The sorption isotherm, the diffusion coefficient and the fluidization parameters were determined experimentally for the solids used.

CONCLUSIONS AND SIGNIFICANCE

Model calculations show (Figure 5) that for particles with diameter larger than 120 μm no difference between the two transfer mechanisms is to be expected. However, for the smaller particles there is a significant difference in drying efficiency to be expected.
The first experiments were carried out with larger particles, 260 and 130 μm, and although results show some experimental scatter a reasonable fit with both model calculations is found. For smaller particles, 85 and 115 μm, the model with the cloud fits better to the experimental findings than the boundary layer model.
This is an important result because it shows that the drying rate will only increase when the particle size is reduced as long as the particles remain larger than 120 μm. Below this size the drying rate will decrease because the cloud will present an extra resistance to mass transfer.

DRYING MODELS

The water transport is supposed to take place in three consecutive steps. First water diffuses from the interior of the particles to the surface. This is a slow process that is governed by molecular diffusion. Throughout this model the diffusion coefficient is supposed to be constant.
From the particle's surface the water is transferred to the gas in the emulsion phase. In this step it is assumed that because of the large surface area of the particles the mass transfer resistance at the particle's surface can be neglected and the concentrations in gas and solid are in equilibrium.

Furthermore it is assumed that the moisture transfer is so fast that all particles will have the same surface concentration, regardless the particle's average water concentration. Because the particles have a relatively long residence time in the bed they will be perfectly mixed. From the gas in the emulsion phase the water is transferred to the bubbles. Here two different mechanisms can be distinguished. In the first mechanism it is assumed that the influence of the cloud which surrounds the bubbles can be neglected and that mass transfer can be described by boundary layer theory (see figure 1). In the second mechanism it is assumed that solids and gas flow co-currently through the cloud which results in a fast drying process by which only the surface of the particles is effected (see figure 2).
In the last step the bubbles take up the water vapour as they rise in plug flow through the bed.

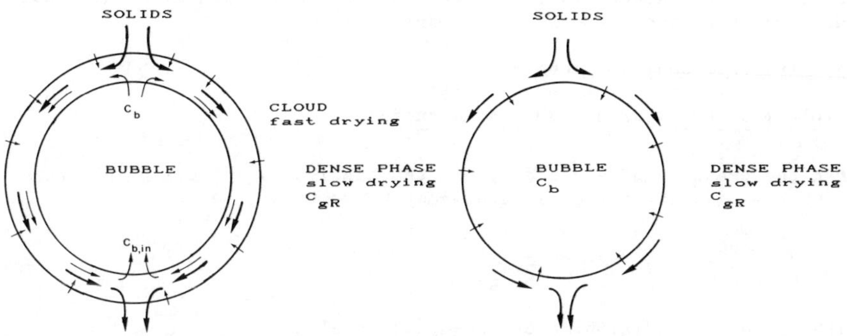

MOISTURE TRANSFER TO BUBBLE

Figure 1. Boundary layer model   Figure 2. Model with cloud

Drying inside the dense phase particle

The drying process inside the particle can be described by the following differential equation:

$$\frac{\partial C_s}{\partial t} = \frac{D}{r^2} \left\{ \frac{\partial}{\partial r} \left(r^2 \frac{\partial C_s}{\partial r}\right)\right\} \tag{1}$$

with initial condition: $t=0$ $0<r<R_p$ $C=C_{so}$

and boundary condition: $t>0$ $r=0$ $\frac{\partial C_s}{\partial r} = 0$

$r=R_p$ $C=C_{sR}$

The solution is well known and gives the average moisture concentration of a particle, C_s, as a function of its residence time in the bed:

$$\frac{\bar{C}_s - C_{sR}}{C_{so} - C_{sR}} = \sum_{n=1}^{\infty} \frac{6}{(n\pi)^2} \exp\left(-(n\pi)^2 \cdot \frac{Dt}{R_p^2}\right) \tag{2}$$

As the solids in the bed are ideally mixed the average moisture concentration of the bed can be calculated using the age distribution.
$I(t/\tau) = \exp(-t/\tau)$

$$C_{se} = \int_0^\infty \bar{C}_s \, I(t/\tau)d(t/\tau) \tag{3}$$

The substitution of (2) in (3) results in:

$$f = \frac{C_{se} - C_{sR}}{C_{so} - C_{sR}} = \sum_{n=1}^{\infty} \frac{6}{(n\pi)^2((n\pi)^2 Fo+1)} \tag{4}$$

in which $Fo=D\tau/R_p^2$.
Equation 4 describes the average moisture concentration of the bed as a function of residence time, τ, with C_{sR}, in equilibrium with the dense phase water concentration C_{gR}, as parameter.

Mass transfer emulsion-bubble

Simple mass transfer at the bubble surface

When the presence of a cloud around the bubbles is not taken into account the mass balance of a bubble may be written:

$$\frac{4}{3}\pi R_b^3 U_b \frac{dC_b}{dh} = -4\pi R_b^2 K_c (C_b - C_{gR}) \tag{5}$$

with boundary condition: $h=0 \quad C_b=C_{go}$ results:

$$Z = \frac{C_{ge} - C_{gR}}{C_{go} - C_{gR}} = s.\exp\left(-\int_0^H \frac{3K_c}{U_b R_b} dh\right) \tag{6}$$

Mass transfer model including the cloud

The moisture concentration in the bubbles is the result of a short term process in which gas from the bubble circulates through the cloud and contacts the solids which pass through the cloud (figure 2). As the slip velocity between the gas and the particles in the cloud is relatively low, the short term process is treated as a co-current plug flow. During this process, which takes only 0.01-0.1 seconds, the particles are exposed to a sudden increased driving force caused by the low moisture concentration in the bubble. Hence during this short contact time the outer shell is dried intensively.
Hoebink [2] has shown that during this process the overall mass transfer coefficient can be described by:

$$\frac{1}{K_{og}} = \frac{R_p + 3K_g t}{K_g R_p} \tag{7}$$

When this time dependent mass transfer coefficient is combined with the mass transfer model for the cloud presented earlier by Hoebink [4] it is found that

$$Z = s.\exp\left(-\int_{o}^{H} \frac{Q_g(1-P)}{U_b V_b}\, dh\right) \tag{8}$$

in which

$$P = \frac{C_{b,in}-C_{gR}}{C_b-C_{gR}} = \left(1 + \frac{2\pi\sqrt{2}(R_c^3 - R_b^3)K_g}{Q_g R_p}\right)^{-0.4(1-\varepsilon)} . \exp\left(\frac{-2\pi\sqrt{2}\ R_c^2 K_c}{Q_g}\right) \tag{8a}$$

Overall moisture balance

By a moisture balance over the whole bed the slow drying process inside the particles can be combined with the mass transfer towards the bubbles. The overall mass balance may be written:

$$Q_s(C_{so}-C_{se}) = Q_a(C_{gR}-C_{go})(1-Z) \tag{9}$$

With the equations (4), (6) or (8) and (9) the water concentration of the particles leaving the bed C_{so} can be calculated. To that end the integrals in equation (6) or (8) must be calculated numerically.
The drying efficiency, G, of the bed is defined as the ratio of the amount of water removed from the particles and the amount of water that would have been removed if the particles were in equilibrium with the drying air. It can be calculated from C_{se}.

EXPERIMENTS

Materials

In all experiments a sub-microporous silicagel was used. Particles of various diameter were obtained from the same initial batch by milling and sieving. The relevant properties were determined experimentally.
Diameter: d_{32} was determined by sieving. Diameters used: 85, 115, 135, 250 µm. Sorption isotherm and diffusion coefficient in the solid phase: determined gravimetrically using a method described by Dengler [5]. The result for the sorption isotherm is given in figure 3. For use in calculations this curve is approximated by 4 straight line pieces.
The diffusion coefficient was found to be a function of the moisture concentration in the solid. Figure 4 gives the result.
Fluidization characteristics: were measured in the same bed as in which the drying experiments were carried out. U_{mf}, ε_{mf} and the bubble hold up were determined. The diameters of the bubbles were determined as a function of the height in the bed. Therefore the bubble frequency was measured with a capacitance probe at various heights. The results fitted very well to Darton's correlation [6].

Drying experiments

Equipment: The drying experiments were performed in a cylindrical fluidized bed with internal diameter of 30 cm. The silicagel was dried with hot air at a constant bed temperature of 50°C. The gas distributor plate was fitted with 197 holes of 2 mm diameter. With a screw feeder the wet powder was continuously fed to the bed, the height of which was maintained constant. A large number of drying experiments were carried

out at widely varying conditions:
τ: 0.6-4.4 hr; U_o: 5.5, 8.7, 11.0 cm/s; H_{bed}: 16, 30 cm; C_{so}: 88-145 g water/kg dry solid.
The drying process was monitored by determining the moisture concentration in silicagel samples drawn from the bed at regular intervals. When steady state was reached the drying efficiency, G, was calculated from C_{so} and C_{se}.

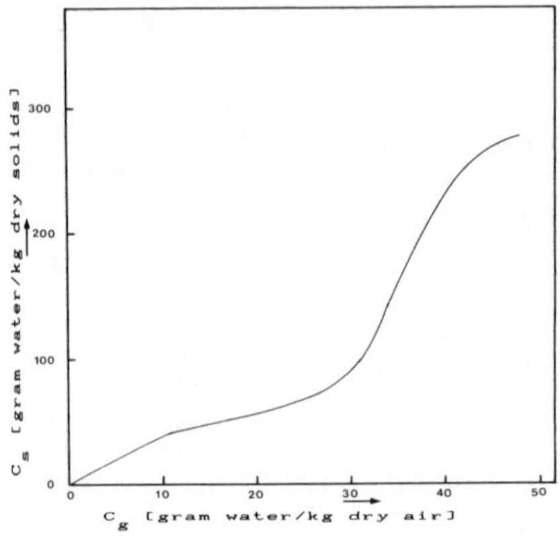

Figure 3. Sorption isotherm of water in silicagel at 50°C

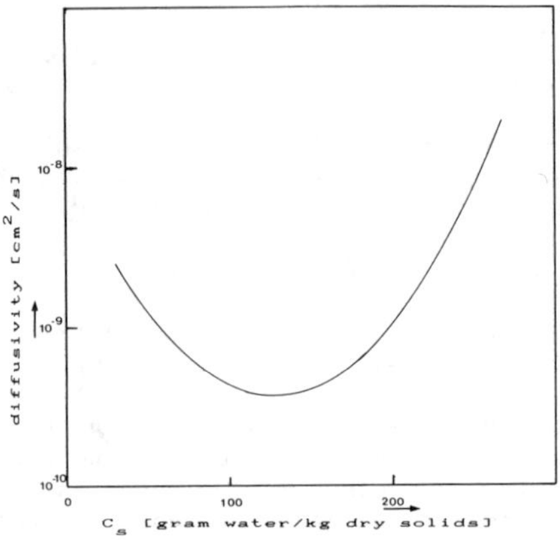

Figure 4. Diffusion coefficient of water in silicagel at 50°C

COMPARISON WITH MODEL CALCULATIONS

In the model a large number of parameters are used. Some of these were measured in separate experiments. For the remaining parameters the following correlations were used:
The mass transfer coefficient K_c is obtained from the Chiba-Kobayashi equation [3]. A correction factor of 2.0 is necessary when the equation is applied to freely bubbling beds because this equation was originally derived for single bubbles only.
The Davidson-Harrison [7] model was used to calculate the cloud diameter R_c and the gas flow from the bubble to the cloud Q_g.
For the particle to gas mass transfer coefficient K_g the Nelson-Galloway [8] equation is used.
In a computer program the drying efficiency is calculated with both the simple mass transfer model and the cloud model using the above-mentioned parameter values from experiment and literature. In figure 5 the influence of the particle diameter on the drying efficiency, G, is given. It shows that for larger particles the results for the two mass transfer models are identical. For smaller particles the cloud forms an extra resistance against mass transfer between the bubble and the dense phase.
In figures 6 and 7 the drying efficiencies measured experimentally are compared with the results from the computer calculations for the conditions of the experiment. Figure 6 gives the results for the larger par-

ticles. Here both models predict the same drying efficiency. Although there is experimental scatter, the model predicts the drying process rather well. In figure 7 the results for the smaller particles are given. Here the two models give different results. The results are very well described with the model that takes the cloud phenomena into account. The simple mass transfer model gives a systematic error of 4%.

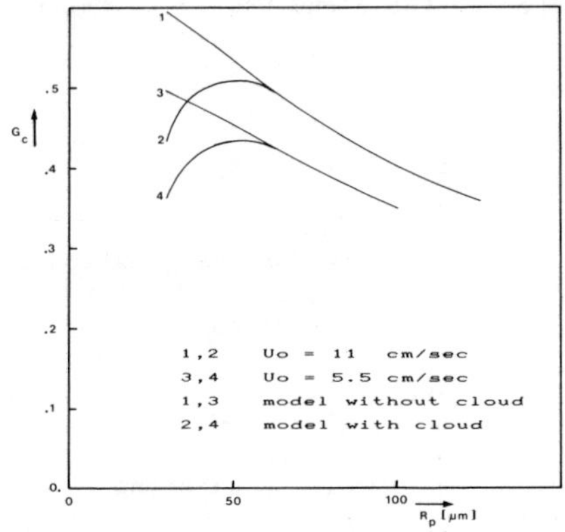

Figure 5. Calculated drying efficiency, G_c, as function of particle radius.
Parameter values:
C_{so} = 110 g water//kg dry solids, H = 16 cm, τ = 3600 sec.

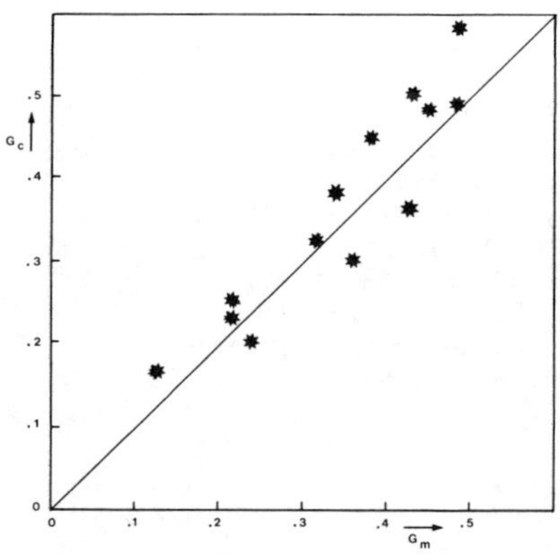

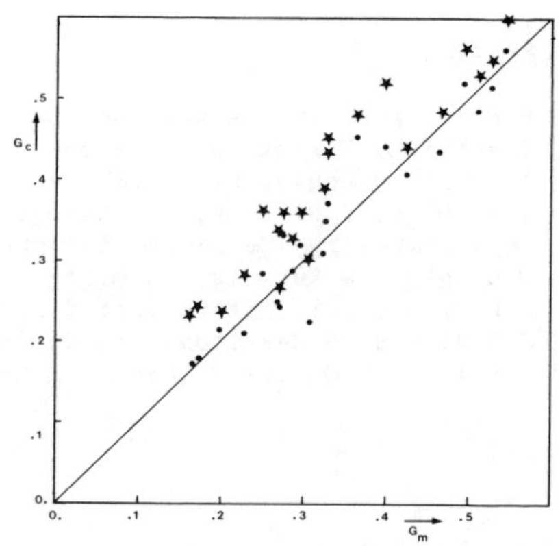

MEASURED DRYING EFFICIENCY, G_m, VERSUS RESULTS OF MODEL CALCULATION, G_c

Figure 6. Large particles
$R_p > 60$ µm

Figure 7. Small particles
$R_p < 60$ µm
* model without cloud
● model with cloud

CONCLUSION

- A model has been developed with which the drying efficiency can be calculated for particles with internal diffusion limitation in a continuously operated fluidized bed.
- Drying of large particles (>120 µm) can be calculated with a simple mass transfer model between bubble and dense phase.
- For smaller particles mass transfer in the cloud has to be taken into account.

SYMBOLS

C	concentration	t	time
D	diffusion coefficient	U	velocity
f	dimensionless solid moisture concentration, equation 4	V	volume
		Z	defined by equation 6
Fo	dimensionless residence time	ε	voidage
G	drying efficiency	τ	residence time
h	vertical coordinate		
H	bed height	subscripts	
K_c	bubble-bed mass transfer coef.	a	air
K_g	particle-gas mass transfer coef.	b	bubble
n	integer number $1<n<\infty$	e	exit conditions
P	bubble's dimensionless moisture concentration, equation 8	g	gas
		mf	minimum fluidization
Q	volumetric flow rate	o	feed
r	radial coordinate	p	particle
R	radius	R	at particle surface
s	bubble fraction in gas flow	s	solid

REFERENCES

[1] B.Palancz, Chem.Eng.Sci. 38, p.1045 (1983).
[2] J.Hoebink, K.Rietema, Chem.Eng.Sci. 35, p.2257, (1980).
[3] T.Chiba, H.Kobayashi, Chem.Eng.Sci. 25, p.1375.
[4] J.Hoebink, K.Rietema, in: H.Angelino, Fluidization and its applications, Ste.Chemie Industrielle, 1974.
[5] W.Dengler, W.Kuckels, Chem.Ing.Techn. 42, p.1258 (1970).
[6] D.Darton et al., Trans.Inst.Chem.Eng. 55, p.274 (1977).
[7] J.Davidson, D.Harrison, Fluidisation, Academic Press, 1971.
[8] P.Nelson, T.Galloway, Chem.Eng.Sc. 30, p.1, (1975).

DRYING UNDER PRESSURE

by

Arne Sloth Jensen

A/S De Danske Sukkerfabrikker (DDS)
(The Danish Sugar Corporation)

Langebrogade 5, DK-1001 Copenhagen K

ABSTRACT

Wet particles are charged into a modified cellular fluid bed under pressure. The particles are carried by an upward superheated steam flow circulated within a closed system. Heat is added to the system and the steam developed by evaporation from the material can be tapped from the system for other purposes.

SCOPE

In beet sugar factories it is usual procedure to dry the beet remains after slicing, extraction of sugar and mechanical pressing to about 25% dry matter.

Up till now, this drying process has been carried out in rotary drums to which is added a 700 to 900°C warm mixture of air and flue gas from combustion of coal or oil. This process has three drawbacks:

1) The added energy of approx. 3000 kJ per kg water evaporated leaves the plant together with the exhaust gas in the form of an air/water-steam mixture with a dew point of abt. 70°C in which energy has only little value.

2) The exhaust gas contains some more or less charred plant parts, and the result is air pollution.

3) Residues from the combustion get in contact with the dried product and thereby pollute it.

In order to avoid these three drawbacks of the traditional drying process, a new drying method as well as equipment for the implementation of the process have been developed at DDS in recent years.

The basic object has been to let the drying process take place in a closed system as indicated in figure 1.

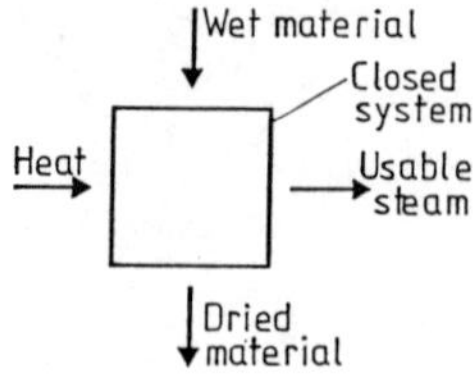

Figure 1

This simple process can have many forms. The efforts of this project has tended to let the drying process take place in a superheated steam which at the same time acts as a carrying agent in a fluid bed. The steam must then be recirculated in a closed system.

CONCLUSION

Flow conditions by fluidization of beet pulp in air have been examined in acrylic plastic models. Experiments have been made with fluidization of beet pulp in a pressure vessel with direct steam from a steam boiler. In this way, information was obtained for the elaboration of a pilot plant in which beet pulp is fluidized in a superheated water steam recirculated in a closed system. The plant has a capacity of 1 ton water evaporation per hour. Tests and reconstructions have made it possible to give this plant a simple form, and experience has been gathered which forms the basis for the design of a larger plant now under construction.

The method has three advantages:

1) Energy saving

2) Less pollution

3) Product improvement

FUNDAMENTAL LABORATORY TESTS

As a start, a ϕ200 mm fluid bed in acrylic plastic was made with air as fluidizing agent. Beet pulp with 25% dry matter could be fluidized at an air velocity of 4.5 m/sec. and dried material (90% dry matter) at about 3 m/sec. Fall tests of single particles in a vertical ϕ75 mm glass tube did, however, show that particles of a maximum length of 1 mm will fall through static air at a rate of under 1 m per second. Since these small particles constitute 5-10% of the material, it is necessary to greatly reduce the velocity in the upper part of the bed or to have a special separating system for these particles.

In another acrylic plastic model different shapes of heating surfaces were tested. Partly tubes and partly plates. Vertical plates could be placed at intervals of about 10 cm without causing a blocking of the bed. If tubes are used as heating surfaces with the same heat transfer per cubic metre bed, the tubes must be placed so closely that the result will be an obstruction of the bed.

Based on this experience, a fluid bed was built in 1982 with steam as fluidizing agent. It was constructed as shown on figure 2.

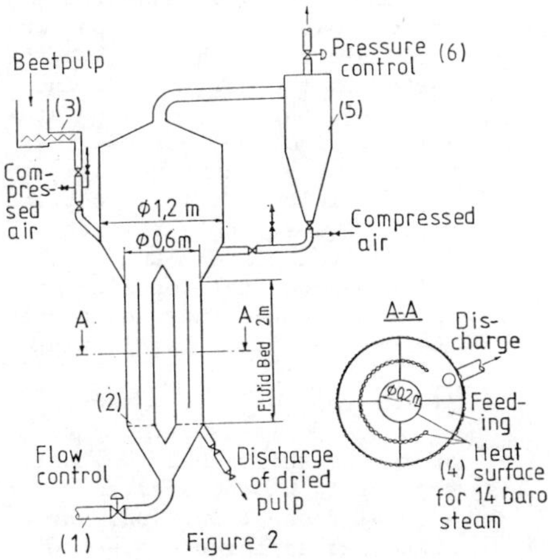

Figure 2

Through the tube (1) steam is added from a 4.8 baro boiler. The flow-controlled steam passes through the perforated plate (2) to the fluid bed to which the material is charged with the screw conveyor (3) and fed discontinuously through the automatically controlled valves.

In the bed itself heat is added through heating surfaces (4) heated with 14 baro steam. The used steam leaves the plant through the cyclone (5). A pressure control (6) is mounted on the outlet pipe.

As it appears from section A-A, the bed is circular and provided with a partition between the places where the material is fed and discharged. It proved necessary to divide the bed into 4 cells with vertical cell walls from about 100 mm above the perforated plate (2) in order to increase the retention time.

The main results from tests with the plant are:

The retention time for beet pulp should be 4-6 min at 2 baro and about 140°C. A division of the bed into more than 4 cells is necessary. A steam velocity of about 4 m/sec. at 2 baro and a temperature of 140°C is adequate for fluidization. Heat-transfer coefficient for the heating surfaces was measured at different pressures and velocities.

At 2 baro and 4 m/sec. it was 96 W/m^2 ^{o}C. As expected it increases with the pressure in the plant and the gas velocity. The quantity of the material in the plant has no significant impact.

PILOT PLANT 1000 kg EVAP./H

Based on the information gained from the above mentioned plant, a plant was built in 1983 in which the big steam flow for fluidization is reheated after the bed and returned to the bed by means of a centrifugal blower. The necessary heat is added to the plant as 14 baro steam.

The Fluid-Bed Apparatus

The bed is ring-shaped with an exterior diameter of 1.8 m and divided into 16 cells by radial partitions emanating from a ϕ0.6 m centre pipe. The cell walls and the ϕ1.8 m cylinder surface are heating surfaces heated by 14 baro steam. Above this the apparatus expands conically to ϕ3 m. Figure 3 shows a vertical section of the apparatus.

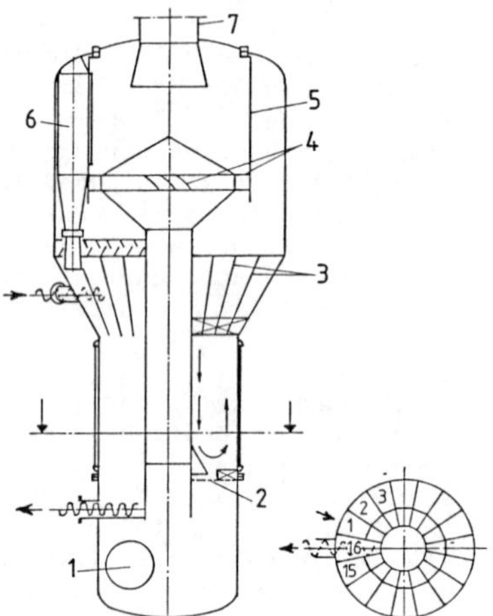

Figure 3

Through (1) a superheated steam of eg. 3.5 baro is added. The steam flows through the perforated plate (2) which forms the bottom of the cells 1-15. In the conic part of the apparatus a set of inclined plates (3) is placed. They contribute to catching particles thrown up between these plates. The particles fall on the plates and slide back on the top side of same. Above the inclined plates there is a set of baffle plates.

At the top there is a separation system for dust. Figure 3 shows the shape that the dust separation has been given after several reconstructions. The steam passes through a set of guide blades (4) and thus creates a vortex in the cylinder which is limited by the plate (5). The dust accumulates along the periphery and is separated through a slot to the side cyclone (6) placed above the last cell (No. 16). The separated dust is led directly down to the bottom of the cell from which it is discharged from the apparatus. The dust particles require only a short drying time and will therefore be dry before they reach cell 16. In order to overcome the pressure difference between the side cyclone and cell 16, an ejector driven by 14 baro steam has been made in the bottom of this cyclone. To obtain a shape with little risk of obstruction, the live steam is added through 12 small nozzles placed in the periphery of the ϕ120 mm hole in the bottom of the side cyclone. From here a mixture of dust and steam is led down through a diffuser. The purified steam is discharged from the apparatus through the tube at the top (7).

The material to be dried is charged into the apparatus through a rotary valve and carried on by a screw conveyor to cell 1. The material then passes on through the cells to cell 16 where no steam flow is blown through the bottom. Instead there is a screw conveyor at the bottom which leads the material to a rotary valve through which the dried material is discharged to atmospheric pressure.

The flow conditions in each cell were studied in a full-scale acrylic plastic model. In beet pulp, for which the plant was originally designed, the size of the particles varies greatly. The majority is 3-8 mm particles, but there is a lot of dust and some larger particles of 8-20 mm. There are also a few particles of more than 30 mm. This makes a traditional fluid bed impossible. The flow in the cells should be regulated so that all the particles are kept moving. Different shapes have been studied in the acrylic plastic model. The final shape is shown in fig.4.

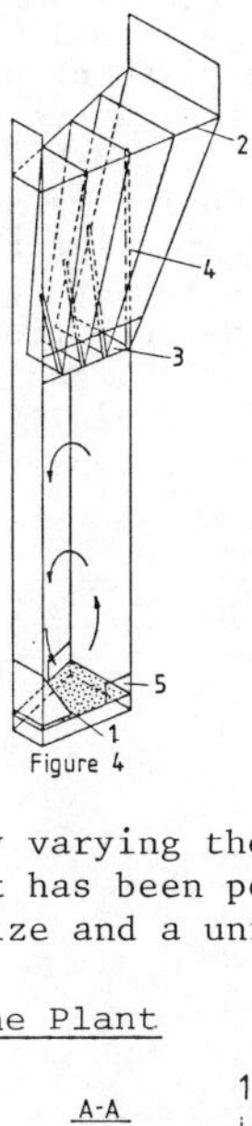

Figure 4

By means of the triangle (1) a downgoing flow of particles above this triangle is secured. The particles slide down the top side of (1) and are then blown horizontally from the lower corner of the triangle by the steam that has passed up through the perforated plate below the triangle. The particles are blown back into the cell by the flow through the perforated plate. A circular motion has thus been created, and it is intensified by the shape of the cell, because the distance between the cell walls is most narrow in the part above the triangle. Thus the lowest vertical steam velocity is here.

The passage from one cell to another can take place through 3 canals. The smallest particles can cross the top edge of the cell wall (2). Other particles can pass through the opening (3) to which some of the material that slides down the slanting plates is led by the guidebars (4). Finally, the heavy particles can pass through the opening (5) at the bottom of the cells. By varying the size of the openings and the length of the guidebars (4) it has been possible to ensure a suitable drying time for each particle size and a uniform filling in each cell.

The Plant

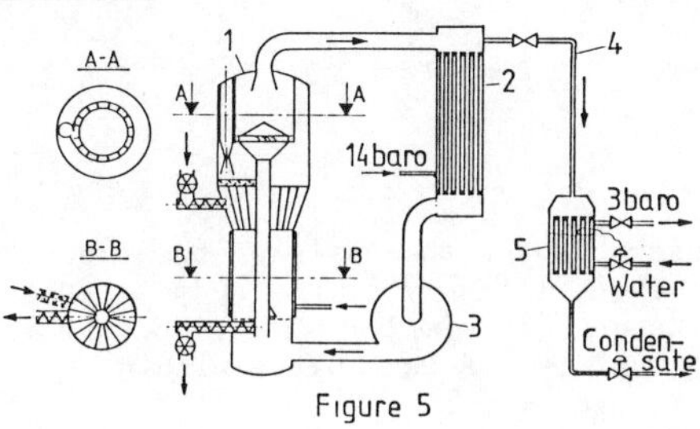

Figure 5

The bed forms part of the complete plant as shown in fig. 5.

Originally, the steam was led from the top of the bed (1) through a cyclone and then through a wet scrubber before it continued to the superheater (2). But the improved dust separation in the top of the bed made these external dust separations superfluous.

After the superheater, which is heated with 14 baro steam, the steam continues to the centrifugal blower (3) from which the steam is blown back to the bed, and the big steam circuit is closed.

The steam quantity coming from evaporation from the material must be compensated by a tapping of steam through the pipe (4). This steam may contain some dust and some acids if it is the result of beet pulp drying. It is replaced by a clean steam by heat exchange in the steam generator (5).

Operating Results

A number of characteristic operating data are stated in the diagram, figure 6. Variations have been made to the parameters: supply steam pressure, pressure in plant and supplied flow of beet pulp. The evaporation capacity depends strongly on the difference between the two steam pressures.

The dry matter content in the dried material depends on the superheating of the steam and the retention time of the material. The retention time and the deviation on same decrease at increasing flow of supplied pulp; however, not inversely proportionally to the quantity supplied, since the filling in the apparatus increases at increasing capacity. Consequently, the temperature at the top of the plant drops when the dosing is increased and the evaporation of the plant increases. At the same time the dry matter content of the dried material decreases. This dry matter content may be controlled by varying the supplied pulp flow. The retention time has been examined at shock dosing of beet pulp to which LiCl has been added. Then samples have been taken currently of dried material, in which the concentration of LiCl has been determined. The result appears in figure 7.

Supply steam pressure	baro	5.2	5.2	12.0	11.8
Pressure in dryer	baro	0.5	2.5	3.0	2.9
Supplied pulpflow	kg/h	1071	1056	1503	2046
Dry substance in wet pulp	%	26.8	26.9	26.9	26.2
Dry substance in dried pulp	%	78.5	58.2	87.0	72.0
Evaporation	kg/h	705	568	1038	1301
Temperature in bottom	°C	158	158	185	185
Temperature in top	°C	130	142	154	150

Figure 6

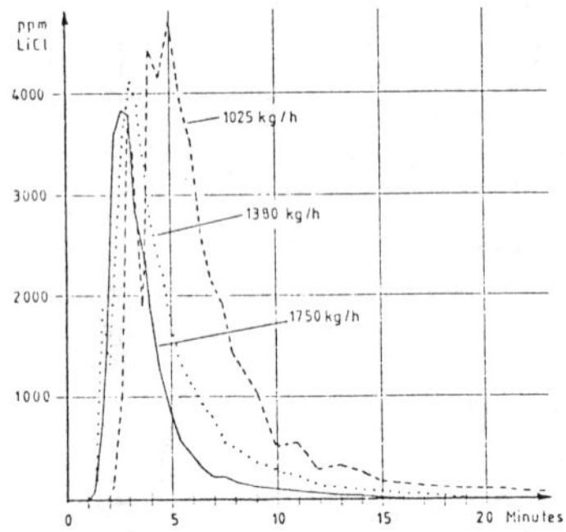

Figure 7

FUTURE PLANTS

A new and bigger plant has been constructed on the basis of the experience gathered from the pilot plant. The relatively simple shape of the pilot plant makes it possible to make an even simpler plant, because the circulation of the great steam quantity can take place within a closed pressure vessel as shown on figure 8.

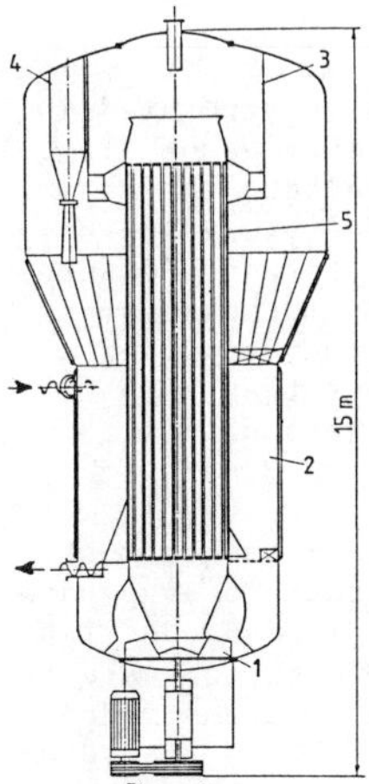

Figure 8

The blower (1) is placed at the bottom of the apparatus. From the impeller, the steam is blown up through the cells (2) to the top of the apparatus where dust is separated in the cyclone (3) and transferred to the side cyclone (4) and thereby to the last cell. At the centre of the cyclone, the steam is led to the superheater (5) where it is heated on its passage down through the pipes which are heated by steam at higher pressure, eg. 14 baro. From the superheater, the steam is led directly into the suction side of the blower, and the big steam circuit is closed.

The excess steam produced in the plant is discharged at the top in the middle of the cyclone section. It is led to a heat exchanger where it is condensated inside vertical tubes. Water is boiled between these tubes and clean steam is thus produced for other purposes.

For this boiling, condensate from the heating surfaces in the drier is added as feed water. Figure 9 shows a mass and energy flow for a plant as shown on figure 8 with an evaporating capacity of 6 tons of water per hour.

A plant like the one described here is under construction at Stege Sugar Factory in Denmark.

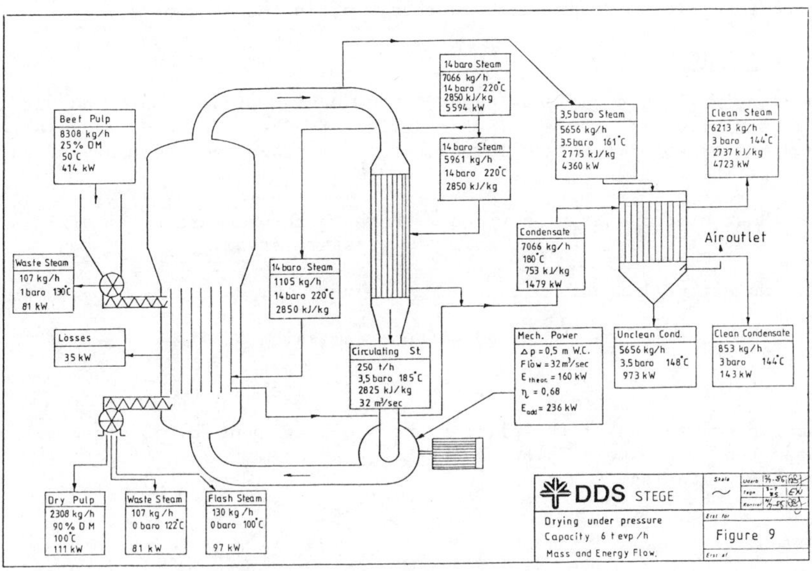

Figure 9

INTEGRATION IN A SUGAR FACTORY

In a sugar factory, the purified sugar juice is normally evaporated in a 4 or 5 step evaporator. 1st step is heated with steam of about 2.5 baro. The steam is drawn from steam boilers of high pressure that may be reduced through a turbine which runs an electric generator. The boiler pressure is often reduced through a reducing valve to the pressure wanted in the evaporator. If a greater or smaller part of the steam is reduced in this way, a pressure drying plant like the one described here can favourably be applied. From figure 7 it appears that 7066 kg 14 bar steam is added to the plant, and the plant delivers 6213 kg 3 bar steam at 144°C. If 7066 kg 14 bar steam is reduced to 3 bar with subsequent water injection cooling with 80°C hot water, the result will be 7411 kg 3 baro steam at 144°C. Consequently, the steam boiler must produce 7411 - 6213 = 1198 kg more steam per hour in order to evaporate 6000 kg water in the pressure drying plant. This will give an additional energy consumption of 3070 kJ for evaporation of 6000 kg equivalent to 512 kJ per kg water evaporated in the plant. To this must be added a power consumption in the pressure drying plant of 236 kW to the rotary blower. This corresponds to 142 kJ per kg water evaporated. The consumption will thus be 512 + 142 = 654 kJ per kg water evaporated. This is only 22% of the energy consumption of 3000 kJ per kg water evaporated in traditional drying plants.

Steam systems in sugar factories are very different so that the incorporation of a pressure drying plant must vary from one plant to another. A factory may not have steam of adequate pressure for the superheater, for example 12 to 20 bar, but instead 40 ato. In that case, it will be natural to run the rotary blower by a turbine which will reduce 40 bar steam to approx. 20 bar which is then used in the superheater.

OTHER APPLICATIONS

This drying method was developed for the drying of beet pulp, but it may be applied advantageously for drying of other products, especially when one or more of the following three advantages of the process can be exploited:

1) Most of the energy provided with the high-pressure steam will leave the plant as a usable low-pressure steam.

2) No air pollution.

3) The material can be dried without admission of air.

The implementation of the pilot plant 1,000 kg evap./h. is financially supported by the Commission of the European Communities, Directorate for Energy Saving.

DESIGNING FLUID BED DRYERS FOR LOW ENERGY OPERATIONS

Ole Kjærgaard
H. Bender Mortensen
Sten Mortensen (speaker)
A/S Niro Atomizer
305, Gladsaxevej
DK-2860 Søborg, Denmark

ABSTRACT

The objective of this paper is the optimization of industrial fluidized bed dryers with regard to energy consumption by a systematic evaluation of dryer design and operating conditions. A hygroscopic polymer is used as an example. The conclusions are valid in principle for many other products.

SCOPE

It is now more than ever before important to improve all industrial processes towards lower energy consumption. The energy required for drying constitutes an essential part of the total energy consumption involved in the production of many materials. Fluidized beds are widely applied as industrial dryers. They exist in many varieties such as continuous or batch dryers, stationary or vibrating, single or multi-stage, with or without contact heating surfaces. Solids flow patterns in continuous units may vary from back mix to plug flow, and relative to the gas flow from cross to cross/counter-current flow.

This paper will demonstrate how to minimize the energy consumption through a systematic evaluation of

- operating parameters
- solids flow patterns
- gas flow patterns
- contact heat input
- heat recovery

One product example is used to illustrate the effect of these general factors, namely polypropylene (PP).

PP produced by suspension polymerization in hexane is an example of a hygroscopic, heat-sensitive thermoplastic material.

CONCLUSIONS AND SIGNIFICANCE

A systematic evaluation of fluidized bed design and operating parameters may lead to a significant improvement in the operating costs as demonstrated in the case of hexane-wet polypropylene. Similar evaluations are used for other products based on their specific drying characteristics.

DESIGNING FLUID BED DRYERS FOR LOW ENERGY OPERATIONS

HEXANE-WET POLYPROPYLENE

Polypropylene polymerized in liquid hexane is fed to the dryer from a continuous decanter centrifuge at a volatiles level of 0.5 kg/kg of dry polymer and a temperature of 60°C. The particle size range is 100-1500 micron, the mean about 500 micron and the bulk density 400 kg/m^3. The feed material is not directly fluidizable. The desired final volatiles level is 500 ppm.

The equilibrium volatiles concentrations in PP as function of the relative gas saturation are shown in fig. 1. PP is truly hygroscopic above and below the final volatiles level of 500 ppm.

Figure 1
Equilibrium volatiles content of PP versus relative gas saturation.

Inert gas:
Nitrogen with PP

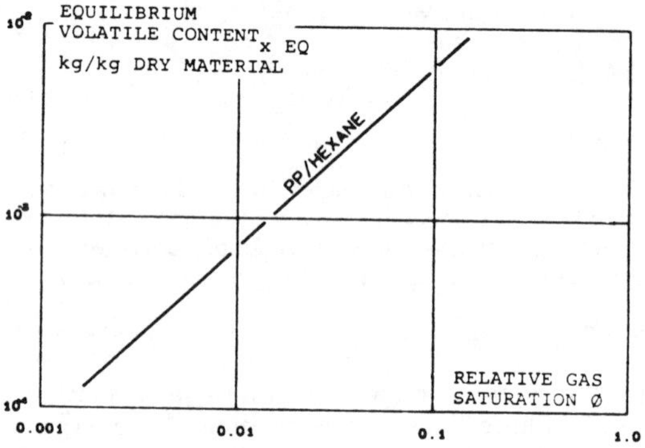

Data from a batch test in a fluidized bed with nitrogen recycle system are shown in fig. 2 - volatile contents and solids temperature versus drying time. The inlet gas had a hexane dew point of -16°C and a temperature of 110°C. The fluidizing velocity was 0.30 m/s. Fluidization starts when the volatile level is below 0.25 kg/kg. The drying takes place at constant rate (zero order) down to a volatiles level of 0.10 and at a decreasing rate (1st order) below 0.10 kg/kg.

Figure 2
Volatiles content and temperature of PP versus time in batch fluid bed

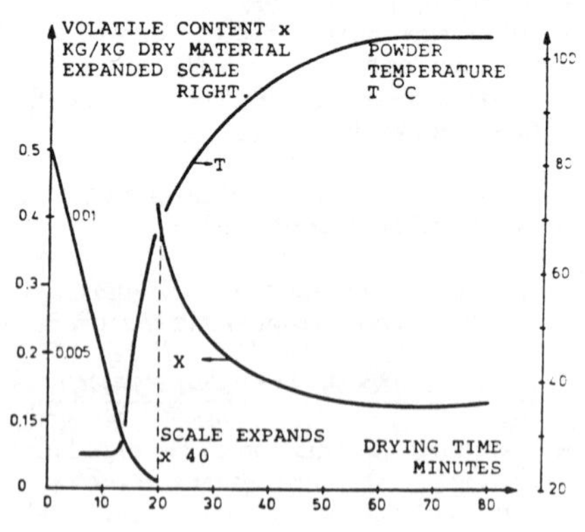

Dryer Design Parameters - a Qualitative Study

- Due to explosion risks, cost of volatiles and polymer oxidation risk, the drying must take place in an inert gas recycle system in which the volatiles are recovered in a condenser.

- Polymer heat-sensitivity restricts the process temperatures. The maximum permissible temperatures are 110°C for the polymer and 130°C for the drying gas.

- The amount of gas required for evaporation of the volatiles and heating of the polymer may be significantly reduced by the introduction of heating surfaces submerged in the fluidized layer. Maximum permissible heating surface temperature is 130°C.

- As a consequence of the hygroscopicity and the heat sensitivity of the polymer, the final drying stage has to operate at a dew point less than -16°C.

- However, most of the solvent may be evaporated at a much higher dew point corresponding to the use of cooling water for the condensation, that is a dew point in the range 30°C to 40°C.

- Consequently, the fluidized bed drying may be performed in either of two systems:

  a) a single cycle system operating at a dew point less than -16°C

  b) a dual cycle system, the first cycle utilizing cooling water and only the second operating at a dew point less than -16°C, see fig. 3

- The capital cost is lower for a single than for a dual cycle system. However, due to the high cost of refrigeration, the latter has been favoured for all industrial installations. Consequently, the calculations and results presented below are restricted to the optimization of this concept.

- The solids flow pattern in a continuous fluidized bed dryer may be designed to be nearly ideal back mix or ideal plug flow or in between the two extremes. The back mix flow variety is particularly useful, e.g. if the feed is not directly fluidizable. In all other cases the plug flow variety is preferred due to better heat utilization and more narrow solids residence time distribution.

- It is readily deduced from the drying curve that the drying gas leaving the final part of the process contains a considerable amount of sensible heat which may be utilized for:

  a) A previous drying stage. This is accomplished by the introduction of multi-deck fluidized beds.

  b) Heat-exchange with the cold gas recycling from the condenser. This is especially relevant in the second cycle thereby saving a considerable amount of refrigeration energy. Consequently, a gas/gas heat-exchanger is incorporated in all the second cycle systems studied below.

Figure 3 Fluid bed drying system for hexane-wet PP

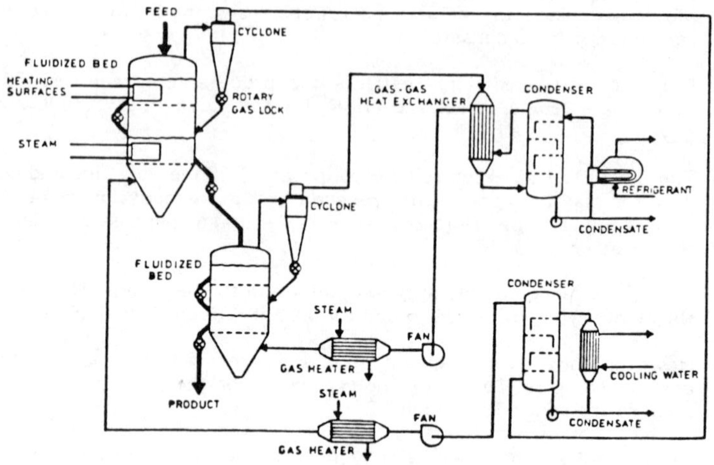

Dryer Design Parameters - a Quantitative Study

The dryer configuration Fig. 3 is used as an example, and the effects of various parameters are calculated for a throughput of 10,000 kg/h dry polymer. Capital costs are not entered into the calculations. Following energy costs in USD/kWh are used:

Steam	:	.026
Electricity	:	.055
Cooling water	:	.0021
Refrigerant;		
-30°C	:	.077
-35°C	:	.103
-40°C	:	.138
-45°C	:	.181

Gas rate versus solids flow pattern. The effect of solids flow pattern on gas rate of the post dryer is shown in fig. 4 (for the initial part of the drying, the effect is much less pronounced). A "tanks in series" model is employed (n = number of tanks). n = 1 corresponds to ideal back mix flow, n = infinite to ideal plug flow. It is seen that if n > 10, the effect of increasing the number of tanks is slight.

It may be concluded that the post dryer -operating in the regime of decreasing drying rate - should be designed for the maximum practical degree of plug flow.

Figure 4
Inert drying gas rate for PP versus solids flow pattern in post dryer

Assumptions:
Solids inlet vol. content : 0.014 kg/kg
Solids inlet temp. : 110°C
Inlet gas dew point : -25°C
One deck
No heating surfaces

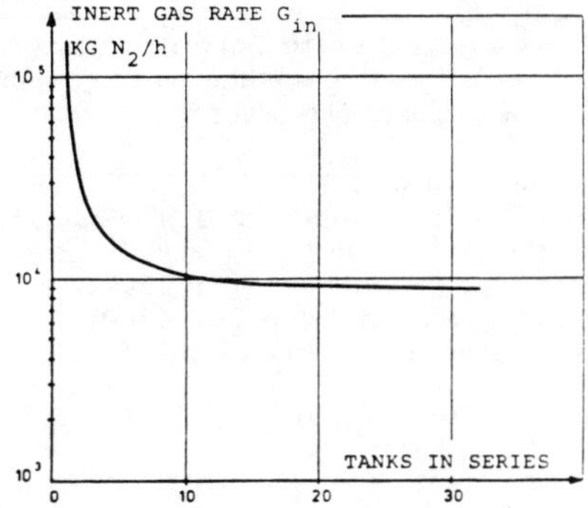

Gas rate versus number of decks in the fluidized bed post dryer. Fig. 5 shows that for the post dryer the gas rate is reduced significantly as the number of decks is increased. In practice, 2 or 3 decks are used.

Figure 5
Inert drying gas rate for PP versus number of decks in the post dryer

Assumptions:
Solids inlet vol. content : 0.014 kg/kg
Solids inlet temp. : 110°C
Inlet gas dew point : -25°C
No heating panels
Max. practical degree of plug flow

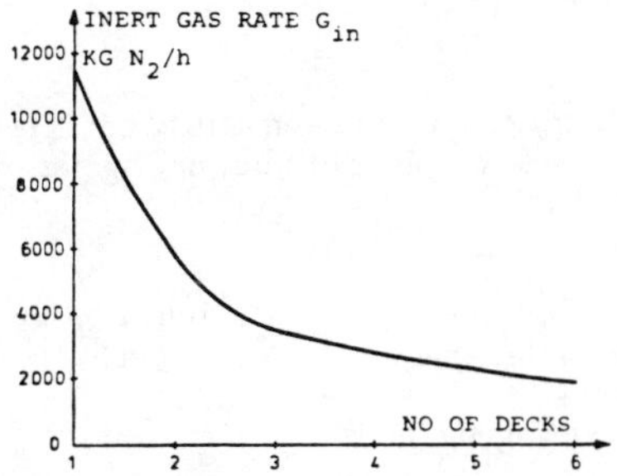

Gas rate and utility costs versus heating surface area and No. of decks of pre-dryer. The effect of heating surfaces is a very marked reduction in utility costs in the pre-dryer, see fig. 6. The product leaving the pre-dryer is very close to the maximum permissible solids temperature for which reason the use of heating surfaces in the post dryer is not worthwhile.

As in the case of the post dryer the amount of circulating inert drying gas and thereby the capital costs and utility costs are reduced especially when going from one to two decks.

Figure 6
Inert drying gas rate and utility costs for PP versus heating surface area and number of decks of pre-dryer

Assumptions:
Solids outlet temp. : 110°C
Inlet gas dew point : 40°C
Inlet gas temp. : 130°C
Heating surface temp. : 130°C
Max. practical degree of plug flow

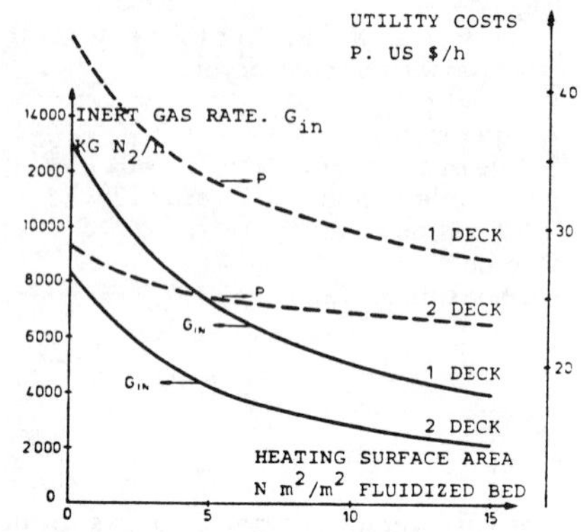

Gas rate and utility costs versus dew point of inlet drying gas of post dryer. Whereas the gas rate decreases with the dew point, fig. 7 shows that, the utility costs exhibit a minimum around -25°C due to the increasing cost of refrigeration at lower temperatures.

Figure 7
Inert drying gas rate and utility costs for PP versus dew point of inlet drying gas of post dryer

Assumptions:
Solids inlet vol. : 0.014 kg/kg
Solids inlet temp. : 110°C
2 decks
No heating surfaces

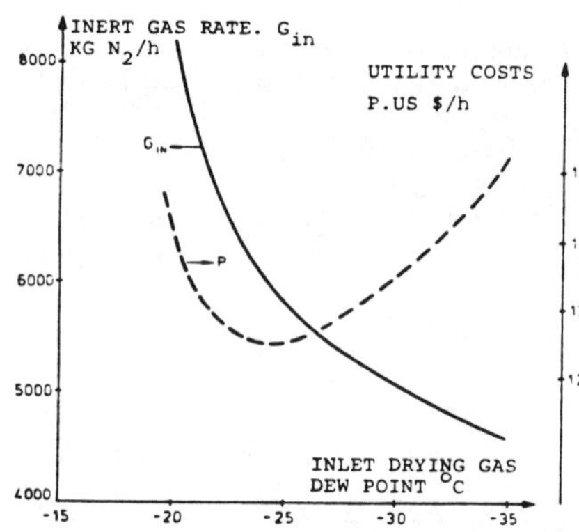

Gas rate and utility costs versus volatiles content exit pre-dryer
- or the effect of apportioning the drying to the two drying circuits.
The minimum possible outlet volatiles content from the pre-dryer at 40°C dew point and 110°C product temperature is approx. 0.007 kg hexane/kg dry material. Consequently, the gas rate and utility costs increase markedly as this value is approached. The reverse effect is seen in the post dryer system.

The minimum compound utility cost is reached at approximately 0.014 kg hexane/kg dry material exit the pre-dryer.

Figure 8

Inert drying gas rate and utility costs for PP versus volatiles content exit pre-dryer

Pre-Dryer
Inlet gas dew point : 40°C
2 decks
Max. 5 m^2 heating surface/m^2 fluid bed

Post Dryer
Inlet gas dew point : -25°C
2 decks
No heating surfaces

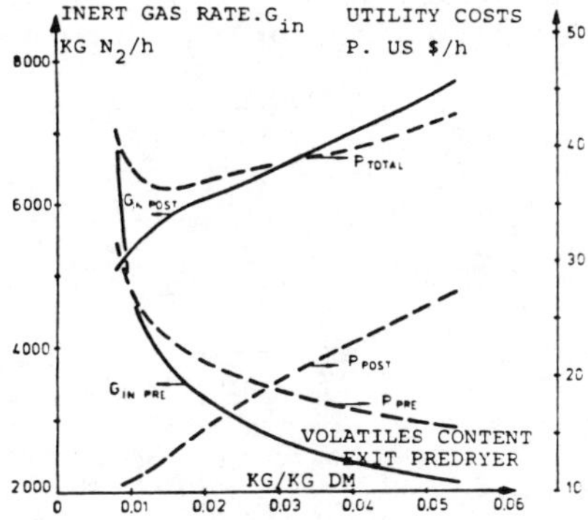

Conclusion

In conclusion, the analysis above has shown how the product parameters (like heat sensitivity, fluidization characteristics, drying rates and equilibrium relationship) combined with fluidized bed characteristica (like solids flow pattern, contact heating surfaces and dew point of drying gas) and utility costs lead to an optimized design of the overall drying system.

THE USE OF A ROTATING FLUIDIZED BED AS A HIGH EFFICIENCY DUST FILTER

by
Robert Pfeffer, Gabriel I. Tardos and Eli Gal
Department of Chemical Engineering
The City College of The City University of New York
New York, New York 10031

ABSTRACT

The filtration efficiency of a laboratory-sized continuously operating rotating fluidized bed filter (RFBF) was measured for different operating variables and conditions including rotating speed, gas velocity, gas humidity, size and type of granules, size and type of aerosol and bed thickness. The data were compared to theoretical models assuming inertial impaction to be the primary filtration mechanism.

SCOPE

This research was carried out to demonstrate the feasibility of using a rotating fluidized bed composed of inert granules as a high efficiency filter to separate aerosols from a contaminated gas stream. The application of classical filtration theory to predict the efficiency and the pressure drop through the RFBF was also demonstrated.

CONCLUSIONS AND SIGNIFICANCE

It was shown that the RFBF can separate gas-borne aerosols in the size range between 0.3 to 10μm with high efficiency at superficial gas velocities of 3 to 4 m/sec; these values are considerably higher than the operational velocities of other granular or fabric filters. The high operating velocities allow for a relatively compact filter design which can still separate large gas throughputs. It was also shown that even at these high gas velocities the pressure drop through the device did not exceed 20 to 30 cm H_2O.

* * *

The idea of using a fluidized bed to act as a dust filter dates back to Meissner and Mickley (1) who established that fluidized beds could serve to remove mists and dusts from air at high efficiencies. The basic concept involves using the gas to be filtered as the fluidizing medium to fluidize a bed of collector particles; the collector particles can then be removed and replaced continuously without difficulty. Care must be taken to prevent plugging of the distributor with dust particles, but otherwise operation of the filter is no different from that of a conventional fluidized bed.

Although it has been found that high collection efficiency could be obtained in a continuously operating fluidized bed filter (Knetting and Beekmans (2), McCarthy et al (3), Patterson and Jackson (4), Gutfinger and Tardos (5)) the fluidized bed dust filter is limited to operation at velocities no higher than 1.2 to 1.5 times the minimum fluidization velocity since higher velocities lead to by-passing of gas in the form of bubbles and consequently to poor collection efficiency (Scott and Guthrie (6), Black and Boubel (7), Tardos (8)). To overcome this limitation, Pfeffer and Hill (9) and Gal et al. (10) proposed and tested the use of a rotating fluidized bed as a filter device so that bubbles can be avoided at high gas flow rates simply by increasing the rotating speed of the bed. High filtration efficiency in the rotating fluidized bed is obtained mainly due to inertial impaction which is caused by the high gas velocity through the bed and by electrostatic effects which are the result of triboelectrification of the bed granules. The RFBF has the advantages of high filtration efficiency, continuous operation and high flow rates of gas per unit area of distributor; thus it appears very attractive for industrial applications.

DESCRIPTION OF APPARATUS AND PROCEDURE

The rotating fluidized bed (Lindauer et al (11), Levy and Chen (12), and Levy et al (13)) is essentially a vertical cylinder with porous walls which rotates around its axis of symmetry. Fluidization material in the form of granules is introduced into the cylinder and is forced to the wall due to the large centrifugal forces produced by the rotation. The wall serves as the gas distributor and the gas flows radially inward through it and through the bed of granules. When the drag forces on the granules balance the centrifugal forces, the bed becomes fluidized and fluidization can be achieved at any gas velocity by changing the rotating speed of the bed.

Figure 1 is a schematic representation of the device. The apparatus consists of a 0.2m diameter, 0.3m high distributor contained in a cylindrical plenum. The cylinder's lateral wall is perforated (43% open area) with 3/8 inch diameter holes. The wall is covered on the inside by 16 to 40 mesh screens depending on the size of granules used. The distributor (cylinder and screen) can be rotated between 340 to 750 RPM corresponding to radial accelerations of 13 to 63 g's. Liquid aerosol of dioctylphtalate (DOP) is introduced into the inlet air using a Royco 258 smoke generator and particle concentration is measured using a Climet particle size and concentration monitoring system; 99% of all liquid aerosol particles (DOP) is in the 0.3 to 3.0 micrometer range. Solid aluminum silicate particles in the size range between 0.3 to 10.0 micrometers, are also used. The solid dust is injected pneumatically into the entering air stream using a home-built generator containing a screw feeder and an ejector.

Two sets of experimental runs were carried out. The first set of experiments were performed batch-wise with respect to the collector granules, i.e., they were not replaced during each run, while the dirty air flow was continuous. A second set of experiments was performed with continuous introduction and removal of the fluidized granules from the bed. The circulation of the granules assured that clean granules are always present in the bed and, therefore, a filtration efficiency decrease due to the accumulation of dirty granules and/or re-entrainment of aerosol will not occur.

A schematic of the continuously operating rotating fluidized bed, also showing electrostatic charge measurement, is shown in Figure 2. The feeding system includes a hopper and a screw feeder. The clean particles are forced pneumatically to the surface of the bed by a high pressure air stream (15 psig), at a rate determined by the screw feeder at the bottom of the hopper. The granule discharge system is composed of a baffle in the bed and an inertial separator. As the radial velocity of the gas through the granules increases above minimum fluidization, particles are forced to the center of the rotating cylinder and out of the filter in the gas stream. Carry-over of particles from the bed is usually very small since the gas vortex will tend to send particles back into the bed but can be increased by the baffle (see Figure 2) which stops the rotation of some granules. The centrifugal force on these granules is small and the drag force can easily carry them downstream to the inertial separator (cyclone) where they are separated from the filtered air. The distance between the edge of the baffle and the screen is adjustable and determines the thickness of the bed. Thus changing the feed rate of granules does not affect the bed thickness but it reduces the average residence time of the granules in the bed.

EXPERIMENTAL RESULTS AND DISCUSSION

Pressure Drop and Minimum Fluidization Velocity

In a relatively shallow rotating fluidized bed operated at high enough rotational speed the bed assumes the shape of a cylindrical annulus (Levy et al (13)). The pressure drop and minimum fluidization velocity in such a bed can be obtained by using classical relationships found in the literature for ordinary fluidized beds, by replacing the acceleration of gravity g by the radial acceleration a_r produced by the rotation of the distributor. It was shown by Gal et al (10) and by Levy and Chen (12) that the pressure drop in a rotating fluidized bed is given by:

$$\Delta p = (1 - \epsilon)\, \rho_g W^2\, (r_0^2 - r_i^2)/2 \qquad (1)$$

while the pressure drop through the granules in the fixed bed regime is given by a modification of the Ergun correlation:

$$\Delta p = \frac{150(1-\epsilon)^2 \mu U_0 r_0 \ln(r_0/r_i)}{\epsilon^3 d_g^2} + \frac{1.75(1-\epsilon)\rho_f U_0^2 r_0^2 [1/r_i - 1/r_0]}{\epsilon^3 d_g} \qquad (2)$$

where W is the angular velocity, ϵ is the porosity, U_0, μ and ρ_f are the gas velocity, viscosity and density, respectively, d_g is the granule diameter and r_0 and r_i are the outer and inner radii of the bed. The minimum fluidization velocity, U_{mf}, can be found by equating the pressure drop in Equations (1) and (2) and replacing U_0 by U_{mf}.

Pressure drop as a function of air velocity for 0.3 cm in diameter polyethylene granules is given in Figure 3 for bed heights of L = 0.52, 1.67, 3.0 cm and rotational speeds of 350 and 450 RPM. The experimental results agree well with the prediction of Equation (2) for the fixed bed region and with those of Equation (1) for the fluidized bed region. As seen, the bed thickness increases as the minimum fluidization velocity decreases from about 240 cm/sec (L = 0.52 cm and W = 350 RPM) to about 190 cm/sec (L = 3.0 cm and W = 350 RPM) and from 300 cm/sec (L = 0.52 cm and W = 450 RPM) to about 250 cm/sec (L = 3.0 cm and W = 450 RPM).

Filtration Efficiency Measurements

Experiments were performed to examine the effect of the following parameters: gas velocity, granule size, aerosol density and aerosol size on the filtration efficiency of the rotating fluidized bed filter. To diminish the effect of electrostatics all experiments were performed at a relative humidity larger than 30% where these effects are small (Gal (14)). Furthermore DOP was used in most of the experiments so that bouncing and re-entrainment of aerosol was minimized.

The filtration efficiency for a certain size of dust particle can be calculated from:

$$\eta = 1 - n_{out}/n_{in} \tag{3}$$

where n_{in} and n_{out} are the number concentration of dust particles of a certain size in a unit volume of gas before and after passing through the filter, respectively. These concentrations were measured simultaneously using a Climet CI-225 dust particle counter with two sensors.

The effect of gas velocity and aerosol particle size on filtration efficiency in the rotating fluidized bed filter is shown in Figures 4. As the velocity increases from 60 cm/sec to 240 cm/sec, filtration efficiency increases from 34% to 98% for the 1 to 2μm DOP particles. There is hardly any change in filtration efficiency, which is practically zero, for the 0.5 to 0.8 μm particles at velocities lower than 180 cm/sec. At higher velocity filtration efficiency increased sharply to 49% at 300 cm/sec.

The bed material may significantly effect filtration efficiency due to surface phenomena and attraction and repulsion forces that different granules may exert on the aerosol particles. However, under conditions of high relative humidity (low electrostatic charge on bed granules) and low rate of bouncing (use of liquid areosol or low velocity), one can isolate the effect of granule size even if different bed materials are used. For example, experiments performed with 0.89 to 1.0μm DOP particles at 180 cm/sec showed the filtration efficiency to be very high-98 to 99% for the 0.03 to 0.04 cm granules and then drop sharply to 92%, and further to 48% as the granule size increases to 0.07 cm and 0.3 cm, respectively. The combined effect of gas velocity and granule size is shown in Figure 5 where filtration efficiency is given as a function of the Reynolds number,

$Re = U_0 d_g \rho_g / \mu$. based on the granule diameter. As seen, the filtration efficiency increases sharply with increasing Re for small values of the granule diameter and moderately for large values.

The inertia of an aerosol particle depends on its density ρ_p. As the density increases the inertial effect becomes stronger and the efficiency increases. The filtration efficiency as a function of velocity for liquid DOP aerosol and for solid aluminum silicate dust is compared in Figure 6. The filtration efficiency of the solid aerosol is higher than that of the DOP at constant velocity, 61% vs. 33% at 69 cm/sec and 85% vs. 63% at 120 cm/sec. Using DOP aerosol it takes a 1.5 to 2.0 times higher velocity to obtain the same filtration efficiency as obtained for the solid dust in this range of velocities. However, as the velocity increases, bouncing of solid dust particles occurs and filtration efficiency drops slightly from 95% at 240 cm/sec to 90% at 300 cm/sec. At velocities higher than 200 cm/sec the filtration efficiency of the DOP and the solid dust is about the same.

Since the RFBF is operated at relatively high gas velocities, inertial impaction characterized by the Stokes number defined as:

$$St = 4C\rho_p U_0 r_p^2 / 9\mu d_g \qquad (4)$$

where C is the Cunningham correction factor and r_p the aerosol radius is the major mechanism which characterizes filtration in a RFBF. Thus the efficiency increases with increasing air velocity, dust particle size and density and decreases with increasing granule size.

The Effect of Rotating Speed and Bed Thickness on the RFBF

In the rotating fluidized bed minimum fluidization may be obtained at almost any given velocity by adjusting the rotating speed of the bed within the mechanical limitations of the system. It was found that in the fixed bed region filtration efficiency is not affected by the rotation. As the rotating speed decreases at constant gas velocity the bed changes from a fixed to a fluidized bed. The filtration efficiency at minimum fluidization conditions is about the same as the efficiency in the fixed bed region and drops slightly as the rotating speed decreases. A further decrease in rotating speed results in bubble formation which reduces filtration efficiency very rapidly.

As the bed thickness increases, dust particles must pass through more layers of bed material and the probability of them being captured increases. Assuming that each granule experiences the same filtration phenomena the relationship between the total bed efficiency and the bed thickness becomes:

$$\eta = 1 - \exp[-1.5E_T(1-\epsilon)L/d_g] \qquad (5)$$

where E_T is the so-called single total particle efficiency and L is the filter thickness. In Figure 7 the total bed penetration, which is defined as $1-\eta$, is shown as a function of bed thickness for the 0.3 cm polyethylene granules. The curves in Figure 7 are linear up to a bed thickness of about 2 to 2.5 cm as they should be according to Equation (5) but the slopes change somewhat as the thickness increases. This phenomena of lower than expected efficiency near the surface of a thicker bed, especially for high velocities and small granules, can be explained by the nonuniformity of the fluidization in the existing rotating fluidized bed apparatus.

Continuously Operated RFBF

A set of experimental runs were performed using the continuously operated RFBF, with clean granules introduced into the system (see Figure 2) and dirty granules removed, to study the effect of granule feeding and removal on pressure drop and filtration efficiency. The system was operated continuously with granules fed and discharged from the system for periods of up to 3.25 hours using DOP and up to 1.75 hours using solid dust as the aerosol to be filtered. In both cases the runs were stopped due to mechanical problems in the feeding system not related to the RFBF operation itself. No problems were observed with the bed itself; the 14 mesh screen remained almost clean and no increase in pressure drop was observed. Bed thickness could be maintained constant at up to 3 cm and filtration efficiencies also remained relatively constant. It was concluded from these experiments that continuous operation does not appear to alter the performance of the RFBF.

COMPARISON OF EXPERIMENTAL DATA TO THEORETICAL MODEL

In order to compare experimental and theoretical results it is customary to recalculate filtration efficiencies based on the concept of the total single particle efficiency, E_T, using Equation (5).

Experiments using 0.07 cm polystyrene granules in the RFBF were compared to the theoretical correlation of Gal et al (15) which was derived for inertial impaction effects only in Figure 8. All the experimental points which are far right of the curve can be explained by experimental error or to bouncing effects for the solid dust. All points which are significantly to the left of the curve represent the effects of filtration mechanisms other than inertia such as diffusion, direct interception and gravity.

Electrostatic effects play a major role in improving filtration efficiency in the RFBF at low Stokes numbers in the range $0.001<St<0.01$. This was recently demonstrated by the authors (Gal (14)) by showing that, due to triboelectrification, caused by fluidization, the granules carry a wide distribution of electric charges and that the aerosol filtered in the bed may sometimes be charged and in other cases be electrically neutral. It was experimentally shown, for example, that the DOP aerosol is neutral while the solid dust generated by atomization is always charged and this charge is highly dependent on the humidity of the carrier gas. These results explain the much higher collection efficiency of solid dust as compared to DOP aerosol(Figure 6).

REFERENCES

1. Meissner, H.P. & Mickley, M.S., Ind. Eng. Chem., 41, 1238 (1949).
2. Knetting, P. & Beekmans, J.M., Can. J. Chem. Eng., 52, 703 (1974).

3. McCarthy, D., Yankel, A.J., Patterson, R.G. and Jackson, M.L., Ind. Eng. Chem., Process Des. & Dev., 15, 266 (1976).
4. Patterson, R.G. & Jackson, M.L., AIChE Symp. Ser. 161, 64 (1977).
5. Gutfinger, C. and Tardos, G.I., Atmos. Envi., 13, 553 (1979).
6. Scott, D.S. & Guthrie, D.A., Can. J. of Chem Eng., 37, 200 (1959).
7. Black, G.H. and Boubel, R.W., IEC Proc. Des. Des., 8, 573 (1969).
8. Tardos, G.I., "The Granular Bed Filter - Theory and Experiments", Ph.D. Dissertation, Technion-Israel Institute of Technology (1977).
9. Pfeffer, R. and Hill, F.B., "A Feasibility Study on the Use of a Rotating Fluidized Bed as a Dust Filter", Brookhaven National Laboratory Report 50990, September (1978).
10. Gal, E., Tardos, G.I. and Pfeffer, R., Proc. World Filtration Congress III, Downingtown, Pa., Sept. (1982).
11. Lindauer, G.C., Tichler, P. & Hatch, L.P., BNL 50013, T-435 (1966).
12. Levy, E.K. and Chen, J.C., Paper presented at Int. Powder and Bulk Solid Handling and Processing Conf., Rosmont, Ill. (1977).
13. Levy, E.K. et al, Fluidization, J. Davidson and D.L. Keairns, Editors, Cambridge University Press (1978).
14. Gal, E., Ph.D. Dissertation, The City College of The City University of New York (1984).
15. Gal, E., Tardos, G.I. and Pfeffer, R., AIChE J., 31, 1093 (1985).

* * *

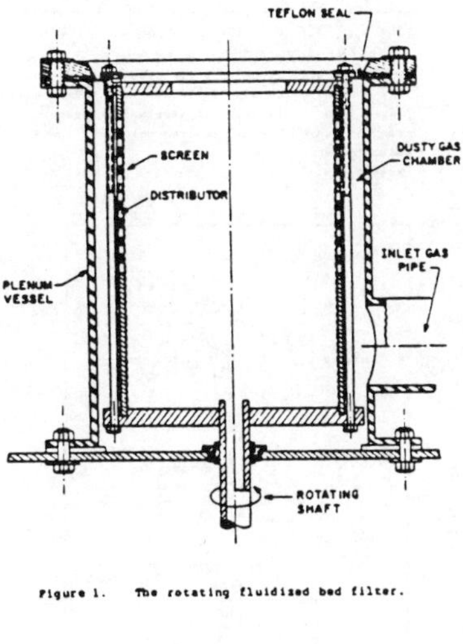

Figure 1. The rotating fluidized bed filter.

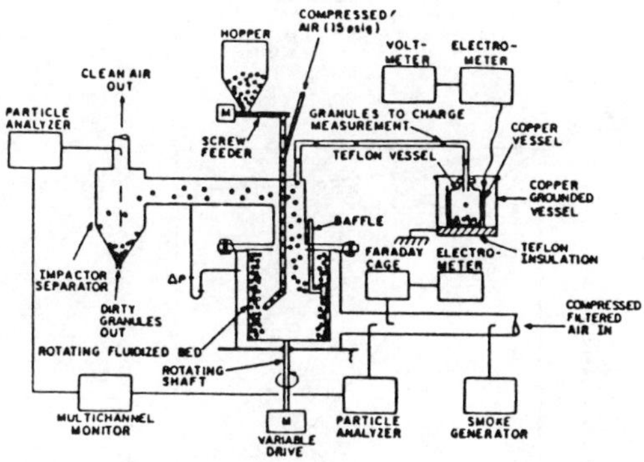

Figure 2. The rotating fluidized bed filter experimental set-up.

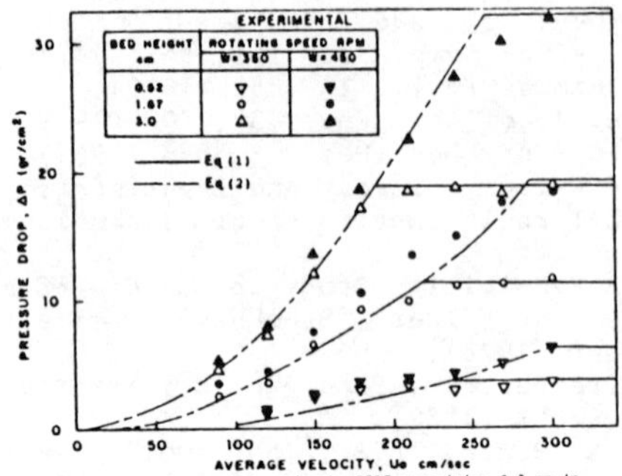

Figure 3. Pressure drop in the RFBF containing 0.3 cm in diameter polyethylene granules as a function of air velocity for different bed heights. Bed Porosity, ε = 0.4

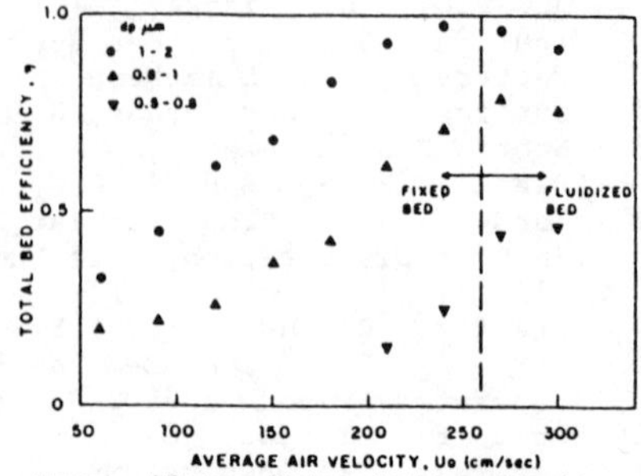

Figure 4. Filtration efficiency of a bed containing 0.3 cm diameter polyethylene granules as a function of average air velocity for different sizes of DOP particles. Relative humidity, RHu = 30%; Bed thickness, L = 3.0 cm; Bed porosity, ε = 0.4; Rotating speed, W = 450 RPM

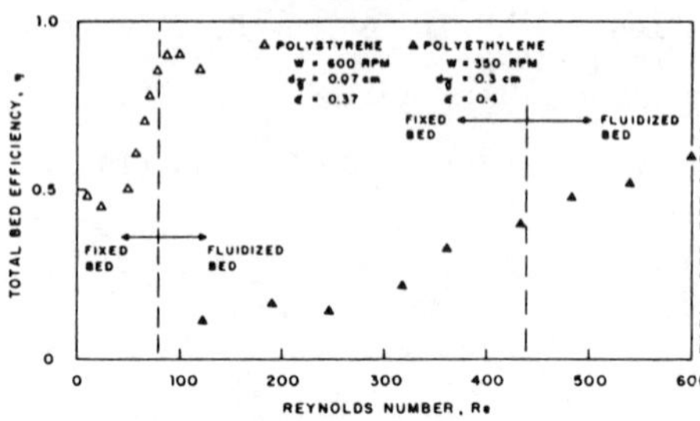

Figure 5. Total bed efficiency in the filtration of 0.8-1.0 μm diameter DOP particles as a function of the Reynolds number, Re. Bed thickness, L = 1.0 cm; Relative humidity, RHu = 30-40%

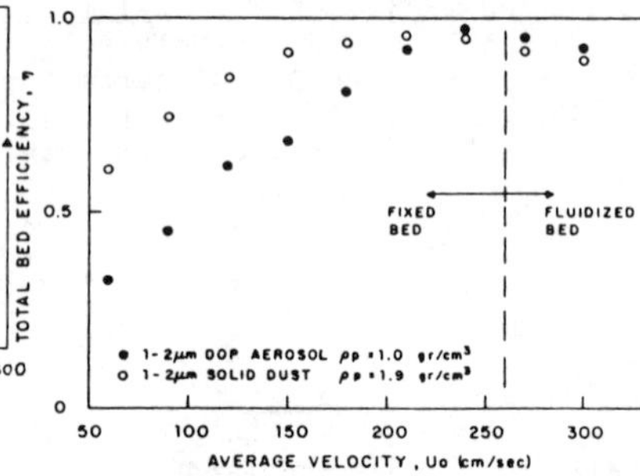

Figure 6. Total bed efficiency as a function of air velocity for DOP and solid dust particles in a RFBF containing 0.3 cm diameter polyethylene granules. Bed thickness, L = 3.0 cm; Bed porosity, ε = 0.4; Rotating speed, W = 450 RPM; Relative Humidity, RHu = 35-40%

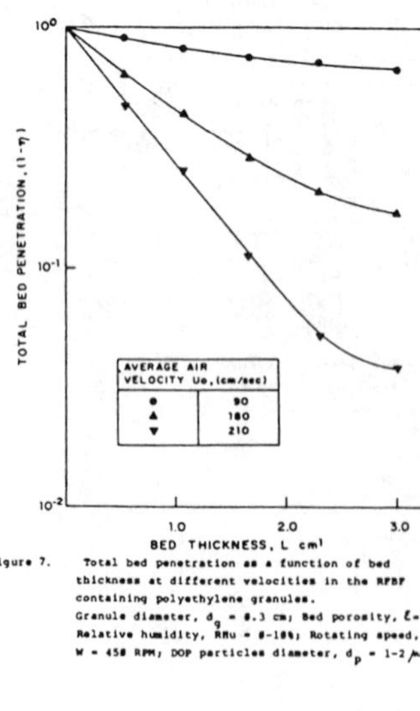

Figure 7. Total bed penetration as a function of bed thickness at different velocities in the RFBF containing polyethylene granules. Granule diameter, d_g = 0.3 cm; Bed porosity, ε = 0.4; Relative humidity, RHu = 0-10%; Rotating speed, W = 450 RPM; DOP particles diameter, d_p = 1-2 μm

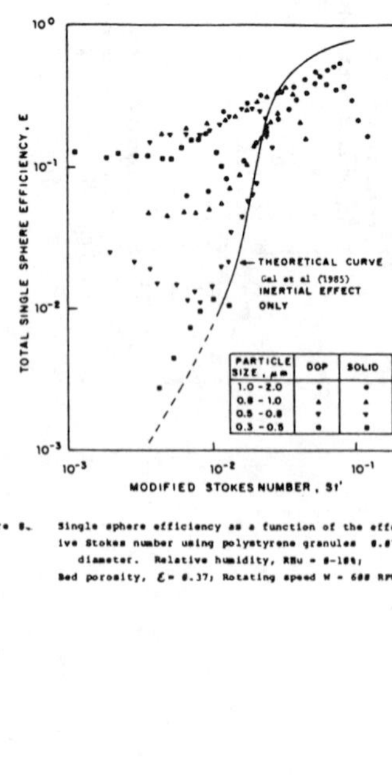

Figure 8. Single sphere efficiency as a function of the effective Stokes number using polystyrene granules 0.07 cm diameter. Relative humidity, RHu = 0-10%; Bed porosity, ε = 0.37; Rotating speed W = 600 RPM

Author Index

List of Referees

Dr. R. Abed
E.I. Dupont De Nemours, Wilmington, Del., USA
Dr. H. Arastoopour
Institute of Gas Technology, Chicago, Ill., USA
Dr. A. Avidan
Mobil Research and Development Corp., Paulsboro, N.J., USA
Professor M. Baerns
Ruhr-Universität Bochum, Bochum-Querenburg, Germany
Dr. J. Baeyens
Fabricom, Bruxelles, Belgium
Dr. C.G.J. Baker
RHM Research Ltd., High Wycombe, England
Professor A.P. Baskakov
Ural Polytechnic Institute, Sverdlovsk, USSR
Professor W. Bauer
Frauenhofer-ILV, München, Germany
Professor L.A. Behie
University of Calgary, Alberta, Canada
Professor M.A. Bergougnou
University of Western Ontario, Canada
Professor R. Bilbao
University of Zaragoza, Spain
Dr. H.J. Bock
Universität Erlangen-Nürnberg, Germany
Dr. V.A. Borodulya
Luikov Heat and Mass Transfer Institute, Minsk, USSR
Dr. J.S.M. Botterill
University of Birmingham, England
Professor J. Bridgwater
University of Birmingham, England
Dr. C.E. Capes
National Research Council, Ottawa, Canada
Professor H.S. Caram
Lehigh University, Bethlehem, Pa., USA
Professor C. Chavarie
Ecole Polytechnique de Montréal, Canada
Dr. D.J. Cheesman
University College London, England
Professor Gan-Tang Chen
Zhejiang University, Hangzhou, China
Professor J.C. Chen
Lehigh University, Bethlehem, Pa., USA
Dr. T. Chiba
Hokkaido University, Sapporo, Japan
Professor R. Clift
University of Surrey, Guildford, England
Professor J.-P. Couderc
Institut National Polytechnique de Toulouse, France

Dr. R.C. Darton
Koninklijke/Shell-Laboratorium, Amsterdam, Holland
Professor J.F. Davidson
University of Cambridge, England
Dr. J. Deinert
Rheinische Braunkohlenwerke Aktiengesellschaft, Köln, Germany
Professor H.I. de Lasa
University of Western Ontario, Canada
Dr. S.A. El-Temtamy
Egyptian Petroleum Research Institute, Cairo, Egypt
Professor N. Epstein
University of British Columbia, Canada
Professor Liang-Shih Fan
Ohio State University, USA
Professor L.T. Fan
Kansas State University, USA
Dr. T.J. Fitzgerald
TRW Inc., Redondo Beach, Ca., USA
Professor P.U. Foscolo
University of L'Aquila, Italy
Professor S. Furusaki
University of Tokyo, Japan
Professor T. Furusawa
University of Tokyo, Japan
Dr. D. Geldart
University of Bradford, England
Dr. B.M. Gibbs
University of Leeds, England
Professor H. Gibert
Institut National Polytechnique de Toulouse, France
Professor D. Gidaspow
Illinois Institute of Technology, USA
Dr. L.R. Glicksman
Massachusetts Institute of Technology, USA
Professor J.R. Grace
University of British Columbia, Canada
Professor R.A. Graff
City College of the City University of New York, USA
Dr. P. Guigon
Université de Technologie de Compiegne, France
Dr. J.E. Gwyn
Shell Development Company, Houston, Texas, USA
Dr. D. Harrison
University of Exeter, England
Professor M. Hasatani
Nagoya University, Japan
Dr. H. Hatano
Tokyo Institute of Technology, Yokohama, Japan
Mr. J. Highley
National Coal Board, Cheltenham, England
Dr. M. Horio
Tokyo University of Agriculture and Technology, Japan
Dr. J.R. Howard
University of Aston in Birmingham, England

Professor Tsung-ting Hu
Tianjin University, China
Professor Jin Yong
Qinghua University, Beijing, China
Dr. G. Jovanovic
Belgrade University, Yugoslavia
Professor M.R. Judd
University of Natal, Durban, South Africa
Professor K. Kato
Gunma University, Kiryu-shi, Japan
Dr. D.L. Keairns
Westinghouse R&D Center, Pittsburgh, Pa., USA
Dr. T.M. Knowlton
Institute of Gas Technology, Chicago, Ill., USA
Professor H.O. Kono
West Virginia University, USA
Dr. J. Korenberg
York-Shipley, Inc., York, Pa., USA
Professor D. Kunii
Yokohama National University, Japan
Professor Mooson Kwauk
Academia Sinica, Beijing, China
Professor C. Laguerie
Institut National Polytechnique de Toulouse, France
Dr. R.D. LaNauze
CSIRO, North Ryde, N.S.W., Australia
Professor J.-F. Large
Université de Technologie de Compiegne, France
Professor B. Leckner
Chalmers University of Technology, Göteborg, Sweden
Professor L.S. Leung
University of Queensland, Australia
Professor E.K. Levy
Lehigh University, Bethlehem, Pa., USA
Professor H. Littman
Rensselaer Polytechnic Institute, Troy, N.Y., USA
Dr. Liu Dalu
Academia Sinica, Beijing, China
Professor Y.A. Liu
Virginia Polytechnic Institute and State University, USA
Dr. J.F. Lottes
Mobil Chemical Company, Edison, N.J., USA
Dr. L. Marcussen
Technical University of Denmark, Lyngby, Denmark
Professor L. Massimilla
Universita de Napoli, Italy
Dr. H. Masson
Iniex, Liège, Belgium
Dr. J. Matsen
Exxon Research & Engineering Co., Florham Park, N.J., USA
Professor O. Molerus
Universität Erlangen-Nürnberg, Germany
Professor M.H. Morgan, III
Rensselaer Polytechnic Institute, Troy, N.Y., USA

Professor S. Mori
Nagoya Institute of Technology, Japan
Professor S. Morooka
Kyushu University, Fukuoka, Japan
Dr. S. Mortensen
A/S Niro Atomizer, Søborg, Denmark
Dr. L. Neuzil
Prague Institute of Chemical Technology, Czechoslovakia
Dr. R.A. Newby
Westinghouse R&D Center, Pittsburgh, Pa., USA
Professor A.W. Nienow
University of Birmingham, England
Dr. I.W. Noordergraaf
Koninklijke/Shell-Laboratorium, Amsterdam, Holland
Dr. T.J. O'Brien
Morgantown Energy Technology Center, Morgantown, W.Va., USA
Professor M.H. Peters
Rensselaer Polytechnic Institute, Troy, N.Y., USA
Professor R. Pfeffer
City College of the City University of New York, USA
Dr. W.W. Poersch
Deutsche Babcock AG, Krefeld-Uerdingen, Germany
Professor O.E. Potter
Monash University, Victoria, Australia
Dr. R.Y. Qassim
Cidade Universitaria, Rio de Janeiro, Brazil
Professor J.F. Richardson
University College Swansea, Wales, UK
Professor K. Rietema
Eindhoven University of Technology, Holland
Dr. G.M. Rios
Université des Sciences et Techniques du Languedoc, France
Professor P.N. Rowe
University College London, England
Dr. L.A. Ruth
Exxon Research & Engineering Co., Florham Park, N.J., USA
Dr. S.C. Saxena
University of Illinois, Chicago, Ill., USA
Professor K. Schügerl
Universität Hannover, Germany
Dr. J.T. Shaw
National Coal Board, Cheltenham, England
Dr. T. Shingles
Sasol Technology (Pty) Limited, Sasolburg, South Africa
Dr. T. Shirai
Science University of Tokyo, Chiba, Japan
Dr. A. Sørensen
Technical University of Denmark, Lyngby, Denmark
Mr. L.M. Southwick
C.F. Braun & Co., Rutherford, N.J., USA
Professor A.M. Squires
Virginia Polytechnic Institute and State University, USA
Dr. F. Staub
General Electric Corp., Schenectady, N.Y., USA

Dr. J.B. Stepanek
University of Salford, England
Dr. J.J. van Deemter
Technische Hogeschool Delft, Holland
Professor P.J. van den Berg
Technische Hogeschool Delft, Holland
Professor W.P.M. van Swaaij
Twente University of Technology, Enschede, Holland
Dr. A.H.M. Verkooijen
Eindhoven University of Technology, Holland
Professor J. Werther
Technical University of Hamburg, Germany
Mr. A.B. Whitehead
Painesville, Australia
Professor Cheng-yi Xia
Nanjing Institute of Chemical Technology, China
Professor R. Yamazaki
Nagoya University, Japan
Dr. Wen-ching Yang
Westinghouse R&D Center, Pittsburgh, Pa., USA
Dr. J.G. Yates
University College London, England
Professor Yu Zhi-qing
Qinghua University, Beijing, China
Dr. V. Zakkay
New York University, Westbury L.I., N.Y., USA
Dr. F.A. Zenz
F.A. Zenz, Inc., Garrison, N.Y., USA